Laboratory Guide to the Methods in Biochemical Genetics

Nenad Blau • Frédéric M. Vaz
Editors

Laboratory Guide to the Methods in Biochemical Genetics

Second Edition

Springer

Editors
Nenad Blau
Division of Metabolism
University Children's Hospital Zürich
Zürich, Switzerland

Frédéric M. Vaz
Laboratory Genetic Metabolic Disease,
Department of Laboratory Medicine
Amsterdam UMC
Amsterdam, The Netherlands

ISBN 978-3-031-58818-1 ISBN 978-3-031-58819-8 (eBook)
https://doi.org/10.1007/978-3-031-58819-8

This Springer imprint is published by the registered company Springer Nature Switzerland AG
The registered company address is: Gewerbestrasse 11, 6330 Cham, Switzerland

Foreword

The field of inborn errors of metabolism is rapidly expanding, now encompassing over 1000 entities that affect approximately one in 500 newborns worldwide. These diseases are often severe and affect multiple systems in the body. Patients present with symptoms that may not clearly indicate a specific biochemical pathway, and the underlying cause of the disease is often not diagnosed until adulthood. However, an accurate diagnosis is crucial for successful, targeted therapies that can significantly improve or even prevent the manifestations of the disease if initiated early.

Advancements in medicine and science often occur at the intersection of different fields. In the case of inherited metabolic diseases, there is a significant gap between medical doctors and biochemical methods, which are constantly evolving. Fortunately, there are established medical and scientific societies for biochemists, molecular geneticists, and medical doctors, which provide official training and quality control programs for laboratories.

An international interdisciplinary group has successfully condensed the current knowledge of *Laboratory Guide to the Methods in Biochemical Genetics* into this book. It will be essential for all laboratories working in this field, providing both experienced professionals and busy doctors with critical information on methods, control ranges, and potential pitfalls. This well-written book offers in-depth knowledge to specialists as well as busy doctors worldwide who care for patients with these diseases as part of their work.

A better understanding of the central basis of diagnosis and specific treatments will enable healthcare professionals to ask crucial questions to the lab when faced with challenging clinical problems that require quick resolution. Additionally, this book should be included in every medical library and medical school worldwide.

Heidelberg, Germany | Georg F. Hoffmann

Preface

Experienced chefs often have a collection of delicious entree recipes stored in their memory. The rest of us, who cook for pleasure or necessity, typically rely on recipes that provide detailed lists of ingredients and instructions for mixing and baking, which are crucial for achieving a successful outcome. We have all likely experienced the disappointment of a "flat" cake due to the omission of a small ingredient like baking powder. It is through these experiences that we learn the importance of following recipes accurately, and this is why "cookbooks" are so valuable. Similarly, the second edition of the *Laboratory Guide to the Methods in Biochemical Genetics* aims to address the current lack of up-to-date laboratory techniques in the field. The previous edition published in 2008 has become outdated due to advancements in methodology and the introduction of high-throughput, high-sensitivity equipment, including metabolomic techniques. This new edition seeks to bridge the methodological gap that has emerged over recent years. Together with the *Physician's Guide to the Treatment and Follow-Up of Inherited Metabolic Diseases*, this book completes on the subject.

In recent years, the field of biochemical genetics has gained significant attention in laboratory settings due to the widespread implementation of expanded newborn screening worldwide. As a result, the number of samples being processed in biochemical genetics laboratories is increasing due to the necessary evaluation of positive newborn screening results. Therefore, it is in the best interest of all parties involved, including physicians, patients, metabolic specialists, and other healthcare providers, that these laboratories have access to the most up-to-date methods for their testing procedures.

The editors of this new book have tried to compile a comprehensive list of laboratory tests that will be beneficial for most laboratories. Each chapter has been meticulously developed by experts with the intention of making them self-contained and independent. This structure ensures that there is no need to refer to other technique papers, allowing for a quicker implementation of these techniques in the laboratory.

It is important to note that most methods used in biochemical genetics are labor-intensive and complex, especially when compared to automated testing laboratories. Therefore, having a compendium specifically designed for laboratory use will prove to be a valuable resource. Our goal in compiling this edition has been to provide a useful tool that can either enhance existing protocols for readers and colleagues who are already well-versed in the biochemical genetics laboratory or fill any gaps in the test menu for those who are just starting out.

We would like to express our gratitude to Dr. Leo Kluijtmans for his significant contributions to the planning of the book. Unfortunately, due to personal circumstances, he had to step down from the editorial board.

To all our readers and colleagues, we hope that this book will serve as a valuable asset in your biochemical genetic endeavors. Happy biochemical cooking!

Zürich, Switzerland Nenad Blau
Amsterdam, The Netherlands Frédéric M. Vaz

Contents

Contributors

Erland Arning Institute of Metabolic Disease, Baylor Scott & White Research Institute, Dallas, TX, USA

Rafael Artuch Clinical Biochemistry Department, Sant Joan de Déu Research Institute and CIBERER-ISCIII, Barcelona, Spain

Jean François Benoist Laboratoire de Biochimie, Hopital Necker Enfants Malades, Paris-Saclay Université France, Paris, France

Caroline Bhairo Neurometabolic Unit, National Hospital, London, UK

Nenad Blau Division of Metabolism, University Children's Hospital Zürich, Zürich, Switzerland

Annet M. Bosch Emma Children's Hospital, Department of Pediatrics, Division of Metabolic Diseases, Amsterdam Gastroenterology Endocrinology Metabolism Institute, University of Amsterdam, Amsterdam, The Netherlands

Claudia Carducci Department of Experimental Medicine, Sapienza University of Rome, Rome, Italy

Peter T. Clayton Genetics and Genomic Medicine Department, UCL GOS Institute of Child Health, London, UK

Christelle Corne Laboratoire des maladies métaboliques et dépistage néonatal, Institut de Biologie et pathologie, Grenoble, France

Alessio Cremonesi Division of Clinical Chemistry and Biochemistry, University Children's Hospital Zürich, University of Zurich, Zürich, Switzerland

Marli Dercksen Biochemistry Department, Faculty of Natural and Agricultural Sciences, North-West University, Potchefstroom, South Africa

Katrina M. Dipple Division of Genetic Medicine, Department of Pediatrics, University of Washington, Seattle, WA, USA

Genetic Medicine, Seattle Children's Hospital, Seattle, WA, USA

Center for Translational and Clinical Research, Seattle Children's Research Institute, Seattle, WA, USA

Marinus Duran Academic Medical Center, Lab Genetic Metabolic Diseases, Amsterdam, The Netherlands

Sarah H. Elsea Department of Molecular and Human Genetics, Baylor College of Medicine, Houston, TX, USA

Baylor Genetics, Houston, TX, USA

Sacha Ferdinandusse Laboratory Genetic Metabolic Diseases, Department of Clinical Chemistry and Pediatrics, Emma Children's Hospital, Amsterdam UMC Location University of Amsterdam, Amsterdam, The Netherlands

Brian Fowler Division of Metabolism and Molecular Pediatrics, University Children's Hospital, Zürich, Switzerland

Maria Fuller Genetics and Molecular Pathology, SA Pathology, Women's and Children's Hospital, North Adelaide, SA, Australia

Adelaide Medical School, University of Adelaide, Adelaide, SA, Australia

K. Michael Gibson Washington State University, Section of Clinical Pharmacology, Spokane, USA

Kevin Glinton Department of Molecular and Human Genetics, Baylor College of Medicine, Houston, TX, USA

Susanna M. I. Goorden Center for Lysosomal and Metabolic Diseases, Department of Clinical Genetics, Erasmus University Medical Center, Rotterdam, The Netherlands

Marne C. Hagemeijer Center for Lysosomal and Metabolic Diseases, Department of Clinical Genetics, Erasmus University Medical Center, Rotterdam, The Netherlands

Patricia L. Hall Department of Laboratory Medicine and Pathology, Mayo Clinic, Rochester, MN, USA

Simon J. R. Heales Neurometabolic Unit, National Hospital, London, UK

UCL Great Ormond Street Institute of Child Health, London, UK

Nastassja Himmelreich Department 1, Centre for Child and Adolescent Medicine, University of Heidelberg, Heidelberg, Germany

M. C. D. G. Huigen Translational Metabolic Laboratory, Department of Laboratory Medicine, Radboud University Medical Centre, Nijmegen, The Netherlands

Edwin H. Jacobs Center for Lysosomal and Metabolic Diseases, Department of Clinical Genetics, Erasmus University Medical Center, Rotterdam, The Netherlands

Cornelis Jakobs Clinical Chemistry, Metabolic Unit, VU University Medical Center, Amsterdam, Netherlands

Erwin E. W. Jansen Clinical Chemistry, Metabolic Unit, VU University Medical Center, Amsterdam, Netherlands

Youssef Khalil Genetics and Genomic Medicine Department, UCL GOS Institute of Child Health, London, UK

Christian Körner Division of Metabolism, University Children's Hospital Zürich, Germany

Dirk J. Lefeber Department of Neurology, Radboud University Medical Centre, Nijmegen, The Netherlands

Department of Human Genetics, Radboud University Medical Centre, Nijmegen, The Netherlands

Jürgen Lübbehusen Department 1, Centre for Child and Adolescent Medicine, University of Heidelberg, Heidelberg, Germany

Daniel Mackenzie Neurometabolic Unit, National Hospital, London, UK

Shayne Mason Biochemistry Department, Faculty of Natural and Agricultural Sciences, North-West University, Potchefstroom, South Africa

Dietrich Matern Laboratory Medicine, Medical Genetics, and Pediatrics, Biochemical Genetics Laboratory, Division of Laboratory Genetics and Genomics, Department of Laboratory Medicine and Pathology, Mayo Clinic, Rochester, MN, USA

Déborah Mathis University Institute of Clinical Chemistry, Inselspital, Bern University Hospital, University of Bern, Bern, Switzerland

Philippa B. Mills Genetics and Genomic Medicine Department, UCL GOS Institute of Child Health, London, UK

Thomas Opladen Division for Child Neurology and Metabolic medicine, Centre for Paediatric and Adolescent Medicine, Heidelberg, Germany

Martin Poms Division of Clinical Chemistry and Biochemistry, University Children's Hospital, Zurich, Switzerland

Ana Pop Laboratory Genetic Metabolic Diseases, Department of Clinical Chemistry, Amsterdam UMC Location University of Amsterdam, Amsterdam, The Netherlands

Simon A. S. Pope Neurometabolic Unit, National Hospital, London, UK

UCL Great Ormond Street Institute of Child Health, London, UK

H. C. M. T. Prinsen Department of Genetics, section Metabolic Diagnostics, Wilhelmina Children's Hospital, University Medical Centre Utrecht, Utrecht, The Netherlands

Richard J. Rodenburg Radboud Center for Mitochondrial Medicine (RCMM), Department of Pediatrics, 830 Translational Metabolic Laboratory (TML), Radboud University Medical Center, HB, Nijmegen, The Netherlands

George J. G. Ruijter (on behalf of the ERNDIM Scientific Advisory Board) Department of Clinical Genetics, Erasmus University Medical Center, GD, Rotterdam, The Netherlands

Center for Lysosomal and Metabolic Diseases, Department of Clinical Genetics, Erasmus University Medical Center, Rotterdam, The Netherlands

Gajja S. Salomons Amsterdam UMC Location AMC, Laboratory Genetic Metabolic Diseases, Amsterdam, The Netherlands

Matthew J. Schultz Department of Laboratory Medicine and Pathology, Mayo Clinic, Rochester, MN, USA

Cristina Sierra Clinical Biochemistry Department, Sant Joan de Déu Research Institute and CIBERER-ISCIII, Barcelona, Spain

Desirée E. C. Smith Laboratory Genetic Metabolic Diseases (F0-132), Department of Clinical Chemistry, Amsterdam UMC Location University of Amsterdam, Amsterdam, The Netherlands

V. Reid Sutton Department of Molecular and Human Genetics, Baylor College of Medicine, Houston, TX, USABaylor Genetics, Houston, TX, USA

Christian Thiel Department 1, Centre for Child and Adolescent Medicine, University of Heidelberg, Heidelberg, Germany

Silvia Tortorelli Department of Laboratory Medicine and Pathology, Mayo Clinic, Rochester, MN, USA

Jiddeke M. van de Kamp Department of Human Genetics, Amsterdam UMC Location VUMC, Amsterdam, The Netherlands

Marie van Dijk Laboratory of Genetic Metabolic Diseases, Departments of Clinical Chemistry, Amsterdam UMC Location University of Amsterdam, Amsterdam, The Netherlands

André B. P. van Kuilenburg Laboratory of Genetic Metabolic Diseases, Departments of Clinical Chemistry, Amsterdam UMC Location University of Amsterdam, Amsterdam, The Netherlands

Amsterdam Gastroenterology Endocrinology Metabolism, Amsterdam, The Netherlands

Henk van Lenthe Laboratory of Genetic Metabolic Diseases, Departments of Clinical Chemistry, Amsterdam UMC Location University of Amsterdam, Amsterdam, The Netherlands

Anne Vassault Laboratoire de Biochimie, Hopital Necker Enfants Malades, Paris-Cité University France, Paris, France

Frédéric M. Vaz Laboratory Genetic Metabolic Diseases, Department of Clinical Chemistry and Pediatrics, Emma Children's Hospital, Amsterdam UMC Location University of Amsterdam, Amsterdam, The Netherlands

Christine Vianey-Saban (on behalf of the ERNDIM Scientific Advisory Board) Division of Inborn Errors of Metabolism, Department of Biochemistry and Molecular Biology, CHU de Lyon, Bron, France

Mirjam M. C. Wamelink Laboratory of Genetic Metabolic Diseases (F0-132), Departments of Clinical Chemistry, Amsterdam UMC Location University of Amsterdam, Amsterdam, The Netherlands

Ronald J. A. Wanders Department of Clinical Chemistry and Pediatrics, Emma Children's Hospital, Laboratory Genetic Metabolic Diseases, Amsterdam University Medical Center, AZ, Amsterdam, The Netherlands

Brandi Wasek Institute of Metabolic Disease, Dallas, TX, USA

Hans R. Waterham Laboratory Genetic Metabolic Diseases, Department of Clinical Chemistry, Amsterdam UMC, University of Amsterdam, Amsterdam, The Netherlands

Ron A. Wevers Department Laboratory Medicine, Translational Metabolic Laboratory, Radboud University Medical Centre, GA, Nijmegen, The Netherlands

Strategies in Biochemical Genetics

1

Ronald J. A. Wanders and Ron A. Wevers

Contents

Abbreviations

CPT2	Carnitine palmitoyltransferase 2
CSF	Cerebrospinal fluid
FUS	Feature of Unknown Significance
Hilic	Hydrophilic interaction liquid chromatography
HMDB	Human Metabolome Data Base
ICIMD	International Classification of Inherited Metabolic Diseases
IEM	Inborn Error of Metabolism
IMD	Inherited Metabolic Disease
LC	Liquid chromatography
LC-MS/MS	Liquid chromatography-tandem mass spectrometry
LC-QTof MS	Liquid chromatography-Time of flight Mass Spectrometry
MCAD	Medium-chain acyl-CoA dehydrogenase
MTP	Mitochondrial trifunctional protein
NGMS	Next generation metabolic screening
NMR	Nuclear Magnetic Resonance
Rt	Retention time
VLCAD	Very long-chain acyl-CoA dehydrogenase
VUS	Variant of Unknown Significance

R. J. A. Wanders (✉)
Department of Clinical Chemistry and Pediatrics, Emma Children's Hospital, Laboratory Genetic Metabolic Diseases, Amsterdam University Medical Center, AZ, Amsterdam, The Netherlands
e-mail: r.j.wanders@amsterdamumc.nl

R. A. Wevers (✉)
Department Laboratory Medicine, Translational Metabolic Laboratory, Radboud University Medical Centre, GA, Nijmegen, The Netherlands
e-mail: ron.wevers@radboudumc.nl

N. Blau, F. M. Vaz (eds.), *Laboratory Guide to the Methods in Biochemical Genetics*,
https://doi.org/10.1007/978-3-031-58819-8_1

1.1 Introduction and Historical Perspective

Archibald Garrod is now rightly regarded as the father of biochemical genetics. He published his key concept of Inborn Errors of Metabolism (IEM) in a series of four Croonian Lectures to the Royal College of Physicians of London which he had developed over the years ever since his work on alkaptonuria [1]. Following his father's footsteps who was a talented physician scientist in his own right, Garrod first obtained a degree in Chemistry followed later by his training in Medicine and received his doctorate from Oxford with a thesis on rheumatoid arthritis (see: Prasad and Galbraith, [2]). Garrod's fascination for pigments and the different colors in patients' urines led him to describe erythropoietic porphyria, but also prompted him to work on alkaptonuria. This condition had already been described many years before in literature in 1859 in fact and the identity of the substance which is responsible for the deep brown colorization of patients' urine upon exposure to air, was resolved as being homogentisic acid. Garrod's realization that much can be gained by studying as many patients as he could get access to, inspired him to set up an improved method for the quantitative analysis of homogentisic acid and collect urines from many patients and their relatives which he obtained by contacting different physicians all over Europe. He quickly realized the high proportion of alkaptonuric versus healthy members in 9 families (see [1, 3]; later extended to 17 families in 1908) and the large proportion of alkaptonurics who are offspring of marriages of first cousins. Having heard of these findings William Bateson—a zoologist with a keen interest in heredity which he studied through plant breeding—began to correspond with Garrod in January 1902 and suggested that recessive inheritance might be at the basis of alkaptonuria and referred to the work of Gregor Mendel in this respect. Interestingly, Bateson himself was introduced to the work of Mendel just 2 years earlier through a letter written to him by Hugo de Vries following the rediscovery of the forgotten work of Gregor Mendel by De Vries and others. Bateson is a true genius in his own right and was the first to introduce the term "genetics" and is rightly recognized today as the principal founder of Mendelism and human genetics (see [3]). Especially since his ideas on the underlying Mendelian basis of alkaptonuria were received with little enthusiasm, Garrod extended his work to other conditions including cystinosis, albinism, and pentosuria which formed the basis of his four subsequent Croonian Lectures delivered on June 18, 23, 25, and 30, 1908 and published in *The Lancet*.

Since these early days, much has happened and the list of inborn errors of metabolism has expanded progressively. This rapid pace of discovery is explained by the marked advances in technology which have led to the identification of many, "new" inborn errors of metabolism over the years. Indeed, after Følling's first description of phenylketonuria using a simple chemical test (ferric chloride test; [4]) which led him to identify phenylpyruvate as the unknown accumulating substance in phenylketonuria (PKU) patients, the number of "new" amino acid disorders quickly increased when amino acid analyzers entered the laboratories in the 1950s. Similarly, the introduction of gas chromatography coupled with flame ionization detection or Mass Spectrometry (MS) allowed the recognition of many organic acidurias. The rapid pace of metabolite analysis was soon followed by the equally rapid development in the field of enzymatic analyses. This not only allowed identification of the precise enzyme defect responsible for the metabolite abnormalities observed in patients but also paved the way toward identification of the mutant gene involved. The successive emergence of all these new laboratory techniques and the rapid increase in knowledge about inborn errors of metabolism quickly led to the institution of specialized laboratories separate from routine clinical chemistry laboratories preferably led by dedicated "metabolic (bio)chemists" acting as the linking pin between the clinician involved and the laboratory.

In more recent years, the amazing progress in the field of gene sequencing has revolutionized the field of inborn errors of metabolism even more with important consequences for the biochemical genetics laboratories now and in the future. At the same time, the rapid expansion of our field has led to a reappraisal of the definition of inborn errors of metabolism. The challenge of the definition of inborn errors of metabolism highlighted by the ever-increasing number of disorders of glycosylation led Morava et al. to propose that an IEM requires only that "an impairment of specific enzymes or biochemical pathways is intrinsic to the pathomechanism" irrespective of whether it is associated with abnormalities in biochemical laboratory tests [5]. More recently, Ferreira and coworkers published an internationally agreed classification (ICIMD) of more than 1400 disorders subdivided into 24 categories comprising 124 groups [6]. In contrast, Saudubray et al. have proposed a different classification primarily based on clinical considerations which only includes three different groups [7]. Yet another view would be to choose a gene-based classification as proposed by David Valle, especially since current knowledge holds that the phenotype associated with each individual IEM broadens over the years (in Blau et al. [8]). Indeed, whereas the initial phenotype of an IEM is usually well-defined in terms of clinical signs and symptoms, this is often followed by the description of patients with incomplete phenotypes sometimes lacking any of the stigmata observed in the classical patients. A case in point is the group of peroxisomal disorders with mutations in the peroxisome biogenesis gene *PEX1* linked to classical Zellweger syndrome with all its clinical stigmata including craniofacial dysmorphia, early death, and multiple organ involvement whereas mutations in *PEX1* are also associated with much milder phenotypes including Usher syndrome [9] and Heimler syndrome [10].

Below we will describe the current state of knowledge regarding biochemical genetics and the laboratories involved. We will also discuss the potential of new technologies and the future of biochemical genetics, especially in these days in which genetic analysis has taken center stage in the identification of inborn errors of metabolism, or rather "inherited metabolic disorders (IMD)" as advocated by Ferreira et al. [6].

1.2 The Biochemical Genetics Laboratory Now and in the Future

Traditionally, the diagnostic work-up of a patient suspected to suffer from a metabolic disease has largely relied on targeted metabolite analyses of the three body fluids plasma, urine, and CSF and this procedure is still operational in many laboratories worldwide. Although initially extremely expensive, holistic genetic tests have become increasingly cheaper and are now comparable in price with testing at the metabolite level. As a consequence, Whole Exome Sequencing (WES) and/or Whole Genome Sequencing (WGS) have slowly taken over the place of metabolite testing especially since the turnaround time of results is almost equal nowadays which leaves the field with the question about the role of metabolomics now and in the (nearby) future. The advantage of holistic genetic tests is that they cover almost each and every gene involved in a metabolic pathway albeit by coding for an enzyme, a transporter or a different protein playing an essential role in the particular metabolic pathway involved. This advantage is only valid when the genetic laboratory has not used a diagnostic gene panel but truly works in an "open the exome/genome mode." In practice, targeted metabolite assays will never enable us to measure all metabolites that act as a substrate or as a product of an enzymatic reaction. Some metabolites will never reach the blood circulation because they cannot pass the cellular membrane (for example, phosphorylated metabolites). Also, a vast array of metabolites will simply not be detected when using the broad range of assays which are available in the toolbox of the metabolic laboratory (see other chapters in this book). What then about untargeted metabolomics? Certainly, the number of diseases that can be diagnosed in a single patient with this technique clearly surpasses the number of diseases covered by any other test at the metabolite level. The technique works best when combining Liquid Chromatography (LC) with Mass Spectrometry (MS). Most laboratories use reversed-phase LC because of its stable performance between runs. But whatever chromatography is used there will always be a significant percentage of molecular species that will elute from the column in—or close by—the void volume leading to ion suppression and loss of sensitivity. Other column types, like Hilic, perform better in separating the more polar metabolites. Other approaches are required when diagnosing the growing group of inborn errors of lipid metabolism because of their poor solubility in water. To that end, untargeted lipidomics techniques might well fill that gap. Also, metabolic diseases concerning posttranslational modifications in proteins, like glycosylation, ubiquitination, sumoylation, and/or phosphorylation require yet other dedicated approaches including proteomics. In conclusion, there is no single analytical technique at the small molecule level that will be able to identify the vast majority of metabolites involved in inborn errors of metabolism. We will not discuss the rapidly growing areas of untargeted lipidomics, glycomics, and proteomics in the diagnostic laboratory in the rest of this chapter simply because these techniques are still very much in development. In the following paragraphs, we will showcase examples of indications for targeted and untargeted metabolomics for the metabolic laboratory now or in the near future. Most of the examples are on untargeted metabolomics and use LC-Qtof Mass Spectrometry as an analytical technique. This approach has been coined as Next Generation Metabolic Screening (NGMS; [11]) and the methodology has been described by Coene et al. [12]. Untargeted metabolomics assays have the characteristic of being hypothesis generating and are in that respect more powerful both in research and in patient care than the conventional techniques that set out to confirm or reject a hypothesis [13].

A. Newborn Screening

Targeted metabolite testing plays a major role in the various Newborn Screening (NBS) programs used worldwide for most of the diseases in the screening panel. The number of IEMs included in NBS programs varies widely even between different states in the USA. Unfortunately, neonatal screening is not perfect and is inevitably associated with the phenomenon of false positives which requires additional investigations either at the level of metabolites and/or enzymatic analyses to resolve whether the neonate is a true-positive or not. At present, first-tier DNA testing is considered in several countries. Accepting larger-scale DNA testing in NBS programs holds the promise of being able to add dozens of treatable diseases in one go for which currently no metabolite test is available. One of the major challenges of this DNA-first approach will be the interpretation of Variants of Unknown Significance (VUS). Most likely such programs will start by reporting only cases with loss of function mutations and known pathogenic mutations. But also diseases where a second-tier metabolite test is or can be made available could be included. In such cases, the result of the metabolite assay can decide whether a VUS is or is not disease causing. Whether or not DNA-first testing in NBS will become a reality, it will always be relevant to confirm the diagnosis at the metabolite level where possible. Furthermore, the follow-up of the neonate after therapy has started will always include reliable and quantitative metabolite and/or enzyme testing.

B. Acute Clinical Situations

As metabolite assays are faster than DNA-based assays targeted or untargeted assays are preferred in acute situations where a metabolic disease is considered. Examples are acute decompensations of organic acidurias, urea cycle defects, or aminoacidopathies in children but also in adult comatic patients due to OTC deficiency.

C. Selective Screening, Follow-Up, and Therapy Control

Over the next few years requests for selective screening will gradually diminish in countries where WES/WGS testing is in place thus endangering the relevance—and maybe even the existence—of dedicated metabolic laboratories that have confined themselves to targeted metabolite assays only. For sure, metabolite testing will remain relevant in the follow-up of diagnosed patients and as well monitor therapy strategies and patient compliance.

D. WES/WGS and Metabolomics in Concert
In WES and WGS techniques, the occurrence of VUSses is a major factor that contributes to diagnoses being missed. A genetic laboratory may decide not to report a class 3 variant or may not be able to decide with certainty about the pathogenicity of a class 4 variant. If this occurs in a "metabolic gene" metabolomics assays can provide functional information and help to decide whether the variant is disease causing or not. The other way around can also occur. When a metabolite is found in abnormally high or abnormally low concentrations in a targeted or untargeted assay this finding may point the way to the mutated gene. An example of the latter situation is the story that led to the discovery of NANS deficiency also referred to as NANS-CDG (see below). Of course, these situations only occur when a gene coding for an enzyme or transporter is involved. Taken together, ideally it would make sense to perform WES/WGS and untargeted metabolomics simultaneously. This should be followed up by the joint interpretation of the results by a biochemical geneticist and a molecular geneticist so that the data of WES/WGS analyses can be combined with the (un)targeted metabolomics data and preferably reported in a single, joint report to the clinician.

1.3 Ways of Applying Untargeted Metabolomics

1.3.1 Diagnosis in a Single Sample of One Patient

The first question that comes up is whether or not untargeted metabolomics can play a role in the diagnostics of a single patient. The 2018 paper of Coene et al. in the special edition of JIMD on metabolomics revealed that the answer to this question unequivocally is "yes!" [12]. Ford et al. have used a metabolomics approach with multiple LC-MS/MS methods and showed this approach as precise and reproducible in measuring biomarkers diagnostic for IEMs [14]. The paper by Liu et al. showed that the diagnostic yield of untargeted metabolomics is approximately sixfold higher compared with the conventional screening techniques used in metabolic laboratories [15]. The paper suggests that broader screening approaches should be considered in the initial evaluation for metabolic disorders. Lists of IEMs that could be detected using untargeted techniques have been provided by Coene and by Odom and coworkers [12, 13]. The output of untargeted metabolomics profiles contains 15–20.000 signals per plasma sample and the majority of these have till now, unfortunately, remained "unknowns." Estimates of the size of this dark matter in the metabolome vary considerably [16]. Such "unknowns" may either come from endogenous human metabolism, may derive from medication, from dietary sources or from contributions of the gut microbiome. They may thus not be relevant for the diagnosis of the patient but they surely cannot be excluded because a specific "unknown" may be the relevant biomarker that will lead to the diagnosis. Although databases like the Human Metabolome Database (HMDB) and KEGG may help to identify an "unknown" it seems easy to get confused at a higher level. So in the case of a single sample from one patient, it does not make sense to spend much time on signals that are significantly higher or lower than in control samples. What then can be done in such a situation? If the laboratory involved has composed a panel of metabolites that connect to specific metabolic diseases and knows their mass-to-charge ratio (m/z value) and Retention time (Rt) an in-house bioinformatics pipeline will allow inspection of these metabolites in the untargeted metabolomics output file. By comparing these data to a set of age-matched control samples this approach will lead the way to the diagnosis. Figure 1.1 gives an example of this: panels 1 and 2 show the increased plasma levels of xanthine and the very low concentration of plasma uric acid in a patient with xanthinuria. Panel 3 in the figure shows the high citrulline concentration in a patient with citrullinemia. You might say that by working with an in-house disease-related metabolite panel you analyze your untargeted data in a targeted way. The paper of Coene et al. 2018 showed that this worked very well in practice. Indeed, 42 of the 45 diseases

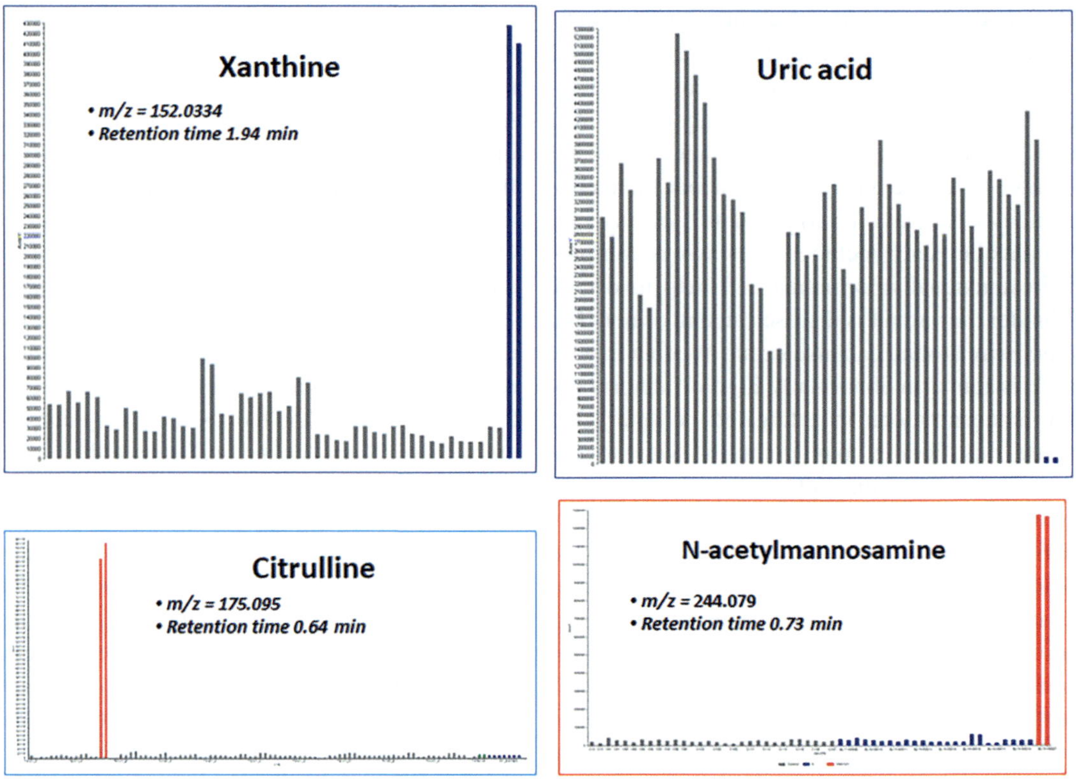

Fig. 1.1 Untargeted metabolomics in plasma in a single patient: All patients and controls in the graphs are in duplicate. Panels 1 and 2 show the high xanthine and low uric acid in this patient leading to the diagnosis of Xanthinuria. In blue the patient sample and in gray the controls. Panel 3 shows in red the high citrulline concentration (chloride adduct) in a patient diagnosed with citrullinemia; controls in gray and blue. Also in red panel 4 shows the high plasma concentration of N-acetylmannosamine (sodium adduct) in a patient; controls in gray and in blue. The feature for N-acetylmannosamine was found via an open metabolome approach in the untargeted metabolomics data. These findings led to the discovery of N-acetylmannosamine synthase deficiency in a novel IEM now also known as NANS-CDG

that were in the panel of the Nijmegen translational metabolic laboratory at that time, were correctly identified using this strategy. The number of metabolites and diseases in an untargeted metabolomics panel will grow over time and thus gain impact.

It is important to emphasize that actual experience with the diagnosis of an individual disease is required before the disease can be added to the diagnostic panel in an accredited laboratory setting. In a research setting there are other options. The laboratory can run relevant metabolites that are commercially available to document their m/z and Rt characteristics and to establish reference ranges. These can even be metabolites that have never been associated with a metabolic disease in literature before. In that way, the laboratory can develop a list of biomarkers that may be useful in diagnosing metabolic diseases that have never been identified before in the formal patient care setting of that particular laboratory. This can be framed as an "open the metabolome" approach.

1.3.2 Diagnosis in Subsequent Samples of a Single Patient

It may be worthwhile to analyze two or three samples of the same patient over a certain time interval of weeks or months. Normally that will exclude the diet as a source of the "unknown." Assuming the clinician has provided information on the medication used by this patient it can occur that in all samples the same feature is flagged as abnormal. Of course, this may well be an "unknown". In that situation, the sample will likely enter the research area of the metabolic laboratory for further identification. At one time one of us (RAW) was involved in a case of two patients in one family with a very serious neurological disease and feeding abnormalities. The sibs did not use any medication. Our untargeted metabolomics assay showed a consistent abnormal signal in several body fluids of both children on three different days. In the end, after a lot of work, this turned out to be a signal from acesulfame-potassium, an artificial no-calorie sweetener used in foods and beverages. The explanation was that both children drank Diet Coke every day. This example shows you to be cautious with this approach but it also shows that the approach may work when the unusual signal would have been caused by a metabolic disease.

1.3.3 Finding Biomarkers in a Known Syndrome

A first example showcasing how untargeted plasma metabolomics can identify relevant metabolic perturbations in a syndrome in monozygotic twins diagnosed with Snyder-Robinson syndrome was published by Abela et al. [17]. This extremely rare X-chromosomal syndrome is clinically characterized by intellectual disability, hypotonia, epilepsy, and gait abnormalities and genetically by mutations in the *SMS* gene. The gene encodes the enzyme spermine synthase that converts spermidine into spermine in polyamine metabolism. Statistical analysis of the metabolomics dataset with the plasma samples of the twins and 19 controls identified several metabolites in polyamine metabolism with abnormal concentrations. These were four metabolites upstream of the metabolic defect: spermidine, N-methylputrescine, isoputreanine, and N8-acetylspermidine. The latter two derive from accumulating spermidine. The four metabolites clearly illustrate the disturbance in spermine synthesis. When used as biomarkers for this disease they provide functional evidence for pathogenicity of variants in the *SMS* gene. The example shows that already a small number of patients having the same disease allows the identification of relevant metabolic abnormalities reflecting a functional disturbance in metabolism. Likewise, Abela et al. were the first to define the characteristic metabolite fingerprint in mitochondrial aconitase-2 deficiency [18]. In plasma they found significantly decreased concentrations of the downstream metabolites cis-aconitate and isocitrate and to a lesser extent alpha-ketoglutarate.

1.3.4 Finding Novel Biomarkers in Well-Known Inborn Errors of Metabolism

Classical metabolite screening largely relies on finding the accumulation of the direct substrate of an enzyme—or cargo of a transporter—and/or the decrease in concentration of the enzyme's product. The actual metabolic effect of an enzyme deficiency, however, may be much broader for instance through alternative metabolism of the accumulating substrate. This may give rise to unwanted and sometimes toxic side products that are clinically relevant. But also more distant effects of the deficiency may be observed in seemingly unrelated pathways. From 21 known metabolic diseases that we have investigated using untargeted LC-Qtof metabolomics analysis on samples from two or more patients per disease we observed so far unknown features representing novel biomarkers in 19 of these 21 IEMs. The findings on PKU described by Václavík and by van Outersterp et al. may serve as an example [19, 20]. Apart from phenylalanine itself, abnormal concentrations were found for six other biomarkers: N-lactoyl-Phe, Glu-Phe, Glu-Glu-Phe, N-acetyl-Phe, Phe-Phe, and Phe-glucose. Even in the most intensively studied inborn error which PKU is, it thus turned out possible to discover novel biomarkers. Further investigations on Phe-glucose revealed that this metabolite was formed non-enzymatically in the circulation of the PKU patients. The sum of these PKU-metabolites theoretically reflects the Phenylalanine status better than the Phenylalanine concentration alone. It remains to be established whether the total Phe-load of the patient or any one of the individual Phe-metabolites will turn out to be a better parameter in the follow-up of PKU patients than the current clinical monitoring based on the phenylalanine concentration.

1.3.5 Identifying Novel Metabolite Classes

The PKU example (point 4) showed that an amino acid can react non-enzymatically with glucose and perhaps as well with other sugars. The paper by van Outersterp et al. showed that these so-called Amadori rearrangement products (so-called ARPs) also occur in other aminoacidopathies like hyperlysinemia type I, methionine adenosyltransferase deficiency I/III, citrullinemia type I, cystathionine beta synthase deficiency and hyperprolinemia II and most likely as well in others [20]. It is expected that also amino acids that accumulate in various other amino acid-related diseases will form ARPs. The example shows how the untargeted metabolome pointed the way toward this as yet unknown group of metabolites.

1.3.6 Defining a Novel Metabolic Disease

A first illustration clarifying that the combined use of untargeted metabolomics with whole exome sequencing in a single patient could lead the way to defining a novel inborn error is the discovery of NANS deficiency, a defect in the CMP-sialic acid biosynthetic pathway. WES analysis in the index patient had been inconclusive. There were no causative mutations in genes associated with known diseases. Instead, the exome of this patient contained biallelic mutations in 19 candidate genes but none of these could be prioritized on the basis of mutational severity scores. Body fluid samples of a patient with this disease showed a significantly increased concentration of an N-acetyl sugar that turned out to be N-acetyl-D-mannosamine, an intermediate of the CMP-sialic acid biosynthetic pathway (Fig. 1.1 panel 4). This matched with the biallelic mutations found in one of the 19 candidate genes, i.e., the *NANS*-gene coding for the enzyme N-acetylneuraminic acid synthase which catalyzes an intermediate step in the CMP-sialic acid biosynthesis pathway. Finding other patients with mutations in the gene further defined the clinical phenotype of this novel metabolic disease that more recently was classified as a CDG subform NANS-CDG [11, 21].

1.3.7 Unraveling Novel Pathomechanisms

Peters et al. set out to apply untargeted metabolomics on CSF samples and focused among others on the glucose transporter defect, caused by mutations in the *GLUT1* gene [22]. The disease is considered to cause an energy deficit in the brain because of the local glucose shortage. The institution of a ketogenic diet has improved the clinical outcome of this disease considerably but still there seemed to be residual disease. The untargeted CSF metabolomics data showed, apart from the low glucose and lactate, unknown features at a lower-than-normal concentration. These features could be identified as the two saccharides Glc-Xyl and Glc-Xyl-Xyl. Existing literature mentioned these glycans as O-glycans on many proteins containing evolutionary conserved epidermal growth factor-like domains (EGF-repeats; see [22]). The published data exemplify that the glucose transporter defect not only results in an energy problem for the brain but that the disease in fact also is an O-glycosylation defect due to the shortage of available glucose. This should lead to reconsidering the therapeutic strategy for GLUT1 deficiency. Taken together, untargeted metabolomics, in this case in CSF, has led to the discovery of a novel pathomechanism involved in this disease.

1.3.8 Diseases in Search of Newborn Screening Biomarkers

The power of untargeted metabolomics to discover novel biomarkers can be applied when searching for a reliable biomarker suited for introduction in new-born screening (NBS) programs. The marker should be stable on the Guthrie cards, stable over time at room temperature, and preferably not give false positives. One of the treatable diseases that lacked such stable biomarkers was Pyridoxine Dependent Epilepsy (PDE), caused by mutations in the *ALDH7A1* gene. The known diagnostic biomarkers pipecolic acid, piperideine-6-carboxylic acid (P6C), and α-aminoadipic semialdehyde do not qualify because of insufficient stability and/or insufficient discriminative power. Untargeted metabolomics identified two novel metabolites that can be used in diagnosing the disease and are sufficiently stable for NBS purposes. These are 2-oxopropylpiperidine-2-carboxylic acid (2-OPP) and 6-oxopipecolic acid (6-oxoPIP; [23–25]). Both biomarkers are currently considered for introduction in NBS programs worldwide. The work in these papers further showed that 2-OPP derives from P6C and acetoacetic acid. This again, just as in the example of Amadori rearrangement products is another example of a non-enzymatic reaction occurring in blood. Such spontaneous reactions apparently occur more often. It seems important to have in mind or to actively investigate such processes in body fluids but as well as inside cells.

1.4 Challenges and Chances for the Field

1. *The dark metabolome.* In eight examples we have demonstrated the power of untargeted metabolomics both in patient care and in research. The issue remains that the vast majority of the features observed with untargeted metabolomics techniques have so far remained unidentified. The Toronto group of Montenegro-Burke has coined these unknown areas of the metabolome as the "dark metabolome" [26]. As yet we do not know the importance of these unknowns and cannot estimate their importance in the context of human health. An international collaboration involving clinicians, laboratory scientists, bioinformaticians as well as machine learning- and artificial intelligence techniques will be required to shed more light on this issue. Of course, database searches, MS/MS fragmentation, and proton-NMR spectroscopy can sometimes help to identify an "unknown." Gauglitz et al. have used a metadata-based source annotation strategy to enhance the interpretation of untargeted MS/MS metabolomics data [27]. Proton-NMR spectroscopy has its inherent drawback of limited sensitivity but still is a cornerstone of metabolite analysis and interpretation. If the concentration of an "unknown" is high enough (micromolar or higher) it certainly is worthwhile to use NMR because the spectra contain a wealth of structural information on a metabolite. Often, however, these strategies do not bring the solution. Therefore, novel tools and methods for metabolite identification need to be developed to advance the field.

A new kid on the block in this respect is infrared spectroscopy which is a factor 1000 more sensitive than NMR and also contains structural information. The FELIX facility in Nijmegen is a unique electron laser facility that enables to record an infrared spectrum of a molecule of your choice in the chamber of the mass spectrometer; FELIX Laboratory (ru.nl). Also, the molecule can be fragmented and IR-spectra of the fragments can be recorded thus contributing to further structure elucidation. FELIX is open for users worldwide and for the purposes of "unknowns" relevant to a specific metabolic disorder it is certainly worthwhile to apply infrared spectroscopy in attempts to resolve the molecular structure of the "unknown." In several of the eight examples in this chapter "unknowns" could be identified with the help of FELIX IR-spectroscopy technology. A tutorial on the technique has been published by Martens et al. [28, 29].

2. *The intracellular metabolome.* Over the last few decades, metabolic laboratories have mainly studied the body fluid metabolome. Plasma and urine are easy to obtain and considered non-invasive. At the same time important and useful information, hidden in patient's tissues or specific cell types, is ignored. Many metabolites will not be able to leave the cell as they cannot pass the membrane(s). Applying targeted and/or untargeted metabolomics to tissue biopsies and cell cultures will help in understanding the pathophysiology of individual metabolic diseases. Working with iPS cells brings the opportunity to study the metabolome of specific cell types relevant to individual diseases and as well to observe the impact of therapeutic interventions and strategies.
3. *MS imaging.* A promising recent development is a possibility of doing untargeted metabolomics on tissues or on cultured cells without any homogenization. The mass spectrometer then works as a sensitive molecular microscope. When working for instance with KO-mice tissues the data will hold spatial information and can for instance show the concentration of the metabolite of your choice in different areas of the brain. The technique can also be applied in an untargeted way. When comparing diseased tissue with control tissue the bioinformatics pipeline will provide the differentials between the two; for instance, illustrating that a metabolite with a specific m/z value is significantly different in the disease state compared to the control. MS imaging can be applied for lipids [30] as well for other small molecule classes. This technique brings a new dimension to the research in metabolic diseases.
4. *The lipidome.* Many metabolic diseases are known in lipid metabolism. Although the analytical techniques to evaluate the body fluid lipidome are available lipidomics is not used in routine metabolic screening. It is a challenge to further uncover this very important part of metabolism and to gain a more complete understanding of the role of the different lipid classes in health and disease.

1.5 Enzymology

The current toolbox available to the biochemical genetics laboratory also includes enzymatic analysis. In most cases, enzyme activities are measured following earlier metabolite and/or genetic testing but the reverse is also true as detailed below. Since the majority of enzymes are localized inside cells, enzymatic analysis is usually performed in blood cells (erythrocytes, lymphocytes, etc.), in tissue biopsies, and/or in cultured skin fibroblasts. In earlier years enzyme activities were usually measured using spectrophotometric and/or fluorometric assays but current knowledge holds that these assays often lack the sensitivity to quantify the precise extent of the enzyme deficiency. Resolution of this question is all the more important since the residual activity of an enzyme almost by definition determines the flux through the metabolic pathway in conjunction with other parameters including the metabolic network the enzyme is embedded in but also the load applied to the metabolic pathway involved as determined by dietary sources among others. Procedures for the precise quantification of residual enzyme activities have been described in literature which includes the use of mass spectrometry for the quantitative determination of the product of an enzyme reaction. In this respect, the work notably by Gelb and coworkers may be mentioned who developed MS/MS-based methods for the quantitative assessment of various lysosomal enzymes and showed that MS/MS-based assays for different mucopolysaccharidoses had a much larger analytical range when compared with fluorescence-based assays [31]. Furthermore, Lin et al. developed an LC-MS/MS-based assay for the quantitative analysis of alpha-glucosidase the enzyme deficient in Pompe disease which showed a 700-fold higher analytical range when compared to the classical, fluorometric assay for alpha-glucosidase [32]. Importantly, this newly developed assay allowed them to discriminate patients suffering from the infantile form of

Pompe disease from patients with the late-onset form [32] (for discussion [33]).

Apart from the issue described above which involves the methods used to measure enzyme activities, there are other concerns that make it difficult to interpret the results of enzyme activity measurements beyond the conclusion that an enzyme is deficient or not. One major concern in this respect is the fact that many enzymes are measured under so-called optimal conditions which may be far away from the conditions an enzyme faces in its proper physiological situation. Indeed, enzymes are usually measured at their optimal pH and not at their proper physiological pH which should be +/– 7.0 for cytosolic enzymes but +/ -8.0 for intramitochondrial enzymes. Secondly, enzyme assays should be performed in a medium mimicking that of the compartment the enzyme is located in (cytosol- and mitochondria-mimicking media). Third, enzymes should be measured using the enzyme's physiological substrate and not artificial substrates as is clear from the work of Gelb and coworkers on lysosomal enzymes described above. Fourth, ideally enzyme activities should not only be measured at saturating substrate concentrations as usually done but also at lower substrate concentrations. The reason is that enzymes rarely encounter substrates at saturating levels and are thus not running at their maximal speed (Vmax) but at reduced rates as dictated by the non-saturating concentrations of substrate. The consequence of this rationale is that it is not so much the Vmax of an enzyme that is important but its catalytic efficiency defined as Vmax/Km in which the Km is the Michaelis-Menten constant defined as the substrate concentration at which the enzyme operates at half-maximal speed. The superiority of Vmax/Km over Vmax as the best parameter to describe the kinetic properties of an enzyme is also very important if it is realized that it is not only the Vmax of an enzyme that may be changed as a consequence of a certain mutation but also the Km. This is especially true for mutations causing amino acid substitutions. The work done by Diez-Fernandez and Haeberle illustrates this point beautifully [34]. They studied a range of mutant Argininosuccinate Synthetase (ASS) enzymes and showed that for each of the mutant ASS proteins the Km values for aspartate and citrulline were much higher compared to the wildtype enzyme. As a consequence, the catalytic efficiency of the different mutant ASS enzymes was much lower when compared with the corresponding Vmax of the mutant enzymes with obviously major consequences for the flux through the urea cycle. It is important to realize this point despite the fact that in daily practice it may be difficult to run enzyme assays at different substrate concentrations.

Although it is important to measure true residual enzyme activities, it remains difficult to extrapolate these results directly to the in vivo situation of patients especially since the ultimate effect of an enzyme deficiency on pathway flux is also very much dependent on the context, that is the biological network surrounding the enzyme especially since enzymes rarely operate in isolation. To this end, flux studies have long been used to study the effects of an enzyme deficiency on pathway flux. This was first done using different radioactively labelled substrates including [14C]—and [3H]—labelled substrates which have now been surpassed by non-radioactive isotopes. Using [3H]-oleate as substrate, we have found that whole-cell (fibroblasts) [3H]-oleate oxidation flux correlates much better with the phenotypic outcome of VLCAD-deficient patients when compared with the results of VLCAD enzyme activity measurements [35]. Ideally, in vitro analysis of residual enzyme activities and flux analysis should be followed up by in vivo studies in patients using stable isotopes. The availability of an increasing number of differentially labelled isotopes now allows such studies to be performed [36, 37].

1.6 Conclusion

Whole exome- and whole genome sequencing have considerably changed and improved the diagnostics of IEMs thereby also vastly expanding the number of metabolic diseases (still counting). Simultaneously, the options for genetic therapies have gained an enormous impetus bringing new opportunities and challenges to the field. These innovations have changed the way we work and have provided answers to the "Quo vadis question" (Morava 2015). It is clear that biochemical genetics will remain an essential player required in the diagnostic process but also to improve our understanding of the biochemical basis of diseases and their pathomechanisms. The development of untargeted metabolomics techniques will assist in the diagnostic process, will shed light on the "dark matter" in the metabolome, and will result in meaningful novel biomarkers for therapy follow-up and for the design and development of novel therapeutic approaches. Similar innovations are to be expected for lipidomics, proteomics, and glycomics thus giving the biochemical genetics laboratories unprecedented relevance.

References

1. Garrod AE (1902) About Alkaptonuria. Med Chir Trans 85:69–78
2. Prasad C, Galbraith PA (2005) Sir Archibald Garrod and alkaptonuria–'story of metabolic genetics'. Clin Genet 68(3):199–203. https://doi.org/10.1111/j.1399-0004.2005.00487.x
3. Harper PS (2005) William Bateson, human genetics and medicine. Hum Genet 118(1):141–151. https://doi.org/10.1007/s00439-005-0010-3
4. Følling A (1934) Excretion of phenylpyruvic acid in urine as a metabolic anomaly in connection with imbecility. Nord Med Tidskr 8:1054–1059
5. Morava E, Rahman S, Peters V, Baumgartner MR, Patterson M, Zschocke J (2015) Quo vadis: the redefinition of "inborn metabolic diseases". J Inherit Metab Dis 38(6):1003–1006. https://doi.org/10.1007/s10545-015-9893-x
6. Ferreira CR, Rahman S, Keller M, Zschocke J, ICIMD Advisory Group (2021) An international classification of inherited metabolic disorders (ICIMD). J Inherit Metab Dis 44(1):164–177. https://doi.org/10.1002/jimd.12348
7. Saudubray JM, Mochel F, Lamari F, Garcia-Cazorla A (2019) Proposal for a simplified classification of IMD based on a pathophysiological approach: a practical guide for clinicians. J Inherit Metab Dis 42(4):706–727. https://doi.org/10.1002/jimd.12086
8. Blau N, Dionisi-Vici C, Ferreira CR, Vianey-Saban C, van Karnebeek CDM (eds) (2022) Foreword in physician's guide to the diagnosis, treatment, and follow-up of inherited metabolic diseases, vol 1, 2nd edn. Springer, pp vii–ix
9. Raas-Rothschild A, Wanders RJ, Mooijer PA, Gootjes J, Waterham HR, Gutman A, Suzuki Y, Shimozawa N, Kondo N, Eshel G, Espeel M, Roels F, Korman SH (2002) A PEX6-defective peroxisomal biogenesis disorder with severe phenotype in an infant, versus mild phenotype resembling usher syndrome in the affected parents. Am J Hum Genet 70(4):1062–1068. https://doi.org/10.1086/339766
10. Ratbi I, Falkenberg KD, Sommen M, Al-Sheqaih N, Guaoua S, Vandeweyer G, Urquhart JE, Chandler KE, Williams SG, Roberts NA, El Alloussi M, Black GC, Ferdinandusse S, Ramdi H, Heimler A, Fryer A, Lynch SA, Cooper N, Ong KR, Smith CE, Inglehearn CF, Mighell AJ, Elcock C, Poulter JA, Tischkowitz M, Davies SJ, Sefiani A, Mironov AA, Newman WG, Waterham HR, Van Camp G (2015) Heimler syndrome is caused by Hypomorphic mutations in the peroxisome-biogenesis genes PEX1 and PEX6. Am J Hum Genet 97(4):535–545. https://doi.org/10.1016/j.ajhg.2015.08.011
11. van Karnebeek CD, Bonafé L, Wen XY, Tarailo-Graovac M, Balzano S, Royer-Bertrand B, Ashikov A, Garavelli L, Mammi I, Turolla L, Breen C, Donnai D, Cormier-Daire V, Heron D, Nishimura G, Uchikawa S, Campos-Xavier B, Rossi A, Hennet T, Brand-Arzamendi K, Rozmus J, Harshman K, Stevenson BJ, Girardi E, Superti-Furga G, Dewan T, Collingridge A, Halparin J, Ross CJ, Van Allen MI, Rossi A, Engelke UF, Kluijtmans LA, van der Heeft E, Renkema H, de Brouwer A, Huijben K, Zijlstra F, Heise T, Boltje T, Wasserman WW, Rivolta C, Unger S, Lefeber DJ, Wevers RA, Superti-Furga A (2016) NANS-mediated synthesis of sialic acid is required for brain and skeletal development. Nat Genet 48(7):777–784. https://doi.org/10.1038/ng.3578. Erratum in: 2017;49(6):969
12. Coene KLM, Kluijtmans LAJ, van der Heeft E, Engelke UFH, de Boer S, Hoegen B, Kwast HJT, van de Vorst M, Huigen MCDG, Keularts IMLW, Schreuder MF, van Karnebeek CDM, Wortmann SB, de Vries MC, Janssen MCH, Gilissen C, Engel J, Wevers RA (2018) Next-generation metabolic screening: targeted and untargeted metabolomics for the diagnosis of inborn errors of metabolism in individual patients. J Inherit Metab Dis 41(3):337–353. https://doi.org/10.1007/s10545-017-0131-6
13. Odom JD, Sutton VR (2021) Metabolomics in clinical practice: improving diagnosis and informing management. Clin Chem 67(12):1606–1617. https://doi.org/10.1093/clinchem/hvab184
14. Ford L, Kennedy AD, Goodman KD, Pappan KL, Evans AM, Miller LAD, Wulff JE, Wiggs BR, Lennon JJ, Elsea S, Toal DR (2020) Precision of a clinical metabolomics profiling platform for use in the identification of inborn errors of metabolism. J Appl Lab Med 5(2):342–356. https://doi.org/10.1093/jalm/jfz026
15. Liu N, Xiao J, Gijavanekar C, Pappan KL, Glinton KE, Shayota BJ, Kennedy AD, Sun Q, Sutton VR, Elsea SH (2021) Comparison of untargeted Metabolomic profiling vs traditional metabolic screening to

identify inborn errors of metabolism. JAMA Netw Open 4(7):e2114155. https://doi.org/10.1001/jamanetworkopen.2021.14155
16. Jones OAH (2018) Illuminating the dark metabolome to advance the molecular characterisation of biological systems. Metabolomics 14(8):101. https://doi.org/10.1007/s11306-018-1396-y
17. Abela L, Spiegel R, Crowther LM, Klein A, Steindl K, Papuc SM, Joset P, Zehavi Y, Rauch A, Plecko B, Simmons TL (2017) Plasma metabolomics reveals a diagnostic metabolic fingerprint for mitochondrial aconitase (ACO2) deficiency. PLoS One 12(5):e0176363. https://doi.org/10.1371/journal.pone.0176363
18. Abela L, Simmons L, Steindl K, Schmitt B, Mastrangelo M, Joset P, Papuc M, Sticht H, Baumer A, Crowther LM, Mathis D, Rauch A, Plecko B (2016) N(8)-acetylspermidine as a potential plasma biomarker for Snyder-Robinson syndrome identified by clinical metabolomics. J Inherit Metab Dis 39(1):131–137. https://doi.org/10.1007/s10545-015-9876-y
19. Václavík J, Coene KLM, Vrobel I, Najdekr L, Friedecký D, Karlíková R, Mádrová L, Petsalo A, Engelke UFH, van Wegberg A, Kluijtmans LAJ, Adam T, Wevers RA (2018) Structural elucidation of novel biomarkers of known metabolic disorders based on multistage fragmentation mass spectra. J Inherit Metab Dis 41(3):407–414. https://doi.org/10.1007/s10545-017-0109-4
20. van Outersterp RE, Moons SJ, Engelke UFH, Bentlage H, Peters TMA, van Rooij A, Huigen MCDG, de Boer S, van der Heeft E, Kluijtmans LAJ, van Karnebeek CDM, Wevers RA, Berden G, Oomens J, Boltje TJ, Coene KLM, Martens J (2021) Amadori rearrangement products as potential biomarkers for inborn errors of amino-acid metabolism. Commun Biol 4(1):367. https://doi.org/10.1038/s42003-021-01909-5
21. den Hollander B, Rasing A, Post MA, Klein WM, Oud MM, Brands MM, de Boer L, Engelke UFH, van Essen P, Fuchs SA, Haaxma CA, Jensson BO, Kluijtmans LAJ, Lengyel A, Lichtenbelt KD, Østergaard E, Peters G, Salvarinova R, Simon MEH, Stefansson K, Thorarensen Ó, Ulmen U, Coene KLM, Willemsen MA, Lefeber DJ, van Karnebeek CDM (2021) NANS-CDG: delineation of the genetic, biochemical, and clinical Spectrum. Front Neurol 12:668640. https://doi.org/10.3389/fneur.2021.668640
22. Peters TMA, Engelke UFH, de Boer S, van der Heeft E, Pritsch C, Kulkarni P, Wevers RA, Willemsen MAAP, Verbeek MM, Coene KLM (2020) Confirmation of neurometabolic diagnoses using age-dependent cerebrospinal fluid metabolomic profiles. J Inherit Metab Dis 43(5):1112–1120. https://doi.org/10.1002/jimd.12253
23. Wempe MF, Kumar A, Kumar V, Choi YJ, Swanson MA, Friederich MW, Hyland K, Yue WW, Van Hove JLK, Coughlin CR 2nd (2019) Identification of a novel biomarker for pyridoxine-dependent epilepsy: implications for newborn screening. J Inherit Metab Dis 42(3):565–574. https://doi.org/10.1002/jimd.12059. Epub 2019 Mar 11
24. Engelke UF, van Outersterp RE, Merx J, van Geenen FA, van Rooij A, Berden G, Huigen MC, Kluijtmans LA, Peters TM, Al-Shekaili HH, Leavitt BR, de Vrieze E, Broekman S, van Wijk E, Tseng LA, Kulkarni P, Rutjes FP, Mecinović J, Struys EA, Jansen LA, Gospe SM Jr, Mercimek-Andrews S, Hyland K, Willemsen MA, Bok LA, van Karnebeek CD, Wevers RA, Boltje TJ, Oomens J, Martens J, Coene KL (2021) Untargeted metabolomics and infrared ion spectroscopy identify biomarkers for pyridoxine-dependent epilepsy. J Clin Invest 131(15):e148272. https://doi.org/10.1172/JCI148272
25. van Outersterp RE, Engelke UFH, Merx J, Berden G, Paul M, Thomulka T, Berkessel A, Huigen MCDG, Kluijtmans LAJ, Mecinović J, Rutjes FPJT, van Karnebeek CDM, Wevers RA, Boltje TJ, Coene KLM, Martens J, Oomens J (2021) Metabolite identification using infrared ion spectroscopy–novel biomarkers for pyridoxine-dependent epilepsy. Anal Chem 93(46):15340–15348. https://doi.org/10.1021/acs.analchem.1c02896
26. Crow JM (2021) A growing body of work. Nature 599:S14–S15
27. Gauglitz JM, West KA, Bittremieux W, Williams CL, Weldon KC, Panitchpakdi M, Di Ottavio F, Aceves CM, Brown E, Sikora NC, Jarmusch AK, Martino C, Tripathi A, Meehan MJ, Dorrestein K, Shaffer JP, Coras R, Vargas F, Goldasich LD, Schwartz T, Bryant M, Humphrey G, Johnson AJ, Spengler K, Belda-Ferre P, Diaz E, McDonald D, Zhu Q, Elijah EO, Wang M, Marotz C, Sprecher KE, Vargas-Robles D, Withrow D, Ackermann G, Herrera L, Bradford BJ, Marques LMM, Amaral JG, Silva RM, Veras FP, Cunha TM, Oliveira RDR, Louzada-Junior P, Mills RH, Piotrowski PK, Servetas SL, Da Silva SM, Jones CM, Lin NJ, Lippa KA, Jackson SA, Daouk RK, Galasko D, Dulai PS, Kalashnikova TI, Wittenberg C, Terkeltaub R, Doty MM, Kim JH, Rhee KE, Beauchamp-Walters J, Wright KP Jr, Dominguez-Bello MG, Manary M, Oliveira MF, Boland BS, Lopes NP, Guma M, Swafford AD, Dutton RJ, Knight R, Dorrestein PC (2022) Enhancing untargeted metabolomics using metadata-based source annotation. Nat Biotechnol 40(12):1774–1779. https://doi.org/10.1038/s41587-022-01368-1
28. Martens J, Berden G, Bentlage H, Coene KLM, Engelke UF, Wishart D, van Scherpenzeel M, Kluijtmans LAJ, Wevers RA, Oomens J (2018) Unraveling the unknown areas of the human metabolome: the role of infrared ion spectroscopy. J Inherit Metab Dis 41(3):367–377. https://doi.org/10.1007/s10545-018-0161-8
29. Martens J, van Outersterp RE, Vreeken RJ, Cuyckens F, Coene KLM, Engelke U, Kluijtmans LAJ, Wevers RA, Buydens LMC, Redlich B, Berden G, Oomens J (2020) Infrared ion spectroscopy: new opportunities for small-molecule identification in mass

spectrometry–a tutorial perspective. Anal Chim Acta 1093:1–15. https://doi.org/10.1016/j.aca.2019.10.043
30. Tortorella S, Tiberi P, Bowman AP, Claes BSR, Ščupáková K, Heeren RMA, Ellis SR, Cruciani G (2020) LipostarMSI: comprehensive, vendor-neutral software for visualization, data analysis, and automated molecular identification in mass spectrometry imaging. J Am Soc Mass Spectrom 31(1):155–163. https://doi.org/10.1021/jasms.9b00034
31. Kumar AB, Masi S, Ghomashchi F, Chennamaneni NK, Ito M, Scott CR, Turecek F, Gelb MH, Spacil Z (2015) Tandem mass spectrometry has a larger analytical range than fluorescence assays of lysosomal enzymes: application to newborn screening and diagnosis of Mucopolysaccharidoses types II, IVA, and VI. Clin Chem 61(11):1363–1371. https://doi.org/10.1373/clinchem.2015.242560
32. Lin N, Huang J, Violante S, Orsini JJ, Caggana M, Hughes EE, Stevens C, DiAntonio L, Chieh Liao H, Hong X, Ghomashchi F, Babu Kumar A, Zhou H, Kornreich R, Wasserstein M, Gelb MH, Yu C (2017) Liquid chromatography-tandem mass spectrometry assay of leukocyte acid α-glucosidase for Post-newborn screening evaluation of Pompe disease. Clin Chem 63(4):842–851. https://doi.org/10.1373/clinchem.2016.259036
33. Gelb MH (2018) Newborn screening for lysosomal storage diseases: methodologies, screen positive rates, normalization of datasets, second-tier tests, and Post-analysis tools. Int J Neonatal Screen 4(3):23. https://doi.org/10.3390/ijns4030023
34. Diez-Fernandez C, Wellauer O, Gemperle C, Rüfenacht V, Fingerhut R, Häberle J (2016) Kinetic mutations in argininosuccinate synthetase deficiency: characterisation and in vitro correction by substrate supplementation. J Med Genet 53(10):710–719. https://doi.org/10.1136/jmedgenet-2016-103937
35. Diekman EF, Ferdinandusse S, van der Pol L, Waterham HR, Ruiter JP, Ijlst L, Wanders RJ, Houten SM, Wijburg FA, Blank AC, Asselbergs FW, Houtkooper RH, Visser G (2015) Fatty acid oxidation flux predicts the clinical severity of VLCAD deficiency. Genet Med 17(12):989–994. https://doi.org/10.1038/gim.2015.22
36. Wanders RJA, Vaz FM, Ferdinandusse S, van Kuilenburg ABP, Kemp S, van Karnebeek CD, Waterham HR, Houtkooper RH (2019) Translational metabolism: a multidisciplinary approach towards precision diagnosis of inborn errors of metabolism in the omics era. J Inherit Metab Dis 42(2):197–208. https://doi.org/10.1002/jimd.12008
37. Reijngoud DJ (2018) Flux analysis of inborn errors of metabolism. J Inherit Metab Dis 41(3):309–328. https://doi.org/10.1007/s10545-017-0124-5

2 Quality Control and Quality Assurance in Biochemical Genetics Laboratory Testing

George J. G. Ruijter and Christine Vianey-Saban

Contents

Abbreviations

EQA	External quality assurance
ERNDIM	European Research Network for the evaluation and improvement of screening, Diagnosis and treatment of Inherited disorders of Metabolism
IMD	Inherited metabolic disorders
IQC	Internal quality control

G. J. G. Ruijter (✉)
(on behalf of the ERNDIM Scientific Advisory Board) Department of Clinical Genetics, Erasmus University Medical Center, Rotterdam, The Netherlands
e-mail: g.ruijter@erasmusmc.nl

C. Vianey-Saban
(on behalf of the ERNDIM Scientific Advisory Board) Division of Inborn Errors of Metabolism, Department of Biochemistry and Molecular Biology, CHU de Lyon, Bron, France

2.1 Introduction

Diagnostic testing is one of the cornerstones in medical decision-making and, hence, high-quality laboratory testing is critical for optimal patient care, i.e. accurate diagnosis and proper therapy. Inaccurate diagnostic results may have serious consequences for patient care, i.e. delay in correct diagnosis, unnecessary additional diagnostic testing, failure to provide the right treatment, or application of unnecessary or even harmful treatment. To achieve high-quality laboratory testing, implemen-

N. Blau, F. M. Vaz (eds.), *Laboratory Guide to the Methods in Biochemical Genetics*,
https://doi.org/10.1007/978-3-031-58819-8_2

tation of quality control is key. Ultimately, accreditation provides a formal acknowledgement of laboratory competence in providing a diagnostic service and reassures physicians that high-quality results will be returned in a timely manner upon ordering a test. The way accreditation is organised differs from country to country, but for clinical laboratories commonly the ISO 15189:2012 standard is applied. In December 2022, an updated version (4th) of the ISO 15189 standard was published. In 2022, 336 of the 402 laboratories (84%) that participated in External Quality Assurance schemes provided by ERNDIM (European Research Network for the evaluation and improvement of screening, Diagnosis, and treatment of Inherited disorders of Metabolism) were accredited to one or more standards, with the majority, 213 laboratories, having ISO 15189 accreditation (source: ERNDIM registration records). In complying with ISO 15189, a laboratory must adopt a quality management system, which involves the whole of activities to safeguard reliable test results. A quality management system includes both managerial and technical aspects of the laboratory procedures, such as (1) optimal resources (e.g. buildings, equipment, and document control), (2) competent personnel, (3) establishment of standardised operating methods (SOPs) ensuring robust laboratory processes and validated tests, and (4) assessments to investigate performance and compliance aiming at continuous improvement. The latter includes Internal Quality Control (IQC) and External Quality Assurance (EQA) which are used to detect false test results, with the intention to correct and prevent them. Although IQC and EQA share an overall objective, major differences exist (Table 2.1). In this chapter, we will discuss how IQC and EQA can be used to assess and optimise precision, accuracy, and correct interpretation of test results in biochemical genetics testing.

2.2 Quality Control in Biochemical Genetics Testing

The purposes of laboratory testing in biochemical genetics are usually either (1) to diagnose or exclude inherited metabolic disorders (IMD) or (2) to monitor therapy in patients who already have been diagnosed. In the first situation, the laboratory would typically perform screening for many different IMD by investigating a large number of different metabolites or metabolite groups using a variety of methods. The tests employed may produce quantitative or qualitative results as long as patients can be differentiated from normal controls. For many IMD, the level of diagnostic metabolites is grossly abnormal, reducing the need for high-precision test results. For example, classic phenylketonuria (PKU), alkaptonuria, or Hurler's disease do not call for high performance of the tests used to diagnose or screen for these disorders. Therapy monitoring, on the other hand, is targeted to one or a few metabolites relevant to a particular treatment and requires quantitative test results to guide the physician or dietician. Obviously, precision is much more important in therapy monitoring as results will be related to target metabolite concentration ranges that are much closer to nor-

Table 2.1 Characteristics of IQC vs EQA

	Internal quality control	External quality assurance
Results (values/ diagnoses)	Known	Unknown
Availability of results	Immediately after test completion	Later, when reports are received
Samples, concentrations	Normal, pathological	Multiple concentration levels or many different diagnoses
Frequency of testing	Every experimental batch	Periodically, e.g. every month or two circulations per year
Assessment	Comparison to previously established values in own laboratory	Comparison to other laboratories' (quantitative) results or diagnostic proficiency testing
Evaluation	Precision, reproducibility	Precision and accuracy

mal. Moreover, new test results obtained for a patient will be compared to previous ones and trends noticed will be used to decide on treatment. Thus, for treatment monitoring measurement uncertainty must be low compared to biological variation. Also, when a target concentration range from published guidelines is used to perform therapy monitoring, high accuracy is a requirement.

Biochemical genetic laboratories are faced with a few specific challenges that call for rigorous quality control, including both IQC and participation in relevant EQA programmes. First, IMD are individually rare, but more than a thousand different disorders are known [1]. Laboratories are challenged to reach and maintain sufficient experience to perform tests and interpret results for a substantial number of different IMD. Moreover, IMD are commonly heterogeneous in the biochemical abnormalities they show as a result of varying residual enzyme activity and dependence on diet and other external factors. Diagnostic Proficiency Testing (see below) is an important tool to educate laboratories on metabolite abnormalities of IMD. Secondly, to perform clinical screening for a large number of different IMD, often, "profiling" assays are used to examine functionally or structurally related compounds, e.g. amino acids, organic acids, and oligosaccharides. To some extent, these tests produce qualitative results, which are intrinsically more difficult to monitor by quality control procedures. Third, patients may only be investigated once for the possibility of an IMD. In case of a negative screening result, an IMD is excluded and may not be investigated further. Obviously, a false negative result would have dramatic consequences for the patients and families and must be avoided at all times. However, at the same time, laboratories need to avoid unnecessary repetition of investigations.

2.3 Internal Quality Control

The main objective of Internal Quality Control (IQC) is to ensure day-to-day stability of an analytical test and to warrant that patient results are reliable and can be released. IQC monitors precision/reproducibility of test results through the use of quality control samples analysed at a certain frequency. It is performed within the laboratory and results are available immediately at test completion, which can then be used for acceptance/rejection of test results (Table 2.1). A correct result typically validates proper functioning of all aspects of the analytical process, including sample handling, reagents, instruments, and data processing. Although the overall objective of IQC is clear, major differences exist in the characteristics of IQC for quantitative and qualitative tests and these will be described separately.

2.3.1 Quantitative Assays

Nowadays, many tests used by biochemical genetics laboratories produce (semi-)quantitative results owing to developments in mass spectrometry. Performing IQC for quantitative tests first requires that the method is able to generate reproducible results under standard conditions, such that test results are amenable to quality control using statistical analysis. For quantitative tests, the following aspects should be taken into consideration during implementation of IQC:

Samples

- The sample matrix should preferably be identical to the patient samples routinely analysed. When different types of sample matrices (urine, plasma etc.) are analysed, also different IQC samples representative of these matrices should be included.
- A steady source of IQC samples is best, as from time to time batches run out and must be replaced by new ones. IQC samples may be prepared from laboratory stocks or purchased. ERNDIM provides IQC samples in collaboration with MCA laboratory (Winterswijk, The Netherlands). The website https://erndimqa.nl/ includes pages to submit and monitor results of the IQC samples delivered by ERNDIM-MCA.

Metabolites and Concentrations

- The metabolites monitored must be stable during the time a batch of IQC samples is used as unstable metabolites will show a trend of

decreasing concentration not related to the test.
- Metabolite concentrations in untreated IMD patient samples are typically grossly abnormal and it is therefore recommended to include two IQC samples, one with metabolite levels in the physiological range and another one in the pathological range. Samples with metabolite concentrations in the pathological range may be produced by adding pure compounds to a suitable sample matrix, taking solubility, stability etc. into account.
- For tests quantifying one or only a few metabolites these need to be monitored in IQC. However, in screening tests with many metabolites being quantified, e.g. amino acids, acylcarnitines and organic acids, a selection of these metabolites may be checked in IQC in order to limit the workload spent on IQC. Metabolites that are critical for clinical reasons should be included in IQC.

Frequency of Analysis
- IQC samples must be included in each experimental run.
- Continuous analysis is not very common in biochemical genetics testing but may be in use with some tests in larger laboratories with high throughput, e.g. with ion exchange chromatography amino acid analysers. When this approach is used, IQC samples may be included after every 20–30 patient samples.

Acceptable Variation in Test Results
- Before starting validation of a test, laboratories should set criteria for acceptable variation in metabolite concentrations. The analytical methodology chosen must be able to meet these precision requirements.
- As described above, acceptable variation is related to expected clinically relevant differences. So, coming back to the example of diagnosing alkaptonuria, a relatively large variation in the result of homogentisic acid is acceptable, as the level in untreated patients is a few orders of magnitude higher than the reference population. On the other hand, deciding on adjustment of the diet in PKU patients does require high precision.
- For the above-mentioned reasons, it is impossible to state a value for acceptable variation that would be valid for any test. The maximum value of the day-to-day variation coefficient (CV; standard deviation/mean value) in a quantitative test result that would be acceptable is around 20%. For tests used to guide therapy acceptable CV should be closer to 10%.
- Another way to estimate acceptable analytical performance is to review EQA programmes, assuming that participants use one or a few robust methods, in which case interlaboratory CV approximates intra-laboratory day-to-day variation. In the 2022 ERNDIM Quantitative Amino Acids in Serum (QTAS) scheme the interlaboratory CV of phenylalanine at the level relevant to therapy monitoring, 351 μmol/l, was 6.3%.
- During validation of an analytical test, the actual day-to-day variation of a metabolite is established experimentally by analysing the IQC sample(s) in a number of different, typically at least 10, test runs. The results are then used to set the target concentration range for subsequent runs (see below "Interpretation of IQC Results").
- The variation coefficient is often also used to estimate measurement uncertainty.

Interpretation of IQC Results
- Results obtained for metabolite concentrations in IQC samples (see above "Acceptable Variation in Test Results") are typically analysed in plots showing the established mean, plus/minus 2–3 standard deviations (Fig. 2.1). These control charts were first described by Levey and Jennings [2] and can be implemented as simple paper graphs or in digital forms, using Excel® or specialised software packages. Consecutive test results are added to the chart, and this will (1) enable evaluation of the last result, and (2) will document the history of test results.
- To monitor the performance of a quantitative test, laboratories must define control rules: a set of conditions that must be met for a run to be accepted or rejected. Ideally, the set of control rules identifies true errors, but does

not result in a high "false rejection rate", which would obviously result in wasting time and effort in the laboratory. In other words, the set of rules defined for a particular test must serve a purpose and it is likely that different assays need different IQC rules. Many laboratories use combinations of Westgard rules [3] (multirule QC, see also https://www.westgard.com/mltirule.htm) to achieve optimal IQC.

- Commonly applied IQC rules are the 3SD rule and the 2x2SD rule (see Fig. 2.1), where a result in excess of 3 SD would lead to run rejection, whilst a result between 2 and 3 SD would be a trigger for closer inspection of the data. A single occasion would usually not result in rejection, but two consecutive results between 2 and 3 SD would, by applying the 2x2SD rule.
- Levey-Jennings charts will also reveal trends in test performance that may not be obvious when evaluating results by single QC rules.
- When an IQC sample is analysed multiple times in one experiment, additional rules might be applied. An example is run rejection when two within-run test results are +2SD and -2SD, respectively, since this would be indicative of a large random error.

2.3.2 Qualitative Assays

For some groups of disorders, qualitative screening tests are still used by the majority of biochemical genetics laboratories, such as urinary organic analysis, mucopolysaccharide electrophoresis, and oligosaccharide thin-layer chromatography (TLC). Reasons for qualitative tests may be lack of (commercially available) standards or lack of an appropriate quantitative test method, although with developments in mass spectrometry, virtually all compounds relevant to IMD can be assayed (semi-) quantitatively. Qualitative tests are more challenging to quality control as statistical analysis cannot be performed on "peaks" or "bands" and interpretation is, to some extent, subjective. This subjectivity is a serious concern and should be addressed by using well-defined test procedures and extensive training. Interpretative EQA schemes have a clear educational value in this regard. For qualitative tests, the following aspects should be taken into consideration to ensure high-quality results:

- Negative control samples from normal individuals and positive control samples from patients with established and relevant IMD diagnoses should be included in an experi-

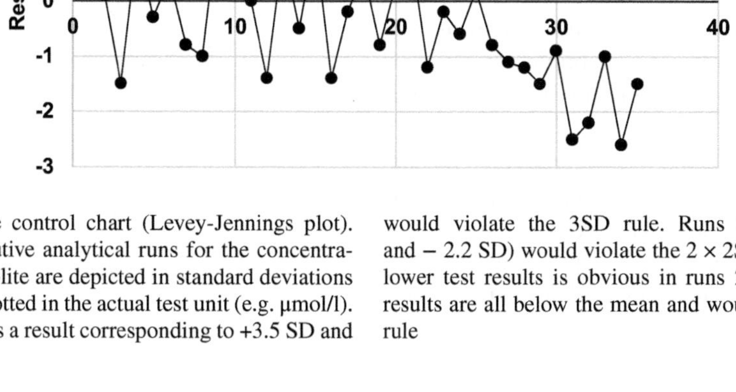

Fig. 2.1 Example control chart (Levey-Jennings plot). Results in consecutive analytical runs for the concentration of one metabolite are depicted in standard deviations but may also be plotted in the actual test unit (e.g. μmol/l). Run number 10 has a result corresponding to +3.5 SD and would violate the 3SD rule. Runs 31 and 32 (−2.5 and − 2.2 SD) would violate the 2 × 2SD rule. A trend to lower test results is obvious in runs 20–35; the last 10 results are all below the mean and would trigger the 10× rule

ment. The results should be as expected and reproducible in qualitative terms. Including these extra samples may lead to a considerably larger workload but ensures that each run is reliable.

- Positive controls may also help to interpret indistinct results in a sample of a new patient. Thus, retention of a chromatographic peak or the retention, shape, and colour of a TLC band may be compared directly between a patient sample and a positive control.
- To reduce interpretative subjectivity of qualitative assays, it is good practice to implement independent interpretation by at least two staff. Interpretations obviously should match and when differences in interpretation occur, results need to be closely looked at.
- Pictures or scans of test results should be stored as digital copies to document outcome of negative and positive controls in consecutive experiments. Similar to a control chart for quantitative tests, such a library serves as reference material to judge results of control samples in a newly performed experiment and will support the evaluation of test results during a longer time to notice trends. For example, staining of a TLC plate may become weaker due to degradation of reagents and eventually lead to a false-negative result. A library of images may also serve as a training resource.
- Organic acid analysis is an example of a test that will produce qualitative results for the majority of metabolites analysed, but semi-quantitative results may be produced for the organic acids that are included in a calibrator sample. These quantified organic acids can be assessed in quantitative quality control procedures as described above and may be used as quality indicators of metabolites that are not quantified. A minimal requirement in organic acid analysis is to include at least one non-naturally occurring compound (e.g. heptanoylglycine or 3,3-dimethylglutaric acid) as an "internal standard". The areas of the resulting peaks can be assessed in quantitative quality control procedures and used as quality indicators as described above.

2.4 External Quality Assurance

EQA compares performance between laboratories and provides information about the accuracy and bias of the results obtained. As opposed to IQC, which is organised within the laboratory and delivers results in "real time", EQA is provided by an external organisation and evaluation of results is received sometime after analysis (Table 2.1, Fig. 2.2). It has been documented that participation in EQA leads to improved quality of analytical and diagnostic performance in biochemical genetics testing [4]. The benefits of EQA include [5]:

1. Comparison of performance between laboratories.
2. Comparison of performance of different analytical methods to guide method selection.
3. Warning to systematic errors calling for improvement.
4. Detection of areas requiring education and training.
5. Providing evidence of quality for participating laboratories.

EQA schemes may address quantitative as well as qualitative/interpretative aspects. Quantitative EQA schemes require stable samples, with metabolite concentrations relevant to the clinical situation, distributed at a frequency supporting regular testing. Results of participating laboratories are

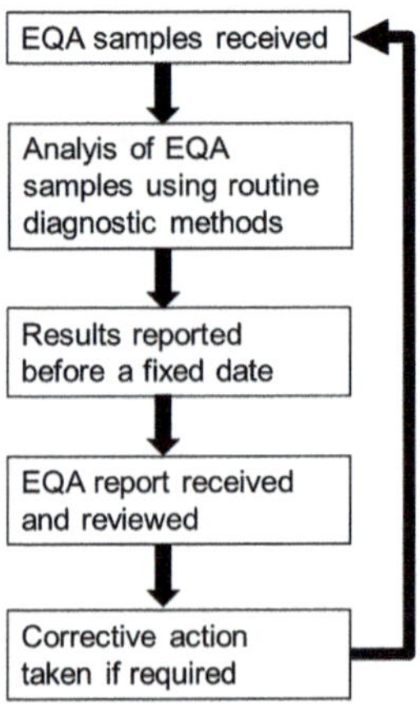

Fig. 2.2 EQA cycle

analysed by the EQA provider to reveal individual variation from a target or consensus value: bias, either across different analytical methods or in a method-specific manner. Bias may be expressed as a percentage of the target/consensus value or as z-score, which takes into account the variation across different laboratories (and methods). When different concentrations of a metabolite are used EQA may also inform on linearity, whereas duplicate samples inform on intra-laboratory precision. EQA scheme organisers must provide clear and useful reports with quantitative assessment of data and explanations of how data is managed. Qualitative EQA schemes often circulate authentic patient samples and require both laboratory testing and interpretation of results, e.g. provide a diagnosis. This type of EQA scheme is often termed Proficiency Testing (PT) and uses predefined scoring systems to assess proficiency. EQA reports for qualitative schemes should include summaries of results submitted and scores. An EQA provider must have a system to identify poor performers, to notify them and to offer support in order to improve competence.

In the field of IMD, the majority of EQA schemes are provided by ERNDIM, and will be detailed in the next section. CAP (College of American Pathologists), in collaboration with ACGM (American College of Medical Genetics), provides a proficiency testing EQA scheme called BGL [6], which consists of two shipments per year of a set of samples for qualitative or quantitative analysis of acylcarnitines (0.1 ml plasma), amino acids (1.0 ml plasma or 2 ml urine), glycosaminoglycans (2 ml urine), and organic acids (7.5 ml urine). CAP additionally provides quantitative EQA for free carnitine (BGL1) and one for amino acids (BGL2): two shipments per year of three 0.3 mL serum specimens and three 1.0 mL liquid specimens, respectively. These CAP EQA schemes are mainly intended for American laboratories. EQA schemes for porphyrins are provided for many years by different organisations: CSCQ (Centre Suisse de Contrôle de Qualité, Geneva, Switzerland), EPNet (European Porphyria Network, Bergen, Norway), RCPA (Royal College of Pathologists of Australasia, Surry Hills, Australia), SKML (Stichting Kwaliteistsbewaking Medische Laboratoriumdiagnostiek, Nijmegen, The Netherlands) and WEQAS (Welsh External Quality Assurance Scheme, Cardiff, UK).

2.5 ERNDIM: General Aspects

ERNDIM is an independent not-for-profit foundation which has been providing EQA schemes in the field of IMD since 1994. Further information is available at: https://www.erndim.org. Whilst originally a European organisation, in 2022 ERNDIM had over 400 participants from 62 countries, with approximately 40% non-European participants. In 2023, ERNDIM is operating 17 EQA schemes which are developed and monitored by the Scientific Advisory Board (SAB) comprising 22 leading scientists from 10 countries.

Four categories of schemes are provided: quantitative (7 schemes), qualitative (5 schemes), hybrid (4 schemes), and interpretative (1 scheme). The main purpose of the quantitative schemes is to evaluate the ability of the participating laboratories to quantitatively analyse the concentrations of the analytes included in each scheme. Qualitative schemes evaluate the ability of the laboratory to establish or exclude a specific diagnosis of an IMD. Hybrid schemes evaluate both quantitative analysis and interpretation of metabolite patterns to establish or exclude diagnosis. The qualitive schemes use clinical samples as the EQA materials, whereas the quantitative and the hybrid schemes (except lysosomal enzymes in fibroblasts) use human or artificial matrices spiked with commercially available analytes. The cognitive scheme "Amino acid interpretation" has been introduced in 2023. Unlike the other schemes, no physical material is circulated for this scheme. Clinical information and quantitative values for amino acids in plasma, urine, and/or CSF (with reference ranges) are provided, and participants should report a diagnosis.

The main characteristics of these schemes are given in Table 2.2. Samples are prepared and dispatched by the scheme organisers following specifications planned by the Scientific

Table 2.2 Main characteristics of ERNDIM EQA schemes

Scheme category	Scheme name	Scheme code	Number of organising centres	Specimen	Type of specimen	Sample preparation	Sample volume	Number of surveys per year	Number of samples per survey	Total number of samples per year	Starting year	Number of participants in 2022
Quantitative	Acylcarnitines in serum	ACS	1	Serum	Spiked[a]	Lyophilised	0.5 mL	8	1	8	2017	126
	Purines and pyrimidines	PPU	1	Urine	Spiked[a]	Lyophilised	2.5 mL	8	1	8	2000	56
	Quantitative amino acids	QTAS	1	Serum	Spiked[a]	Lyophilised	1.0 mL	8	1	8	1993	269
	Quantitative organic acids	QTOU	1	Urine	Spiked[a]	Lyophilised	5 mL	8	1	8	1993	127
	Special assays in dried blood spots	SADB	1	Blood on filter paper	Spiked[a]	Dried blood spots	75 µL	8	1	8	2019	104
	Special assays in serum	SAS	1	Serum	Spiked[a]	Lyophilised	5 mL	8	1	8	1993	237
	Special assays in urine	SAU	1	Urine	Spiked[a]	Lyophilised	5 mL	8	1	8	1993	192
Hybrid	Cystine in white blood cells (WBC)	CWBC	1	Supernatant WBC + Protein solution	Spiked[a]	Lyophilised	5 mL	8	1	8	2005	38
	Lysosomal enzymes	LEFB	1	Cultured skin fibroblasts	Clinical[b]	Lyophilised	0.5 mg protein	2	3	6	2011	69
	Neurotransmitters in cerebrospinal fluid (CSF)	NCSF	1	CSF	Spiked[a]	Lyophilised	0.5 mL	8	1	8	2016	37
	Pterins in urine	PTU	1	Urine	Spiked[a]	Lyophilised	1.0 mL	8	1	8	2017	37

Qualitative	Acylcarnitines in dried blood spots	ACDB	3	Blood on filter paper	Clinical[b]	Dried blood spots	30–50 μL	2	3	6	2003	137
	Congenital disorders of glycosylation	CDG	1	Plasma / serum	Clinical[b]	Lyophilised	25 μL	2	3	6	2010	62
	Diagnostic proficiency testing	DPT	5	Urine	Clinical[b]	Heat treated samples	5–10 mL	2	3	6	1990	98
	Qualitative organic acids	QLOU	3	Urine	Clinical[b]	Heat treated samples	2–3 mL	2	3	6	1992	224
	Urine mucopolysaccharides	UMPS	1	Urine	Clinical[b]	Lyophilised	5 mL	2	3	6	2012	88
Interpretative	Amino acid interpretation	AAI	1	–	–	**Data only** (plasma, urine, or CSF)	–	2	3	6	2023	105[1]

[a]Matrix of human origin, spiked with commercially available analytes
[b]Real clinical samples donated by patients
[1] Pilot scheme

Advisor of each scheme and agreed upon by the SAB. These scheme organisers are two accredited EQA providers subcontracted by ERNDIM: MCA (Winterswijk, The Netherlands) for the quantitative and hybrid schemes, and CSCQ (Geneva, Switzerland) for the qualitative schemes. Clinical samples are obtained by the Scientific Advisors or by participants with the appropriate patient consent and sent to the scheme organiser for aliquoting and dispatch.

The dispatch of samples and the deadlines for the submission of results are scheduled according to a calendar established by the SAB. Participants have access to the websites of the scheme organisers to submit results and to view clinical information, when it is relevant, during a fixed period of time for each survey. Results are evaluated by each Scientific Advisor according to ERNDIM policies and procedures, and a customised report is made available on the scheme organiser website. The principles used for the evaluation of results for each scheme are outlined below (paragraph 6). For the qualitative and hybrid schemes any error which would be unacceptable to the majority of laboratories is separately assessed as Critical Error. Absence of Critical Error is required for satisfactory performance; hence a Critical Error will automatically lead to poor performance regardless of the overall score in the EQA scheme. The critical errors for each scheme year are proposed by the Scientific Advisors for the individual EQA schemes and ratified at the SAB meeting held after the completion of the EQA scheme year.

Participating laboratories that failed to reach a satisfactory performance, because of a low score, a critical error or both, receive a Performance Support Letter (PSL) prepared by the Scientific Advisor stating the details of this poor performance. The aim of this PSL is to begin a dialogue between the Scientific Advisor and the participating laboratory to attempt to give clues for improvement. Procedures are in place to appeal against classification as a poor performer. Customised Annual Reports for each scheme and an Annual Certificate of participation are issued after completion of the scheme year.

Figure 2.3 gives the percentage of poor performance per scheme in 2022, and whether critical errors have been made for hybrid and qualitative schemes.

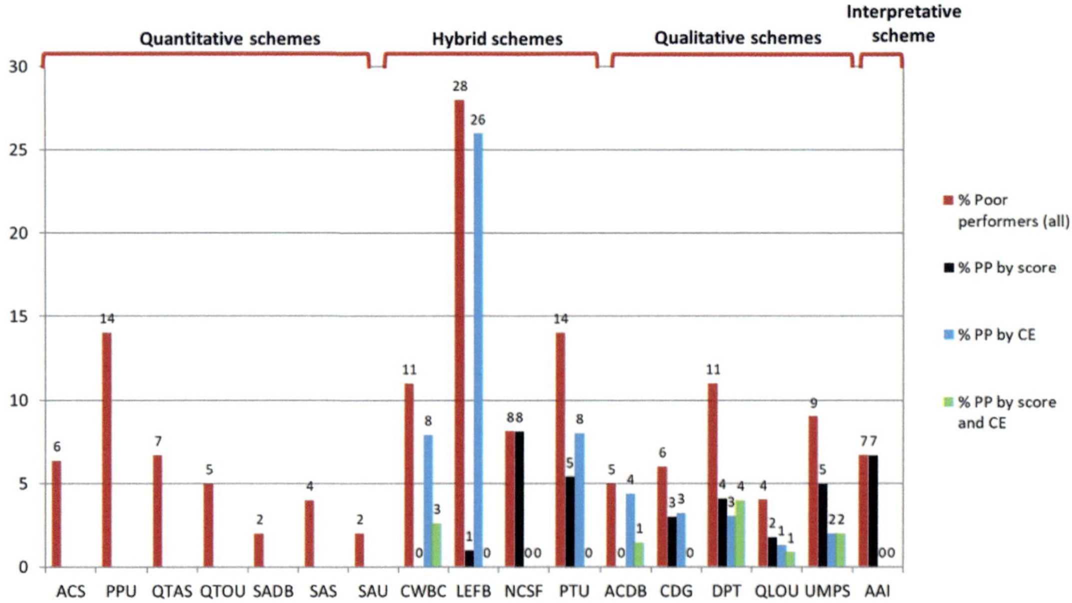

Fig. 2.3 Percentage of poor performers by EQA scheme. Scheme codes are given in Tables 2.2 and 2.3. *CE* critical error

2.6 ERNDIM Schemes

2.6.1 Quantitative Schemes

The details of the analytes included in the quantitative schemes are given in Table 2.3. For each quantitative scheme, eight lyophilised samples are provided per year, all prepared from the same matrix to which various amounts of analytes are added. The added quantities are identical in pairs of samples (4 pairs). From March to October one sample is analysed per month and results are submitted by participants on the ERNDIM website which supplies short- and long-term reports. Short-term reports on the eight individual specimens are available 2 weeks after the submission deadline and provide up-to-date information on analytical performance. A characteristic of the website is that it allows individual laboratories the choice of access to fully detailed and/or summarised reports of results. The "Analyte in Detail" is the most detailed report and shows results of a specific analyte in a specific sample. A more condensed report is the "Cycle Review" which summarises the performance of all analytes in a specific sample. The Annual Report summarises all results of a complete scheme year and gives an indication of the overall performance for all analytes in all 8 samples (Fig. 2.4). Four parameters are evaluated for each analyte: accuracy, precision, linearity, and recovery. When one of these 4 parameters falls in the category "5% worst performances of all labs", it will be marked by a "red flag" for this analyte (see Fig. 2.4). Satisfactory performance for a quantitative scheme is achieved when a participant laboratory obtains less than 25% of "red flags" ([number of red flags] / [number of analytes × 4] × 100%).

Table 2.3 Details of the analytes included in the ERNDIM quantitative and hybrid schemes in 2023

Scheme	Scheme code	Specimen	Analytes	
Acylcarnitines in serum	ACS	Serum	Free carnitine (C_0) Acetylcarnitine (C_2) Propionylcarnitine (C_3) Malonylcarnitine ($C_{3\text{-}DC}$) Butyrylcarnitine (C_4) 3-hydroxybutyrylcarnitine ($C_{4\text{-}OH}$) Methylmalonylcarnitine ($C_{4\text{-}DC}$) Tiglylcarnitine ($C_{5:1}$) Isovalerylcarnitine (C_5) 3-Hydroxyisovalerylcarnitine ($C_{5\text{-}OH}$) Glutarylcarnitine ($C_{5\text{-}DC}$)	Hexanoylcarnitine (C_6) Octanoylcarnitine (C_8) Decanoylcarnitine (C_{10}) Dodecanoylcarnitine (C_{12}) Cis-5-Tetradecenoylcarnitine ($C_{14:1}$) Palmitoylcarnitine (C_{16}) 3-hydroxypalmitoylcarnitine ($C_{16\text{-}OH}$) Oleoylcarnitine ($C_{18:1}$) Stearoylcarnitine (C_{18}) 3-Hydroxystearoylcarnitine ($C_{18\text{-}OH}$)
Purines & Pyrimidines	PPU	Urine	5-aminoimidazole 4-carboxamide ribonucleoside (AICAr) Adenine Adenosine Creatinine[a] Cytidine Deoxyadenosine Deoxyguanosine Deoxyinosine Deoxyuridine Dihydrothymine Dihydrouracil Guanosine 5-Hydroxymethyluracil Hypoxanthine Inosine	Orotic acid Orotidine Pseudouridine Succinyladenosine Succinyl-5-aminoimidazole-4-carboxamide-1-ribonucleotide (SAICAr) Thymidine Thymine Uracil 3-Ureidoisobutyric acid 3-Ureidopropionic acid Uric acid[a] Uridine Xanthine

(continued)

Table 2.3 (continued)

Scheme	Scheme code	Specimen	Analytes	
Quantitative amino acids	QTAS	Serum	Alanine Allo-isoleucine 2-aminobutyric acid Arginine Argininosuccinic acid Asparagine Aspartic acid Aspartylglucosamine Citrulline Cystathionine Cystine Glutamic acid Glutamine Glycine Histidine Homocitrulline Hydroxyproline	Isoleucine Leucine Lysine Methionine Ornithine Phenylalanine Pipecolic acid Proline Saccharopine Sarcosine Serine Sulphocystine Taurine Threonine Tryptophan Tyrosine Valine
Quantitative organic acids	QTOU	Urine	N-Acetylaspartic acid Adipic acid Creatinine[a] Ethylmalonic acid Fumaric acid Glutaric acid Hexanoylglycine 3-Hydroxybutyric acid 4-Hydroxybutyric acid 2-Hydroxyglutaric acid 3-Hydroxyglutaric acid 3-Hydroxyisovaleric acid 3-Hydroxy-3-methylglutaric acid 3-Hydroxypropionic acid	Isovalerylglycine 2-Ketoglutaric acid 2-Methylcitric acid 3-Methylglutaconic acid 3-Methylglutaric acid Methylmalonic acid Mevalonic acid Pyroglutamic acid Sebacic acid Suberic acid Suberylglycine Tiglylglycine Vanillactic acid
Special assays in dried blood spots	SADB	Blood on filter paper	Allo-isoleucine Carnitine free (C0) Homocystine Isoleucine Leucine Methionine	NTBC (nitisone) Phenylalanine Succinylacetone Tyrosine Valine
Special assays in serum	SAS	Serum	Behenic acid (C22:0) Biotinidase[a] Carnitine free (C0) Cerotic acid (C26:0) Cholestane-3β,5α,6β-triol Cholestanol Cholesterol[a] Coenzyme Q10 Creatine 7-Dehydrocholesterol Galactose Glycosylsphingosine Guanidinoacetic acid Homocystine 3-Hydroxybutyrate	7-Ketocholesterol Lactic acid Lignoceric acid (C24:0) Lyso-globotriaosylcéramide (Lyso-Gb3) C26:0-Lysophosphatidylcholine (C26:0-LPC) Lysosphingomyeline Methylmalonic acid Non-esterified fatty acids (NEFA)[a] Phytanic acid Pipecolic acid Pristanic acid Pyruvic acid Succinylacetone

(continued)

Table 2.3 (continued)

Scheme	Scheme code	Specimen	Analytes	
Special assays in urine	SAU	Urine	5-Aminolevulinic acid Carnitine free (C0) Creatine Creatinine[a] L-Cystine Galactitol D,L-Glyceric acid Glycolic acid Guanidinoacetic acid Homocitrulline Homogentisic acid Homovanillic acid	4-Hydroxyglutamic acid 5-Hydroxyindolacetic acid Lactic acid Mucopolysaccharides (chondroitin sulphate) Orotic acid Oxalic acid Pipecolic acid Sialic acid Succinylacetone Sulphocysteine
Cystine in WBC	CWBC	White blood cells	Cystine Protein	
Neurotransmitters in CSF	NCSF	CSF	5-Hydroxyindoleacetic acid (5-HIAA) 5-Hydroxytryptophan (5-HTTP)	3-Methyl dopa (3-MD) Homovanillic acid (HVA)
Pterins in urine	PTU	Urine	Biopterin Creatinine[a]	Neopterin Primapterin

Unless mentioned and when relevant, analytes are in the L-form

[a]These analytes are present in the matrix, so results are recorded for comparison between labs, but the concentrations do not vary. They are not included in the Individual Online Annual Reports or Certificates of participation

2.6.1.1 Acylcarnitines in Serum (ACS)

All samples are prepared from the same basic serum but with various amounts of added analytes at 3 levels, thus 4 concentration levels. Acylcarnitines can be quantified either as butylated esters or underivatized. When measuring underivatized acylcarnitines, some analytes will be isobaric: malonylcarnitine ($C_{3\text{-DC}}$) and 3-hydroxybutyrylcarnitine (C_{4OH}), as well as methylmalonylcarnitine ($C_{4\text{-DC}}$) and 3-hydroxyisovalerycarnitine ($C_{5\text{-OH}}$). Therefore, their accurate measurement can only be achieved using LC-MS/MS instead of the commonly used FIA (flow injection analysis) method which does not allow their separation. This probably explains the poorer interlab CV for these compounds in 2022 (up to 58.7%), compared to the interlab CV of e.g. free carnitine (13.9%).

2.6.1.2 Purines and Pyrimidines in Urine (PPU)

All samples are prepared from the same native urine sample. The compounds included in the scheme are added at three levels, and with the basal level, giving four concentration levels. The aim of this scheme is to include as many purines and pyrimidines as possible, however, this is constrained by availability, cost, and solubility. For this reason, some analytes are included at two levels. The inclusion of 2,8-dihydroxyadenine is not feasible as this compound proved to be impossible to dissolve when in solid state. Over the first 10 years of the scheme, the interlab CV has significantly decreased for many of the compounds included, gradually levelling off towards 20% in recent years, and in 2022 it is even below 20%. This success confirms the educational relevance of the scheme, even though the percentage of poor performers was relatively high in 2022.

2.6.1.3 Quantitative Amino Acids in Serum (QTAS)

The same basic human serum, which has been dialysed to remove most of the amino acids present and to which various amounts of analytes are added, is used for all samples. Unusual amino acids are also added (e.g. aspartylglucosamine, sarcosine, and argininosuccinic acid) but can be changed from year to year. It is the most successful ERNDIM scheme. For most amino acids, the

ERNDIMQA - ANNUAL REPORT

Quant.org.acids [redacted]

[redacted]

Methodset : **GC-MS**

	Accuracy (mean)		Precision (CV% duplicates)		Linearity (r)		Recovery (%added analyte)		Data All Labs	
Analyte	Your Lab	All Labs	Your Lab	All Labs	Your Lab	All Labs	Your Lab	All Labs	n	Interlab CV
2-OH Glutaric acid	MP	152	32.0%	20.8%	0.877	0.980	89%	93%	77	35.6%
3 methylglutaric acid	MP	70.4	26.3%	15.0%	0.914	0.985	117%	101%	77	23.6%
3-OH-3 methylglutaric acid	FR	66.9	FR	24.7%	FR	0.983	FR	77%	65	62.4%
3-OH-Isobutyric acid	OR	136	OR	20.4%	OR	0.979	OR	71%	53	37.0%
3-OH-Isovaleric acid	FR	78.1	FR	39.3%	FR	0.940	FR	112%	73	55.5%
4-OH-Butyric acid	FR	72.8	FR	31.2%	FR	0.982	FR	70%	63	56.7%
Adipic acid	MP	178	31.6%	12.6%	0.950	0.993	85%	101%	85	26.4%
Creatinine		3258		3.4%		0.000		0%	67	5.25%
D,L- Glyceric acid	FR	278	FR	22.0%	FR	0.985	FR	83%	64	47.0%
Ethylmalonic acid	ORFR	50.0	ORFR	17.1%	ORFR	0.992	ORFR	99%	89	36.3%
Fumaric acid	MP	60.0	19.5%	14.5%	0.991	0.991	75%	100%	78	28.5%
Glutaric acid	MP	112	44.0%	13.1%	0.878	0.994	86%	101%	90	21.8%
Glycolic acid	OR	236	OR	20.6%	OR	0.963	OR	89%	77	38.0%
Hexanoylglycine	MP	17.0	29.3%	29.9%	0.949	0.950	111%	100%	64	37.2%
Keto-glutaric acid		198		21.4%		0.992		103%	73	40.1%
Malic acid		117		19.8%		0.995		81%	53	47.4%
Methylmalonic acid	305	330	31.2%	11.4%	0.995	0.996	95%	102%	104	27.6%
Mevalonic acid	FR	132	FR	21.7%	FR	0.992	FR	82%	56	46.1%
Pyroglutamic acid	353	322	61.6%	23.0%	0.879	0.980	106%	85%	73	44.2%
Sebacic acid	MP	55.4	15.2%	19.4%	0.987	0.987	91%	101%	83	35.9%
Suberic acid	123	172	11.5%	16.3%	0.992	0.990	64%	105%	84	32.4%
Tiglylglycine	MP	83.6	17.0%	24.9%	0.975	0.987	54%	85%	57	48.0%
Vanillactic acid		38.5		19.4%		0.975		98%	38	33.0%
Overall	260	270	29.0%	20.1%	0.944	0.940	88%	89%	71	37.7%

Fig. 2.4 Example of an Annual Report for a quantitative scheme. The Annual Report is based on all data submitted during the annual cycle for one scheme. Outliers (results exceeding the mean +/− 7SD) are removed before the calculations. *Analytes*: The first column lists all analytes in the scheme. *Accuracy*: the mean of 8 results of the participant lab is compared to the mean of all labs. *Precision*: the mean coefficient of variation (CV) of 4 duplicates is compared to the median CV of all labs. *Linearity*: for each analyte, the regression coefficient r is calculated of test results as a function of added amounts. The lab's r and the median r of all labs are listed. *Recovery*: the slope of linearity calculation is multiplied with 100% and is the recovery of added amounts. Again, recovery of the lab and the median recovery of all labs is shown. *Data all lab*: the last column shows general data of all labs: the number of labs that submitted results (n) and the Interlab CV. *Green flag* (analyte name in green box): marks satisfactory performance of an analyte and applies when at least 3 of the 4 analytical parameters show satisfying results. *Red flag* with a numeric result (example Suberic acid): marks a result in the category "5% worst performances of all labs". *Empty white box* (example: Malic acid): No results submitted. *FR* in a white box (example: Mevalonic acid): Few Results (defined as <6 out of 8 results). *OR* in a purple box (example Glycolic acid): Outlying Results (defined as 2 or more outliers). *ORFR* in a yellow box (example Ethylmalonic acid): Outlying Results and Few Results (defined as <6 results including at least one outlier). *MP* in an amber box (example Adipic acid): Missing Pair (defined as the situation in which both results of a sample pair are missing or outlier).

interlab CV is <15%, which is satisfying and allows the comparison of amino acid levels for diagnosis and monitoring of treatment for one patient in different hospitals and for the use of shared reference values. Several amino acid calibrators are commercially available but do not contain asparagine, glutamine and some unusual amino acids, and this probably explains the higher interlab CV for these compounds (e.g. 20.9% for asparagine).

2.6.1.4 Quantitative Organic Acids (QTOU)

All samples are prepared from the same basic human urine but with various amounts of added analytes, at final concentrations covering the physiological range and the typical pathological range. This EQA scheme often encounters a high interlab CV, up to 60%, whereas overall reported linearity is excellent for all compounds: this highlights the major problem in the analysis of organic acids, the lack of standardisation. To accurately quantify an organic acid, a stable isotope-labelled form of the analyte should be used as an internal standard, since most organic acids are not optimally extracted. Since this would be too cumbersome to achieve, most laboratories do not use stable isotope dilution methods. Other difficulties are that labs need to prepare their own calibrator(s) by weighing and diluting solid compounds and the lack of commercial sources for some organic acids. An interesting initiative has been developed by the Amsterdam University (AUMC) that offers for sale a liquid calibrator containing 33 different organic acids. The wider use of this calibrator is likely to improve the performance of this scheme. In the meantime, it is important that follow-up of a patient affected with an organic aciduria, for monitoring the treatment, is carried out in the same centre.

2.6.1.5 Special Assays in Dried Blood Spots (SADB)

Since 2022, the scheme consists of eight dried blood spots, four duplicates, all prepared from the same basic whole blood with various amounts of added analytes. Blood is depleted to create low basal concentrations but not to a zero level. Thus, concentrations of analytes in the samples are the remaining endogenous concentrations plus the spiked amounts. Most laboratories submit results for phenylalanine whereas only a few submit results for NBTC. In 2022, the intralab variation was acceptable for all analytes (mean variance 11.2%), but the interlab CV ranged from 17.2% for phenylalanine to 59.5% for homocysteine. This interlab imprecision highlights the challenges associated with the measurement of some of these analytes in dried blood spots, and there is evidently a need to improve the performance of these analytes.

2.6.1.6 Special Assays in Serum (SAS)

The scheme consists of eight lyophilized samples, four duplicates, prepared from the same basic serum but with various amounts of added analytes, except for biotinidase, NEFA, and total cholesterol, which are present as endogenous activity/concentration in the lyophilized samples and are not added (results are the mean values reported by the different laboratories). It is also a successful scheme, with 237 participating labs in 2022. Except for coenzyme Q10 (CoQ) (49%), good recoveries were observed, and the overall precision of 10.4% is quite satisfying. The interlab CV ranged from 5.5% for lactic acid to 115% for CoQ. To be more precise, 13 metabolites displayed a CV <15%, 4 between 15–20% and 6 > 20% (CoQ and succinylacetone showed very high CV: > 50%). It is difficult to draw general conclusions concerning this scheme because of the large number of metabolites analysed with many different analytical methods, some with commercially available calibrators, others with calibrators prepared in-house.

2.6.1.7 Special Assays in Urine (SAU)

As for the other quantitative schemes, this scheme consists of eight lyophilised samples, four duplicates, prepared from the same basic urine but with various amounts of added analyte. In 2022, the interlab CV varied from 5.7% for creatinine to 54.2% for homogentisic acid. In this EQA scheme as well, the analytes are quantified using different methods.

2.6.2 Hybrid Schemes

The details of the analytes included in the hybrid schemes are available in Table 2.3. In these hybrid schemes, the quantitative results are assessed like in quantitative schemes, but in addition, participants are asked to interpret the results and to give a diagnosis. For all hybrid schemes, a wrong diagnosis can be assessed as a critical error if the majority of participants concluded the correct diagnosis and therefore a laboratory can be evaluated as a poor performer even if the number of points required for satisfactory performance is reached.

2.6.2.1 Cystine in White Blood Cells (CWBC)

Participants receive eight liquid samples of 250 μL equivalent to extracts from 5 mL whole blood samples for cystine measurement and eight lyophilised samples for measurement of protein. Participants should analyse one pair (cystine sample + corresponding protein sample) every month and report the cystine result as nmol ½ cystine/mg protein. In addition, they must interpret the result by choosing the correct diagnosis from a list box. A limited number of laboratories participate in this scheme (38 in 2022). In 2022, the mean recovery of cystine (nmol/aliquot) was 99% and recovery of protein was 92%. The lower recovery in the protein assay may reflect the lack of standardisation in the protein assays concerning the method and calibrator used. The interlab CV for cystine (nmol ½ cystine/mg protein) was relatively high, 35.2%, but is partly explained by the precision being the cumulated precision of both cystine and protein assays. Poor performance is generally by both score and critical error since a poor measurement of cystine and/or protein can lead to a wrong interpretation.

2.6.2.2 Lysosomal Enzymes in Fibroblasts (LEFB)

Four pellets of lyophilised cultured skin fibroblasts from patients affected with a lysosomal storage disorder, together with two pellets of lyophilised cultured skin fibroblasts from control patients, are sent at the beginning of the year, each pellet being equivalent to 0.5 mg protein. Participants analyse the samples in two surveys of 3 pellets (2 affected patients and 1 control) by measuring the activity of a maximum of 10 lysosomal enzymes. The measurement of all 10 enzyme activities is not mandatory, and the enzyme activities to be measured can change from year to year. No clinical data are provided. Participants must express the enzyme activities of the affected patients as a percentage of the control cell line and interpret the results by giving the diagnosis of the lysosomal storage disorder using a multi-choice box with a free text box for comments. Since participants may not measure all enzyme activities, performance is evaluated for each enzyme separately, which probably explains the worst performance amongst all EQA schemes.

2.6.2.3 Neurotransmitters in Cerebrospinal Fluid (NCSF)

The matrix used is either a pool of CSF from patients not affected with a disorder of neurotransmitter metabolism or an artificial matrix. Four pairs of samples, spiked with four different analytes at concentrations that mimic neurotransmitter disorders or a non-affected patient, are circulated. Quantitative data are evaluated, and participant laboratories must also interpret the quantitative results and give a diagnosis using a multi-choice box with a free text box for comment. Despite a relatively high percentage of unsatisfactory quantitative performance for this scheme, interpretation is commonly correct.

2.6.2.4 Pterins in Urines

A pool of control urine is split into four aliquots, which are spiked with four different concentrations of biopterin, neopterin, and primapterin. Four pairs of samples are circulated. Results must be expressed as mmol/mol creatinine. Quantitative data are evaluated as well as interpretation using a multi-choice box. In 2022, the overall interlab CV was 26.9% ranging from 23.5% for biopterin to 32.3% for primapterin, but the latter compound had been measured by only two-third of the laboratories. The consequence is that the overall unsatisfactory performance is relatively high in this scheme, but it is almost equally due to poor quantitative results and interpretation error(s).

2.6.3 Qualitative Schemes

The main purpose of the qualitative schemes is to evaluate the ability of the laboratory to establish or exclude a specific diagnosis of an Inherited Metabolic Disease (IMD). Participants are expected to obtain correct analytical results, to recognise the characteristic diagnostic patterns, to make a diagnostic conclusion, and to suggest additional test(s) necessary to confirm the diagnosis. These schemes use clinical samples as the EQA materials. All qualitative schemes are method specific, except the DPT scheme. Two elements are evaluated: 1) analytical performance and 2) interpretative proficiency, also considering recommendations for further investigation (Table 2.4).

The total score is calculated as a sum of these two criteria. The maximum to be achieved is four points per sample. The scores are calculated only for laboratories submitting results. The minimum score for satisfactory performance is decided by the Scientific Advisory Board. In 2022, the minimum score has been set at 17 points out of a maximum of 24 (70%) for all qualitative schemes. For all qualitative schemes, a wrong diagnosis can be assessed as a critical error. With particularly challenging samples with low overall proficiency, the SAB may decide to classify it as educational, excluding it from performance assessment.

2.6.3.1 Acylcarnitines in Dried Blood Spots

This scheme offers dried blood spots obtained from patients with confirmed diagnoses to enable laboratories to gain or maintain experience to identify mitochondrial fatty acid oxidation defects and organic acidurias. Due to the difficulty in obtaining sufficient samples, this programme is managed by three centres: Heidelberg,

Table 2.4 Scoring of qualitative schemes

A	Analytical performance	Correct results of the appropriate tests	2
		Partially correct or non-standard methods	1
		Unsatisfactory or misleading	0
I	Interpretative proficiency and recommendations	Good (diagnosis was established)	2
		Helpful but incomplete	1
		Misleading or wrong diagnosis	0

London, and Rome. Each participant receives six dried blood spots per year. One of these is distributed to all labs participating in this scheme as a common sample. In 2022, the common sample was obtained from a patient affected with 3-hydroxy-3-methylglutaryl-CoA lyase deficiency. The key metabolites to diagnose this disorder are 3-methylglutarylcarnitine ($C_{6\text{-DC}}$) and 3-hydroxyisovalerylcarnitine ($C_{5\text{-OH}}$). A critical error was assigned to the five participant laboratories that failed to identify an increase of these metabolites and did not recommend performing urinary organic acids, the other test that enables to diagnose this disorder. Proficiency is generally satisfactory, ranging from 75 to 100% in 2022 (median 98% on 18 samples in 3 schemes).

2.6.3.2 Congenital Disorders of Glycosylation (CDG)

Each participant receives six samples per year of 25 µl lyophilised plasma or serum from patients affected with CDG or from control patients. The aim of this EQA scheme is the qualitative interpretation of sialotransferrin profiles in the screening for Congenital Disorders of Glycosylation (CDG). Proficiency is generally good. The major problem of this scheme is to obtain clinical samples in sufficient amounts.

2.6.3.3 Diagnostic Proficiency Testing (DPT)

The aim of this EQA is to assess test selection, analysis, interpretation, and advice related to the detection of inherited metabolic disorders. DPT schemes are the ultimate EQA challenge for biochemical genetics laboratories coming closest to routine diagnostics. In practice, 6 samples of 7–10 ml of heat-treated urine (20 min at 59 °C) from patients affected with aminoacidopathies, organic acidurias, lysosomal storage disorders, purine or pyrimidine disorders, or normal controls are sent to participants with some clinical information. Each participant decides what test(s) to carry out according to the clinical presentation from a minimal test panel including amino acids, organic acids, oligosaccharides, mucopolysaccharides, and purines and pyrimidines. Additionally, specific assays such as orotic acid and sialic acid can be performed. If a participant is not performing one of these tests, it is allowed to outsource the analysis to another lab, called a cluster lab, but the participant is responsible for the results provided by this cluster laboratory. The use of cluster labs is strictly forbidden in the other qualitative schemes which are method specific. Due to the difficulty in obtaining sufficient samples, this programme is managed by five centres: Czech Republic (Prague), France (Lyon), Switzerland (Bern), The Netherlands (Rotterdam), and the UK (Sheffield). Every year, generally before the Annual Meeting of SSIEM (Society for the Study of Inborn Errors of Metabolism), the participants of each DPT centre meet the Scientific Advisor(s) of the scheme to discuss the results. This is an important part of the proficiency testing, since the goal of the programme is to improve the competence of the participating laboratories, which includes the critical review of all results and discussing possible improvements. The percentage of poor performers is one of the highest amongst the ERNDIM EQA schemes, but this can be explained by the fact that DPT scheme investigates a large panel of methods, and the investigation to be carried out is the decision of the laboratory staff. Proficiency ranged from 63% to 100% in 2022 (median 93% of 29 samples in 5 schemes). A retrospective study on all DPT surveys over the last 15 years demonstrated that, when a sample with the same disorder was redistributed in a subsequent survey (115 samples), performance improved for 75 of them. Further, an overall trend in improvement of proficiency in ERNDIM DPT schemes over the years has been observed, although poor performance for some samples remains, all of which suggesting that ERNDIM diagnostic proficiency testing is most probably a valuable activity [7].

2.6.3.4 Qualitative Organic Acids

The aim of this EQA is to educate and assess the ability of laboratories to detect inherited disorders resulting in recognisable patterns of organic acid excretion. Urine samples are from affected or control patients. Three centres are involved: Barcelona, Heidelberg, and Sheffield. Six samples of 2–3 ml heat-treated urine samples are distributed every year, with some clinical information. The overall performance is generally good (67–100% in 2022; median 96% of 17 samples in 3 schemes).

2.6.3.5 Urine Mucopolysaccharides

The participants of this scheme must perform the quantitative (related to creatinine) and qualitative analysis of mucopolysaccharides (GAG) with interpretation of results obtained. No clinical information is provided. To achieve satisfactory results in this programme, it is mandatory to perform a qualitative analysis of mucopolysaccharides. But, despite the recommendations of the Scheme Advisor, some laboratories persist in registering when they do not perform qualitative GAG analysis, and this partly explains the relatively high unsatisfactory performance.

2.6.4 Interpretative Scheme: Amino Acid Interpretation

This is a relatively new scheme, which was introduced as a full scheme since 2023. The aim is to assess the ability of laboratories to detect inherited disorders resulting in recognisable patterns of amino acids in plasma and other specimens. No physical samples are circulated for this scheme, but instead, cases with data only are circulated: quantitative data with the corresponding reference range, and clinical information. Participating laboratories must carry out the interpretation of the EQA sample data. In the 2022 pilot scheme, overall proficiency of the 6 samples ranged from 68 to 92%. An asset of an interpretative EQA scheme is that it allows a review of the competence of individuals, which is demanded in laboratory accreditation.

2.7 Sample Exchange Programmes When EQA Is Lacking

In case a diagnostic metabolite, a group of metabolites or an enzyme activity is measured by only a limited number of laboratories, e.g. for very rare disorders or newly identified analytes, EQA cannot be set up by ERNDIM. As an alternative, interlaboratory sample exchange programmes may be developed by laboratories. Ideally, blinded samples are exchanged between laboratories to assess proficiency. Although such programmes usually are relatively small, the design will be similar to formal EQA schemes. Thus, the organising laboratory should consider: sample selection or preparation, stability of analytes in the samples, sample dispatch, forms or a website to enable result submission by participants, statistical analysis of results, and evaluation of results. Often, in sample exchange programmes, compiled results are circulated amongst the participants without formal evaluation, whilst review of results is done in a collaborative manner.

For laboratories that are considering organising sample exchange, ERNDIM can assist by including in the annual survey a question on the interest amongst its participants in a certain sample exchange programme and then linking them by providing contact details.

2.8 The Future of Biochemical Genetics EQA Provided by ERNDIM

In a laboratory specialised in the diagnosis of IMD, more analytes than those included in the ERNDIM EQA schemes may be measured. Each year, ERNDIM sends a questionnaire to all participants asking for their opinion on the existing EQA schemes, but also on the analytes they would like to see added (or possibly removed) and on new EQA schemes they would like to have set up. The ERNDIM policy is that a new analyte or a new EQA scheme cannot be introduced if less than 30 participant laboratories would register for it. Likewise, new EQA will not be developed if it is already available from another provider, such as an EQA scheme for porphyrins. The ERNDIM SAB has ratified the development for 2025 of one new scheme: Complex Lipids in Serum. This programme will include some of the lipids already included in the Special Assay in Serum scheme. It remains to be decided what other lipids will be added. The recent development of untargeted metabolomics in an increasing number of laboratories prompted the SAB to investigate the possibility of a new metabolomics EQA scheme. However, this project was postponed since the annual questionnaire revealed that there were not enough labs interested to participate and it was not clear which specimen would be useful: plasma or urine.

Because of the difficulty of obtaining clinical samples in sufficient amounts, especially plasma or CSF samples, ERNDIM will put effort into the development of new interpretative schemes. This type of programmes has the advantage that members of laboratory staff can register and participate individually: this would demonstrate the competence of individuals which is a real plus in the context of laboratory accreditation.

Although performance is improving year on year, the final goal of ERNDIM would be that all participants in biochemical genetics EQA schemes measure accurately the analytes and interpret reliably the quantitative and qualitative results. This would allow the comparison of analyte levels for diagnosis and monitoring of treatment for one patient in different hospitals and to use of common reference values. Although things seem to be moving in the right direction, this is not yet the case for all analytes. There are several explanations for sub-optimal diagnostics: Standard Operating Procedures (SOP) are not harmonised amongst laboratories and commercially available standard materials are often lacking. That is why ERNDIM's tasks in the future will also be to provide consensus guidelines for diagnostic testing and to encourage suppliers to develop standard material for the highly specialised analytes that biochemical genetics laboratories need to quantify.

2.9 Conclusion

In biochemical genetics laboratories, quality management, including IQC and EQA, is of fundamental importance to ensure reliable and timely diagnostic results. Besides implementing IQC, it is important for laboratories to investigate which EQA schemes are available for implementation in their structure. By providing such EQA schemes, ERNDIM assists biochemical genetics laboratories in assessing the quality of their diagnostic tests. Besides, proficiency testing has a significant educational value providing labs with the necessary knowledge to optimise diagnostic testing for these rare disorders.

Acknowledgements The authors thank all the Scientific Advisors of the different ERNDIM schemes and their Deputies, the Administration Office of ERNDIM and the Scheme Organisers, without whom this organisation could not function.

References

1. Ferreira CR, Rahman S, Keller M, Zschocke J (2021) An international classification of inherited metabolic disorders (ICIMD). J Inherit Metab Dis 44:164–177
2. Levey S, Jennings ER (1950 Nov) The use of control charts in the clinical laboratory. Am J Clin Pathol 20(11):1059–1066. https://doi.org/10.1093/ajcp/20.11_ts.1059
3. Westgard JO, Barry PL, Hunt MR, Groth T (1981 Mar) A multi-rule Shewhart chart for quality control in clinical chemistry. Clin Chem 27(3):493–501
4. Fowler B, Burlina A, Kozich V, Vianey-Saban C (2008) Quality of analytical performance in inherited metabolic disorders: the role of ERNDIM. J Inherit Metab Dis 31:680–689
5. Libeer JC (2001 Jul 20) Role of external quality assurance schemes in assessing and improving quality in medical laboratories. Clin Chim Acta 309(2):173–177. https://doi.org/10.1016/s0009-8981(01)00518-6
6. Oglesbee D, Cowan TM, Pasquali M et al (2018) CAP/ACMG proficiency testing for biochemical genetics laboratories: a summary of performance. Genet Med 20:83–90
7. Mathis D, Croft J, Chrastina P, Fowler B, Vianey-Saban C, Ruijter GJG (2022 Sep) The role of ERNDIM diagnostic proficiency schemes in improving the quality of diagnostic testing for inherited metabolic diseases. J Inherit Metab Dis 45(5):926–936. https://doi.org/10.1002/jimd.12523

Simple Tests

3

K. Michael Gibson and Marinus Duran

Contents

3.1 Cyanide Nitroprusside Test

3.1.1 Principle

The cyanide nitroprusside test determines the presence of free sulfhydryl or disulfide compounds in urine samples [1–3]. During the first step of the assay, cyanide reduces any disulfides that are present in free sulfhydryl compounds. In the second step, a reddish color reaction results when the free sulfhydryl groups complex with nitroprusside. A positive result is most usually due to cystine in the urine. Familial cystinuria is among the most common aminoacidurias. Disulfides are also excreted in other metabolic disorders such as homocystinuria and β-mercaptolactate-cysteine disulfiduria. Both will also produce positive results according to the following reaction: RSH + $Na_2Fe(CN)_5NO$ (sodium nitroprusside) → chromophore + NO.

3.1.2 Specimen Requirements

The specimen requirements for this test are as follows:

1. 1.0 ml of clear urine.
2. Store and transport specimen frozen.
3. Gloves and a laboratory coat should be worn. Use the proper technique to avoid aerosols.

3.1.3 Reagents

1. Sodium cyanide (NaCN), 5% in H_2O. Weigh out 25 g of NaCN carefully; avoid creating dust. Add NaCN to 500 ml H_2O and stir. Store the

K. M. Gibson (✉)
Washington State University, Section of Clinical Pharmacology, Spokane, USA
e-mail: mike.gibson@wsu.edu

M. Duran
Academic Medical Center, Lab Genetic Metabolic Diseases, Amsterdam, The Netherlands

N. Blau, F. M. Vaz (eds.), *Laboratory Guide to the Methods in Biochemical Genetics*,
https://doi.org/10.1007/978-3-031-58819-8_3

reagent in a plastic bottle at room temperature under a fume hood. This reagent is stable for up to 1 year. The reagent bottle must be stored in a container large enough to hold the bottle's contents. Dispose of any unused 5% solution appropriately. Always perform the assay in a fume hood; protect the eyes and skin with glasses and gloves. Discard any completed assay mixture (liquid in tubes) down the drain in the fume hood after adding 70/30 (v:v water) bleach solution into each tube and allowing it to stand for 1 h. Avoid inhalation, contact with skin/mucous membranes, and ingestion.

2. Sodium nitroprusside (sodium nitroferricyanide), 0.5% in H_2O. Carefully weigh out 0.5 g of the reagent; avoid creating dust. Add the reagent to 100 ml H_2O. The reagent is light sensitive, so store it in a dark bottle at +4 °C. The solution is stable for up to 4 months. Dispose of any unused standard reagent appropriately. Always perform the assay in a fume hood; protect the eyes and skin with glasses and gloves. Discard any completed assay mixture (liquid in tubes) down the drain in the fume hood after adding 70/30 (v:v water) bleach solution into each tube and allowing it to stand for 1 h.

3.1.4 Procedure

1. Set up one 12 × 100 mm test tube per patient. Label the tubes with the patient number/name. Thaw one tube each of pre-made positive and negative controls.
2. Aliquot 1.0 ml of clear patient urine to the corresponding test tubes. If the urine is not clear, spin at ~500 × *g* for 3 min.
3. In the fume hood, add 0.4 ml NaCN to each patient's specimen and control tube and vortex.
4. Incubate at room temperature for 20 min.
5. After 20 min, add 0.2 ml sodium nitroprusside to each patient's specimen and control tube one at a time, vortex, and read the results immediately.
6. Note that color develops immediately and fades quickly. A pink to red, or beet color, indicates a positive. Grade the positives on a scale of trace to 4 + .
7. Record the results.
8. Add 70/30 bleach solution into each tube and allow to stand for 1 h to ensure that the sodium cyanide will not be reactive when disposed of down the drain.

3.1.5 Linearity/Calculations

Not applicable for this test.

3.1.6 Interference

The urine sample used should be at a neutral pH. If the sample has been preserved with acid, a false positive may occur. Drugs such as N-acetylcysteine, 2-mercaptoethanesulfonate, 2-mercaptopropionylglycine, captopril, penicillamine, and large amounts of synthetic penicillin metabolites and acetoacetate will give positive results. Bacterial contamination may also give a false positive.

3.1.7 Quality Control

Positive and negative controls (pooled urine supplemented with and without cysteine, respectively; 50 mg/dl) are run with each batch of patient samples.

3.1.8 Normal/Therapeutic Ranges

A slight pink color may be detected from a normal urine that is concentrated; however, the detection of any positive result warrants further investigation.

3.1.9 Notes

Cystathionine, methionine, and taurine are not detected by this assay. In addition, this assay may be helpful for the detection of heterozygotes of cystinuria types 2 and 3.

3.2 Ehrlich's Aldehyde Reagent

3.2.1 Principle

Ehrlich's test is used to identify the presence of porphobilinogen (PBG) or urobilinogen in the urine [1, 2, 4]. PBG in urine reacts with Ehrlich's aldehyde reagent to give a red color in samples with acute porphyria. Except for PBG synthase deficiency, acute porphyria is generally associated with increased urinary excretion of PBG. Urobilinogen, a component of heme degradation, results when bilirubin derivatives are secreted in the bile and further degraded by bacterial enzymes in the large intestine. In the normal patient, some of the urobilinogen is reabsorbed and transported to the kidneys where it is converted to urobilin (yellow) and excreted. However, the majority of urobilinogen is converted microbially to sterobilin (deep red-brown). Urobilin is responsible for the characteristic color of urine, while sterobilin is the main pigment of feces.

3.2.2 Specimen Requirements

The specimen requirements for this test are as follows:

1. 1.0 ml of urine.
2. Store and transport the specimen frozen.
3. Gloves and a laboratory coat should be worn. Use the proper technique to avoid aerosols.

3.2.3 Reagents

1. 2% (w/v) p-dimethylaminobenzaldehyde (p-DABA) in 2 N HCL. Mix 2 g p-DABA in 100 ml 2 N HCl. Store in a brown bottle at 4 °C. Note: this reagent is an irritant and is corrosive; avoid inhalation and contact with skin/mucous membranes. Dispose of the reagent in the drain at a concentration below 10% in an aqueous solution.
2. Reagent-grade chloroform and butanol.

3.2.4 Procedure

1. Set up one 12 × 100 mm test tube per patient and a control. Label each tube with the patient's number/name or control name.
2. Aliquot 1.0 ml each of well-mixed patient and control urine to the corresponding tubes.
3. Add 0.1 ml p-DABA reagent to each tube and vortex.
4. Incubate at room temperature for 10 min.
5. After 10 min, examine the color of tube mixture. A deep orange, pink, or red color is considered positive.
6. If positive, add 1.0 ml chloroform to the tube, vortex, and allow to settle for a few minutes.
7. After it has settled, if the bottom layer is pink, the sample is urobilinogen positive and PBG negative. Grade the color intensity (trace, 1 +, 2 +, 3 +, 4 +) and record the results.
8. If the top layer is pink, pipette it out and place it into a clean tube.
9. Extract by adding 1.0 ml butanol to the removed layer (top layer). Vortex and allow the suspension to settle for a few minutes.
10. Upon settling, if the bottom layer in the tube is pink, the sample is PBG positive. Grade the color intensity (trace, 1 +, 2 +, 3 +, 4 +) and record the results.

3.2.5 Linearity/Calculations

Not applicable to this assay.

3.2.6 Interference

Clinical samples often contain complex mixtures of different porphyrins. Porphyrin solubility in organic solvents is inversely proportional to the number of acid side chains. Indoles in the urine may also give a positive result.

3.2.7 Quality Control

A negative and positive control sample (pooled urine, spiked with urobilinogen/PBG) is run with each batch of patient samples.

3.2.8 Normal/Therapeutic Ranges

The range is determined visually by observing the mixture color. A deep orange, pink, or red color is considered positive and is graded on a scale of trace to 4 +. A grade of 3–4 + should be acted on immediately. The range for normal tube test results may include results of negative to 1 + potentially due to the natural occurrence of small amounts of urobilinogen/PBG in urine, and/or interfering substances.

3.2.9 Notes

Individuals affected with porphyria present with acute attacks, skin lesions, or both. The onset of these attacks rarely occurs before puberty. An attack usually consists of severe abdominal pain and often neurological sequelae. During and after such attacks, excessive amounts of aminolevulinic acid and PBG are excreted in the urine. The most common porphyria is PBG deaminase deficiency (acute intermittent porphyria), which primarily affects liver function. A positive result coupled with a clinical indication of hepatosplenomegaly suggests that evaluation for tyrosine metabolites in the urine should be pursued (using the nitrosonaphthol test).

3.3 Dinitrophenylhydrazine Assay

3.3.1 Principle

The Dinitrophenylhydrazine (DNPH) assay detects α-keto acids in the urine [1, 3]. 2,4-DNPH reacts with α-keto acids to form hydrazones, which precipitate out of solution. Several metabolic disorders can be detected, including phenylketonuria (PKU), maple syrup urine disease (MSUD), tyrosinosis, tyrosyluria, histidinemia, and methionine malabsorption (Oasthouse syndrome). The presence of acetone also gives a positive result and may suggest a metabolic disorder. A positive result due to acetone is consistent with the diagnosis of disorders such as hyperglycinemia, isovaleric acidemia, and glycogen storage diseases. A 2 N HCl test is performed on all DNPH positives to determine the presence of false-positive results.

3.3.2 Specimen Requirements

For this assay, 1.0 ml of random, void urine (morning void preferable) is required, stored, and transported frozen.

3.3.3 Reagents

1. 0.2% 2,4-DNPH in 2 N HCL. Store at 4 °C. Stable for 1 year. This reagent is corrosive.
2. 2 N HCl. Store at room temperature in an acid storage area. Use in a fume hood.
3. 2-Ketoacid standard (3-methyl-2-oxobutyric acid, 2-ketoisovaleric); mix 50 mg in 25 ml (2 mg/ml) deionized water. Store the solution at 4 °C.

3.3.4 Procedure

1. Label one 13 × 100 mm test tube with the patient's number or name. Add 1.0 ml of clear urine. If the urine is turbid, spin at 500 × *g* for 3 min.
2. Prepare a positive control of 20 mg/dl. For this, add to a 13 × 100 mm test tube, 100 µl of 2-ketoacid standard and 900 µl of H_2O. The amount of precipitate that will occur in this positive control is considered to be approximately 2+. This aids in judging the amount of precipitate in patient samples.

3. Add 1 ml DNPH reagent to each patient and control tube and vortex. Incubate at room temperature for 10 min.
4. After 10 min, examine the samples. The presence of a yellow or whitish precipitate indicates a positive result.
5. If no precipitate occurs, the patient sample is recorded as negative. Grade the samples with a precipitate on a trace–4 + scale and record the results.
6. For positive samples, test to ensure that the reaction did not produce a false positive. Label a clean 13 × 100 mm test tube with the patient's name or number and place 1 ml of the patient's clear urine into the tube. Add 1 ml of 2 N HCl, vortex, and allow to incubate at room temperature for 10 min.
7. The presence of a precipitate after 10 min indicates a positive result. This suggests that the original positive DNPH result observed was probably a false positive.
8. Record the 2 N HCl results as either negative or grade on a trace–4 + scale if a precipitate is present. Generally, results of 2 + or more are considered a concern.

3.3.5 Interfering Substances

Substances with low solubility in acid can interfere with the DNPH assay. Mandelamine (methenamine mandelate), an antibacterial medication, and radiopaque contrast material will form a precipitate immediately upon addition of the DNPH reagent. The color and immediacy of the formation of this precipitate distinguishes it from the yellowish precipitate of α-keto acid hydrazones. It is very important to have information on patient medications prior to evaluation of the DNPH test.

3.3.6 Quality Control

Control urines are run with each batch of patient samples, consisting of pooled negatives (no appreciable precipitate) in addition to spiked pool urine containing significant exogenous α-keto acid standard.

3.3.7 Affected/Unaffected Values

An unaffected patient will have a negative DNPH result. Mildly affected or partially treated patients may also yield negative results. Patients with a blood phenylalanine level (indicative of PKU) over 1 mmol/l should generate positive DNPH results. Patients with a blood leucine level (indicative of MSUD) of 0.8 mmol/l or higher usually show a positive DNPH result. Patients with pyruvate metabolic disorders may also give positive results, as will patients with true and transient disorders of tyrosine degradation.

3.4 Multistix Reagent Strips for Urinalysis

3.4.1 Principle

Multistix 10 SG and Multistix PRO 11 (Bayer HealthCare, Elkhart, IN, USA) are reagent test strips that provide qualitative colorimetric analysis of several intermediates in urine, including protein, blood, leukocytes, nitrite, glucose, ketones (acetoacetic acid), pH, specific gravity, creatinine, bilirubin, and urobilinogen. The methodology, interpretation, and characteristics of the approximate linearity of these procedures are provided in the package insert with each reagent bottle. In addition, for each test, sensitivity parameters, expected values, limitations, and interfering substances in urine are described. Results obtained from this screening procedure can provide insight into the status of a patient's kidney and liver function, acid-base balance, and/or carbohydrate metabolism.

3.4.2 Specimen Requirements

Fresh void urine is the sample of choice for this analysis, enough to saturate all reagent patches on one strip; the minimum requirement is 1.0 ml of urine. The sample should be stored and transported frozen.

3.4.3 Reagents

Reagent strips for urinalysis, manufactured by Bayer HealthCare. Each strip contains reagent patches for the testing of glucose, bilirubin, ketone (acetoacetic acid), blood, pH, and protein. See the package insert for exact reagent compositions.

3.4.4 Procedure

1. Dip one strip into well-mixed, thawed urine to ensure saturation of reagent patches, then remove the strip from the urine. The strip should remain in the urine only briefly to reduce the potential for leaching of reagents from the patches.
2. Place the strip onto a paper towel (reagent patch side facing up) and allow it to develop for an appropriate time period; the development times of the specific patches are listed on the package insert.

3.4.5 Interferences

Certain medications may induce discoloration of the urine sample, which may interfere with the results. The colorimetric reaction may be affected by riboflavin or drugs containing azo dyes, or by nitrofurantoin. Other interferences may include:

1. Glucose: False negatives may occur when the urine contains small amounts of glucose (75–125 mg/dl) and/or ascorbic acid concentrations of 50 mg/dl or more. Sensitivity may also be decreased by the presence of ketone bodies.
2. Bilirubin: A false negative may occur when the sample contains ascorbic acid concentrations of 25 mg/dl or more. The yellow/orange-to-red color produced by the presence of indican (indoxyl sulfate) is also problematic.
3. Ketone: False-positive results may occur in highly pigmented samples, samples containing levodopa metabolites, or samples containing significant amounts of sulfhydryls.
4. Blood: False positives may arise from oxidizing contaminants (such as hypochlorite), or microbial peroxidase (e.g., as a result of a urinary tract infection).
5. Protein: False positives may occur with highly buffered or alkaline samples or from contamination with quaternary ammonium compounds (antiseptics, detergents).

3.4.6 Quality Control

A control urine is run with each batch of patient samples. In most instances, this is a pooled control urine that has been aliquoted.

3.4.7 Normal/Therapeutic Ranges

1. Glucose: The small amount of glucose excreted normally by the kidneys generally produces a result considered negative to trace.
2. Bilirubin: The detection of even trace amounts of bilirubin in the urine is abnormal.
3. Ketone: A negative result is typical. Positive results can indicate physiological stress (e.g., fasting and pregnancy). Ketoacidosis and other abnormalities of carbohydrate, lipid, and amino acid metabolisms may yield high levels of ketones in the urine and positive strip results.
4. Blood: The development of non-hemolyzed blood (intact erythrocytes) or a green color (free hemoglobin/myoglobin) on the strip is not typically normal and may warrant further investigation.
5. pH: Urinary pH may range from 5 to 9 for both normal and abnormal samples.
6. Protein: A result beyond trace amounts is suggestive of proteinuria and renal dysfunction.

3.4.8 Notes

Refer to the package insert for details concerning the chemical principles of each reagent patch. If samples are allowed to sit at room temperature for prolonged periods before testing, microbial proliferation may occur and alter the results.

3.5 Tyrosine Metabolites: The Nitrosonaphthol Test

3.5.1 Principle

In the presence of nitric acids, 4-hydroxylated phenolic acids (and to some extent also hydroxylated indoles from tryptophan) conjugate with 1-nitroso-2-naphthol to yield orange/red chromophores [3]. The structure of 1-nitroso-2-naphthol is shown in Fig. 3.1.

3.5.2 Specimen Requirements

This test requires 1 ml of clear voided urine that is free of particulate matter. If present, any precipitate is removed by centrifugation for 5 min at 3000 × *g*. The urine sample is stored frozen.

3.5.3 Reagents

1. Reagent A: 2.63 N nitric acid (one volume of concentrated nitric acid in five volumes of deionized H_2O). Store the solution at 4 °C; it is stable for 1 year. Avoid inhalation and contact with skin or mucous membranes.
2. Reagent B: 2.5 g sodium nitrite dissolved in 100 ml H_2O. Store at 4 °C. Avoid inhalation and contact with skin or mucous membranes.
3. Reagent C: 100 mg 1-nitroso-2-napthol dissolved in 100 ml 95% ethanol. Store at 4 °C in a brown glass bottle; it remains stable for 3 months. Avoid inhalation and contact with skin or mucous membranes. This reagent is an irritant and possible carcinogen.
4. Standard: 0.01 M N-acetyl-L-tyrosine in 1 M HCl: 40 mg/100 ml 1 M HCl. Use only L-tyrosine (human form). Store at 4 °C.

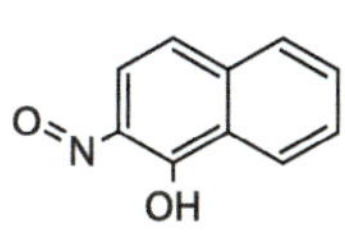

Fig. 3.1 Structure of 1-nitroso-2-naphthol

3.5.4 Procedure

1. Label a 13 × 100 mm test tube with the patient's number or name. Add three drops of their urine to the test tube.
2. Prepare the positive controls. Label four 13 × 100 mm test tubes as "10", "20", "40", and "60", respectively. These correspond to the amount of standard in microliters added to the tubes and provide a range of positive controls from trace to 4 + .
3. Add 10 μl standard to the tube labeled "10", 20 μl to that labeled "20" tube, 40 μl to the "40" tube, and 60 μl to the "60" tube.
4. Add three drops of water to each control tube.
5. Add 1 ml of reagent A to each patient and control tube.
6. Quickly add 1 drop of reagent B to each tube.
7. Quickly add 200 μl of reagent C to each tube.
8. Vortex all tubes and incubate them at room temperature for 15 min.
9. After 15 min, observe the color in each tube. The "10" control tube yields the color intensity of a trace positive. Any patient tube color that is less intense than the "10" control tube is a negative result. Grade any positives according to the range given by the control tubes. Positive controls should yield results as listed in Table 3.1.

3.5.5 Quality Control

A control urine is run with each batch of patient samples. As for other control samples in this section, this is generally a pooled urine sample that has been aliquoted and stored frozen.

Table 3.1 Nitrosonaphthol test results

Tyrosine control	Qualitative grade	Color
40	Trace	Light orange-red
80	1 +	Orange-red
120	2 +	Deep orange-red
240	3 +/4 +	Red/immediate red

3.5.6 Normal/Therapeutic Ranges and Interpretations

An orange-red color suggests the presence of hydroxylated tyrosine analogs (e.g., 4-hydroxyphenylpyruvate, 4-hydroxyphenyllactate, and 4-hydroxyphenylacetate), as well as tyrosine. A key limitation is the occurrence of transient tyrosinemia of the newborn, a not uncommon finding as the hepatic enzymes involved in tyrosine metabolism are not well developed in the newborn. 4-Hydroxyphenylacetate levels may be elevated due to intestinal bacterial metabolism, malabsorption, and other disorders, and may lead to false positives. Some disorders of carbohydrate metabolism, and unassociated liver dysfunction, may alter tyrosine metabolism with artifactual results. Patients undergoing parenteral nutrition are often supplemented with tyrosine analogs, which can lead to difficulties in the interpretation of this test. Patients with tumors may excrete increased homovanillic acid and/or 5-hydroxyindoleacetic acids (end products of dopamine and serotonin metabolism, respectively), which produce pink and purple colorimetric reactions in the nitrosonaphthol reaction. Any positive should be correlated with clinical history and/or followed-up with more specialized testing (blood amino acids/urine organic acids).

References

1. Buist NR (1968) Set of simple side-room urine tests for detection of inborn errors of metabolism. Br Med J 22:745–749
2. O'Brien D, Ibbott FA, Rodgerson DO (1968) Laboratory manual of pediatric Microbiochemical techniques (4th edn). Harper and Row, New York, p 70
3. Shih VE, Mandell R, Sheinhait I (1991) General metabolic screening tests. In: Hommes FA (ed) Techniques in diagnostic human biochemical genetics. Wiley-Liss, New York, pp 45–68
4. Elder GH, Smyth SJ (1991) Porphyrins and porphobilinogen. In: Hommes FA (ed) Techniques in diagnostic human biochemical genetics. Wiley-Liss, New York, pp 327–344

Amino Acids

4

H. C. M. T. Prinsen and M. C. D. G. Huigen

Contents

Abbreviations

ASA	Argininosuccinic acid
CSF	Cerebrospinal Fluid
DBS	Dried Blood Spot
EQA	External Quality Assessment
ERNDIM	European Research Network for evaluation and improvement of screening, Diagnosis and treatment of Inherited disorders of Metabolism
HILIC	Hydrophilic Interaction Liquid Chromatography
HPLC	High-Performance Liquid Chromatography

H. C. M. T. Prinsen (✉)
Department of Genetics, section Metabolic Diagnostics, Wilhelmina Children's Hospital, University Medical Centre Utrecht, Utrecht, The Netherlands
e-mail: b.prinsen@umcutrecht.nl

M. C. D. G. Huigen
Translational Metabolic Laboratory, Department of Laboratory Medicine, Radboud University Medical Centre, Nijmegen, The Netherlands
e-mail: marleen.huigen@radboudumc.nl

N. Blau, F. M. Vaz (eds.), *Laboratory Guide to the Methods in Biochemical Genetics*,
https://doi.org/10.1007/978-3-031-58819-8_4

IEM	Inborn Error of Metabolism
IUPAC	International Union of Pure and Applied Chemistry
MRM	Multiple Reaction Monitoring
NKH	Non-Ketotic Hyperglycinemia
OPA	ortho-phthalaldehyde
pI	Isoelectric point
RP	Reversed Phase

4.1 Introduction

Amino acids are important metabolites for the human body. They are building blocks of our proteins and peptides and serve as sources of energy, are precursors for different metabolites such as neurotransmitters or can act as neurotransmitters themselves.

The synthesis of proteins predominantly takes place in the liver and this organ has a high constant amino acid composition. Proteins consist of 20 different DNA-coded amino acids and the amino acids that are incorporated into proteins are termed proteinogenic amino acids. Some of these proteinogenic amino acids are not synthesized endogenously and need to be taken up from the diet. These amino acids (methionine, threonine, valine, phenylalanine, isoleucine, leucine, lysine and tryptophan) are termed essential amino acids. For infants, the amino acids histidine, cysteine and arginine can be essential as well. Amino acids that are not incorporated into peptides and proteins are termed non-proteinogenic amino acids. In the human body, amino acids or amino acid-derived metabolites exist and are the product of amino acid biosynthesis, the enzymatic degradation of proteins and peptides or are introduced via the diet [1–3].

For a long time, the detection of amino acids in biological fluids was dependent on its reaction with ninhydrin. The coloured amino acid ninhydrin reaction product could be visualized on filter paper or quantified by photometry at 400–600 nm [4]. The experiments performed by Moore and Stein [5] resulted in the introduction of the amino acid analyzer, a prototype of the product that is still used today in many genetic metabolic laboratories. The introduction of the amino acid analyzer accelerated the large-scale investigation in the field of Inborn Error of Metabolism (IEM), since it was now possible to monitor dietary interventions in patients with IEM.

In literature abundant methods are described for analysis of amino acids. Frequently used strategies for amino acid analysis included High-Performance Liquid Chromatography (HPLC), cation exchange HPLC with post-column ninhydrin derivatization, GC-MS and LC-MS/MS related methods (with or without derivatization). The choice of a particular method is dependent of several factors. Here, we describe different approaches for amino acid analysis, that according to the authors, are used nowadays frequently in a metabolic setting. With the selection of the chosen methods, we considered the costs, the equipment, the applicability of the method and analysis time.

4.2 Clinical Significance

The clinical presentation of defects in amino acid metabolism are variable and are often non-specific. Onset of symptoms may occur in the neonatal period to adolescence or patients may be asymptomatic (e.g., newborn screening). Clinical indications for amino acid analysis are diverse and are summarized by Sharer et al. [6]. Amino acid analysis is commonly performed in biological specimens like plasma, serum, dried blood spot (DBS), cerebrospinal fluid (CSF) and urine. It is evident that the physiological abundance of amino acids varies for these different matrices. In the genetic metabolic laboratory, the analysis of amino acids is performed to diagnose inborn errors in amino acid metabolism or transport or for diagnosis of inborn errors in the urea cycle [6–8]. However, disturbances in amino acid concentrations are not only related to IEM. Amino acids can also serve as sensitive markers for nutritional status or reflect the function of various organs such as the liver, kidney, intestine, or muscle. Therefore, the amino acid analysis is also frequently performed for diet monitoring in patients with a known IEM or for nutritional assessment of patients with non-metabolic conditions (e.g., short bowel syndrome or proximal tubular dysfunction) due to medication [7, 9].

4.3 Properties of Amino Acids

Amino acids or 2-amino carboxylic acids are classified into different categories. The subdivision is based on the chemical structure of the sidechain and their polarity. Amino acids are low-molecular weight mono- or dicarboxylic acids containing at least one amino group. The carboxylic acid group contains a proton that is loosely bound and thus tends to have a slightly negative charge. On the other hand, the nitrogen atom of the amino group has a free electron pair and is prone to bind a proton, resulting in a positively charged molecule. The pH at which the amino acid has no net charge, despite two ionized groups is called the isoelectric point (pI). At the pI, the amino acid is a zwitterion (or dipolar ion) since it has both anionic and cationic properties. Amino acids with two carboxylic groups have a low pI and those with two amino groups have a high pI. Since amino acids differ in composition, it is obvious that each amino acid has a different pI (Table 4.1). The differences in pI and

Table 4.1 Some properties of the most relevant amino acids. IUPAC name: Name of the amino acid according to the nomenclature of the International Union of Pure and Applied Chemistry, *MW* Molecular weight, *pI* isoelectric point

Amino acid	IUPAC name	MW	pI
Alanine	2-Aminopropanoic acid	89.1	6,00
Allo-isoleucine	2-Amino-3-methylpentanoic acid	131.2	
Arginine	2-Amino-5-carbamimidamidopentanoic acid	174.2	10.76
Argininosuccinic acid	(N-4-Amino-4-carboxybutyl) carbamimidoyl-aminobutanedioic acid	290.3	
Asparagine	2-Amino-3-carbamoylpropanoic acid	132.1	5.41
Aspartic acid	2-Aminobutanedioic acid	133.1	2.77
Citrulline	2-Amino-5-(carbamoylamino)pentanoic acid	175.2	5.91
Cystine	2-Amino-3-[(2-amino-2-carboxyethyl)disulfanyl]propanoic acid	240.3	
Figlu	2-Methanimidamidopentanedioic acid	174.2	
Glutamic acid	2-Aminopentanedioic acid	147.1	3.22
Glutamine	2-Amino-4-carbamoylbutanoic acid	146.1	5.65
Glycine	2-Aminoacetic acid	75.1	5.97
Histidine	2-Amino-3-(1H-imidazol-5-yl)propanoic acid	155.2	7.59
Homoarginine	2-Amino-6-carbamimidamidohexanoic acid	188.2	
Homocitrulline	2-Amino-6-(carbamoylamino)hexanoic acid	189.2	
Homocystine	2-Amino-4-[(3-amino-3-carboxypropyl)disulfanyl]butanoic acid	268.4	
Hydroxyproline	4-Hydroxypyrrolidine-2-carboxylic acid	131.1	
Isoleucine	(2S,3S)-2-Amino-3-methylpentanoic acid	131.2	6.02
Leucine	2-Amino-4-methylpentanoic acid	131.2	5.98
Lysine	2,6-Diaminohexanoic acid	146.2	9.74
Methionine	2-Amino-4-(methylsulfanyl)butanoic acid	149.2	5.74
Ornithine	2,5-Diaminopentanoic acid	132.2	
Phenylalanine	2-Amino-3-phenylpropanoic acid	165.2	5.48
Phosphoethanolamine	(2-Aminoethoxy)phosphonic acid	141.1	
Pipecolic acid	Piperidine-2-carboxylic acid	129.2	
Proline	Pyrrolidine-2-carboxylic acid	115.1	6.30
Saccharopine	2-[(5-Amino-5-carboxypentyl)amino]pentanedioic acid	276.3	
Serine	2-Amino-3-hydroxypropanoic acid	105.1	5.68
Sulphocysteine	3-Sulfanyl-2-(sulfoamino)propanoic acid	201.2	
Taurine	2-Aminoethane-1-sulphonic acid	125.1	
Threonine	2-Amino-3-hydroxybutanoic acid	119.1	5.60
Tryptophan	2-Amino-3-(1H-indol-3-yl)propanoic acid	204.2	5.89
Tyrosine	2-Amino-3-(4-hydroxyphenyl)propanoic acid	181.2	5.66
Valine	2-Amino-3-methylbutanoic acid	117.1	5.96

length of the amino acid chain can be used to separate amino acids.

4.4 Specimens for Analysis

The appropriate sample for investigation aminoacidopathies is plasma, but for the analysis of transport defects, renal amino acidurias or amino acid biosynthesis defects an additional urine sample or CSF is required. Most laboratories prefer heparinized plasma, but EDTA plasma or serum can also be analyzed. Usually, 100 μL–300 μL plasma, depending on the method that is used, is sufficient for analysis. To prevent haemolysis, it is important to separate plasma from the blood cells as fast as possible. Due to a high intracellular level of specific amino acids in erythrocytes (e.g., glutamic acid, aspartic acid and taurine), haemolysis can result in an increase of these amino acids in plasma. Furthermore, the degradation of erythrocytes will result in the release of the enzyme arginase, which will convert arginine into ornithine. Sulphur-containing amino acids cysteine, homocysteine and tryptophan tend to bind plasma proteins. Therefore, care should be taken in the deproteinization of biological samples in order to preserve these amino acids.

Although a fasting plasma is preferred, many blood samples are taken in daily practice during the morning or afternoon. This may result in mildly increased levels of (essential) amino acids, which is non-specific. Therefore, caution should be taken in the diagnostics of an amino acid biosynthesis defect that necessitates a fasting plasma sample or CSF.

The analysis of amino acids is also routinely performed in DBS. DBS sampling provides advantages compared to plasma, which is less invasive compared to whole blood sampling. DBS sampling can be assessed by patients or caretakers and DBS transport requires less resources for shipping, storage and archiving. One should be aware that DBS and plasma concentrations are influenced by several factors including, differences in distribution of the amino acids between plasma and erythrocytes, extraction efficiency from DBS, sample preparation, derivatization of sample and methodological test biases [10–14], implying that for therapy control, plasma and DBS amino acid concentrations cannot easily be exchanged. Significant differences in amino acid concentrations (e.g., phenylalanine) are reported for plasma and DBS [11, 14]. For PKU patients, maintaining adequate phenylalanine concentrations is the strongest determinant for cognitive outcome and advised age-dependent ranges of blood phenylalanine are presented in European guidelines. For patients below 12 years of age and for patients during pregnancy, the advised range of blood phenylalanine concentration is 120–360 μmol/L, whilst for patients older than 12 years the advised blood phenylalanine range from 120 to 600 μmol/L [15, 16]. Since no preferred sample type (DBS or plasma) is defined in these European guidelines, the clinicians should be aware that treatment outcomes can be different for PKU patients when monitored by plasma or DBS [11, 14]. For uniformity, it is advisable to use a DBS-plasma correction factor [17]. DBS is not only used for monitoring therapy control for PKU patients but also for other IEM [18].

For the diagnosis of neurometabolic aminoacidopathies, amino acid profiling in CSF is essential. CSF contains few potential interfering compounds in contrast to plasma and urine. However, the sampling of CSF is much more complex since a lumbar puncture is time consuming and invasive for the patient. For many metabolites, there is a rostro-caudal gradient present [19] and sampling of CSF should preferably be performed according to a standardized protocol. Usually, the first 0.5–1 mL is used for analysis of biogenic amines, whilst the second mL is used for amino acid analysis. Amino acid concentrations in CSF are lower compared to plasma, except for glutamine. Therefore, one must be aware that the CSF is not contaminated with blood, the result of a traumatic lumbar puncture or impaired blood–brain barrier permeability resulting from epilepsy or asphyxia. Release of amino acids from erythrocytes results in a non-specific increase of multiple amino acids masking diagnostic abnormalities. Therefore, it is highly recommended to centrifuge the CSF prior to analysis. In addition, it is meaningful to analyze amino acids that are normally not present in CSF such as proline, cystine and tryptophan. These amino acids serve as quality control, in order to detect blood contamination biochemically, since blood contamination is not always clearly visible [20, 21].

When Non-Ketotic Hyperglycinemia (NKH) or glycine encephalopathy (OMIM # 605899) is clinically suspected, plasma amino acids should be analyzed as a first step. If an isolated increased concentration of glycine is observed, and no abnormal metabolites are found in urine organic acid analysis (excluding propionic, methylmalonic aciduria or use of anti-epileptic drugs (valproic acid)), glycine concentrations should be analyzed in CSF and plasma. CSF/plasma amino acid ratios are more informative than an isolated CSF sample and therefore, it is highly recommended to obtain a paired plasma sample preferably within 2 h of CSF sampling. The diagnosis of NKH is biochemically suspected based on an elevated CSF/plasma glycine ratio. The analysis of amino acids in CSF and plasma is not only appropriate for NKH, but also applicable for diagnosis of amino acid biosynthesis defects. It is important to realize that IEM presenting with a neurometabolic phenotype do not always express a biochemical phenotype and therefore diagnosis can be challenging [22].

Urine amino acid analysis can be performed in a random urine sample (100–1000 μL). In the kidney nephron, the proximal tubule is the major site of nutrient absorption and under normal circumstances about 95–99% of all amino acids are reabsorbed. Serine, glycine and histidine are the major components lost in urine, whilst other amino acids are only in low amounts available. Due to the non-invasiveness of its collection and the possibility of home sampling, urine represents an attractive biological matrix for metabolic investigation. Proper collection and storage are essential to avoid misleading conclusions. Despite the hydrophilic nature of amino acids, some of them are poorly soluble in water. Cystine (dimer of cysteine) and tyrosine are poorly soluble, and care should be taken to fully dissolve these amino acids in urine prior to amino acid analysis. In daily practice, it is often hard to obtain a 24-h urine sample and therefore a spot urine is frequently used for metabolic profiling with amino acid excretion normalized per creatinine. The preferred sample is a first-morning urine specimen to minimize confounders from dietary intake. The spot urine should be stored deep frozen or preservatives can be added to avoid bacterial growth and metabolite degradation [6, 7].

Abnormal excretion of amino acids, either specific or non-specific gives rise to further investigation. Disruptive functioning of renal tubules results in impaired transport as characterized by a generalized aminoaciduria or by other renal tubular defects that affect the reabsorption of phosphate, bicarbonate, uric acid and glucose. In general, an increased excretion of amino acids in combination with near-normal plasma amino acids is suggestive of an inherited or acquired renal transporter defect. Therefore, for correct interpretation of the amino acid profile, it is evident to evaluate both amino acids in plasma and urine. Specimens for amino acid analyses that cannot be analyzed immediately should be kept frozen at −20 °C or preferably −80 °C for CSF.

To summarize, each matrix has its (dis)advantages and the matrix choice for amino acids analysis is dependent on the question that needs to be answered.

4.5 High-Performance Liquid Chromatography (HPLC)

In literature, several methods are described for the measurement of amino acids by HPLC [23–27].

4.5.1 Principle of the Assay

HPLC is an effective type of column chromatography. In HPLC chromatography, the column is important for the separation of amino acids because it contains the stationary phase. In general, for separation of amino acid analysis a Reversed Phase (RP) column is used. However, it is noteworthy that separation by RP-HPLC alone is not specific and sensitive enough for all amino acids. Therefore, alternatives were developed and a technique that is still used today is the analysis of amino acids by precolumn fluorescent derivatization coupled to RP-HPLC. The most commonly used agent for derivatization of amino acids is Ortho-phthalaldehyde (OPA). OPA is a derivatization agent for amino acids because it reacts with amino groups to form highly fluorescent products in the presence of excess thiols. These compounds can be visualized and measured. The method described below is an example of precolumn fluorescent derivatization coupled to RP-HPLC and in literature several alternatives are available.

4.5.2 Reagents and Chemicals

All chemicals and solvents should have the highest purity available for HPLC grade. The following reagents and chemicals are required.

Ortho-phthaldialdehyde 97% and potassium borate buffer 1.0 M were obtained from Sigma-Aldrich (Zwijndrecht, the Netherlands). 2-Mercapto-ethanol from BioRad (Lunteren, the Netherlands) and Lithium citrate buffer B from Pfizer (Capelle aan den IJssel, the Netherlands). The methanol 99.8% Sodium acetate p.a., acetic acid 100%, HCl 30%, Isopropanol, EDTA, and 5-sulphosalicylic acid were purchased from Merck (Amsterdam, the Netherlands).

Preparation of reagents, internal standard and calibrators was performed as follows. The OPA reagent is prepared by dissolving 20 mg in 500 μL methanol. To this solution 3 ml of the potassium borate buffer and 2 mL H_2O is added, followed by the addition 30 μL of mercapto-ethanol. This reagent is stable for 1 week when stored in the dark. Stock solution sodium acetate (1 M) is prepared by dissolving 82 g of sodium acetate and 375 mg EDTA in 1000 mL H_2O. Eluent A: 0.01 M NaAc +9.1% methanol, pH 5.75 is prepared by diluting 100 mL of the stock NaAc to 1000 mL and 100 ml of methanol and the pH of the eluent was adjusted to 5.75 with acetic acid. The eluent is stable at room temperature under helium for 2 weeks. Eluent B was prepared as follows. To 1000 mL methanol, 20 mL isopropanol and 100 mL H_2O was added. Eluent B is stable for 4 weeks. Five percent sulphosalicylic acid (SSA) was prepared by dissolving 5 g SSA in 100 mL H_2O. The internal standard is carboxymethylcysteine. Ten milligrams of carboxymethylcysteine was dissolved in 10 mL 5% SSA and was, prior to use diluted 1:20. Labile amino acids such as tryptophan, glutamine, asparagine and homocarnosine should be prepared fresh at each calibration step but can be stored at −20 °C for at least 4 weeks.

4.5.3 Instrumentation

A regular HPLC instrument with an ordinary reversed-phase column and a fluorescence detector is sufficient for proper analysis. The sensitivity of the detector makes it ideally suited for CSF analysis, although most of the reported applications are for plasma. Because of the limited stability of the OPA derivatives, the derivatization procedure should be carried out in an intelligent autosampler such as the Gilson 231XL sampler, equipped with a temperature-controlled sample holder (Gilson 832, Den Haag, the Netherlands). HPLC separation is achieved on an Alltech OPA-HR LC-8 column (5 micron; 150 × 4.6 mm,

Grace, Columbia, USA) protected by a Supelco LC-18 pellicular guard column (20 micron; 20 × 4.6 mm, Merck, Amsterdam, the Netherlands). Fluorescence detection (Jasco FP-1520, Jasco, de Meern, the Netherlands) comprises an excitation wavelength of 330 nm and an emission wavelength of 450 mm.

4.5.4 Analytical Procedure

In brief, a 200-μL sample is deproteinized with 20 μL sulphosalicylic acid 5% containing internal standard. Following precipitation and centrifugation, 150 μL supernatant is diluted with 150 μL of the lithium citrate buffer and placed in the autosampler. The derivatization step takes 40 μL of sample and 20 μL of the OPA reagent which are mixed and left to react for 30 sec. An amount of 20 μL is injected. Separation is achieved at 21 °C using the elution programme shown in Table 4.2.

4.5.5 Calculation

Amino acids are detected by a fluorescence detector and the electric signal is recorded. Amino acids are detected and identified based on the retention time that is determined for every amino acid under defined experimental conditions. The calculation was based on the area under peak for a given amino acid of known concentration.

Table 4.2 HPLC elution programme

Step	*Time (min)*	*Flow (ml/min)*	*%A*	*%B*
0	*0*	*0.5*	*95*	*5*
Injection				
1	*8*	*1.0*	*95*	*5*
2	*17*	*1.0*	*65*	*35*
3	*15*	*0.6*	*20*	*80*
4	*1*	*1.0*	*10*	*90*
5	*2*	*1.0*	*0*	*100*
6	*8*	*1.0*	*95*	*5*

4.5.6 Post-analytical

Amino acid profiles in CSF for three different IEM are depicted in Fig. 4.1. The plots demonstrate HPLC chromatograms from patients with argininosuccinate lyase deficiency, NKH and maple syrup urine disease, respectively.

4.5.7 Limitations of the Method

The instrumentation and material for HPLC analysis are more cost-effective than the traditional ion-exchange systems or mass spectrometers. However, one should be aware that this equipment is less/not suitable for separation of isobaric amino acids and interfering compounds (e.g., artefacts from the diet and medication) that can be present in physiological fluids. Furthermore, problems with instability and reproducibility of derivatization agents are reported. For instance, OPA is frequently used for derivatization, despite its inability to react with all relevant amino acids such as proline, hydroxyproline and sulphuric-containing amino acids. The instability of the reaction products, make an automated derivatization system coupled with an automated injector and constant retention times an absolute necessity. Considering these considerations, the HPLC analysis is useful for biological fluids other than urine. The use of this equipment for amino acid analysis is therefore not the first line of choice in many genetic metabolic laboratories, since more sophisticated equipment is needed.

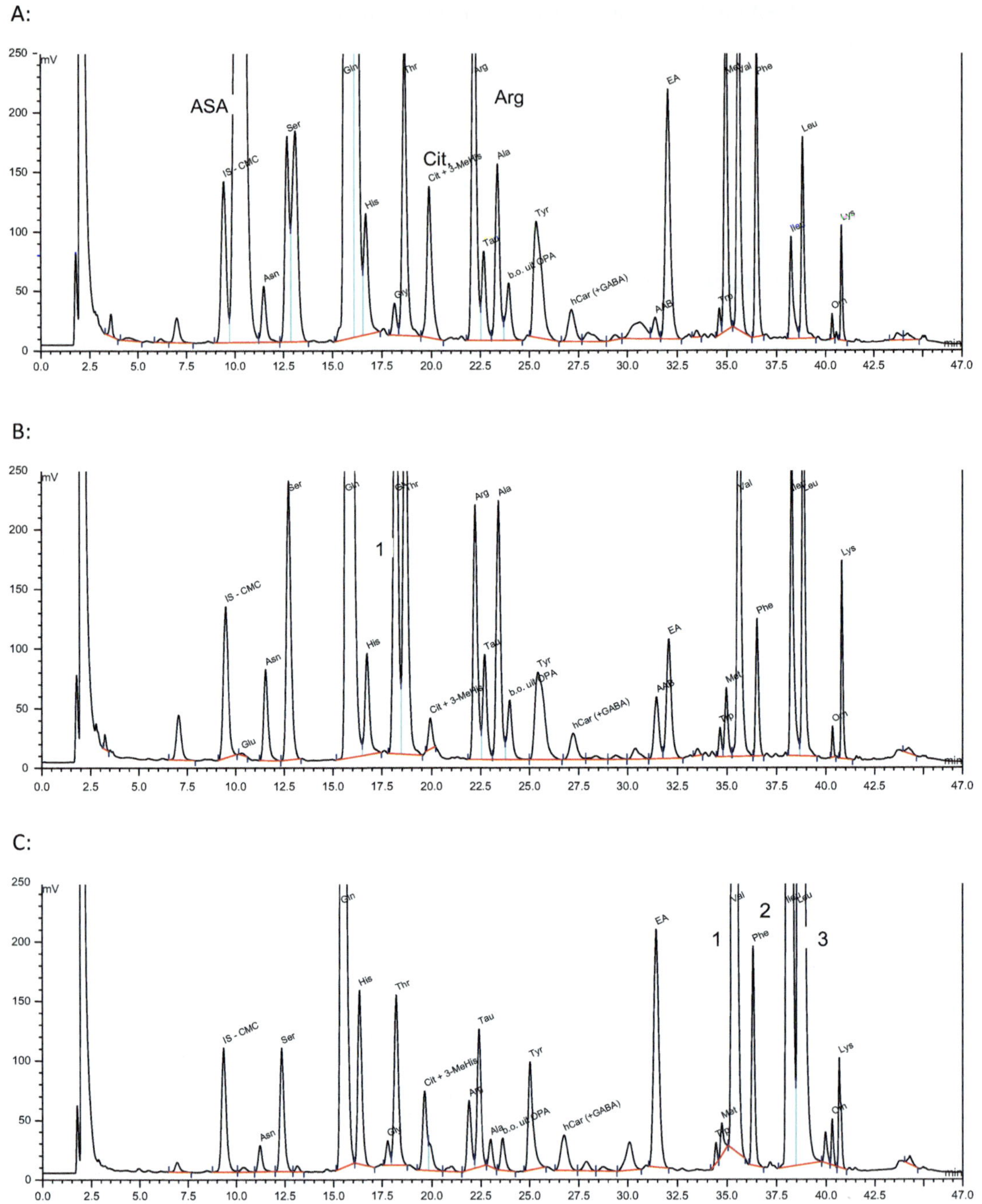

Fig. 4.1 (**a**) Cerebrospinal fluid amino acids by HPLC in a patient with argininosuccinate lyase deficiency. Note the extremely high level of argininosuccinate (ASA), which is higher than the plasma level. In addition, the concentration of citrulline—precursor of ASA in the urea cycle—was increased. The patient was on arginine supplementation treatment, hence an increased Arg level in this fluid. (**b**) Cerebrospinal fluid amino acids by HPLC in a patient with non-ketotic hyperglycinemia (NKH). Glycine (peak 1) was elevated, no other changes were observed. CSF is the fluid of choice for the diagnosis of NKH. (**c**) Cerebrospinal fluid levels of amino acids by HPLC in a patient with maple syrup urine disease (untreated). High levels of leucine (peak 3), isoleucine (peak 2) and valine (peak 1) are the hallmarks of this disorder. The HPLC approach is unable to separate isoleucine and allo-isoleucine

4.6 Amino Acid Analyzer (Ion Exchange Chromatography)

4.6.1 Principle of the Assay

Amino acids can be separated without prior derivatization using ion exchange chromatography. The basic principle of this separation technique is the interaction of acids, that are present in the elution buffer (e.g., amino acids), with the stationary phase that consists of small sulphonated polystyrene particles. The negative charge of these sulphonated acid residues is balanced by the lithium cations that are present in the elution buffer. Since the separation process takes place in a mild acidic environment, resulting in protonation of the carboxylic residues, the interaction with the stationary phase of the column is achieved by the protonated (and thus positively charged) amino group(s) of the amino acids. This indicates that amino acids with a high pI have a strong interaction with the stationary phase and consequently have a longer retention time. This principle is the basic mode of separation of the amino acid analyzer. In daily practice, the separation of amino acids is achieved using a combination of buffers with different pH, lithium citrate concentrations and a temperature gradient. Quantification of the amino acids occurs upon the post-column reaction of amino acids with ninhydrin at a temperature of approximately 135 °C. Ninhydrin interacts with the free amide group of the amino acid and this interaction produces a purple complex for the primary amino acids and a yellow complex for the secondary amino acids. The purple complex is detected at an absorbance of 570 nm, whilst the yellow complex is detected at an absorbance of 440 nm. It is important to realize that every metabolite with a free amide group can react with ninhydrin and can therefore co-elute with amino acids and interfere with the amino acid chromatogram. In general, the order of elution of the amino acids is acidic amino acids, followed by neutral amino acids and finally basic amino acids. Amino acids are identified by comparing the retention time of the amino acid and the 570/440 ratio of the peaks that are standardized with certified reference material.

4.6.2 Reagents and Chemicals

All chemicals and solvents should have the highest purity available for HPLC grade. The following reagents and chemicals are required. Amino acid standard solution (AAS18), S-aminoethylcysteïne (SAEC), asparagine, and 2-propanol were obtained from Sigma-Aldrich (Zwijndrecht, the Netherlands). Glutamine and SSA were purchased from VWR (Amsterdam, the Netherlands). Homocitrulline was obtained from Santa Cruz Biotechnology (Heidelberg, Germany). Biochrom Lithium buffer 1 pH = 2.20, Lithium buffer 2 pH = 3.00, Lithium buffer 3 pH = 3.15, Lithium buffer 4 pH = 3.50, Lithium buffer 5 pH = 3.55, Lithium regeneration buffer 6 and ready to use ninhydrin (EZ Nin) were obtained from Isogen Life Science (De Meern, the Netherlands).

Preparation of reagents and calibrators was performed as follows. 12.5 g SSA was added to 40 mg SAEC, and the mixture was dissolved in milliQ-water to a final concentration of 800 μM for SAEC and 5% for SSA. Preparation of a one-point calibrator for asparagine, glutamine and homocitrulline was achieved by dissolving 16.5 mg asparagine, 18.2 mg glutamine and 23.6 mg homocitrulline in 250 milliQ-water, resulting in final concentrations of 500 μmol/L for these amino acids. 2-Propanol solution (20%) was prepared by adding 200 mL 2-propanol to 800 milliQ-water.

4.6.3 Instrumentation

Amino acid analyzers are offered by different companies and the equipment should give comparable results. In general, the instrument is equipped with software that controls the inlet of the buffers and the temperature of the column. The manufacturer will supply a basic separation programme for physiological fluid that can be adapted depending on the amino acids of interest.

The response of the system is calibrated using a commercial standard solution (AAS18) to which known amounts of freshly prepared (unstable) amino acids (asparagine, glutamine and homocitrulline) are administered. To obtain reproducible results, it is advisable to check the quality of the ninhydrin solution regularly. Therefore, an aqueous amino acid standard mixture should be analyzed before every run and adjustment of the response factors is necessary when the amino acid concentrations do not fulfil the quality controls. Reduced ninhydrin tends to be unstable and therefore all kinds of precautions can be taken (e.g., storage in the dark and cold, administration of a reducing agent and oxygen reduction) to prevent deterioration of the ninhydrin solution. Nowadays, ready-to-use ninhydrin that guarantees no degradation with time or oxygen exposure is commercially available.

4.6.4 Analytical Procedure

All samples that are analyzed need to be deproteinized prior to application onto the column, since the resin is sensitive to protein contamination. The procedure that is used most frequently combines both the deproteinization step with SSA and the addition of the internal standard. The procedure for analysis of amino acids in plasma, CSF and urine is described here. Hundred µL plasma, CSF or urine was mixed with 100 µL SAEC in 5% SSA in an Eppendorf tube and the sample was stored for 30 min at 4 °C. The sample was centrifuged for 5 min at 17000 × g in an Eppendorf centrifuge at room temperature. The supernatant was transferred to a 0.2 µm GHP filter plate (VWR, Amsterdam). The filtrate was collected in a 96-wells plate. The sample was subsequently transferred to glass vials, whereafter the sample was ready for analysis. The injection volume of the prepared samples is 40 µL.

4.6.5 Calculation

After analysis, amino acid peaks are integrated. For the amino acid analyzer, peak area and concentration of each specific amino acid are linear over a certain range. Therefore, it is essential to determine the linear range of each amino acid during validation studies to ascertain that the calculated amino acid concentration is in the linear range. At the end of each series, the AAS18 mixture is analyzed again to control the performance of the system. Periodically, recalibration of the system is necessary, when a new ninhydrin batch is used.

4.6.6 Post-analytical

Identification of the amino acids is assessed by comparing the retention time and 570/440 ratio with reference substances.

Figure 4.2 shows the chromatogram of an amino acid standard solution (500 µmol/L for each amino acid), that is analyzed by an amino acid analyzer. The absorbance of the amino acids at 570 nm is shown in blue and the absorbance of the amino acids at 440 nm is shown in red. A characteristic amino acid profile in the urine of a patient with cystinuria is shown in Fig. 4.3. The "COLA" amino acids (cystine, ornithine, lysine and arginine) are clearly increased.

4.6.7 Limitations of the Method

Although the amino acid analyzer is excellent in terms of reproducibility for the analysis of amino acids, main disadvantages are the long analysis time (approximately 3 h per sample including column regeneration) and that the equipment is exclusively applicable for the analysis of amino acids.

Some laboratories developed more specific and sensitive methods for amino acids that could not be reliably determined using an amino acid analyzer, such as pipecolic acid in CSF. However, with the introduction of tandem mass spectrometry in most laboratories, these assays are hardly performed anymore.

As described above, ninhydrin reacts with the amide group of amino acids to visualize amino acids by optical detection. Iatrogenic compounds such as aminoglycosides, antiepileptic drugs and other medications (e.g., acetaminophen) with a free amide group can interfere with the amino acid chromatogram.

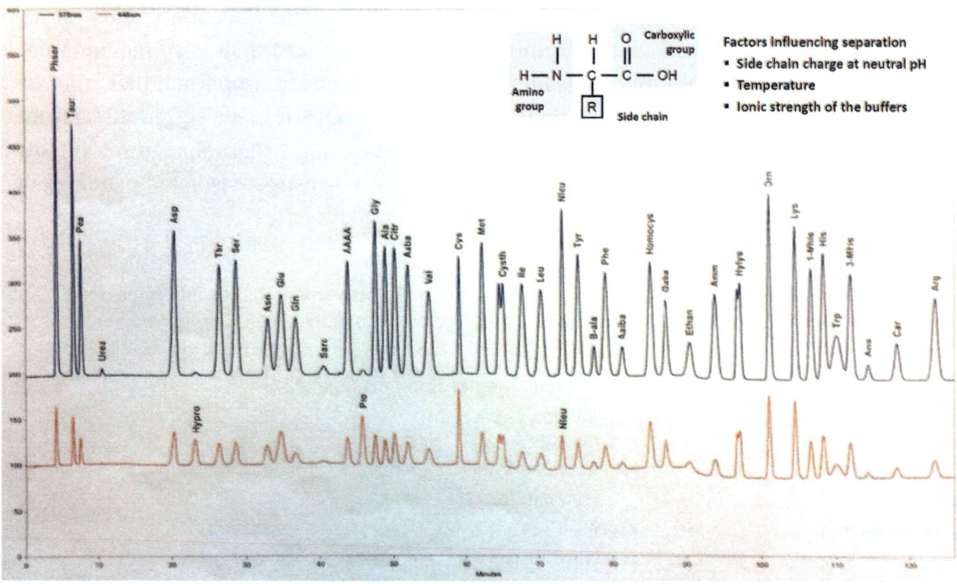

Fig. 4.2 Amino acid profile of an amino acid standard solution analyzed by an amino acid analyzer. The concentration of each amino acid is 500 μmol/L. The absorbance of the amino acids at 570 nm is shown in blue and the absorbance of the amino acids at 440 nm is shown in red. Norleucine is used as an internal standard. *Phser* phosphoserine, *Tau* taurine, *Pea* phosphoethanolamine, *Asp* aspartic acid, *Hypro* hydroxyproline, *Thr* threonine, *Ser* serine, *Asn* asparagine, *Glu* glutamic acid, *Gln* glutamine, *Sarc* sarcosine, *AAAA* alpha-amino-adipic acid, *Pro* proline, *Gly* glycine, *Ala* alanine, *Citr* citrulline, *Aaba* alpha-amino-butyric acid, *Val* valine, *Cys* cystine, *Met* methionine, *Cysth* cystathionine, *Ile* isoleucine, *Leu* leucine, *Nleu* norleucine, *Tyr* tyrosine. *B-ala* beta-alanine, *Phe* phenylalanine, *Aaiba* alpha-amino-isobutyric acid, *Homocys* homocystine, *Gaba* gamma-aminobutyric acid, *Ethan* Ethanolamine, *Amm* ammonia, *Hylys* hydroxylysine, *Orn* ornithine, *Lys* lysine, *1-Mhis* 1-methylhisitidine, *His* histidine, *Trp* tryptophan, *3-Mhis* 3-methylhistidine, *Ans* anserine, *Car* carnosine and *Arg* arginine

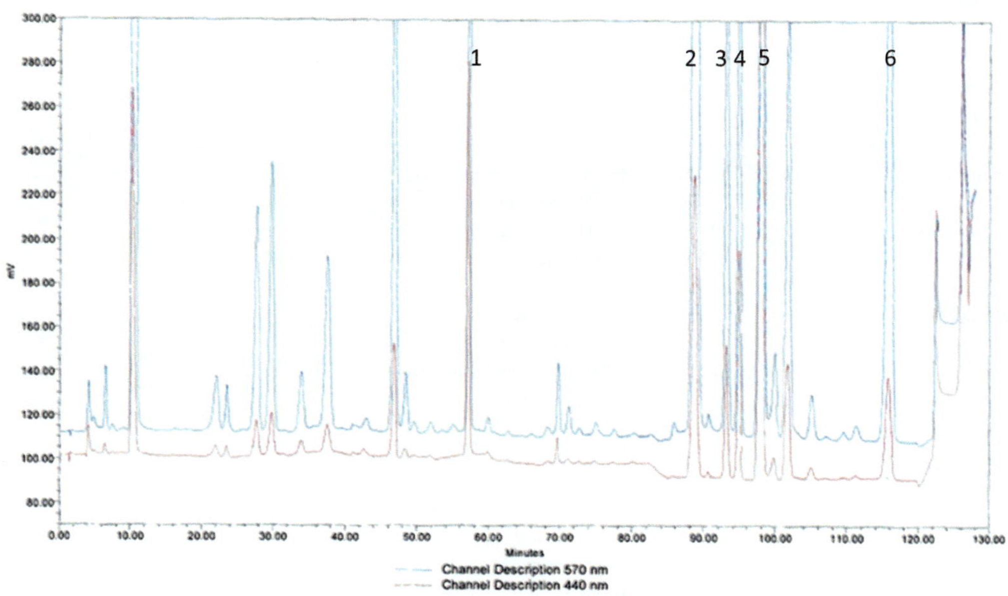

Fig. 4.3 Urinary amino acid profile of a patient with cystinuria analyzed by an amino acid analyzer. Cystinuria is characterized by an elevated excretion of cystine, ornithine, lysine and arginine. The indicated peaks are (1) cystine, (2) ammonia, (3) S-aminoethylcysteïne (SAEC), (4) ornithine, (5) lysine and (6) arginine

Furthermore, Kavanagh et al. [28] reported that pregabalin co-chromatographed with arginine and Wiencek et al. [29] reported that zonisamide co-chromatographed with allo-isoleucine on ion-exchange chromatography resulting in false-positive results for these amino acids. In Table 4.3, unusual amino acids and related metabolites in the region of known amino acids are depicted. For correct interpretation of the amino acid profile, it is therefore important that laboratory specialists be aware of these pitfalls and that other techniques (e.g., mass spectrometry) should be applied to investigate isolated increases of amino acids in these cases.

Table 4.3 Unusual amino acids and related substances which may be encountered in inborn errors of metabolism. Note: This table is only relevant for the amino acid analyzer. The position of the neighbouring components is not always well defined and therefore the laboratory should try and collect as many reference substances as possible

Usual component	Neighbouring/coeluting component(s)
Cysteic acid	Sulphocysteine Phosphoserine Cysteine-2-mercaptoethanesulphonic acid disulphide
Taurine	Hypotaurine
Phosphoethanolamine	
Urea	Hawkinsin AADG (aspartylglucosaminuria) Reduced glutathione
Aspartic acid	Δ^1-Pyrroline-3-hydroxy-5-carboxylic acid
4-Hydroxyproline	Methionine-sulphoxide Methionine-sulphone
Threonine	
Serine	Phospho-hydroxylysine Cysteine-β-mercapto-lactate disulphide
Asparagine	
Glutamic acid	
Glutamine	Δ^1-Pyrroline-5-carboxylic acid S-Carboxy-isopropylcysteine (isobuteine)
Sarcosine	
2-Aminoacidpic acid	Oxidized glutathione Cysteine
Proline	Penicillamine
Glycine	
Alanine	
Citrulline	Formiminoglutamic acid Galactosamine
2-aminobutyric acid	N^{ε}-acetyllysine Lanthionine
Valine	
Cystine	Saccharopine Pipecolic acid γ-Glutamylphenylalanine
Methionine	Homocitrulline Allo-isoleucine Penicillamine-cysteine disulphide
Isoleucine	Dihydroxyphenylalanine N^{ε}-carboxylmethyllysine
Leucine	Argininosuccinic acid Glycylproline
Tyrosine	Cysteine-homocysteine disulphide 3-Methoxytyrosine Erythromycin glucoheptonate

(continued)

Table 4.3 (continued)

Usual component	Neighbouring/coeluting component(s)
Phenylalanine	N^{α}-Acetyllysine Acetaminophen metabolite
β-Alanine	Δ-Aminolevulinic acid
β-Aminoisobutyric acid	GABA ASA anhydride 1 ASA anhydride 2 Homocystine Ethanolamine
Ammonia	Kynurenine γ-Vinyl-GABA Cysteinylglycine
Hydroxylysine	
Ornithine	
1-Me-histidine	
Histidine	Methyllysine
Lysine	
3-Me-histidine	
Tryptophan	Carnosine Anserine Homocarnosine
Arginine	Homoarginine 3-Aminopiperidone Methylarginine(s)

4.7 Tandem Mass Spectrometry

Although the post-column ninhydrin derivatization method is very accurate and reproducible, the method is less suitable, with respect to time, for modern genetic metabolic laboratories. Since 2005, numerous LC-MS/MS and GC-MS alternatives for amino acids analysis have developed each with their own advantages/disadvantages. Abundant literature is available and the applicability of the method that is used is dependent on the purpose.

Reversed-phase liquid chromatography, a commonly used separation mode in LC-MS applications is not ideal for amino acid analysis. Due to the hydrophilicity of these metabolites, amino acids retain a low retention on these columns. Therefore, many laboratories have introduced derivatization procedures to improve retention and separation. With the introduction of ion-pairing agents, the separation of amino acids on a C18-reversed-phase column improved. Complex derivatization procedures were no longer necessary and acceptable separation of metabolites was achieved. Unfortunately, problems with retention time stability and mass spectrometric problems with the detector in electrospray negative mode limited the use of ion-pairing reagents [30, 31]. With the introduction of Hydrophilic Interaction Liquid Chromatography (HILIC) columns for UPLC several applications (e.g., amino acid analysis) became with reach [32]. In this paragraph, mass spectrometric analysis using derivatization of amino acids is not discussed. For some recent publications about amino acid derivatization procedures see [33, 34]. An extensive review of various amino acid separation strategies using mass spectrometry is provided by Ferré et al. [35].

4.7.1 Principle of the Assay

The separation principle of HILIC is based on the strong interaction of polar compounds (amino acids and carbohydrates) with the hydrophilic phase. These compounds obtain good retention by HILIC and a benefit of HILIC is the use of aqueous organic solvent eluent that enhances ionization. In this paragraph, a method for the analysis and quantification of amino acids in several human matrices without derivatization using HILIC tandem mass spectrometry is described. Determination of amino acids in plasma is described in more detail in [36].

Plasma, CSF or urine are mixed with an internal standard solution. After vortexing, a solvent consisting of ammoniumformiate, acetonitrile and formic acid is added to precipitate protein and ionize the amino acids. The specimen is vortexed again, centrifuged and filtrated using a 96-wells plate, whereafter the sample is ready for analysis. Amino acids are separated using HILIC and analyzed by LC-MS/MS using Multiple Reaction Monitoring (MRM).

4.7.2 Reagents and Chemicals

All chemicals and solvents should preferably be UPLC grade. The following reagents and chemicals are required: 2,3,3,3-D_4-alanine, guanido-$^{15}N_2$-arginine:HCl, 2,3,3-D_3-aspartic acid, 5-^{13}C, 4,4,5,5-D_4-citrulline, 3,3,3,3-D_4-cystine, 2,4,4-D_3-glutamic acid, 2,3,3,4,4-D_5-glutamine, 1,2-$^{13}C_2$-glycine, U-$^{13}C_6$-histidine:HCl.H_2O, 3,3,3,3,4,4,4,4-D_8-homocystine, U-$^{13}C_6$,U-D_{10}, ^{15}N-isoleucine, 5,5,5-D_3-leucine, 4,4,5,5-D_4-lysine:2HCl, methyl-D_3-methionine, 5,5-D_2-ornithine:HCl, ring-D_5-phenylalanine, 2,3,3,4,4,5,5-D_7-proline, 1,2,3-$^{13}C_3$-serine, 1,2-^{13}C-taurine, U-$^{13}C_4$, U-D_5,^{15}N-threonine, indole-D_5-tryptophan, ring-D_4-tyrosine and 1-^{13}C-valine were purchased from Cambridge Isotope Laboratories (Massachusetts, USA). D_9-pipecolic acid, 2,3,3-D_3-S-Sulpho-cysteine were purchased from CDN-isotopes (Quebec, Canada). D_8-homocystine was purchased from Buchem (Apeldoorn, the Netherlands), 3-amino-piperidone, $^{13}C_4$-$^{15}N_2$-asparagine, alanine, delta-aminolevulinic acid, arginine, asparagine, aspartic acid, argininosuccinic acid (ASA), citrulline, cysteine, formiminoglutaminic acid (Figlu), histidine, homocitrulline, homocystïne, hydroxyproline, isoleucine, leucine, lysine, methionine, methionine-sulphoxide, methionine-sulphone, phenylalanine, phosphoethanolamine, pipecolic acid, proline, saccharopine, serine, s-sulpho-cysteine, taurine, threonine, tryptophan, tyrosine, valine, ammoniumformiate and DTT ((*2S, 3S*)-1,4-bis(sulfanyl)butane-2,3-diol) were purchased from Sigma-Aldrich (Zwijndrecht, the Netherlands). Homocitrulline and homoarginine were obtained from Santa Cruz Biotechnology (Heidelberg, Germany). Glutamic acid, glutamine, glycine, ornithine, sulphosalicylic acid (SSA), hydrochloric acid (HCl) and formic acid were purchased from VWR (Amsterdam, the Netherlands). Acetonitrile (ACN) and methanol both UPLC grade were purchased from Biosolve (Valkenswaard, the Netherlands).

4.7.3 Instrumentation

A Xevo-TQ MS triple quadrupole mass spectrometer with an Electrospray Ionization (ESI) source and an Acquity UPLC system (Waters, Manchester, United Kingdom) were used. MassLynx software (Waters, Manchester, United Kingdom) was used for instruments' control and data acquisition. The mass spectrometer operated in ESI-positive mode, capillary voltage 1.00 kV, desolvation temperature 550 °C, source temperature 150 °C, cone gas flow 50 L/h, and desolvation gas flow was 1000 L/h. Collision energy and cone voltage were optimized for each amino acid and the compounds of interest were analyzed using Multiple Reaction Monitoring (MRM). The dwell time was set automatically. The MRM transitions for the amino acid internal standards and compounds of interest are shown in Table 4.4, respectively. The amino acids are tabulated by increasing mass.

The chromatographic separation was carried out on an Acquity UPLC BEH Amide column (2.1 mm × 100 mm, 1.7 μm particle size) including a Van Guard™ UPLC BEH Amide precolumn (2.1 × 5 mm, 1.7 μm particle size) (Waters, Milford, USA). The column was maintained at a temperature of 35 °C and the sample volume injected was 2 μL. Optimal chromatographic separation was achieved at a flowrate of 0.4 mL/min using a gradient with solvent A (10 mM ammoniumformiate in 85% acetonitrile containing 0.15% formic acid) and solvent B (10 mM ammoniumformiate in milliQ-water containing 0.15% formic acid pH 3.0). The solvent gradient used for separation of the amino acids is described in Table 4.5. The total run time was 18 min includ-

Table 4.4 Multiple Reaction Monitoring (MRM) of amino acids and their corresponding internal standards for quantification by LC-MS/MS

Amino acid	MRM	Internal standard	MRM
Glycine	76.0 > 30.1	Glycine - 1,2-$^{13}C_2$	78.0 > 31.1
Alanine_62	90.0 > 62.0	DL-Alanine - 2,3,3,3-D_4	94.0 > 66.0
Alanine	90.1 > 44.1	DL-Alanine - 2,3,3,3-D_4	94.1 > 48.1
Serine	106.0 > 60.0	L-Serine - $^{13}C_3$	109.0 > 62.0
Proline	116.0 > 70.1	DL-Proline - 2,3,3,4,4,5,5-D_7	123.1 > 77.1
Valine	118.1 > 72.1	L-Valine - $^{13}C_5$	123.1 > 76.0
Threonine	120.0 > 74.1	L-Threonine - $^{13}C_4$, D_5, ^{15}N	130.0 > 83.1
Taurine	126.0 > 108.0	Taurine - 1,2-^{13}C	128.0 > 110.0
Pipecolic acid	130.1 > 84.1	Pipecolic acid - D_9	139.1 > 93.1
d-Aminolevulinic acid	132.0 > 86.0	Pipecolic acid - D_9	139.1 > 93.1
Hydroxyproline	132.1 > 68.0	L-Threonine - $^{13}C_4$, D_5, ^{15}N	130.0 > 83.1
Isoleucine	132.1 > 86.1	L-Isoleucine - $^{13}C_6$, D_{10}, ^{15}N	149.1 > 102.2
Leucine	132.1 > 86.1	L-Leucine - 5,5,5-D_3	135.1 > 89.1
Asparagine	133.0 > 74.0	Asparagine - $^{13}C_4$-$^{15}N_2$	139.0 > 77.0
Ornithine	133.1 > 70.1	L-Ornithine:HCl - 5,5-D_2	135.1 > 72.2
Aspartic acid	134.0 > 74.0	DL-Aspartic acid- 2,3,3-D_3	137.0 > 75.1
Phosphoethanolamine	142.0 > 44.0	L-Lysine:2HCl - 4,4,5,5-D_4	151.1 > 88.1
Glutamine	147.1 > 84.1	L-Glutamine - 2,3,3,4,4-D_5	152.1 > 88.1
Lysine	147.1 > 84.1	L-Lysine:2HCl - 4,4,5,5-D_4	151.1 > 88.1
Glutamic acid	148.0 > 84.0	DL-Glutamic acid - 2,4,4-D_3	151.1 > 87.0
Methionine	150.0 > 104.0	L-Methionine - methyl-D_3	153.0 > 107.1
Histidine	156.1 > 110.0	L-Histidine:HCl:H_2O - $^{13}C_6$	162.1 > 115.1
Phenylalanine	166.1 > 120.1	DL-Phenylalanine - ring-D_5	171.1 > 125.1
Glycylproline	173.0 > 116.0	L-Threonine - $^{13}C_4$, D_5, ^{15}N	130.0 > 83.1
Formiminoglutamic acid	175.0 > 84.0	L-Glutamine - 2,3,3,4,4-D_5	152.1 > 88.1
Arginine	175.1 > 70.1	L-Arginine:HCl - Guanido-$^{15}N_2$	177.1 > 70.2
Citrulline	176.1 > 159.0	L-Citrulline - 5-^{13}C, 4,4,5,5,-D_4	181.2 > 75.1
Tyrosine	182.1 > 136.1	L-Tyrosine - ring-D_4	186.1 > 140.1
Homoarginine	189.1 > 172.1	L-Arginine:HCl - Guanido-$^{15}N_2$	177.1 > 70.2
Homocitrulline	189.9 > 172.9	L-Citrulline - 5-^{13}C, 4,4,5,5,-D_4	181.2 > 75.1
S-sulphocysteine	202.0 > 120.0	S-sulpho-DL-cysteine - 2,3,3-D_3	205.0 > 123.0
Tryptophan	205.1 > 146.0	Indole-D_5-tryptophan	210.0 > 150.2
Cystine	241.0 > 152.0	DL-Cystine - 3,3,3,3-D_4	245.1 > 154.0
ASA anhydride II	255.0 > 70.2	L-Citrulline - 5-^{13}C, 4,4,5,5,-D_4	181.2 > 75.1
Homocystine	269.0 > 136.0	DL-Homocystine - 3,3,3,3,4,4,4,4-D_8	277.2 > 140.1
ASA anhydride I	273.2 > 70.2	L-Citrulline - 5-^{13}C, 4,4,5,5,-D_4	181.2 > 75.1
Saccharopine	277.4 > 84.0	DL-Homocystine - 3,3,3,3,4,4,4,4-D_8	277.2 > 140.1
ASA	291.2 > 70.2	L-Citrulline - 5-^{13}C, 4,4,5,5,-D_4	181.2 > 75.1
Aspartylglucosamine	336.0 > 126.0	L-Arginine:HCl - Guanido-$^{15}N_2$	177.1 > 70.2

ing column equilibration. The column was coupled to the mass spectrometer.

4.7.4 Analytical Procedure

Forty μL plasma was mixed with 40 μL internal standard solution. For more details on the preparation of the internal standard solution see [36]. After vortexing this mixture, 280 μL solvent A (10 mM ammoniumformiate (final concentration) in 85% acetonitrile +0.15% formic acid) was added. The sample was vortexed again, subsequently centrifuged for 5 min at 17000 × g in an Eppendorf centrifuge and all the supernatant was transferred to a 96-wells plate (Waters, Etten-Leur). After filtration with a 0.2-μm GHP filter (VWR, Amsterdam) the sample was ready for analysis. Preparation of the calibrators is described in detail in [36]. The prepared samples can be stored for 3 days at 4 °C. Injection volume of the prepared plasma samples is 2 μL. For CSF, 20 μL was mixed with 20 μL internal standard volume and after vortexing this mixture 200 μL solvent A was added. The sample preparation is similar as described above. Injection volume of the prepared CSF samples is 3 μL. For urine 50 μL was mixed with 50 μL internal standard solution. The sample preparation procedure is similar as described above. The injection volume of the prepared urine samples was also 2 μL.

Table 4.5 Gradient elution for separation of amino acids by LC-MS/MS

Time (minutes)	Flow (mL/min)	Rate solvent A (%)	Rate solvent B (%)	Curve
0.00	0.400	100.0	0.0	0
6.00	0.400	100.0	0.0	Linear (6)
6.10	0.400	94.1	5.9	Linear (6)
10.00	0.400	82.4	17.6	Linear (6)
12.00	0.400	70.6	29.4	Linear (6)
12.10	0.400	100.0	0.0	Linear (6)
18.00	0.400	100.0	0.0	Linear (6)

4.7.5 Calculation

Amino acid peaks were integrated using the software programme TargetLynx (Waters, Manchester, United Kingdom). Calculation of the ratio (peak area amino acid/peak area IS-amino acid) for known concentrations of the calibrators resulted for each amino acid in the generation of linear regression coefficients. Calculation of the amino acid concentrations in patient samples was performed by determination of the ratio (peak area amino acid/peak area IS-amino acid) and results were calculated with the linear regression model.

4.7.6 Post-analytical

In Fig. 4.4, the results of a plasma amino acid profile using tandem mass spectrometry are depicted. A plasma profile of a patient with

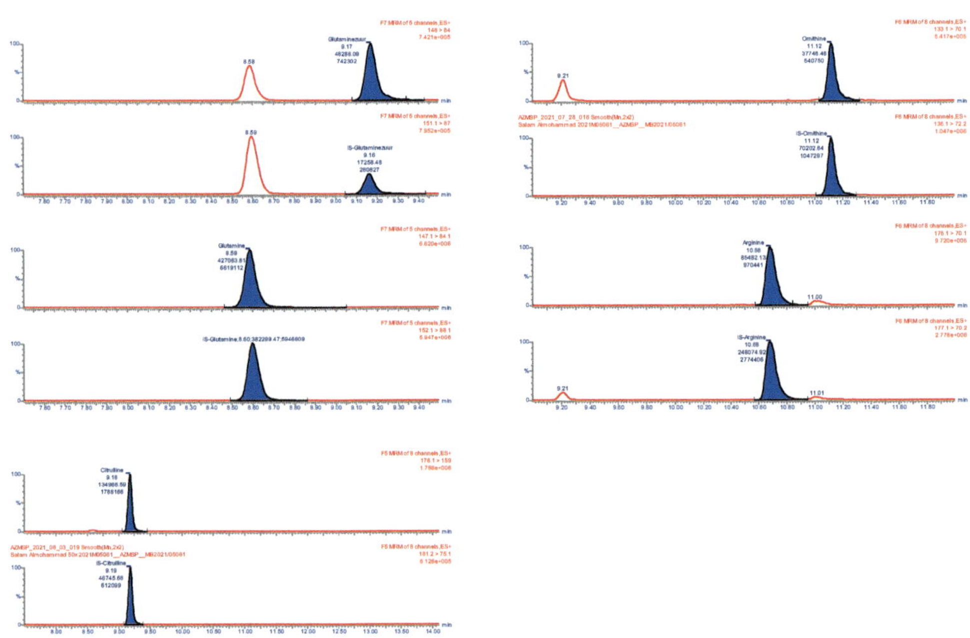

Fig. 4.4 Plasma amino acid profile of a patient with argininosuccinate synthetase 1 (ASS1) deficiency (citrullinemia type I) by tandem mass spectrometry. Plasma concentrations of glutamic acid, glutamine and citrulline were strongly elevated. Plasma ornithine was reduced, and arginine was normal

argininosuccinate synthetase 1 (ASS1) deficiency is shown.

4.7.7 Limitations of the Method

This method described here cannot chromatographically separate allo-isoleucine from the isomeric amino acid isoleucine. In addition, ASA carry-over is noticed when a HILIC amide column is used. These limitations were resolved by the introduction of mixed-mode columns. Baseline separation of allo-isoleucine can be achieved from the isobaric isoleucine and no carry-over of ASA is observed [37, 38]. Since the sample preparation is similar as described above, these methods can be easily combined to quantify the full range of clinically important amino acids using a single sample [38].

4.8 Additional Assays for Amino Acid Analysis

We have described three different approaches for amino acid analysis: HPLC (paragraph 5), amino acid analyzer (paragraph 6) and LC-MS/MS (paragraph 7). We realize that this is not the complete overview of all available techniques for amino acid analysis. In addition, several laboratories possess high-end mass spectrometers, such as Orbitrap or QTOF equipment. These analyzers are very sensitive and accurate but also very expensive. However, the methods that are being developed are less quantitative but rather qualitative and are used as a screening and are less appropriate for therapy control [39, 40].

Nowadays, ready-to-use kits are commercially available for amino acid analysis as well, but they are expensive and, similar to other derivatization methods, also increase possible errors, imprecision, sample complexity and sample preparation time and are not superior compared to existing derivatization methods or ion-exchange chromatography [41]. Therefore, a procedure without derivatization for analysis of amino acids would be preferable for many laboratories for simplicity and cost savings. Manufacturers are aware of these problems and alternatives (without derivatization) are developed.

4.9 Amino Acid Interpretation and Reference Values

It is evident that the necessity to analyze amino acids is dependent on the demand and interpretation of amino acid concentrations is difficult without appropriate reference values. Therefore, pre-analytical requirements are frequently provided by the laboratory to prevent these issues. Reference values for plasma, urine and CSF are shown in Tables 4.6, 4.7 and 4.8. One should be aware that generated reference values cannot be easily exchanged between equipment [6]. Comparability of results is necessary to test the validity of the obtained values for your own equipment. Besides, several factors influence the levels of amino acids. Pasquali et al. [7] reported possible artefacts that can arise due to inappropriate collection, handling and storage of samples. In addition, age, clinical status (fever, infections, liver and kidney dysfunction) of the patient, nutritional status and treatment with medication should be considered when interpreting results [6, 7].

As mentioned before, amino acids are preferably analyzed in fasting plasma, but in daily practice this is a challenge. Nutrition has a considerable effect on amino acid concentrations. A high protein intake or administration of parenteral amino acid solutions will result in an increase of most plasma amino acids. For other amino acids including homocitrulline and taurine, it is known that elevated urinary excretion of these amino acids is frequently of dietary origin (e.g., canned formula, breastmilk and energy drinks), whilst consumption of white meat can result in increased levels of carnosine, anserine and 3-methylhistidine, respectively. During prolonged fasting gluconeogenesis from the muscle amino acids will be upregulated and this results in a decline of plasma amino acids, except for the branched-chain amino acids that can rise. (Severe) protein malnutrition can lead to decreased levels of essential amino acids in

Table 4.6 Reference values for amino acids in plasma (µmol/L) obtained by amino acid analyzer

Plasma							
Amino acid	1st wk	1wk-1mo	1-4mo	4mo-2 yr	2-10 yr	10-18 yr	>18 yr
Taurine	14–238	9–201	19–139	19–139	24–92	10–162	6–126
Aspartate	1–21	5–37	6–30	0–16	3–15	4–28	1–9
Hydroxyproline	18–66	20–72	22–54	6–30	0–16	0–44	0–33
Threonine	53–141	55–187	66–182	36–136	48–140	72–192	102–190
Serine	62–206	60–240	85.221	42.174	77–169	75–175	68–160
Asparagine	38–114	22–94	28–80	26–70	24–64	32–64	35–63
Glutamate	32–104	0–109	29–81	24–68	11–51	11–59	1–57
Glutamine	198–886	178–670	368–652	337–709	373–709	396–740	435–721
Proline	120–344	65–457	98–254	53–201	93–221	75–307	88–290
Glycine	101–317	20–356	73–241	74–290	113–261	148–324	123–319
Alanine	108–448	116–376	182–378	144–348	158–314	192–508	182–552
Citrulline	5–33	10–50	16–32	6–34	18–50	17–49	16–46
Valine	65–201	59–199	96–228	79–267	133–273	142–278	144–269
Cystine	20–60	28–52	33–57	43–67	24–52	32–64	35–63
Methionine	6–50	13–53	19–51	9–29	11–27	16–36	12–32
Isoleucine	22–82	22–94	28–84	32–92	31–83	38–94	34–84
Leucine	47–175	47–167	55–155	53–149	64–164	76–168	78–160
Tyrosine	38–178	40–160	31–115	14–114	34–82	40–92	35–84
Phenylalanine	21–85	21–93	29–85	26–70	35–67	38–78	39–74
Ornithine	31–207	10–214	51–123	9–105	24–64	20–84	27–98
Histidine	25–113	33–121	55–115	30–110	54–106	58–106	67–109
Lysine	67–291	42–242	56–172	56–148	77–181	105–221	111–248
Tryptophan	16–72	25–65	32–92	27–59	5–57	27–75	30–95
Arginine	12–116	12–104	38–102	18–78	38–98	45–125	46–128

Table 4.7 Reference values for amino acids in urine (µmol/mmol creatinine) obtained by amino acid analyzer

Urine									
Amino acid	1st wk	1 wk–1 mo	1–4mo	4 mo-2 yr	2–4 yr	4–6 yr	2–10yr	10–18yr	>18yr
Taurine	0–638	2–61	0–79	0–79	0–79	0–79	0–79	0–79	22–80
Phosphoethanolamine	10–51	10–51	10–5	10–51	5–27	5–27	4–10	4–10	0–5
Aspartate	0.43	8–46	2–23	0–53	4–21	4–21	4–21	3–7	3–5
Hydroxyproline	0–282	36–478	2–384	0–41	0–3	0–3	0–3	0–3	0–3
Threonine	6–55	10–139	19–140	9–100	10–89	1–73	2–45	0–36	1–48
Serine	21–204	23–308	41–288	30–191	30–148	19–112	1–95	12–78	5–69
Asparagine	0–47	1–60	0–67	0–54	0–53	0–74	0–29	0–27	3–9
Glutamate	0–52	0–55	0–51	0–69	0–48	0–38	0–31	1–6	1–3
Glutamine	0–182	41–216	40–239	28–253	23–187	13–151	7–137	18–98	19–57
2-Aminoadipate	0–21	3–59	4–49	3–55	0–47	5–23	1–23	0–16	0–16
Proline	4–142	2–233	3–341	0–34	0–26	0–13	0–7	0–6	0–7
Glycine	0–1046	78–1259	105–796	40–616	76–516	24–397	20–201	18–252	12–199
Alanine	0–135	41–308	58–297	34–189	18–175	8–111	8–80	10–85	5–59
Citrulline	0–2	0–2	0–2	0–8	0–1	0–1	4–18	0–5	1–2
Valine	9–15	2–28	6–27	4–32	6–20	4–11	1–12	2–11	2–7
Cystine	0–65	2–52	7–54	0–36	5–25	4–22	3–17	5–16	1–19
Homocitrulline	0–135	0–15	0–15	0–4	1–26	1–11	0–8	0–8	0–8
Methionine	5–25	0–8	2–20	2–20	2–9	2–9	0–8	0–8	0–8
Isoleucine	3–9	1–18	0–26	0–10	4–14	2–12	0–7	0–7	1–5
Leucine	3–24	1–25	0–25	3–21	2–21	2–12	1–12	1–9	2–6
Tyrosine	0–29	9–55	13–75	10–72	0–71	8–51	6–42	6–37	5–27
Phenylalanine	0–34	3–34	3–40	5–37	0–41	2–25	1–20	1–17	2–11

(continued)

Table 4.7 (continued)

Urine									
Ornithine	2–15	0–24	1–39	0–20	0–9	0–8	0–7	1–4	4–17
Histidine	0–206	43–312	76–344	59–373	99–267	36–277	13–185	6–196	13–148
Lysine	0–82	10–172	18–148	0–182	0–92	0–52	0–64	0–20	12–52
Arginine	2–8	0–20	0–19	0–19	1–13	1–7	0–9	0–8	1–7

Table 4.8 Reference values for amino acids in CSF (μmol/L) obtained by amino acid analyzer

Cerebrospinal fluid amino acids	0–1 yr	1–18 yr	>18 yr
Taurine	4–14	4–14	4–14
Aspartate	3.7–10.5	2–4.4	1–5
Threonine	15–130	14–50	14–50
Serine	29.6–86	21–44	21–44
Asparagine	4.5–16.4	3–9	3–17
Glutamate	0.8–11	1.2–11	1.2–11
Glutamine	390–824	352–680	352–680
Glycine	3–8.3	3–8.3	1–14
Alanine	13.-48	13–48	13–48
Citrulline	1.2–7.8	1–6	1–6
Valine	12–29	9–18	5–38
Methionine	1.6–7.4	1–5	1–9
Isoleucine	5–12	2.5–6.5	3.4–13.4
Leucine	12.4–20	6.5–16	4.2–27
Tyrosine	7–25	8–14	8–14
Phenylalanine	5–22.5	4–14	24–19
Ornithine	2.8–19	2.4–9	2.4–9
Histidine	10.5–30	9–22	9–22
Lysine	12–36	10–28	15–43
Tryptophan	2.1–7.5	1–4	1–4
Arginine	12–32	13–35	13–35
Homocarnosine	5.5–12	2.5–10.6	2.5–10.6

plasma and an increase in glycine, serine and proline concentrations.

Amino acid concentrations change with age and therefore appropriate reference ranges are essential for proper interpretation of the results. In the neonatal period, the function of liver and kidney of premature neonates are not fully matured, and increased plasma concentrations of methionine and tyrosine can reflect this immaturity. In addition, an elevated excretion of specific amino acids such as cystine and glycine can be supportive. Due to maturation of the renal tubular reabsorption system, the reference values in urine are much higher in the neonatal period compared to adulthood (Table 4.7). The reference values in CSF do not show a clear dependence on age (Table 4.8). Most amino acids are lower in concentration compared to plasma, except glutamine. Caution should be taken when analyzing a traumatic lumbar puncture, that will result in an increase of amino acid levels in CSF (see paragraph 4). No ventral/dorsal gradient for amino acids is observed.

A correct interpretation of quantitative amino acid profiles involves more than just comparing concentrations with reference values. Most disorders result in clearly abnormal amino acid profiles, but some disorders present with more subtle biochemical abnormalities. Pattern recognition is therefore of utmost importance and attention must be paid to *relatively* high or low concentrations of specific amino acids. Especially biosynthesis defects can be hard to distinguish in a non-fasting plasma and putative differences in amino acid concentrations in other matrices can be helpful. At present, approximately 60 primary inborn errors in amino acid metabolism are known, and the number is still growing. These disorders reveal defects in catabolic or anabolic pathways but can also be caused by a defective transport system. Besides a primary transport defect, several other IEM (e.g., Lowe syndrome, classical galactosemia and mitochondrial disorders) are accompanied by generalized aminoaciduria. It is important to realize that obvious amino acid changes can be secondary as well and are not always the result of a primary inborn error of metabolism [6, 7]. Remarkably enough, not all defects in amino acid metabolism are associated with clinical symptoms. For instance, histidinemia (OMIM #235800), which is caused by mutations in the enzyme histidine ammonia-lyase and results in clearly increased histidine levels, is generally considered to be a non-disease. Recent overviews of pathological conditions that are associated with abnormal amino acid concentrations (primary and secondary causes) are shown in [7, 8, 42].

4.10 Quality Control

As is clear, the analysis of amino acids is routinely performed in a genetic metabolic laboratory. Nowadays, a requisite is that these methods are validated. Medical laboratories follow the ISO15189 standard in which requirements for quality and competence including validation are described. Parameters that should be tested in the validation process are intra- and interassay variation, measurement uncertainty, recovery, limit of detection, limit of quantitation, linearity, stability, carry-over, interferences, selectivity, reference values and a clinical validation should be performed.

Quality control samples need to be included during analysis and in general and for amino acid analysis two or three quality controls (QC-low, (QC-middle) or QC-high) are prepared for internal quality control. The obtained values of the QC's are in the range of the calibrators and are analyzed in each run. Normally, all amino acids that are quantified are displayed in a quality control card (e.g., Shewart card) for internal quality control monitoring. For amino acids, an external quality assessment scheme exists that is proved by the European Research Network for evaluation and improvement of screening, Diagnosis and treatment of Inherited disorders of Metabolism (ERNDIM). External Quality Assessment (EQA) schemes use consensus values and provide information on accuracy, precision, linearity and recovery of your method compared to others. This enables a laboratory to adapt its procedure if necessary. More recently, ERNDIM added an amino acid interpretation scheme to the catalogues which focusses on education and assesses the ability to recognize patterns of amino acids in plasma.

4.11 Conclusion

In the screening for an inborn error of metabolism (IEM), the analysis of amino acids is an important tool. Adequate interpretation of amino acid profiles in which patterns are recognized is required for a correct diagnosis. The interpretation of urine profiles can be complicated due to the interference of medication and dietary components. However, with the use of mass spectrometry analyses, interference will not play a major role anymore in these laboratories.

In this chapter, we have described three different methods for amino acid analyses. Considering the cost-effectiveness, the HPLC analysis is a cheaper alternative than the classic amino acid analyzer and mass spectrometers. However, this method is less suitable for separation of isobaric amino acids and suffers a lot from interfering components.

Many laboratories use the classical amino acid analyzer which has great reproducibility. The disadvantages, however, are the long analysis time and the interference with other components that react with ninhydrin. Also, the equipment is exclusively applicable for the analysis of amino acids. Since several laboratories have purchased tandem mass spectrometry equipment, it is logical that mass spectrometry-based methods for amino acid analysis are increasingly used. The main advantages are the relatively short analysis time and there is hardly any interference with other components in the specimens.

Acknowledgements We are thankful for Dr. M. Duran who wrote the first version of this chapter in the first edition (2008). Ing. R. Versteeg was thanked for preparing the tables and carefully reading part of the manuscript.

References

1. Aliu E, Kanungo S, Arnold GL (2018) Amino acid disorders. Ann Transl Med 6(24):471. https://doi.org/10.21037/atm.2018.12.12
2. Lopez MJ, Mohiuddin SS (2023) Biochemistry, essential amino acids. In: Disclosure: Shamim Mohiuddin declares no relevant financial relationships with ineligible companies. StatPearls, Treasure Island (FL). ineligible companies
3. Wu G (2009) Amino acids: metabolism, functions, and nutrition. Amino Acids 37(1):1–17. https://doi.org/10.1007/s00726-009-0269-0
4. Moore S, Stein WH (1948) Photometric ninhydrin method for use in the chromatography of amino acids. J Biol Chem 176(1):367–388
5. Moore S, Stein WH (1954) A modified ninhydrin reagent for the photometric determination of amino acids and related compounds. J Biol Chem 211(2):907–913

6. Sharer JD, De Biase I, Matern D, Young S, Bennett MJ, Tolun AA, Committee ALQA (2018) Laboratory analysis of amino acids, 2018 revision: a technical standard of the American College of Medical Genetics and Genomics (ACMG). Genet Med 20(12):1499–1507. https://doi.org/10.1038/s41436-018-0328-6
7. Pasquali MLN (2022) Amino acids. In: Physician's guide to the diagnosis, treatment, and follow-up of inherited metabolic diseases, 2nd edn. Springer, pp 41–50
8. Yahyaoui R, Perez-Frias J (2019) Amino acid transport defects in human inherited metabolic disorders. Int J Mol Sci 21(1). https://doi.org/10.3390/ijms21010119
9. Wu G (2021) Amino acids in nutrition, health, and disease. Front Biosci (Landmark Ed) 26(12):1386–1392. https://doi.org/10.52586/5032
10. Carling RS, Barclay Z, Cantley N, Emmett EC, Hogg SL, Finezilber Y, Schulenburg-Brand D, Murphy E, Moat SJ (2023) Investigation of the relationship between phenylalanine in venous plasma and capillary blood using volumetric blood collection devices. JIMD Rep 64(6):468–476. https://doi.org/10.1002/jmd2.12398
11. Coene KLM, Timmer C, Goorden SMI, Ten Hoedt AE, Kluijtmans LAJ, Janssen MCH, Rennings AJM, Prinsen H, Wamelink MMC, Ruijter GJG, Korver-Keularts I, Heiner-Fokkema MR, van Spronsen FJ, Hollak CE, Vaz FM, Bosch AM, Huigen M (2021) Monitoring phenylalanine concentrations in the follow-up of phenylketonuria patients: an inventory of pre-analytical and analytical variation. JIMD Rep 58(1):70–79. https://doi.org/10.1002/jmd2.12186
12. Holub M, Tuschl K, Ratschmann R, Strnadova KA, Muhl A, Heinze G, Sperl W, Bodamer OA (2006) Influence of hematocrit and localisation of punch in dried blood spots on levels of amino acids and acylcarnitines measured by tandem mass spectrometry. Clin Chim Acta 373(1–2):27–31. https://doi.org/10.1016/j.cca.2006.04.013
13. Lawson AJ, Bernstone L, Hall SK (2016) Newborn screening blood spot analysis in the UK: influence of spot size, punch location and haematocrit. J Med Screen 23(1):7–16. https://doi.org/10.1177/0969141315593571
14. Moat SJ, Schulenburg-Brand D, Lemonde H, Bonham JR, Weykamp CW, Mei JV, Shortland GS, Carling RS (2020b) Performance of laboratory tests used to measure blood phenylalanine for the monitoring of patients with phenylketonuria. J Inherit Metab Dis 43(2):179–188. https://doi.org/10.1002/jimd.12163
15. van Wegberg AMJ, MacDonald A, Ahring K, Belanger-Quintana A, Blau N, Bosch AM, Burlina A, Campistol J, Feillet F, Gizewska M, Huijbregts SC, Kearney S, Leuzzi V, Maillot F, Muntau AC, van Rijn M, Trefz F, Walter JH, van Spronsen FJ (2017) The complete European guidelines on phenylketonuria: diagnosis and treatment. Orphanet J Rare Dis 12(1):162. https://doi.org/10.1186/s13023-017-0685-2
16. van Spronsen FJ, van Wegberg AM, Ahring K, Belanger-Quintana A, Blau N, Bosch AM, Burlina A, Campistol J, Feillet F, Gizewska M, Huijbregts SC, Kearney S, Leuzzi V, Maillot F, Muntau AC, Trefz FK, van Rijn M, Walter JH, MacDonald A (2017) Key European guidelines for the diagnosis and management of patients with phenylketonuria. Lancet Diabetes Endocrinol 5(9):743–756. https://doi.org/10.1016/S2213-8587(16)30320-5
17. van Vliet K, van Ginkel WG, van Dam E, de Blaauw P, Koehorst M, Kingma HA, van Spronsen FJ, Heiner-Fokkema MR (2020) Dried blood spot versus venous blood sampling for phenylalanine and tyrosine. Orphanet J Rare Dis 15(1):82. https://doi.org/10.1186/s13023-020-1343-7
18. Moat SJ, George RS, Carling RS (2020a) Use of dried blood spot specimens to monitor patients with inherited metabolic disorders. Int J Neonatal Screen 6(2):26. https://doi.org/10.3390/ijns6020026
19. Crawford PM, Lloyd KG, Chadwick DW (1988) CSF gradients for amino acid neurotransmitters. J Neurol Neurosurg Psychiatry 51(9):1193–1200. https://doi.org/10.1136/jnnp.51.9.1193
20. Akiyama T, Kobayashi K, Higashikage A, Sato J, Yoshinaga H (2014) CSF/plasma ratios of amino acids: reference data and transports in children. Brain and Development 36(1):3–9. https://doi.org/10.1016/j.braindev.2012.12.001
21. Jones CM, Smith M, Henderson MJ (2006) Reference data for cerebrospinal fluid and the utility of amino acid measurement for the diagnosis of inborn errors of metabolism. Ann Clin Biochem 43(Pt 1):63–66. https://doi.org/10.1258/000456306775141759
22. de Koning TJ (2017) Amino acid synthesis deficiencies. J Inherit Metab Dis 40(4):609–620. https://doi.org/10.1007/s10545-017-0063-1
23. Cunningham G, Zhou Y, Summar M (2022) Development of a robust 30-minute reverse-phase high pressure liquid chromatography method to measure amino acids using widely available equipment and its comparison to current clinical ion-exchange chromatography measurement. Mol Genet Metab Rep 31:100868. https://doi.org/10.1016/j.ymgmr.2022.100868
24. Dolowy M, Pyka A (2014) Application of TLC, HPLC and GC methods to the study of amino acid and peptide enantiomers: a review. Biomed Chromatogr 28(1):84–101. https://doi.org/10.1002/bmc.3016
25. Frank MP, Powers RW (2007) Simple and rapid quantitative high-performance liquid chromatographic analysis of plasma amino acids. J Chromatogr B Analyt Technol Biomed Life Sci 852(1–2):646–649. https://doi.org/10.1016/j.jchromb.2007.01.002
26. Leah JM, Palmer T, Griffin M, Wingad CJ, Briddon A, Oberholzer VG (1986) Urine amino-acid-analysis by Hplc in the investigation of inborn-errors of metabolism. J Inherit Metab Dis 9:250–253. https://doi.org/10.1007/Bf01799720
27. Nolasco DM, Fortes ICP, Valadares ER (2020) Quantitative analysis of amino acids by HPLC in

dried blood and urine in the neonatal period: establishment of reference values. Biomed Chromatogr 34(11):e4931. https://doi.org/10.1002/bmc.4931
28. Kavanagh D, Holmes M, Appleton M, Bartlett K (2012) Interference of urinary amino acid analysis by pregabalin. Clin Chim Acta 413(9–10):953–954. https://doi.org/10.1016/j.cca.2012.02.006
29. Wiencek JR, Dietzen DJ, Murray T, Dawling S, Colby JM, Nichols JH (2018) Persistently increased Alloisoleucine in a patient with seizures. Clin Chem 64(10):1433–1437. https://doi.org/10.1373/clinchem.2017.278804
30. de Person M, Chaimbault P, Elfakir C (2008) Analysis of native amino acids by liquid chromatography/electrospray ionization mass spectrometry: comparative study between two sources and interfaces. J Mass Spectrom 43(2):204–215. https://doi.org/10.1002/jms.1287
31. Piraud M, Vianey-Saban C, Bourdin C, Acquaviva-Bourdain C, Boyer S, Elfakir C, Bouchu D (2005) A new reversed-phase liquid chromatographic/tandem mass spectrometric method for analysis of underivatised amino acids: evaluation for the diagnosis and the management of inherited disorders of amino acid metabolism. Rapid Commun Mass Spectrom 19(22):3287–3297. https://doi.org/10.1002/rcm.2197
32. Guo S, Duan JA, Qian D, Tang Y, Qian Y, Wu D, Su S, Shang E (2013) Rapid determination of amino acids in fruits of Ziziphus jujuba by hydrophilic interaction ultra-high-performance liquid chromatography coupled with triple-quadrupole mass spectrometry. J Agric Food Chem 61(11):2709–2719. https://doi.org/10.1021/jf305497r
33. Ozturk Er E, Erarpat S, Bodur S, Gunkara OT, Ozbek B, Bakirdere S (2022) Accurate determination of amino acids by quadruple isotope dilution-reverse phase liquid chromatography-tandem mass spectrometry after derivatization with 2-Naphthoyl chloride. J Chromatogr A 1667:462870. https://doi.org/10.1016/j.chroma.2022.462870
34. Roper SM, Weindel AL, Dietzen DJ (2022) Comprehensive determination of amino acids for diagnosis of inborn errors of metabolism. Methods Mol Biol 2546:55–64. https://doi.org/10.1007/978-1-0716-2565-1_6
35. Ferre S, Gonzalez-Ruiz V, Guillarme D, Rudaz S (2019) Analytical strategies for the determination of amino acids: past, present and future trends. J Chromatogr B Analyt Technol Biomed Life Sci 1132:121819. https://doi.org/10.1016/j.jchromb.2019.121819
36. Prinsen H, Schiebergen-Bronkhorst BGM, Roeleveld MW, Jans JJM, de van der Velden MGM, Visser G, van Hasselt PM, Verhoeven-Duif NM (2016) Rapid quantification of underivatized amino acids in plasma by hydrophilic interaction liquid chromatography (HILIC) coupled with tandem mass-spectrometry. J Inherit Metab Dis 39(5):651–660. https://doi.org/10.1007/s10545-016-9935-z
37. DeArmond PD, Bunch DR (2022) Quantitation of non-derivatized free amino acids for detecting inborn errors of metabolism by incorporating mixed-mode chromatography with tandem mass spectrometry. J Mass Spectrom Adv Clin Lab 25:1–11. https://doi.org/10.1016/j.jmsacl.2022.05.002
38. Griffin C, Ammous Z, Vance GH, Graham BH, Miller MJ (2019) Rapid quantification of underivatized alloisoleucine and argininosuccinate using mixed-mode chromatography with tandem mass spectrometry. J Chromatogr B Analyt Technol Biomed Life Sci 1128:121786. https://doi.org/10.1016/j.jchromb.2019.121786
39. Coene KLM, Kluijtmans LAJ, van der Heeft E, Engelke UFH, de Boer S, Hoegen B, Kwast HJT, van de Vorst M, Huigen M, Keularts I, Schreuder MF, van Karnebeek CDM, Wortmann SB, de Vries MC, Janssen MCH, Gilissen C, Engel J, Wevers RA (2018) Next-generation metabolic screening: targeted and untargeted metabolomics for the diagnosis of inborn errors of metabolism in individual patients. J Inherit Metab Dis 41(3):337–353. https://doi.org/10.1007/s10545-017-0131-6
40. Liu H, Zhu J, Li Q, Wang D, Wan K, Yuan Z, Zhang J, Zou L, He X, Miao J (2021) Untargeted metabolomic analysis of urine samples for diagnosis of inherited metabolic disorders. Funct Integr Genomics 21(5–6):645–653. https://doi.org/10.1007/s10142-021-00804-w
41. Carling RS, McDonald BA, Austin D, Burden D, Correia J, Leung J, Mayers B, John C (2020) Challenging the status quo: a comparison of ion exchange chromatography with liquid chromatography-mass spectrometry and liquid chromatography-tandem mass spectrometry methods for the measurement of amino acids in human plasma. Ann Clin Biochem 57(4):277–290. https://doi.org/10.1177/0004563220933303
42. Kandasamy P, Gyimesi G, Kanai Y, Hediger MA (2018) Amino acid transporters revisited: new views in health and disease. Trends Biochem Sci 43(10):752–789. https://doi.org/10.1016/j.tibs.2018.05.003

Organic Acids

5

Patricia L. Hall and Silvia Tortorelli

Contents

Abbreviations

2OHGA	2-Hydroxyglutaric acid
3OHGA	3-Hydroxyglutaric acid
BSTFA + TMCS	N,O-Bis(trimethylsilyl) trifluoroacetamide
GC-MS	Gas chromatography-mass spectrometry
IVA	Isovaleric acidemia
MMA	Methylmalonic acidemia
PA	Propionic acidemia
PDA	Pentadecanoic acid
UOA	Urine organic acid analysis

P. L. Hall (✉) · S. Tortorelli
Department of Laboratory Medicine and Pathology, Mayo Clinic, Rochester, MN, USA
e-mail: Hall.patricia@mayo.edu; Tortorelli.silvia@mayo.edu

5.1 Organic Acid Analysis

Quantitative and qualitative organic acid analyses can be used for the identification of many disorders.

5.1.1 Introduction

Organic acid analysis in urine (UOA) is the first line test for the diagnosis of organic acidemias and serves as a valuable screening tool for other inborn errors of metabolism, including fatty acid oxidation defects, ketone body metabolism disorders, aminoacidopathies, neurotransmitter defects, mitochondrial disorders and cofactor deficiencies. At least one-third of the algorithms developed by the American College of Medical Genetics and Genomics to follow up abnormal newborn screens recommend organic acid analysis as part of the workup to confirm an abnormal result (ACMG, accessed 02/06/2023) [1]. UOA is an early part of laboratory workups for many

N. Blau, F. M. Vaz (eds.), *Laboratory Guide to the Methods in Biochemical Genetics*,
https://doi.org/10.1007/978-3-031-58819-8_5

Table 5.2 Characteristic urine organic acid profiles of disorders not commonly included on newborn screening panels

Disease	Key analytes
3-Hydroxyisobutyryl-CoA hydrolase deficiency	2-Hydroxyisovaleric acid, methacrylic acid conjugates (HIBCH)
3-Hydroxyisobutyrate dehydrogenase deficiency	2-Hydroxyisovaleric acid, 3-hydroxybutyric acid (HIBADH)
Methylmalonate semialdehyde dehydrogenase deficiency	3-Hydroxybutyric acid, ethylmalonic acid
Malonic aciduria[a]	Malonic acid (methylmalonic acid)
Combined malonic and methylmalonic aciduria acetyl-CoA-synthase 3 deficiency	Malonic acid, methylmalonic acid
Ethylmalonic encephalopathy[a]	Ethylmalonic acid, methylsuccinic acid, 2-methylbutyrylglycine, isovalerylglycine
Crotonase deficiency	2-Methyl-2,3-dihydrobutyrate
Canavan disease, Aspartoacylase deficiency	N-Acetylaspartic acid
Aminoacylase 1 deficiency	N-Acetylalanine, N-acetylglutamic acid, N-acetylglycine, N-acetylisoleucine, N-acetylleucine, N-acetylmethionine, N-acetylvaline
Alkaptonuria	Homogentisic acid
Hawkinsinuria	4-Hydroxycyclohexylacetate
2-Aminoadipic 2-oxoadipic aciduria	2-Hydroxyadipic acid, 2-ketoadipic acid, C6-C10 dicarboxylic acids
5-Oxoprolinase deficiency	5-Oxoproline
Glutathione synthetase deficiency	5-Oxoproline
Primary hyperoxaluria type 1, AGT alanine-glyoxylate aminotransferase deficiency (peroxisomal)	Oxalic acid, glycolic acid
Primary hyperoxaluria type 2, D-glycerate dehydrogenase deficiency	Oxalic acid, glyceric acid
Glycolate oxidase 1 deficiency, isolated glycolic aciduria Hydroxyacid oxidase 1 deficiency	Oxalic acid, glycolic acid
Oxalate transporter deficiency	Oxalic acid
Enteric hyperoxaluria and nephrolithiasis, SLC26A6 deficiency	Oxalic acid, citric acid (↓)
Primary hyperoxaluria type 3, mitochondrial 4-hydroxy-2-oxoglutarate aldolase 1 deficiency	Oxalic acid
Aromatic L-amino acid decarboxylase deficiency	Vanillactic acid
Succinic semialdehyde dehydrogenase deficiency	4-Hydroxybutyric acid + erythroen threo-4,5-dihydroxyhexanoic acid lactone
2-Oxoglutaric aciduria, Amish lethal microcephaly	2-Ketoglutaric acid
Fumaric aciduria	Fumaric acid
Primary lactic acidemia	Lactic acid, pyruvic acid (↑↑↑)
Combined D-2- and L-2-hydroxyglutaric aciduria	2-Hydroxyglutaric acid
L-2-Hydroxyglutaric aciduria L-2-hydroxyglutarate dehydrogenase deficiency	2-Hydroxyglutaric acid
D-2-Hydroxyglutaric aciduria type I D-2-hydroxyglutarate dehydrogenase deficiency	2-Hydroxyglutaric acid

[a]denotes disorders that have variable detection by NBS programs

Table 5.3 Disorders with abnormal excretion of 3-methylglutaconic acid

Disease	Gene	OMIM
Primary 3-methylglutaconic acidurias		
3-Methylglutaconyl-CoA hydratase deficiency	*AUH*	250950
HMG-CoA lyase deficiency[a]	*HMGCL*	613898
Historical secondary 3-methylglutaconic acidurias		
Barth syndrome	*TAZ*	302060
Costeff syndrome	*OPA3*	258501
MEGDEL syndrome—3-MEthylGlutaconic aciduria, deafness, encephalopathy, and Leigh-like syndrome	*SERAC1*	614725
DCMA syndrome—dilated CardioMyopathy with ataxia	*DNAJC*	610198
TMEM70 deficiency	*TMEM70*	612418
Newly described secondary 3-methylglutaconic acidurias		
Sengers syndrome	*AGK*	212,50
Complex V deficiency, nuclear type 1	*ATPAF2*	604273
Complex V deficiency, nuclear type 3	*ATP5E*	614053
MGCA7B—3-methylglutaconic aciduria type VIIB	*CLPB*	616271
CPS1 (carbamoyl phosphatase synthase I) deficiency	*CPS1*	237300
Short-chain enoyl-CoA hydratase deficiency	*ECHS1*	602292
SPPRS—spastic paraplegia and psychomotor retardation with or without seizures	*HACE1*	616756
MGCA8—3-methylglutaconic aciduria type VIII	*HTRA2*	617248
COXPD19—combined oxidative phosphorylation deficiency 19	*LYRM4*	615595
MELAS syndrome—mitochondrial myopathy, encephalopathy, lactic acidosis, and stroke	*m.3243A > G tRNA*	540000
COXPD52—combined oxidative phosphorylation deficiency 52	*NFS1*	619386
POLG1 deficiency	*POLG1*	174763
COXPD37—combined oxidative phosphorylation deficiency 37	*C19orf70 (MICOS13)*	618329
MTDPS5—mitochondrial DNA depletion syndrome 5	*SUCLA2*	612073
MGCA9—3-methylglutaconic aciduria type IX	*TIMM50*	617698
SLOS—smith-Lemli-Opitz syndrome	*DHCR7*	270400
Statin-induced myopathy		

on the expertise of the laboratory personnel involved at all stages [4].

5.1.2.2 Properties of Metabolites

Organic acids are water-soluble compounds containing one or more carboxyl groups and other non-amino functional groups (Fig. 5.1). They originate from the intermediary metabolism of all major groups of organic cellular components. Primary precursors are amino acids, lipids, carbohydrates, nucleic acids, and steroids. UOA is not limited to the detection of these endogenous organic acids. Hundreds of different compounds, including medications and dietary artifacts can be found in an UOA profile. Identification of this many analytes relies on the combination of mass spectra and retention time. A full table of analytes detected by UOA would take up much of the space dedicated to this chapter. Local expertise in mass spectrometry and intermediate metabolism is essential for the identification of these analytes and mapping them to an appropriate interpretation. For single-stage mass spectrometry, three points of identification are usually required, most commonly a retention time match and two characteristic ions. Several libraries are commercially available and are a good starting point for establishing an in-house library. It is important to recognize that parent molecules may not be the only species detected for endogenous compounds, intermediate metabolites are also excreted in urine, and it is important to recognize and identify them appropriately.

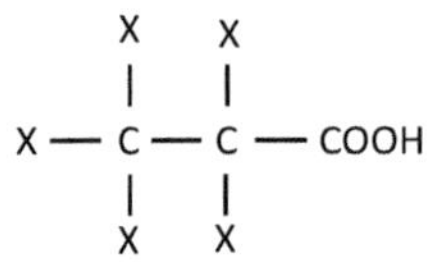

Fig. 5.1 Generic structure of organic acids. A wide variety of functional groups (hydrogen, keto-, hydroxy-, carboxyl-) and side chains (methyl, ethyl, propyl, butyl) can be present at each X position, creating a massive number of organic acids that can be detected

5.2 Organic Acid Analysis by Gas Chromatography-Mass Spectrometry

5.2.1 Methods

5.2.1.1 Principle

Urine samples corresponding to a known amount of creatinine are acidified and then a liquid-liquid extraction with ethyl acetate is performed. The organic residue is evaporated and sialylated with N,O-bis-(trimethylsilyl) trifluoroacetamide with 1% trimethylchlorosilane. Samples are analyzed using electron impact gas chromatography-mass spectrometry.

If desired, UOA analysis can include an oximation step with hydroxylamine hydrochloride, for the improved detection of keto acids, such as pyruvic acid, succinylacetone, for the detection of tyrosinemia type 1, and small abnormal metabolites (2-ketoisocaproic acid and 2-ketoisovaleric acid) characteristic of a diagnosis of maple syrup urine disease. This step is described in the sample preparation that follows. It is a local decision whether to oximate all samples prior to initial analysis, or to repeat targeted analyses with oximation based upon review of a suspicious profile. Our laboratory does not oximate all samples for initial analysis.

5.2.1.2 Pre-analytical

1. *Specimens*

Random urine specimens should be collected, stored in a plastic urine tube with no preservatives and shipped frozen. The recommended volume for analysis is 10 mL, however, smaller volumes can be analyzed. If possible, samples should be collected during an acute episode.

If desired, urine spotted on filter paper can be validated as an appropriate specimen type. An appropriate filter paper should be soaked entirely in urine (~ 2–3 mL), and allowed to dry at room temperature for at least 3 hours. If possible, the sample should be collected during an acute episode.

2. *Reagents and Chemicals*
   - Pentadecanoic acid (PDA; internal standard; Sigma)
     - Dissolve 100 mg of PDA in 70 mL methanol and make up to volume in a 100 mL volumetric flask.
   - Sodium hydroxide (Fisher)
     - Dissolve 48 g of NaOH in 150 mL of water and make up to volume in a 200-mL volumetric flask to provide 6 N NaOH.
   - Hydrochloric acid (EMD)
     - Add 250 mL concentrated HCl to 250 mL water in a glass bottle and mix well to provide 6 N HCl.
   - Hydroxylamine-HCl (Sigma)
     - Add 25 g of Hydroxylamine-HCl to 700 mL of water and make up to volume in a 1-L amber volumetric flask to provide a solution of 2.5% Hydroxylamine-HCl. Protect from light.
   - Sodium chloride (Fisher)
     - Used as provided.
   - Ethyl acetate (EMD)
     - Used as provided.
   - Methanol (EMD)
     - Used as provided.
   - Sodium sulfate (Sigma)
     - Used as provided.
   - BSTFA + TMCS (ThermoFisher Scientific)
     - Used as provided.
   - Carrier gas (Hydrogen/Helium)
3. *Instrumentation*
   - Gas Chromatograph—Mass spectrometer (Agilent 6890 N GC with Agilent 5973 N / 5975 mass spectrometer, or equivalent)
   - HP-5 (Cross-linked 5% phenylmethyl silicone; 25 m; ID 0.20 mm; film thickness 0.33 μm) or DB-5MS (Cross-linked 5%

phenylmethyl silicone; 20 m; ID 0.18 mm; film thickness 0.18 μm)
- Evaporator
- Heating Block
- Vortexer
- Centrifuge
- Liquid Handler (Optional—dependent on volume)

4. *Calibration*

Calibration requirements for organic acid analysis can vary based on reporting requirements and method setup. Fully qualitative interpretations may only quantify a small number of key analytes (methylmalonic acid, orotic acid, glutaric acid), while fully quantitative organic acid profiles may quantify or provide relative abundance values for 50 or more unique organic acids.

If quantification of specific organic acids is desired, standards should be selected and calibration curves prepared for clinically relevant ranges. These curves should be run in full at least every 6 months or when instrument parameters change (new column installation). Calibration verification using a single point per daily analysis should be performed as well. All calibration procedures should be verified to comply with any local regulatory requirements, which can vary. Our calibration materials and methods are provided here as an example.

A 6-point calibration curve is run for 19 targeted analytes (orotic acid, succinic acid, fumaric acid, malic acid, cis-aconitic acid, citric acid, adipic acid, suberic acid, sebacic acid, 3-methylglutaric acid, pyroglutamic acid, ethylmalonic acid, methylsuccinic acid, homogentisic acid, malonic acid, glutaric acid, 2-hydroxyglutaric acid, 2-hydroxyglutaric lactone, methylmalonic acid, and 2-ketoglutaric acid) each time a new internal standard is prepared or every 6 months, whichever is more frequent. Eighteen of these analytes are analyzed unoximated and 1 is analyzed after oximation (2-ketoglutaric acid).

Each calibration curve must be reviewed for acceptability prior to being put into use. A new curve should provide acceptable results (most commonly +/− 2 SD for both levels of quality control included on the run) and appropriate linearity over the measured range ($R^2 \geq 0.98$).

5.2.1.3 Analytical procedure

1. Sample Preparation
   - Thaw all specimens and quality control samples and mix gently.
   - Calculate the required urine volumes for each patient specimen to correspond to the known amount of creatinine required for analysis. A verified, locked spreadsheet or worksheet may be useful for this purpose. Note any short or dilute specimens.
   - Pipet the calculated volume for each patient sample and control into a labeled 16x100 mm screw cap glass tube. Add water to make the final volume 3.0 mL if needed.
   - Add 100 μL of 1 mg/mL PDA (internal standard) to each vial.
   - Oximation *(for the identification of 2-keto acids—optional).*
     - *Check pH and adjust to 7.0 with 6 N HCl or NaOH as needed.*
     - *Add 0.5 mL of hydroxylamine-HCl to each tube.*
     - *Cap the tubes, vortex to mix, and place on heating block (60 °C) for 30 minutes.*
     - *Remove tubes from rack and return to room temperature before proceeding.*
   - Add 1 g NaCl to each tube, vortex to mix.
   - Add 250 μL of 6 N HCl to each tube and vortex to mix (pH should be <2).
   - Add 2.5 mL of ethyl acetate to each tube and shake for ~10 seconds. Centrifuge tubes at 1769 RCF for 1 minute.

     Note: If needed, preparation of samples can be paused at this step.
   - Use a Pasteur pipet to transfer the organic ethyl acetate layer (upper layer) into a labeled 16 x 100 mm glass tube. If the aqueous (lower) layer is disturbed, return any transferred liquid to the screw top tube and re-centrifuge. Discard this Pasteur pipet and use a clean one for transfer of the specimen after re-centrifugation is com-

plete. If the ethyl acetate layer has solidified, tap the side of the tube several times to loosen the layer and re-centrifuge. This step should take place in a fume hood.

- Repeat this liquid-liquid extraction (addition of 2.5 mL of ethyl acetate and removal of organic layer) 3 more times. All four ethyl acetate layers for a given specimen should be combined in one tube. The remainder of the extracted specimen (aqueous phase) can be discarded.
- Add 1 g Na_2SO_4 to each tube containing the ethyl acetate layers to remove any water. Wait 1 minute and then centrifuge for 1 minute at 1769 RCF.
- Transfer all ethyl acetate to a labeled screw top vial, avoiding contact with the sodium sulfate pellet. If the pellet is disturbed, re-centrifuge.
- Evaporate the ethyl acetate at 60 °C using a gentle flow of nitrogen gas. Samples should be evaporated just to dryness.
- Add 100 μL of BSTFA+TMCS to each tube. Cap with a fresh cap, vortex to mix, and place in 80 °C heating block for 30 minutes. This step must take place in a fume hood.
- Remove the tubes from the heating block and allow them to reach room temperature.
- Transfer the contents of each tube to a corresponding labeled GC vial and cap.
- All GC vials should be loaded on the instrument in a known sequence for the day's run, including all controls, calibrators, and patient samples.

2. Quality Control

Two levels of quality control should be analyzed with each run. Acceptance criteria will vary based on calibration and reporting practices. Quality control specimens should be selected to ensure the assay is performing appropriately for relevant qualitative and quantitative parameters. Depending on analyte selection for routine QC samples, an additional control may need to be added for runs where a specimen is oximated.

- The low control is made by spiking small amounts (3–8 μg/0.25 mg creatinine) of abnormal (tiglylglycine, 3-hydroxyglutaric acid, succinylacetone, 2-methylcitric acid, hexanoylglycine, and suberylglycine) analytes into approximately 1 L of normal urine. This control is only monitored for the presence of these abnormal analytes, rather than quantitatively. If any analyte in the low QC is not able to be detected, the run should be rejected.
- The high control is prepared by spiking 25 μg/0.25 mg creatinine of 2-hydroxyglutaric acid and 100 μg/0.25 mg creatinine of suberic acid, sebacic acid, methylmalonic acid, ethylmalonic acid, glutaric acid, and adipic acid in approximately 1 L of donor urine. The exact amounts added will vary based on the creatinine value of the donor urine and will need to be calculated each time a new control is prepared. These analytes are quantified and standard multianalyte QC rules are used to determine acceptance of the run.
- The oximate control is prepared by spiking 100 μg/0.25 mg creatinine of 2-ketoglutaric acid into 1 L of normal urine. This control is analyzed quantitatively when oximated samples are included in the analysis.

5.2.1.4 Calculation and laboratory data analysis

1. *Laboratory Data Analysis*

Analysis of urine organic acid profiles, more than almost any other test performed in a biochemical genetics laboratory, varies greatly between performing sites. Any written document can serve as a guideline for best practices and cannot replace experience and practice with samples analyzed by the method performed in an individual's laboratory. Variations such as carrier gas can dramatically change the appearance of an organic acid profile (Fig. 5.2).

Commercial software packaged with most GC-MS systems can handle the basic functions required for organic acid analysis—capture of scanned mass range, basic library matching, and providing quantitation for desired analytes. The use of such software requires development and verification of appropriate methods. If quantitative results are generated, they should be verified according to appropriate local regulations and imported into the laboratory information management system after review is complete.

Urine organic acid analysis cannot always be completed using a simple quantitative method and library matches. Not all targeted physiological compounds are present in commercially available libraries, particularly downstream metabolites of medications and other endogenous compounds. Procedures should exist for adding verified compounds to the library of a laboratory to ensure proper identification going forward.

Similarly, not all relevant physiological and / or pathological compounds, even those that are diagnostic for specific metabolic disorders, are available as standards that allow for the construction of standard curves. Indeed, it is not always necessary to provide quantification of a metabolite that is abnormal simply by its presence for a diagnostic assay.

Data analysis of organic acid profiles in the laboratory requires significant training and experience. Given the rarity of some organic acid conditions, even experienced laboratories may only encounter some conditions infrequently. Laboratories should use these specimens as training opportunities for staff, including the analysis of previously analyzed profiles in routine procedures for training new technologists.

Technologists reviewing organic acid profiles for qualitative identification of peaks should use documented criteria for matching spectra, including both retention time and relevant mass fragments. Peaks that are labeled manually should use standard naming and abbreviation if necessary. Mixed peaks on the profile (see Fig. 5.3) should be labeled as such, and for highly clinically relevant compounds, it may be necessary to print / save the overlay for review during interpretation. Unknown compounds should be labeled with key masses. Notations about reason for referral, clinical notes, and other testing performed or pending in the laboratory are also helpful. After review of a profile is complete, the

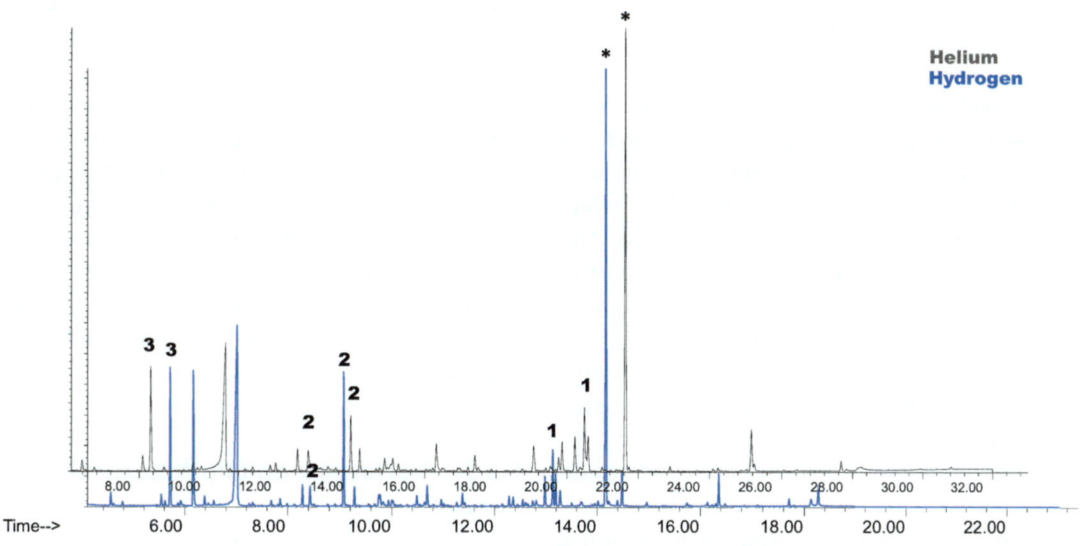

*: pentadecanoic acid (PDA); 1: double peak of 2-methylcitric acid; 2: propionylglycine; 3: 3-hydroxypropionic acid

Fig. 5.2 Urine sample from a known patient with propionic acidemia analyzed using two different carrier gases (Hydrogen and Helium). The characteristic peaks required for the diagnosis are present in both profiles, however, the resolution and retention time differences are an example of how method changes can alter the appearance of urine organic acid testing

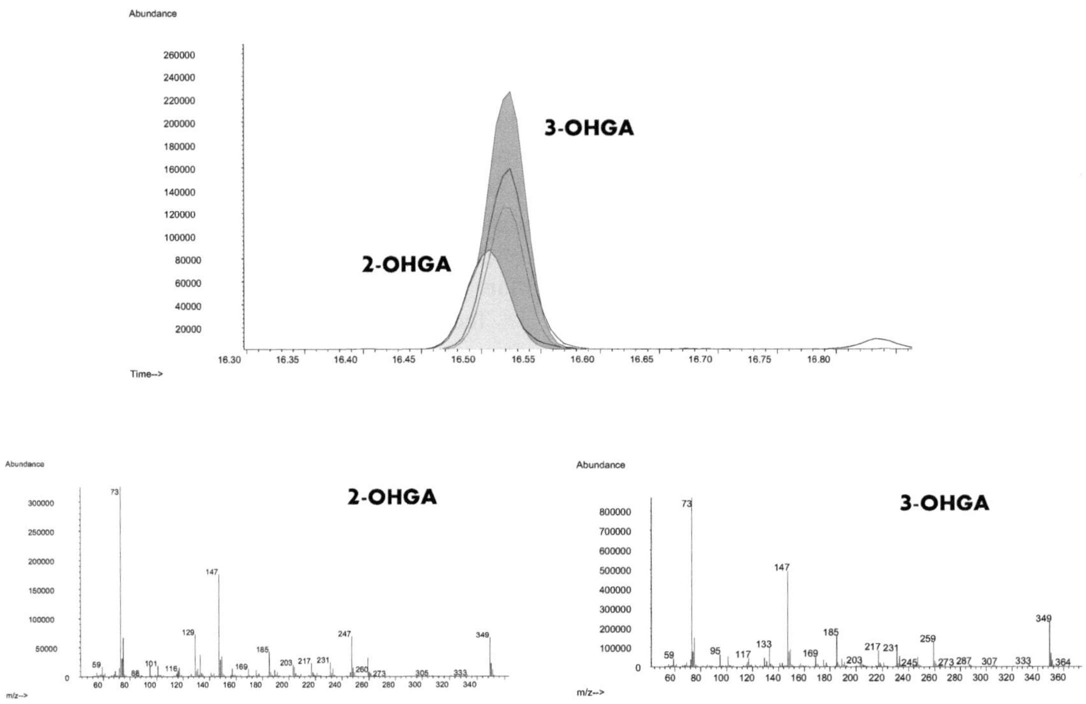

Fig. 5.3 Overlay of 2-hydroxyglutaric acid and 3-hydroxyglutaric acid peaks in a urine sample from a patient with glutaric acidemia type I. The mass spectra for each compound are shown, highlighting the difficulties in distinguishing the two species

performing technologist should be identified on any accompanying paperwork.

8. *Interpretation of Urine Organic Acid Results*

A broad spectrum of conditions can be identified using urine organic acid analysis, and it is important for a laboratory director to be familiar with the full scope, even for conditions which they may never have personally identified. This can involve literature reviews, proficiency testing specimens, or saved samples from previous analyses. Professional guidelines from the American College of Medical Genetics and Genomics exist to provide a starting point for this complex interpretation [8].

Interpretation of urine organic acid results should take into consideration all relevant information provided for a specimen, including other testing and a reason for referral when available. When a specific reason for referral is provided, and a suspected condition is NOT detected by the analysis, an explanation of the criteria used to rule this out is beneficial, such as "3-hydroxyglutaric acid was not detected" when a clinical suspicion for glutaric acidemia type I is documented. Follow-up testing recommendations should also be included in any report where abnormalities are detected. The variability in how laboratories perform and report UOA analysis is very broad. Our laboratory only reports a limited number of quantitative values for abnormal patients (most commonly methylmalonic acid, glutaric acid, orotic acid, and homogentisic acid), and does not extensively quantify a broad scope of analytes.

5.2.2 Reference and Pathological Values

Recognizing the need for reference ranges for laboratories that may be evaluating performance or establishing methods, a basic list of organic acid reference ranges has been adapted from a previous version of this chapter and provided as Table 5.4.

Table 5.4 Selected reference ranges for organic acid analysis. All values provided in mmol/mol creatinine. Adapted from the previous version of this chapter with permission

	Age			
Analyte	<1 month	1 month–6 months	6 months–5 years	>5 years
Cis-aconitic acid	5–31	10–97	10–97	3–44
Adipic acid	9–37	9–37	<15	<5
Azelaic acid	<1	<1	<1	<1
Citric acid	<1045	104–268	<656	87–639
Ethylmalonic acid	<17	<17	<17	<17
Fumaric acid	10–45	4–45	1–27	2–4
Glutaric acid	<13	<13	<13	<13
Homogentisic acid	<10	<10	<10	<10
Homovanillic acid	<22	<22	<8	<7
3-Hydroxybutyric acid	<5	<5	<5	<10
2-Hydroxyglutaric acid	<15	<15	<15	<15
4-Hydroxyphenyllactic acid	<50	<10	<10	<10
4-Hydroxyphenylpyruvic acid	<20	<5	<5	<5
Isocitric acid	<368	<67	<77	16–99
2-Ketoglutaric acid	22–567	63–552	36–103	41–82
Lactic acid	46–348	57–346	21–38	20–101
Malic acid	<52	8–73	4–57	17–47
Methylmalonic acid	<4	<4	<4	<4
Methylsuccinic acid	<12	<12	<12	<12
Orotic acid	1–5	1–3	1–3	<1
Pyroglutamic acid	<61	<61	<61	<61
Sebacic acid	3–16	3–16	<8	<8
Suberic acid	4–20	4–20	<8	<8
Succinic acid	35–547	34–157	16–118	29–87
Vanillylmandelic acid	<15	<10	<7	<5

Due to the primarily diagnostic nature of the urine organic acid profile, and the presence of abnormal metabolites in extremely large amounts for some disorders, laboratories may choose not to reject samples that fall short of minimum volume or creatinine requirements. Interpretation of the profiles should be handled with caution, as while some disorders have extremely abnormal profiles that would likely be detected with a high dilution factor (such as methylmalonic acidemia of a primary lactic acidemia), others would be missed (low excretor glutaric acidemia, some variant forms of propionic acidemia). In cases of a seemingly normal profile with a high dilution, recommending the collection of another sample is appropriate.

Urine organic acid reports vary greatly between laboratories, and thus reporting criteria and report appearance will as well. It is our practice to report a "normal" urine organic acid profile in a strictly qualitative manner, while other laboratories will still report a list of quantified analytes for both normal and abnormal profiles. Regardless of these differences in methodology, any report should include a cohesive summary of the abnormal findings in the submitted specimen, clear identification of the suspected underlying disorder or cause, including common name, enzyme defect, and gene(s) involved where appropriate. If confirmatory testing is required to clarify UOA findings, these details should be included. Whenever possible, UOA interpretations should make an overall interpretation and recommendation of complex profiles, rather than separate comments for each abnormality without proper context.

5.3 Conclusion

UOA is surely the most challenging assay to interpret in the biochemical genetics laboratory. The complete scope of endogenous and exogenous analytes, combined with the wide variability in normal profiles results in a near infinite number of combinations. The thought that "no two profiles look alike" has no doubt occurred to many trainees working through their first attempts at UOA profiles. In addition to this variation in normal, the number of disorders that could be detected by UOA far outpaces most biochemical genetics testing, particularly when compared to core assays such as plasma acylcarnitine or amino acid analysis. Outside of the more common disorders, particularly those included in newborn screening programs where ascertainment is very high, many biochemical geneticists may only see some disorders once or twice in their career, however, it is still necessary to understand their potential profiles and be able to accurately interpret these profiles when they do present themselves.

Acknowledgment The authors would like to acknowledge Dr. Piero Rinaldo [9] for his permission to reference and use materials from his previous version of this chapter and for his comprehensive training in organic acid analysis.

References

1. ACMG ACT Sheets and Algorithms. https://www.acmg.net/ACMG/Medical-Genetics-Practice-Resources/ACT_Sheets_and_Algorithms.aspx
2. Tanaka K, Budd MA, Efron ML, Isselbacher KJ (1966) Proc Natl Acad Sci USA 56:236
3. Ferreira CR, Rahman S, Keller M, Zschocke J, ICIMD Advisory Group (2021) An international classification of inherited metabolic disorders (ICIMD). J Inherit Metab Dis 44:44
4. Kumps A, Duez P, Mardens Y (2002) Metabolic, nutritional, iatrogenic, and Artifactual sources of urinary organic acids: a comprehensive table. Clin Chem 48(5):708–717. https://doi.org/10.1093/clinchem/48.5.708
5. Wortmann SB, Duran M, Anikster Y, Barth PG, Sperl W, Zschocke J, Morava E, Wevers RA (2013) Inborn errors of metabolism with 3-methylglutaconic aciduria as discriminative feature: proper classification and nomenclature. JIMD 36:923–928
6. Jones DE, Klacking E, Ryan RO (2021) Inborn errors of metabolism associated with 3-methylglutaconic aciduria. CCA 522:96–104
7. Vamecq J, Papegay B, Nuyens V, Boogaerts J, Leo O, Kruys V (2020) Mitochondrial dysfunction, AMPK activation and peroxisomal metabolism: a coherent scenario for non-canonical 3- methylglutaconic acidurias. Biochimie 168:53–82
8. Gallagher RC, Pollard L, Scott AI, Huguenin S, Goodman S, Sun Q (2018) ACMG biochemical genetics Subcommittee of the Laboratory Quality Assurance Committee. Laboratory analysis of organic acids, 2018 update: a technical standard of the American College of Medical Genetics and Genomics (ACMG). Genet Med 20(7):683–691. https://doi.org/10.1038/gim.2018.45. Epub 2018 Mar 15
9. Rinaldo P (2008) Organic Acids. In: Blau N, Duran M, Gibson K (eds) Laboratory guide to the methods in biochemical genetics. Springer, Berlin, Heidelberg. https://doi.org/10.1007/978-3-540-76698-8_9

Acylcarnitines

6

Dietrich Matern

Contents

Abbreviations

ESI	Electrospray ionization
FAO	Mitochondrial fatty acid oxidation
LC	Liquid chromatography
MS	Mass spectrometry
MS/MS	Tandem mass spectrometry

D. Matern (✉)
Biochemical Genetics Laboratory, Division of Laboratory Genetics and Genomics, Department of Laboratory Medicine and Pathology, Mayo Clinic, Rochester, MN, USA
e-mail: matern@mayo.edu

6.1 Clinical Significance

Carnitine and its esters are physiologically present in all biological fluids, but carnitine is most abundant in tissues with high energy requirements, particularly skeletal and cardiac muscle [1]. Carnitine was discovered early in the last century by Gulewitsch and Krimberg when studying meat extracts [2]. Because its function was unknown and considering the source of this compound, they suggested naming it *Carnitine*. The chemical structure was determined in 1927 [3] and in 1952 it was categorized as a vitamin B when studies in the meal worm *Tenebrio molitor*

N. Blau, F. M. Vaz (eds.), *Laboratory Guide to the Methods in Biochemical Genetics*,
https://doi.org/10.1007/978-3-031-58819-8_6

suggested a role of carnitine as a growth factor (Vitamin B_T) [4, 5].

In 1955, Fritz determined that carnitine plays an essential role in fatty acid β-oxidation (FAO) [6] and in 1973 the first two clinically relevant disorders affecting this pathway were described: primary carnitine deficiency and carnitine palmitoyltransferase II deficiency [7, 8]. To date, more than 20 different enzyme and transporter deficiency states affecting fatty acid mitochondrial β-oxidation and FAD synthesis and transport have been described [9–13].

Carnitine is not only involved in the metabolism of long-chain fatty acids but also in branched-chain amino acid metabolism. Most of the classic organic acidurias resulting from defects in the latter metabolic pathway are associated with secondary carnitine deficiency. In fact, the clinical utility of acylcarnitine analysis was first identified for organoacidopathies such as propionic, methylmalonic, and isovaleric acidemias [14–19]. While urine initially was the preferred specimen, plasma became the specimen of choice because acylcarnitine profiles even in healthy probands are generally less complex in plasma than in urine, and because the sensitivity of acylcarnitine analysis is higher when plasma is analyzed as opposed to urine, especially for the diagnosis of long-chain FAO disorders [20]. However, urine acylcarnitine analysis can add diagnostic value to the metabolic work up of patients with organic acidemias but inconclusive or borderline abnormal urine organic acid and plasma acylcarnitine profiles [21–25].

Acylcarnitine analysis for the diagnosis of organic acidemias and particularly of FAO disorders plays a prominent role in all venues of clinical biochemical genetics: prenatal diagnosis, newborn screening, evaluation of symptomatic patients, and postmortem screening. Almost exclusively performed by tandem mass spectrometry (MS/MS), plasma/serum is the primary specimen type in diagnostic settings. Blood dried on filter paper is analyzed for newborn screening, and together with bile in the postmortem evaluation of cases of sudden and unexpected death [12, 26]. Cell-free supernatant of amniotic fluid can be used for the prenatal diagnosis of select inborn errors of metabolism [27–30].

Cultured fibroblasts or amniocytes can be probed with FAO substrates and carnitine. Cell cultures deficient of an FAO enzyme will accumulate specific acylcarnitine species when incubated with substrates such as palmitate, allowing for the diagnosis of FAO disorders [31–40]. Modifications of this assay system have also been developed for the diagnosis of defects affecting the metabolism of branched-chain amino acids [21, 35, 36]. This approach was also adapted for the study of peripheral blood mononuclear cells [41] and whole blood [42]. The latter, however, requires incubation of EDTA-blood within 24 h of collection.

Acylcarnitine analysis using stable isotope-labeled internal standards provides quantitative data for acylcarnitine species [15]. However, to provide meaningful results to referring health care professionals, it is critical to complement analytical proficiency with in-depth interpretation of results as is true for many other examples of complex metabolic profiles [43].

6.2 Properties of Carnitine and Acylcarnitines

Carnitine, L-3-hydroxy-4-(trimethylammonium) butyrate, is a water-soluble, trimethylammonium derivative of γ-amino-β-hydroxybutyric acid which is formed from trimethyllysine via γ-butyrobetaine [44]. Carnitine originates in about 75% of dietary intake of meat, fish, and dairy products containing proteins with trimethyllysine residues. Under normal conditions, endogenous synthesis from lysine and methionine plays a minor role but can be stimulated by a diet low in carnitine [45]. Carnitine is not further metabolized and is excreted in urine and bile as free carnitine or as conjugated carnitine esters [1, 46, 47]. Adequate intracellular levels of carnitine are therefore maintained by mechanisms that modulate dietary intake, endogenous synthesis, reabsorption, and cellular uptake.

Under physiologic conditions, carnitine is primarily required to shuttle long-chain fatty acids across the inner mitochondrial membrane for fatty acid β-oxidation and products of peroxisomal β-oxidation to the mitochondria for further metabolism in the citric acid cycle [44, 48]. Acylcarnitines are formed by conjugating acyl-CoA moieties to carnitine which in the case of activated long-chain fatty acids is accomplished by carnitine palmitoyltransferase I (CPT-I) [10]. The acyl group of the activated fatty acid (fatty acyl-CoA) is transferred by CPT-I from the sulfur atom of CoA to the hydroxyl group of carnitine (Fig. 6.1). Carnitine-acylcarnitine translocase (CACT) then transfers the long-chain acylcarnitines across the inner mitochondrial membrane, where CPT-II reverses the action of CPT-I by formation of acyl-CoA and release of free unesterified carnitine.

In pathologic conditions, such as FAO disorders or organic acidemias due to acyl-CoA dehydrogenase deficiencies, the functions of carnitine as a regulator of substrate flux and energy balance across cell membranes and as a modulator of intracellular concentrations of free CoA become crucial. In such conditions acyl-CoAs accumulate within the mitochondrial matrix and carnitine is utilized to shuttle these compounds out of the mitochondria as acylcarnitines, providing free CoA at the same time.

Carnitine and its esters are present in all biological fluids and depending on the enzyme defect, a particular acylcarnitine pattern becomes apparent where those acylcarnitine species serving as direct substrates for the defective enzyme accumulate disproportionately to the down- and upstream metabolites (Table 6.1).

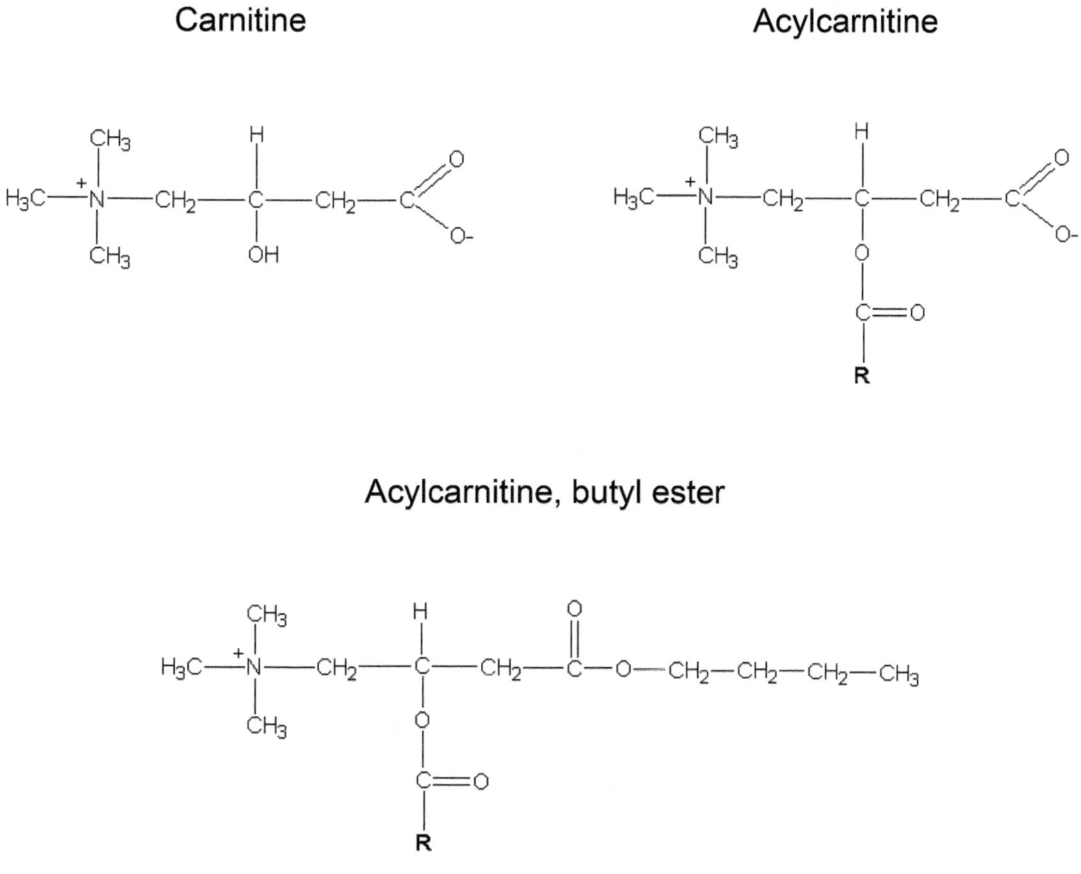

Fig. 6.1 Structures of carnitine, acylcarnitine, and butylated acylcarnitine. The R represents the acylcarnitine species with up to 18 carbons, which are typically the targets of an acylcarnitine analysis

Table 6.1 Clinically relevant acylcarnitine species included in a typical (butylated) acylcarnitine analysis and their relevance when abnormally elevated (unless otherwise noted)

[M + H]+	Acylcarnitine species		Disorder
218	C0	Free carnitine	Carnitine supplementation (deficiency if low)
260	C2	Acetyl-	Carnitine supplementation or ketosis (deficiency if low)
274	C3	Propionyl-	PA, MMA, MCD, SUCLA2, treatment with heptanoic acid
287		FIGLU	Glutamate formimino transferase deficiency
288	C4	Butyryl–/ Isobutyryl-	SCAD, IBDH, MADD, EE
300	C5:1	Tiglyl-/3-Methylcrotonyl-	BKT, MCC, MHBD, MCD
302	C5	Isovaleryl–/ 2-Methylbutyryl–/ Pivaloyl-	IVA, SBCAD, MADD, EE, antibiotics/cream derived artifact, treatment with heptanoic acid
304	C4-OH	3-Hydroxy butyryl-	SCHAD, ketosis
314	C6:1	3-Methylglutaconyl-	MGA
316	C6	Hexanoyl-	MCAD, MKAT, MADD
318	C5-OH	3-Hydroxy isovaleryl-	MCC, HMG, MCD, MGA, BIOT[a]
		2-Methyl3-hydroxybutyryl-	BKT, MHBD
322		Benzoyl-	Treatment with Na-benzoate
325		Dextrose (fragment)	Sample contamination with dextrose
330	C7	Heptanoyl-	Treatment with heptanoic acid
336		Phenylacetyl-	Treatment with phenylacetic acid
338		Salicylyl-	Treatment with salicylic acid
344	C8	Octanoyl-	MCAD, M/SCHAD, MKAT, MADD
360	C3-DC	Malonyl-	Malonic aciduria
	C8-OH	3-hydroxy octanoyl-	MKAT
368	C10:2	Decadienoyl-	DCR
370	C10:1	Decenoyl-	MCAD
372	C10	Decanoyl-	MCAD, MADD
374	C4-DC	Methylmalonyl–/succinyl-	MMA[#], SUCLA2
388	C5-DC	Glutaryl-	GA-1
	C10-OH	3-Hydroxy decanoyl-	M/SCHAD, MKAT
399		Dextrose (fragment)	Sample contamination with dextrose
400	C12	Dodecanoyl-	MADD
402	C6-DC	3-Methylglutaryl-	HMG
416	C12-OH	3-Hydroxy dodecanoyl-	LCHAD/TFP
426	C14:1	Tetradecenoyl-	CACT, CPT II, MADD, VLCAD, LCHAD/TFP
428	C14	Tetradecanoyl- (myristoyl-)	CACT, CPT II, MADD, VLCAD, LCHAD/TFP
444	C14-OH	3-Hydroxytetradecanoyl-	LCHAD/TFP
456	C16	Hexadecanoyl- (palmitoyl-)	CACT deficiency, CPT II deficiency, VLCAD deficiency, LCHAD/TFP deficiency; CPT I[§]
470	C16:1-OH	3-Hydroxyhexadecenoyl-	Antibiotics derived artifact
472	C16-OH	3-Hydroxyhexadecanoyl-	LCHAD/TFP
473		Dextrose (fragment)	Sample contamination with dextrose
480	C18:2	Octadecadienoyl- (linolyl-)	CACT, CPT II, VLCAD, LCHAD/TFP
482	C18:1	Octadecenoyl- (oleyl-)	CACT, CPT II, VLCAD, LCHAD/TFP; CPT type I[§]
484	C18	Octadecanoyl- (stearyl-)	CACT, CPT II, VLCAD, LCHAD/TFP
498	C18:1-OH	3-Hydroxy octadecenoyl-	LCHAD/TFP
500	C18-OH	3-Hydroxy octadecanoyl-	LCHAD/TFP

[a]Rarely associated with abnormal acylcarnitine profiles; [#]Respective analyte is not consistently elevated in MMA; [§]CPT I is suggested by low concentrations of long-chain acylcarnitine species and relatively high free carnitine; *BIOT* biotini-

(continued)

Table 6.1 (continued)

dase deficiency; *BKT* β-ketothiolase deficiency; *CACT* carnitine-acylcarnitine translocase deficiency; *CPT* carnitine palmitoyltransferase deficiency; *DCR* 2,4-dienoyl-CoA reductase deficiency [49]; *EE* ethylmalonic encephalopathy; *FIGLU* formiminoglutamate; *GA-1* glutaric acidemia type I (glutaryl-CoA dehydrogenase deficiency); *HMG* 3-hydroxy 3-methylglutaryl-CoA lyase deficiency; *IBDH* isobutyryl-CoA dehydrogenase deficiency; *IVA* isovaleric acidemia (isovaleryl-CoA dehydrogenase deficiency); *LCHAD* long-chain 3-hydroxy acyl-CoA dehydrogenase deficiency; $[M + H]^+$, molecular weight of $[M + H]^+$ ions corresponding to the butyl ester derivatives of acylcarnitine species; *MADD* multiple acyl-CoA dehydrogenase deficiency (glutaric acidemia type II and related conditions of riboflavin transport and metabolism); *MCAD* medium-chain acyl-CoA dehydrogenase deficiency; *MCC* 3-methylcrotonyl-CoA carboxylase deficiency; *MCD* multiple carboxylase (holocarboxylase) deficiency; *MGA* 3-methylglutaconic aciduria type I (3-methylglutaconyl-CoA hydratase deficiency); *MMA* methylmalonic acidemias; *MHBD* 2-methyl 3-hydroxy butyryl-CoA dehydrogenase deficiency; *PA* propionic acidemia (propionyl-CoA carboxylase deficiency); *SCAD* short-chain acyl-CoA dehydrogenase deficiency; *SCHAD* short-chain 3-hydroxy acyl-CoA dehydrogenase deficiency; *SUCLA2* ADP-forming succinyl-CoA synthetase deficiency [50]; *TFP* mitochondrial trifunctional protein deficiency; *VLCAD* very long-chain acyl-CoA dehydrogenase deficiency

6.3 Methods

Several techniques have been described to differentiate and eventually quantify specific carnitine esters. In 1977, Bieber and Choi demonstrated the presence of short-chain acylcarnitine species in several animal tissues by use of gas chromatography and mass spectrometry (MS) [51]. This approach was improved over time and as recently as 1997 new protocols for quantitative gas chromatography-MS analysis of acylcarnitines were published [52]. Additional methods described and which do not require volatile samples include thin layer chromatography and radioisotopic exchange/HPLC [53], as well as liquid chromatography (LC) MS. HPLC thermospray MS and fast atom bombardment MS were first applied to acylcarnitine analysis by Alfred Yergey, David Millington, Charles Roe and colleagues in the early 1980s [17, 54, 55]. This group later applied tandem mass spectrometry (MS/MS) to the analysis of selected amino acids [56], then to acylcarnitine analysis [57]. The advantage of MS/MS over other methods lies in its sensitivity which allows for efficient preparation procedures of small sample volumes and fast analytical times, therefore providing for rapid throughput of even large numbers of samples. However, the significant cost and sophistication of MS/MS instruments delayed the adoption of this technology in other than highly specialized, often research laboratories. Only in the late 1990s instruments became commercially available that were affordable, robust, computer-driven, and increasingly user friendly [58]. Today then, acylcarnitines are analyzed by a large number of laboratories and for the most part by MS/MS. The primarily used instruments are triple quadrupole analyzers combined with electrospray ionization (ESI) sources. With ESI-MS/MS, the liquid sample is introduced into the ion source through a capillary tube and exposed to a strong electric field and a counter flow of nitrogen gas, which in combination produces the electrospray. The solvent evaporates and eventually desorption of charged ions occurs into the mass analyzer. ESI occurs under atmospheric pressure, allowing efficient ionization, and the mass analyzers operate at very low pressure. The combination is possible by introducing the ions into the analyzers through a very small orifice (<1 mm) and using powerful vacuum pumps.

This chapter focuses on acylcarnitine analysis in various sample types following derivatization to butyl or methyl esters and flow injection ESI-MS/MS. Capillary electrophoresis MS/MS and LC-MS/MS methods have also been described with the benefits of isobar separation and further simplification of the sample preparation steps; however, at the cost of larger sample volume requirements and longer analytical times [59–61].

6.4 Acylcarnitine Analysis of Plasma or Serum by Tandem Mass Spectrometry (MS/MS)

6.4.1 Principle

Acylcarnitines are extracted from the sample by mixing with methanol or an acidified acetonitrile solution containing isotopically labeled acylcarnitines of various chain lengths at defined concentrations as internal standards. Following centrifugation, the supernatant is evaporated and the residue is derivatized with either n-butanol HCl or n-methanol HCl yielding the acylcarnitines for analysis by flow injection ESI-MS/MS. Following analysis, a graphical acylcarnitine profile is generated by the software provided by the MS/MS vendor which can be interpreted qualitatively. Additional vendor-provided software allows semiquantitative calculation of the concentration of each individual acylcarnitine species based on the abundance of the assigned internal standard.

6.4.2 Pre-analytical

(i) Specimen

Plasma or serum is obtained from blood collected in sodium heparin (lithium heparin and EDTA are also acceptable) or in clot activator tubes, respectively. Most informative results are generally achieved when samples are obtained during acute illness. Because inborn errors of metabolism are traditionally not entered early into differential diagnostic considerations, sample collection should alternatively be timed before a meal, preferably after an overnight fast. A prolonged fasting challenge, however, should not routinely be undertaken as these require close surveillance typically not possible in an outpatient setting.

Samples should be kept frozen until analysis. Reliable results, particularly for short-chain acylcarnitine species cannot be achieved following several days at ambient temperatures [62].

(ii) Reagents and Chemicals

– Standards

The following standards are used in our laboratory:

- Acetylcarnitine (C_2)
- Propionylcarnitine (C_3)
- Butyrylcarnitine (C_4)
- Isovalerylcarnitine (C_5)
- Hexanoylcarnitine (C_6)
- Octanoylcarnitine (C_8)
- Decanoylcarnitine (C_{10})
- Dodecanoylcarnitine (C_{12})
- Tetradecanoylcarnitine (C_{14})
- Hexadecanoylcarnitine (C_{16})
- Octadecanoylcarnitine (C_{18})
- Isotopically labeled standards (internal standards)

  – Acetyl-d_3-carnitine (C_2),
  – Propionyl-d_3-carnitine (C_3),
  – Butyryl-d_3-carnitine (C_4),
  – Octanoyl-d_3-carnitine (C_8),
  – Dodecanoyl-d_3-carnitine (C_{12}),
  – Palmitoyl-d_3-carnitine (C_{16}).

  The internal standards are dissolved in methanol. The final working internal standard solution is prepared in acetonitrile at the following concentrations and stored refrigerated:

- 1.330 μmol/L acetyl-d_3-carnitine
- 0.266 μmol/L propionyl-d_3-carnitine
- 0.266 μmol/L butyryl-d_3-carnitine
- 0.266 μmol/L octanoyl-d_3-carnitine
- 0.266 μmol/L dodecanoyl-d_3-carnitine
- 0.543 μmol/L palmitoyl-d_3-carnitine
- Additional or different internal standards can be used. Labeled and unlabeled acylcarnitine standards are available from the Organic Synthesis Laboratory at the Amsterdam University Medical Centers, Amsterdam, The Netherlands (https://www.vumc.com/departments/clinical-chemistry/metabolic-laboratory/organic-synthesis-laboratory.htm) or from Cambridge Isotope Laboratories, Tewksbury, Massachusetts, USA (http://www.isotope.com).

- 3 N HCl in n-butanol (butanolic-HCl) Regis Chemical Co. (Morton Grove, IL) # 201009
- Methanol
- 80% Acetonitrile:Water[1]
- Nitrogen gas, in-house, scrubbed with BHT-1 hydrocarbon trap Chrom Tech (Apple Valley, MN)
- Air, compressed
- Air, zero grade (filtered by Balston filter unit)
- Formic Acid, 98–100%
- Equipment and Supplies Samples can be prepared using multi-well plates or in per-sample vials. Described is the procedure using 96-well plates (for procedure using per-sample vials see previous edition of this chapter [63]).
- 96-well, polypropylene, U-bottom microplate, Greiner #650201
- 96-deep-well (2 mL), polypropylene plate, Chrom Tech #96–6009
- Microplate aluminum sealing tape, Fisher Scientific #07200684
- 96-Well Pattern Blue PTFE Sealing Film, Chrom Tech BST-9790
- Integra 1250 μL Standard V96 Inserts, Integra #6442
- Viaflo 96 Pipettor, Integra Biosciences Co.
- SPE Dry 96 Dual Microplate Evaporator, Agronaut
- Fisherbrand Isotemp General Purpose Heating and Drying Oven, Fisher Scientific
- Thermo Max Q2000 Orbital Shaker, Thermo Scientific
- Fisherbrand MultiTube Vortexer, Fisher Scientific
- Fisherbrand Accuspin Microcentrifuges with 24 × 1.5/2.0 mL Rotor (Thermo FisherScientific #75003524).
- AB SCIEX API 3200 tandem mass spectrometry system with Turbo V IonSpray interface source
- Shimadzu LC-20 AB Binary Pump
- Leap Technologies HTC PAL Autosampler
- 0.5 U peek filter end fitting, Chromtech A-428X
- Back pressure column, BDS Hypersil C18, 3 μm, 50 × 1, Keystone Scientific, Inc., #053–46-1

(iii) Calibration

Mass calibration of the quadrupole mass spectrometers, Q1 and Q3, is based on poly(propylene)glycol in methanol as supplied by the instrument manufacturer and following instructions on calibration and tuning provided in the instrument operating manual.

(iv) Quality Control (QC)

- QC samples:
  - A QC standard mixture, a normal control, and an abnormal control are run with each batch. The normal control is made by aliquoting 100 μL portions of a verified normal-range plasma obtained from a healthy volunteer donor into screw-cap vials which are stored at −70 °C.
  - The abnormal control is made by spiking 50 mL of normal plasma with defined concentrations of unlabeled acylcarnitine standards spanning the mass range covered by the analysis (see Reagents and Chemicals). The concentration for each standard should be chosen based on the upper limit of the reference range for the respective acylcarnitine species.

[1] Use of reverse osmosis or HPLC grade water in every step requiring water is recommended.

100 μL portions are aliquotted into screw-cap vials and stored at −70 °C.
- The mean and standard deviation for each QC sample are calculated with a minimum of 20 between run values.
- In our laboratory, control values that fall within ±2 standard deviations of the mean are considered acceptable and require no further action. Any control values that are either greater than 2 standard deviations or trends require review by a laboratory director. The laboratory director decides whether the analysis of the complete batch or of specific samples needs to be repeated.

6.4.3 Analytical

i. Sample Preparation

Formic acid is added to the working internal standard solution to achieve a concentration of 0.3% formic acid. 300 μL are then pipetted into each well of a 96-deep well plate before 20 μL of thawed and vortexed plasma sample is added. The plate is then sealed with adhesive aluminum tape and a plastic plate cover. After thorough vortexing and centrifugation for 5 min at 5180 x g, the supernatant is carefully transferred to corresponding wells of a U-bottom polypropylene plate and dried under a gentle stream of warm nitrogen. Working in a hood, the samples are derivatized to butyl esters by the addition of 100 μL 3 M butanol-HCl and, following the application of cling wrap and a plate cover, heating at 65 °C in an Isotemp oven for 25 min. The plate is then allowed to cool before the samples are again dried down under warm nitrogen. The dried samples are redissolved in 100 μL 80% acetonitrile:water, the plate is covered with Blue PTFE Sealing Film and rotated for at least 1 min at 120 rpm. The samples are now ready for MS/MS analysis which should proceed promptly. If necessary, prepared samples can be stored refrigerated for up to 4 days prior to analysis.

ii. Instrument Setup

LC system:
Controlled by the MS/MS software, the Shimadzu LC-20 AB binary pump operates isocratic using 80% acetonitrile:water at 40 μL/min. A 50 x 1 mm C18 column (Keystone) is used between the pump and autosampler to provide backpressure. The syringe/system flush solution is the 80% acetonitrile:water that is used as mobile phase. The autosampler is connected directly to the MS/MS Turbo V Ion Spray source. The injected sample volume is 10 μL.

A 0.5-U Peek filter is installed between the autosampler and Turbo V Ion Spray electrospray interface source. The filter end fitting is discarded and replaced after approximately 200 injections.

MS/MS system:
Acylcarnitine analysis is performed as a precursor scan in positive ion mode. Q1 is set to scan a mass range from m/z 200 to 505, while Q3 is set to determine a precursor ion of m/z 85. The method is optimized for a mixture of butyl esters containing d_3-C_2, d_3-C_3, d_3-C_4, d_3-C_8, d_3-C_{12}, and d_3-C_{16}-carnitine.

The sample queue is entered into the MS/MS instrument's software (i.e., Analyst for SCIEX instruments) according to the manufacturer's instructions. Samples can be given specific designations that allow for correct QC and reference range assignment.

Prior to analysis, the method should be equilibrated, and the system purged if necessary. The LC line should be attached to the sprayer after proper equilibration for approximately 30 min.

iii. Calculation

- A datafile is generated for each sample and presented as a mass profile. All acylcarnitine species in a mass range from m/z 200 to 505 that generate a product m/z of 85 are represented in this profile based on their detected abundance.
- Calculation of the molecular weights of the n-butyl esters of acylcarnitines, expressed as $(M + H)^+$ molecular ions,

Table 6.2. Reference range for acylcarnitine species (as butylesters) in plasma (nmol/mL) (from [64])

	≤ 7 days (n = 1551)			8 days–7 years (n = 26,701)			> 7 years (n = 12,783)		
Acylcarnitine species	1st %ile		99th %ile	1st %ile		99th %ile	1st %ile		99th %ile
C0	4.82	–	24.28	6.13	–	32.30	5.92	–	30.61
C2	1.86	–	17.50	2.40	–	27.62	1.77	–	24.10
C3		<	1.17		<	1.60		<	1.46
FIGLU		<	0.10		<	0.04		<	0.04
C4		<	0.81		<	0.81		<	0.78
C5:1		<	0.05		<	0.04		<	0.05
C5		<	0.71		<	0.54		<	0.39
C4-OH		<	0.16		<	0.30		<	0.27
C6:1		<	0.05		<	0.05		<	0.05
C6		<	0.17		<	0.22		<	0.19
C5-OH		<	0.21		<	0.15		<	0.08
C7		<	0.05		<	0.05		<	0.05
C6-OH		<	0.06		<	0.07		<	0.05
C8:1		<	0.73		<	0.75		<	0.70
C8		<	0.41		<	0.41		<	0.45
C3-DC / C8-OH		<	0.11		<	0.13		<	0.16
C10:2		<	0.10		<	0.10		<	0.10
C10:1		<	0.34		<	0.43		<	0.46
C10		<	0.50		<	0.56		<	0.67
C4-DC		<	0.07		<	0.06		<	0.07
C10:1-OH		<	0.06		<	0.04		<	0.05
C5-DC / C10-OH		<	0.10		<	0.08		<	0.10
C12:1		<	0.15		<	0.18		<	0.19
C12		<	0.21		<	0.20		<	0.19
C6-DC		<	0.13		<	0.13		<	0.13
C12:1-OH		<	0.05		<	0.06		<	0.07
C12-OH		<	0.07		<	0.08		<	0.11
C14:2		<	0.10		<	0.13		<	0.13
C14:1		<	0.19		<	0.25		<	0.24
C14		<	0.12		<	0.12		<	0.11
C14:1-OH		<	0.03		<	0.04		<	0.04
C14-OH		<	0.03		<	0.02		<	0.03
C16:1		<	0.09		<	0.08		<	0.09
C16		<	0.29		<	0.24		<	0.22
C16:1-OH		<	0.37		<	0.04		<	0.03
C16-OH		<	0.05		<	0.03		<	0.03
C18:2		<	0.16		<	0.17		<	0.19
C18:1		<	0.29		<	0.32		<	0.35
C18		<	0.11		<	0.11		<	0.12
C18:2-OH		<	0.02		<	0.02		<	0.03
C18:1-OH		<	0.02		<	0.03		<	0.03
C18-OH		<	0.02		<	0.02		<	0.02

allows designation of each peak (Table 6.1).

- Quantitation software provided by the instrument vendor (i.e., ChemoView for SCIEX instruments) allows calculation of quantitative and semiquantitative concentrations for these acylcarnitines by comparison of the detected abundance of each acylcarnitine versus that of a designated internal standard with known concentration.
- The results for each sample are further compared to age-appropriate reference ranges (Table 6.2).

6.4.4 Post-analytical: Interpretation and Reference Ranges

The laboratory director, typically a board certified Clinical Biochemical Geneticist, reviews all profiles, and provides an interpretation based on pattern recognition and not simply on single abnormal values (Table 6.2). The laboratory director also determines the need for sample dilutions and carryover checks.

Simple reporting of numeric results is not appropriate because most physicians are not familiar with pattern recognition. A comprehensive interpretation takes into consideration any available clinical and dietary information and other laboratory results, provides possible differential diagnoses, recommendations for additional biochemical testing and confirmatory studies if indicated, as well as contact information for the laboratory director in case the referring physician has additional questions.

6.4.5 Pitfalls

Acylcarnitine profiles are dependent on the clinical status of the patient at the time of sample collection [65, 66]. It is, therefore, important to provide the biochemical genetics laboratory with information regarding the clinical context during which the sample was collected. The laboratory must be conscientious of the fact that carnitine deficiency states can be associated with acylcarnitine profiles that lack any acylcarnitine species that are elevated above the reference range. Consequently, it is essential that the complete profiles are reviewed and even borderline elevated acylcarnitines should prompt further follow-up in the presence of abnormally low free and acetylcarnitine (Fig. 6.2). If clinically indicated, a repeat sample should be collected at least 24 h after L-carnitine supplementation.

While MS/MS allows for unequivocal identification of most metabolites, there are a few exceptions. In particular, the short-chain acylcarnitines of 4 and 5 carbons represent more than one analyte. C_4-Carnitine is known to be a mixture of butyrylcarnitine derived from fatty acid metabolism and isobutyrylcarnitine derived from the metabolism of valine (Fig. 6.3) [23, 67]. C_5-carnitine is a mixture of isovalerylcarnitine and 2-methylbutyrylcarnitine derived from leucine and isoleucine degradation, respectively (Fig. 6.4) [21, 68]. Samples of patients treated with antibiotics containing pivalic acid (including newborns whose mothers were treated with such antibiotics prenatally) or exposed to neopentanoate may contain pivaloylcarnitine, another C_5 species [69–71]. Differentiation is of clinical importance and is usually achieved by urine acylglycine and organic acid analyses. Patients with SCAD deficiency, for example, will reveal an abnormal excretion of ethylmalonic acid suggesting that an elevated C_4-carnitine represents primarily butyrylcarnitine. Patients with isobutyryl-CoA dehydrogenase deficiency on the other hand will demonstrate an elevated C_4-carnitine in plasma and urine and either normal acylglycines or an elevated excretion of isobutyrylglycine [23]. The detection of an overexcretion of either isovalerylglycine or 2-methylbutyrylglycine will determine the origin of a C_5-carnitine elevation as either due to isovaleric acidemia or 2-methylbutyrylglycinuria as opposed to an artifact [68, 70]. Alternatively, separation of isobars can be achieved by LC-MS/MS analysis [61].

Several other metabolites are also nonspecific markers for several disorders. For example, C5-OH-carnitine, which represents 3-hydroxy isovalerylcarnitine and 2-methyl 3-hydroxybutyrylcarnitine, can be elevated in seven different organic acidemias (Table 6.1 and Fig. 6.5). The interpretation of elevated C5-OH-carnitine in newborns or breast-fed infants is further complicated by the fact that it can indicate maternal 3-methylcrotonylglycinuria while the infant is only an unaffected carrier [72].

The long-chain FAO disorders of CACT deficiency and CPT-II deficiency cannot be differentiated because both cause accumulation of the same long-chain acylcarnitine species which is explained by the fact that neither enzyme is involved in the chain-shortening action of FAO (Table 6.1 and Fig. 6.6e and f). Isolated LCHAD deficiency and complete mitochondrial TFP defi-

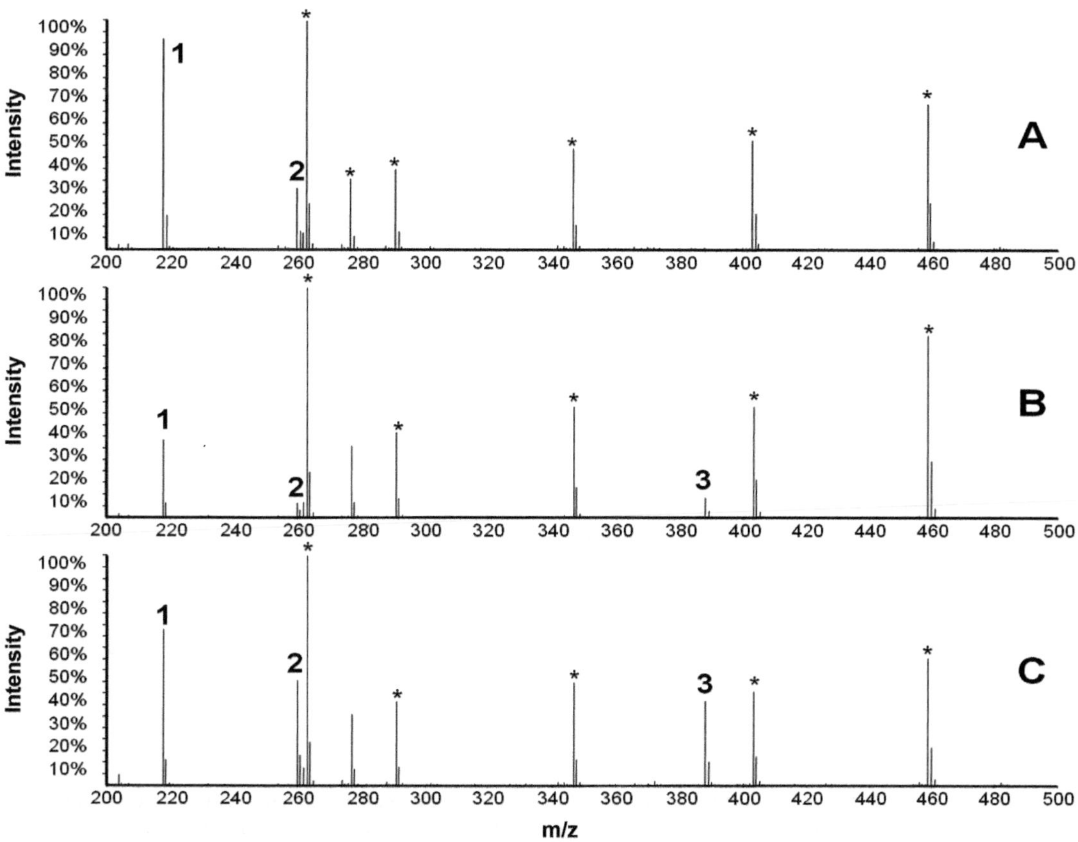

Fig. 6.2 Profiles of acylcarnitines as their butyl esters in plasma (precursor of m/z 85 scan) of normal control (**a**), a patient with glutaric acidemia type I and low carnitine levels (**b**), and a patient with glutaric acidemia type I and normal carnitine status (**c**). Peak #1, free carnitine (m/z 218); peak #2, acetylcarnitine (C_2; m/z 260); peak #3, glutarylcarnitine (C_5-DC; m/z 388). The asterisks represent the internal standards (from left to right): d_3-acetylcarnitine (C_2; m/z 263), d_3-propionylcarnitine (C_3; m/z 277), d_3-butyrylcarnitine (C_4; m/z 291), d_3-octanoylcarnitine (C_8; m/z 347), d_3-dodecanoylcarnitine (C_{12}; m/z 403), and d_3-palmitoylcarnitine (C_{16}; m/z 459)

ciency also cannot be differentiated by routine acylcarnitine analysis [62, 66, 75]. When such profiles are encountered, delineation of the correct defect is only possible by either specific enzyme assay or molecular genetic analysis of the relevant genes. Isolated elevations of propionylcarnitine (C_3) are not specific for propionic acidemia but also observed in methylmalonic acidemias of various etiologies (Table 6.1 and Fig. 6.5b and c). Because methylmalonylcarnitine (C_4-DC) is not consistently elevated in methylmalonic acidemias, elucidation of the correct diagnosis requires at a minimum urine organic acid analysis.

Another antibiotic that may cause problems in the interpretation of butylated acylcarnitines is cefotaxime (Fig. 6.5d) [76]. This antibiotic or metabolites thereof, reveals itself by acylcarnitine analysis at m/z 470 which otherwise is considered to represent the monounsaturated form of 3-hydroxy hexadecenoylcarnitine ($C_{16:1}$-OH). In poorly resolved scans, this may be difficult to differentiate from m/z 472 which is a marker for LCHAD and TFP deficiencies. However, whereas m/z 472 (C_{16}-OH) is more abundant than $C_{16:1}$-OH in these FAO disorders, the profile of a patient treated with cefotaxime usually reveals an m/z 470 to m/z 472 ratio that is greater than 1. Furthermore and in contrast to cefotaxime treatment, both LCHAD and TFP deficiencies are usually accompanied by elevations of other long-chain species (Table 6.1) [66].

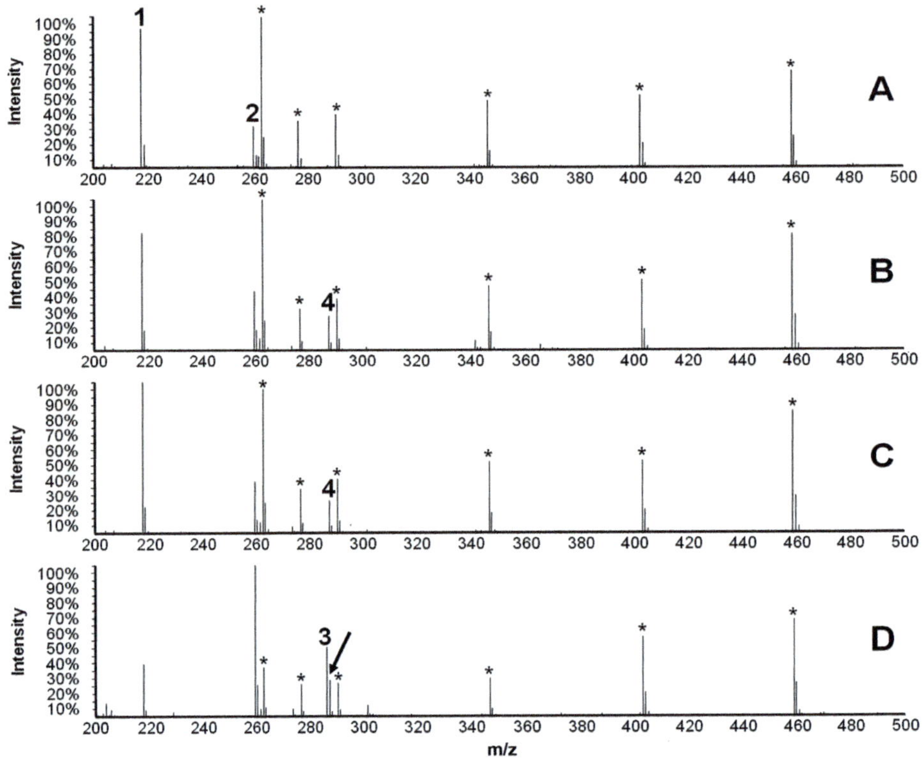

Fig. 6.3 Profiles of acylcarnitines as their butyl esters in plasma (precursor of m/z 85 scan) of a normal control (**a**) and three patients with elevated C_4-carnitine (m/z 288; peak #4 indicated by arrow in panel D) that represents primarily butyrylcarnitine in a patient with SCAD deficiency (**b**), isobutyrylcarnitine in a patient with isobutyryl-CoA dehydrogenase (IBDH) deficiency (**c**), and a natural isotope of formiminoglutamate (FIGLU; m/z 287; peak #3) in a patient with glutamate formimino-transferase deficiency (**d**). Peak #1, free carnitine (m/z 218); peak #2, acetylcarnitine (C_2; m/z 260). The asterisks represent the internal standards (from left to right): d_3-acetylcarnitine (C_2; m/z 263), d_3-propionylcarnitine (C_3; m/z 277), d_3-butyrylcarnitine (C_4; m/z 291), d_3-octanoylcarnitine (C_8; m/z 347), d_3-dodecanoylcarnitine (C_{12}; m/z 403), and d_3-palmitoylcarnitine (C_{16}; m/z 459)

Formiminoglutamate (FIGLU), a marker for glutamate formimino-transferase deficiency, is also detectable by acylcarnitine analysis; represented as a peak with m/z 287 (Fig. 6.3d) [77]. In poorly resolved acylcarnitine profiles, this peak may be confused with iso-/butyrylcarnitine (m/z 288). To avoid the incorrect interpretation of acylcarnitine profiles, we recommend performing the analysis in product scan mode as opposed to Multiple Reaction Monitoring (MRM) mode. For example, the FIGLU peak at m/z 287 would not be identified correctly in MRM mode because the transition of 287–85 is often not selected. However, the 288/85 transition would reveal abnormal results but in fact, does not represent either butyryl- or isobutyrylcarnitine, but the natural isotope of FIGLU.

6.5 Acylcarnitine Analysis in Urine by MS/MS

6.5.1 Principle

While acylcarnitine analysis was first performed in urine specimens for the evaluation of patients with organic acidemias, plasma was found to be more informative for the diagnosis of FAO disorders [20]. However, urine acylcarnitine analysis still has a role in the diagnostic evaluation of patients with organic acidurias but uninformative or borderline abnormal results of plasma acylcarnitine and urine organic acid analyses, which was first described in patients with the low excretor form of glutaric acidemia type I [21–25].

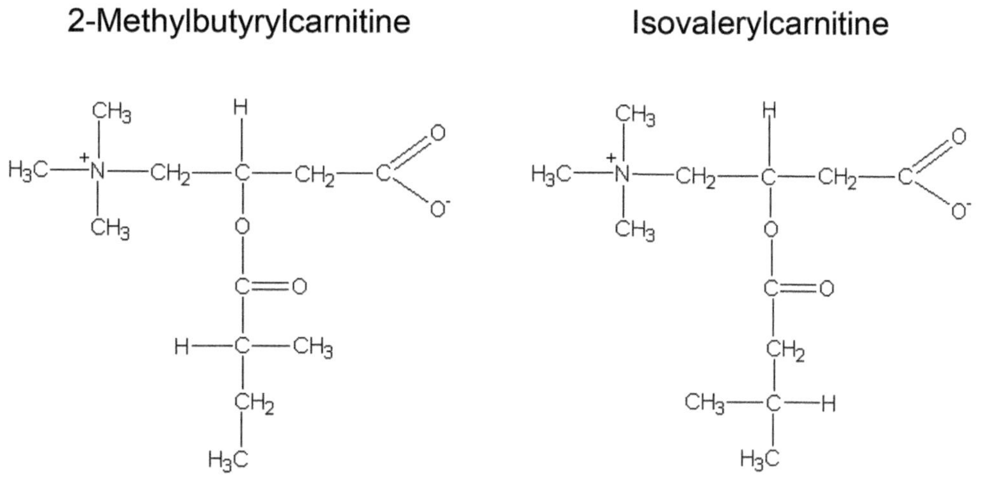

Fig. 6.4 Structures of the isomers 2-methylbutyryl- and isovalerylcarnitine. These isomers cannot be differentiated by acylcarnitine analysis as described here because both appear as butylated C_5-carnitine esters at m/z 302. Differentiation would require an additional chromatography step prior to MS/MS analysis

In our laboratory, sample preparation and analysis are identical to that of plasma once a urine aliquot has been prepared that is based on the creatinine concentration.

6.5.2 Specimen

Urine is collected from patients suspected to have an organic acidemia preferably during an acute metabolic decompensation. As this is often not possible, an early morning specimen should be collected. The sample should be sent frozen and without preservatives.

6.5.3 Procedure and Calculation

The creatinine concentration is measured in the sample using routine methods (i.e., Jaffe reaction). A urine volume equivalent to 0.25 mg creatinine is diluted to 300 μL with deionized water (if the creatinine equivalent exceeds 300 μL, no dilution is made). 20 μL of the diluted or undiluted urine is then analyzed following the procedure described above for plasma acylcarnitine analysis (Sect. 4). The final result is expressed in mmol/mol of creatinine.

6.5.4 Interpretation

As is true for plasma acylcarnitine analysis, the interpretation of urine acylcarnitine profiles is based on quantitative reference ranges and pattern recognition. The reference ranges used in our laboratory for urine are provided in Table 6.3.

6.5.5 Pitfalls

Aside from the above-mentioned potential problems (see 6.4.5.) such as the inability to discriminate isobars and interfering metabolites of antibiotics, the interpretation of urine acylcarnitine profiles is inherently more complex and overinterpretation must be avoided. Therefore, urine acylcarnitine analysis should not be included in the first line of screening investigations but be targeted to specific diagnostic considerations, in particular, organic acidemias such as glutaric acidemia type I. Given that long-chain acylcarnitines are not excreted in urine there is, apart from establishing (hydroxy-)dicarboxylcarnitinuria, no established role of urine acylcarnitine analysis for the diagnosis of FAO disorders.

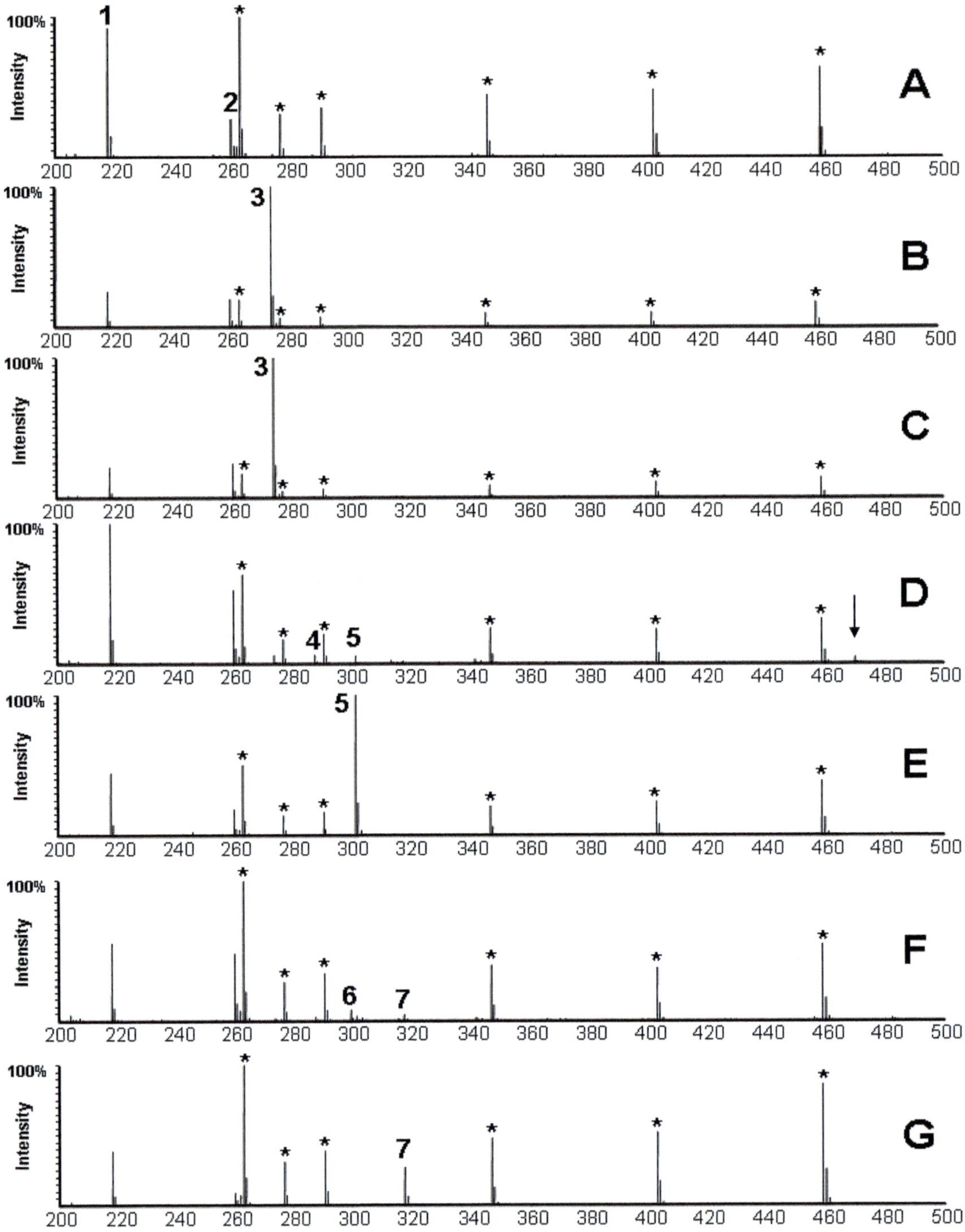

Fig. 6.5 Profiles of acylcarnitines as their butyl esters in plasma (precursor of m/z 85 scan) of a normal control (**a**) and patients with various organic acidemias. Propionylcarnitine (C_3; m/z 274; peak #3) is the primary marker for both Propionic acidemia (**b**) and Methylmalonic acidemias (**c**). Note that an elevation of methylmalonylcarnitine (C_4-DC; m/z 374) is not typically found in patients with Methylmalonic acidemias. In the several cases of Ethylmalonic Encephalopathy (**d**) analyzed in our laboratory, elevations of C_4-(m/z 288; peak #4) and

(continued)

Table 6.3 Reference range for urine acylcarnitines (as butylesters) based on samples with normal organic acid results ($n = 40$) (from [63]). Note that the concentrations are determined (and named here) by the m/z values of the corresponding acylcarnitine species found in plasma/serum. Whether the non-hydroxylated long-chain species are indeed the same acylcarnitine species as in plasma/serum was not verified

Acylcarnitine species	(mmol/mol creatinine)		
C0	0.35	–	31.60
C2		<	16.46
C3		<	1.20
C4		<	2.74
C5:1		<	0.34
C5		<	1.53
C4-OH		<	0.26
C6		<	0.16
C5-OH		<	0.52
C6-OH		<	0.32
C8:1		<	4.30
C8		<	0.61
C3-DC		<	0.50
C10:2		<	0.48
C10:1		<	0.65
C10		<	0.21
C4-DC		<	0.57
C10:1-OH		<	0.26
C5-DC (C10-OH)		<	0.37
C12:1		<	0.07
C12		<	0.19
C6-DC		<	0.81
C12:1-OH		<	0.27
C12-OH		<	0.16
C14:2		<	0.02
C14:1		<	0.21
C14		<	0.39
C14:1-OH		<	0.14
C14-OH		<	0.09
C16:1		<	0.04
C16		<	0.18
C16:1-OH		<	0.02
C16-OH		<	0.05
C18:2		<	0.02
C18:1		<	0.02
C18		<	0.05
C18:2-OH		<	0.01
C18:1-OH		<	0.03
C18-OH		<	0.02

6.6 Acylcarnitine Analysis in Dried Blood and Bile Spots by MS/MS

6.6.1 Applications

Dried blood spots are the most frequently used specimen for acylcarnitine analysis for the simple reason that it has become part of newborn screening worldwide. However, in this setting, the acylcarnitine analysis is combined with the measurement of amino acids, succinylacetone, and an increasing number of analytes such as adenosine, lysophosphatidylcholines, guanidinoacetate, creatine, and creatinine. Because analytical kits are available that have been approved by the US Food and Drug Administration (FDA), newborn screening will not be further considered in this chapter (refer to package inserts and relevant standards available from Clinical and Laboratory Standards Institute, CLSI; clsi.org).

In the clinical laboratory, dried blood spots can be used as a convenient specimen type for acylcarnitine analysis. However, in the author's laboratory experience, clinics and hospitals

C_5-carnitine (m/z 302; peak #5) species were noted. Isolated C_5-carnitine elevations are encountered in patients with Isovaleric acidemia (**e**) where it represents isovalerylcarnitine. C_5-Carnitine is also elevated in patients with Short/branched-chain acyl-CoA dehydrogenase (SBCAD) deficiency where it represents 2-methylbutyrlcarnitine (see Fig. 6.4), and in patients treated with antibiotics containing pivalic acid (including newborns whose mothers were treated with such antibiotics prenatally) or exposed to neopentanoate where it represents pivaloylcarnitine. Patients with β-Ketothiolase deficiency (**f**) present with elevations of tiglylcarnitine ($C_{5:1}$; m/z 300; peak #6) and C_5-OH-carnitine (m/z 318; peak #7). In most cases of 3-Methylcrotonyl-CoA carboxylase deficiency (**g**) C_5-OH-carnitine is the only abnormal acylcarnitine species present. The differential diagnosis of C_5-OH-carnitine elevations includes eight different conditions (Table 6.1). Also note that C_5-OH-carnitine represents 3-hydroxy isovalerylcarnitine in 3-Methylcrotonyl-CoA carboxylase deficiency (**g**), and 2-methyl 3-hydroxy butyrylcarnitine in β-Ketothiolase deficiency (**f**). The arrow in panel D represents an artifact at m/z 470 (the same mass as $C_{16:1}$-OH-carnitine) secondary to the patient being treated with cefotaxime. The asterisks represent the internal standards (from left to right): d_3-acetylcarnitine (C_2; m/z 263), d_3-propionylcarnitine (C_3; m/z 277), d_3-butyrylcarnitine (C_4; m/z 291), d_3-octanoylcarnitine (C_8; m/z 347), d_3-dodecanoylcarnitine (C_{12}; m/z 403), and d_3-palmitoylcarnitine (C_{16}; m/z 459)

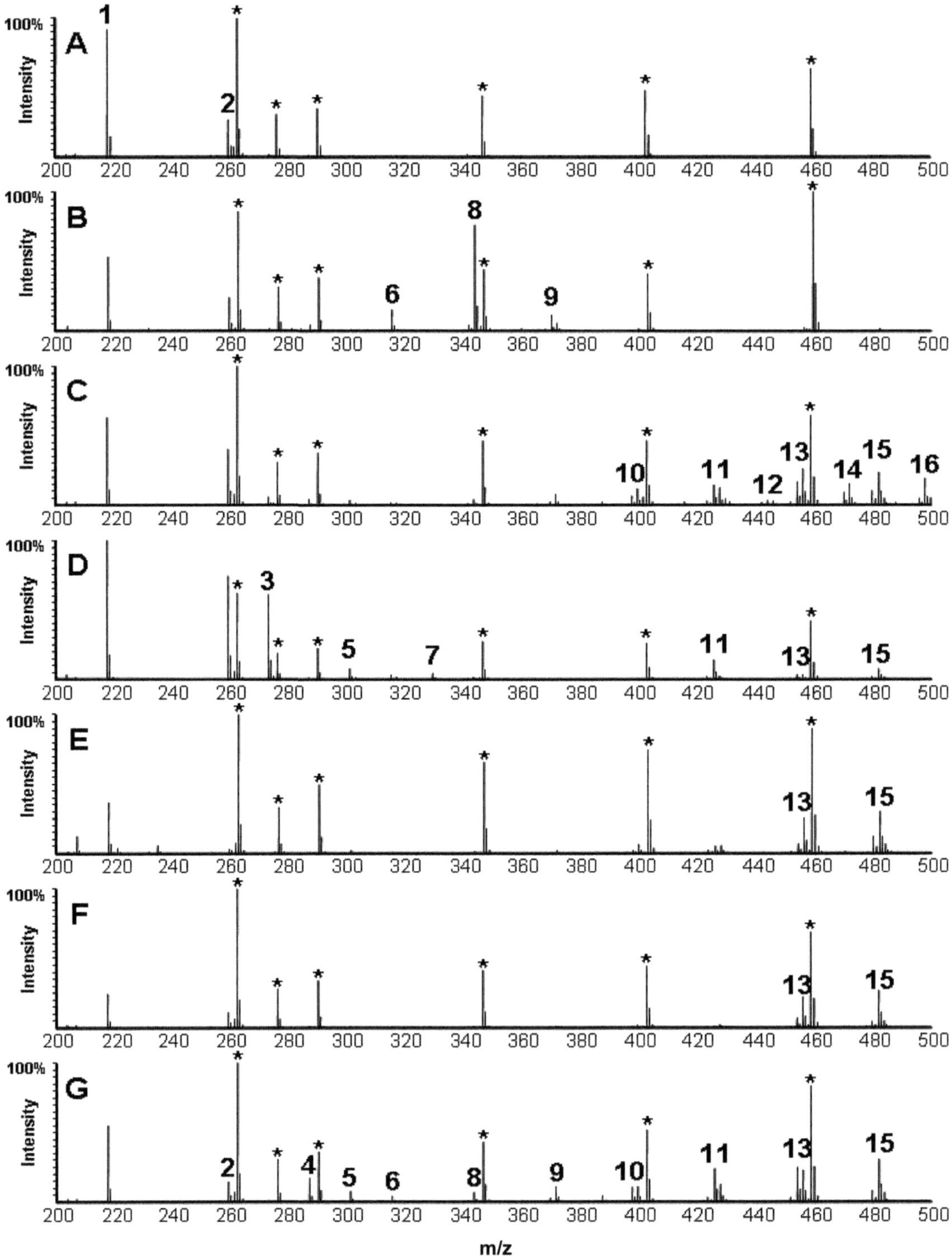

Fig. 6.6 Profiles of acylcarnitines as their butyl esters in plasma (precursor of m/z 85 scan) of a normal control (**a**) and patients with various fatty acid β-oxidation disorders. A typical profile observed in patients with MCAD deficiency is shown in panel B where octanoylcarnitine (C_8; m/z 344; peak #8) is greater than both hexanoylcarnitine (C_6; m/z 316; peak #6) and decenoylcarnitine ($C_{10:1}$; m/z 370; peak #9). In LCHAD (**c**) and VLCAD (**d**) deficiencies long-

(continued)

(except newborn nurseries) prefer venous blood draws where blood can be contained and no drying on filter paper is required. However, for postmortem metabolic screening, dried blood, and bile spots offer an inexpensive specimen type that could also be used for molecular genetic confirmation or additional diagnostic workup.

6.6.2 Principle

Acylcarnitine analysis of dried blood or bile spots is very similar to the analysis of plasma. A small disk (diameter typically 5 mm or less) is punched out of both the blood or bile spot and the acylcarnitines extracted by the addition of methanol and known concentrations of isotopically labeled acylcarnitines which function as internal standards. The extract is dried under a stream of nitrogen, and derivatized by the addition of either n-butanol HCl or n-methanol HCl. The acylcarnitines are measured as their butyl or methyl esters by MS/MS. The concentrations of the analytes are established by computerized comparison of ion intensities of these analytes to that of the internal standards.

6.6.3 Pre-analytical

i. Specimen

For postmortem analysis, blood and bile are collected at the latest at the time of autopsy. Otherwise, as is true for plasma samples, the most informative results are obtained when blood samples are collected during acute illness or at least prior to a meal. Blood should be obtained by capillary stick of well-perfused skin (heels in young infants or fingers) and free dripping of a few drops of blood directly on the filter paper card. Following complete drying at room temperature for at least 2 h, the sample can be sent ambient. Analysis should be completed within a few days. While diagnostic results can be obtained in most cases of medium- and long-chain FAO defects even after prolonged storage time at room temperature, samples should be stored frozen because particularly short-chain acylcarnitine species are not reliably measurable several months after collection [62].

ii. Materials

The reagents used for filter paper analysis are the same as for plasma acylcarnitine analysis except for the internal standard/extraction solution which is prepared using methanol as opposed to acetonitrile.

The samples are typically collected on Whatman 903, Revvity 226, or Ahlstrom 226 filter paper and punched using automatic (i.e., DBS Puncher, Revvity, Waltham, MA) or manual punchers (i.e., M.C. Mieth, Inc., Port Orange, FL) depending on testing volume.

iii. QC samples

A QC standard, normal blood and normal bile spots are run with each assay. The controls are made by pooling normal blood or bile and spotting it on filter paper.

6.6.4 Sample Preparation and Calculation

Into each well of a 96-deep well plate one 5 mm blood or 1.5 mm bile spot is punched and 300 µl

chain acylcarnitines (C_{12} [m/z 400; peak #10], $C_{14:1}$ [m/z 426; peak #11], C_{16} [m/z 456; peak #13], $C_{18:1}$ [m/z 482; peak #15]) are elevated including long-chain 3-hydroxy species (C_{14}-OH [m/z 444; peak #12], C_{16}-OH [m/z 472; peak #14], and $C_{18:1}$-OH [m/z 498; peak #16]) in LCHAD deficiency. CPT II (**e**) and CACT (**f**) deficiencies are indistinguishable by acylcarnitine analysis. Patients with severe glutaric acidemia type II (**g**) reveal elevations of most acylcarnitine species with chain-length of C_4 (m/z 288; peak #4) and longer. Note that the samples represented in panels E, F, and G have low levels of free carnitine (m/z 218; peak #1) and acetylcarnitine (C_2; m/z 260; peak #2) indicating relative carnitine deficiency. Also, note that the patient with VLCAD deficiency (**d**) reveals only moderate elevations of long-chain acylcarnitine species while free carnitine (m/z 218) is relatively high suggesting ongoing L-carnitine supplementation. Furthermore, the odd-chain species C_3- (m/z 374; peak #3), C_5-(m/z 302; peak #5), and C_7-carnitine (m/z 330; peak #7) are elevated which represent dietary artifacts secondary to the patient being supplemented with triheptanoin [73, 74]. The asterisks represent the internal standards (from left to right): d_3-acetylcarnitine (C_2; m/z 263), d_3-propionylcarnitine (C_3; m/z 277), d_3-butyrylcarnitine (C_4; m/z 291), d_3-octanoylcarnitine (C_8; m/z 347), d_3-dodecanoylcarnitine (C_{12}; m/z 403), and d_3-palmitoylcarnitine (C_{16}; m/z 459) internal standard/extraction solution is added. The plate is sealed with adhesive aluminum tape and agitated for 30 min on a plate rotator. Then the liquid is transferred to corresponding wells of a U-bottom polypropylene plate and dried under a gentle stream of warm nitrogen. Butylesters are then formed and the analysis is performed as described for plasma samples (see 5.c.).

When only a few specimens need to be analyzed that do not justify the use of 96-well plates or automation, consider per-sample vials as described in the previous edition of this chapter [63].

6.6.5 Interpretation

As noted for acylcarnitine analysis of other specimen types, the interpretation of results for blood and bile spots is based on quantitative reference ranges and ultimately pattern recognition by a trained biochemical geneticist. The reference ranges currently used in our laboratory for postmortem blood and bile spots represent prospectively updated percentile ranks (Table 6.4). There are significant differences among all specimen types, particularly in the abundance of long-chain acylcarnitine species as has been well documented in the context of carnitine palmitoyltransferase 1 (CPT-1) deficiency [78, 79].

6.7 In Vitro Probe Assay for the Diagnosis of FAO Disorders and Selected Organic Acidemias in Fibroblast Cultures Utilizing MS/MS

6.7.1 Principle

Most FAO disorders have a rather similar presentation and their biochemical diagnosis can at times be difficult. Commonly used metabolite screens, such as urine organic acids, plasma acylcarnitines, and fatty acids are influenced by dietary factors and the clinical status of the patient [66]. This often leads to incomplete diagnostic information or even false negative results [65]. Molecular genetic analysis targeted to specific genes, gene panels, or application of large-scale parallel sequencing is now available and sufficiently diagnostic if pathogenic variants are identified. However, when genotypes involve variants of uncertain significance, additional studies may be required. Enzyme assays can determine a final diagnosis and are preferred, especially when they can be performed on blood cells for the condition under consideration. The purpose of the in vitro probe assay is to offer screening for several defects of FAO and organic acid metabolism under controlled laboratory conditions using fibroblast cultures (Table 6.5) when the above-mentioned testing is inconclusive. However, this assay is not informative if the deficient enzyme is physiologically not expressed in skin fibroblasts.

As indicated in the Introduction to this chapter, several different approaches to the in vitro probing of mitochondrial fatty acid β-oxidation have been published and are in use in several biochemical genetics laboratories around the world [31, 34–42]. The method described here is based on Shen et al. [38] and was further modified to allow for the diagnosis of several defects of branched-chain amino acid metabolism [21, 23].

The principle of this assay relies on the assumption that skin fibroblasts of patients affected with relevant disorders will accumulate certain species of acyl-CoA esters reflecting the metabolic defect when the cell medium is supplemented with a long-chain fatty acid. This in turn leads to the accumulation of acylcarnitine species through the action of carnitine acyltransferases. The formation of acylcarnitines from acyl-CoA esters is an equilibrium process and can be driven by the addition of excess L-carnitine in the medium. Metabolites of branched-chain amino acids can be differentiated by the addition of isotopically labeled branched-chain amino acids to the incubation medium (Fig. 6.7). Acylcarnitine profiles in the cell medium following incubation are easily obtained by tandem mass spectrometry as for the other specimen types described above.

Table 6.4 Percentile ranks of acylcarnitine species (as butylesters) in postmortem dried blood and bile spots (in nmol/mL) (from Matern [64])

	Blood spots ($n = 9788$)			Bile spots ($n = 8848$)		
Acylcarnitine species	5th percentile		95th percentile	5th percentile		95th percentile
C0	29	–	203	57	–	371
C2	26.8	–	194	36	–	311
C3		<	10.9		<	11.6
C4		<	17.2		<	7.5
C5:1		<	0.23		<	0.71
C5		<	2.10		<	3.13
C4-OH		<	7.36		<	3.05
C6		<	2.19		<	2.64
C5-OH		<	0.96		<	1.24
C7		<	0.16		<	0.92
C6-OH		<	0.56		<	1.22
C8:1		<	0.77		<	26.1
C8		<	0.78		<	5.20
C3-DC		<	0.45		<	1.35
C10:2		<	0.11		<	2.53
C10:1		<	0.17		<	6.66
C10		<	0.52		<	4.33
C4-DC		<	1.10		<	1.21
C10:1-OH		<	0.15		<	1.36
C5-DC		<	0.30		<	1.13
C12:1		<	0.11		<	5.19
C12		<	0.54		<	5.47
C12:1-OH		<	0.13		<	1.96
C12-OH		<	0.17		<	1.00
C14:2		<	0.56		<	4.31
C14:1		<	0.24		<	5.26
C14		<	0.46		<	2.28
C14:1-OH		<	0.09		<	1.34
C14-OH		<	0.09		<	0.57
C16:1		<	0.27		<	1.76
C16		<	2.04		<	3.07
C16:1-OH		<	0.11		<	1.21
C16-OH		<	0.11		<	0.68
C18:2		<	0.69		<	2.15
C18:1		<	1.98		<	3.38
C18		<	1.54		<	3.58
C18:2-OH		<	0.13		<	0.58
C18:1-OH		<	0.14		<	0.70
C18-OH		<	0.11		<	0.61

6.7.2 Pre-analytical

(i) Specimen

Fibroblasts are grown from a small skin biopsy collected during an outpatient visit or as part of a planned surgical procedure following routine culturing techniques. Cell cultures may also derive from umbilical cord or, for prenatal diagnostic purposes, from amniocytes obtained by amniocentesis. In our laboratory, the analysis for each patient is performed in triplicate.

(ii) Reagents and Chemicals

Table 6.5 Disorders detectable by the In Vitro probe assay

Fatty acid β-oxidation disorders	Short-chain acyl-CoA dehydrogenase (SCAD) deficiency
	Medium-chain acyl-CoA dehydrogenase (MCAD) deficiency
	Long-chain 3-hydroxyacyl-CoA dehydrogenase (LCHAD) deficiency
	Mitochondrial trifunctional protein (TFP) deficiency
	Very long-chain acyl-CoA dehydrogenase (VLCAD) deficiency
	Carnitine palmitoyl transferase II (CPT-II) deficiency
	Carnitine-acylcarnitine translocase (CACT) deficiency
Organic acid disorders	Electron transfer flavoprotein (ETF) deficiency, ETF-dehydrogenase deficiency (multiple acyl-CoA dehydrogenase deficiency, MADD; glutaric acidemia type II)[a]
	Isovaleric Acidemia
	Short branched-chain acyl-CoA dehydrogenase deficiency (SBCAD deficiency; 2-methylbutyrylglycinuria)
	Isobutyryl-CoA dehydrogenase (IBD) deficiency

[a]excluding riboflavin responsive forms of MADD

– Isotopically labeled standards (internal standards):

- Acetyl-d_3-carnitine (d_3-C_2)
- Isovaleryl-d_9-carnitine (d_9-C_5)
- Octanoyl-d_3-carnitine (d_3-C_8)
- Dodecanoyl-d_3-carnitine (d_3-C_{12})
- Palmitoyl-d_3-carntitine (d_3-C_{16})
  The internal standards are dissolved in methanol and a final working internal standard solution is prepared in methanol at the following concentrations and stored refrigerated:
- 131.1 nmol/L acetyl-d_3-carnitine
- 25.3 nmol/L isovaleryl-d_9-carnitine
- 25.0 nmol/L octanoyl-d_3-carnitine
- 24.6 nmol/L dodecanoyl- d_3-carnitine
- 48.5 nmol/L palmitoyl-d_3-carnitine

– Standards (unlabeled)

- Acetylcarnitine (C_2)
- Isovalerylcarnitine (C_5)
- Octanoylcarnitine (C_8)
- Dodecanoylcarnitine (C_{12})
- Palmitoylcarntitine (C_{16})

Standards are dissolved in methanol at the same concentrations as the internal standards (see above).

Additional or different labeled and unlabeled standards can be used. However, note that the incubation as described here includes isotopically labeled L-valine and L-isoleucine and that the metabolites of these branched-chain amino acids can interfere with isotopically labeled internal standards for butyrylcarnitine or propionylcarnitine.

– Methanol
– 80% Acetonitrile:Water[2]
– Nitrogen gas, in-house, scrubbed with BHT-1 hydrocarbon trap; Chromtech (Apple Valley, MN)
– Minimal Essential Medium (MEM) with Earl's Salts and L-glutamine
Sigma #M4655
Note: Complete MEM is used during fibroblast culturing, but the In Vitro Probe Medium (see below) is used during the 72-h incubation described below.
Palmitic Acid (16:0) 200 mmol/L (MW = 256.4); Sigma P-5585
Dissolve 104 mg of palmitic acid into 2 mL of absolute ethanol.
– L-carnitine Inner Salt 400 mmol/L (MW = 161.2); Sigma C-0158.
Dissolve 128 mg of L-carnitine into 2 mL water.
– 18% Bovine Serum Albumin (BSA) Solution Essentially Fatty Acid Free (MW = 66,200); Sigma A-6003
To make 5 mL of BSA solution, 0.9 g BSA is added to a 50-mL conical tube containing 5 mL of PBS (Dulbecco's 1x

[2]Use of reverse osmosis or HPLC grade water in every step requiring water is recommended.

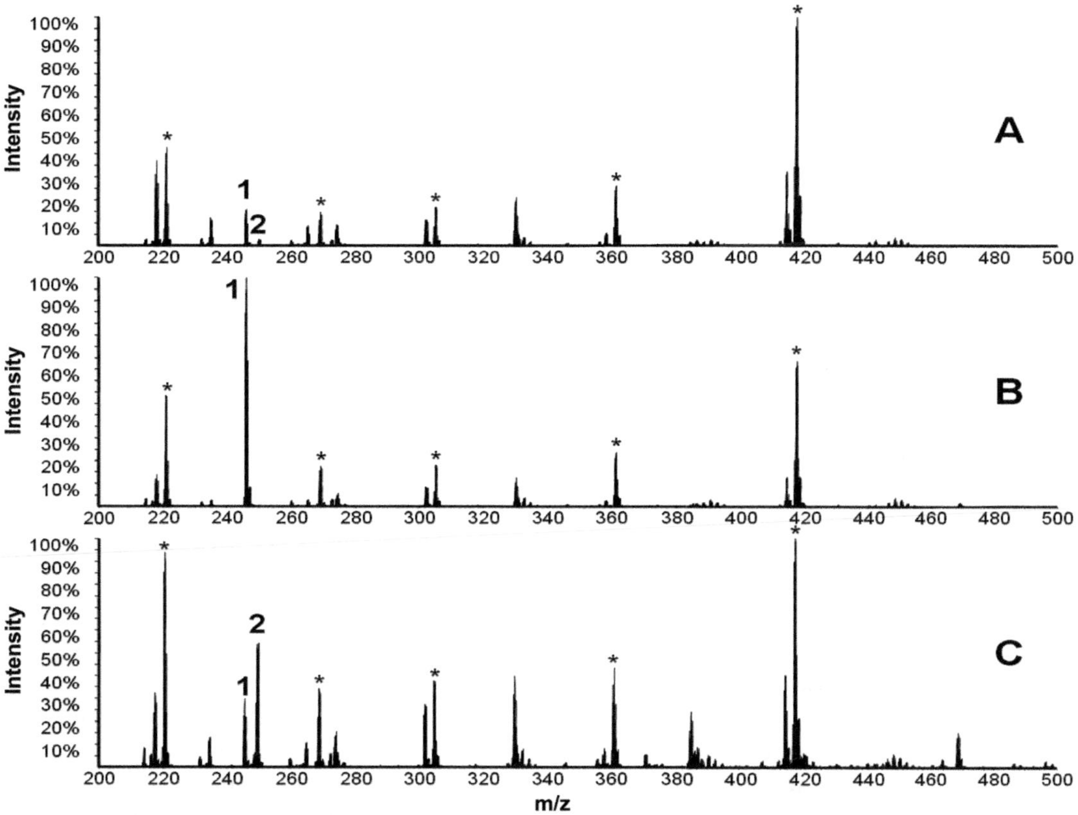

Fig. 6.7 Profiles of acylcarnitines as their methyl esters in cell culture medium (precursor of m/z 99 scan) following the in vitro probe assay in fibroblast cultures of a normal control (**a**) and a patient with SCAD deficiency (**b**) and Isobutyryl-CoA dehydrogenase deficiency (**c**). Peak #1, unlabeled C_4-carnitine (m/z 246) representing butyrylcarnitine accumulating in panel B due to a SCAD deficiency causing a block in fatty acid β-oxidation of unlabeled palmitate. Peak #2, isotopically labeled (^{13}C) C_4-carnitine (m/z 250) representing isobutyrylcarnitine accumulating in panel C due to IBDH deficiency causing a block in the degradation of isotopically labeled L-valine. The asterisks represent the internal standards (from left to right): d_3-acetylcarnitine (C_2; m/z 221), d_9-isovalerylcarnitine (C_5; m/z 269), d_3-octanoylcarnitine (C_8; m/z 305), d_3-dodecanoylcarnitine (C_{12}; m/z 361), and d_3-palmitoylcarnitine (C_{16}; m/z 419)

Phosphate Buffered Saline, Gibco Laboratories, 14,190–144).

- Antibiotic Antimycotic Solution 100X; Invitrogen 15,240–062
  5 mL of Gentamicin Reagent Solution (10 mg/mL; Sigma G1272) is added to a 100-mL bottle of antibiotic solution, which is then sterilized using a 0.2 μ Nalgene filtration lask.
- L-Valine-U-$^{13}C_5$; Cambridge Isotope Laboratories (CLM-2249)
  48 mg are added to a 10-mL volumetric flask, which is then filled to 10 mL with water.
- L-Isoleucine-U-$^{13}C_6$; Cambridge Isotope

Laboratories (CLM-2248)
54 mg are added to a 10-mL volumetric flask, which is then filled to 10 mL with water.

- Unlabeled amino acid solution (10x Stock):
  To a 100 mL volumetric flask, add:
  - 0.031 g L-Cystine-2HCl; MP Biomedicals Inc. 101,454
  - 0.052 g L-Tyrosine; MP Biomedicals Inc. 103,183
  - ~50 mL water

  Adjust pH to >8.0 by adding 2 M NaOH drop wise while stirring until reagent dissolves.
  Add remaining amino acids to the 100-mL volumetric flask:
- 0.126 g L-Arginine-HCl; Sigma, A6969
- 0.042 g L-Histidine HCl-H_2O; MP Biomedicals Inc. 101,957
- 0.052 g L-Leucine; MP Biomedicals Inc. 102,158
- 0.073 g L-Lysine-Dihydrochloride; Acros Organics, AC413360250
- 0.015 g L-Methionine; MP Biomedicals Inc. 102,291
- 0.032 g L-Phenylalanine; MP Biomedicals Inc. 102,623
- 0.048 g L-Threonine; MP Biomedicals Inc. 103,053
- 0.010 g L-Tryptophan; Sigma, T0254
- 10 mL L-Glutamine; Invitrogen, 25,030–081 (must be in solution before use)

  Mix until the solution is dissolved. Bring to volume with water.
  10 mL aliquots are prepared using 15 mL conical screwtop tubes and stored at −80 °C.
- In Vitro Probe Medium (incomplete):
  For each 100 mL of medium needed:

70.8 mL	water
10.0 mL	Earle's Balanced Salt Solution 10X; Sigma E-7510
1.0 mL	MEM Vitamin Solution 100X; Sigma M-6895
2.9 mL	Sodium Bicarbonate Solution 7.5% (w/v); Invitrogen 25,080–094
2.0 mL	L-Valine-U-$^{13}C_5$ Solution (see above)
2.0 mL	L-Isoleucine-U-$^{13}C_6$ Solution (see above)
10.0 mL	Unlabeled amino acid solution (see above)
1.2 mL	Complete antibiotic antimycotic solution 100x (see above)
0.1 mL	L-carnitine (see above)

  The above components are mixed and the pH is adjusted to 7.0–7.2 using 1 M NaOH or 1 M HCl as needed. The medium is then filter sterilized using a 0.2 μ Nalgene filter flask, which causes an increase of the final pH by approximately 0.18.
- In Vitro Probe Medium (Complete; made fresh each time assay is performed).
  For 50 mL of medium, first 1.0 mL of BSA solution (see above) is added to a 15x85-mm glass test tube and pre-warmed to 37 °C. While vortexing, 50 μl palmitic acid (see above) is added and this solution is then, also under vortexing, added to the 50 mL of incomplete In Vitro Probe Medium (see above) which has been pre-warmed to 37 °C. The complete In Vitro Probe Medium is mixed well and sterilized via a 0.2 μ filtration flask.

(iii) Equipment and Supplies

- Sonicator/Cell Disruptor; Misonix Inc., Farmingdale, NY
- SPE Dry-96, Plate Sample Concentrator; Argonaut, Charlottesville, VA
- 6-well cell culture plates; Corning Life Sciences, Acton, MA
- Cluster Tubes 96 tubes per rack; Corning Life Sciences, Acton, MA; #4410
- Filter Paper—Schleicher and Schuell #900
- 96 M-03-96 Round Sealing Silicone Mat; Chrom Tech #96 M-03
- Adhesive Free Film Mat; Chrom Tech, Inc. #ZAF-PE—50
- 96-Well Clear Flat Bottom Cell Culture Microplate; Falcon #351172
- 96-Well U-bottom Microplate; Falcon, #351177
- 6-Well Sterile Culture Cluster Flat Bottom Plate with Lid; Costar, #3516
- Sterile 50 mL conical tube; Falcon, #352098
- Nalgene 0.90 mm 0.2 μm filter flask (500 mL)
- AB SCIEX API 5500 tandem mass spectrometry system
- Shimadzu LC-20AB pump
- Leap PAL HTS-xt LC autosampler with 96-well plate sample tray insert

(iv) Calibration (see 6.4.2.iii.)

(v) QC

- QC samples:
  - Cell medium is pooled after incubation of several negative cell lines and added to the acylcarnitine analysis of each batch as controls. The mean and standard deviation are calculated with a minimum of 20 between run values.
  - A standard solution made up of equimolar amounts of standard and internal standard solution will undergo acylcarnitine analysis with each batch (Standard Control). The expectation is that the standard and corresponding internal standard peaks are of equal abundance in the acylcarnitine profile and that no other acylcarnitine peaks are present.
  - In addition, at least one cell line from a proband not affected with a FAO disorder will be analyzed with each run as a normal control. At least two cell lines from patients with different, previously established FAO disorders are also analyzed simultaneously. The latter cell lines act as abnormal controls.
- Control values for the pooled QC sample that fall within ±2 standard deviations of the mean are considered acceptable and require no further action. Any control values that are either greater than 2 standard deviations or trends require review by a laboratory director.
- The results for the normal and abnormal control cell lines and the control samples are reviewed by the laboratory director who determines whether the analysis of the complete batch or of specific samples needs to be repeated.

6.7.3 Analytical

i. Cell Culture and Pre-incubation

Fibroblast cultures for each patient and control are grown to confluency in T-75 culture flasks. The assay is then performed in triplicate using 6-well plates. First 0.5 mL of 0.25% Trypsin-EDTA solution (Sigma T4049) is added to each T-75 flask. After the cells have lifted from the flask, they are resuspended in 6 mL complete MEM. 1 mL of the cell suspension is transferred to a 50-mL conical tube containing 14 mL of MEM complete medium. Following further resuspension, 4 mL are plated into each of three wells in a 6-well culture plate and placed in the incubator. Once the cells have attached to the seeded plates and are approximately 90% to 100% confluent (3–4 days), complete MEM is replaced with 1.5 mL of Complete In Vitro Probe Medium (see above).

ii. Incubation and Harvesting

After 72-h incubation at 37 °C, remove plate from incubator and view each well under the microscope. Check each well for cell growth, cell death, and possible contamination. Gently rotate the plate to mix medium. Remove all medium from each well using a transfer pipette. Transfer approximately 1 mL of the medium for acylcarnitine analysis into a corresponding labeled cluster tube for each patient. It can be spotted on filter paper immediately or stored frozen for later acylcarnitine analysis. The remaining medium is discarded.

Then, 2.0 mL of PBS is added to each well and each flask is gently washed. The saline solution is then replaced by 0.2 mL of 0.25% Trypsin-EDTA solution (Sigma T4049). After 3–5 min, the cells will lift off the plate bottom which should be confirmed by light microscopy. To bring the cells back into suspension, 1.0 mL PBS is added to each trypsinized well and cell clumps are broken up by careful pipetting. The cell suspension is then transferred to 1.5 mL microfuge tubes and centrifuged at 21,100 x g for 30 seconds. Afterward, the supernatant is carefully removed without disrupting the cell pellet to which 75 μL of water is now added. Without further manipulation, cell pellet and water are stored frozen at −20 ° C until protein analysis.

iii. Protein Analysis of Cell Pellet

Cell pellets are thawed at room temperature. Then each tube is sonicated twice for at least 30 seconds with cooling on ice between each 30-second burst. The protein concentration is determined by a routine method (i.e., Lowry method) and the result is recorded for each cell pellet.

iv. Preparation of Cell Medium for Acylcarnitine Analysis

Into each well of a 96-well flat bottom plate, one 6 mm filter paper disk (Schleicher & Schuell #900) is placed using a paper puncher (i.e., M.C. Mieth Manufacturing, Inc., Port Orange, FL). For each patient and control, 50 μL of collected cell culture medium is dripped on a filter disk. In addition, 50 μL of Complete In Vitro Probe Medium is added to one disk as a blank. After the samples have dried overnight, 300 μL labeled internal standard/extraction solution (see above) is added to each well and the plate is agitated on an orbital mixer for 30 min. Then, 100 μL of the extract is transferred to a clean U-bottom 96-well plate and dried under nitrogen. To derivatize the acylcarnitines to methyl esters, 100 μL of 3 M methanolic-HCl (TCI America X0041) is added to each well and the covered plate is heated at 50 °C for 15 min before the samples are dried again under warm nitrogen. After reconstitution of the dried residue in 100 μL of 80:20 acetonitrile:water (mobile phase) the samples are ready for acylcarnitine analysis.

v. Instrument Setup

LC System:

Controlled by the MS/MS software, the Shimadzu LC-20AB pump operates isocratic using 80% acetonitrile:water at 40 μL/min. A 50 x 1 mm C18 column (Keystone) is used between the pump and autosampler to provide backpressure. The syringe/system flush solution is the 80% acetonitrile:water that is used as mobile phase. The autosampler is connected directly to the MS/MS Turbo V Ion Spray source. The injected sample volume is 10 μL.

A 0.5-U Peek filter is installed between the autosampler and Turbo V Ion electrospray interface source. The filter end fitting is discarded and replaced after approximately 200 injections.

MS/MS System:

Acylcarnitine analysis is performed as a precursor scan in positive ion mode. Q1 is set to scan a mass range from m/z 200 to 500 m/z, while Q3 is set to determine a precursor ion of m/z 99. The method is optimized for a mixture of methyl esters containing d_3-C_2, d_9-C_5, d_3-C_8, d_3-C_{12}, and d_3-C_{16}-carnitine.

The sample queue is entered into the MS/MS instrument's software (Analyst for SCIEX instruments) according to the manufacturer's instructions. Samples can be given specific designations that allow for correct QC and reference range assignment.

Prior to analysis, the method should be equilibrated, and the system purged if necessary. The

LC line should be attached to the sprayer after proper equilibration for approximately 30 min.

vi. Calculation

- A datafile is generated for each sample and presented as a mass profile. All acylcarnitine species in a mass range from 200 to 500 Da that generate a product m/z of 99 are represented in this profile based on their detected abundance.
- Calculation of the molecular weights of the n-methyl esters of acylcarnitines, expressed as $(M + H)^+$ molecular ions, allows designation of each peak (Table 6.6).
- Quantitation software provided by the instrument vendor (i.e., ChemoView for SCIEX instruments) allows calculation of quantitative and semiquantitative concentrations for these acylcarnitines by comparison of the detected abundance of each acylcarnitine versus that of a designated internal standard with known concentration. The final results for each acylcarnitine species are normalized to the protein concentration determined in the cell pellet.
- The results for each sample and analyte are compared to previously established reference ranges (Table 6.6).

Table 6.6 Mass assignment and reference range (n = 128[a]) for acylcarnitine species (as methylesters) accumulating in fibroblast cultures following incubation with palmitate, $^{13}C_5$-valine, $^{13}C_6$-isoleucine, and L-carnitine

$[M + H]^+$	Acylcarnitine species		(μmol/g protein)
218	C_2	Acetyl-	0.100–1.785
221	d_3-C_2	*Internal standard*	
232	C_3	Propionyl-	<0.075
235[#]	$^{13}C_3$	Propionyl-	<0.517
246	C_4	Butyryl-	<0.129
250[#]	$^{13}C_4$	Isobutyryl-	<0.059
260	C_5	Isovaleryl-	<0.052
265[#]	$^{13}C_5$	2-Methylbutyryl-	<0.175
269	d_9-C_5	*Internal standard*	
274	C_6	Hexanoyl-	<0.094
302	C_8	Octanoyl-	<0.099
304	C_5-DC	Glutaryl-	<0.029
305	d_3-C_8	*Internal standard*	
326	$C_{10:2}$	Decadienoyl-	<0.004
328	$C_{10:1}$	Decenoyl-	<0.009
330	C_{10}	Decanoyl-	<0.148
358	C_{12}	Dodecanoyl-	<0.035
361	d_3-C_{12}	*Internal standard*	
374	C_{12}-OH	3-Hydroxydodecanoyl-	<0.007
384	$C_{14:1}$	Tetradecenoyl-	<0.021
386	C_{14}	Myristoyl-	<0.035
402	C_{14}-OH	3-Hydroxytetradecanoyl-	<0.006
456	C_{16}	Palmitoyl-	<0.270
459	d_3-C_{16}	*Internal standard*	
430	C_{16}-OH	3-Hydroxyhexadecanoyl-	<0.014
438	$C_{18:2}$	Linoleyl-	<0.004
440	$C_{18:1}$	Oleyl-	<0.019
442	C_{18}	Stearoyl-	<0.019
458	C_{18}-OH	3-Hydroxystearoyl-	<0.009

[a]Controls with confirmed wild type for the common ACADS variants 625G > A and 511C > T. [#]Mass corresponds to analyte derived from incubation with isotopically labeled valine and isoleucine. $[M + H]^+$, molecular weight of $[M + H]^+$ ions corresponding to the methyl ester derivatives of acylcarnitine species

6.7.4 Post-analytical: Interpretation and Reference Ranges

As always, the interpretation of metabolite profiles is aided by comparison of analytes to quantitative reference ranges established during implementation and validation of the assay. Ultimately, however, the interpretation is based on pattern recognition by a trained biochemical geneticist.

6.7.5 Pitfalls

The In Vitro probe assay is performed under standardized conditions and is independent of the patient's status at the time the skin biopsy is obtained. The accumulation of acylcarnitines in this assay system appears to be dependent on potentially present residual enzyme activity and therefore provides information regarding the severity of an enzyme deficiency state for some disorders. This has been well documented for VLCAD deficiency (Fig. 6.8) [36]. In our experience, this seems also to hold true for the different SCAD genotypes. Cells lines homozygous or compound heterozygous for variants that abolish

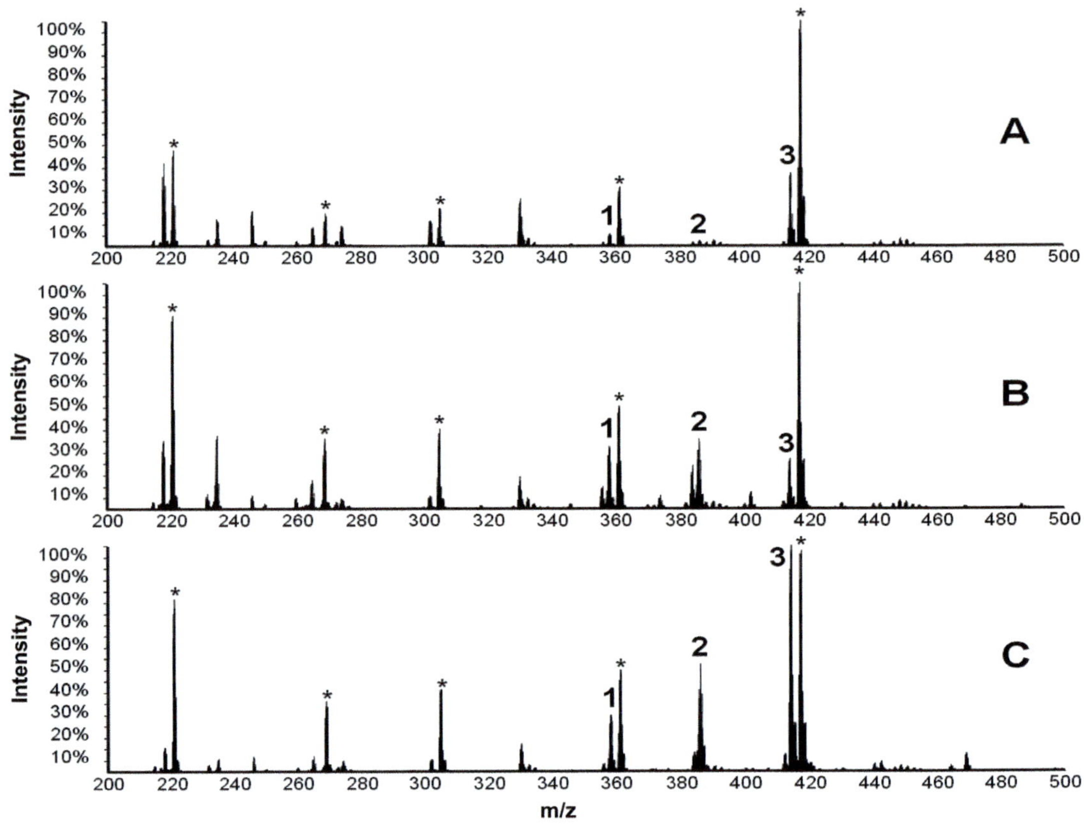

Fig. 6.8 Profiles of acylcarnitines as their methyl esters in cell culture medium (precursor of m/z 99 scan) following the In Vitro probe assay in fibroblast cultures of a normal control (**a**) and a patient with the milder (**b**) and the severe variant (**c**) of VLCAD deficiency. Note the more prominent elevation of dodecanoyl-(C_{12}; m/z 358; peak #1) and myristoylcarnitine (C_{14}; m/z 386; peak #2) compared to a relatively normal accumulation of palmitoylcarnitine (C_{16}; m/z 416; peak #3) in the milder VLCAD variant compared to the severe variant where palmitoylcarnitine is markedly elevated. The asterisks represent the internal standards (from left to right): d_3-acetylcarnitine (C_2; m/z 221), d_9-isovalerylcarnitine (C_5; m/z 269), d_3-octanoylcarnitine (C_8; m/z 305), d_3-dodecanoylcarnitine (C_{12}; m/z 361), and d_3-palmitoylcarnitine (C_{16}; m/z 419)

SCAD activity accumulate larger amounts of butyrylcarnitine than those that are homozygous or compound heterozygous for only the common, non-pathogenic SCAD variants, 625G > A and 511C > T. Furthermore, as is true for acylcarnitine analysis of other sample types, CPT-II and CACT deficiencies cannot be differentiated using this assay system. However, Roe et al. suggested that specific metabolite ratios may facilitate this differentiation [80]. Finally, the analysis of fibroblasts does not allow for the diagnosis of enzyme deficiency states that are not expressed in this tissue (for example, the liver-specific SCHAD deficiency). Moreover, riboflavin-responsive multiple acyl-CoA dehydrogenase deficiencies (Glutaric acidemia type II) are typically not identified because the incubation medium includes riboflavin in the vitamin solution. In conclusion, the ordering physician must be aware of the limitations of this assay system and should consider involving the biochemical genetics laboratory in discussions regarding the most appropriate diagnostic work up of patients, preferably before a skin biopsy is taken.

6.8 Conclusion

Since its introduction into clinical laboratory practice, plasma acylcarnitine analysis has evolved into one of the most commonly ordered tests when an inborn error of energy metabolism is suspected, either based on clinical grounds or in follow up of an abnormal newborn screening result. Applications of acylcarnitine analysis have been suggested for common conditions from heart disease, cancer, and psychiatric disorders to infectious diseases such as HIV and COVID-19, however, have not entered routine practice. Accordingly, 85% of patients whose nearly 200,000 samples were received in the author's laboratory since 2000, were less than 18 years old (age range: 1 day to 103 years old; mean: 3 years, median 9.8 years). A certain diagnosis or differential diagnosis (such as CACT deficiency vs. CPT-II deficiency) was provided for ca. 6000 patients over that period with MCAD and VLCAD deficiencies the most common.

References

1. Bremer J (1983) Carnitine–metabolism and functions. Physiol Rev 63(4):1420–1480
2. Gulewitsch WI, Krimberg R (1905) Zur Kenntnis der Extraktivstoffe der Muskeln. II. Mitteilung. Ueber das Carnitin. Hoppe Seylers Z Physiol Chem 45:326–330
3. Tomita M, Sendju Y (1927) Über die Oxyaminverbindungen, welche die Biuretreaktion zeigen. III. Spaltung der gamma-Amino-beta-oxybuttersäure in die optisch-aktiven Komponenten. Hoppe Seylers Z Physiol Chem 169:263–277
4. Bhattacharyya PK, Carter HE, Fraenkel G, Weidman KR (1952) The identity of vitamin BT with carnitine. Arch Biochem 35(1):241–242
5. Carter HE, Bhattacharyya PK, Weidman KR, Fraenkel G (1952) Chemical studies on vitamin BT isolation and characterization as carnitine. Arch Biochem Biophys 38:405–416
6. Fritz I (1955) The effect of muscle extracts on the oxidation of palmitic acid by liver slices and homogenates. Acta Physiol Scand 34(4):367–385. https://doi.org/10.1111/j.1748-1716.1955.tb01256.x
7. DiMauro S, DiMauro PM (1973) Muscle carnitine palmityltransferase deficiency and myoglobinuria. Science 182(115):929–931
8. Engel AG, Angelini C (1973) Carnitine deficiency of human skeletal muscle with associated lipid storage myopathy: a new syndrome. Science 179(76):899–902
9. Ensenauer R, He M, Willard JM, Goetzman ES, Corydon TJ, Vandahl BB, Mohsen AW, Isaya G, Vockley J (2005) Human acyl-CoA dehydrogenase 9 (ACAD-9): characterization of a novel role in the mitochondrial beta -oxidation of unsaturated fatty acids. J Biol Chem 280(37):32309–32316
10. Knottnerus SJG, Bleeker JC, Wust RCI, Ferdinandusse S, Wijburg FA, Wanders RJA, Visser G, Houtkooper RH (2018) Disorders of mitochondrial long-chain fatty acid oxidation and the carnitine shuttle. Rev Endocr Metab Disord 19(1):93–106. https://doi.org/10.1007/s11154-018-9448-1
11. Prasun P (1993) Multiple acyl-CoA dehydrogenase deficiency. In: Adam MP, Everman DB, Mirzaa GM et al (eds) GeneReviews((R)), Seattle (WA)
12. Rinaldo P, Matern D, Bennett MJ (2002) Fatty acid oxidation disorders. Annu Rev Physiol 64:477–502
13. Zhang J, Zhang W, Zou D, Chen G, Wan T, Zhang M, Cao X (2002) Cloning and functional characterization of ACAD-9, a novel member of human acyl-CoA dehydrogenase family. Biochem Biophys Res Commun 297(4):1033–1042
14. Chalmers RA, Roe CR, Stacey TE, Hoppel CL (1984) Urinary excretion of l-carnitine and acylcarnitines by patients with disorders of organic acid metabolism: evidence for secondary insufficiency of l-carnitine. Pediatr Res 18(12):1325–1328
15. Millington DS, Roe CR, Maltby DA (1984) Application of high resolution fast atom bombardment and constant B/E ratio linked scanning to the identification and analysis of acylcarnitines in metabolic disease. Biomed Mass Spectrom 11(5):236–241
16. Roe CR, Bohan TP (1982) L-carnitine therapy in propionicacidaemia. Lancet 1(8286):1411–1412
17. Roe CR, Hoppel CL, Stacey TE, Chalmers RA, Tracey BM, Millington DS (1983) Metabolic response to carnitine in methylmalonic aciduria. An effective strategy for elimination of propionyl groups. Arch Dis Child 58(11):916–920
18. Roe CR, Millington DS, Maltby DA, Bohan TP, Hoppel CL (1984a) L-carnitine enhances excretion of propionyl coenzyme a as propionylcarnitine in propionic acidemia. J Clin Invest 73(6):1785–1788
19. Roe CR, Millington DS, Maltby DA, Kahler SG, Bohan TP (1984b) L-carnitine therapy in isovaleric acidemia. J Clin Invest 74(6):2290–2295
20. Millington DS, Terada N, Chace DH, Chen YT, Ding JH, Kodo N, Roe CR (1992) The role of tandem mass spectrometry in the diagnosis of fatty acid oxidation disorders. Prog Clin Biol Res 375:339–354
21. Ensenauer R, Vockley J, Willard JM, Huey JC, Sass JO, Edland SD, Burton BK, Berry SA, Santer R, Grunert S, Koch HG, Marquardt I, Rinaldo P, Hahn S, Matern D (2004) A common mutation is associated with a mild, potentially asymptomatic phenotype in patients with Isovaleric Acidemia diagnosed by newborn screening. Am J Hum Genet 75(6):1136–1142
22. Guenzel AJ, Hall PL, Scott AI, Lam C, Chang IJ, Thies J, Ferreira CR, Pichurin P, Laxen W, Raymond K, Gavrilov DK, Oglesbee D, Rinaldo P, Matern D,

Tortorelli S (2021) The low excretor phenotype of glutaric acidemia type I is a source of false negative newborn screening results and challenging diagnoses. JIMD Rep 60(1):67–74. https://doi.org/10.1002/jmd2.12217
23. Oglesbee D, He M, Majumder N, Vockley J, Ahmad A, Angle B, Burton B, Charrow J, Ensenauer R, Ficicioglu CH, Keppen LD, Marsden D, Tortorelli S, Hahn SH, Matern D (2007) Development of a newborn screening follow-up algorithm for the diagnosis of isobutyryl-CoA dehydrogenase deficiency. Genet Med 9(2):108–116. https://doi.org/10.1097/gim.0b013e31802f78d6
24. Pitt JJ, Eggington M, Kahler SG (2002) Comprehensive screening of urine samples for inborn errors of metabolism by electrospray tandem mass spectrometry. Clin Chem 48(11):1970–1980
25. Tortorelli S, Hahn SH, Cowan TM, Brewster TG, Rinaldo P, Matern D (2005) The urinary excretion of glutarylcarnitine is an informative tool in the biochemical diagnosis of glutaric acidemia type I. Mol Genet Metab 84(2):137–143
26. Millington DS, Kodo N, Norwood DL, Roe CR (1990) Tandem mass spectrometry: a new method for acylcarnitine profiling with potential for neonatal screening for inborn errors of metabolism. J Inherit Metab Dis 13(3):321–324
27. Braida L, Crovella S, Boniotto M, Luchesi A, de Vonderweid U, Casetta B, Amoroso A (2001) A rapid and quantitative mass spectrometry method for determining the concentration of acylcarnitines and aminoacids in amniotic fluid. Prenat Diagn 21(7):543–546
28. Morel CF, Watkins D, Scott P, Rinaldo P, Rosenblatt DS (2005) Prenatal diagnosis for methylmalonic acidemia and inborn errors of vitamin B12 metabolism and transport. Mol Genet Metab 86(1–2):160–171
29. Rinaldo P, Studinski AL, Matern D (2001) Prenatal diagnosis of disorders of fatty acid transport and mitochondrial oxidation. Prenat Diagn 21(1):52–54
30. Van Hove JL, Chace DH, Kahler SG, Millington DS (1993) Acylcarnitines in amniotic fluid: application to the prenatal diagnosis of propionic acidaemia. J Inherit Metab Dis 16(2):361–367
31. Giak Sim K, Carpenter K, Hammond J, Christodoulou J, Wilcken B (2002) Quantitative fibroblast acylcarnitine profiles in mitochondrial fatty acid beta-oxidation defects: phenotype/metabolite correlations. Mol Genet Metab 76(4):327
32. Nada MA, Rhead WJ, Sprecher H, Schulz H, Roe CR (1995) Evidence for intermediate channeling in mitochondrial beta-oxidation. J Biol Chem 270(2):530–535
33. Nada MA, Vianey-Saban C, Roe CR, Ding JH, Mathieu M, Wappner RS, Bialer MG, McGlynn JA, Mandon G (1996) Prenatal diagnosis of mitochondrial fatty acid oxidation defects. Prenat Diagn 16(2):117–124
34. Okun JG, Kolker S, Schulze A, Kohlmuller D, Olgemoller K, Lindner M, Hoffmann GF, Wanders RJ, Mayatepek E (2002) A method for quantitative acylcarnitine profiling in human skin fibroblasts using unlabelled palmitic acid: diagnosis of fatty acid oxidation disorders and differentiation between biochemical phenotypes of MCAD deficiency. Biochim Biophys Acta 1584(2–3):91
35. Roe CR, Roe DS (1999) Recent developments in the investigation of inherited metabolic disorders using cultured human cells. Mol Genet Metab 68(2):243–257
36. Roe DS, Vianey-Saban C, Sharma S, Zabot MT, Roe CR (2001) Oxidation of unsaturated fatty acids by human fibroblasts with very-long-chain acyl-CoA dehydrogenase deficiency: aspects of substrate specificity and correlation with clinical phenotype. Clin Chim Acta 312(1–2):55–67
37. Schmidt-Sommerfeld E, Bobrowski PJ, Penn D, Rhead WJ, Wanders RJA, Bennet MJ (1998) Analysis of carnitine esters by radio-high performance liquid chromatography in cultured skin fibroblasts from patients with mitochondrial fatty acid oxidation disorders. Pediatr Res 44(2):210–214
38. Shen JJ, Matern D, Millington DS, Hillman S, Feezor MD, Bennett MJ, Qumsiyeh M, Kahler SG, Chen YT, Van Hove JL (2000) Acylcarnitines in fibroblasts of patients with long-chain 3-hydroxyacyl-CoA dehydrogenase deficiency and other fatty acid oxidation disorders. J Inherit Metab Dis 23(1):27–44
39. Ventura FV, Costa CG, Struys EA, Ruiter J, Allers P, Ijlst L, de Almeida T, Duran M, Jakobs C, Wanders RJA (1999) Quantitative acylcarnitine profiling in fibroblasts using U-C-13 palmitic acid: an improved tool for the diagnosis of fatty acid oxidation defects. Clin Chim Acta 281(1–2):1–17
40. Young SP, Matern D, Gregersen N, Stevens RD, Bali D, Liu HM, Koeberl DD, Millington DS (2003) A comparison of in vitro acylcarnitine profiling methods for the diagnosis of classical and variant short chain acyl-CoA dehydrogenase deficiency. Clin Chim Acta 337(1–2):103–113
41. Schulze-Bergkamen A, Okun JG, Spiekerkotter U, Linder M, Haas D, Kohlmuller D, Mayatepek E, Schulze-Bergkamen H, Greenberg CR, Zschocke J, Hoffmann GF, Kolker S (2005) Quantitative Acylcarnitine profiling in peripheral blood mononuclear cells using in vitro loading with palmitic and 2-Oxoadipic acids: biochemical confirmation of fatty acid oxidation and organic acid disorders. Pediatr Res
42. Janzen N, Hofmann AD, Schmidt G, Das AM, Illsinger S (2017) Non-invasive test using palmitate in patients with suspected fatty acid oxidation defects: disease-specific acylcarnitine patterns can help to establish the diagnosis. Orphanet J Rare Dis 12(1):187. https://doi.org/10.1186/s13023-017-0737-7
43. American College of Medical Genetics and Genomics (ACMG) Technical Standards for Clinical Genetics Laboratories (2021) ACMG
44. Vaz FM, Wanders RJ (2002) Carnitine biosynthesis in mammals. Biochem J 361(Pt 3):417–429
45. Ringseis R, Wege N, Wen G, Rauer C, Hirche F, Kluge H, Eder K (2009) Carnitine synthesis and

uptake into cells are stimulated by fasting in pigs as a model of nonproliferating species. J Nutr Biochem 20(11):840–847. https://doi.org/10.1016/j.jnutbio.2008.07.012
46. Hamilton JJ, Hahn P (1987) Carnitine and carnitine esters in rat bile and human duodenal fluid. Can J Physiol Pharmacol 65(9):1816–1820
47. Rashed MS, Ozand PT, Bennett MJ, Barnard JJ, Govindaraju DR, Rinaldo P (1995) Inborn errors of metabolism diagnosed in sudden death cases by acylcarnitine analysis of postmortem bile. Clin Chem 41(8 Pt 1):1109–1114
48. Jakobs BS, Wanders RJ (1995) Fatty acid beta-oxidation in peroxisomes and mitochondria: the first, unequivocal evidence for the involvement of carnitine in shuttling propionyl-CoA from peroxisomes to mitochondria. Biochem Biophys Res Commun 213(3):1035–1041
49. Houten SM, Denis S, Te Brinke H, Jongejan A, van Kampen AH, Bradley EJ, Baas F, Hennekam RC, Millington DS, Young SP, Frazier DM, Gucsavas-Calikoglu M, Wanders RJ (2014) Mitochondrial NADP(H) deficiency due to a mutation in NADK2 causes dienoyl-CoA reductase deficiency with hyperlysinemia. Hum Mol Genet 23(18):5009–5016. https://doi.org/10.1093/hmg/ddu218
50. Carrozzo R, Dionisi-Vici C, Steuerwald U, Lucioli S, Deodato F, Di Giandomenico S, Bertini E, Franke B, Kluijtmans LA, Meschini MC, Rizzo C, Piemonte F, Rodenburg R, Santer R, Santorelli FM, van Rooij A, Vermunt-de Koning D, Morava E, Wevers RA (2007) SUCLA2 mutations are associated with mild methylmalonic aciduria, Leigh-like encephalomyopathy, dystonia and deafness. Brain 130(Pt 3):862–874
51. Bieber LL, Choi YR (1977) Isolation and identification of aliphatic short-chain acylcarnitines from beef heart: possible role for carnitine in branched-chain amino acid metabolism. Proc Natl Acad Sci USA 74(7):2795–2798
52. Costa CG, Struys EA, Bootsma A, ten Brink HJ, Dorland L, Tavares de Almeida I, Duran M, Jakobs C (1997) Quantitative analysis of plasma acylcarnitines using gas chromatography chemical ionization mass fragmentography. J Lipid Res 38(1):173–182
53. Schmidt-Sommerfeld E, Penn D, Duran M, Rinaldo P, Bennett MJ, Santer R, Stanley CA (1992) Detection and quantitation of acylcarnitines in plasma and blood spots from patients with inborn errors of fatty acid oxidation. Prog Clin Biol Res 375:355–362
54. Millington DS, Bohan TP, Roe CR, Yergey AL, Liberato DJ (1985) Valproylcarnitine: a novel drug metabolite identified by fast atom bombardment and thermospray liquid chromatography-mass spectrometry. Clin Chim Acta 145(1):69–76
55. Yergey AL, Liberato DJ, Millington DS (1984) Thermospray liquid chromatography/mass spectrometry for the analysis of L-carnitine and its short-chain acyl derivatives. Anal Biochem 139(2):278–283
56. Millington DS, Maltby DA, Roe CR (1986) Rapid detection of argininosuccinic aciduria and citrullinuria by fast atom bombardment and tandem mass spectrometry. Clin Chim Acta 155(2):173–178
57. Millington DS, Norwood DL, Kodo N, Roe CR, Inoue F (1989) Application of fast atom bombardment with tandem mass spectrometry and liquid chromatography/mass spectrometry to the analysis of acylcarnitines in human urine, blood, and tissue. Anal Biochem 180(2):331–339
58. Matern D, Magera MJ (2001) Mass spectrometry methods for metabolic and health assessment. J Nutr 131(5):1615S–1620S
59. Ghoshal AK, Guo T, Soukhova N, Soldin SJ (2005) Rapid measurement of plasma acylcarnitines by liquid chromatography-tandem mass spectrometry without derivatization. Clin Chim Acta 358(1–2):104–112
60. Heinig K, Henion J (1999) Determination of carnitine and acylcarnitines in biological samples by capillary electrophoresis-mass spectrometry. J Chromatogr B Biomed Sci Appl 735(2):171–188
61. Luna C, Griffin C, Miller MJ (2022) A clinically validated method to separate and quantify underivatized acylcarnitines and carnitine metabolic intermediates using mixed-mode chromatography with tandem mass spectrometry. J Chromatogr A 1663:462749. https://doi.org/10.1016/j.chroma.2021.462749
62. Matern D, Strauss AW, Hillman SL, Mayatepek E, Millington DS, Trefz FK (1999) Diagnosis of mitochondrial trifunctional protein deficiency in a blood spot from the newborn screening card by tandem mass spectrometry and DNA analysis. Pediatr Res 46(1):45–49
63. Matern D (2008) Acylcarnitines, including in vitro loading tests. In: Blau N, Duran M, Gibson KM (eds) Laboratory guide to the methods in biochemical genetics. Springer Verlag, Heidelberg, pp 171–206
64. Matern D (2022) Acylcarnitines. In: Blau N, Dionisi-Vici C, Ferreira C, Vianey Saban C, van Karnebeek CDM (eds) Physician's guide to the diagnosis, treatment, and follow-up of inherited metabolic diseases, vol 1, 2nd edn. Springer Verlag, pp 65–74
65. Browning MF, Larson C, Strauss A, Marsden DL (2005) Normal acylcarnitine levels during confirmation of abnormal newborn screening in long-chain fatty acid oxidation defects. J Inherit Metab Dis 28(4):545–550
66. Van Hove JL, Kahler SG, Feezor MD, Ramakrishna JP, Hart P, Treem WR, Shen JJ, Matern D, Millington DS (2000) Acylcarnitines in plasma and blood spots of patients with long-chain 3-hydroxyacyl-coenzyme a dehydrogenase deficiency. J Inherit Metab Dis 23(6):571–582
67. Koeberl DD, Young SP, Gregersen NS, Vockley J, Smith WE, Benjamin DK Jr, An Y, Weavil SD, Chaing SH, Bali D, McDonald MT, Kishnani PS, Chen YT, Millington DS (2003) Rare disorders of metabolism with elevated butyryl- and isobutyryl-carnitine detected by tandem mass spectrometry newborn screening. Pediatr Res 54(2):219–223
68. Matern D, He M, Berry SA, Rinaldo P, Whitley CB, Madsen PP, Van Calcar SC, Lussky RC, Andresen BS,

Wolff JA, Vockley J (2003) Prospective diagnosis of 2-Methylbutyryl-CoA dehydrogenase deficiency in the Hmong population by newborn screening using tandem mass spectrometry. Pediatrics 112(1):74–78
69. Abdenur JE, Chamoles NA, Guinle AE, Schenone AB, Fuertes AN (1998) Diagnosis of isovaleric acidaemia by tandem mass spectrometry: false positive result due to pivaloylcarnitine in a newborn screening programme. J Inherit Metab Dis 21(6):624–630
70. Boemer F, Schoos R, de Halleux V, Kalenga M, Debray FG (2014) Surprising causes of C5-carnitine false positive results in newborn screening. Mol Genet Metab 111(1):52–54. https://doi.org/10.1016/j.ymgme.2013.11.005
71. Murko S, Aseman AD, Reinhardt F, Gramer G, Okun JG, Mutze U, Santer R (2023) Neonatal screening for isovaleric aciduria: reducing the increasingly high false-positive rate in Germany. JIMD Rep 64(1):114–120. https://doi.org/10.1002/jmd2.12345
72. Gibson KM, Bennett MJ, Naylor EW, Morton DH (1998) 3-Methylcrotonyl-coenzyme a carboxylase deficiency in Amish/Mennonite adults identified by detection of increased acylcarnitines in blood spots of their children. J Pediatr 132(3 Pt 1):519–523. https://doi.org/10.1016/s0022-3476(98)70032-0
73. Karunanidhi A, Van't Land C, Rajasundaram D, Grings M, Vockley J, Mohsen AW (2022) Medium branched chain fatty acids improve the profile of tricarboxylic acid cycle intermediates in mitochondrial fatty acid beta-oxidation deficient cells: a comparative study. J Inherit Metab Dis 45(3):541–556. https://doi.org/10.1002/jimd.12480
74. Roe CR, Sweetman L, Roe DS, David F, Brunengraber H (2002) Treatment of cardiomyopathy and rhabdomyolysis in long-chain fat oxidation disorders using an anaplerotic odd-chain triglyceride. J Clin Invest 110(2):259–269
75. Hintz SR, Matern D, Strauss A, Bennett MJ, Hoyme HE, Schelley S, Kobori J, Colby C, Lehman NL, Enns GM (2002) Early neonatal diagnosis of long-chain 3-hydroxyacyl coenzyme a dehydrogenase and mitochondrial trifunctional protein deficiencies. Mol Genet Metab 75(2):120–127
76. Vianey-Saban C, Boyer S, Levrat V, Cheillan D, Piraud M, Guffon N, Maire I (2004) Interference of cefotaxime in plasma acylcarnitine profile mimicking an increase of 3-hydroxypalmitoleylcarnitine (C16:1-OH) using butyl esters. J Inherit Metab Dis 27(Suppl. 1):94
77. Majumdar R, Yori A, Rush PW, Raymond K, Gavrilov D, Tortorelli S, Matern D, Rinaldo P, Feldman GL, Oglesbee D (2017) Allelic spectrum of formiminotransferase-cyclodeaminase gene variants in individuals with formiminoglutamic aciduria. Mol Genet Genomic Med 5(6):795–799. https://doi.org/10.1002/mgg3.333
78. de Sain-van der Velden MG, Diekman EF, Jans JJ, van der Ham M, Prinsen BH, Visser G, Verhoeven-Duif NM (2013) Differences between acylcarnitine profiles in plasma and bloodspots. Mol Genet Metab 110:116. https://doi.org/10.1016/j.ymgme.2013.04.008
79. Heiner-Fokkema MR, Vaz FM, Maatman R, Kluijtmans LA, van Spronsen FJ, Reijngoud DJ (2017) Reliable diagnosis of carnitine Palmitoyltransferase type IA deficiency by analysis of plasma Acylcarnitine profiles. JIMD Rep 32:33–39. https://doi.org/10.1007/8904_2016_564
80. Roe DS, Yang BZ, Vianey-Saban C, Struys E, Sweetman L, Roe CR (2005) Differentiation of long-chain fatty acid oxidation disorders using alternative precursors and acylcarnitine profiling in fibroblasts. Mol Genet Metab 87:40

7 Purines and Pyrimidines

Marie van Dijk, Henk van Lenthe,
Susanna M. I. Goorden, Mirjam M. C. Wamelink,
and André B. P. van Kuilenburg

Contents

M. van Dijk · H. van Lenthe · M. M. C. Wamelink
Laboratory of Genetic Metabolic Diseases,
Departments of Clinical Chemistry, Amsterdam UMC
Location University of Amsterdam,
Amsterdam, The Netherlands

S. M. I. Goorden
Department of Clinical Genetics, Erasmus Medical
Centre, Rotterdam, The Netherlands

A. B. P. van Kuilenburg (✉)
Laboratory of Genetic Metabolic Diseases,
Departments of Clinical Chemistry, Amsterdam UMC
Location University of Amsterdam,
Amsterdam, The Netherlands

Amsterdam Gastroenterology Endocrinology
Metabolism, Amsterdam, The Netherlands
e-mail: a.b.vankuilenburg@amsterdamumc.nl

Abbreviations

β-AIB	β-Aminoisobutyrate
β-UP	β-Ureidopropionase
ADA	Adenosine deaminase
ADSL	Adenylosuccinate lyase
AICAR	5-amino-4-imidazolecarboxamide ribotide
AICAriboside	5-amino-4-imidazolecarbox amide riboside
AICAR-TF	AICAR transformylase
AMPD1	Adenosine monophosphate deaminase 1

N. Blau, F. M. Vaz (eds.), *Laboratory Guide to the Methods in Biochemical Genetics*,
https://doi.org/10.1007/978-3-031-58819-8_7

APRT	Adenine phosphoribosyltransferase
CAD	Carbamoyl-phosphate synthetase, aspartate transcarbamylase, dihydroorotase
CDA	Cytidine deaminase
CTPS	CTP synthetase
dCK	Deoxycytidine kinase
dGK	Deoxyguanosine kinase
DHODH	Dihydroorotate dehydrogenase
DHP	Dihydropyrimidinase
DPD	Dihydropyrimidine dehydrogenase
HGPRT	Hypoxanthine-guanine phosphoribosyltransferase
IMP-CH	Inosine monophosphate (IMP) cyclohydrolase
IS	Internal standard
N-C-β-AIB	*N*-carbamyl-β-aminoisobutyric acid
N-C-β-alanine	*N*-carbamyl-β-alanine
P5N-1	Pyrimidine 5′-nucleotidase type-I
PNP	Purine nucleoside phosphorylase
PRPS1	Phosphoribosyl pyrophosphate synthetase 1
RRM2B	Ribonucleotide reductase regulatory TP53 inducible subunit M2B
SAICAR or Succinyl-AICAR	Succinyl-5-amino-4-imidazolecarboxamide ribotide
SAICAriboside	Succinyl-5-amino-4-imidazolecarboxamide riboside
TK1	Thymidine kinase 1
TK2	Thymidine kinase 2
TP	Thymidine phosphorylase
UCK1,2	Uridine-cytidine kinase 1,2
UMPS	UMP synthase
XDH	Xanthine dehydrogenase

7.1 Introduction

Inherited disorders of purine and pyrimidine metabolism have a wide variety of clinical presentations which include, among others, anaemia, immunodeficiency, kidney stones, convulsions, mental retardation, autism and growth retardation. In contrast to the well-known defects of purine metabolism, most of the defects of pyrimidine metabolism have only recently been discovered [1–3]. Genetic enzyme defects are not only biological curiosities but they can also have serious clinical consequences in more frequently occurring situations such as adverse reactions to medication. Defects in the degradation of pyrimidines have been associated with a variable clinical phenotype whereas the same defects can lead to severe life-threatening toxicities when (partially) deficient individuals are treated with the pyrimidine analogue 5-fluorouracil or capecitabine [4–6].

Because of this broad clinical spectrum, rapid and specific screening methods are desired for the diagnosis of these disorders in patients at risk to enable appropriate treatment, genetic counseling and prenatal diagnosis. Targeted and general screening methods for purine and pyrimidine analyses in liquid urine or urine-soaked filter paper strips have been developed using UPLC-electrospray tandem mass [7–11]. The availability of these methods provides powerful tools to (pre)screen patients at risk.

7.2 Purines and Pyrimidines

Purine and pyrimidine nucleotides are essential for a vast number of biological processes such as the synthesis of RNA, DNA, phospholipids, glycogen and the sialylation and glycosylation of proteins. Both purines and pyrimidines can be synthesized de novo in mammalian cells through multistep processes (Figs. 7.1 and 7.2). In addition to the de

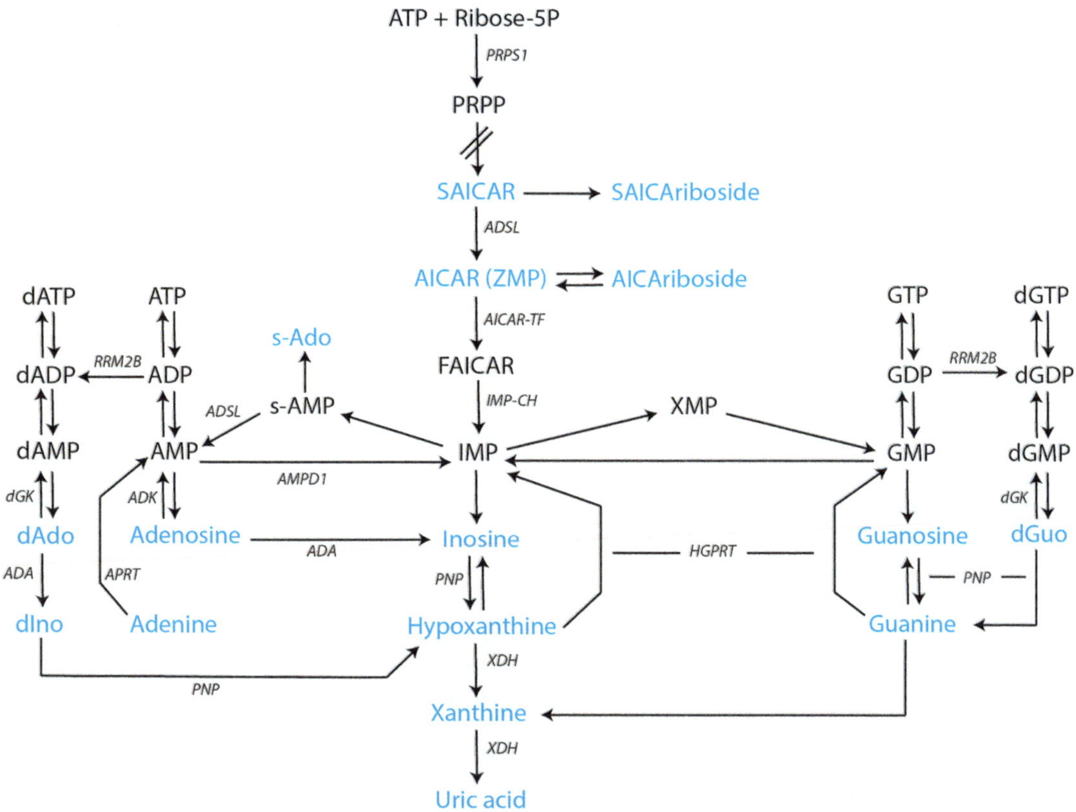

Fig. 7.1 Schematic representation of purine metabolism. The enzymes of inherited disorders of purine metabolism are depicted in italics and characteristic diagnostic metabolites are depicted in blue. ADA, adenosine deaminase; ADSL, adenylosuccinate lyase; AICAR, 5-amino-4-imidazolecarboxamide ribotide; AICAriboside, 5-amino-4-imidazolecarboxamide riboside; AICAR-TF, AICAR transformylase; AMPD1, adenosine monophosphate deaminase 1; APRT, adenine phosphoribosyltransferase; dGK, deoxyguanosine kinase; HGPRT, hypoxanthine-guanine phosphoribosyltransferase; IMP-CH, inosine monophosphate (IMP) cyclohydrolase; PRPS1, phosphoribosyl pyrophosphate synthetase 1; PNP, purine nucleoside phosphorylase; RRM2B, ribonucleotide reductase regulatory TP53 inducible subunit M2B; SAICAR, succinyl-5-amino-4-imidazolecarboxamide ribotide; SAICAriboside, succinyl-5-amino-4-imidazolecarboxamide riboside; XDH, xanthine dehydrogenase

novo synthesis, purine nucleotides can also be synthesized via the salvage of the purine bases adenosine, guanosine and hypoxanthine. In contrast, the salvage pathway for pyrimidine nucleotides occurs at the level of the nucleosides uridine and cytidine. Patients with a deficiency of the purine or pyrimidine metabolism usually present with altered levels of the nucleoside bases and/or (deoxy)ribonucleosides. The structure of the naturally occurring purine and pyrimidine nucleoside bases and (deoxy)ribonucleosides is shown in Figs. 7.3 and 7.4, respectively.

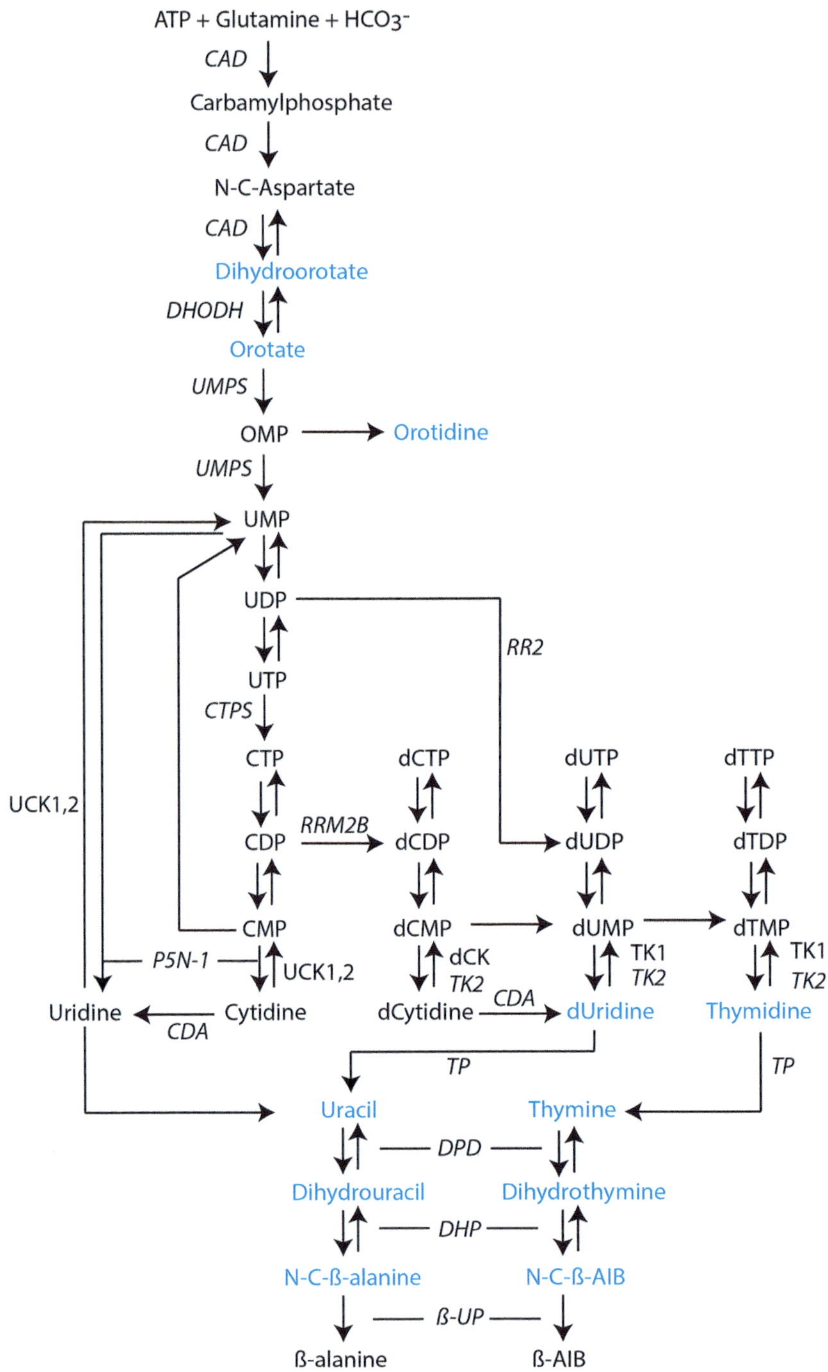

Fig. 7.2 Schematic representation of pyrimidine metabolism. The enzymes of inherited disorders of pyrimidine metabolism are depicted in italics and characteristic diagnostic metabolites are depicted in blue. CAD, carbamoyl-phosphate synthetase, aspartate transcarbamylase, dihydroorotase; CDA, cytidine deaminase; CTPS, CTP synthetase; dCK, deoxycytidine kinase; DPD, dihydropyrimidine dehydrogenase; DHODH, dihydroorotate dehydrogenase; DHP, dihydropyrimidinase; *N*-C-β-alanine, *N*-carbamyl-β-alanine; *N*-C-β-AIB, *N*-carbamyl-β-aminoisobutyric acid; P5N-1, pyrimidine 5′-nucleotidase type-I; TK1, thymidine kinase 1; TK2, thymidine kinase 2; TP, thymidine phosphorylase; RRM2B, ribonucleotide reductase regulatory TP53 inducible subunit M2B, ribonucleotide reductase; UCK1,2, uridine-cytidine kinase 1,2; UMPS, UMP synthase; β-UP, β-ureidopropionase; β-AIB, β-aminoisobutyrate

Purine nucleoside bases

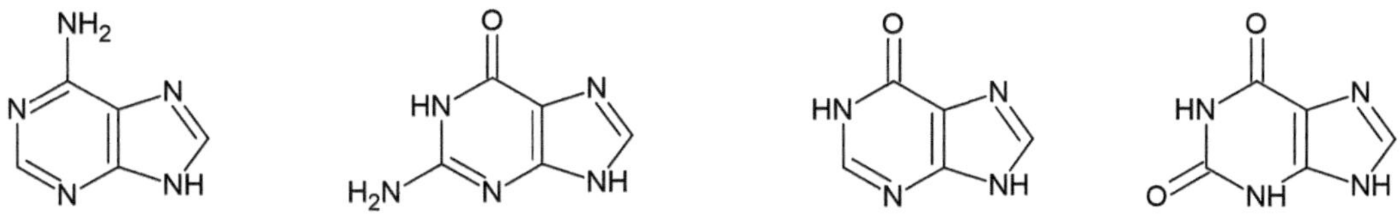

Adenine **Guanine** **Hypoxanthine** **Xanthine**

Purine ribonucleosides

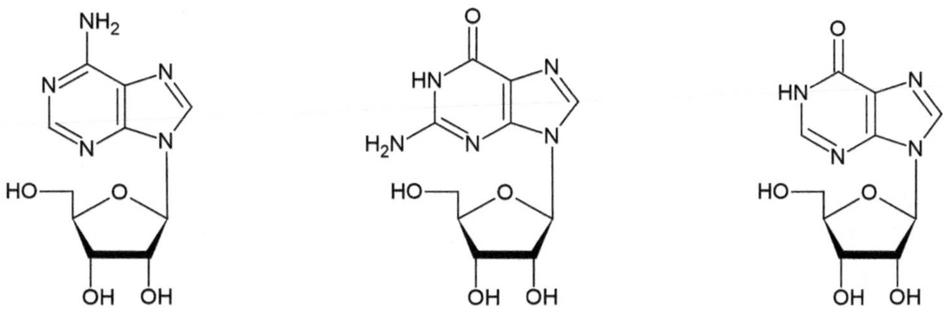

Adenosine **Guanosine** **Inosine**

Purine deoxyribonucleosides

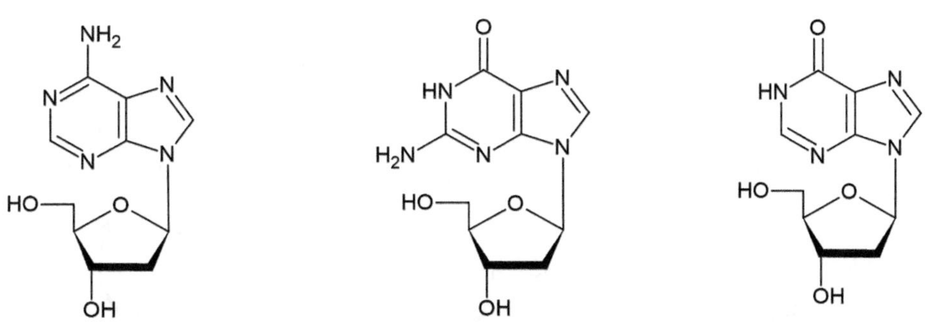

Deoxyadenosine **Deoxyguanosine** **Deoxyinosine**

Fig. 7.3 Structure of the naturally occurring purine nucleoside bases and (deoxy)ribonucleosides

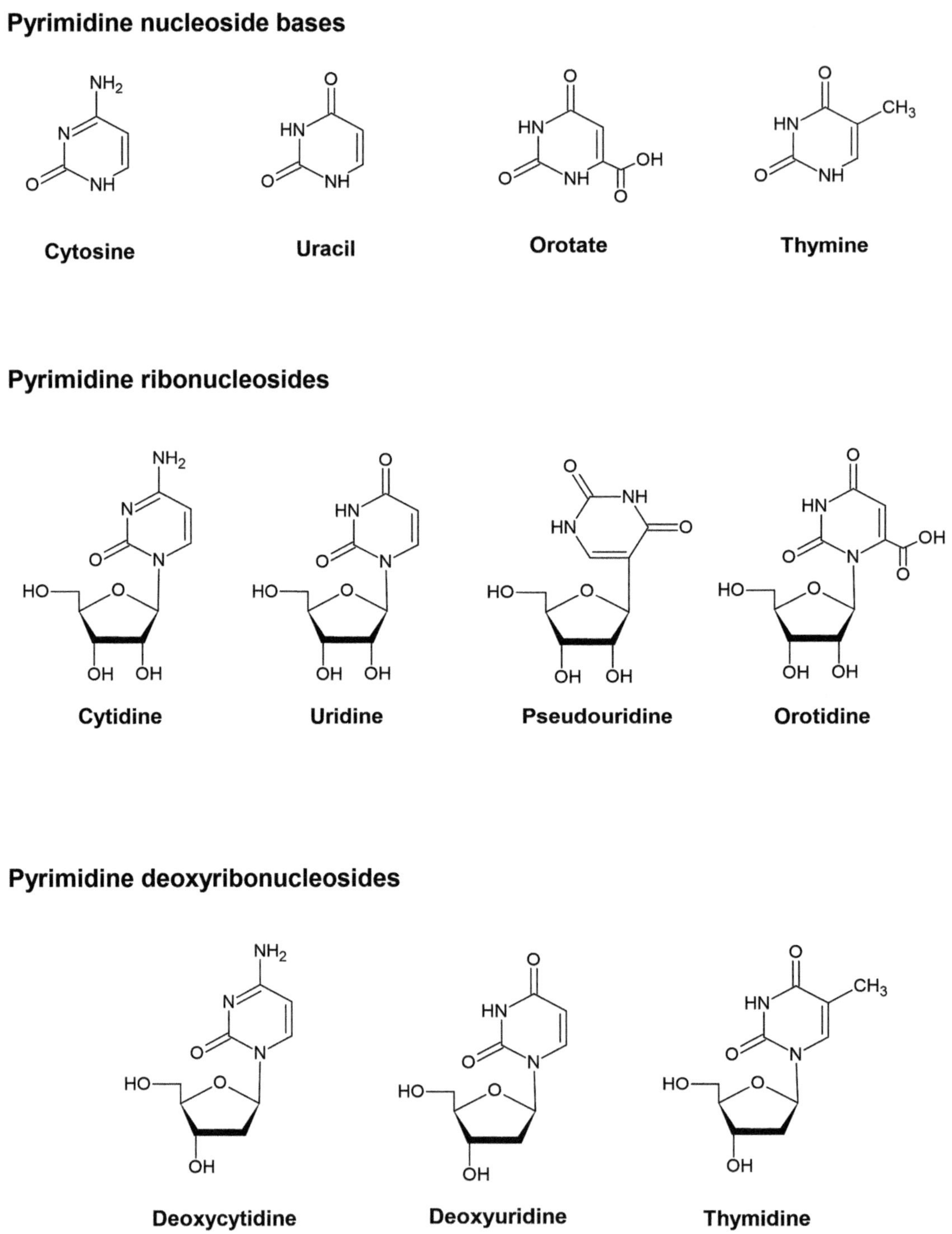

Fig. 7.4 Structure of the naturally occurring pyrimidine nucleoside bases and (deoxy)ribonucleosides

7.3 Methods

7.3.1 Principle

The metabolites of interest in urine are separated using reversed-phase UPLC combined with electrospray ionization tandem mass spectrometry and detection is performed using multiple-reaction monitoring. Stable-isotope-labelled reference compounds are used as internal standards.

7.3.2 Pre-analytical

Urine samples are stored at –20 °C, until analysis. Prior to analysis, the liquid urine samples are vortexed. Samples with much sediment can be left standing until the majority has sunk to the bottom after which a sample can be taken from the supernatant above.

For urine-soaked filter paper strips: Dip filter paper strips (12 × 40 mm; type 2992; Schleicher & Schuell) completely into urine and remove the excess urine by wiping it off along the wall of the test tube. Dry the strips at room temperature. The strips can be stored at room temperature until analysis.

7.3.3 Reagents and Chemicals

Standards and internal standards can be found in Tables 7.1 and 7.2. All standards (Table 7.1) were obtained from Sigma, apart from thymine and hypoxanthine which were obtained from Calbiochem, orotate which was obtained from Acros, and succinyl-adenosine and succinyl-AICAR which were obtained from the University of Prague. Succinyladenosine can also be prepared according to the method of Jaeken and Van den Berghe [12]. In short, 0.5 ml of a solution of 0.8 mM adenylosuccinate and 5 U of Crotalus adamanteus venom 5′-nucleotidase, both obtained from Sigma are mixed in 40 mM Tris-HCl buffer, pH 8.0, containing 4 mM Mg^{2+} and incubated at 37 °C for 2 h. Reversed-phase HPLC is used to check hydrolysis efficiency and to calculate the amount of succinyladenosine formed during hydrolysis [13].

Table 7.1 Standards

Standards
Uracil
Thymine
Dihydrouracil
Dihydrothymine
N-carbamyl-beta-alanine
N-carbamyl-beta-isobutyric acid
5-hydroxymethyluracil
Adenine
Adenosine
AICAriboside
Cytidine
Deoxy-adenosine
Deoxy-guanosine
Deoxy-inosine
Deoxy-uridine
Guanine
Guanosine
Hypoxanthine
Inosine
Orotate
Pseudouridine
Succinyl-adenosine
Succinyl-AICAriboside (SAICAriboside)
Thymidine
2,8-dihydroxyadenine
Uridine
Xanthine

All internal standards (Table 7.2) were obtained from Cambridge Isotope Laboratories, except 5-hydroxymethyl-D2-6-D1-uracil and D2-deoxyuridine which were obtained from CDN isotopes, and 13C4-succinyl-adenosine and 13C4-SAICAR which were obtained from the University of Prague. D6-N-carbamyl-β-aminoisobutyric acid and 13C4;15N2-N-carbamyl-beta-alanine were prepared in house: D6-N-carbamyl-β-aminoisobutyric acid is prepared by the alkaline hydrolysis of 5,6,6D3-MeD3-dihydrothymine, essentially as described before [14]. Dissolve 5,6,6D3-MeD3-dihydrothymine in 0.1 M NaOH and hydrolyse at 37 °C for 1 h. Neutralize the solution with HCl. 13C4;15N2-N-carbamyl-beta-alanine is prepared in the same manner from 13C4;15N2-dihydrouracil.

Ammonia solution 25% and acetic acid are purchased from Merck. Methanol is obtained from Biosolve. Water was passed through a MilliQ Lab system (Millipore).

Table 7.2 Internal standards

Internal standards	Internal standard concentration (μM)	Internal standard mix 1 or 2
1,3-$^{15}N_2$-uracil	150	IS1
α,α,α,6-D_4-thymine	200	IS1
$^{13}C_4$;$^{15}N_2$-dihydrouracil	1000	IS1
5,6,6D_3-MeD_3-dihydrothymine	1000	IS1
5-hydroxymethyl-D_2-6-D_1-uracil	250	IS1
D_6-*N*-carbamyl-β-aminoisobutyric acid	1000	IS1
$^{13}C_4$;$^{15}N_2$-*N*-carbamyl-beta-alanine	300	IS1
$^{15}N_5$-adenine	150	IS2
$^{13}C_5$-adenosine	150	IS2
$^{15}N_3$-cytidine	150	IS2
$^{15}N_5$-2′-deoxy-adenosine	150	IS2
$^{13}C_{10}$,$^{15}N_5$-deoxy-guanosine	150	IS2
D_2-deoxy-uridine	150	IS2
^{13}C,$^{15}N_2$-guanine	25	IS2
$^{15}N_5$-guanosine	50	IS2
$^{13}C_5$-hypoxanthine	150	IS2
$^{15}N_4$-inosine	150	IS2
$^{15}N_2$-orotate	150	IS2
$^{13}C_4$-succinyl-adenosine	25	IS2
$^{13}C_4$-SAICAriboside	25	IS2
$^{13}C_5$-AICAriboside	5	IS2
$^{15}N_2$-thymidine	150	IS2
$^{15}N_2$-uridine	150	IS2
$^{15}N_2$-xanthine	150	IS2

7.3.4 Instrumentation

A Waters Acquity H-Class—XEVO TQ-S Micro UPLC-mass spectrometer system in combination with:

IS1 metabolites

- A Phenomenex Aqua C18 analytical column (250 × 4.6 mm; particle size, 5 μm) and C18 ODS 4 × 3 mm guard cartridge

IS2 metabolites

- A Waters Acquity HSS-T3 C18 column (100 × 2.1 mm; particle size, 1.8 μm)

7.3.5 MS Settings and Chromatographical Conditions

The mobile phases for IS1 metabolites on the Phenomenex column are as follows: 50 mM acetic acid, adjusted to pH 4.0 with 25% ammonia solution (eluent A) and methanol (eluent B). The elution gradient is as follows (flow rate, 1 ml/min): 0–1.30 min, 100% A; 1.30–11.30 min, 100% A-50% A; 11.30–11.40 min, 50% A-100% A; 11.40–15.90 min, equilibration with 100% A. All gradient steps are linear and the total analysis time, including equilibration, is 15.9 min. For the IS2 metabolites on the Waters column the following mobile phases are used: 25 mM ammonium acetate, adjusted to pH 5.0 with 25% ammonia solution (eluent A) and methanol (eluent B). The elution gradient is as follows (flow rate, 0.4 ml/min): 0–1.00 min, 100% A; 1.00–4.00 min, 100% A-60% A; 4.00–5.00 min, 60% A; 5.00–5.01 min, 60% A-100% A; 5.01–8.00 min equilibration with 100% A. The gradient steps are linear and the total analysis time, including equilibration, is 8 min. A XEVO TQ-S Micro mass spectrometer is used in the positive ESI mode for the general purine and pyrimidine screening application. Only orotate and 5-hydroxymethyluracil are measured in the negative ESI mode. The detector is used in MS/MS mode using multiple-reaction monitoring to detect a specific transition of precursor to fragment ion for each analyte. The transition, cone voltage and collision energy established for each compound are listed in Tables 7.3 and 7.4.

7.3.6 Calibration

Calibration curves are obtained for each analyte for the metabolite/IS peak-area ratio versus the concentration of the metabolite in the calibration mixture. The calibration curves are linear for all compounds up to 0.5 mM, except for orotate (0.35 mM), uridine (0.2 mM), thymidine (0.25 mM), deoxyadenosine (0.1 mM), succinyl-adenosine (0.05 mM) and deoxyuridine (0.3 mM).

7.3.7 Quality Control

- A pooled urine and a pooled urine spiked with all relevant metabolites are analysed with each batch of patient samples.
- The response of the internal standards is monitored during a series and should be constant. In case of strong deviations in the response of an internal standard, there might be quenching of the signal and the sample should be re-analysed after dilution.
- In case the concentration of a metabolite is higher than the highest concentration of the calibration curve, the sample will be analysed again after dilution.

Table 7.3 MS settings for IS1 metabolites in the positive ESI mode, 5-hydroxymethyluracil in the negative ESI mode

Compound	Parent ion m/z	Daughter ion m/z	Cone voltage V	Collision energy eV
Uracil	113	70	35	13
1,3-$^{15}N_2$-uracil	115	71	35	13
Thymine	127	110	30	8
α,α,α,6-D_4-thymine	131	114	30	8
Dihydrothymine	129	69	35	12
5,6,6D_3-MeD_3-dihydrothymine	135	74	35	12
Dihydrouracil	115	73	30	13
$^{13}C_4$;$^{15}N_2$-dihydrouracil	121	77	30	13
N-carbamyl-β-alanine	133	90	20	6
$^{13}C_4$;$^{15}N_2$-*N*-carbamyl-beta-alanine	139	94	20	6
5-hydroxymethyluracil	141	123	20	14
5-hydroxymethyl-D_2-6-D_1-uracil	144	125	20	14
N-carbamyl-β-aminoisobutyric acid	147	129	19	6
D_6-*N*-carbamyl-β-aminoisobutyric acid	153	135	19	6
Dihydrouridine	247	115	20	9

Table 7.4 MS settings for IS2 metabolites in the positive ESI mode, orotate in the negative ESI mode

Compound	Parent ion m/z	Daughter ion m/z	Cone voltage V	Collision energy eV
Adenine	136	119	40	18
$^{15}N_5$-adenine	141	123	40	18
Hypoxanthine	137	110	32	17
$^{13}C_5$-hypoxanthine	142	114	32	17
Guanine	152	135	30	16
^{13}C,$^{15}N_2$-guanine	155.1	138	30	16
Xanthine	153	110	35	17
$^{15}N_2$-xanthine	155	111	35	17
Deoxyuridine	229	113	10	10
D_2-deoxy-uridine	231	115	10	10
Thymidine	243	127	10	8

(continued)

Table 7.4 (continued)

Compound	Parent ion m/z	Daughter ion m/z	Cone voltage V	Collision energy eV
$^{15}N_2$-thymidine	245.1	129	10	8
Cytidine	244	112	10	11
$^{15}N_3$-cytidine	247.1	115.01	10	11
Uridine	245	113	20	9
D_2-deoxy-uridine	247.1	115	20	9
Pseudouridine	245	209	22	10
Deoxyadenosine	252	136	20	14
$^{15}N_5$-2′-deoxy-adenosine	257.1	141	20	14
Deoxyinosine	253	137	13	8
AICAriboside	259	127	22	12
$^{13}C_5$-AICAriboside	264	127	22	12
Adenosine	268	136	30	15
$^{13}C_5$-adenosine	273.1	136	30	15
Deoxyguanosine	268	152	10	8
$^{13}C_{10}$,$^{15}N_5$-deoxy-guanosine	283.1	162	10	8
Inosine	269	137	15	9
$^{15}N_4$-inosine	273.1	141	15	9
Guanosine	284	152	15	12
$^{15}N_5$-guanosine	289.1	157	15	12
SAICAriboside	375	243	10	11
$^{13}C_4$-SAICAriboside	379.1	247.1	10	11
Succinyladenosine	384	252	35	17
$^{13}C_4$-succinyl-adenosine	388.1	256.1	35	17
2,8-dihydroxyadenine	168	80	30	30
Orotate	155	111	20	9
$^{15}N_2$-orotate	157	113	20	9

7.4 Analytical Procedure

7.4.1 Liquid Urine

- Transfer 100 μl (IS1 metabolites) or 25 μl (IS2 metabolites) urine, calibration standard or control into an autosampler vial (Gilson)
- Add 10 μl of IS1-mix or IS2-mix and mix thoroughly
- Centrifuge 2 min at full speed (>10,000 g)
- Inject 25 μl (IS1) or 3 μl (IS2) into the UPLC-tandem MS system

7.4.2 Urine-Soaked Filter Paper

- Add 20 μl of IS1-mix or IS2-mix on the filter paper strip which has been soaked in urine
- After drying, transfer the filter paper strip to a 2 ml Eppendorf tube
- Add 1.5 ml of 75% (v/v) methanol and sonicate for 10 min, or perform the extraction twice using 750 μl per extraction and combine the extracts
- Transfer the extract to a glass tube
- Evaporate the extract at 40 °C under a constant stream of nitrogen
- Dissolve the dried extract in 200 μl 50 mM acetic acid (pH 4.0) for IS1 metabolites or 200 μl 100 mM ammonia (IS2 metabolites) and sonicate for 5 min
- Centrifuge the sample at full speed (>10,000 g) for 2 min
- Transfer the supernatant to an autosampler vial
- Save 100 μl of the supernatant for determination of the creatinine concentration
- Inject 25 μl (IS1 metabolites) or 3 μl (IS2 metabolites) into the UPLC-tandem MS system

7.4.3 Calculation

The concentration of each analyte was determined by use of the slope and intercept of the calibration curve that is obtained by the analyte/

IS peak-area ratio versus the concentration of the analyte in the calibration mixture.

7.5 Post-analytical

7.5.1 Interpretation

Concentrations of metabolites outside the reference ranges may constitute a typical pattern indicating the presence of an inborn error of purine or pyrimidine metabolism. However, the altered excretion of purine and pyrimidines may also be a secondary phenomenon due to the presence of other metabolic disorders, such as a deficiency of the urea cycle [10]. Increased concentration of a single metabolite or combinations of metabolites may also result from bacterial contamination, sample conditions, medication or dietary compounds [13].

7.6 Chromatograms (Figs. 7.5 and 7.6)

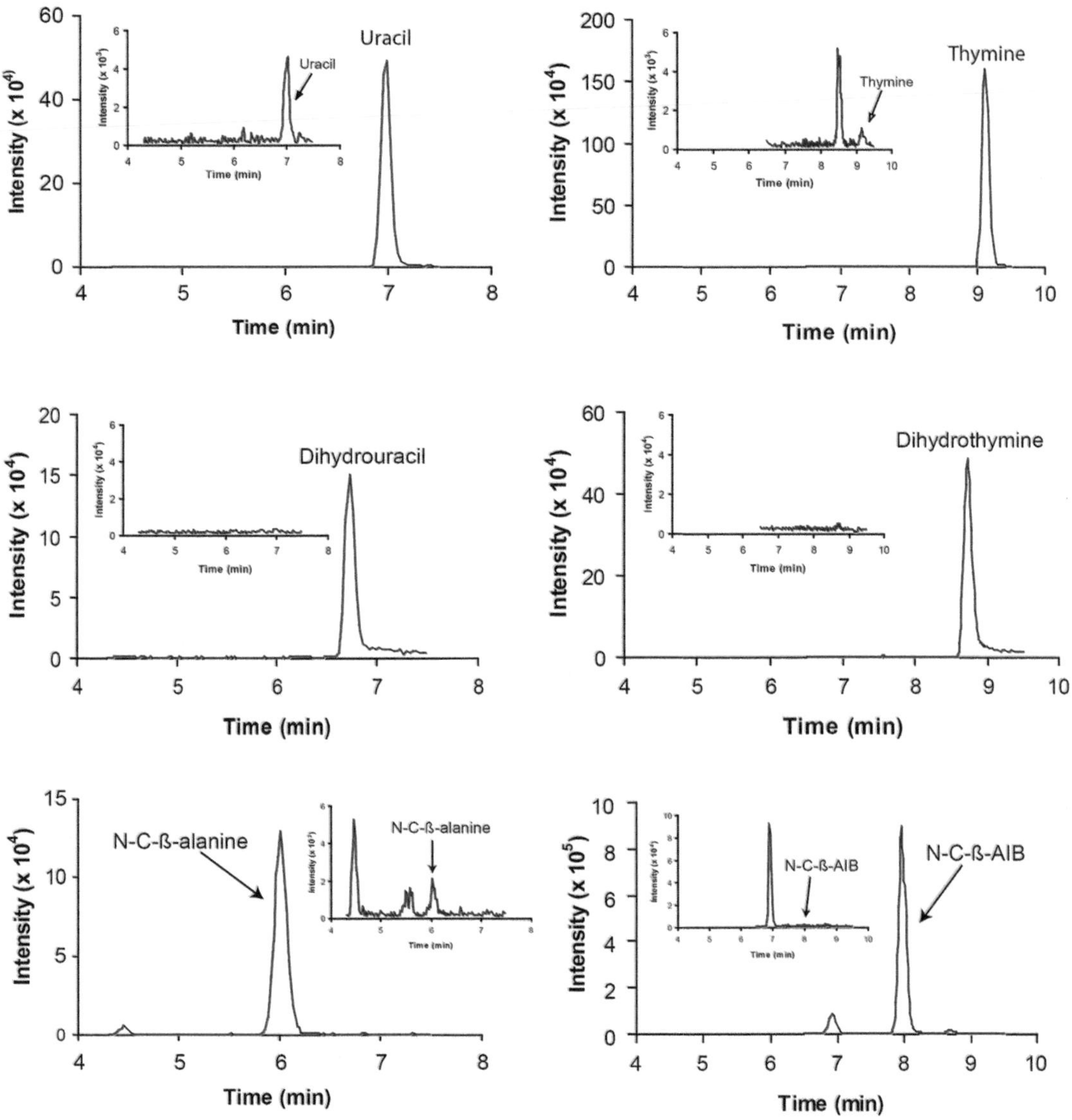

Fig. 7.5 HPLC/ESI tandem-MS profiles of urines from patients with a dihydropyrimidine dehydrogenase (upper panel), dihydropyrimidinase (middle panel) and a β-ureidopropionase deficiency (lower panel). The inserts show the profiles of a control urine, note the differences in scale. *N*-C-β-alanine, *N*-carbamyl-β-alanine; *N*-C-β-AIB, *N*-carbamyl-β-aminoisobutyric acid

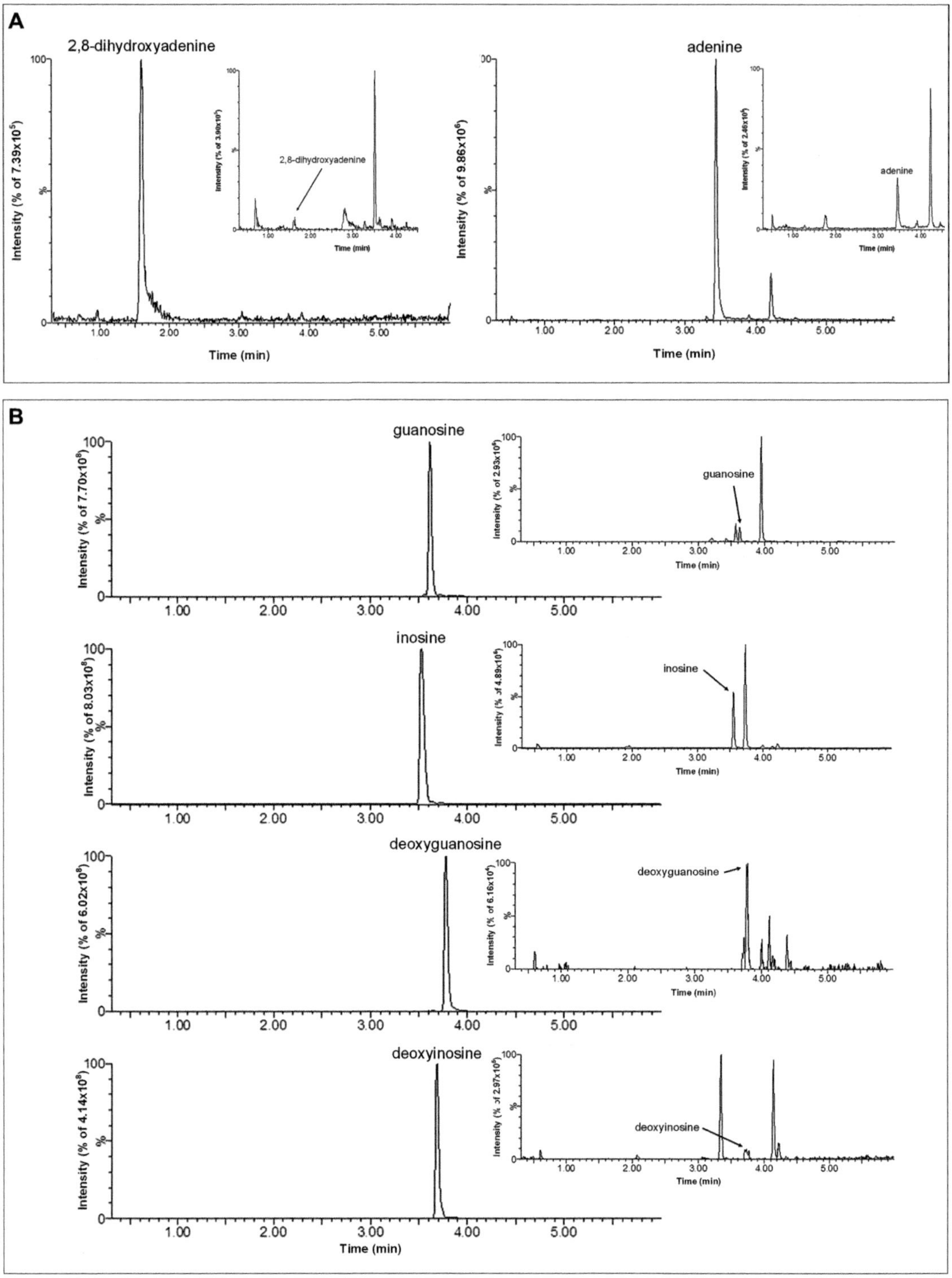

Fig. 7.6 HPLC/ESI tandem-MS profiles of urines from patients with an adenine phosphoribosyltransferase (**a**) and a purine nucleoside phosphorylase deficiency (**b**). The inserts show the profiles of a control urine, note the differences in scale

7.7 Reference Values

Reference values for purine and pyrimidine nucleosides and bases in urine (≤3 or ≤4 years) are shown in Table 7.5.

7.8 Typical Pathological Values (Tables 7.6 and 7.7)

Table 7.5 Reference values of purine and pyrimidine metabolites in urine

Compound	Min. ref. value in µmol/mmol creatinine	Max. ref. value in µmol/mmol creatinine	Method used	Age range (years)	N
Uracil	0	30	Mean ± 2SD	0–3	104
Thymine	0	1.7	Mean ± 2SD	0–3	104
Dihydrouracil	0	16.9	Mean ± 2SD	0–3	104
Dihydrothymine	0	7.3	Mean ± 2SD	0–3	104
N-carbamyl-beta-alanine	0	29.5	Mean ± 2SD	0–3	104
N-carbamyl-beta-isobutyric acid	0	6.4	Mean ± 2SD	0–3	104
5-hydroxymethyluracil	0	2	Mean ± 2SD	0–3	104
Adenine	0	34	Mean ± 2SD	0–1	26
Adenine	0	6.4	Mean ± 2SD	1–4	27
Adenosine	0	4.9	Mean ± 2SD	0–4	53
AICAriboside	0	3.5	97.5 percentile	0–4	>450
Cytidine	0	2	Mean ± 2SD	0–4	53
Deoxy-adenosine	0	0.1	Mean ± 2SD	0–4	53
Deoxy-guanosine	0	0.2	Mean ± 2SD	0–4	53
Deoxy-inosine	0	0.1	Mean ± 2SD	0–4	53
Deoxy-uridine	0	1.3	Mean ± 2SD	0–4	53
Guanine	0	4.2	Mean ± 2SD	0–4	53
Guanosine	0	1.6	Mean ± 2SD	0–4	53
Hypoxanthine	0	48	Mean ± 2SD	0–4	53
Inosine	0	4.6	Mean ± 2SD	0–4	53
Orotate	0	3.9	Mean ± 2SD	0–4	53
Pseudouridine	1	109	Mean ± 2SD	0–4	53
Succinyl-adenosine	1	7.5	2.5–97.5 percentile	0–4	>450
SAICAriboside	0	1.8	97.5 percentile	0–4	>450
Thymidine	0	0.5	Mean ± 2SD	0–4	53
Uridine	0	2.1	Mean ± 2SD	0–4	53
Xanthine	4	39	Mean ± 2SD	0–1	26
Xanthine	0	65	Mean ± 2SD	1–4	27

Table 7.6 Concentration range of characteristic purine metabolites in urine (μmol/mmol creatinine) from patients

Compound	ADA deficiency [15]	PNP deficiency [15]	HGPRT deficiency [15]	APRT deficiency [15]	PRPP superactivity [15]	ADSL deficiency [15]	ATIC deficiency [15]	Adenosine kinase deficiency [16]	XDH/molybdenum cofactor deficiency [15]
AICAriboside			2–21				280 ($n = 1$)		
SAICAriboside						9–802			
Adenine				8–41					
Hypoxanthine			90–270		>50				20–660
Guanine									
Xanthine			25–109						134–2900
2,8-dihydroxyadenine				23–32					
Deoxyadenosine	10–266								
Deoxyinosine		100–650							
Deoxyguanosine		100–400							
Inosine		500–1900							
Guanosine		286–405							
Adenosine								<4–45	
Succinyladenosine						69–2603			

ADA, adenosine deaminase; PNP, purine nucleoside phosphorylase; HGPRT, hypoxanthine-guanine phosphoribosyltransferase; APRT, adenine phosphoribosyltransferase; ADSL, adenylosuccinate lyase

Table 7.7 Concentration range of characteristic pyrimidine metabolites in urine (μmol/mmol creatinine) from patients

Compound	DHODH deficiency [17]	UMPS deficiency [18]	TP deficiency [19]	DPD deficiency [20]	DHP deficiency [21, 22]	β-UP deficiency [23–25]
Orotate	33	200–8930				
Dihydroorotate	39					
Orotidine		149–465				
Thymidine			125			
Deoxyuridine			150			
Uracil			77	92–888	11–135	1–167
Thymine			48	59–628	4–225	0.5–83
5-hydroxymethyluracil				3–66		
Dihydrouracil					100–849	8–2531
Dihydrothymine					46–622	15–1375
N-carbamyl-β-alanine						130–1523
N-carbamyl-β-AIB						21–1204

DHODH, dihydroorotate dehydrogenase; UMPS, uridine monophosphate synthase; TP, thymidine phosphorylase; DPD, dihydropyrimidine dehydrogenase; DHP, dihydropyrimidinase; β-UP, β-ureidopropionase

7.9 Pitfalls

Bacterial contamination of the urine may result in strongly increased levels of uracil due to the bacterial degradation of pseudouridine. Thymine-uraciluria which is indicative for a dihydropyrimidine dehydrogenase or dihydropyrimidinase deficiency may also result from increased tissue degradation. However, the latter situation is also characterized by hyper-β-aminoisobutyric aciduria and hyper-β-alanineuria [13]. Under alkaline conditions, due to the presence of bacterial contamination, the deoxynucleosides may be hydrolysed towards their corresponding nucleoside bases and dihydrouracil/dihydrothymine may be hydrolysed towards *N*-carbamyl-β-alanine/*N*-carbamyl-β-AIB.

7.10 Follow-Up Enzyme Assays

- The activity of dihydropyrimidine dehydrogenase can be measured in peripheral blood mononuclear cells, fibroblasts and liver using radiolabeled thymine followed by separation of radiolabeled thymine and the radiolabeled reaction products with reversed-phase HPLC combined with online detection of the radioactivity [26].
- The activity of dihydropyrimidinase or β-ureidopropionase can only be measured in liver or kidney. The activity of dihydropyrimidinase is determined using a radiochemical assay with subsequent separation of radiolabelled dihydrouracil from radiolabelled *N*-carbamyl-β-alanine with reversed-phase HPLC combined with detection of $^{14}CO_2$ by liquid scintillation counting [27]. The activity of β-ureidopropionase can be determined using radiolabeled *N*-carbamyl-β-alanine followed by separation of radiolabeled *N*-carbamyl-β-alanine from radiolabeled β-alanine by reversed-HPLC [14, 28].
- The activity of thymidine phosphorylase can be detected in leukocytes using a nonradiochemical assay in which thymine is detected at 265 nm after separation with reversed-phase HPLC [29].
- The activity of hypoxanthine-guanine phosphoribosyltransferase (HPRT), adenine phosphoribosyltransferase (APRT), adenosine deaminase (ADA) and purine nucleoside phosphorylase (PNP) can be determined in dried blood spots using HPLC linked assays [30–32].

7.11 Conclusion

The detection of purine and pyrimidine nucleosides and bases, using state-of-the-art reversed-phase UPLC hyphenated with electrospray tandem mass spectrometry, enables the diagnosis of patients suffering from inherited metabolic diseases, affecting the synthesis or degradation pathways of purine and pyrimidine metabolites.

References

1. Balasubramaniam S, Duley JA, Christodoulou J (2014) Inborn errors of purine metabolism: clinical update and therapies. J Inherit Metab Dis 37(5):669–686
2. Balasubramaniam S, Duley JA, Christodoulou J (2014) Inborn errors of pyrimidine metabolism: clinical update and therapy. J Inherit Metab Dis 37(5):687–698
3. Kamatani N, Jinnah HA, Hennekam RCM, van Kuilenburg ABP (2021) Purine and pyrimidine metabolism. In: Pyeritz RE, Korf BR, Grody WW (eds) Emery & Rimoin's principles and practice of medical genetics and genomics; metabolic disorders. Elsevier Academic, New York, pp 183–234
4. van Kuilenburg ABP (2004) Dihydropyrimidine dehydrogenase and the efficacy and toxicity of 5-fluorouracil. Eur J Cancer 40(7):939–950
5. van Kuilenburg AB, Meijer J, Maurer D, Dobritzsch D, Meinsma R, Los M, Knegt LC, Zoetekouw L, Jansen RL, Dezentje V, van Huis-Tanja LH, van Kampen RJ, Hertz JM, Hennekam RC (2017) Severe fluoropyrimidine toxicity due to novel and rare DPYD missense mutations, deletion and genomic amplification affecting DPD activity and mRNA splicing. Biochim Biophys Acta 1863(3):721–730
6. Henricks LM, Siemerink EJM, Rosing H, Meijer J, Goorden SMI, Polstra AM, Zoetekouw L, Cats A, Schellens JHM, van Kuilenburg ABP (2018) Capecitabine-based treatment of a patient with a novel DPYD genotype and complete dihydropyrimidine dehydrogenase deficiency. Int J Cancer 142(2):424–430
7. Steinbusch LKM, Wang P, Waterval H, Stassen F, Coene KLM, Engelke UFH, Habets DDJ, Bierau J, Körver-Keularts I (2021) Targeted urine metabolomics with a graphical reporting tool for rapid diagnosis of inborn errors of metabolism. J Inherit Metab Dis 44(5):1113–1123
8. Ito T, van Kuilenburg ABP, Bootsma AH, Haasnoot AJ, van Cruchten AG, Wada Y, van Gennip AH (2000) Rapid screening of high-risk patients for disorders of purine and pyrimidine metabolism using HPLC-electrospray tandem mass spectrometry of liquid urine or urine-soaked filter paper strips. Clin Chem 46(4):445–452
9. van Lenthe H, van Kuilenburg ABP, Ito T, Bootsma AH, van Cruchten AG, Wada Y, van Gennip AH (2000) Defects in pyrimidine degradation identified by HPLC-electrospray tandem mass spectrometry of urine specimens or urine-soaked filter paper strips. Clin Chem 46(12):1916–1922
10. van Kuilenburg ABP, van Lenthe H, Loffler M, van Gennip AH (2004) Analysis of pyrimidine synthesis "de novo" intermediates in urine and dried urine filter-paper strips with HPLC-electrospray tandem mass spectrometry. Clin Chem 50(11):2117–2124
11. Hartmann S, Okun JG, Schmidt C, Langhans CD, Garbade SF, Burgard P, Haas D, Sass JO, Nyhan WL, Hoffmann GF (2006) Comprehensive detection of disorders of purine and pyrimidine metabolism by HPLC with electrospray ionization tandem mass spectrometry. Clin Chem 52(6):1127–1137
12. Jaeken J, Van den Berghe G (1984) An infantile autistic syndrome characterised by the presence of succinylpurines in body fluids. Lancet 2(8411):1058–1061
13. van Gennip AH, Busch S, Elzinga L, Stroomer AE, van Cruchten A, Scholten EG, Abeling NG (1993) Application of simple chromatographic methods for the diagnosis of defects in pyrimidine degradation. Clin Chem 39(3):380–385
14. van Kuilenburg ABP, van Lenthe H, van Gennip AH (1999) A radiochemical assay for beta-ureidopropionase using radiolabeled N-carbamyl-beta-alanine obtained via hydrolysis of [2-(14)C]5, 6-dihydrouracil. Anal Biochem 272(2):250–253
15. Bierau J, Šebesta I (2022) Purine and pyrimidine disorders. In: Blau N, Vici CD, Ferreira CR, Vianey-Saban C, Karnebeek CDMV (eds) Physician's guide to the diagnosis, treatment, and follow-up of inherited metabolic diseases. Springer Nature, Cham, pp 191–212
16. Staufner C, Lindner M, Dionisi-Vici C, Freisinger P, Dobbelaere D, Douillard C, Makhseed N, Straub BK, Kahrizi K, Ballhausen D, la Marca G, Kölker S, Haas D, Hoffmann GF, Grünert SC, Blom HJ (2016) Adenosine kinase deficiency: expanding the clinical spectrum and evaluating therapeutic options. J Inherit Metab Dis 39(2):273–283
17. Duley JA, Henman MG, Carpenter KH, Bamshad MJ, Marshall GA, Ooi CY, Wilcken B, Pinner JR (2016) Elevated plasma dihydroorotate in Miller syndrome: Biochemical, diagnostic and clinical implications, and treatment with uridine. Mol Genet Metab 119(1-2):83–90
18. Bailey CJ (2009) Orotic aciduria and uridine monophosphate synthase: a reappraisal. J Inherit Metab Dis 32(Suppl 1):S227–S233
19. Fairbanks LD, Marinaki AM, Carrey EA, Hammans SR, Duley JA (2002) Deoxyuridine accumulation in urine in thymidine phosphorylase deficiency (MNGIE). J Inherit Metab Dis 25(7):603–604
20. van Kuilenburg ABP, Dobritzsch D, Meinsma JR, Haasjes J, Waterham HR, Nowaczyk MJM, Maropoulos GD, Hein G, Kalhoff H, Kirk JM, Baaske H, Aukett A, Duley JA, Ward KP, Lindqvist Y, van Gennip AH (2002) Novel disease-causing mutations in the dihydropyrimidine dehydrogenase gene interpreted by analysis of the three-dimensional protein structure. Biochem J 364(Pt 1):157–163
21. van Kuilenburg ABP, Dobritzsch D, Meijer J, Meinsma R, Benoist JF, Assmann B, Schubert S, Hoffmann GF, Duran M, de Vries MC, Kurlemann G, Eyskens FJM, Greed L, Sass JO, Schwab KO, Sewell

AC, Walter J, Hahn A, Zoetekouw L, Ribes A, Lind S, Hennekam RCM (2010) Dihydropyrimidinase deficiency: Phenotype, genotype and structural consequences in 17 patients. Biochim Biophys Acta 1802(7-8):639–648
22. Nakajima Y, Meijer J, Dobritzsch D, Ito T, Zhang C, Wang X, Watanabe Y, Tashiro K, Meinsma R, Roelofsen J, Zoetekouw L, van Kuilenburg ABP (2017) Dihydropyrimidinase deficiency in four East Asian patients due to novel and rare DPYS mutations affecting protein structural integrity and catalytic activity. Mol Genet Metab 122(4):216–222
23. van Kuilenburg AB, Dobritzsch D, Meijer J, Krumpel M, Selim LA, Rashed MS, Assmann B, Meinsma R, Lohkamp B, Ito T, Abeling NG, Saito K, Eto K, Smitka M, Engvall M, Zhang C, Xu W, Zoetekouw L, Hennekam RC (2012) β-ureidopropionase deficiency: phenotype, genotype and protein structural consequences in 16 patients. Biochim Biophys Acta 1822(7):1096–1108
24. Nakajima Y, Meijer J, Dobritzsch D, Ito T, Meinsma R, Abeling NG, Roelofsen J, Zoetekouw L, Watanabe Y, Tashiro K, Lee T, Takeshima Y, Mitsubuchi H, Yoneyama A, Ohta K, Eto K, Saito K, Kuhara T, van Kuilenburg AB (2014) Clinical, biochemical and molecular analysis of 13 Japanese patients with beta-ureidopropionase deficiency demonstrates high prevalence of the c.977G > A (p.R326Q) mutation [corrected]. J Inherit Metab Dis 37(5):801–812
25. Dobritzsch D, Meijer J, Meinsma R, Maurer D, Monavari AA, Gummesson A, Reims A, Cayuela JA, Kuklina N, Benoist JF, Perrin L, Assmann B, Hoffmann GF, Bierau J, Kaindl AM, van Kuilenburg ABP (2022) β-Ureidopropionase deficiency due to novel and rare UPB1 mutations affecting pre-mRNA splicing and protein structural integrity and catalytic activity. Mol Genet Metab 136(3):177–185
26. van Kuilenburg ABP, van Lenthe H, Tromp A, Veltman PCJ, van Gennip AH (2000) Pitfalls in the diagnosis of patients with a partial dihydropyrimidine dehydrogenase deficiency. Clin Chem 46(1):9–17
27. van Kuilenburg ABP, van Lenthe H, van Gennip AH (1999) Radiochemical assay for determination of dihydropyrimidinase activity using reversed-phase high-performance liquid chromatography. J Chromatogr B Biomed Sci Appl 729(1–2):307–314
28. van Kuilenburg ABP, Meinsma R, Beke E, Assmann B, Ribes A, Lorente I, Busch R, Mayatepek E, Abeling NGGM, van Cruchten A, Stroomer AEM, van Lenthe H, Zoetekouw L, Kulik W, Hoffmann GF, Voit T, Wevers RA, Rutsch F, van Gennip AH (2004) {beta}-Ureidopropionase deficiency: an inborn error of pyrimidine degradation associated with neurological abnormalities. Hum Mol Genet 13(22):2793–2801
29. van Kuilenburg ABP, Zoetekouw L (2005) Determination of thymidine phosphorylase activity by a non-radiochemical assay using reversed-phase high-performance liquid chromatography. J Chromatogr B Analyt Technol Biomed Life Sci 820(2):269–273
30. Jacomelli G, Micheli V, Peruzzi L, Notarantonio L, Cerboni B, Sestini S, Pompucci G (2002) Simple non-radiochemical HPLC-linked method for screening for purine metabolism disorders using dried blood spot. Clin Chim Acta 324(1–2):135–139
31. van Kuilenburg AB, Zoetekouw L, Meijer J, Kuijpers TW (2010) Determination of adenosine deaminase activity in dried blood spots by a nonradiochemical assay using reversed-phase high-performance liquid chromatography. Nucleosides Nucleotides Nucleic Acids 29(4–6):461–465
32. van Kuilenburg AB, Zoetekouw L, Meijer J, Kuijpers TW (2010) Identification of purine nucleoside phosphorylase deficiency in dried blood spots by a non-radiochemical assay using reversed-phase high-performance liquid chromatography. Nucleosides Nucleotides Nucleic Acids 29(4–6):466–470

Peroxisomal Biomarkers

8

Frédéric M. Vaz and Ronald J. A. Wanders

Contents

List of Abbreviations

ACOX1/2/3	Acyl-CoA oxidase 1/2/3
AMACR	α-methylacyl-CoA racemase
BCFA	Branched-chain fatty acids
BHT	2,6-di-*tert*-butyl-4-hydroxy-toluene
DBP	D-bifunctional protein
DHCA	Dihydroxycholestanoic acid 3α,7α-dihydroxy-5β-cholestanoic acid

F. M. Vaz (✉) · R. J. A. Wanders
Laboratory Genetic Metabolic Diseases, Department of Clinical Chemistry and Pediatrics, Emma Children's Hospital, Amsterdam UMC Location University of Amsterdam,
Amsterdam, The Netherlands
e-mail: f.m.vaz@amsterdamumc.nl; r.j.wanders@amsterdamumc.nl

N. Blau, F. M. Vaz (eds.), *Laboratory Guide to the Methods in Biochemical Genetics*,
https://doi.org/10.1007/978-3-031-58819-8_8

ESI	Electrospray ionization
GC	Gas chromatography
LC	Liquid chromatography
m/z	Mass over charge ratio
MRM	Multi-reaction monitoring
MS	Mass spectrometry
MTBSTFA	*N*-methyl-*N*-(tert-butyldimethylsilyl) trifluoroacetamide
PBD	Peroxisome biogenesis disorders
RCDP	Rhizomelic chondrodysplasia punctata
SCPx	Sterol-carrier protein X
THCA	Trihydroxycholestanoic acid = 3α,7α,12α-trihydroxy-5β-cholestanoic acid
UPLC	Ultra-performance liquid chromatography
VLCFA	Very long-chain fatty acids
ZSDs	Zellweger spectrum disorders

8.1 Introduction

Peroxisomes, like mitochondria and the endoplasmic reticulum, are subcellular organelles that are involved in many degradative and synthetic processes. Peroxisomes are made in a process called peroxisome biogenesis which encompasses the formation of peroxisomal membranes, the import of peroxisomal membrane and matrix proteins, proliferation of peroxisomes, and finally degradation of peroxisomes. The key features of peroxisome biogenesis have been resolved in the last decades which include the identity and functional role of the so-called PEX proteins defined as proteins directly involved with peroxisome biogenesis [1]. Peroxisomal metabolism involves different metabolic pathways localized in human peroxisomes which are catalyzed by a distinct set of enzyme proteins unique to peroxisomes now amounting to more than 100 [1]. Relevant for this chapter (also see Figs. 8.1 and 8.2):

1. *Fatty acid β-oxidation*: oxidation of very long-chain fatty acids (VLCFA), 2-methyl branched-chain fatty acids including pristanic acid, and the C_{27}-bile acid intermediates di- and trihydroxycholestanoic acid (DHCA and THCA).
2. *Ether lipid synthesis*: synthesis of 1-alkyldihydroxyacetone phosphate as the building block for ether lipids, including plasmalogens.
3. *Fatty acid α-oxidation*: oxidation of 3-methyl branched-chain fatty acids, notably phytanic acid by shortening fatty acids by one carbon.
4. *Glyoxylate detoxification*: glyoxylate is a toxic molecule because of its oxidation to oxalate. The peroxisomal enzyme alanine glyoxylate aminotransferase (AGT) converts glyoxylate into glycine.

Like other organelles, the peroxisome contains different transport systems to transport substrates, co-factors, and products in and out peroxisomes. These include ABCD1 for the transport of VLCFA and ABCD3 for branched-chain fatty acids, DHCA and THCA. Recently, tethering proteins including acyl-CoA binding domain containing 5 (ACBD5) have been identified to play an important role in metabolite transfer by bringing organelles physically closer together, making these relevant for peroxisomal pathology.

In this chapter we describe the measurement of different peroxisomal biomarkers, namely very long-chain fatty acids, phytanic and pristanic acid, the relatively new biomarker C26:0-lysoPC, and plasmalogens. We focus on the plasmalogen analysis and do not include the measurement of other long-chain fatty acids as described in the previous version of this book as these are not valuable as peroxisomal biomarkers [2]. For the measurement of two other relevant peroxisomal biomarkers, bile acids and pipecolic acid, we refer the reader to elsewhere in this book Chaps. 14 and [3].

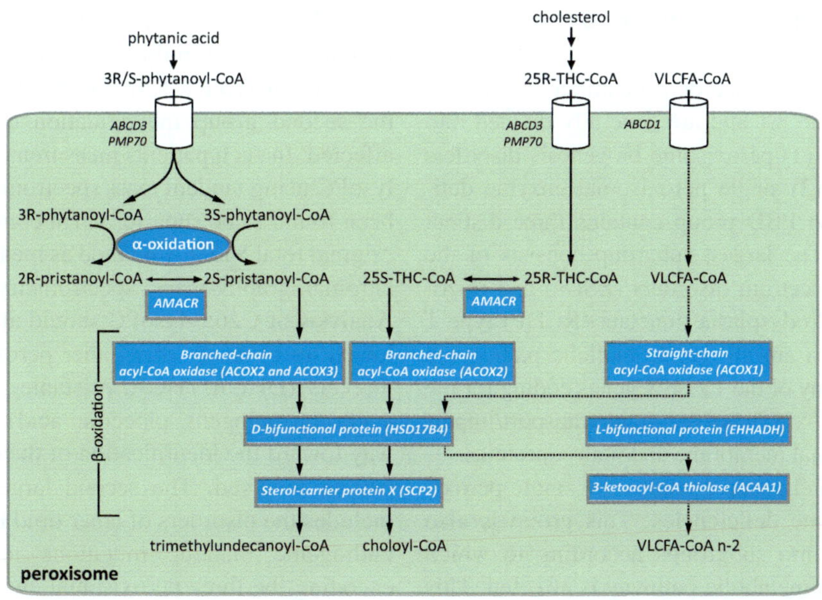

Fig. 8.1 Peroxisomal transport and α- and β-oxidation. Transport of branched-chain fatty acids into the peroxisomes is mediated by ABCD3 (PMP70) whereas VLCFA are transported via ABCD1. Oxidation is mediated by three different acyl-CoA oxidases (ACOX 1, 2, and 3), two bifunctional enzymes (EHHADH and HSD17B4) catalyzing the second and third step of peroxisomal β-oxidation, and finally two thiolases (ACAA1 and SCP2). All these enzymes have different substrate specificities depending on the type of fatty acids that is metabolized. α-methylacyl-CoA racemase (AMACR) is needed for the oxidation of 2-methyl branched-chain fatty acids which converts activated (2R)-pristanic acid (pristanoyl-CoA) and (2R)-C_{27}-bile acid intermediates (THC-CoA) into the corresponding (2S)-fatty acyl-CoAs, which is important because (2R)-acyl-CoAs are substrates for ACOX1, 2, and 3 whereas (2R)-acyl-CoAs are not

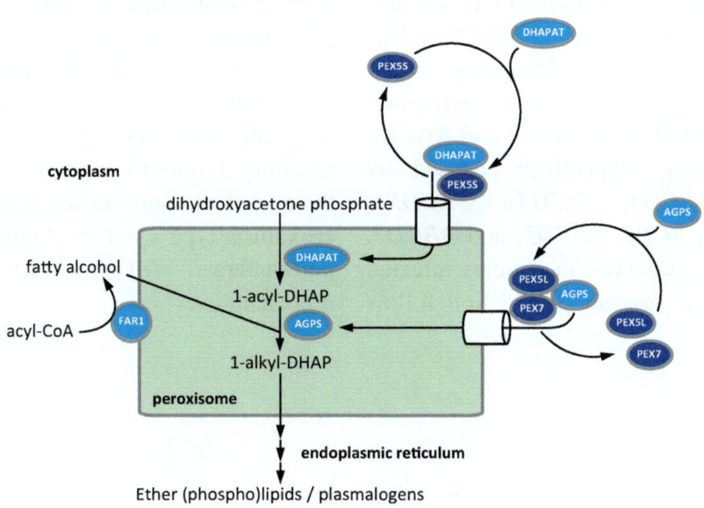

Fig. 8.2 Ether lipid biosynthesis in the peroxisome: dihydroxyacetone phosphate (DHAP)-acyltransferase (DHAPAT) binds via its peroxisome targeting signal (PTS) 1 to PEX5S which is the soluble receptor that delivers DHAPAT to the transport machinery at the peroxisomal membrane for import into the peroxisomal matrix. The import of the PTS2 protein alkylglycerone phosphate synthase (AGPS) needs both PEX5L and the PTS2 receptor PEX7 for import into the peroxisome. In the peroxisome, DHAPAT esterifies DHAP with a long-chain fatty acyl-CoA yielding 1-acyl-DHAP. Next, AGPS replaces the acyl unit by a fatty alcohol, produced by reduction of an acyl-CoA by fatty acyl-CoA reductase 1 (FAR1), resulting in the formation of 1-alkyl-DHAP. 1-Alkyl-DHAP is further converted to ether lipids, including plasmalogens, in the endoplasmic reticulum

8.2 Clinical Significance

The peroxisomal disorders identified so far are listed in Table 8.1 and are generally divided into two groups: (1) peroxisome biogenesis disorders (PBD) and (2) single peroxisomal enzyme deficiencies. The PBD group contains three distinct subgroups. The largest subgroup consists of the Zellweger spectrum disorders (ZSDs) and rhizomelic chondrodysplasia punctata (RCDP) type 1 and 5, which are all due to biallelic pathogenic variants in any of the 13 PEX genes coding for the different peroxins involved in the transport/import of peroxisomal membrane and matrix proteins.

The second group contains the single peroxisomal enzyme deficiencies. This group is also subdivided into subgroups according to which peroxisomal metabolic pathway is affected. This includes the disorders of peroxisomal β-oxidation, transport and tethering, ether lipid synthesis, phytanic acid α-oxidation, and glyoxylate detoxification.

X-linked adrenoleukodystrophy is most frequent among the single peroxisomal enzyme deficiencies followed by the peroxisomal β-oxidation disorders D-bifunctional protein deficiency, acyl-CoA oxidase 1 deficiency, α-methylacyl-CoA racemase (AMACR) deficiency, acyl-CoA oxidase 2 deficiency, sterol-carrier protein X deficiency, and the tethering/transport defects ACBD5 deficiency and ABCD3 (PMP70) deficiency, respectively, caused by mutations in *ABCD1*, *HSD17B4*, *ACOX1*, *AMACR*, *ACOX2*, *SCP2*, *ACBD5*, and *ABCD3*. The abnormalities observed in patients affected by a defect in peroxisomal β-oxidation follow logically from the functional role of each of the proteins involved in this degradative process and are listed in Table 8.1. In 5 of the 13 disorders in the second group the oxidation of VLCFA is affected. In such patients measurement of C26:0-lysoPC using tandem mass spectrometry has now been found to be superior when compared to the original total VLCFA method as measured by gas chromatography-mass spectrometry (GC-MS). Analysis of C26:0-lysoPC should always be followed up by analysis of other peroxisomal biomarkers (DHCA/THCA, pristanic and phytanic acid, plasmalogens, pipecolic acid) to guide the way toward the identification of the defect in the patient involved. The second largest subgroup includes the disorders of ether lipid biosynthesis. Pathogenic biallelic mutations in the genes encoding the three peroxisomal enzymes FAR1, GNPAT, and AGPS, which catalyze the first phase of ether lipid synthesis, have been described in patients affected by rhizomelic chondrodysplasia punctata and are called RCDP types 2, 3, and 4. Ether lipid synthesis is deficient in all three disorders as exemplified by the deficient plasmalogen levels in erythrocytes. The fourth subgroup includes the disorders of peroxisomal α-oxidation with Refsum disease due to phytanoyl-CoA 2-hydroxylase deficiency as the only disorder identified so far. The key pathognomonic biomarker for Refsum disease is phytanic acid which is usually extremely high in patients with levels reaching 1 mmol/l. The last subgroup involves the disorders of glyoxylate metabolism with hyperoxaluria type 1 due to alanine glyoxylate aminotransferase (*AGXT*) deficiency as sole representative.

Table 8.1 Metabolite alterations in different peroxisomal disorders

	Gene	VLCFA C26:0-lysoPC	Pristanic	Phytanic	Plasmalogens	Pipecolic acid	DHCA/THCA
Group 1: Peroxisome biogenesis disorders							
ZSDs [4]	*PEX*[a]	↑	N-↑[b]	N-↑[b]	N-↓	↑	N-↑
RCDP type 1 [5, 6]	*PEX7*	N	N-↑[b]	N-↑[b]	↓	N	N
RCDP type 5 [7]	*PEX5L*	N	N-↑[b]	↑	↓	N	N
Group 2: Single-enzyme deficiencies							
Peroxisomal β-oxidation							
Acyl-CoA oxidase 1 deficiency [8]	*ACOX1*	↑	N	N	N	N	N
Acyl-CoA oxidase 2 deficiency [9–11]	*ACOX2*	N	N	N	N	N	↑
D-bifunctional protein deficiency [12]	*HSD17B4*	↑	N-↑[b]	N-↑[b]	N	N	N-↑[b]
SCPx deficiency [13]	*SCP2*	N	N-↑[b]	N	N	N	N-↑[b,c]
AMACR deficiency [14]	*AMACR*	N	↑	N-↑[b]	N	N	↑
Transport and tethering							
X-linked adrenoleukodystrophy [15]	*ABCD1*	↑	N	N	N	N	N
70-kDa peroxisomal membrane protein (PMP70) [16]	*ABCD3*	N	N	N	N	N	↑
Acyl-CoA binding domain containing 5 [17]	*ACBD5*	↑	N	N	N	N	N
Ether lipid biosynthesis							
RCDP type 2 [5, 6]	*GNPAT*	N	N	N	↓	N	N
RCDP type 3 [5, 6]	*AGPS*	N	N	N	↓	N	N
RCDP type 4 [18]	*FAR1*	N	N	N	↓	N	N
Phytanic acid α-oxidation							
Refsum disease [19]	*PHYH*	N	N	↑	N	N	N
Glyoxylate detoxification							
Hyperoxaluria type 1 [20]	*AGXT*	N	N	N	N	N	N

[a]ZSDs can be caused by variants in 13 different *PEX* genes
[b]Levels may vary from normal to (markedly) elevated, which depends on the diet and/or age
[c]Abnormal bile alcohol glucuronides

8.3 Properties of Metabolites

Table 8.2 lists the properties of the peroxisomal biomarkers described in this chapter.

8.3.1 VLCFA/BCFA

VLCFA and BCFA are very hydrophobic molecules that are practically insoluble in water. VLCFA are defined as straight-chain fatty acids with 22 or more carbon atoms. VLCFA are either absorbed from the diet or synthesized endogenously by chain-elongation of long-chain fatty acids. VLCFA are incorporated in membrane phospholipids, for example in the skin in acylceramides where they are needed for skin-barrier formation, but also in glycerophospholipids in myelin in the nervous system [21].

BCFA relevant for peroxisomal disorders are phytanic acid (3,7,11,15-tetramethylhexadecanoic acid) and pristanic acid (2,6,10,14-tetramethylpenta decanoic acid) and are obtained primarily from the diet as phytanic acid (or formed from ingested phytol by oxidation) and pristanic acid, where the latter can also be formed by peroxisomal α-oxidation of phytanic acid [1, 19]. In the circulation, VLCFA and BCFA are not present as free fatty acids but are part of triglycerides, ceramides, and phospholipids and therefore not excreted in the urine.

8.3.2 C26:0-lysoPC

C26:0-lysoPC is a lysophosphatidylcholine with a hexacosanoic acid as side-chain and is one of the best biomarkers of VLCFA accumulation in the lipid fraction.

8.3.3 Plasmalogens

Plasmalogens are a subclass of glycero(phospho) lipids that contain a vinyl-ether bond at the *sn*-1 position. Plasmalogens play a crucial role in the functioning of the nervous system as part of the myelin sheath, regulate membrane fluidity, are involved in signal transduction, and have antioxidant properties to protect against oxidative stress [5, 6]. In circulation, the main lipid classes that contain these vinyl-ethers are phosphatidylethanolamine ether lipids but they also are present at lower levels in phosphatidylcholine ether lipids and neutral ether lipids. The majority of the radyl units on the *sn*-1 position of plasmalogens are C16:0 and C18:0 fatty alcohols (the vinyl-ether is not counted as a double bound as it is considered part of the backbone of the ether lipid).

Table 8.2 Properties of metabolites

Metabolite	Structural formula [monoisotopic mass (Da)]	Systematic name, trivial name	HMBD ID
C22:0	$C_{22}H_{44}O_2$ [340.33]	Docosanoic acid, behenic acid	HMDB0000944
C24:0	$C_{24}H_{48}O_2$ [368.37]	Tetracosanoic acid, lignoceric acid	HMDB0002003
C26:0	$C_{26}H_{52}O_2$ [396.40]	Hexacosanoic acid, cerotic acid	HMDB0002356
C26:0-lysoPC	$C_{34}H_{70}NO_7P$ [635.49]	1-O-hexacosanoic-*sn*-glycero-3-phosphocholine	HMDB0029205
Phytanic acid	$C_{20}H_{40}O_2$ [312.30]	3,7,11,15-tetramethylhexadecanoic acid	HMDB0000801
Pristanic acid	$C_{19}H_{38}O_2$ [298.29]	2,6,10,14-tetramethylpentadecanoic acid	HMDB0000795
C16:0/18:0-plasmalogens	Many	Ether (phospho)lipids composed of many different species with a 2-vinyl-ether bond on the *sn*-1 position of the glycerol backbone with a C16:0 or C18:0 radyl unit	Many

8.4 Methods

8.4.1 VLCFA, BCFA in Plasma/Serum or Cellular Homogenate

Quantitative analysis of VLCFA and BCFA using stable isotope dilution gas chromatography-mass spectrometry (GC-MS) analysis after derivatization with *N*-methyl-*N*-(tert-butyldimethylsilyl) trifluoroacetamide (MTBSTFA) is a robust and reliable method [22]. The addition of stable isotopes for C22:0, C24:0, C26:0, phytanic acid, and pristanic acid as internal standards allows the measurement of the total pool of VLCFA, phytanic acid, and pristanic acid after both acidic and alkaline hydrolysis, followed by hexane extraction. The hexane phase is washed once more, after which the sample is dried under nitrogen followed by addition of pyridine and MTBSTFA and heating of the samples at 80 °C for 30 min. The sample is subsequently dried again under nitrogen and taken up in hexane, followed by GC-MS analysis.

8.4.2 C26:0-lysoPC in Plasma, Blood Spot, or Cellular Homogenate

Measuring C26:0-lysoPC by stable isotope dilution UPLC-MS/MS is relatively straightforward. In the presence of a stable isotope-labeled internal standard, plasma or an aliquot of a cellular homogenate is deproteinized using methanol. Similarly, blood spots are extracted with methanol. After removal of protein precipitates by centrifugation, the extract is taken to dryness under a nitrogen stream and resuspended in methanol, followed by UPLC-MS/MS analysis.

8.4.3 Plasmalogens in Erythrocytes

The analysis of plasmalogens in erythrocytes uses the same sample preparation as that for the analysis of fatty acids as described in the previous edition of the book [2] and [23]. We focus solely on the analysis of total C16:0- and C18:0-plasmalogens relative (in %) to their corresponding fatty acids palmitic acid (C16:0) and stearic acid (C18:0) as biomarkers for peroxisomal ether lipid synthesis. To determine the plasmalogen levels, lipids in an erythrocyte hemolysate are hydrolyzed in methanolic HCl at 90 °C resulting in the cleavage of the vinyl-ether bond and concomitant formation of dimethyl acetals. Fatty acids are transmethylated to form fatty acid methyl esters. Both fatty acid methyl esters and dimethyl acetals (C16:0- and C18:0-dimethylacetal and C16:0 and C18:0 fatty acids) are extracted with hexane, concentrated, and analyzed using GC with flame ionization detection.

8.5 Pre-analytical

8.5.1 VLCFA/BCFA

Plasma is prepared by centrifugation for 10 min at 1000 × g of whole blood and stored at −20 °C until analysis. Both EDTA and heparin plasma can be used. Needed for analysis: 50 μl. Serum is not frequently used but is also possible if reference ranges have been determined for this matrix. For adequate and meaningful measurement of VLCFA, samples should be collected before the first meal of the day after an overnight fast. As peanut or peanut food derivatives contain high levels of VLCFA, they cause elevated VLCFA levels up to 12 h after the meal (especially around 3–4 h) [24], peanut-containing food should be avoided if samples are taken without an overnight fast. Analyses of VLCFA in plasma are usually combined with that of BCFA, which is less diet-dependent. No particular precautions need to be taken for the analysis of BCFA and the same sample amount (50 μl of plasma) is needed for the assay.

Both VLCFA and BCFA are stable compounds that are not likely lost by oxidation given their lack of desaturations and therefore freezing of the plasma is sufficient until analysis. The low levels of especially C26:0 and the presence of VLCFA in waxes, (skin)grease, and skin-crèmes do make it necessary to meticulously clean reusable glassware and make sure that clean disposables are used throughout the analysis.

8.5.2 C26:0-lysoPC

C26:0-lysoPC is a cellular metabolite of VLCFA (of cerotic acid, or hexacosanoic acid, C26:0), but this biomarker is not sensitive to short-term effects of the diet. For C26:0-lysoPC both plasma (needed for analysis: 20 µl), blood spot (needed for analysis: one 1/4 in. blood spot, 1/8 in. possible) or cellular homogenates (100–200 µg protein equivalent in homogenate) can be used.

Contamination of external sources as stated above for C26:0 is less relevant—if at all—for C26:0-lysoPC as this molecule is measured without hydrolysis and is not present in aforementioned possible contaminants.

8.5.3 Plasmalogens

One milliliter of EDTA-blood is sufficient for the analysis. If only heparinized blood is available, this can be used too. Blood should be transported to the laboratory within 1 day to allow separation and washing of erythrocytes. Yet, as the C16:0- and C18:0-plasmalogen content is calculated as a ratio over the corresponding fatty acids, even older blood or even hemolyzed blood can be used for an (less precise because of possible plasma contribution where dietary fatty acids can be present!) estimation of the plasmalogen levels. In whole or centrifuged (EDTA) blood, plasmalogens are relatively stable. In our hands, centrifuged blood samples are kept at 4 °C for 3 weeks and can still be used for plasmalogen determination/estimation if required during this period. Erythrocyte plasmalogen levels are not influenced by dietary intake which explains why dietary measures nor fasting samples are not required to perform the analysis. The EDTA blood is centrifuged at 1000 × g after which the plasma and buffy coat is removed (or used for other analyses) and the erythrocytes pellet is washed twice with 0.9% NaCl. After removal of the saline, cells are diluted 1:1 with 0.9% NaCl and stored in a freshly 2,6-di-*tert*-butyl-4-hydroxytoluene (BHT)-treated Eppendorf vial (see Sect. 8.6) at least overnight to make the hemolysate.

8.6 Reagents and Chemicals

All reagents are minimally of analytical grade and preferably ultrapure LC-MS grade, especially for the tandem MS determination of C26:0-lysoPC.

8.6.1 VLCFA/BCFA

See Table 8.3 for the standards/stable isotope-labeled internal standards.

Other materials/chemicals needed: acetonitrile, 25% and 37% HCl, sodium hydroxide (NaOH),

Table 8.3 VLCFA/BCFA standards/stable isotope-labeled internal standards

Metabolite[a]	Name/label	HMDB ID
C22:0	Docosanoic acid	HMDB0000944
C24:0	Tetracosanoic acid	HMDB0002003
C26:0	Cerotic acid	HMDB0002356
Phytanic acid	3,7,11,15-tetramethylhexadecanoic acid	HMDB0000801
Pristanic acid	2,6,10,14-tetramethylpentadecanoic acid	HMDB0000795
$^{2}H_{4}$-C22:0	3,3,5,5-$^{2}H_{4}$-docosanoic acid	
$^{2}H_{4}$-C24:0	3,3,5,5-$^{2}H_{4}$-tetracosanoic acid	
$^{2}H_{4}$-C26:0	3,3,5,5-$^{2}H_{4}$-hexacosanoic acid	
$^{2}H_{3}$-Phytanic acid	3-methyl-$^{2}H_{3}$-7,11,15-trimethyl pentadecanoic acid	
$^{2}H_{3}$-Pristanic acid	2-methyl-$^{2}H_{3}$-6,10,14-trimethyl pentadecanoic acid	

[a]Standards/stable isotope-labeled internal standards can be obtained from The Organic acid Synthesis Laboratory VU Medical Centre, Amsterdam, The Netherlands (https://www.vumc.com/departments/clinical-chemistry/metabolic-laboratory/organic-synthesis-laboratory.htm) or from other sources

Table 8.4 Stock and work solutions of (stable isotope-labeled) standards

Metabolite	Stock (mM)	Dilution factor	Work solution (μM)
($^2H_4$-)C22:0	1	20 ×	50
($^2H_4$-)C24:0	1	20 ×	50
($^2H_4$-)C26:0	0.5	500 ×	1
($^2H_3$-)Phytanic acid	0.25	60 ×	4
($^2H_3$-)Pristanic acid	0.25	250 ×	1

potassium hydroxide (KOH), *n*-Hexane, pyridine, toluene, MTBSTFA (Aldrich), ultrapure water (such as Millipore), nitrogen gas, 0.9% NaCl.

Solutions to be made:

- Stock and work solutions of (stable isotope-labeled) standards made according to Table 8.4.
- Acid hydrolysis reagent: Mix 1 part of 37% HCl with 23 parts of acetonitrile. Prepare freshly before each series of analyses (about 1 ml/sample) and keep at room temperature.
- Alkaline hydrolysis reagent: Dissolve 40 g of NaOH in approximately 50 ml water and dilute to 1000 ml with methanol. The solution is turbid (this is normal, it will precipitate to the bottom upon standing) and is stable for 1 year when stored at room temperature.
- 1 M KOH: Dissolve 56 g of KOH in 1 l of ultrapure water. The solution is kept at room temperature and can be used for 1 year.

8.6.2 C26:0-lysoPC

See Table 8.5 for reagents/chemicals needed for the C26:0-lysoPC assay.

Solutions to be made:

- Standard: 0.2 μM C26:0-lysoPC in methanol. Stability: 5 years at –20 °C
- Internal standard: 1 μM $^2H_4$-C26:0-lysoPC in methanol. Stability: 5 years at –20 °C
- Eluent A: 0.1 % (v/v) formic acid in water. Stability: 3 months at room temperature
- Eluent B: 0.1 % (v/v) formic acid in methanol. Stability: 1 year at room temperature

Table 8.5 Reagents/chemicals for C26:0-lysoPC determination

Metabolite/reagent	Vendor, product number
Methanol	Biosolve, BIS 136835
Formic acid (98-100%)	Merck, M264
C26:0-lysoPC	Avanti Polar Lipids, 855810P
$^2H_4$-C26:0-lysoPC	Avanti Polar Lipids, 860389P

Table 8.6 Reagents/chemicals for the plasmalogen assay

Metabolite/reagent	Vendor, product number
2,6-di-*tert*-butyl-4--hydroxytoluene (BHT)	Merck, 822021
18-methyl-C19:0 methyl ester	Larodan, 21-1918
Methanolic HCl (3M)	Merck 90964
C16:0-dimethylacetal	Sigma, H7391
Chloroform	Biosolve, 3480601
n-Hexane	Biosolve, 8290601
Ethanol	Supelco, 100983

8.6.3 Plasmalogens

Plasmalogen analysis can be combined with that of polyunsaturated fatty acids and for this we refer the reader to the previous version of this book [2]. We have chosen to only focus on peroxisomal markers and hence only describe the analysis of the plasmalogens. See Table 8.6 for reagents/chemicals needed for plasmalogen assay.

- Internal standard stock solution: weigh 33 mg of the 18-methyl-C19:0 methyl ester and dissolve it in 50 ml of chloroform (approximately 2 mmol/l, store at 6 °C).
- Internal standard working solution: dilute the stock solution 28-fold with chloroform. The final concentration is 72 μmol/l.
- BHT solution: dissolve 1.0 g BHT in 100 ml ethanol. Store the solution in a dark bottle at 6 °C. Stable for 1 year.
- BHT-treated Eppendorf vials: Prepare BHT-treated vials immediately before aliquoting the erythrocyte suspension. Pipet 100 μl of BHT solution into an Eppendorf vial and take to dryness under a nitrogen flow at 40 °C. These vials can be used to make and store the erythrocyte hemolysate at –20 °C.

8.7 Instrumentation

8.7.1 VLCFA/BCFA

- GC-MS: Agilent 7890B GC and MSD 5977A
- GC column: CP-sil 19CB, 25 m x 0.25 mm, film thickness 0.20 μm (Varian, CP7712)
- Glass 5 ml tube:
  - DWK LifeSciences DURAN® 5 ml glass tubes with GL14 Thread (VWR, 391-4022)
  - DWK LifeSciences DURAN® Screw Cap 12 mm diameter with PTFE protected seal (VWR, 201-0000)
- Centrifuge: Hettich Rotina 148s (for large tubes)
- Glass 4 ml vials:
  - Vial (BGB Analytik 130400)
  - Black screw cap with silicone protected seal (BGB Analytik, 130801)
- Sample concentrator with heating (N_2 as drying gas) (Techne, FCS400D, Techne, DB.3D)
- Oven (for acid/alkalic hydrolysis and derivatization): Memmert UN30 or equivalent
- GC vials:
  - Glass crimp neck vial, 1.5 ml (BGB Analytik, 110400)
  - 150 μl Conical Glass Insert (total volume 200 μl), 31 × 5 mm, Tip: 15 mm (BGB Analytik 110500)
  - 11 mm Crimp Cap (silver) with Septa Natural Rubber/TEF (BGB Analytik, 11031201)
  - 11 mm Electronic crimper for aluminum caps (BGB Analytik, 297041)

8.7.2 C26:0-lysoPC

- Vortex mixer (Labo Tech Mix TM 01)
- Ultrasonic bath (Branson 3510)
- Sample concentrator with heating (N_2 as drying gas) (Techne, FCS400D, Techne, DB.3D)
- UPLC-MS/MS vial 500 μl (Gilson, 450830) + cap (Gilson, 450831)
- Glass HPLC crimp neck vial, 1.5 ml (BGB Analytik, 110400)
- UPLC-MS/MS analytical column: Kinetex C8, 50 × 2.1 mm, 2.6 micron (Phenomenex, 00B-4497-AN)
- High speed Eppendorf centrifuge: Hettich 220R or equivalent
- UPLC with column oven
- Triple-quad MS: Waters Xevo TQ-S micro or equivalent

8.7.3 Plasmalogens

- Oven (for methylation): Memmert UN30 or equivalent
- GC: Agilent 7890B or equivalent with flame ionization detector
- GC columns:
  - Capillary pre-column: DB-17, 15 m × 0.25 mm, film thickness 0.25 μm (Agilent: J&W Scientific 122-1712)—*length used: 1.90 m*
  - Capillary column: J&W HP-FFAP, 25 m × 0.2 mm, film thickness 0.33 μm (Agilent, 19091F-102)
- Sample concentrator with heating (N_2 as drying gas) (Techne, FCS400D, Techne, DB.3D)
- Centrifuge: Hettich Rotina 148s (for large tubes)
- VWR Scientific DVX-2500 Digital Multi Tube Vortexer
- Glass 5 ml tube:
  - DWK LifeSciences DURAN® 5 ml glass tubes with GL14 Thread (VWR, 391-4022)
  - DWK LifeSciences DURAN® Screw Cap 12 mm diameter with PTFE protected seal (VWR, 201-0000)
- Glass 10 ml conical tube: 90 × 16 mm (Fisher Emergo, 350307)
- GC vials:
  - Glass crimp neck vial, 1.5 ml (BGB Analytik, 110400)
  - 150 μl Conical Glass Insert (total volume 200 μl), 31 × 5 mm, Tip: 15 mm (BGB Analytik, 110500)
  - 11 mm Crimp Cap (silver) with Septa Natural Rubber/TEF (BGB, 11031201)
  - 11 mm Electronic crimper for aluminum caps (BGB Analytik, 297041)

8.8 Analytical Procedure

8.8.1 VLCFA/BCFA in Plasma/Serum or Cellular Homogenate

Cellular Homogenate

1. Resuspend cellular pellet in appropriate volume of 0.9% NaCl to achieve a protein concentration of about 4 mg/ml and place on ice
2. Sonicate using a tip-sonicator 1 × 20 seconds at 8 watt on ice
3. Determine the protein concentration (we use the bicinchoninic acid assay)
4. Pipet 100 µl of the homogenate into a 5 ml glass vial and proceed with step 5 of the plasma/serum method

Plasma/Serum

- *Sample preparation*
  - Thaw plasma or serum and internal standard mixtures on ice
  - Set the oven at 110 °C
  - Collect and label 5 ml glass tubes equipped with screw caps and Teflon liner
  - Transfer 50 µl plasma/serum into a 5 ml glass tube
  - Add 50 µl internal standard work solution
- *Acid hydrolysis*
  - Add 1 ml acid hydrolysis reagent. Cap the tubes tightly, mix carefully, and put the tubes in the oven for 45 (±5) min
  - Remove the tubes from the oven and allow to cool down to room temperature.
- *Alkaline hydrolysis*
  - Remove the screw caps and add 1 ml of alkaline hydrolysis reagent
  - Cap the tubes tightly, mix carefully, and put the tubes in the oven again for 45 (±5) min at 110 °C
  - Remove the tubes from the oven and allow to cool down to room temperature.
- *Extraction of fatty acids—1*
  - Remove the screw caps and add 250 µl of 25% hydrochloric acid
  - Add 2 ml hexane, screw the tubes tightly, and mix for about 60 s by hand
  - Centrifuge for 10 min at 1000 × g at room temperature to allow phase separation
  - Collect the upper hexane layer with the aid of a glass Pasteur pipette and transfer the sample into another glass 5 ml tube with screw cap while not disturbing the layer between the two phases
- *Alkaline wash step*
  - Add 1.75 ml 1 M KOH to the tubes with the hexane layer, screw the tubes tightly, and mix the samples by hand for 1 min
  - Centrifuge for 10 min at 1000 × g at room temperature to allow phase separation
  - Remove the upper hexane phase with the aid of a glass pipette attached to a liquid suction filtration vacuum pump and keep the water phase
- *Extraction of fatty acids—2*
  - Add 300 µl of 25% hydrochloric acid (HCl)
  - Add 2 ml hexane and mix the samples by hand
  - Centrifuge for 10 min at 1000 × g at room temperature to allow phase separation
  - Transfer the upper hexane layer in a glass 4 ml vial with the aid of a glass Pasteur pipette
  - Now also include the calibration standards with internal standards prepared as described below in glass 4 ml vials
  - Take the sample to dryness under a nitrogen stream in a heating block set at 40 °C
- *Derivatization*
  - Dissolve the residue in 25 µl pyridine and add 25 µl MTBSTFA reagent
  - Cap the tubes tightly and allow derivatization at 80 °C for 30 min in the oven
  - Take the samples to dryness under nitrogen in a heating block set at 40 °C
  - Dissolve the residue in 100 µl hexane and transfer the contents into a 1 ml crimp top vial with insert and cap
  - Samples are now ready for analysis by GC-MS

Calibration Standards

Prepare the calibration curve according to Table 8.7 in a 4 ml glass vial. These are to be included in step 23 onward of the plasma/serum procedure listed above.

Table 8.7 Preparation of the standard curve for VLCFA/BCFA

Standard	Work solution (# μl)	Internal standard (# μl)
1	0	50
2	25	50
3	50	50
4	100	50

GC-Analysis

Inject 1 μl into the GC-MS system in the splitless mode, use helium as carrier gas with a constant flow of 1.5 ml/min. Use the following program GC program:

1. Starting temperature is 60 °C which is maintained for 1 min after injection
2. Increase temperature to 240 °C at a rate of 30 °C/min (6 min)
3. Increase temperature to 270 °C at a rate of 10 °C/min (3 min)
4. Increase temperature to 300 °C at a rate of 4 °C/min (7 min 30 s)
5. Maintain temperature at 300 °C for 5 min

Total runtime will be around 25 min.

Figure 8.3 shows the mass fragmentogram of a control, a Refsum patient, and a patient with Zellweger syndrome.

MS Settings

See Table 8.8 for MS settings of the GC-MS for the VLCFA/BCFA assay.

8.8.2 C26:0-lysoPC in Plasma, Blood Spot, or Cellular Homogenate

Important: Use a positive displacement pipette when indicated. Using air displacement pipettes will severely affect your precision/accuracy.

- To prepare the standard curve pipet the following amount of 0.2 μM C26:0-lysoPC solution directly in 2 ml glass HPLC vials using a positive displacement pipet. The amount of pmol that is pipetted from the 0.2 μM C26:0-lysoPC standard is chosen such that the virtual "concentration" in 20 μl corresponds to the amounts listed in the last column of Table 8.9 and to the amount of plasma that is added.
- Add 20 μl of the 1 μM $^{2}H_{4}$-C26:0-lysoPC (20 pmol) using a positive displacement pipette
- Add 0.5 ml of methanol and mix by vortexing for 10 s
- Take to dryness under nitrogen flow at 40 °C
- Resuspend residue in 100 μl Methanol
- Transfer to UPLC-MS vial

For each matrix use the appropriate procedure listed below.

Plasma (or Serum)

1. Make sure the samples have thawed properly and are at room temperature and mix by vortexing for 10 s
2. Pipette 20 μl of plasma in a 1.5 ml Eppendorf cup
3. Add 20 μl of the 1 μM $^{2}H_{4}$-C26:0-lysoPC (20 pmol) using a multipipette or positive displacement pipette
4. Add 0.5 ml Methanol and mix by vortexing for 10 s
5. Incubate for 5 min in the ultrasonic bath
6. Centrifuge for 5 min at 16000 × g at room temperature
7. Transfer the extract in a 2 ml glass HPLC vial
8. Take to dryness under nitrogen flow at 40 °C
9. Resuspend residue in 100 μl Methanol
10. Transfer to UPLC-MS vial

Blood Spots

1. Punch a ¼ in. blood spot and add to 2 ml Eppendorf cup. This represents 10 μl of blood
2. Add 20 μl of the 1 μM $^{2}H_{4}$-C26:0-lysoPC (20 pmol) using a multipipette or positive displacement pipette
3. Add 0.5 ml Methanol and mix by vortexing for 10 s
4. Incubate for 5 min in the ultrasonic bath
5. Transfer the extract in a 2 ml glass HPLC vial
6. Take to dryness under nitrogen flow at 40 °C
7. Resuspend residue in 100 μl Methanol
8. Transfer to UPLC-MS vial

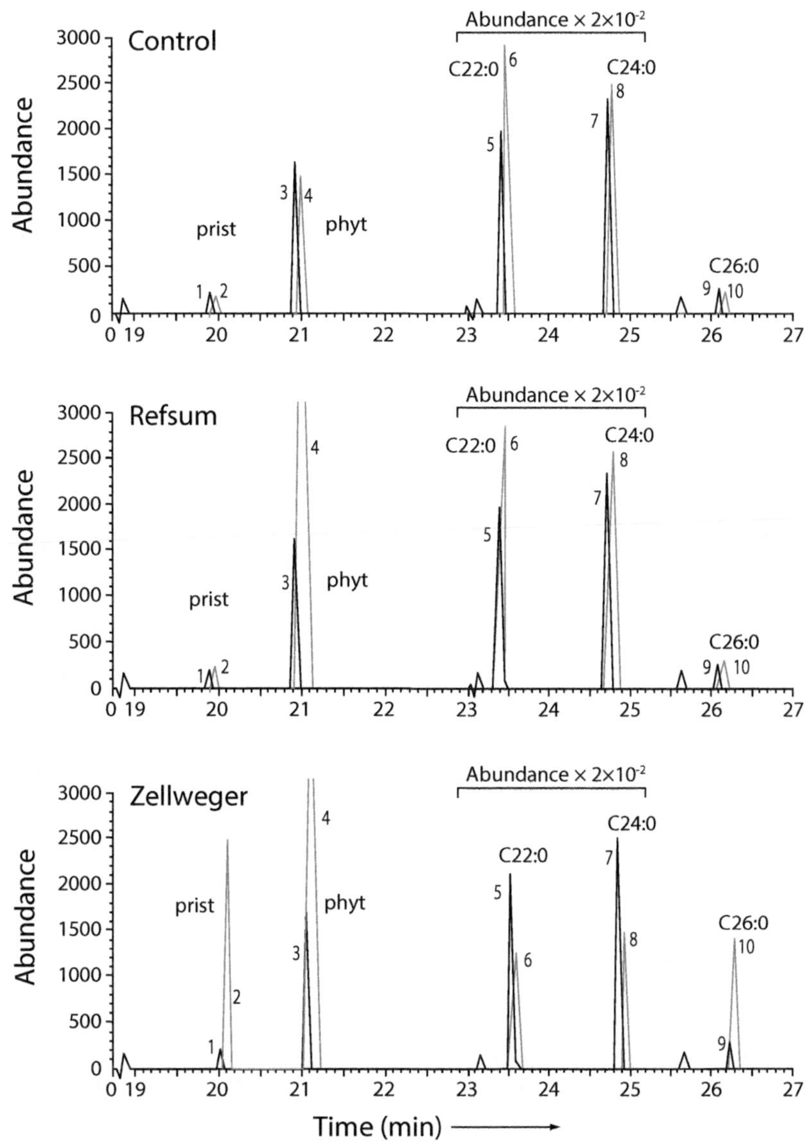

Fig. 8.3 Mass fragmentogram of the plasma very long-chain fatty acids, phytanic acid, and pristanic acid in a control, a patient with Refsum disease, and a patient with a peroxisome biogenesis defect (Zellweger syndrome). Plots represent the (m-57)-fragments of the tertiarybutyl-dimethylsilyl derivatives. Numbered peaks represent: (1) $^2H_3$-pristanic acid; (2) native pristanic acid; (3) $^2H_3$-phytanic acid; (4) native phytanic acid; (5) $^2H_4$-C22:0 fatty acid; (6) native C22:0 fatty acid; (7) $^2H_4$-C24:0 fatty acid; (8) native C24:0 fatty acid; (9) $^2H_4$-C26:0 fatty acid; (10) native C26:0 fatty acid. prist, pristanic acid; phyt, phytanic acid

Cellular Homogenate

1. Determine the protein concentration of the protein homogenate (we use the bicinchoninic acid assay)
2. Pipette 50 μl of homogenate (100–200 μg protein equivalent) in a 1.5 ml Eppendorf cup
3. Proceed as for plasma form step 3 onward (see above)

Liquid Chromatography for UPLC-MS/MS

Two μl of the extract is injected into the UPLC system. Oven temperature is 50 °C. Below is the gradient table for the Waters Acquity used in our laboratory (Table 8.10). This should be fine-tuned to the system that is used.

Table 8.8 GC-MS settings for the VLCFA/BCFA assay

Metabolite	M-57$^+$ ion (m/z)	Other parameters
C22:0	397.40	• M-57$^+$ ion is monitored using SIM mode • Electron impact mode • Dwell time: 100 msec • Injector temperature 300°C • Interface temperature 290°C
($^2H_4$-)C22:0	401.40	
C24:0	425.40	
($^2H_4$-)C24:0	429.40	
C26:0	453.40	
($^2H_4$-)C26:0	457.40	
Phytanic acid	369.30	
($^2H_3$-)Phytanic acid	372.30	
Pristanic acid	355.30	
($^2H_3$-)Pristanic acid	358.30	

Table 8.9 Preparation of the standard curve for plasma, blood spot, and homogenate

Standard	0.2 µM C26:0-lysoPC (# µl)	Amount C26:0-lysoPC (pmol)	Final concentration in 20 µl (nM)
1	0	0	0
2	5	1.0	50
3	10	2.0	100
4	25	5.0	250
5	50	10.0	500
6	100	20.0	1000

Table 8.10 Gradient program for UPLC-MS/MS

Time	%A	%B	Flow (ml/min)	Curve
Initial	50	50	0.400	Initial
0.10	24	76	0.400	0.6
4.00	0	100	0.400	0.6
9.00	0	100	0.400	0.6
9.10	50	50	0.400	0.6
10.00	50	50	0.400	0.6

Mass Spectrometry for UPLC-MS/MS

ESI in the positive ion mode is used. Settings in Table 8.11 are settings for the Waters Xevo TQ-S micro.

8.8.3 Plasmalogens

1. One day before the analysis: wash erythrocytes as described in Sect. 8.5
2. Aliquot a minimum of 300 µl of the washed erythrocyte suspension in a BHT-treated Eppendorf vial and freeze at –20 °C for at least 16 h (overnight)

   If analysis needs to be performed the same day: freeze at –80 °C for at least 30 min
3. Thaw the erythrocyte hemolysates and control samples to room temperature
4. Preheat the stove at 90 °C
5. Prepare the 5 ml glass tubes and screw caps with Silicone/PTFE septa
6. Pipet 100 µl of the internal standard work solution into the 5 ml glass tube
7. Pipet 50 µl of (control) erythrocyte hemolysate into the 5 ml glass tube
8. Add 1 ml of 3M Methanolic HCl and firmly screw on the cap
9. Vortex for 5 s (multivortex can be used)
10. Incubate in the oven for 4 h at 90 °C
11. Let cool to room temperature (high inside pressure when hot!)
12. Add 2 ml of hexane, recap
13. Vortex for 10 s (multivortex can be used)

Table 8.11 MRM transitions and other mass spectrometric parameters

Metabolite	Parent (m/z)	Daughter (m/z)	Dwell (s)	Collision energy (eV)	Cone voltage (V)
C26:0-lysoPC	636.50	104.10	0.164	29.0	20.0
$^2H_4$-C26:0-lysoPC	640.50	104.10	0.64	29.0	20.0
Additional parameters: • Capillary 3.50 kV • Cone 10.50 V • Source temperature 150 °C • Desolvation temperature 350 °C • Cone gas flow 50 l/h • Desolvation gas flow 650 l/h					

14. Pipet the hexane layer into a clean glass 10 ml conical tube
15. Take to dryness under nitrogen flow WITHOUT heating
16. Dissolve the residue in 175 μl of hexane and pipet into a glass crimp neck vial with insert and cap
17. Inject 1 μl into the GC system in the splitless mode, use helium as carrier gas with an inlet pressure of 2.5 Bar (250 kPa), and use the following program:
    (a) Starting temperature is 50 °C which is maintained for 1.5 min after injection
    (b) Increase temperature to 190 °C at a rate of 30 °C/min (4 min and 40 s)
    (c) Maintain temperature at 190 °C for 5 min
    (d) Increase temperature to 230 °C at a rate of 8 °C/min (5 min)
    (e) Maintain this temperature until high-boiling compounds (cholesterol!) have eluted (about 30 min). Total runtime will be around 45 min.

An example chromatogram is provided in Fig. 8.4. GC specifics:

- The glass wool in the liner of the GC system should be replaced prior to each series and as dimethylacetals derived from plasmalogens are sensitive to decomposition in "active spots" in the injection system, the inertness of the injection system should be checked with every series by analyzing a "plasmalogen extract." In practice, we only analyze ten samples for plasmalogens and then change the glass wool for a new set of ten samples and repeat this until all samples have been measured.
- Injection of a blank hexane sample will show the presence of any high-boiling artifacts that have remained on the column since the previous series and emerge with the hexane blank.

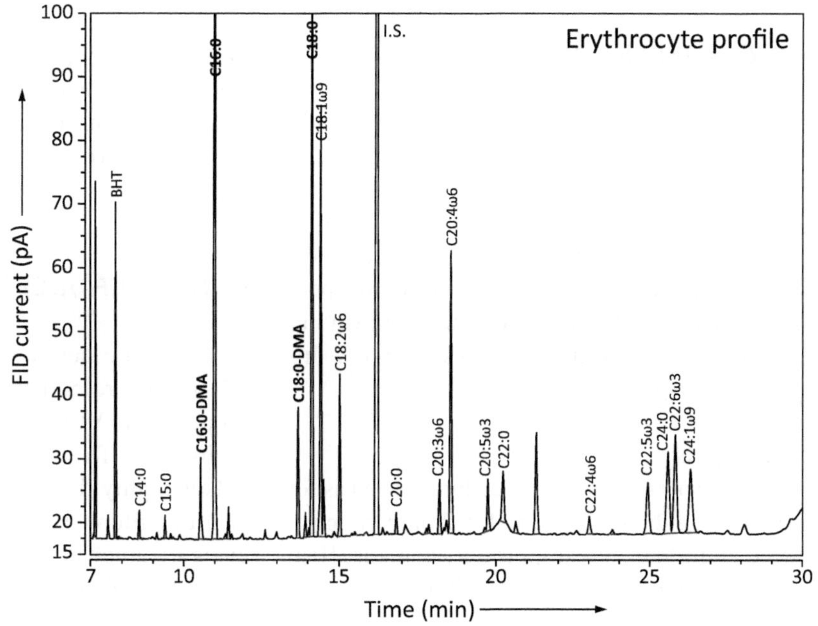

Fig. 8.4 Gas chromatography trace of essential fatty acid methyl esters and plasmalogen dimethylacetals in control erythrocytes. C16:0/C18:0-DMA elute before their corresponding fatty acid methyl esters (bold). I.S., internal standard

8.9 Calculation

8.9.1 VLCFA/BCFA

The abundance of the different analytes in the sample, controls, and standard curve is divided by the abundance of the corresponding internal standard and this "response" is interpolated on the standard curve using the concentrations mentioned in Table 8.4. Because of the possible contribution of endogenous third isotope of phytanic and pristanic acid to the abundance of ($^2H_3$-)Phytanic acid and ($^2H_3$-)Pristanic acid, respectively, the abundance of the BCFA internal standards is corrected according to the following formula:

- Abundance of ($^2H_3$-)BCFA—1.2 abundance of endogenous BCFA

GC-MS vendor software (in our case ChemStation or MassHunter, Agilent) always supports this type of calculation.

8.9.2 C26:0-lysoPC

Plasma (or Serum)
The abundance of the C26:0-lysoPC in the sample, controls, and standard curve is divided by the abundance of the internal standard $^2H_4$-C26:0-lysoPC and this "response" is interpolated on the standard curve using the concentrations mentioned in Table 8.9. MS vendor software (in our case TargetLynx, Waters) always supports this type of calculation.

Blood Spots
For blood spots, the same procedure is used as for plasma but a dilution factor of 2 should be introduced as the ¼ in. blood spots only contains 10 µl of blood instead of the assumed 20 µl in the standard curve.

Cellular Homogenate
For a cellular homogenate the penultimate column "Amount C26:0-lysoPC (pmol)" of Table 8.9 should be used on the x-axis of the standard curve to calculate the amount of pmol present in the 50 µl of homogenate that was used. Using the protein concentration the amount of C26:0-lysoPC per mg protein can be calculated.

8.9.3 Plasmalogens

The abundance of C16:0-DMA and C18:0-DMA is divided by C16:0 and C18:0, respectively, and multiplied by 100% to obtain the percentage of C16:0- and C18:0-plasmalogens.

8.10 Post-analytical (e.g., Chromatograms of Pathological Samples)

8.11 Reference Values

Reference ranges should be determined in the representative population using your own laboratory method, ranges below are provided as example based on our methods. For pipecolic acid and bile acid intermediates, see [3] and Chap. 14, respectively.

8.11.1 VLCFA/BCFA

As BCFA are derived from exogenous sources only (food), reference values increase with age as opposed to VLCFA where no age-dependency is observed. The reference range of VLCFA in cells is highly dependent on the medium that is used to culture the cells. In media with higher levels of glucose (like DMEM) VLCFA levels are higher, likely due to enhanced endogenous fatty acid synthesis and elongation. Reference ranges should therefore be determined for a specific medium and be checked with new batches of—especially—fetal calf serum (Tables 8.12 and 8.13).

Table 8.12 Reference ranges VLCFA/BCFA in plasma

Age range (days)	Phytanic acid (μM)	Pristanic acid (μM)	C22:0 (μM)	C24:0 (μM)	C26:0 (μM)	C24:0/C22:0	C26:0/C22:0
0–120	0.04–5.28	0–0.20	40–119	33–84	0.45–1.32	0.57–0.92	0.00–0.02
121–240	0.21–5.70	0–0.20					
241–365	0.38–4.40	0–0.50					
366–730	1.34–8.62	0–1.72					
731–43,800	0.49–9.88	0–1.60					

Table 8.13 Reference ranges VLCFA/BCFA in fibroblasts

Source (medium)	C22:0 (pmol/mg protein)	C24:0 (pmol/mg protein)	C26:0 (pmol/mg protein)	C24:0/C22:0	C26:0/C22:0
Fibroblasts (HAM-F10)	2.46–6.59	6.37–13.87	0.16–0.41	1.39–2.31	0.04–0.09
Fibroblasts (DMEM)	4.12–8.81	8.41–15.17	0.25–0.56	1.68–2.92	0.03–0.10

Table 8.14 Reference ranges (all ages) C26:0-lysoPC

Matrix	Concentration range
Plasma	29–88 nM
Blood spot	29–72 nM
Fibroblasts (cultured on HAM-F10)	2–14 pmol/mg protein

Table 8.15 Reference ranges (all ages) plasmalogens in erythrocytes (Table 8.15)

Metabolite ratio	DMA/FA × 100%
C16:0-DMA/C16:0	5.5–8.1
C18:0-DMA/C18:0	11.8–19.4

8.11.2 C26:0-lysoPC

The reference range of C26:0-lysoPC in cells is, like that of VLCFA, highly dependent on the medium use to culture the cells, see comment about this in previous VLCFA paragraph (Table 8.14).

8.11.3 Plasmalogens

8.12 Pathological Values

See Table 8.1 for accumulating or deficient metabolites in the different disorders. For interpretation and pathological values of bile acid intermediates, see Chap. 14. It is important to realize that depending on the disorder, the profile can vary considerably and in some disorders may even be without obvious abnormalities, especially for milder peroxisome biogenesis disorders and older patients where the biochemical abnormalities have been shown to (partially) resolve in time [25]. Hence, a strong clinical suspicion of peroxisomal disease should always be verified by genetic and fibroblast investigation, regardless of the outcome of plasma analyses.

8.12.1 VLCFA and C26:0-lysoPC

To screen for peroxisomal disorders it is not sufficient to only measure VLCFA and BCFA, as also mentioned in the discussion of the previous version of this book [26]. Ideally, in addition to VLCFA/BCFA, peroxisomal bile acid intermediates in plasma (see Chap. 14), pipecolic acid in plasma (see [3]), and plasmalogens in erythrocytes are measured but especially C26:0-lysoPC in plasma/blood spot should be measured and evaluated together. C26:0-lysoPC has become an invaluable parameter to ascertain possible accumulation of VLCFA in the lipid fraction and has proven to be a superior biomarker when compared to the classic VLCFA measurement [27]. This is especially relevant for ALD carriers where in the VLCFA analysis C26:0 and the C26:0/C22:0 ratio are normal in 15–20% of the cases [28]. In peroxisome biogenesis disorders, Zellweger spectrum disorders, C26:0-lysoPC also is elevated but in a very limited amount of cases borderline normal values were measured.

Table 8.16 Pathological values of VLCFA in ALD and Zellweger in plasma

Disorder	C22:0 (μM)	C24:0 (μM)	C26:0 (μM)	C24:0/C22:0	C26:0/C22:0
ALD carrier	41–107	47–102	1.11–2.93	0.84–1.33	0.02–0.04
ALD	32–76	44–115	2.00–4.12	1.06–1.83	0.04–0.10
Zellweger	9–33	20–45	2.90–12.22	1.29–2.13	0.11–0.63

Historical values from own laboratory, also see more recent comparison in [27]

Table 8.17 Pathological values of C26:0-lysoPC in ALD and Zellweger in plasma and blood spot

	C26:0-lysoPC (nM)	
Disorder	Plasma	Blood spot
ALD carrier	118–576	80–497
ALD	190–1004	224–1208
Zellweger	74–2485	28–3133
		124–2881

See [27] and [29]

See indication of pathological values in plasma for VLCFA and C26:0-lysoPC in ALD and Zellweger in Tables 8.16 and 8.17.

8.12.2 BCFA

Phytanic and pristanic acid results should always be interpreted together. In general, patients with a peroxisome biogenesis defect/Zellweger spectrum disorders or a single-enzyme deficiency of the peroxisomal β-oxidation system will show increases of both BCFA in variable ratios, except for ALD or acyl-CoA oxidase 1 deficiency having normal branched-chain fatty acid levels.

Patients with Refsum disease may have extremely high phytanic acid levels (higher than 200 μmol/l up to more than 1000 μmol/l), whereas pristanic acid is low (<1 μmol/l) because of the phytanoyl-CoA hydroxylase deficiency, which is also expressed in a very low phytanic/pristanic ratio (<0.05, reference range: 0.05–0.4). Less pronounced phytanic acid elevations are observed in RCDP type 1 patients, which applies to both the classical form and the variant forms. Values may range from 200 to 900 μmol/l, somewhat depending on age. In neonatal RCDP patients of less than 1 week old, plasma phytanic acid levels (0.7–5.8 μmol/l) were normal. This increases in 2–3-week-old RCDP patients to phytanic acid levels of 9.1–13.2 μmol/l but this obviously depends on the intake of BCFA and its precursors.

In AMACR deficiency, the phytanic acid levels are marginally elevated (around 20 μmol/l) but pristanic acid levels are 15–50-fold elevated, as well as the pristanic/phytanic ratio (>1.3, reference range: 0.05–0.4). These patients also have clearly elevated levels of the C_{27}-bile acid intermediates DHCA and THCA [14].

In SCPx deficiency, phytanic acid levels can be elevated somewhat, but pristanic acid, the substrate for the branched-chain β-oxidation system, is clearly elevated [13, 30]. As SCPx is also involved in the oxidation of THCA and DHCA, these can also accumulate as well as bile alcohol glucuronides (see Chap. 14).

When patients with elevated levels of BCFA undergo a phytol/phytanic acid restricted diet, the assay can be used for therapy control.

8.12.3 Pipecolic Acid

Pipecolic acid is a biomarker for generalized loss of peroxisomal function in peroxisome biogenesis disorders that sometimes is the only elevation in plasma that is detectable. Yet, the biomarker is not specific for peroxisomal disorders as it also is elevated in pyridoxine-responsive seizures (see [3] for methodology and reference ranges).

8.12.4 Plasmalogens

Erroneously low red cell levels can be encountered when the transmethylation process has not been completed. Breaking the ether lipid bond of the plasmalogens requires more energy than hydrolysis of the fatty acid esters. Evaluation of the plasmalogen levels should not be done after a blood transfusion. Donor erythrocytes will be present for up to 120 days following a transfusion.

Plasmalogen levels are low in PBD and in single-enzyme deficiencies affecting enzymes involved in ether lipid synthesis (see Table 8.1). In milder deficiencies, C16:0-plasmalogens are frequently more affected than C18:0-plasmalogens. In DBP deficiency plasmalogens can be low-normal. Although biochemically not well understood, this is possibly caused by accumulating β-oxidation intermediates inside the peroxisome that interfere with ether lipid synthesis. There is a new ether lipid disorder where *FAR1* variants specifically mutate the arginine residue at position 480, which results in a disorder reminiscent of FAR1 deficiency but where uncontrolled ether lipid synthesis results in elevated levels of plasmalogens [31].

8.13 Quality Control

8.13.1 VLCFA/BCFA

Three different internal quality control samples are used. These are:

1. Two different pools of a large number of left-over plasma samples of patients are thoroughly mixed. Aliquots of 100 µl are put into Eppendorf vials and stored at –20 °C. These samples are stable for 2 years. Each analytical run includes two pool plasma sample, one around the reference range and one at elevated levels (pool of peroxisome biogenesis patients).
2. A so-called VLCFA/BCFA extract. This is a pool of extracted and derivatized samples. It is used solely to differentiate between machine- or sample workup-related problems.
3. External quality control samples are provided by European Research Network for the evaluation and improvement of screening, diagnosis, and treatment of Inherited disorders of Metabolism in its Special Assays Serum scheme. Eight lyophilized samples are distributed each year, consisting of four pairs of samples with increasing fatty acid levels, from physiological to pathological concentrations. This gives a good indication of the precision, linearity, and recovery of the laboratory's analytical performance. Comparison of the results with the mean of all participating laboratories gives an indication of the accuracy.

8.13.2 C26:0-lysoPC

- Plasma: Two controls are made from plasma (pool) of patients with C26:0-lysoPC within the reference range and pool of patients with high C26:0-lysoPC levels (e.g. peroxisome biogenesis disorders, adrenoleukodystrophy)
- Blood spot: Two controls are made from blood of an individual with C26:0-lysoPC concentration within the reference range and that of a patient with adrenoleukodystrophy with elevated levels. These can be prepared as regular blood spots (we spot 40 µl of EDTA blood on filter paper, air-dry at room temperature, and store in Ziplock bags at 4 °C.
- Cellular homogenate: we use fibroblast pellets prepared in bulk from controls and peroxisome biogenesis/adrenoleukodystrophy patients which are kept as pellets at –80 °C.

8.13.3 Plasmalogens

Two different internal quality control samples are used. These are:

1. A pool of erythrocyte samples. This is made up of erythrocytes from left-over EDTA blood of a large number of different patients. Aliquots of 300 µl suspension are stored in BHT-treated Eppendorf vials at –20 °C for a maximum period of 1 year. One of these samples is included in each weekly series of analyses. The results of C16:0-DMA, C18:0-DMA, C16:0 and C18:0 and the C16:0-DMA/C16:0 and C18:0-DMA/C18:0 can be monitored.
2. A so-called plasmalogen extract. This is a pool of extracted and derivatized erythrocyte samples. It is used solely to check for the detector response of the plasmalogens, thereby identifying "active spots" in the injection system. Also very useful to differentiate between machine- or sample workup-related problems.

No external quality control scheme is available for this analysis.

8.14 Conclusion

In this chapter we have described the methods used to measure VLCFA, BCFA, C26:0-lysoPC, and plasmalogens in different matrices, which is suitable for the identification/confirmation of peroxisomal disorders.

Acknowledgments The authors gratefully acknowledge Ries Duran for expert advice and contribution from the previous edition of this book. Sacha Ferdinandusse, Albert Bootsma, Henk van Lenthe, Henny Rusch, Elise van der Sluijs, and Lia van Lint are acknowledged for the technical expertise and advice.

References

1. Wanders RJA, Baes M, Ribeiro D, Ferdinandusse S, Waterham HR (2023) The physiological functions of human peroxisomes. Physiol Rev 103:957–1024
2. Duran M, Wanders RJA (2008) Plasmalogens and polyunsaturated fatty acids. In: Blau N, Duran M, Gibson KM (eds) Laboratory guide to the methods in biochemical genetics. Springer, Berlin, pp 207–220
3. Duran M, Wanders RJA (2008) Pipecolic acid. In: Blau N, Duran M, Gibson KM (eds) Laboratory guide to the methods in biochemical genetics. Springer, Berlin, pp 129–135
4. Klouwer FCC, Huffnagel IC, Ferdinandusse S, Waterham HR, Wanders RJA, Engelen M et al (2016) Clinical and biochemical pitfalls in the diagnosis of peroxisomal disorders. Neuropediatrics 47:205–220
5. Braverman NE, Moser AB (2012) Functions of plasmalogen lipids in health and disease. Biochim Biophys Acta 1822:1442–1452. Elsevier
6. Brites P, Waterham HR, Wanders RJA (2004) Functions and biosynthesis of plasmalogens in health and disease. Biochim Biophys Acta 1636:219–231
7. Barøy T, Koster J, Strømme P, Ebberink MS, Misceo D, Ferdinandusse S et al (2015) A novel type of rhizomelic chondrodysplasia punctata, RCDP5, is caused by loss of the PEX5 long isoform. Hum Mol Genet 24:5845–5854
8. Ferdinandusse S, Denis S, Hogenhout EM, Koster J, van Roermund CWT, IJlst L et al (2007) Clinical, biochemical, and mutational spectrum of peroxisomal acyl–coenzyme A oxidase deficiency. Hum Mutat 28:904–912
9. Ferdinandusse S, Denis S, van Roermund CWTT, Preece MA, Koster J, Ebberink MS et al (2018) A novel case of ACOX2 deficiency leads to recognition of a third human peroxisomal acyl-CoA oxidase. Biochim Biophys Acta 1864:952–958. Elsevier
10. Vilarinho S, Sari S, Mazzacuva F, Bilgüvar K, Esendagli-Yilmaz G, Jain D et al (2016) *ACOX2* deficiency: A disorder of bile acid synthesis with transaminase elevation, liver fibrosis, ataxia, and cognitive impairment. Proc Natl Acad Sci USA 113:11289–11293
11. Monte MJ, Alonso-Peña M, Briz O, Herraez E, Berasain C, Argemi J et al (2017) ACOX2 deficiency: An inborn error of bile acid synthesis identified in an adolescent with persistent hypertransaminasemia. J Hepatol 66:581–588. European Association for the Study of the Liver
12. Ferdinandusse S, Denis S, Mooyer PAW, Dekker C, Duran M, Soorani-Lunsing RJ et al (2006) Clinical and biochemical spectrum of D-bifunctional protein deficiency. Ann Neurol 59:92–104
13. Ferdinandusse S, Kostopoulos P, Denis S, Rusch H, Overmars H, Dillmann U et al (2006) Mutations in the gene encoding peroxisomal sterol carrier protein X (SCPx) cause leukencephalopathy with dystonia and motor neuropathy. Am J Hum Genet 78:1046–1052
14. Ferdinandusse S, Denis S, Clayton PT, Graham A, Rees JE, Allen JT et al (2000) Mutations in the gene encoding peroxisomal alpha-methylacyl-CoA racemase cause adult-onset sensory motor neuropathy. Nat Genet 24:188–191
15. Honey MIJ, Jaspers YRJ, Engelen M, Kemp S, Huffnagel IC (2021) Molecular biomarkers for adrenoleukodystrophy: An unmet need. Cells 10:1–18
16. Ferdinandusse S, Jimenez-Sanchez G, Koster J, Denis S, Van Roermund CW, Silva-Zolezzi I et al (2015) A novel bile acid biosynthesis defect due to a deficiency of peroxisomal ABCD3. Hum Mol Genet 24:361–370
17. Ferdinandusse S, Falkenberg KD, Koster J, Mooyer PA, Jones R, van Roermund CWT et al (2017) ACBD5 deficiency causes a defect in peroxisomal very long-chain fatty acid metabolism. J Med Genet 54:330–337
18. Buchert R, Tawamie H, Smith C, Uebe S, Innes AM, Al Hallak B et al (2014) A peroxisomal disorder of severe intellectual disability, epilepsy, and cataracts due to fatty acyl-CoA reductase 1 deficiency. Am J Hum Genet 95:602–610. The American Society of Human Genetics
19. Wierzbicki AS (2007) Peroxisomal disorders affecting phytanic acid alpha-oxidation: a review. Biochem Soc Trans 35:881–886
20. Cochat P, Hulton S, Acquaviva C, Danpure CJ, Daudon M, De Marchi M et al (2012) Primary hyperoxaluria Type 1: indications for screening and guidance for diagnosis and treatment. Nephrol Dial Transplant 27:1729–1736

21. Kihara A (2012) Very long-chain fatty acids: Elongation, physiology and related disorders. J Biochem 152:387–395
22. Vreken P, van Lint AE, Bootsma AH, Overmars H, Wanders RJ, van Gennip AH (1998) Rapid stable isotope dilution analysis of very-long-chain fatty acids, pristanic acid and phytanic acid using gas chromatography-electron impact mass spectrometry. J Chromatogr B Biomed Sci Appl Netherlands 713:281–287
23. Dacremont G, Vincent G (1995) Assay of plasmalogens and polyunsaturated fatty acids (PUFA) in erythrocytes and fibroblasts. J Inherit Metab Dis 18:84–89
24. Lam C, Wong D, Cederbaum S, Lim B, Qu Y (2012) Peanut consumption increases levels of plasma very long chain fatty acids in humans. Mol Genet Metab 107:620–622. Elsevier
25. Berendse K, Engelen M, Ferdinandusse S, Majoie CBLM, Waterham HR, Vaz FM et al (2016) Zellweger spectrum disorders: clinical manifestations in patients surviving into adulthood. J Inherit Metab Dis 39:93–106
26. Wanders RJA, Duran M (2008) Very-long-chain fatty acids and phytanic acid. In: Blau N, Duran M, Gibson KM (eds) Laboratory guide to the methods in biochemical genetics. Springer, Berlin, pp 221–231
27. Jaspers YRJ, Ferdinandusse S, Dijkstra IME, Barendsen RW, van Lenthe H, Kulik W et al (2020) Comparison of the diagnostic performance of C26:0-lysophosphatidylcholine and very long-chain fatty acids analysis for peroxisomal disorders. Front Cell Dev Biol 8:690
28. Moser AB, Kreiter N, Bezman L, Lu S, Raymond GV, Naidu S et al (1999) Plasma very long chain fatty acids in 3,000 peroxisome disease patients and 29,000 controls. Ann Neurol United States 45:100–110
29. Klouwer FCC, Ferdinandusse S, van Lenthe H, Kulik W, Wanders RJA, Poll-The BT et al (2017) Evaluation of C26:0-lysophosphatidylcholine and C26:0-carnitine as diagnostic markers for Zellweger spectrum disorders. J Inherit Metab Dis 40:875–881
30. Horvath R, Lewis-Smith D, Douroudis K, Duff J, Keogh M, Pyle A et al (2015) SCP2 mutations and neurodegeneration with brain iron accumulation. Neurology 85:1909–1911
31. Ferdinandusse S, McWalter K, te Brinke H, IJlst L, Mooijer PM, Ruiter JPN et al (2021) An autosomal dominant neurological disorder caused by de novo variants in FAR1 resulting in uncontrolled synthesis of ether lipids. Genet Med 23:740–750

Mucopolysaccharides

9

Maria Fuller

Contents

Abbreviations

CE	Collision energy
CS	Chondroitin sulphate
CXP	Collision exit potential
DMB	1,9-dimethylmethylene blue
DP	Declustering potential
DS	Dermatan sulphate
GAG	Glycosaminoglycans
GalNAc	*N*-acetylgalactosamine
HA	Hyaluronan
Hex	Hexose
HN	Hexosamine
HNAc	*N*-acetylhexosamine
HS	Heparan sulphate
IS	Internal standard
KS	Keratan sulphate
MPS	Mucopolysaccharidoses
MRM	Multiple reaction monitoring
MSD	Multiple sulphatase deficiency
RT	Retention time
S	Sulphate
UA	Uronic acid

M. Fuller (✉)
Genetics and Molecular Pathology, SA Pathology, Women's and Children's Hospital, North Adelaide, SA, Australia

Adelaide Medical School and School of Biological Sciences, University of Adelaide, Adelaide, SA, Australia
e-mail: maria.fuller@adelaide.edu.au

9.1 Introduction and Clinical Significance

The mucopolysaccharidoses (MPS) comprise a family of 11 reported lysosomal storage disorders with a collective incidence of about 1 in

N. Blau, F. M. Vaz (eds.), *Laboratory Guide to the Methods in Biochemical Genetics*,
https://doi.org/10.1007/978-3-031-58819-8_9

33,000, with MPS I and IIIA the most common at 1 in 137,000 [1], and MPS IX reporting only five cases heretofore [2]. As single gene disorders, MPS adhere to autosomal recessive mode of inheritance, except for MPS II, which is X-linked. Pathogenic variants in genes encoding five sulphatases, five hydrolases and one transporter required for the lysosomal degradation of glycosaminoglycans (GAG) lead to the accumulation of incompletely degraded GAG within lysosomes of affected cells. This results in the chronic and progressive deterioration of cells, tissues and organs, and a constellation of progressive clinical features complicit with MPS. There are five categories of GAG affected in MPS; chondroitin sulphate (CS), dermatan sulphate (DS), heparan sulphate (HS), hyaluronan (HA) and keratan sulphate (KS), and the type of GAG often correlates with specific symptoms. For example, accumulation of partially degraded HS is characteristic of neurological impairment, and CS and KS are typically associated with skeletal disease and corneal clouding. Other notable clinical manifestations include coarse facial features, organomegaly and cardio-pulmonary insufficiency. However, there is a wide spectrum of symptomology and severity not just across the MPS disorders, but also within each MPS type suggesting that GAG accumulation alone is not the sole cause of the clinical picture [3]. Table 9.1 provides an overview of the cardinal features pertaining to each MPS—noting there is often broad clinical heterogeneity within each type—and Fig. 9.1 shows the GAG substrate specificity for each enzyme.

The diversity in clinical picture can be exemplified in MPS I, in which a deficiency of the enzyme, α-L-iduronidase (EC 3.2.1.76), second-

Table 9.1 Characteristics of each MPS

MPS subtype	Gene	Enzyme	GAG	Major features	Incidence[a]
MPS I (Hurler/Scheie)	*IDUA*	α-L-iduronidase	DS, HS	Corneal clouding, dysostosis multiplex, organomegaly, intellectual disability in Hurler	137,000
MPS II (Hunter)	*IDS*	Iduronate 2-sulphatase	DS, HS	Dysostosis multiplex, organomegaly, intellectual disability in severe form	176,000
MPS IIIA (Sanfilippo A)	*SGSH*	*N*-sulpho glucosamine sulphohydrolase	HS	Profound intellectual disability, mild somatic disease	148,000
MPS IIIB (Sanfilippo B)	*NAGLU*	α-*N*-acetyl glucosaminidase	HS	Same as MPS IIIA	369,000
MPS IIIC (Sanfilippo C)	*HGSNAT*	Acetyl CoA:α-glucosaminide *N*-acetyltransferase	HS	Same as MPS IIIA	616,000
MPS IIID (Sanfilippo D)	*GNS*	*N*-acetyl glucosamine 6-sulphatase	HS	Same as MPS IIIA	No data
MPS IVA (Morquio A)	*GALNS*	*N*-acetyl galactosamine 6-sulphatase	CS, KS	Corneal clouding, skeletal disease, odontoid hypoplasia	217,000
MPS IVB (Morquio B)	*GLB1*	β-galactosidase	KS	Same as MPS IVA	1,847,000
MPS VI (Maroteaux-Lamy)	*ARSB*	*N*-acetyl galactosamine 4-sulphatase	DS	Corneal clouding, dysostosis multiplex	923,000
MPS VII (Sly)	*GUSB*	β-glucuronidase	DS, HS, CS	Dysostosis multiplex, hepatosplenomegaly, (foetal hydrops and neonatal)	3,694,000
MPS IX[b]	*HYAL1*	Hyaluronidase	HA	Joint inflammation and swelling	No data
MPS X[b]	*ARSK*	Arylsulphatase K	HS, DS, CS	Short stature, dysostosis multiplex, skeletal dysplasia	No data

[a]Australian incidence (2009–2020) [1]
[b]No biochemical assays are described herein for MPS IX or X

Heparan sulphate

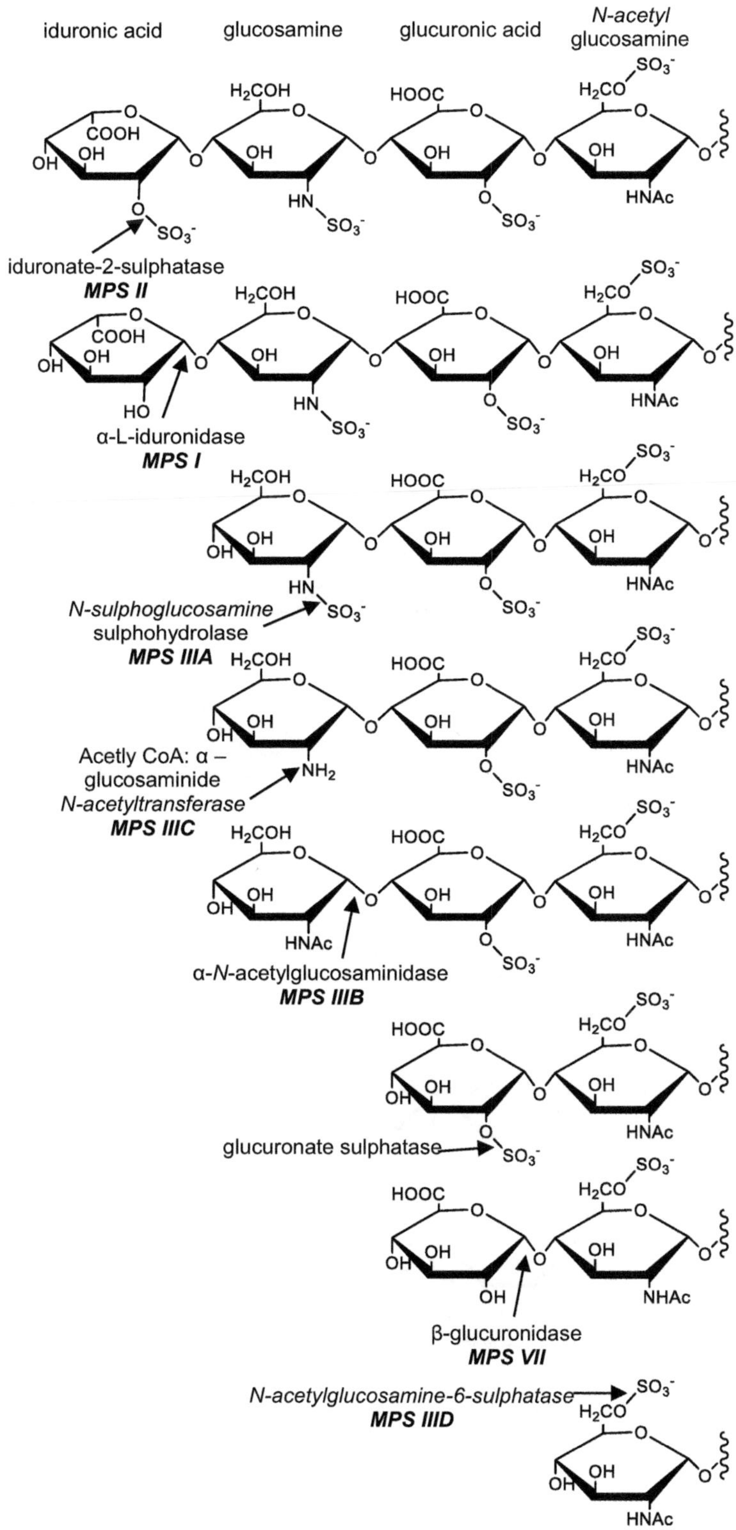

Fig. 9.1 GAG structures with MPS enzyme cleavage sites

Dermatan sulphate

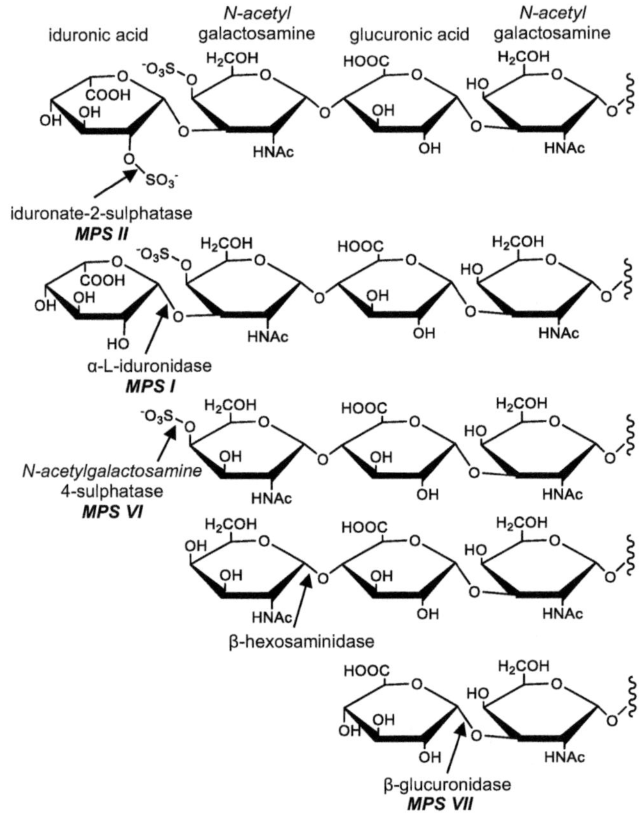

Keratan sulphate

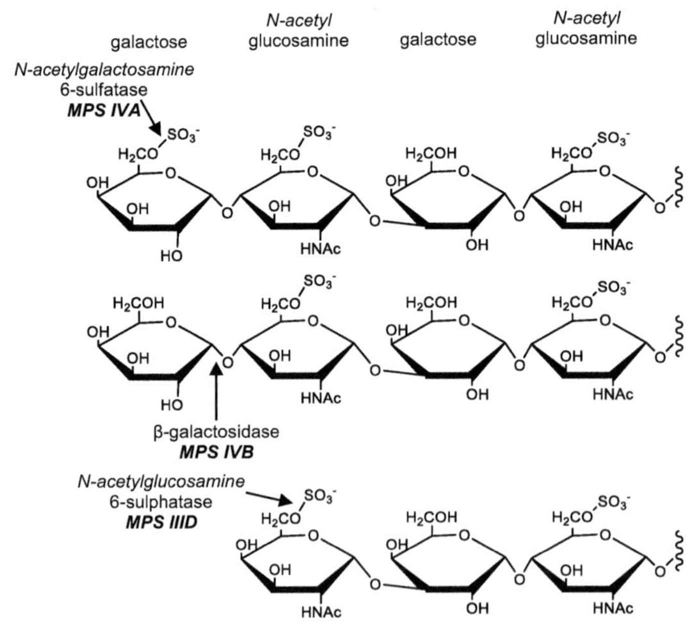

Fig. 9.1 (continued)

Chondroitin sulphate

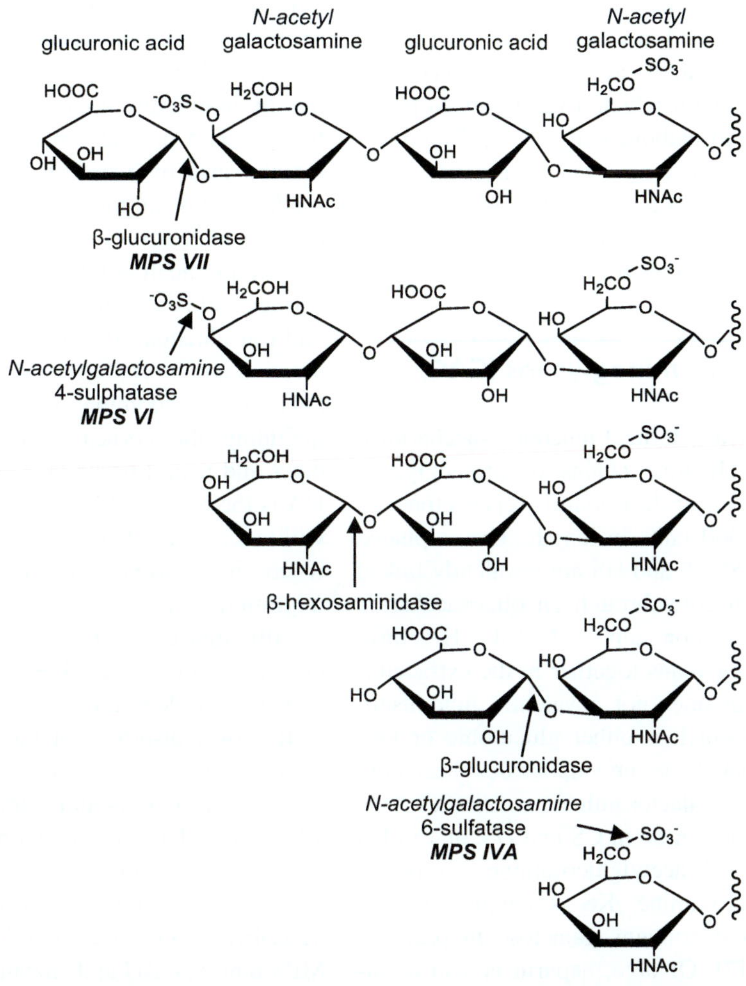

Fig. 9.1 (continued)

ary to biallelic loss-of-function *IDUA* variants, can cause a rapidly progressing disease course with intellectual disability (Hurler syndrome) and a slower progressing phenotype (Scheie syndrome) with normal intelligence. Although some genetic variants that produce no enzyme activity, e.g., c.1205G>A and c.208C>T, typically associate with severe disease and other alleles affording residual enzyme activity manifest a more attenuated phenotype, the relationship between genetic variants, residual enzyme activity and GAG accumulation remains poorly defined [4].

In the 1990s the genes encoding the MPS enzymes were cloned and the years following witnessed a plethora of activity identifying disease-causing variants. The most common genetic variants are indeed missense, but nonsense, splice site, deletions, insertions, frameshifts and rearrangements have all been reported. Genotype–phenotype predictions have been difficult beyond broad, as although null variants are expected to produce severe disease, the interpretation of missense variants is almost impossible without prior experience of the particular variant. More recently, and coinciding with the explosion in molecular genetic technologies, there is increasing awareness of pseudodeficiency alleles and variants of uncertain significance in the MPS genes. The introduction of newborn screening programs for MPS has highlighted the number of pseudodeficiency alleles, where reduced enzyme activity is identified in vitro, but does not reflect the in vivo situation, and has also uncovered genetic variants for which it has not been possible to ascertain pathogenicity [5, 6]. This presents diagnostic challenges when using the genetic and enzymatic tools available, but improvements in contemporary laboratory determinations of GAG are likely to now fill this void.

9.2 Glycosaminoglycans (GAG)

GAG (historically termed mucopolysaccharides) are the carbohydrate chains of proteoglycans consisting of linear disaccharide repeats that can be up to 20,000 units in length. The sulphated GAG, CS, DS, KS and HS are covalently linked to the protein core through an oligosaccharide linker, and the non-sulphated GAG, HA links several proteoglycans together in the extracellular matrix but does not covalently bind itself. Uronic acid which is either glucuronic or iduronic—sulphated or un-sulphated—alternating with *N*-acetylgalactosamine constitutes the repeating disaccharide in CS and DS, and in HS and HA, *N*-acetylglucosamine replaces *N*-acetylgalactosamine. KS is unique amongst the GAG as it contains galactose in place of uronic acid [7]. Of note, heparin is also a sulphated GAG made up of repeating disaccharides of iduronic acid and *N*-acetylglucosamine, but as there is no known MPS arising from impaired catabolism it will not be further mentioned. Although in theory GAG may appear simple structures with repeating units of uronic acid (or galactose in KS) and *N*-acetylhexosamine, there is a high degree of heterogeneity primarily determined by specific glycosyltransferases and sulphotransferases that modify the polymer [8]. For example, the amino group of *N*-acetylglucosamine may be sulphated or unsubstituted instead of *N*-acetylated, and it may also be sulphated on the C6-hydroxyl and occasionally on the C3-hydroxyl. Similarly, *N*-acetyl galactosamine may be sulphated on the C4 and/or C6 position.

Present in almost all tissues, GAG are the foundation of the extracellular matrix, and as participants in so many biological processes they are of fundamental physiological importance. For example, CS is a key structural player of cartilage and by interacting with a variety of growth factors CS is a driving force for the function and maintenance of ligaments and tendons [9]. DS predominates in the skin, cartilage and aorta playing central roles in blood vessel maintenance and wound healing [10]. HS features on the cell surface and is actively involved in cell signalling and transduction affording intercellular communication and morphogenesis of developing tissue [11]. KS predominates in connective tissue including the cornea, cartilage, intervertebral discs and neural tissue [12]. The non-sulphated, HA is the major GAG in the extracellular matrix within connective, epithelial and neural tissue where it contributes to cell proliferation and migration [13].

Although disorders of GAG degradation are described within the MPS family of lysosomal storage disorders, there are also around 20 less well-known disorders of their synthesis. These include defects in genes encoding glycosyltransferases, transporters and sulphotransferases, and all disorders have the common feature of skeletal pathologies. Details of the specific genes, enzymatic defects and clinical manifestations are described in Sasarman et al. [14], and unlike the MPS family, GAG and enzymatic measurements do not yet play a role in the diagnosis of this group of disorders.

9.3 Diagnostic Trajectory

In the absence of a family history, and following clinical suspicion, the laboratory diagnosis of MPS typically begins with the measurement of urinary GAG, and if above reference intervals precipitates secondary testing to identify the type of GAG [15]. However, sensitivity and specificity remain problematic and now with increased knowledge of GAG structure and rapidly advancing mass spectrometry technologies, improved methods of GAG determinations are infiltrating the fabric of laboratory MPS diagnostic testing. This chapter will describe the traditional and contemporary approaches for GAG quantification as well as the methods available to measure enzyme activity that are typically used to provide a laboratory diagnosis of MPS. Although the final step in the diagnostic trajectory is the identification of pathogenic variants in disease-causing genes, molecular genetic testing is beyond the scope of this chapter.

9.4 GAG Screening Tests

9.4.1 Specimen Requirements

Spot urine samples. Urine can be refrigerated, or frozen (for long-term storage) for shipping to testing laboratory to minimize bacterial growth. Urine should be thawed to room temperature prior to testing, mixed before use and the creatinine determined in the automated clinical chemistry/core laboratory.

Urine samples dried onto filter paper. Urine can be eluted from the filter paper by cutting the filter paper into small pieces and placing in a 50 mL centrifuge tube. Cover the filter paper with H_2O and refrigerate overnight. Transfer as much supernatant as possible into a 10 mL centrifuge tube and then determine the creatinine before use.

9.4.2 Qualitative GAG Spot Tests

Principle

Qualitative spot tests have been around since 1960 when the Berry spot test was first reported and its applicability as a urinary screening test for MPS demonstrated in 1987 [16, 17]. Other dyes such as Azure A [18] have also been used. They are based on metachromatic staining of urinary hexuronic acid present in GAG revealing a blue/purple stain on the filter paper. Although very simple, cost effective and fast, they suffer from both false negative and positive results [19].

Reagents

1. CS (Sigma-Aldrich C4384) is used as a positive control typically at three concentrations of 50, 100 and 500 mg/L prepared in H_2O.
2. A 0.4 mM toluidine blue (Sigma-Aldrich T3260) solution is prepared by dissolving 1.22 g in 0.8 L of acetone with stirring, and then made up to a final volume of 1 L with H_2O.
3. Wash solution is 1.8 M acetic acid (Merck).

Analytical Procedure

Ten microlitres of each of the three CS standards and patient urine samples are spotted onto Whatman filter paper strips and allowed to air-dry at room temperature. The filter paper is then immersed in the 0.4 mM toluidine blue solution for 2 min, removed and air-dried. Excess dye is removed with 3 × 5 min washes in 1.8 M acetic acid and a final 5 min wash in H_2O.

Post-analytical and Interpretation

A positive urine sample will be stained dark blue/purple and related to the intensity of the three standards.

Urine samples yielding a strong positive signal should be followed up with further laboratory investigations. However, a negative result does not exclude MPS due to the low sensitivity, particularly for HS and KS, hence MPS III and MPS IVA are typically missed in spot tests. Additionally, false positives are common, especially in neonates due to physiological fluctuations in GAG levels during growth and development [20].

9.4.3 Semi-Quantitative Dye-Spectrometric Methods

Principle

Several methods are available that exploit the properties of GAG to form complexes detectable by spectrophotometry. For example, the acid albumin turbidity test employs the binding of uronic acid residues within GAG to albumin and following acid precipitation the resulting turbidity is measured at 600 nm [21]. Given KS does not contain uronic acid, MPS IVA patients will not be detected, and like the qualitative spot tests, this assay is limited in sensitivity and specificity [20]. The cationic dye, alcian blue, forms ionic interactions with the sulphate groups in GAG, but also lacks specificity due to the interference with other negatively charged substances in the urine and has limited sensitivity for non-sulphated GAG like HA [22]. Currently, the most widely used spectrophotometric method for GAG determinations uses the cationic dye, 1,9-dimethylmethylene blue (DMB), that binds to sulphated GAG present in urine producing a colourimetric complex that absorbs at 520 nm [23] and is described below.

Reagents

1. CS (Sigma-Aldrich C4384) is used as the reference material. A stock standard is prepared by dissolving 10 mg of CS in 100 mL of H_2O. This stock is further diluted in H_2O to give 5 working standards according to Table 9.2.
2. Formate buffer (55 mM, pH 3.3) is prepared by dissolving 3.74 g HCOONa (MW 68.01, Sigma-Aldrich 107603) into 800 mL H_2O. The pH is adjusted to 3.3 with HCl and made up to a final volume of 1 L. Store at 4 °C.
3. Acetate buffer (0.5 M, pH 5.8) is prepared by dissolving 41 g CH_3COONa (anhydrous, MW 82.03) into 900 mL H_2O. The pH is adjusted to 5.8 with glacial acetic acid and made up to a final volume of 1 L. Store at 4 °C.
4. Stock DMB (1,9-dimethylene blue chloride, Polysciences 03610) solution is prepared by dissolving 1.1 mg DMB in 10 mL of 55 mM formate buffer in a glass vial, mix and protect from light with aluminium foil.
5. A fresh working DMB solution is prepared on day of use by diluting the stock DMB solution 1 in 10 with 0.5 M acetate buffer and must also be protected from light.

Table 9.2 Standard curve preparation

Working standard (mg/L)	mL of CS stock (10 mg/100 mL)	mL of H_2O
0	0	10
10	1	9
20	2	8
50	5	5
100	10	0

Quality Control

A positive sample is included which should contain a mixture of MPS patient urine to ensure anonymity, typically a mix of MPS II and MPS IVA to include relevant GAG. Aliquots of the mixture are then stored frozen.

External quality control samples, e.g., ERNDIM should be included as required by proficiency testing (see Sect. 1.2).

Instrumentation

Spectrophotometer for cuvettes, or microtitre plates that can measure wavelengths between 500 and 700 nm.

Analytical Procedure

Mix urines including QC thoroughly and then centrifuge for 5 min at 3100 *g*. Into plastic test tubes, pipette 30 μL of formate buffer (blank), standards, QC or patient samples and add 120 μL of H_2O. Add 825 of working DMB solution and mix thoroughly. Either transfer 0.2 mL into wells of a microtiter plate or transfer 0.5 mL into a cuvette and read absorbance at 520 nm.

Post-analytical and Interpretation

Create a calibration curve by plotting the absorbance at 520 nm against the concentration of each standard. The r^2 should be >0.99. Any samples with absorbances above the highest standard (100 mg/L) should be diluted accordingly with H_2O and repeated. Results can be expressed as g/mol of creatinine to normalize for urine output.

Age-related reference intervals are required as GAG concentrations change with growth and development. Laboratories should calculate their own reference intervals. An example of age-related normal ranges is provided as a guide in Table 9.3.

Urine samples yielding a positive signal should be followed up both clinically and with

Table 9.3 Reference intervals

Age (years)	g/mol creatinine
<1	<23
4	<10
4–9	<7
9–17	<5
>17	<3

further laboratory investigations. This typically involves electrophoresis to identify the pattern of GAG and rule out false positives (Sect. 9.5), or mass spectrometry methods (Sect. 9.6). This also facilitates subsequent enzymatic testing to confirm the MPS type. As with other spectrophotometric methods, GAG determinations are susceptible to false positive and negative results due to low sensitivity and specificity, especially for MPS IVA and MPS patients with attenuated phenotypes [24]. Therefore, all results must be interpreted in conjunction with the clinical picture of the patient, and normal results should not be considered to rule out a diagnosis of MPS when there is strong clinical suspicion.

9.5 Electrophoretic Differentiation of GAG

Principle

GAG are precipitated from urine and electrophoretic separation is based primarily on the size and charge of the GAG. This allows distinct GAG patterns to be visualized that can be compared with normal urine and is typically performed following GAG elevations revealed by the DMB spectrophotometric method (see Sect. 9.4.3). Each separate GAG band can be semi-quantified by densitometry providing an estimate of the amount of specific GAG type present in the urine sample [25, 26].

Reagents

1. CPC citrate is prepared by dissolving 1 g of cetylpyridinium (Sigma-Aldrich C0732) in 1 L of 0.2 M citrate buffer pH 4.8. Citrate buffer (0.2 M, pH 4.8) is prepared by dissolving 58.8 g trisodium citrate · $2H_2O$ (MW 294.1, Sigma-Aldrich S4641) in 800 mL H_2O. The pH is adjusted to 4.8 with citric acid and made up to a final volume of 1 L.
2. 2 M LiCl (Sigma-Aldrich L9650) is prepared by dissolving 8.5 g in 100 mL H_2O.
3. Absolute ethanol (Merck 100986).
4. Barium acetate (1 M, pH 5.0) is prepared by dissolving 255.4 g (MW 255.42, Sigma-Aldrich 243671) in 800 mL H_2O. The pH is adjusted to 5.0 with glacial acetic acid and made up to a final volume of 1 L.
5. Barium acetate (0.1 M, pH 5.0) is prepared by dissolving 25.5 g of barium acetate in 800 mL H_2O. The pH is adjusted to 5.0 with glacial acetic acid and made up to a final volume of 1 L.
6. Barium acetate 15% ethanol buffer (0.1 M, pH 5.0) is prepared as per 5 above using 15% v/v ethanol instead of H_2O.
7. Barium acetate 50% ethanol buffer (0.1 M, pH 5.0) is prepared as per 5 above using 50% v/v ethanol instead of H_2O.
8. Alcian blue (Sigma-Aldrich A9186) solution is prepared by dissolving 0.25 g in 100 mL H_2O. The solution must be made at least 24 h prior to use.
9. Phenol red (Sigma-Aldrich 114529) is prepared by dissolving 5 mg in 10 mL H_2O.
10. Acetic acid 1 % (v/v) is prepared by diluting 20 mL glacial acetic acid (Merck 13700) with 1980 mL H_2O.
11. Whatman 3 MM chromatography paper (Cat # 3030672) is cut into 100 mm square wicks.

12. Whatman No. 1 chromatography paper (Cat # 3001690) is cut into rectangular squares for blotting.

Instrumentation

Electrophoresis unit inclusive of tank and power supply, e.g., LKB Bromma 2219 Multitemp II thermostatic circulator.

Cellulose acetate, e.g., Titan III 94 × 76 mm.

Scanner for densitometry.

Quality Control

The same positive MPS urine mix is described in Sect. 9.4.3. External quality control samples such as the ERNDIM scheme should be included as required by proficiency testing.

Analytical Procedure

Urines are centrifuged at 1500 *g* for 10 min and 0.3 mg of GAG equivalent (from DMB result, Sect. 9.4.3) is mixed with an equal volume of CPC citrate and incubated for 30 min at 37 °C. Repeat centrifugation, decant supernatant and leave inverted for about 10 min to fully drain all liquid. Dissolve pellet in 150 μL 2 M LiCl, vortex and then add 0.8 mL ethanol and vortex again. Transfer the solutions to 4 mL conical tubes and centrifuge at 1500 *g* for 10 min. Aspirate the supernatant and dry the pellet under a gentle stream of nitrogen. Dissolve the pellet in 20 μL of phenol red, vortex to dissolve pellet and then centrifuge at 1500 *g* for 10 min prior to electrophoresis. Cut a piece of cellulose acetate suitable for the size of the tank and immerse slowly in 0.1 M barium acetate for 10 min then blot with Whatman No. 1 to remove excess liquid. Pipette 3 μL of each sample onto the applicator plate and electrophorese for 5 min at 200 V in 1 M barium acetate buffer. The cellulose acetate must be covered with a sheet of plastic, insulated with foam, and then glass slides/plates for pressure (see Fig. 9.2). Following electrophoresis, remove the cellulose acetate and immerse in barium acetate buffer 0.1 M 15% ethanol for 2 min and blot with Whatman No. 1 to remove excess liquid. The cellulose acetate is returned to the tank, covered as described and electrophoresed for another 25 min. The cellulose acetate is then immersed in

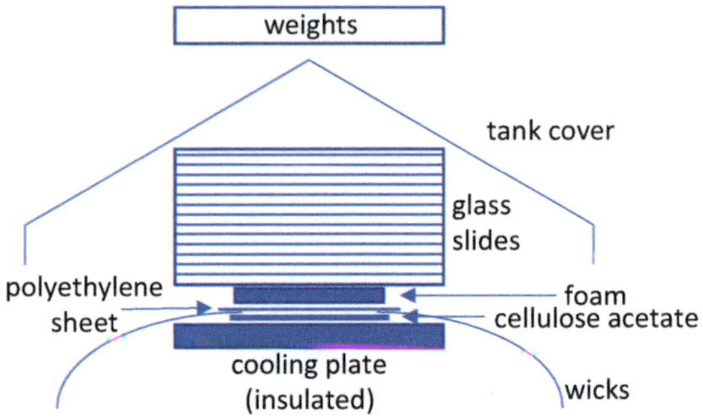

Fig. 9.2 Diagramatic setup for electrophoresis

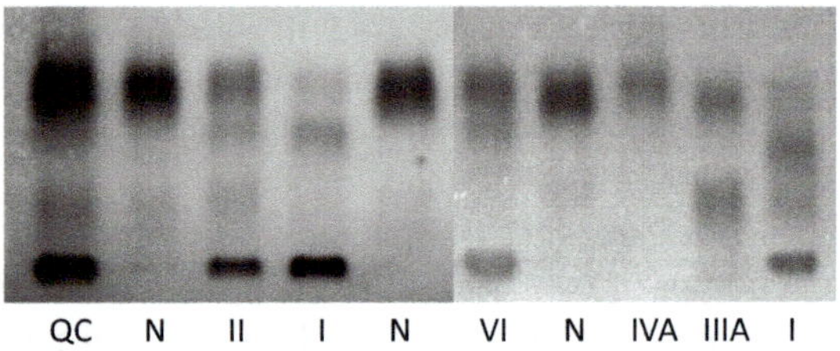

Fig. 9.3 Electrophoresis patterns for different MPS subtypes. QC, quality control; N, normal urine, MPS II (II), MPS I (I), MPS VI (VI), MPS IVA (IVA) and MPS IIIA (IIIA) are shown

barium acetate buffer 0.1 M 50% ethanol for 2 min, blotted and returned to the electrophoresis tank for another 15 min at 200 V. The cellulose acetate is removed and stained with alcian blue for 15 min with gentle mixing, followed by 3 × washes with 1% (v/v) acetic acid and a final wash in H_2O. Following air drying of the cellulose acetate, it can be visualized under transmitted light and bands can be evaluated by densitometry.

Post-analytical and Interpretation

Figure 9.3 shows an example of the electrophoretic patterns seen in some MPS urine samples. Although electrophoresis is able to determine the type of GAG, offering a higher degree of specificity than spectrophotometric methods, it is unable to distinguish the subtypes. The patterns for MPS I, II and VI are the same, and so are all subtypes of MPS III. KS elevations are modest and MPS IVA is often missed due to the appearance of normal electrophoretic patterns [24]. Additionally, abnormal patterns can also be witnessed in other lysosomal disorders, such as mucolipidoses, Sandhoff disease and multiple sulphatase deficiency patients. Care must be taken when interpreting the electrophoretic patterns and enzymatic testing (Sect. 9.6) must follow to confirm any abnormalities.

9.6 Mass Spectrometry of GAG

There has been an explosion of mass spectrometry approaches to analyse GAG over recent years. Primarily dedicated to the elucidation of the fine structure of these complex polymers, these methods have aimed to uncover more about their biological activity [27]. However, these technological advances have also provided a framework for the diagnostic service laboratory and paved the way to measure the accumulated GAG in MPS. High-performance liquid chromatography coupled to electrospray ionization mass spectrometry (LC-ESI-MS/MS) has afforded high sensitivity and specificity for the laboratory diagnosis of MPS compared with standard colourimetric, spectrophotometric and electrophoretic techniques. The most common procedures involve chemical or enzymatic depolymerization of the long chain GAG to smaller oligosaccharides that fall within the mass range for typical MS/MS detection. Chemical hydrolysis, by butanolysis [28], methanolysis [29] or nitrous acid [30] depolymerizes the GAG into simple, uniform disaccharides, with no structural details. This is because the original epimeric nature of the uronic acid and information relating to *N*-acetylation or *N*-sulphation are lost during acid hydrolysis.

In contrast enzymatic digestion preserves structural heterogeneity, depolymerizing GAG into oligosaccharides with an MW of 350–600 Da carrying Δ-4,5-unsaturated hexuronic acid. Heparinase I, II and III and chondroitinase B are used to release HS and DS disaccharides from the GAG polysaccharides, molar enriching the disaccharides, which are then derivatized with [$^{12}C_6$] analine by reductive amination. The analine tag improves chromatographic resolution of the disaccharides followed by MS/MS detection [31]. There are several different disaccharides produced, but a specific method, known as Sensi-Pro, makes use of the characteristic non-reducing end residue to identify each type of MPS [31]. A separate method that quantifies disaccharides generated from KS has been described for MPS IVA [32].

Detailed standard operating procedures for these methods are provided in the supplementary information of Herbst et al. [33, 34]. These two articles also systematically compared current methods of GAG analysis in a retrospective cohort of newborn dried blood spots from patients already diagnosed with MPS I and II. In both studies, a different approach that does not involve depolymerization of the GAG—rather measures the endogenous non-reducing end oligosaccharide [35]—gave greater differentiation between MPS I and MPS II patients and unaffected newborns [33, 34]. Given the simplicity of this approach, its ability to identify all MPS subtypes in one assay (inclusive of MPS IVA), and the performance in the diagnostic laboratory, this method is detailed below.

Principle
In each MPS the catabolic block produces partially degraded GAG fragments in which the non-reducing end residue will be the substrate for the enzyme deficiency. These fragments range in size from very large polysaccharides—too large to be measured—to di- to pentasaccharides that are small enough to be measured using MS/MS. As exo-digestion is sequential, these oligosaccharides are specific for each MPS. Derivatization of the oligosaccharides adds a hydrophobic element allowing conventional reverse phase chromatography and subsequent detection by MS/MS [35].

Specimen
Spot urine samples as described above in Sect. 9.4.1, CSF, plasma and dried blood spots. The method below is provided for urine [35] as this is the most common sample received by the biochemical genetics' laboratory. Plasma and CSF preparation is described in Saville et al. [36] and blood spots in Herbst et al. [33, 34].

Reagents
All chemicals should be of the highest purity available or HPLC grade.

1. Internal standard (IS, 2 mM) is prepared by dissolving 5 mg of chondroitin disaccharide Δdi-4S sodium salt (Sigma-Aldrich C4045, MW 503.3) in 5 mL of H_2O. Store frozen in 0.5 mL aliquots. A working IS stock to give 200 pmol/sample is then prepared by further diluting 10 μL with 90 μL of H_2O, which is sufficient for 90 samples.
2. Prepare on day of use, 250 mM PMP (1-phenyl 3-methyl 5-pyrazolone, TCI Tokyo Kasei Kogyo) by dissolving 436 mg in 10 mL of 0.4 M NH_4OH. Solution can be heated at 70 °C to help dissolve PMP. Check pH and adjust to pH 9.5–10 with NH_4OH if required.
3. Dilute 770 μL formic acid (Merck) with 99.23 mL of H_2O to give a 0.2 M solution.
4. Chloroform ($CHCl_3$) with 1% ethanol (v/v) (Merck, analytical grade).
5. Mobile phase A, add 0.2 mL formic acid to 0.2 L of HPLC grade H_2O in a 1 L glass bottle. Loosen lid and sonicate for 30 min.
6. Mobile phase B, add 1 mL 98% formic acid (Emsure) to 1 L of acetonitrile (LiChrosolv, Merck) in a 1 L glass bottle. Loosen lid and sonicate for 30 min.

Instrumentation

Lyophilizer.

Oven/heating block at 70 °C.

Triple quadrupole mass spectrometer coupled to a high-pressure liquid chromatography system with an autosampler. The author's laboratory uses a Shimadzu Nexera X2 UHPLC and ABSciex 6500 mass spectrometer system. Standard operating procedures for the Waters Acquity coupled to the Xevo TQ-S have also been described [33, 34].

Quality Control

A normal control sample is composed of remaindered unaffected urine samples left over from analyses, thereby ensuring the anonymity of the pooled sample. Individual pooled positive QC urine samples are also prepared; MPS I/II QC, by mixing volumes equating to 0.5 μmoles of creatinine each of known positive MPS I and II urine; MPS III QC, by mixing volumes equating to 0.5 μmoles of creatinine each of known positive MPS IIIA, IIIB and IIIC urine; MPS VI QC, use 0.5 μmoles of creatinine of a known MPS VI patient. ERNDIM proficiency testing should be included as required.

Analytical Procedure

Spot urine samples are centrifuged at 1500 *g* for 5 min and the creatinine is determined in the clinical chemistry/core laboratory. The volume of urine equivalent to 0.5 μmoles of creatinine (μL = 500/creatinine) is lyophilized in 1.5 mL Eppendorf tubes, along with the pooled normal control urine, MPS I/II, MPS III and MPS VI QC. For every nine samples, add 10 μL IS working stock to 1 mL of 250 mM PMP. Add 0.1 mL to each tube of lyophilized urine to give 200 pmoles of IS per sample and to one blank tube as the IS only. Mix well to resuspend and incubate at 70 °C for 90 min. Allow tubes to cool for 10 min at room temperature and acidify with 0.5 mL of 0.2 M formic acid and mix. To each acidified sample add 0.5 mL $CHCl_3$ to wash PMP from the aqueous layer and shake vigorously by hand for 1 min. For large numbers of samples, place tubes in a rack, place another rack on top to hold them in place to shake together. Centrifuge for 5 min in an Eppendorf microfuge on maximum speed and using a glass pipette pre-wetted with $CHCl_3$, remove and discard the bottom $CHCl_3$ layer. Repeat the $CHCl_3$ wash three more times with a further 3 × 0.5 mL $CHCl_3$. After the fourth (final) centrifugation, do not remove the lower $CHCl_3$ phase, rather pipette 0.1 mL of the upper aqueous phase into a well of a microtitre plate or a HPLC glass vial. Cap the vials or seal the plate with heat-sealing foil and place in the autosampler of the LC-ESI/MS/MS system. Include blanks at the beginning and end of the run that only contain 0.1 mL of mobile phase A.

Partial separation of the derivatized oligosaccharides is performed on a Pursuit 3 pentafluorophenyl (PFP) column (2.0 × 100 mm; 3 μm; Agilent) maintained at room temperature in a Shimadzu Nexera*X*2 LC system equipped with an autosampler. The oligosaccharides are eluted over a 12 min LC gradient of mobile phase A and B at a flow rate of 0.3 mL/min following the injection of 4 μL of each sample onto the column. The column is equilibrated at 20% mobile phase B before a linear ramp to 22.5% B at 5 min and up to 90% B at 5.5 min. This is maintained for 2 min before returning to 20% B at 8 min for equilibration (4 min) prior to injection of the next sample. The first 2.2 min (containing salt) and the last 7.5 min (containing PMP) of column flow are diverted to waste.

Table 9.4 provides the chemical formulae, the retention time (RT), multiple reaction monitoring (MRM) pair and mass spectrometry conditions for the oligosaccharides that are elevated in each MPS subtype. Individual MS/MS conditions should be optimized on the mass spectrometer being used in the laboratory with the listed MRM pairs.

Post-analytical and Interpretation

The relative concentration of each oligosaccharide is determined by dividing the peak area of each oligosaccharide by that of the IS. Check there are no contaminating peaks in the blank wells. Begin with the IS and use the retention time (RT) of the IS as the anchor for the samples. Then systematically go through each transition using the RT in the respective QC as to guide each sample according to Table 9.4.

MPS I/II QC contains UA-HNAc (1S) at both an early (MPS I) and late (MPS II) RT, the latter due to the extra sulphate that is believed to be lost in the ESI source, as well as UA-HN-UA (1S). MPS III ABC QC contains HN-UA (1S), $(HNAc\text{-}UA)_2$ (1S) and $(HN\text{-}UA)_2$-HNAc (2S). MPS VI QC contains HNAc (1S), HNAc (2S), HNAc-UA (1S) and $(HNAc\text{-}UA)_2$ (1S).

There is no QC for $(HNAc\text{-}UA)_2$ (2S) but this peak should not be present in the MPS VI QC. There is also no QC for $(Hex\text{-}HNAc)_2$ (2S). For both analytes, if there is a double peak ensure the left peak is integrated. Examples of mass spectrometry traces for each MPS type are depicted in Fig. 9.4 as exceptional care must be taken when integrating to ensure the correct peak is chosen.

Laboratories need to establish their own QC ranges based on the urine samples used in their preparation and should aim to achieve values within 2 SD of the mean. Diagnosing each MPS subtype follows the flow chart in Fig. 9.5 based on reference intervals (shown for each oligosaccharide) calculated on non-MPS urine samples. Again, laboratories should determine their own reference intervals using unaffected MPS urine samples. It is important to follow the path set out in Fig. 9.5 because occasionally samples from young infants may have slightly raised concentrations of multiple analytes and, importantly, due to β-*N*-acetylhexosaminidase activity removing HNAc (1S) residues from the non-reducing end of GAG in MPS IIID, IVA and VI, these three subtypes will also show elevated HNAc (1S) [35]. Even though the MPS subtype can be determined using this method, it is recommended that confirmation be undertaken by

Table 9.4 LC-ESI-MS/MS parameters

Oligosaccharide	RT	MRM	DP	CE	CXP	MPS subtype
ΔUA-GalNAc-4S IS	2.8	788/534	−70	−35	−25	
UA-HNAc (1S)	2.9/3.2	806/295	−84	−45	−25	I, II
UA-HN-UA(1S)	3.3	940/331	−80	−44	−25	I, VII
HNAc (1S)	3.6	630/256	−80	−45	−25	IIID, IVA, VI
HNAc (2S)	3.0	710/256	−80	−45	−25	VI
HNAc-UA (1S)	3.8	806/331	−84	−45	−25	IVA, VI
$(HNAc\text{-}UA)_2$ (1S)	3.0	1185/931	−80	−45	−25	IIIB
$(HNAc\text{-}UA)_2$ (2S) $[M\text{-}H]^{2-}$	3.2	632/298	−80	−45	−25	IIID, IVA, VI
HN-UA (1S)	3.5	764/331	−80	−45	−25	IIIA
$(HN\text{-}UA)_2$-HNAc(2S)$[M\text{-}H]^{2-}$	2.6	691/605	−40	−24	−25	IIIC
$(Hex\text{-}HNAc)_2$ (2S)	2.4	1240/256	−87	−41	−25	IVB

RT, retention time; MRM, multiple reaction monitoring; CE, collision energy; CXP, collision exit potential; DP, declustering potential; GalNAc, *N*-acetylgalactosamine; Hex, hexose; HN, hexosamine; HNAc, *N*-acetylhexosamine; IS, internal standard; S, sulphate; UA, uronic acid

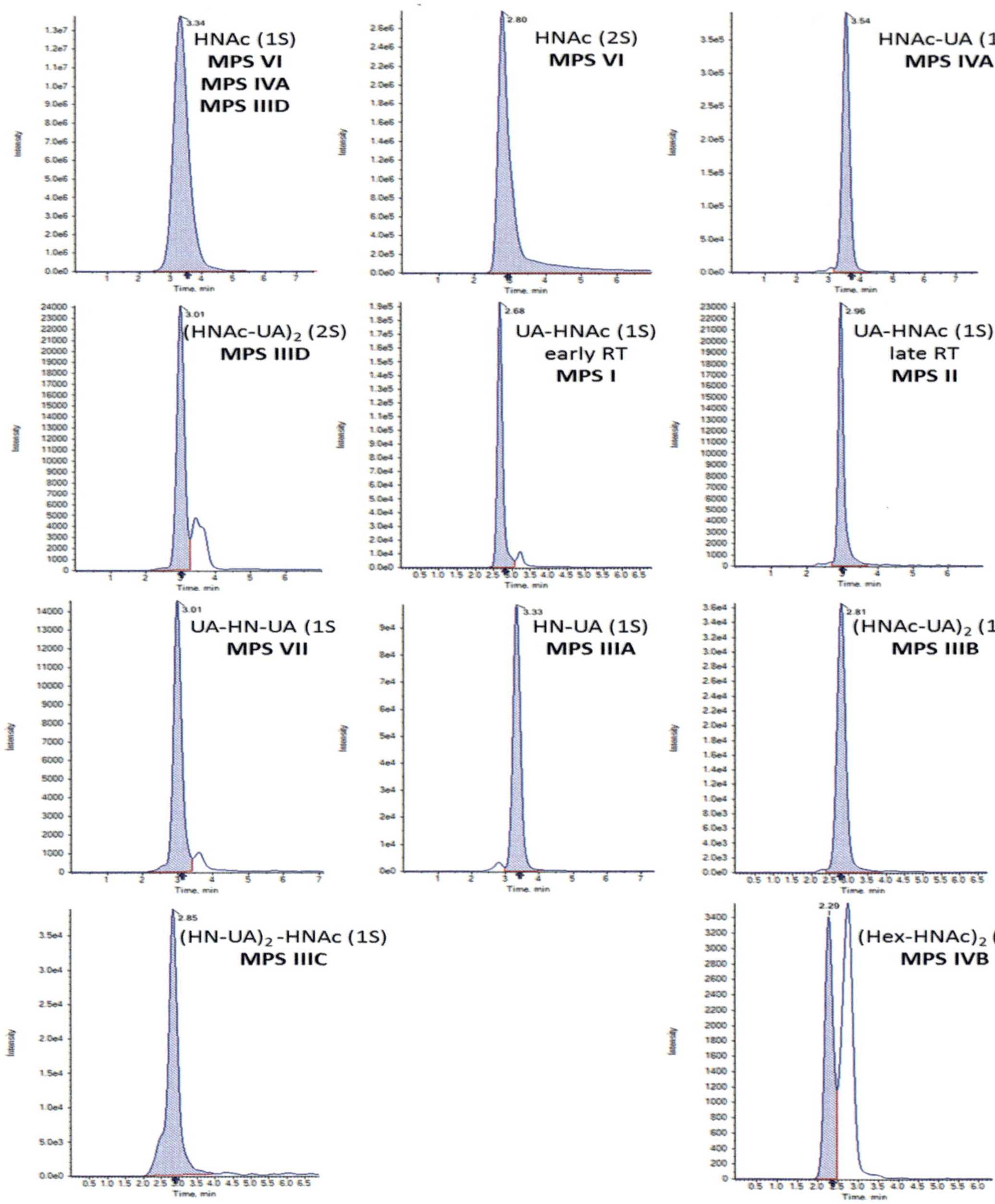

Fig. 9.4 Chromatography for each MRM pair and the respective MPS

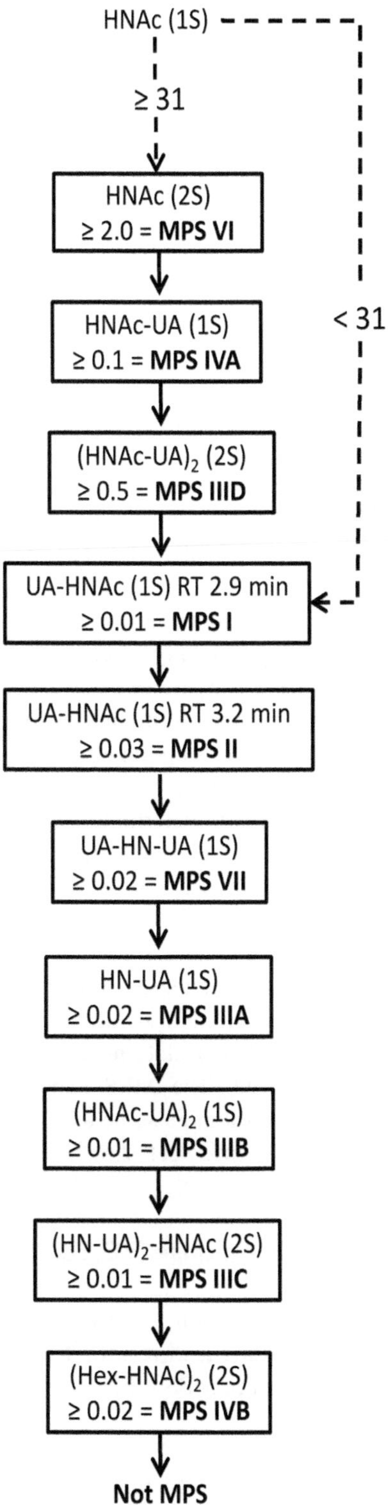

Fig. 9.5 MPS subtype diagnostic flowchart

measuring the requisite enzyme (see Sect. 9.7). The advantage is that only one enzyme, along with a control enzyme to assess integrity of the sample, and in the case of a sulphatase, rule out multiple sulphatase deficiency, needs to be tested to confirm the urine MPS subtype result.

9.7 Enzyme Activity Determinations

Enzyme activity determination for the MPS have evolved over the decades from labour intensive assays using radiolabelled substrates on invasive biopsy samples, to the use of artificial, fluorogenic samples on blood samples. Although MPS enzymes are now often measured in dried blood spots, primarily for newborn screening programs, the typical diagnostic sample is a leukocyte pellet, with fibroblasts remaining the gold standard used in ERNDIM proficiency testing. Therefore, this section will outline methodology for leukocytes as is undertaken in the author's laboratory but are easily adapted for dried blood spots. A multiplex enzymology method for MPS enzyme determinations in dried blood spots, with the exception of MPS IIID is described in Khaledi and Gelb [37] with reference to the synthesis of the substrates in the Supporting Information and their related articles (referenced in the Materials section).

Principle

Leukocytes are separated from erythrocytes by differential sedimentation at unit gravity in a dextran/saline solution [38]. The resultant cell pellet is homogenized and total protein determined by the method of Lowry [39]. The fluorogenic assays are based on the removal of 4-methylumbelliferyl (4-MU) from the substrate by the enzyme at acidic pH. The addition of alkali buffer terminates the enzymatic reaction and enhances the fluorescence of the released 4-MU.

Sample Requirements and Preparation

At least 3 mL of fresh, whole EDTA blood is required and the leukocytes should be isolated within 48 h of blood draw. The blood should be kept at room temperature in the intervening time. Transfer blood from the collection tube into a 10 mL plastic conical tube. Use a separate tube for every 4 mL of blood. Centrifuge at 1700 *g* for 5 min. Remove plasma leaving about 5 mm above the buffy coat. To each tube of packed cells add 0.9% w/v NaCl solution to give a final volume of 8 mL. Add 2 mL of dextran/saline (prepared by dissolving 5 g dextran (MW 250 kDa, Pharmacosmos) and 0.7 g NaCl in 100 mL H_2O) then mix tubes by gentle inversion and allow the red cells to settle for about 45 min. Transfer the leukocyte-containing supernatants into new tubes and sediment the leukocytes by centrifugation at 1000 *g* for 2 min. Discard supernatant and resuspend pellet in 4 mL of 0.2% w/v NaCl solution. Leave the cells for 45 s, but no longer than 1 min, before adding 3.2 mL of 0.2% w/v NaCl solution and mix by inversion. Pellet the cells again by centrifugation, and it may be necessary to repeat the NaCl washing procedure if significant red cells remain in the pellet. The leukocyte pellets can be stored at –20 °C until required.

Before the MPS enzyme assay, thaw leukocyte pellet on ice and resuspend in 0.1–0.3 mL (depending on pellet size) of 0.1% Triton X-100 (BDH Laboratories 30632) in H_2O and sonicate for 15 s on ice. Protein is determined by the method of Lowry et al. [39] and the sample diluted to 1 mg/mL.

General Reagents

1. 4-MU standard (0.142 mM) is prepared by dissolving 25 mg 4-MU (MW 176.2, Toronto Research Chemicals M33300) in 1 mL 100% ethanol and then make up to a final volume of 1 L with H_2O. Store at –20 °C in 0.5 mL aliquots.
2. Glycine buffer, 0.2 M pH 10.7, is prepared by dissolving 15.03 g glycine (Sigma-Aldrich 410225), 13.25 g Na_2CO_3 and 5.85 g NaOH in 800 mL H_2O. Adjust pH to 10.7 with NaOH and make up to a final volume of 1 L with H_2O and store at room temperature.

Instrumentation

Fluorometer for tubes or microtitre plates, e.g., PE Victor NIVO plate reader.

Sonicator, e.g., Misonix Micro XL ultrasonic cell disruptor.

Quality Control

Leukocyte pellets are prepared from unaffected samples to include as normal controls. Prepare a large batch of leukocyte pellets from pooled fresh EDTA blood and store as individual homogenates at 1 mg/mL in single aliquots to be used in each assay. ERNDIM samples are also included when required for proficiency testing.

For all enzyme determinations another enzyme must be measured on the leukocyte homogenate to ensure sample integrity. This is typically another MPS enzyme, but with the exception of MPS IVA, β-galactosidase is also commonly used. Multiple sulphatase deficiency (MSD) must be eliminated when a sulphatase MPS enzyme is deficient unless the diagnosis has been confirmed genetically. Measurement of a second sulphatase enzyme (e.g., *N*-acetylgalactosamine 6-sulphatase, iduronate 2-sulphatase, *N*-sulphoglucosamine sulphohydrolase, glucosamine 6-sulphatse or arylsulphatase A is required.

Laboratories should determine their own reference intervals. The normal cell pellet should yield activity within the normal range and the control enzyme measured in the test samples should also be within the normal range.

Pseudodeficiency of these enzymes can give low activity and result in false diagnoses. In these situations, clinical correlation and molecular analysis is required. The MPS subtype determined in the urine should also be used to confirm enzymatic findings (see Sect. 9.6).

9.7.1 α-L-Iduronidase (EC. 3.2.1.76)

α-L-Iduronidase is responsible for removing the terminal α-L-iduronic acid residue from HS and DS and is deficient in MPS type 1 (Fig. 9.1).

Reagents

1. Substrate buffer (0.1 M formate) is prepared by dissolving 0.67 g of HCOONa (MW 68.01, Sigma-Aldrich 107603) in 80 mL H_2O. If required, adjust pH to 3.6 with formic acid

(Merck F0507) and make up to a final volume of 100 mL with H_2O.

2. Substrate is prepared by dissolving 5 mg of 4-MU-α-L-idopyranosiduronic acid (MW 374.27, Toronto Research Chemicals M334701) in 7 mL substrate buffer to give a final substrate concentration of 2 mM.

Analytical Procedure

The method using the artificial 4-MU substrate was first reported in 1978 [40]. Include an assay blank of 50 μL substrate and 50 μL of 0.9% w/v NaCl. For each sample and the normal control, combine 50 μL substrate with 50 μL of the leukocyte homogenate. Mix and incubate all tubes at 37 °C for 2 h. Terminate the reaction with the addition of 1.5 mL glycine buffer. For the standard, combine 20 μL 4-MU standard with 80 μL 0.9% w/v NaCl and 1.5 mL glycine buffer. Either place tubes in the fluorometer or transfer 0.2 mL of each sample into a microtitre plate and read fluorescence at the excitation wavelength of 355 and emission at 460 nm.

Calculation

α-L-iduronidase activity (nmol/min/mg) is calculated with the formula

$$= \frac{(\text{fluorescence of test - assay blank}) \times 2.84(\text{4MU of standard})}{\text{fluorescence of 4-MU standard} \times 120\,\text{min} \times \text{mg protein in assay}}$$

Interpretation

Normal enzyme activities range from 0.3 to 1.9 nmol/min/mg. MPS I patients with Hurler syndrome (the neurological from of MPS I) typically have undetectable enzyme activity, and patients with a more attenuated disease and no neurological involvement (Scheie syndrome) can have residual activity up to 0.1 nmol/min/mg.

9.7.2 Iduronate 2-Sulphatase (EC. 3.1.6.13)

Iduronate 2-sulphatase is responsible for removing the sulphate group from the 2-position of L-iduronic acid in HS and DS and is deficient in MPS II (Fig. 9.1).

Reagents

1. Substrate buffer is prepared by adding 3 mL of 0.1 M acetic acid to 7 mL 0.1 HCOONa. To 9 mL of this solution, add 37.9 mg of lead acetate (MW 325.29) and pH to 5 with 0.1 M acetic acid and then add the remaining volume to 10 mL.
2. Substrate is prepared by dissolving 5 mg MU-αIdoA-2S (Biosynth EM03201, MW 477) in 8.33 mL of substrate buffer. Aliquot into 1.5 mL Eppendorf tubes and store at – 80 °C. Prepared substrate is stable for at least 3 years.
3. McIlvain's phosphate citrate buffer (Pi/Ci) is prepared by combining 45.43 mL of 0.2 M Na_2HPO_4 (MW 141.96) with 54.48 mL of 0.1 M citric acid monohydrate ($C_6H_8O_7 \cdot H_2O$, MW 210.14). Adjust pH to 4.5 using either solution to a final volume of 100 mL. Store in 10 mL aliquots at –20 °C.
4. Alpha-L-iduronidase solution (Biosynth FI179383).

Analytical Procedure

The protocol is based on the method described by Voznyi et al. [41]. Include an assay blank of 20 μL substrate and 10 μL of 0.1% Triton X-100. For each sample and the normal control, combine 20 μL of substrate with 10 μL of the leukocyte homogenate. Mix and incubate all tubes at 37 °C for 4 h. Add 20 μL Pi/Ci to each tube, mix then add 10 μL alpha-L-iduronidase solution, mix again and incubate for 24 h at 37 °C. Care must be taken when pipetting this enzyme to avoid cross contamination and the use of filter tips is recommended. Terminate the reaction with the addition of 0.2 mL glycine buffer. For the standard, combine 10 μL 4-MU standard with 50 μL H_2O and 0.2 mL glycine buffer. Transfer

0.2 mL of each sample into a microtitre plate and read fluorescence at the excitation wavelength of 355 and emission at 460 nm.

Calculation

Iduronate 2-sulphatase activity (nmol/4 h/mg) is calculated with the formula

$$= \frac{(\text{fluorescence of test - assay blank}) \times 1.42(\text{4MU of standard})}{\text{fluorescence of 4-MU standard} \times \text{mg protein in assay}}$$

Interpretation

Normal enzyme activities range from 23 to 80 nmol/4 h/mg.

MPS II patient range 0 to 0.3 nmol/4 h/mg.

9.7.3 *N*-Sulphoglucosamine Sulphohydrolase (EC. 3.1.6.13)

N-sulphoglucosamine sulphohydrolase is responsible for removing the sulphate groups (sulphamate) attached to the amino groups of glucosamine in HS and is deficient in MPS IIIA (Fig. 9.1).

Reagents

1. Na barbital/acetate (stock) is prepared by dissolving 2.95 g Na barbital (BDH 27283, MW 206.18) and 1.17 g HCOONa (MW 82.03) in 90 mL H_2O. Make up to 100 mL once dissolved.
2. Substrate buffer is prepared by combining 100 mL H_2O, 20 mL of 8.5% (w/v) NaCl and 50 mL of stock. The pH is adjusted to 6.5 with 0.1 M HCl (about 70 mL is required) and then make up to a final volume of 250 mL. Final concentration is 29 mM Na barbital/acetate, 0.68% (w/v) NaCl. Aliquot into 50 mL pots and store at –20 °C.
3. Substrate (20 mM) is prepared by dissolving 17.56 mg 4-MU-α-D-*N*-sulphoglucosaminide (MU-αGLcNS, Biosynth EM06602, MW 439.37) in 2 mL of substrate buffer.
4. Prepare fresh on day of use, 0.9 mg/mL Pefabloc (Roche 11429868001) by dissolving 0.5 mg in 0.555 mL substrate buffer.
5. McIlvain's (2×) phosphate citrate buffer (2×Pi/Ci) is prepared by combining 75 mL of 0.4 M Na_2HPO_4 (MW 141.96) with 25 mL of 0.2 M citric acid monohydrate ($C_6H_8O_7 \cdot H_2O$, MW 210.14). Adjust pH to 6.7 with 5 M NaOH if required. Store at –20 °C.
6. α-glucosidase from *Saccharomyces cerevisiae* (Sigma-Aldrich G5003) is resuspended in H_2O to 0.1 U/μL. For a 26.5 U/mg purchased powder, resuspend 37.74 mg in 10 mL H_2O and store aliquots at –20 °C. Use filter tips to avoid cross contamination.

Analytical Procedure

The protocol is based on the method described by Karpova et al. [42]. Include an assay blank of 10 μL substrate, 10 μL of 0.1% Triton X-100, 5 μL Pefabloc and 5 μL substrate buffer. For each sample and the normal control, combine 10 μL of substrate, 5 μL substrate buffer and 5 μL Pefabloc with 10 μL of the leukocyte homogenate. Mix and incubate all tubes at 47 °C for 17 h. Add 6 μL 2×Pi/Ci to each tube, mix then add 10 μL α-glucosidase, mix again and incubate for 24 h at 37 °C. Care must be taken when pipetting this enzyme to avoid cross contamination and the use of filter tips is recommended. Terminate the reaction with the addition of 0.2 mL glycine buffer. For the standard, combine 10 μL 4-MU standard with 36 μL H_2O and 0.2 mL glycine buffer. Transfer 0.2 mL of each sample into a microtitre plate and read fluorescence at the excitation wavelength of 355 and emission at 460 nm.

Calculation

Enzyme activity (nmol/17 h/mg) is calculated with the formula

$$= \frac{(\text{fluorescence of test - assay blank}) \times 1.42(\text{4MU of standard})}{\text{fluorescence of 4-MU standard} \times \text{mg protein in assay}}$$

Interpretation
Normal enzyme activities range from 3.2 nmol/17 h/mg.

MPS IIIA patient range <0.1 nmol/17 h/mg.

9.7.4 α-*N*-Acetylglucosaminidase (EC. 3.2.1.50)

α-*N*-Acetylglucosaminidase is responsible for removing the *N*-acetylglucosamine residues in HS and is deficient in MPS IIIB (Fig. 9.1).

Reagents
1. Substrate buffer is prepared by dissolving 1.64 CH_3COONa (MW 82.03) in 70 mL H_2O. Adjust pH to 4.3 using glacial acetic acid (MW 60.05) and make up to a final volume of 100 mL. Store at –20 °C.
2. Substrate is prepared by dissolving 7.6 mg of 4-MU 2-acetamido-2-deoxy α-D-glucopyranoside (Toronto Research Chemicals M333800, MW 379.4) in 10 mL of substrate buffer.

Analytical Procedure
The method using this 4-MU substrate was first reported in 1985 [43]. Include an assay blank of 50 μL substrate and 50 μL of 0.9% w/v NaCl. For each sample and the normal control, combine 50 μL substrate with 50 μL of the leukocyte homogenate. Mix and incubate all tubes at 37 °C for 4 h. Terminate the reaction with the addition of 1.5 mL glycine buffer. For the standard, combine 20 μL 4-MU standard with 80 μL 0.9% w/v NaCl and 1.5 mL glycine buffer. Either place tubes in the fluorometer or transfer 0.2 mL of each sample into a microtitre plate and read fluorescence at the excitation wavelength of 355 and emission at 460 nm.

Calculation
Enzyme activity (pmol/min/mg) is calculated with the formula

$$= \frac{(\text{fluorescence of test - assay blank}) \times 2840(\text{4MU of standard})}{\text{fluorescence of 4-MU standard} \times 240\text{min} \times \text{mg protein in assay}}$$

Interpretation
Normal enzyme activities range from 10 to 33 pmol/min/mg.

MPS IIIB patient range 0–2.5 pmol/min/mg.

9.7.5 Acetyl CoA:α-Glucosaminide *N*-Acetyltransferase (EC. 2.3.1.3)

Acetyl CoA:α-glucosaminide *N*-acetyltransferase is responsible for the acetylation of the glucosamine amino groups in HS that are generated from the activity of *N*-sulphoglucosamine sulphohydrolase and is deficient in MPS IIIC (Fig. 9.1).

Reagents
1. Substrate buffer is prepared by combining 28.4 mL of 0.2 M Na_2HPO_4 (MW 141.96) with 21.6 mL of 0.1 M citric acid monohydrate ($C_6H_8O_7 \cdot H_2O$, MW 210.14). Adjust pH to 5.5 if needed. Then add 125 μL Triton X-100 (final 0.25%) and 50 μL 20% Na azide (final 0.02%), the latter is toxic and must be added in a fume hood. Store at –20 °C.
2. Substrate is prepared by dissolving 1.01 mg 4-MU-β-D-glucosaminide (Biosynth EM31025, MW 337.32) in 1 mL substrate buffer. Aliquot into 0.2 mL and store at –80 °C. Stable for at least 3 years.
3. AcetylCo (8 mM) is prepared by reconstituting 1 mg vial of acetyl coenzyme A trisodium salt (Sigma-Aldrich A2056, MW 809.6) in 0.145 mL H_2O. Stable at –20 °C.
4. 0.2% BSA is prepared by dissolving 0.2 g (Sigma-Aldrich A7030) in 100 mL H_2O.

Analytical Procedure
The protocol is based on the method described by Voznyi et al. [44]. Include an assay blank of 10 μL substrate, 10 μL of 0.2% (w/v) BSA and 10 μL AcetylCo. For each sample and the normal control, add 10 μL of substrate and 10 μL AcetylCo to 10 μL of the leukocyte homogenate.

Mix and incubate all tubes at 37 °C for 17 h. Terminate the reaction with the addition of 0.2 mL glycine buffer. For the standard, combine 10 μL 4-MU standard with 20 μL H_2O and 0.2 mL glycine buffer. Transfer 0.2 mL of each sample into a microtitre plate and read fluorescence at the excitation wavelength of 355 and emission at 460 nm.

Calculation
Enzyme activity (nmol/h/mg) is calculated with the formula

$$= \frac{(\text{fluorescence of test - assay blank}) \times 1.42\ (\text{4MU of standard})}{\text{fluorescence of 4-MU standard} \times 10 \times \text{mg protein in assay}}$$

Total β-hexosaminidase (see below) must be measured as the control enzyme, as if this enzyme is low false positive results may arise. This is because the transferase converts the substrate to a non-fluorescing intermediate and β-hexosaminidase is needed to release free 4 MU from this compound.

Interpretation
Normal enzyme activities range from 0.5 to 1.5 nmol/h/mg.

MPS IIIC patient range <0.01 nmol/h/mg.

9.7.6 Total β-Hexosaminidase

Reagents

1. Citrate-phosphate buffer (Ci/Pi) is prepared by combining 266 mL of 0.1 M citric acid monohydrate ($C_6H_8O_7 \cdot H_2O$, MW 210.14) with 246.5 mL of 0.2 M Na_2HPO_4 (MW 141.96). Adjust pH to 4.6 with 5 M NaOH if required. Store at –20 °C.
2. Substrate (0.66 mM) is prepared by dissolving 10 mg 4-MU 2-acetamido 2-deoxy-β-D-glucopyranoside (Melford Lab M1089, MW 379.4) in 10 mL Ci/Pi buffer. Aliquot into 1 mL and store at –20 °C.

Analytical Procedure
The protocol is based on the method described by Leaback and Walker [45] and measures both major isoforms, A and B. Include an assay blank of 0.2 mL H_2O, 0.1 mL of 0.9% (w/v) NaCl, and for the samples and the normal control add 0.2 mL H_2O, 90 μL of 0.9% (w/v) NaCl and 10 μL leukocyte homogenate. Add 0.1 mL substrate to all tubes, mix and incubate at 37 °C for 30 min. Terminate the reaction with the addition of 1.5 mL glycine buffer. For the standard, combine 10 μL 4-MU standard with 20 μL H_2O and 0.2 mL glycine buffer. Transfer 0.2 mL of each sample into a microtitre plate and read fluorescence at the excitation wavelength of 355 and emission at 460 nm.

Calculation
Enzyme activity (nmol/min/mg) is calculated with the formula

$$= \frac{(\text{fluorescence of test - assay blank}) \times 1.42(\text{4MU of standard})}{\text{fluorescence of 4-MU standard} \times 30 \times \text{mg protein in assay}}$$

Normal enzyme activities range from 10 to 50 nmol/min/mg.

9.7.7 *N*-Acetylglucosamine 6-Sulphatase (EC. 3.1.6.14)

N-Acetylglucosamine 6-sulphatase is responsible for desulphating the 6-sulphated *N*-acetylglucosamine residues within HS and is deficient in MPS IIID. This enzyme can also desulphate analogous residues in KS (Fig. 9.1).

Reagents

1. Substrate buffer is prepared by mixing 0.758 g lead (II) acetate trihydrate (Sigma-Aldrich 316512, MW 379.33) with 91 mL 0.2 M CH_3COONa and 9 mL 0.2 M acetic acid. Adjust pH to 5.6 if required with either solu-

tion. To the 100 mL buffer, add 0.1 mL 20% Na-azide and 0.25 mL Triton X-100.
2. Substrate (10 mM) is prepared by dissolving 4.81 mg 4-MU 2-acetamido-2-deoxy α-D-glucopyranoside 6-sulphate sodium salt (Biosynth EM13027, MW 481.41) in 1 mL of substrate buffer. Store at −80 °C.
3. McIlvain's (2×) phosphate citrate buffer (2×Pi/Ci) is prepared by combining 50 mL of 0.4 M Na_2HPO_4 (MW 141.96) with 50 mL of 0.2 M citric acid monohydrate ($C_6H_8O_7 \cdot H_2O$, MW 210.14). Adjust pH to 4.7 with 5 M NaOH if required. Store at −20 °C.
4. LEBT-M3D (lysosome enzymes purified from bovine testis) vials from Biosynth are no longer available at the time of writing. An alternative is being sought. Each vial is typically reconstituted in 2.1 mL H_2O.

Analytical Procedure
The protocol is based on the method described by Wang et al. [46] and has been adapted for leukocytes. Include an assay blank of 10 μL substrate and 10 μL of 0.1% Triton X-100. For each sample and the normal control, combine 10 μL of substrate with 10 μL of the leukocyte homogenate. Mix and incubate all tubes at 37 °C for 17 h. Add 70 μL 2×Pi/Ci to each tube, mix then add 20 μL of reconstituted LEBT-M3D, mix again and incubate for 24 h at 37 °C. Care must be taken when pipetting to avoid cross contamination and the use of filter tips is recommended. Terminate the reaction with the addition of 0.2 mL glycine buffer. For the standard, combine 10 μL 4-MU standard with 70 μL H_2O and 0.2 mL glycine buffer. Transfer 0.2 mL of each sample into a microtitre plate and read fluorescence at the excitation wavelength of 355 and emission at 460 nm.

Calculation
Enzyme activity (nmol/17 h/mg) is calculated with the formula

$$= \frac{(\text{fluorescence of test - assay blank}) \times 1.42(\text{4MU of standard})}{\text{fluorescence of 4 - MU standard} \times \text{mg protein in assay}}$$

Interpretation
Normal enzyme activities range from 7.7 to 19 nmol/17 h/mg.

Reference interval for MPS IIID not available for leukocytes but for cultured skin fibroblast homogenates is 0.2 to 1.2 nmol/17 h/mg.

9.7.8 *N*-Acetylgalactosamine 6-Sulphatase (EC. 3.1.6.4)

N-Acetylgalactosamine 6-sulphatase (arylsulphatase B) is responsible for removing the 6-sulphate from both the galactose residues in KS and the *N*-acetylgalactosamine residues in CS and is deficient in MPS IVA (Fig. 9.1).

Reagents
1. Substrate buffer is prepared by adding 0.29 g NaCl and 0.095 g lead (II) acetate trihydrate (Sigma-Aldrich 316512, MW 379.33) to 31.5 mL of 0.1 M acetic acid and 17 mL 0.1 M HCOONa. Adjust pH to 5 with 0.1 M acetic acid/HCOONa to a final volume of 50 mL. Ensure this is fully dissolved, gentle heating may be required, and then add 0.05 mL of 20% Na-azide.
2. Substrate is prepared by dissolving 1.4 mg 4-MU-β-D-galactose 6-sulphate.Na (Biosynth EM05134, MW 440) in 1 mL substrate buffer. Aliquot into 0.1 mL and store at −80 °C. Stable for at least 3 years.
3. Phosphate buffer is prepared by combining 48 mL of 0.9 M Na_2HPO_4 (MW 141.96) with 1 mL of 0.9 M NaH_2PO_4 (MW 156.01). Adjust pH to 4.3 using either solution to a final volume 50 mL. Store in 10 mL aliquots at −20 °C.
4. Beta-galactosidase stock solution is prepared by adding 10 mL H_2O to a vial of 2500 U of β-D-galactoside galactohydrolase (from *Aspergillus oryzae*, Sigma-Aldrich G1560). This is then further diluted by adding 4980 μL of 0.2% (w/v) BSA to 10 μL of the stock, to give a working concentration of 10 U/mL.

5. 0.2% BSA is prepared by dissolving 0.2 g (Sigma-Aldrich A7030) in 100 mL H_2O.

Analytical Procedure

The protocol is based on the method described by van Diggelen et al. [47]. Include an assay blank of 20 µL substrate and 10 µL of 0.2% (w/v) BSA. For each sample and the normal control, combine 20 µL of substrate with 10 µL of the leukocyte homogenate. Mix and incubate all tubes at 37 °C for 17 h. Add 5 µL phosphate buffer to each tube, mix then add 10 µL of β-galactosidase (10 U/mL), mix again and incubate for 2 h at 37 °C. Care must be taken when pipetting this enzyme to avoid cross contamination and the use of filter tips is recommended. Terminate the reaction with the addition of 0.2 mL glycine buffer. For the standard, combine 10 µL 4-MU standard, 25 µL H_2O, 10 µL of 0.2 % (w/v) BSA and 0.2 mL glycine buffer. Transfer 0.2 mL of each sample into a microtitre plate and read fluorescence at an excitation wavelength of 355 and emission at 460 nm.

Calculation

Enzyme activity (nmol/17 h/mg) is calculated with the formula

$$= \frac{(\text{fluorescence of test - assay blank}) \times 1.42(\text{4MU of standard})}{\text{fluorescence of 4-MU standard} \times \text{mg protein in assay}}$$

Interpretation

Normal enzyme activities range from 90 to 3600 nmol/17 h/mg.

MPS IVA patient range 0–14 nmol/17 h/mg.

9.7.9 β-Galactosidase (EC. 3.2.1.23)

β-galactosidase removes β-linked galactose residues in KS and G_{M1} ganglioside and is deficient in MPS IVB (Fig. 9.1) and G_{M1} gangliosidosis.

Reagents

1. Citrate-phosphate buffer containing NaCl (Ci/PiNaCl) is prepared by dissolving 7 g NaCl in a 500 mL solution made up of 307.5 mL of 0.2 M citric acid monohydrate ($C_6H_8O_7 \cdot H_2O$, MW 210.14) and 192.5 mL of 0.4 M Na_2HPO_4 (MW 141.96). Adjust pH to 4.4 with 5 M NaOH if required. Store at –20 °C.
2. Substrate (0.72 mM) is prepared by dissolving 12.18 mg 4-MU β-D-galactopyranoside (Melford Lab M1095, MW 338.3) in 50 mL Ci/PiNaCl buffer. Aliquot into 1 mL and store at –20 °C.

Analytical Procedure

The protocol is based on the method described by Kolodny and Mumford [48]. Include an assay blank of 20 µL 0.9% (w/v) NaCl, and for all samples and the normal control add 10 µL of 0.9% (w/v) NaCl and 10 µL leukocyte homogenate. Add 80 µL substrate to all tubes, mix and incubate all tubes at 37 °C for 1 h. Terminate the reaction with the addition of 1.5 mL glycine buffer. For the standard, combine 20 µL 4-MU standard with 80 µL H_2O and 1.5 mL glycine buffer. Transfer 0.2 mL of each sample into a microtitre plate and read fluorescence at the excitation wavelength of 355 and emission at 460 nm.

Calculation

Enzyme activity (nmol/min/mg) is calculated with the formula

$$= \frac{(\text{fluorescence of test - assay blank}) \times 2.84(\text{4MU of standard})}{\text{fluorescence of 4-MU standard} \times 60 \times \text{mg protein in assay}}$$

Interpretation

Normal enzyme activities range from 1 to 6 nmol/min/mg.

MPS IVB/G_{M1} gangliosidosis patient range <0.2 nmol/h/mg.

The distinction between G_{M1} gangliosidosis and MPS IVB must be made clinically, given the enzyme is deficient in both disorders. β-D-galactosidase is also reduced in patients with galactosialidosis but not to the same extent, and again clinical distinction is required.

9.7.10 *N*-Acetyl-Galactosamine 4-Sulphatase (Arylsulphatase B; EC 3.1.6.1)

N-acetyl-galactosamine 4-sulphatase is responsible for removing the 4-sulphate from the *N*-acetyl-galactosamine residues in DS and is deficient in MPS VI (Fig. 9.1).

Reagents

1. Acetate buffer (0.5 M) is prepared by dissolving 20.5 g of CH_3COONa (MW 82.03) and 1.28 g of barium acetate (MW 255.42) in 495 mL H_2O. Adjust the pH to 6.0 with NaOH and then make up to 0.5 L. Store at 4 °C.
2. Substrate (0.05 M) is prepared by dissolving 310 mg *p*-nitrocatechol sulphate dipotassium salt (Sigma-Aldrich N7251, MW 311.35) in 10 mL acetate buffer. Adjust the pH to 6.0 and make up to 20 mL in acetate buffer. Store at 4 °C in aluminium foil to protect from light.
3. 1 N NaOH prepared by dissolving 4 g NaOH (MW 40) in 100 mL H_2O.

Analytical Procedure

The protocol is based on the method described by Baum et al. [49]. Four tubes are needed for each sample, labelled 'Reaction', '0 min', '30 min' and '90 min'. In tubes labelled 'Reaction' combine 200 μL H_2O, 100 μL leukocyte homogenate and 450 μL substrate, mix and aliquot 200 μL into tubes labelled '0 min', '30 min' and '90 min'. Add 0.5 mL 1 N NaOH to '0 min' tubes and leave at room temperature. Cover the '30 min' and '90 min' tubes with parafilm and incubate for 30 and 60 min at 37 °C, respectively. At the end of each incubation time add 0.5 mL 1 N NaOH. Read absorbance at 515 nm.

Calculation

Firstly, subtract the absorbance reading for '0 min' from both the '30 min' and '90 min' to give corrected absorbance, then

$$= \frac{\left[\text{corrected absorbance90min - 30min - } 0.2 \times \text{absorbance0min}\right] \times 705.6}{\text{mg / mL protein}}$$

Note that this is based on the molar extinction coefficient of 12,400 for *p*-nitrocatechol to determine enzyme activity in nmol/h/mg of leukocyte homogenate protein

Interpretation

Normal enzyme activities range from 55 to 495 nmol/h/mg.

MPS VI patient range 0–25 nmol/h/mg.

9.7.11 β-D-Glucuronidase (EC. 3.2.1.31)

β-D-glucuronidase is responsible for removing the glucuronic acid residues in DS, HS, CS and hyaluronan and is deficient in MPS VII (Fig. 9.1).

Reagents

1. Substrate buffer is prepared by combining 227 mL of 0.4 M Na_2HPO_4 (MW 141.96) with 273 mL of 0.2 M citric acid monohydrate ($C_6H_8O_7 \cdot H_2O$, MW 210.14). Adjust pH to 5.5 if needed. Store at −20 °C.
2. Substrate (5 mM) is prepared by dissolving 44 mg 4-MU-β-D-glucuronide (Melford M1003, MW 352.3) in 25 mL substrate buffer. Aliquot into 0.5 mL and store at −20 °C. Stable for at least 3 years.

Analytical Procedure

The protocol is based on the method described by Kolodny and Mumford [48]. Include an assay blank of 50 μL 0.9% (w/v) NaCl and 50 μL substrate. For each sample and the normal control, add 50 μL of substrate to 50 μL of the leukocyte homogenate. Mix and incubate all tubes at 37 °C for 60 min. Terminate the reaction with the addition of 1.5 mL glycine buffer. For the standard, combine 20 μL 4-MU standard with 80 μL H_2O and 1.5 mL glycine buffer. Transfer 0.2 mL of each sample into a microtitre plate and read fluorescence at the excitation wavelength of 355 and emission at 460 nm.

Calculation

Enzyme activity (nmol/min/mg) is calculated with the formula

$$= \frac{(\text{fluorescence of test - assay blank}) \times 2.84(\text{4MU of standard})}{\text{fluorescence of 4 - MU standard} \times 60 \times \text{mg protein in assay}}$$

Interpretation

Normal enzyme activities range from 0.5 to 3.5 nmol/min/mg.

MPS VII patient range <0.03 nmol/min/mg.

References

1. Chin S, Fuller M (2021) Prevalence of lysosomal storage disorders in Australia from 2009 to 2020. Lancet Reg Health West Pac 19:100344
2. Natowicz MR, Short MP, Wang Y, Dickersin GR, Gebhardt MC, Rosenthal DI, Sims KB, Rosenberg AE (1996) Clinical and biochemical manifestations of hyaluronidase deficiency. N Eng J Med 335:1029
3. Cyske Z, Gaffke L, Pierzynowska K (2022) Complex changes in the efficiency of the expression of many genes in monogenic diseases, mucopolysaccharidoses, may arise from significant disturbances in the levels of factors involved in the gene expression regulation processes. Genes 13:593
4. Clarke LA, Giugliani R, Guffon N, Jones SA, Keenan HA, Munoz-Rojas MV, Okuyama T, Viskochil D, Whitley CB, Wijburg FA, Muenzer J (2019) Genotype-phenotype relationships in mucopolysaccharidosis type I (MPS I): insights from the international MPS I registry. Clin Genet 96:281–289
5. Burton BK, Hickey R, Hitchens L (2020) Newborn screening for mucopolysaccharidosis type II in Illinois: an update. Int J Neonatal Screen 6:73
6. Hopkins PV, Klug T, Vermette L, Raburn-Miller J, Kiesling J, Rogers S (2018) Incidence of 4 lysosomal storage disorders from 4 years of newborn screening. JAMA Pediatr 172:696–697
7. Lindahl U, Couchman J, Kimata K, Esko JD (2015) Proteoglycans and sulphated glycosaminoglycans. In: Varki A, Cummings RD, Esko JD, Stanley P, Hart GW, Aebi M, Darvill AG, Kinoshita T, Packer NH, Prestegard JH, Schnaar RL, Seeberger PH (eds) Essentials of glycobiology, 3rd edn. Cold Spring Harbor Laboratory, New York, pp 207–221
8. Gottschalk J, Elling L (2021) Current state on the enzymatic synthesis of glycosaminoglycans. Curr Opin Chem Biol 61:71–80
9. Halper J (2014) Proteoglycans and diseases of soft tissues. Adv Exp Med Biol 802:49–58
10. Trowbridge JM, Gallo RL (2002) Dermatan sulphate: new functions from an old glycosaminoglycan. Glycobiology 12:117R–125R
11. Bernfield M, Götte M, Park PW, Reizes O, Fitzgerald ML, Lincecum J, Zako M (1999) Functions of cell surface heparan sulfate proteoglycans. Annu Rev Biochem 68:729–777
12. Caterson B, Melrose J (2018) Keratan sulfate, a complex glyosaminoglycan with unique functional capability. Glycobiology 28:182–206
13. Abatangelo G, Vindigni V, Avruscio G, Pandis L, Brun P (2020) Hyaluronic acid: redefining its role. Cells 9:1743
14. Sasarman F, Maftei C, Campeau PM, Brunel-Guitton C, Mitchell GA, Allard P (2016) Biosynthesis of glycosaminoglycans: associated disorders and biochemical tests. J Inherit Metab Dis 39:173–188
15. Fuller M (2020) Laboratory diagnosis of lysosomal diseases: Newborn screening to treatment. Clin Biochem Rev 41:53–66
16. Berry HK (1987) Screening for mucopolysaccharide disorders with the Berry spot test. Clin Biochem 20:365–371
17. Berry HK, Berry MA, Spinanger J (1960) A paper spot test useful in the study of Hurler syndrome. J Lab Clin Med 55:136–138
18. Berman ER, Vered J, Bach G (1971) A reliable spot test for mucopolysaccharidoses. Clin Chem 17:886–890
19. De Jong JGN, Hasselman JJF, Landeghem AAJ, Vader HL, Wevers RA (1991) The spot test is not a reliable screening procedure for the mucopolysaccharidoses. Clin Chem 37:572–575
20. Chih-Kuang C, Shuan-Pei L, Shyue-Jye L, Tuen-Jen W (2002) MPS screening methods, the Berry spot and acid turbidity tests, cause a high incidence of false-negative results in Sanfilippo and Morquio syndromes. J Clin Lab Anal 16:253–258
21. Carter CH, Wan AT, Carpenter DG (1968) Commonly used tests in the detection of Hurler's syndrome. J Pediatr 73:217–221
22. Björnsson S (1998) Quantification of proteoglycans as glycosaminoglycans in biological fluids using an alcian blue dot blot analysis. Anal Biochem 256:229–237

23. De Jong JGN, Wevers RA, Liebrand-van Sambeek R (1992) Measuring urinary glycosaminoglycans in the presence of protein: an improved screening procedure for mucopolysaccharidoses based on dimethylene blue. Clin Chem 38:803–807
24. Chin SJ, Saville JT, McDermott BK, Zankl A, Fletcher JM, Fuller M (2020) Chondroitin sulphate disaccharide is a specific and sensitive biomarker for mucopolysaccharidosis type IVA. JIMD Rep 55:68–74
25. Cappelletti R, del Rosso M, Chiarugi VP (1979) A New electrophoretic method for the complete separation of all known animal glycosaminoglycans in a monodimensional run. Anal Biochem 99:311–315
26. Hopwood JJ, Harrison JR (1982) High-resolution electrophoresis of urinary glycosaminoglycans: an improved screening test for the mucopolysaccharidoses. Anal Biochem 119:120–127
27. Pepi LE, Sanderson P, Stickney M, Amster IJ (2021) Developments in mass spectrometry for glycosaminoglycan analysis. Mol Cell Proteomics 20:100025
28. Forni G, Malvagia S, Funghini S, Scolamiero E, Mura M, Della Bona M, Villanelli F, Damiano R, la Marca G (2019) LC-MS/MS method for simultaneous quantification of heparan sulphate and dermatan sulphate in urine by butanolysis derivatization. Clin Chim Acta 488:98–103
29. Auray-Blais C, Lavoie P, Tomatsu S, Valayannopoulos V, Mitchell JJ, Raiman J, Beaudoin M, Maranda B, Clarke JTR (2016) UPLC-MS/MS detection of disaccharides derived from glycosaminoglycans as biomarkers of mucopolysaccharidoses. Anal Chim Acta 936:139–148
30. Gill VL, Aich U, Rao S, Pohl CA, Zaia J (2013) Disaccharide analysis of glycosaminoglycans using hydrophilic interaction chromatography and mass spectrometry. Anal Chem 85:1138–1145
31. Lawrence R, Brown JR, Al-Mafraji K, Lamanna WC, Beitel JR, Boons GJ, Esko JD, Crawford BE (2012) Disease-specific non-reducing end carbohydrate biomarkers for mucopolysaccharidoses. Nat Chem Biol 8:197–204
32. Martell LA, Cunico RL, Ohh J, Fulkerson W, Furneaux R, Foehr ED (2011) Validation of an LC-MS/MS assay for detecting relevant disaccharides from keratan sulphate as a biomarker for Morquio A syndrome. Bioanalysis 3:1855–1866
33. Herbst ZM, Urdaneta L, Klein T, Fuller M, Gelb MH (2020) Evaluation of multiple methods for quantification of glycosaminoglycan biomarkers in newborn dried blood spots from patients with severe and attenuated mucopolysaccharidosis I. Int J Neonatal Screen 6:69
34. Herbst ZM, Urdaneta L, Klein T, Burton BK, Basheeruddin K, Liao HC, Fuller M, Gelb MH (2022) Evaluation of two methods for quantification of glycosaminoglycan biomarkers in newborn dried blood spots from patients with severe and attenuated mucopolysaccharidosis type II. Int J Neonatal Screen 8:9
35. Saville JT, McDermott BK, Fletcher JM, Fuller M (2019) Disease and subtype specific signatures enable precise diagnosis of the mucopolysaccharidoses. Genet Med 21:753–757
36. Saville JT, Flanigan KM, Truxal KV, McBride KL, Fuller M (2019) Evaluation of biomarkers for Sanfilippo syndrome. Mol Genet Metab 128:68–74
37. Khaledi H, Gelb MH (2020) Tandem mass spectrometry enzyme assays for multiplex detection of 10-mucopolysaccharidoses in dried blood spots and fibroblasts. Anal Chem 92:11721–11727
38. Kampine JP, Brady RO, Kanfer JN, Feld M, Shapiro D (1966) Diagnosis of Gaucher's disease and Niemann-Pick disease with small samples of venous blood. Science 155:86–88
39. Lowry OH, Roseborough NJ, Farr AL, Randall RJ (1951) Protein measurements with Folin phenol reagent. J Biol Chem 193:265–275
40. Stirling JL, Robinson D, Fensom AH, Benson PF, Baker JE (1978) Fluorimetric assay for prenatal detection of Hurler and Scheie homozygotes or heterozygotes. Lancet 1:147
41. Voznyi YV, Keulemans JLM, van Diggelen OP (2001) A fluorimetric assay for the diagnosis of MPS II (Hunter's disease). J Inherit Metab Dis 24:675–680
42. Karpova EA, Voznyi YV, Keulemans JLM, Hoogeveen AT, Winchester B, Tsvetkova IV, van Diggelen OP (1996) A fluorimetric enzyme assay for the diagnosis of Sanfilippo disease type A (MPS IIIA). J Inherit Metab Dis 19:278–285
43. Marsh J, Fensom H (1985) 4-methylumbelliferyl α-N-acetylglucosaminidase activity for diagnosis of Sanfilippo B disease. Clin Genet 27:258–262
44. Voznyi YV, Karpova EA, Dudukina TV, Tsvetkova IV, Boer AM, Janse HC, van Diggelen OP (1993) A fluorimetric enzyme assay for the diagnosis of Sanfilippo disease C (MPS III C). J Inherit Metab Dis 16:465–472
45. Leaback DH, Walker PG (1961) Studies on glucosaminidase. The fluorimetric assay of N-acetyl-β-glucosaminidase. Biochem J 78:151–156
46. Wang H, Ya V, Boer A, Kleijer W, van Diggelen OP (1993) A fluorimetric enzyme assay for the diagnosis of Sanfilippo disease type D. J Inherit Met Dis 16:935–941
47. Van Diggelen OP, Zhao H, Kleijer WJ, Janse HC, Poorthuis BJHM, van Pelt J, Kamerling JP, Galjaard H (1990) A fluorimetric enzyme assay for the diagnosis of Morquio disease type A (MPS IV A). Clin Chim Acta 187:131–140
48. Kolodny EH, Mumford RA (1976) Human leukocyte acid hydolases: characterisation of eleven lysosomal enzymes and study of reaction conditions for their automated analysis. Clin Chim Acta 70:247–257
49. Baum H, Dodgson KS, Spencer B (1959) The assay of arylsulphatases A and B in human urine. Clin Chim Acta 4:453–458

Oligosaccharides and Sialic Acid

10

Marne C. Hagemeijer, Edwin H. Jacobs,
and George J. G. Ruijter

Contents

List of Abbreviations

Asn	Asparagine
CMAS	CMP-sialic acid synthetase
CMP	Cytidine-5'-monophosphate
CV	Coefficient of variation
Fuc	Fucose
Gal	Galactose
GalNAc	*N*-acetylgalactosamine
Glc	Glucose
GlcNAc	*N*-acetylglucosamine
GM1	Monosialotetrahexosylganglioside
GM2	Disialotetrahexosylganglioside
GNE	UDP-GlcNAc-2-epimerase/ ManNAc kinase
GSD	Glycogen storage disease
Hex	Hexose
HRAM	High-resolution accurate mass
IMD	Inherited metabolic disease
IEM	inborn errors of metabolism
ISSD	Infantile sialic acid storage disorder
LC	Liquid chromatography
Man	Mannose
ManNAc	*N*-acetylmannosamine
ML II	Mucolipidosis type II
ML III	Mucolipidosis type III
MPS	Mucopolysaccharidosis
MS	Mass spectrometry
NAGA	α-*N*-acetylgalactosaminidase
NANA	*N*-acetylneuraminic acid
NANP	*N*-acetylneuraminic acid phosphatase
NANS	*N*-acetylneuraminic acid phosphate synthase
NGLY1	*N*-glycanase 1

M. C. Hagemeijer (✉) · E. H. Jacobs
G. J. G. Ruijter
Center for Lysosomal and Metabolic Diseases, Department of Clinical Genetics, Erasmus University Medical Center, Rotterdam, The Netherlands
e-mail: m.hagemeijer@erasmusmc.nl

N. Blau, F. M. Vaz (eds.), *Laboratory Guide to the Methods in Biochemical Genetics*,
https://doi.org/10.1007/978-3-031-58819-8_10

NPL	*N*-acetylneuraminic acid pyruvate lyase
OS	Oligosaccharide
QC	Quality control
SASD	Free sialic acid storage disorder
SD	Standard deviation
Ser	Serine
SLC	Solute carrier
Thr	Threonine
TLC	Thin-layer chromatography
UDP	Uridine diphosphate
UHPLC	Ultra-high performance liquid chromatography

10.1 Introduction

10.1.1 Oligosaccharides

Oligosaccharides (glycans) are a class of carbohydrates that consist of short repeating units of monosaccharides (usually ~3–12 residues) that are covalently linked via glycosidic bonds and generally exist as glycoconjugates with lipids (e.g., glycosphingolipids and glycoglycerolipids) or proteins (e.g., *N*- or *O*-linked glycosylated proteins). This chapter will focus on the latter type of glycoconjugates, whereas the glycolipids are discussed in another chapter 12.

The most common type of glycoproteins contains *O*-linked or *N*-linked sugar-peptide bonds in which the oligosaccharides are either attached via a glycosidic bond to the hydroxyl group of serine (Ser) or threonine (Thr) or via an *N*-glycosyl bond of the amide nitrogen of asparagine (Asn), respectively [1]. An enormous variety of glycan structures with different lengths and composition can be generated via (1) linear glycosidic linkage of monosaccharides between the anomeric carbons of each sugar but also by (2) anomeric carbon linkage to unmodified hydroxyl groups of monosaccharides in an existing linear chain, which results in oligosaccharide branching [2]. These so-called high-mannose, hybrid, and complex forms are only present on *N*-linked glycoproteins (an example of each form is depicted in Fig. 10.1a). Notably, these *N*- and *O*-linked glycan side-chains encode important biological information.

Due to the huge diversity of protein-bound oligosaccharides these macromolecules are important for maintaining various physiological functions, which include among others (1) structural and modulatory roles, (2) extrinsic (interspecies) glycan recognition, (3) intrinsic (intraspecies) glycan recognition, and (4) molecular mimicry of host glycans (reviewed by Varki [4]). As such, conserving proper cellular homeostasis of glycosylation is essential as deficient activities in any of the enzymes involved in the degradation of glycoproteins result in accumulation of incompletely degraded lysosomal glycan substrates and eventually in severe pathological conditions.

10.1.2 Sialic Acid

A variety of different sialic acids exists with *N*-acetylneuraminic acid (NANA) being the major type of sialic acid in humans. In this chapter, *sialic acid* and *NANA* will be used interchangeably. The majority of this monocarboxylic *N*-substituted monosaccharide is part of oligosaccharides (i.e., "conjugated" sialic acid) and only a small fraction is present as "free" sialic acid. Sialic acid originates from glucose, which is converted in a series of enzymatic reactions to uridine diphosphate (UDP)-*N*-acetyl-glucosamine (UDP-GlcNAc) by the hexosamine biosynthetic pathway (HBP). UDP-GlcNAc-2-epimerase/ManNAc kinase (GNE) catalyzes the next step to *N*-acetylmannosamine (ManNAc) by its epimerase function and is negatively regulated by cytidine-5'-monophosphate (CMP)-NANA, which is the end product of the pathway. ManNAc is converted to ManNAc-6-phosphate, also by UDP-GlcNAc-2-epimerase/ManNAc kinase but by its kinase activity, and subsequently to NANA-9-phosphate (NANA-9P) by *N*-acetylneuraminic acid phosphate synthase (NANS), and finally to NANA by *N*-acetylneuraminic acid phosphate phosphatase (NANP). In the nucleus, NANA is converted to CMP-NANA by CMP-sialic acid synthetase (CMAS) after which it is transported

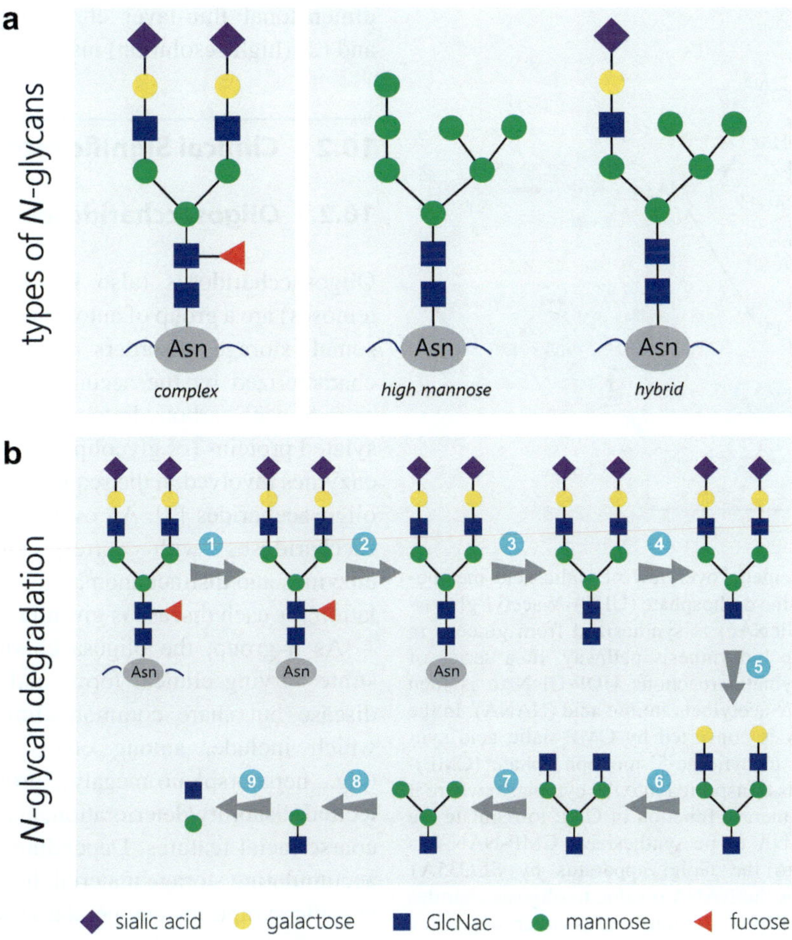

Fig. 10.1 *N*-linked glycan structures and lysosomal degradation. (**a**) A schematic representation of the three most common types of *N*-linked glycoproteins are depicted demonstrating (1) a *complex form* with a fucosylated GlcNAc residue and GlcNAc-substituted mannose residues together with terminal sialic acid residues, (2) a *high-mannose form* that possesses unsubstituted terminal mannose residues attached to the GlcNAc core structure, and (3) a *hybrid form* with both unsubstituted and substituted terminal mannose residues. (**b**) *N*-linked glycoprotein degradation of a representative *complex N*-linked glycan. Each arrow represents an enzymatic reaction. In the first three reactions the polypeptide chain is degraded by proteases (1), followed by removal of the fucose residue by α-fucosidase (2) and hydrolysis of asparagine by *N*-aspartyl-β-glucosaminidase (3). The remaining monosaccharides are released in six sequential enzymatic steps by various exoglycosidases until the complete glycan structure is catabolized: (4) endo-NAc-β-hexosaminidase, (5) α-neuraminidase, (6) β-galactosidase, (7) NAc-β-hexosaminidase, (8) α-mannosidase, and (9) β-mannosidase. *Asn* asparagine, *GlcNAc* N-acetylglucosamine. The figure is adapted from Conte et al. [3] and Michalski and Klein [36]

into the Golgi apparatus by SLC35A1 where it is added to oligosaccharides to generate sialyl-oligosaccharides on glycoproteins and glycolipids. Sialyl-oligosaccharides are degraded mainly in the lysosome by specific lysosomal enzymes (see Fig. 10.1b) with neuraminidase (sialidase) removing the terminal NANA residue. This free NANA is subsequently transported out of the lysosome into the cytoplasm by the lysosomal membrane protein sialin (SLC17A5) where it can either be reused or converted to ManNAc and pyruvate by *N*-acetylneuraminic acid pyruvate lyase (NPL) [5, 6]. A schematic representation of sialic acid metabolism is depicted in Fig. 10.2.

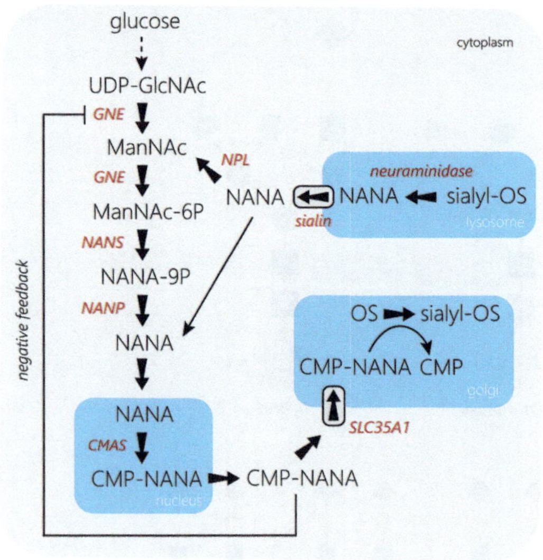

Fig. 10.2 Schematic overview of sialic acid metabolism. First, uridine diphosphate (UDP)-*N*-acetyl-glucosamine (UDP-GlcNAc) is synthesized from glucose in the hexosamine biosynthesis pathway. In a series of sequential enzymatic reactions UDP-GlcNAc is then converted into *N*-acetylneuraminic acid (NANA). In the nucleus NANA is converted by CMP-sialic acid synthase (CMAS) to cytidine-5'-monophosphate (CMP)-NANA, which is transported into the cytoplasm where it inhibits the epimerase function of GNE to regulate the amount of NANA to be synthesized. CMP-NANA is transported into the Golgi apparatus by SLC35A1 where it donates the NANA residue to oligosaccharides (OS), catalyzed by sialyltransferases, after which the sialyl-OS are exported into the cytoplasm. Degradation of sialyl-OS occurs by neuraminidase in lysosomes. The liberated NANA is transported by sialin into the cytoplasm, where it can be re-utilized or degraded into ManNAc and pyruvate (the latter not shown) by *N*-acetylneuraminic acid pyruvate lyase (NPL). *ManNAc* *N*-acetylmannosamine, *ManNAc-6P* *N*-acetylmannosamine-6-phosphate, *NANA-9P* *N*-acetylneuraminic acid-9-phosphate, *GNE* UDP-GlcNAc-2-epimerase/ManNAc kinase, *NANS* *N*- acetylneuraminic acid phosphate synthase, *NANP* *N*-acetylneuraminic acid phosphatase. This figure is adapted from Verheijen [6]

10.1.3 Diagnostic Screening Approaches

Screening for oligosaccharidoses and disorders of sialic acid metabolism is performed by investigating abnormal urinary excretion of partially degraded oligosaccharides and sialic acid. In this chapter, two diagnostic approaches are described to investigate urine samples for the presence of excess storage products: (1) conventional one-dimensional thin-layer chromatography (TLC) and (2) (high-resolution) mass spectrometry.

10.2 Clinical Significance

10.2.1 Oligosaccharidoses

Oligosaccharidoses (also known as glycoproteinoses) are a group of autosomal recessive lysosomal storage disorders (LSDs), which are characterized by the accumulation and urinary excretion of carbohydrate side-chains of glycosylated proteins (or glycolipids) due to defects in enzymes involved in the sequential breakdown of oligosaccharides [7]. An overview of the oligosaccharidoses with corresponding defective enzymes and distinct anomalous glycan accumulation for each disease is given in Table 10.1.

As a group, the oligosaccharidoses demonstrate varying clinical forms and expression of disease but share common clinical symptoms which include, among others, organomegaly (e.g., hepato(spleno)megaly), progressive intellectual disability/deterioration, short stature, and coarse facial features. Depending on the type of accumulating storage material, however, disease-specific symptoms can be observed as well, e.g. corneal clouding, cherry red spots, foam cells, and vacuolated lymphocytes in galactosialidosis, deafness/hearing loss for α- and β-mannosidosis (with hernias and immunodeficiency being characteristic in α-mannosidosis as well), angiokeratoma in α-*N*-acetylgalactosaminidase deficiency, spasticity in fucosidosis (during childhood and adulthood) and hypotonia in the neonatal period in aspartylglucosaminuria [8, 9].

Degradation of all glycans on glycoproteins occurs in the lysosomes and impaired breakdown of *N*-linked glycoproteins specifically results in severe disease. *N*-linked glycan catabolism occurs as a sequential process in which (1) cellular proteases first digest the polypeptide chain of the glycoprotein, followed by (2) enzymatic removal of fucose residues present on the glycan side-chain, and (3) hydrolysis of the asparagine-*N*-acetylglucosamine linkage by an endoglycosidase, after which (4) monosaccharides from the non-reducing end of the glycan

Table 10.1 Overview of the oligosaccharidoses and disorders of sialic acid metabolism

	Gene	Enzyme/protein deficiency	Accumulated products	OMIM #
Oligosaccharidoses				
Galactosialidosis	*CTSA*	Cathepsin A	Sialyl-OS and Galactosyl-OS	256540
α-Mannosidosis	*MAN2B1*	α-D-mannosidase	Mannosyl-OS	248500
β-Mannosidosis	*MANBA*	β-D-mannosidase	GlcNAc-Man and derivatives	248510
α-*N*-Acetylgalactosaminidase deficiency	*NAGA*	α-*N*-acetylgalactosaminidase	GalNAc-Ser, GalNAc-Thr, Glycosphingolipids	609241
Fucosidosis	*FUCA1*	α-L-fucosidase	Fucosyl-OS	230000
Aspartylglucosaminuria	*AGA*	*N*-aspartyl-β-glucosaminidase	GlcNAc-Asn and glycoasparaginyl-OS	208400
Sialidosis (Mucolipidosis I)[a]	*NEU1*	Neuraminidase	Sialyl-OS	256550
Disorders of sialic acid metabolism				
ISSD	*SLC17A5*	Sialin	Free sialic acid	269920
Salla disease	*SLC17A5*	Sialin	Free sialic acid	604369
Sialuria	*GNE*	UDP-GlcNAc-2-epimerase / ManNAc kinase	Free sialic acid	269921
Spondyloepimetaphyseal dysplasia	*NANS*	*N*-acetylneuraminic acid phosphate synthase	ManNAc (with normal free sialic acid)	610442
NPL deficiency	*NPL*	*N*-acetylneuraminic acid pyruvate lyase	Free sialic acid	–

ISSD infantile sialic acid storage disorder, *OS* oligosaccharide, *ManNAc* *N*-acetylmannosamine, *Ser* serine, *Thr* threonine, *Asn* asparagine

[a]Sialidosis is both classified as an oligosaccharidosis and as a disorder of sialic acid metabolism

are sequentially removed by various exoglycosidases (reviewed by Winchester [10]). A schematic representation of the catabolic process of a typical *N*-linked glycan is depicted in Fig. 10.1b (see Table 10.1 for an overview of the disease-specific pathological oligosaccharide accumulations).

10.2.2 Disorders of Sialic Acid Metabolism

One way to classify the sialic acid metabolism disorders that have been described until now is on the basis of whether sialic acid is increased as free sialic acid or as conjugated sialic acid in sialyl-oligosaccharides. The latter group belongs to the oligosaccharidoses and as such only the disorders in which free sialic acid accumulates will be described in this section.

Free sialic acid storage disorder (SASD) is an autosomal recessive disorder in which sialin malfunctions. Sialin is a membrane transporter that exports sialic acid, after being cleaved from sialyl-oligosaccharides, from the lysosomes into the cytoplasm [11]. In SASD there are increased levels of free sialic acid in urine and in cultured fibroblasts. SASD is divided into three clinical forms, depending on the severity of the symptoms. The infantile form (ISSD) is the most severe form, with failure to thrive, mental and motor retardation, hepatosplenomegaly, ascites, dysostosis multiplex, and early death. The milder adult form is *Salla disease*, which originated in the Finnish region of Salla, but patients have also been described outside of Finland. Ataxia and mild mental retardation are symptoms of the mild form, but neurodegeneration, mental retardation, hepatosplenomegaly, and growth retardation can be seen in the intermediate form. In ISSD the urinary excretion of sialic acid is higher than in Salla disease [6]. Not all clinical symptoms may be due to the storage of sialic acid alone, as sialin is also responsible for exocytosis of aspartate and glutamate in synaptic vesicles [12].

Sialuria (non-lysosomal sialuria, French-type sialuria) results from autosomal dominant variants of the *GNE* gene causing increased epimerase activity of the *N*-terminal domain of UDP-GlcNAc-2-epimerase/ManNAc kinase

[13]. These variants disrupt the feedback inhibition by CMP-NANA thereby causing increased cytosolic free sialic acid. Clinical symptoms consist of mild mental retardation, motor retardation, hepatomegaly, prolonged neonatal jaundice, and coarse facial features. Free sialic acid is massively increased in urine; in cultured fibroblasts it is increased as well. Disruption of the *C*-terminal domain of UDP-GlcNAc-2-epimerase/ManNAc kinase by *GNE* variants disrupts its kinase activity [14]. This results in the autosomal recessive disorder *Nonaka myopathy* (hereditary inclusion body myopathy) with progressive muscle dystrophy and muscle wasting. In contrast to sialuria, in Nonaka myopathy there is a defect in the synthesis of sialic acid. In practice, however, a decrease in urinary sialic acid excretion is difficult to observe [6].

Spondyloepimetaphyseal dysplasia is another defect in the sialic acid biosynthesis pathway, in which impaired function of *N*-acetylneuraminic acid phosphate synthase (NANS) results in mental and motor retardation, short stature skeletal dysplasia, hypotonia, and coarse facial features [5]. In urine the excretion of sialic acid is normal, but that of *N*-acetylmannosamine is increased.

N-acetylneuraminic acid pyruvate lyase-deficiency leads to impaired degradation of sialic acid into *N*-acetylmannosamine and pyruvate, thereby increasing the cytosolic concentration of sialic acid. The clinical symptoms are dilated cardiomyopathy, myopathy, and hearing loss. In urine (but not in fibroblasts) free sialic acid is increased [15].

10.3 Properties of Metabolites

The glycan side-chains present on the glycoproteins are composed of various monosaccharides. The most common sugar molecules present in oligosaccharide side-chains are *N*-acetylglucosamine (GlcNAc), galactose (Gal), *N*-acetylgalactosamine (GalNAc), mannose (Man), fucose (Fuc) and the most common form of sialic acid *N*-acetylneuraminic acid [1].

Depending on the specific enzyme deficiency different types of abnormal oligosaccharides will accumulate in the different oligosaccharidoses, which ranges from simple disaccharides like GlcNAc-Fuc in α-L-fucosidase deficiency [16, 17] to large branched oligosaccharides like GlcNAc-Man-(Man-GlcNAc-Gal-NANA)-Man-GlcNAc-Gal-NANA when cathepsin A is defective [18, 19]. Additionally, glycans linked to amino acids like Ser-GalNAc or Thr-GalNAc in NAGA deficiency can be observed [20, 21]. The monocarboxylic *N*-substituted monosaccharide NANA can be detected in pathological urine samples either as increased free sialic acid or increased sialyl-oligosaccharides ("conjugated sialic acid").

An overview of the disease-specific pathological metabolites for the oligosaccharidoses and sialic acid metabolism disorders that can be detected by the screening approaches described in this chapter is given in Table 10.2.

Table 10.2 Oligosaccharide-species detected by UHPLC/HRAM mass spectrometry (negative mode) in patient urine samples

IEM	Rt (min)	Component	m/z	Adduct	Chemical formula	Oligosaccharides	ID	REF
Fucosidosis	2.66	366_fuco (sec)	366.1400	[M-H]-	C14H25NO10	GlcNAc-Fuc	3	16
	3.39	480_fuco	480.1835	[M-H]-	C18H31N3O12	Asn-GlcNAc-Fuc	3	30
	3.37	1055_fuco	1055.3784	[M-H]-	C40H68N2O30	GlcNAc-Man-Man-GlcNAc(-Fuc)-Gal	3	30
GM1 Gangliosidosis MPS IV B	3.27	945_GM1	945.2972	[M+Cl]-	C34H58N2O26	GlcNAc-Man-Man-GlcNAc-Gal	3	30
	3.29	909_GM1	909.3205	[M-H]-	C34H58N2O26	GlcNAc-Man-Man-GlcNAc-Gal	3	30
	3.38	1071_GM1	1071.3812	[M-H]-	C40H68N2O31	GlcNAc-Man-Man-GlcNAc-Gal-Gal	3	30
	3.39	1274_GM1	1274.4527	[M-H]-	C48H81N3O36	GlcNAc-Man-Man-Man-GlcNAc-GlcNAc-Gal	3	30
	3.45	1436_GM1	1436.5055	[M-H]-	C54H91N3O41	GlcNAc-Man(-Man-GlcNAc-Gal)-Man-GlcNAc-Gal	3	30
	3.45	1472_GM1	1472.4822	[M+Cl]-	C54H91N3O41	GlcNAc-Man(-Man-GlcNAc-Gal)-Man-GlcNAc-Gal	3	30
GM2 Gangliosidosis	3.19	747_GM2 (sec)	747.2677	[M-H]-	C28H48N2O21	GlcNAc-Man-Man-GlcNAc	3	30
	3.19	783_GM2 (sec)	783.2444	[M+Cl]-	C28H48N2O21	GlcNAc-Man-Man-GlcNAc	3	30
	3.33	1112_GM2	1112.3999	[M-H]-	C42H71N3O31	GlcNAc-Man(-Man-GlcNAc)-Man-GlcNAc	3	30
	3.33	1148_GM2	1148.3766	[M+Cl]-	C42H71N3O31	GlcNAc-Man(-Man-GlcNAc)-Man-GlcNAc	3	30
	3.36	1315_GM2	1315.4793	[M-H]-	C50H84N4O36	GlcNAc-Man(-Man-GlcNAc)-Man-(GlcNAc)-GlcNAc	3	30
	3.36	1351_GM2	1351.4559	[M+Cl]-	C50H84N4O36	GlcNAc-Man(-Man-GlcNAc)-Man-(GlcNAc)-GlcNAc	3	30
α-NAGA deficiency	3.25	307_alfa-NAG	307.1147	[M-H]-	C11H20N2O8	Ser-GalNAc	3	20
	3.42	321_alfa-NAG	321.1303	[M-H]-	C12H22N2O8	Thr-GalNAc	3	20
Aspartylglucosaminuria NGLY1 deficiency	3.38	334_aspartylglucos	334.1256	[M-H]-	C12H21N3O8	Asn-GlcNAc	3	30
	3.47	496_aspartylglucos	496.1784	[M-H]-	C18H31N3O13	Asn-GlcNAc-Gal	3	30
	3.18	787_aspartylglucos	787.2738	[M-H]-	C29H48N4O21	Asn-GlcNAc-Gal-NANA	3	30
α-Mannosidosis	3.11	544_α-mann	544.1883	[M-H]-	C20H35NO16	GlcNAc-Man-Man	3	30
	3.11	580_α-mann	580.1650	[M+Cl]-	C20H35NO16	GlcNAc-Man-Man	3	30
	3.35	868_α-mann	868.2940	[M-H]-	C32H55NO26	GlcNAc-Man-Man-Man-Man	3	20
	3.34	904_α-mann	904.2706	[M+Cl]-	C32H55NO26	GlcNAc-Man-Man-Man-Man	3	20
	3.44	1030_α-mann	1030.3462	[M-H]-	C38H65NO31	GlcNAc-Man-Man-Man-Man-Man	3	30
β-Mannosidosis	2.85	382_β-mann	382.1355	[M-H]-	C14H25NO11	GlcNAc-Man	3	20
	2.83	418_β-mann	418.1122	[M+Cl]-				
Sialidosis Galactosialidosis	3.16	1008_(galacto)sia	1008.8445	[M-2H]2-	C76H125N5O57	GlcNAc-Man-(Man-GlcNAc-Gal-NANA)-Man-GlcNAc-Gal-NANA	3	19
Mucolipidosis type II	3.11	1200_(galacto)sia	1200.4159	[M-H]-	C45H75N3O34	GlcNAc-Man-Man-GlcNAc-Gal-NANA	3	30

(continued)

Table 10.2 (continued)

IEM	Rt (min)	Component	m/z	Adduct	Chemical formula	Oligosaccharides	ID	REF
Sialic acid metabolism disorders	2.80	*N*-acetylneuraminic acid	308.0987	[M-H]-	C11H19NO9	NANA	1	27
GSD type 2 (M. Pompe)	3.31	665_Hex4	665.2146	[M-H]-	C24H42O21	Tetraglucoside	3	35
	3.44	989_Hex6 (sec)	989.3202	[M-H]-	C36H62O31	Hex6	4	
	3.50	1151_Hex7	1151.3731	[M-H]-	C42H72O36	Hex7	4	29
Sialyllactose (no IEM)	3.52	Sialyllactose	632.2044	[M-H]-	C23H39NO19	Gal-Glc-NANA	4	29

GM1 monosialotetrahexosylganglioside, *GM2* disialotetrahexosylganglioside, *MPS* mucopolysaccharidosis, *NAGA* α-*N*-acetylgalactosaminidase, *NGLY1 N*-glycanase 1, *ML* mucolipidosis, *GSD* glycogen storage disease, *GlcNAc N*-acetylglucosamine, *Fuc* fucosidose, *Asn* asparagine, *Man* mannose, *Gal* galactose, *Ser* serine, *Thr* threonine, *NANA N*-acetylneuraminic acid, *Hex* hexose, *IEM* inborn error of metabolism, *sec* secondary marker, *REF* reference, *ID* confidence level of metabolite annotation according to MSI initiative reporting standard [22]

10.4 Methods

10.4.1 One-Dimensional Thin-Layer Chromatography

10.4.1.1 Principle of the Method

Qualitative screening of oligosaccharidoses is performed by loading of native urine samples on silica gel plates, separation by thin-layer chromatography (TLC), and visualization by sulfuric orcinol staining. TLC plates are investigated for the presence of abnormal oligosaccharide patterns. The protocol outlined here is based on the publication by Humbel and Collart [23]. TLC can also be used for the detection of free sialic acid or sialyl-oligosaccharides. However, given the modest increase in sialic acid in some patients, and the variability between samples [6], it is recommended that sialic acid is measured by one of the mass spectrometric methods described below rather than by TLC.

10.4.1.2 Pre-analytical Phase

Specimen

A urine sample of 0.5 ml is sufficient to perform the TLC analysis. The collected samples should be frozen as soon as possible and stored at –20 °C prior to analysis.

Materials, Reagents, and Chemicals

- TLC eluent: 75 ml butanol + 37.5 ml acetic acid + 37.5 ml water
- TLC eluent (β-mannosidosis): 127.5 ml 1-propanol + 1.5 ml acetic acid + 22.5 ml water
- Oligosaccharidose reference standard mixture (0.5 mg/ml): 10 mg D(+)-lactose-monohydrate, 10 mg maltotetraose, 10 mg D(+)-galactose, and 10 mg D(+)-glucose-monohydrate dissolved in 20 ml of water. Optionally, partly hydrolyzed dextran can be used as a reference standard
- Sialic acid reference sample: 1 mg/ml NANA (optional, see remark in Sect. 10.4.1.1)
- Ninhydrin staining solution for aspartylglucosaminuria pattern
- 3,5-dihydroxytoluolmonohydrate (orcinol) dip reagent: 80 mg orcine monohydrate dissolved in 160 ml acetone + 8 ml of concentrated sulfuric acid (H_2SO_4)
- Bial's reagent: 50 mg orcine monohydrate dissolved in 12.5 ml water + 20 ml HCl (37%) + 0.5 ml $FeCl_3·6H_2O$ (1%)

The eluents can be stored for 1 week at room temperature and the orcinol dip reagent should be prepared freshly before use. The oligosaccharidose reference standard mixture and sialic acid reference sample should be stored at –20 °C.

Instrumentation

- Silica 60 TLC plates, 200 × 200 mm (Merck 1.05721), or half TLC plates 100 × 200 mm (Merck 1.05626)
- CAMAG® Automatic TLC Sampler 4 (ATS 4 robot; CAMAG, Switzerland) with software
- Chromatography tray (inside dimensions 20.5 × 8 cm)
- Benchtop Eppendorf centrifuge
- Laboratory stove at 100 °C

10.4.1.3 Analytical Phase

Procedure

- Determine the creatinine concentration by a separate quantitative measurement (e.g., Roche Modular P Chemistry Analyzer)
- Centrifuge urine samples in Eppendorf tubes for 5 min at 10,000 rpm to remove debris
- Transfer 0.5 ml supernatant into Crimp Vials and seal the vials using Crimp Caps
- Calculate the amount of supernatant that should be applied onto the TLC plates (a minimum of 10 µl is required with a maximum of 90 µl of supernatant that can be applied):

$$\text{volume of the sample}(\mu l) = (\text{age} - \text{dependent factor}^{a}) / [\text{creatinine}(\text{mmol}/\text{l})] \quad (10.1)$$

[a]Age factor = 75, 100, 150, and 200 for the age groups 0–1 years, 1–2 years, 2–8 years, and >8 years, respectively

- Spot the samples onto the TLC plates using an automatic TLC sampler device (CAMAG® ATS 4 robot) using the following settings: band width: 15.0 mm, first application position X: 20.0 mm, application position Y: 15.0 mm, distance between tracks: 25.0 mm, the last lane should be 20.0 mm from the right side of the plate. Manual urine spotting on the TLC plate(s) is possible using a pipet or capillary. However, application by a robot is preferred since it is less time-consuming and delivers a narrower application area and increases resolution
- Spot 10 μl of the oligosaccharidose reference standard mixture (0.5 mg/ml) or (optionally) 10 μl of the sialic acid reference sample (1 mg/ml). Use one lane to apply the standard or apply it overlapping with the last lane
- Fill the chromatography tray with eluent. Ensure that the liquid is below the application line (~1 cm high)

  Note: when considering β-mannosidosis as possible diagnosis use β-mannosidosis-specific eluent
- The inside of the chromatography tray should be covered with filter paper soaked in eluent
- Place a maximum of four TLC plates in the tank (use small notched rods to stabilize the plates)
- Close the chromatography tray and air-seal the lid by applying silicone grease between the lid and the tank
- Once the eluent almost reaches the top of the TLC plates (5–8 h) remove the plates and let them air-dry in a fume hood or laboratory stove
- Pre-heat a laboratory stove to 100 °C
- Stain the TLC plates by dipping in a flat reagent tray filled with (fresh) orcinol dip reagent and heating for approximately 10 min at 100 °C (until colors have developed) and allow them to cool down in the fume hood

  Note: when considering sialic acid metabolism defects or aspartylglucosaminuria as possible diagnosis Bial's reagent and ninhydrin, respectively, can be used in this step
- Once dried, the plates are ready for interpretation and can be scanned and stored digitally

10.4.1.4 Post-analytical Phase

Interpretation of the Silica Gel Plates

Interpretation of the TLC plates occurs qualitatively by analyzing patient urine samples for specific patterns of oligosaccharide bands. Each plate contains an oligosaccharide standard reference mixture (or optionally a sialic acid standard reference sample). The order of compounds included in the mixture described in section "Materials, Reagents, and Chemicals" under Sect. 10.4.1.2 is from the solvent front to the origin: glucose, galactose, lactose, and maltotetraose (see Figs. 10.3 and 10.4a). Optionally, normal urine and/or known patient samples can be included in the analysis to aid in interpretation of the screening results.

Figure 10.3 gives an overview of representative urinary oligosaccharide excretion patterns associated with several of the diseases listed in Table 10.1. As can be seen, a specific series of mannose-containing glycans migrating slower than lactose can be observed for α-mannosidosis, which are not excreted in β-mannosidosis samples. The β-mannosidosis storage product GlcNAc-Man co-elutes with lactose (see Fig. 10.4a and section "Pitfalls and Limitations of the Screening Method" under Sect. 10.4.1.4). Fucosidosis, sialidosis, galactosialidosis, and GM1 gangliosidosis samples demonstrate a specific dense band closely to the origin of the TLC plate. Sialidosis and galactosialidosis samples show an additional closely migrating band. In the fucosidosis and GM1 gangliosidosis staining patterns two additional faster migrating intense bands are present at different positions. GM2 gangliosidosis (Sandhoff disease) samples exhibit two bands migrating closely together. MPS IV B shows an oligosaccharide pattern identical to GM1 gangliosidosis, but generally with weaker-stained bands (data not shown). Free or conjugated sialic acid bands can be visualized by Bial's reagent. Aspartylglucosaminuria and

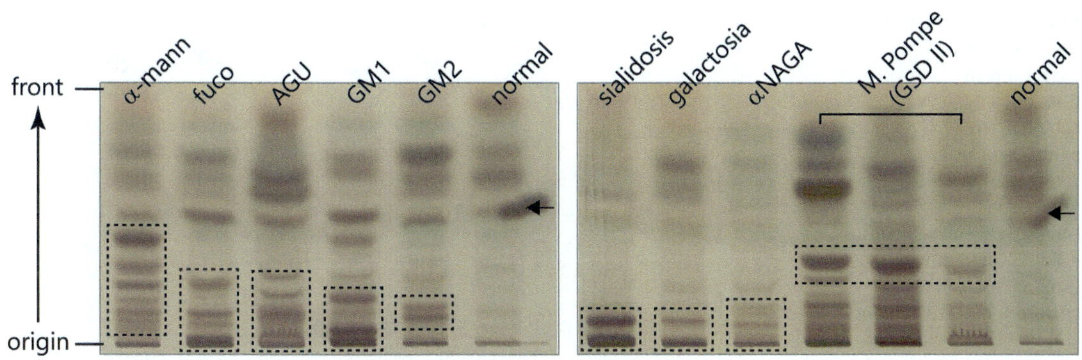

Fig. 10.3 Thin-layer chromatography analysis of oligosaccharidosis urine samples. Examples of oligosaccharide accumulations of various storage disorders detected by TLC are depicted The dashed boxes indicate the disease-specific pathological oligosaccharide excretion patterns. The small arrows indicate the reference compound lactose. *Normal* refers to a control urine sample. *α-mann* α-mannosidosis, *fuco* fucosidosis, *AGU* aspartylglucosaminuria, *GM1* GM1 gangliosidosis, *GM2* GM2 gangliosidosis, *galactosia* galactosialidosis, *αNAGA* α-*N*-acetylgalactosaminidase deficiency, *GSD* glycogen storage disease (three different GSD II samples with varying glucose tetrasaccharide (Glc4) accumulations are presented)

N-acetylgalactosaminidase deficiency samples demonstrate several bands migrating closely together near the origin of the plate. Alternatively, ninhydrin staining (or aspartylglucosamine amino acid analysis) can be applied for characterization of the GlcNAc-Asn and glycoasparaginyl-oligosaccharides storage products that accumulate in aspartylglucosaminuria. Mucolipidosis type II (ML II) can be detected by TLC analysis as well and usually shows a pattern similar to sialidosis. Elevated levels of a tetrasaccharide band are observed in urine samples of Pompe disease patients (GSD type II) of which the intensity seems to correlate to severity of the phenotype (more intense bands in the classic infantile form compared to the late-onset phenotype). In some Pompe urine samples larger glucose-oligosaccharides can be observed as well (Fig. 10.3).

Reference and Pathological Values

Analysis of (pathological) urinary oligosaccharide and free sialic acid excretion by TLC is a qualitative screening method. As such, no pathological values can be established. Interpretation relies on pattern recognition of the various disorders (see section "Interpretation of the Silica Gel Plates" under Sect. 10.4.1.4). Disease-specific reference samples, if available, can be included in the analysis to aid interpretation of the TLC plates.

Quality Control

At least one positive control sample from a patient with an established diagnosis should be included in each experiment. The results should be as expected and reproducible in qualitative terms. Positive controls may also help to interpret indistinct results in a sample of a new patient. Thus, retention of a chromatographic peak or the retention, shape and color of a TLC band may be compared directly between a patient sample and a positive control. To reduce interpretative subjectivity, TLC plates should be interpreted independently by at least two trained individuals. Currently, no specific external quality assurance (EQA) scheme exists for this group of disorders. However, in the ERNDIM (European Research Network for the evaluation and improvement of screening, Diagnosis and treatment of Inherited disorders of Metabolism) Diagnostic Proficiency Testing (DPT) schemes oligosaccharidosis urine samples are included from time to time. In addition, ERNDIM provides an educational oligosaccharide kit with samples that can serve as positive

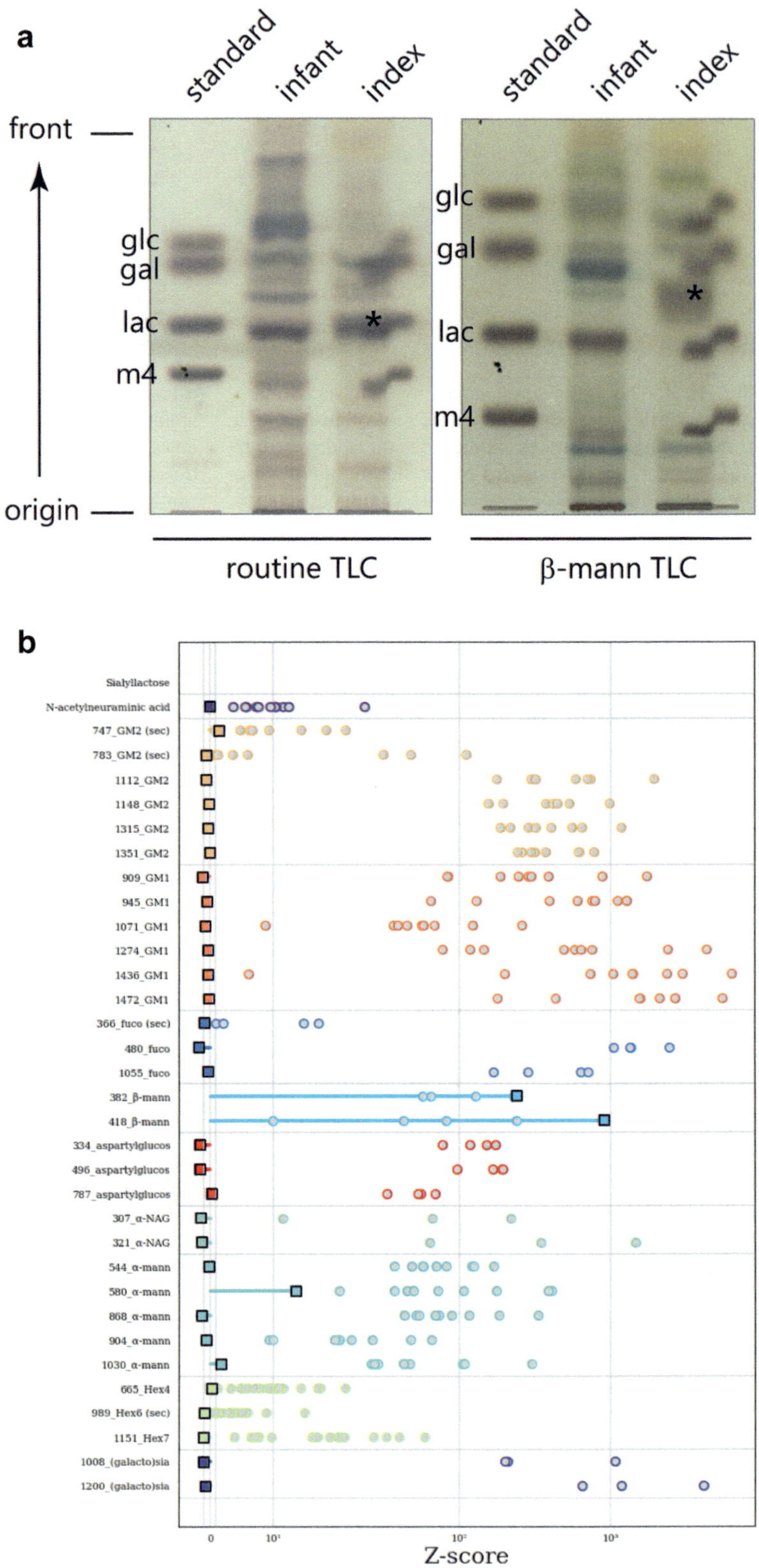

Fig. 10.4 Analysis of oligosaccharide excretions in β-mannosidosis urine samples. (**a**) TLC analyses of urine samples of an *infant* without oligosaccharidosis and an *index* patient with β-mannosidosis using conventional TLC eluent (*routine TLC*) and (β-mannosidosis-specific eluents (*β-mann TLC*). The asterisks indicate the pathological GlcNAc-Man storage product that co-elutes with lactose using the conventional TLC approach, but which is separated from it when using disease-specific eluent. The reference standard consists of glucose (*glc*), galactose (*gal*), lactose (lac), and maltotetraose (*m4*). (**b**) A representative *Z*-score plot obtained by UHPLC/HRAM-MS analysis of a urine sample of a β-mannosidosis patient. Both β-mannosidosis biomarkers (squares) are elevated in the range of the disease-specific reference values (dots) without any interference of non-specific (nutritional) biomarkers

controls for the screening of oligosaccharidoses (see ERNDIMQA—Educational Panels: https://www.erndimqa.nl/Information.aspx?I=1068).

Pitfalls and Limitations of the Screening Method

Interpretation of oligosaccharide excretion patterns by TLC is challenging and requires extensive experience by individuals who analyze these results [24, 25]. This is mainly because of the qualitative nature of the screening method with its limited chromatographic resolution.

A general difficulty associated with this method is that urinary oligosaccharide excretion decreases with age, which makes interpretation of screening results in older patients potentially more difficult than in children. However, the most commonly observed artifact in this analysis is due to nutritional interference in newborn patient urines. These so-called "neonate" patterns consist of milk oligosaccharides received from the mother via breast feeding or infant formula, which may obscure pathological oligosaccharides and renders interpretation of samples demonstrating such interference difficult. Similarly, the main β-mannosidosis-specific storage product GlcNAc-Man co-elutes with lactose. As such, when considering β-mannosidosis as potential diagnosis a specific eluent should be used to separate these two components (see Fig. 10.4a). Samples of galactosialidosis and sialidosis (mucolipidosis type I) patients demonstrate similar excretion patterns and cannot be discriminated [26]. The sensitivity of TLC analysis is insufficient to detect mucolipidosis type III (ML III), a phenotypically milder form of ML II.

The modest increase of free sialic acid in some patients, and the high variability between samples, makes TLC not very reliable for the detection of this metabolite and as such other screening methods (e.g., mass spectrometry) are preferred [6, 27, 28]. When increased free sialic acid is detected, bacterial contamination of the sample or bacterial infection, such as in hemolytic uremic syndrome (HUS) should be considered as a possible cause, as well as diabetes mellitus type II, renal failure or bladder tumors [28].

TLC analysis of oligosaccharide and free sialic acid excretion is a screening method and as such suspected diagnoses should always be confirmed by (a combination of) complementary diagnostic tests that include genetic and enzyme activity analyses.

10.4.2 UHPLC/HRAM Mass Spectrometry

10.4.2.1 Principle of the Method

Underivatized urinary oligosaccharides and free sialic acid are chromatographically separated by ultra-high performance liquid chromatography (UHPLC). Oligosaccharides and sialic acid are detected by high-resolution accurate mass (HRAM) mass spectrometry in negative ionization mode (for analytical details see Table 10.2). Semi-quantitative concentrations of individual components are calculated using an open-source analysis workflow from which *Z*-scores are determined and plotted as comprehensive biomarker profiles. The protocol described here has been published by Hagemeijer et al. [29].

10.4.2.2 Pre-analytical Phase

Specimen

A 200 μl urine sample is sufficient to perform the UHPLC/HRAM-MS analysis. The collected samples should be frozen as soon as possible and stored at −20 °C prior to analysis.

Materials, Reagents, and Chemicals

- Control sample 1: Combine urine samples of α-mannosidosis, sialic acid storage disease, sialidosis, GM1 and GM2 patients to obtain a total volume of 8 ml. Store as 80 μl aliquots in Eppendorf tubes at −20 °C
- Control sample 2: Combine urine samples of aspartylglucosaminuria, fucosidosis, α-NAGA deficiency, and Pompe disease patients to obtain a volume of 8 ml. Store as 80 μl aliquots in Eppendorf tubes at −20 °C

- Control sample 3: An 8 ml urine sample of a β-mannosidosis patient. Store as 80 μl aliquots in Eppendorf tubes at −20 °C

  Note: depending on the type of patient samples that are available composition of the control sample(s) might vary
- Mobile phase A: 5 mM ammonium acetate pH 9, 10:90 water/acetonitrile + 0.1% NH_4OH (0.1925 g ammonium acetate) + 50 ml water + 60 μl ammonium hydroxide (25%) + 450 ml acetonitrile. The solution should be sonicated ("degassed") for at least 30 min
- Mobile phase B: 5 mM ammonium acetate pH 9, 70:30 water/acetonitrile + 0.1% NH_4OH (0.1925 g ammonium acetate) + 350 ml water + 60 μl ammonium hydroxide (25%) + 150 ml acetonitrile). The solution should be sonicated ("degassed") for at least 30 min
- Solvent C: 10:90 (v/v) water/acetonitrile (450 ml acetonitrile + 50 ml water)
- Solvent D: 10:90 water/acetonitrile + 0.1% NH_4OH [50 ml water + 450 ml acetonitrile + 2 ml ammonium hydroxide (25%)]. The solution should be sonicated ("degassed") for at least 30 min
- Internal standard stock solution: 200 μl [$^{13}C_3$]-*N*-acetylneuraminic acid (2.5 mM), 0.9 ml acarbose (500 μM), and 1.45 ml hexaacetyl-chitohexaose (0.5 mM) in a total of 25 ml of water

Instrumentation and Calibration

Here, measurements are performed on a HRAM-MS coupled to an UHPLC system. However, other MS systems can also be used to perform the analysis [20, 30–32]. Importantly, the data generated on these systems can be processed by the open-source bioinformatics pipeline provided that the data complies with the required format of the output file (see section "Data Analysis Using the Bioinformatics Pipeline" under Sect. 10.4.2.3).

The UltiMate 3000 UHPLC RS System with autosampler, pump, and thermostatted column compartment (Thermo Scientific) is used for chromatographic separation of oligosaccharides. At the start of the analysis the injection needle is rinsed using solvent D, followed by injection of 2.5 μl sample onto the ACQUITY UPLC BEH Amide Column 2.1 × 50 mm; 1.7 μm (Waters). Gradient conditions for elution started with 99% mobile phase A with a linear decrease to 50% (0–1.5 min) and 50–5% (2.5–4.0 min) by mixing mobile phase B. After 2 min the column is equilibrated to the starting conditions for the next injection. Once all samples have been analyzed the analytical column is conditioned using solvent C for 15 min. The temperature of the column is set at 40 °C with a flow rate of 0.45 ml/min.

A Q Exactive Plus Hybrid Quadrupole-Orbitrap mass spectrometer (Thermo Fisher) is used in negative ionization and full scan mode (mass range of 150–2000 m/z with a mass accuracy of 5 ppm) with a spray voltage of 3.50 kV, maximum injection time of 200 ms, and resolution of 70,000 (m/Δm, FWHM @ 200 m/z). The nitrogen sheath gas flow and auxiliary gas rates are set at 40 and 15 arbitrary units with capillary and auxiliary gas heater temperatures of 350 °C and 250 °C, respectively. The S-lens value is set at 80. Prior to each analysis, a mass calibration is performed.

10.4.2.3 Analytical Phase

Procedure

- Determine the creatinine concentration by a separate quantitative measurement (e.g., Roche Modular P Chemistry Analyzer)
- Thaw the internal standard stock solution and vortex this solution vigorously once it reaches room temperature
- For each sample add 30 μl of urine to 270 μl internal standard stock solution and vortex the mixture briefly
- Centrifuge the mixtures for 5 min at 13,000 rpm
- Transfer 100 μl of supernatant into 2 ml Crimp Vials with insert and seal the vials using Crimp Caps
- Use 2.5 μl of this mixture for UHPLC/HRAM-MS analysis (see section "Instrumentation and Calibration" under Sect. 10.4.2.2)

Note: these samples can be stored at 4 °C for 2 weeks. Briefly vortex stored samples prior to the analysis

Data Analysis Using the Bioinformatics Pipeline

- Analyze the raw data batch file obtained from the MS using TraceFinder software (Thermo Fisher; or similar mass spectrometry analysis software from other systems)
- Perform (automatic) peak integration of the oligosaccharide storage products using the *m/z* values and retention times as listed in Table 10.2. Retention times of pathological oligosaccharides should be verified using positive control samples (QC samples)
- Calculate the semi-quantitative concentrations of each component using TraceFinder:

$$\left(\frac{\text{peak area analyte}}{\text{peak area IS}}\right)\times[\text{IS}] \qquad (10.2)$$

IS: internal standard (use acarbose for components < 800 m/z and hexaacetyl-chitohexaose for components > 800 m/z, respectively)

- *N*-acetylneuraminic acid concentrations are calculated using [$^{13}C_3$]-NANA as internal standard
- Export the analyzed data as a CSV or XLXS file and save it in a designated *input folder*
- Generate a CSV file that contains patient sample IDs with corresponding age (in days) and creatinine concentration (mmol/l) and store this file in a designated *database folder*
- Load both files in the bioinformatics pipeline for analysis

  Note: the code to run the OLIGO bioinformatics data analysis pipeline can be downloaded from the GitHub repository https://github.com/mbongaerts/OLIGO_pipeline, together with a user manual for operation and/or adjustment of the OLIGO pipeline
- The OLIGO bioinformatics pipeline will analyze the data and comprises the following processing steps:
  - Normalization of the (semi-quantitative) concentration of each component to the corresponding creatinine concentration (values in mmol/mol creatinine)
  - Merging of results from samples of different (historical) experiments indicated as "references batches" to generate a large (age-dependent) reference dataset
  - Determination of patient-specific *Z*-scores of individual components using two possible approaches:

    Fixed approach: for components < 100 non-zero measurements with the mean and standard deviation taken from all measurements (including measurements equal to zero). *Z*-scores are calculated by subtracting the mean and dividing by the standard deviation

    Regression analysis: for components > 100 non-zero measurements by applying a (polynomial) regression model that corrects for age:

$$\hat{y}_i = \hat{\beta}_{\text{intercept}} + \sum_{p=1}^{3} \hat{\beta}_{\text{p}}^{\text{Age}} \left(x_i^{\text{Age}}\right)^{\text{p}} + \hat{\epsilon}_i \qquad (10.3)$$

    $\hat{y}_i$ is the estimated mean for sample i, $\hat{\beta}_{\text{intercept}}$ is the intercept coefficient, $\hat{\beta}_{\text{p}}^{\text{Age}}$ are the slopes for each polynomial term, x_i^{Age} indicates the age of sample, i and $\hat{\epsilon}_i$ are error terms

    To fit the model parameters ordinary least square estimation is used. The standard deviation is determined as described by Bongaerts et al. [33]. Mean and standard deviation from the regression analysis are used for *Z*-score calculations. Prior to fitting the model parameters, values in the reference batches with |*Z*-score| > 5 will be removed. Note that the latter *Z*-scores, i.e. used to remove the outliers, are determined from the mean and standard deviation calculated from all reference samples
  - QC criteria are monitored in parallel for each batch run (see section "Quality Control" under Sect. 10.4.2.4)
  - Disease-specific *Z*-score plots are generated (see Fig. 10.5 for example output files). *Z*-score plots (PNG files), statistical data (CSV files) and QC plots are exported in the dedicated *output folder*

10.4.2.4 Post-analytical Phase

Interpretation of the *Z*-Score Plots

The output file generated by the analysis pipeline consists of a biomarker profile in which disease-specific components are plotted on the *Y*-axis and the calculated *Z*-scores on the *X*-axis. Disease-specific pathological values are depicted in the background of the *Z*-score plot to assist in interpretation of the results. Examples of representative *Z*-score plots are given in Fig. 10.5.

All of the disease-specific biomarkers are elevated (majority of *Z*-scores ± 10^2–10^3) in α-mannosidosis, β-mannosidosis GM1 gangliosidosis, and aspartylglucosaminuria samples, which is also observed for NAGA deficiency but with *Z*-scores of 321_α-NAG being slightly higher than 307_α-NAG.

Markers for (galacto)sialidosis are elevated with Z-scores > 10^2 but these are generally threefold and fivefold higher in sialidosis samples compared to galactosialidosis, which allows for discrimination between these two diseases. These markers are also elevated in ML II/ML III samples but together with a subset of GM1 gangliosidosis and α-mannosidosis markers. This is in contrast to the (galacto)sialidosis samples and distinguishes these diseases.

In fucosidosis and GM2-gangliosidosis samples not all the biomarkers are consistently elevated. Depending on the sample, 366_fuco (fucosidosis) and 747_GM2 and 783_GM2 (GM2 gangliosidosis) may be (close to) normal. All of the remaining biomarkers for both diseases are clearly elevated as a group (*Z*-scores of 10^2–10^3; see section "Pitfalls and Limitations of the Screening Method" under Sect. 10.4.2.4).

Additional LSDs in which aberrant oligosaccharides accumulate can also be detected by UHPLC/HRAM-MS [29]. MPS type IV B biomarker profiles demonstrate consistently elevated GlcNAc-Man2-GlcNAc-Gal (909_GM1) with variable expression of the remaining GM1 gangliosidosis components. Moderately elevated

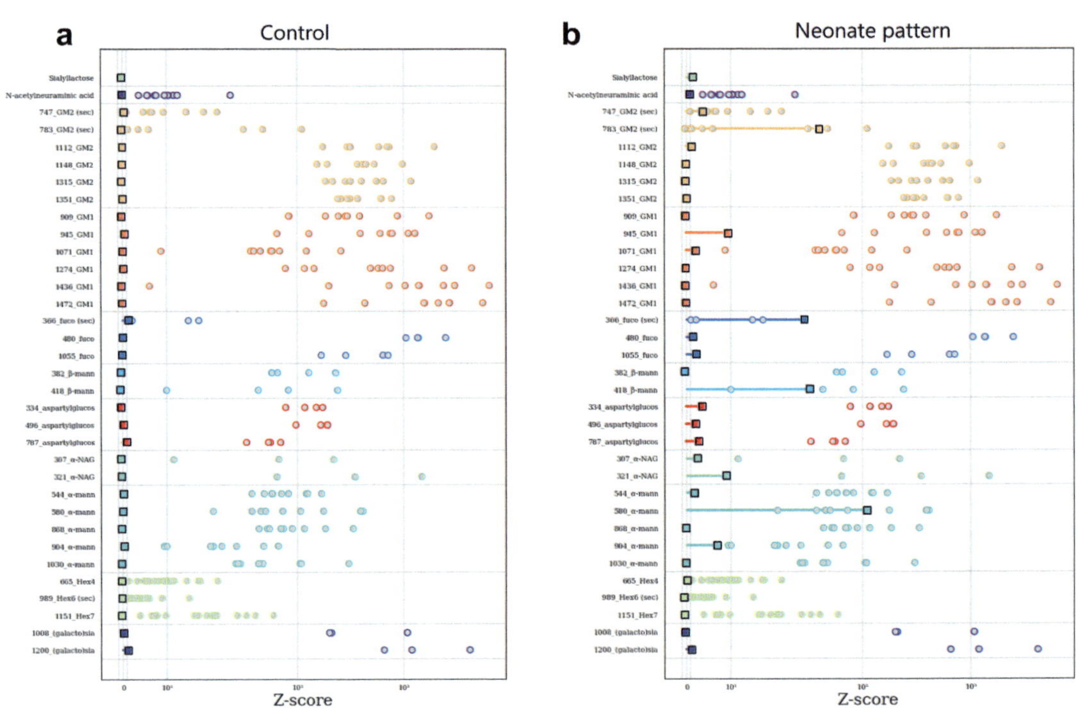

Fig. 10.5 Representative *Z*-score plots of urine samples analyzed by UHPLC/HRAM-MS. *Z*-score plots of a control (non-IMD) sample (**a**) and of a sample with a "neonate pattern" (**b**) are depicted. None of the disease-specific biomarkers are elevated in the control urine sample. The neonate expression profile demonstrates various (non-specific) elevations of several individual biomarkers but not as a disease-specific group of biomarkers. Disease-specific biomarkers are depicted as squares and pathological values of these markers as dots

Z-scores of all aspartylglucosaminuria biomarkers are observed for NGLY-1 but not to the extent as in patients with aspartylglucosaminuria. Finally, excretions of Hex7, Hex4, and Hex6 glycans can be detected (in descending order) in Pompe disease (GSD type II) urine samples.

Reference and Pathological Values

(Age)-dependent reference and pathological values for each component are determined from a user-generated reference data set that consists of measurements of (1) control samples obtained from individuals not diagnosed with an inherited metabolic disease (IMD) and (2) samples of known patients with the disorders listed in Table 10.2 (see section "Data Analysis Using the Bioinformatics Pipeline" under Sect. 10.4.2.3 for additional information). Average semi-quantitative (pathological) reference and patient-specific values for each component are calculated and plotted as *Z*-scores that function as a visual reference set for data interpretation (see Fig. 10.5). It is not possible to list absolute reference and pathological values as the test is semi-quantitative, but approximate *Z*-score values are given in section "Interpretation of the *Z*-Score Plots".

Importantly, the data analysis pipeline operates iteratively, which allows for expansion of the reference data set by simply marking analyzed batches exhibiting normal biomarker profiles as *reference batches* or samples with abnormal profiles as being *disease-specific*. This optimizes and refines the reference data set over time and improves the sensitivity and accuracy of the screening approach. It is also possible to add (novel) biomarkers to the bioinformatics analysis pipeline.

N-acetylneuraminic acid is measured quantitatively; reference values for free sialic acid are given in Table 10.3 and pathological values in

Table 10.3 Age-dependent reference ranges of free and conjugated sialic acid excretion in urine

Age	Free sialic acid (mmol/mol creatinine)	Conjugated sialic acid (mmol/mol creatinine)	*n*
0–6 months	0–97	0–230	28
6–12 months	0–69	0–142	11
1–2 years	0–48	0–132	17
2–5 years	0–29	0–78	21
5–12 years	0–19	0–56	36
>12 years	0–12	0–39	34

Note: these reference values have been established in our laboratory

Table 10.4 Pathological values of free and conjugated sialic acid excretion in urine

		Free sialic acid (mmol/mol creatinine)		Conjugated sialic acid (mmol/mol creatinine)	
	Age (y)	Patient	Reference	Patient	Reference
Salla disease	1.1	124	0–48	40	0–132
	1.9	260	0–48	67	0–132
	4.3	70	0–29	n.d.	n.a.
	11.9	54	0–19	0	0–39
Sialuria	0.4	617	0–97	116	0–230
	0.8	466	0–69	28	0–142
	1.7	351	0–48	0	0–132
	2.8	71	0–29	132	0–78
	3.5	119	0–29	0	0–78
	3.8	46	0–29	n.d.	n.a.
Sialidosis	35.6	9	0–12	232	0–39
Galactosialidosis	0.5	48	0–69	320	0–142

y year, *n.d.* not detected, *n.a.* not applicable

Note: these reference values have been established in our laboratory. Due to the rare occurrence of these diseases the number of analyzed samples is small. Additional pathological values have been published by others [28, 34]

Table 10.4. Additional reference and pathological values have been reported by others [28, 34].

Quality Control

For each batch run three positive quality control (QC) samples are included that consist of a mixture of urines from previously diagnosed patients (see section "Materials, Reagents, and Chemicals" under Sect. 10.4.2.2 for the composition of these samples). As mentioned in section "Quality Control" under Sect. 10.4.1.4, the samples of the ERNDIM educational kit can also be used as positive oligosaccharidosis controls (see ERNDIMQA—Educational Panels: https://www.erndimqa.nl/Information.aspx?I=1068). The internal standards ([$^{13}C_3$]-NANA, acarbose, and hexaacetyl-chitohexaose), which are included in each batch run, are also evaluated for QC purposes (see below).

The following QC parameters are monitored automatically by the bioinformatics pipeline and must meet the following requirements:

- The intra-batch coefficient of variation (CV) < 60% for the IS
- The *Z*-scores of the IS peak areas across all batches < 3 standard deviations (SD)
- $2 < |Z| < 3$ for *Z*-scores of the IS peak areas cannot occur in more than two successive batches
- The expected biomarker patterns have to be obtained in the positive QC samples

  Note: if any of these four QC criteria is not fulfilled, the analysis will be rejected and the experiment must be repeated

Pitfalls and Limitations of the Screening Method

The bioinformatics pipeline can easily be implemented in any laboratory as it functions irrespective of the utilized MS platform (see the GitHub repository). It needs, however, a sufficient amount of user-generated reference data to function optimally (>100 non-zero measurements per biomarker) before it can use age-dependent reference values to calculate *Z*-scores. This is important as it increases the accuracy of the method because glycan urinary excretions decrease with age. Similarly, inclusion of sufficient positive control samples is critical for establishing biomarkers-specific pathological *Z*-score values. However, this might be challenging for new adopters of the pipeline and might delay implementation of this screening approach into a diagnostic setting. The ERNDIM educational kit (see ERNDIMQA—Educational Panels: https://www.erndimqa.nl/Information.aspx?I=1068) might serve as a good initial starting point for laboratories that do not have access to such pathological samples.

All the biomarkers listed in Table 10.2 have been published previously by others [16, 19, 20, 27, 30, 35] but some of them can be inconsistently elevated when using the MS-based screening approach, i.e. 366_fuco, 747_GM2, 783_GM2, and 989_Hex6 (labeled as *secondary markers*). As such, it is important to interpret patterns of disease-specific *primary biomarkers* as a group to obtain a potential diagnosis. Secondary biomarkers can aid the interpretation of the screening results but are not required per se.

Similar to the TLC screening method, inconclusive biomarker profiles can be observed in the UHPLC/HRAM-MS approach, such as the previously mentioned "neonate" patterns (see section "Pitfalls and Limitations of the Screening Method" under Sect. 10.4.1.4). In these profiles subsets of several of the disease-specific biomarkers can be abnormal. These elevations of (non-specific) subsets of biomarkers can vary between patients. However, both (galacto)sialidosis markers have been detected frequently in such samples, which might interfere with detection of galactosialidosis or sialidosis. The inclusion of sialyllactose as a nutritional marker might indicate that milk oligosaccharides have been consumed by the patient but this marker is not always consistently elevated in such cases. The interference in the TLC analysis that might obscure detection of β-mannosidosis is not present in the UHPLC/HRAM-MS approach (compare Fig. 10.4a to Fig. 10.4b).

Analysis of the various oligosaccharidoses by UHPLC/HRAM-MS revealed that sialyl-oligosaccharide biomarker excretion is not exclusively elevated in patients suffering from galactosialidosis or sialidosis but can result from secondary storage in other lysosomal storage disorders [20, 29]. This is also observed in ML II/III samples with additional mannosyl- and galactosyl-biomarker excretion [32]. However, it is not possible to distinguish between mucolipidosis type II or III with UHPLC/HRAM-MS, which requires additional diagnostic testing.

When increased free sialic acid is detected, bacterial contamination of the sample or bacterial infection, such as in HUS, should be considered as a possible cause, as well as diabetes mellitus type II, renal failure, or bladder tumors [28].

Finally, similar to the TLC analysis discussed in Sect. 10.4.1 UHPLC/HRAM-MS analysis is a screening approach and also requires follow-up confirmatory testing for potential oligosaccharidoses or sialic acid metabolism disorders by genetic and enzyme activity testing.

10.4.3 LC-MS/MS to Analyze Sialic Acid

10.4.3.1 Principle of the Method

To identify and differentiate between different forms of sialic aciduria both free and total (free plus conjugated) sialic acid needs to be measured. The total amount of sialic acid can be determined after hydrolysis of conjugated sialic acid. Underivatized sialic acid is chromatographically separated from other compounds by liquid chromatography (LC). Sialic acid is detected by electrospray ionization tandem mass spectrometry (ESI-MS/MS) in the negative ion mode. Part of this protocol has been published previously [27, 28].

10.4.3.2 Pre-analytical Phase

Specimen

A 200 μl urine sample is sufficient to perform the LC-MS/MS analysis. The collected samples should be frozen as soon as possible and stored at –20 °C prior to analysis.

Materials, Reagents, and Chemicals

- Control sample 1: A urine sample of a patient with about 50 μM free sialic acid, and about 300 μM total sialic acid. If such a sample is not available, a urine sample can be spiked with *N*-acetylneuraminic acid for free sialic acid and sialyllactose for conjugated sialic acid. Store as 80 μl aliquots in Eppendorf tubes at –20 °C
- Control sample 2: A urine sample of a patient with about 300 μM free sialic acid, and about 1500 μM total sialic acid. If such a sample is not available, a urine sample can be spiked with *N*-acetylneuraminic acid for free sialic acid and sialyllactose for conjugated sialic acid. Store as 80 μl aliquots in Eppendorf tubes at –20 °C
- Calibration curve samples: a serial dilution of *N*-acetylneuraminic acid (0, 5, 10, 50, 100, 200, 425, and 850 μM)
- Mobile phase A: 50 mM ammonium formate pH 3. The solution should be sonicated ("degassed") for at least 30 min
- Mobile phase B: acetonitrile (100%). The solution should be sonicated ("degassed") for at least 30 min
- Internal standard (IS) solution: [$^{13}C_3$]-*N*-acetylneuraminic acid (0.85 mM)
- Sulfuric acid (H_2SO_4; 63 mM)

Instrumentation and Calibration

The TSQ Quantum triple quadrupole mass spectrometer (MS), interfaced with an electron spray ionization source (Thermo Scientific) connected to a Surveyor autosampler and Surveyor MS Pump Plus (Thermo Scientific) is used to quantify sialic acid. For chromatographic separation an Atlantis dC18 analytical column (3 μm, 3.9 × 150 mm) with dC18 guard column (3 μm, 3.9 × 10 mm, Waters) is used. The temperature of the column is set at 35 °C with a flow rate of 0.3 ml/min. The gradient conditions for elution starts with 100% mobile phase A with a linear increase to 100% mobile phase B in 4 min, followed by 2 min of 100% mobile phase B after which mobile phase A increases linearly to 100% in 6 min. The next injection can occur after this final step. Between 0 and 2.5 min and 6.13–12 min the buffer waste is dispensed to avoid unnecessary introduction of solvents in the source.

Tandem MS analysis is performed in the negative ionization mode with a capillary voltage of 5.0 kV, nitrogen dry gas flow (sheath gas and auxiliary) of 50 and 30 l/min, respectively, source temperature of 325 °C, tube lens settings of 88 V for NANA and 93 V for the internal standard with the collision cell operating at 28 eV at a gas pressure of 1.4 mTorr argon. Ions with *m/z* 308.0 > 86.9 (NANA) and *m/z* 311.0 > 89.9 ([$^{13}C_3$]-NANA) are detected with a mass resolution of 0.7 Da for both quadrupole 1 and 3. Xcalibur software (Thermo Scientific) is used for instrument control and data acquisition/processing.

10.4.3.3 Analytical Phase

Procedure

- Determine the creatinine concentration by a separate quantitative measurement (e.g., Roche Modular P Chemistry Analyzer)
- Thaw the internal standard working solution and calibration curve samples and vortex these vigorously once they reach room temperature
- Thaw the frozen urine and urine control samples
- Centrifuge the urine samples for 10 min at 13,000 rpm
- For the determination of total sialic acid, add 25 µl of each of the urine samples, urine control samples or HPLC-grade water (blank sample) to 25 µl IS solution in a glass vial and add 100 µl of H_2SO_4 (63 mM)
- After vortexing the samples incubate the vials at 80 °C for 1 h in a laboratory stove
- Remove the samples from the stove and let them cool down to room temperature
- Transfer the contents of the vials into 2 ml polypropylene (PP) vials with insert and seal the vials using (polytetrafluoroethylene) PTFE caps
- For the determination of free sialic acid, add 125 µl of each of the urine, urine control or calibration curve samples or HPLC-grade water (blank sample) to 25 µl IS solution into 2 ml PP vials with insert and seal the vials using PTFE caps
- Use 10 µl of processed samples for LC-MS/MS analysis (see section "Materials, Reagents, and Chemicals" under Sect. 10.4.3.2)

  Note: these samples can be stored at 4 °C for 5 days. Briefly vortex stored samples prior to the analysis

Data Analysis

- Analyze the raw data obtained from the MS using Xcalibur software (Thermo Scientific) (or similar mass spectrometry analysis software from other systems)
- Perform (automatic) peak integration of sialic acid using the *m/z* values 308.0 > 86.9 (sialic acid) and 311.0 > 89.9 ([$^{13}C_3$]-*N*-acetylneuraminic acid). The retention time is around 5 min but should be determined exactly using solutions of pure *N*-acetylneuraminic acid
- Calculate the free (or total) sialic acid concentrations using [$^{13}C_3$]-*N*-acetylneuraminic acid as internal standard and the calibration curve
- Calculate the concentration of conjugated sialic acid by subtracting the value obtained for free sialic acid in a sample from the value obtained for total sialic acid in that sample

  Note: the correlation coefficient of the calibration curve should be ≥ 0.98. If not, the experiment should be repeated

10.4.3.4 Post-analytical Phase

Reference and Pathological Values

Reference values for free and conjugated sialic acid excretion in urine are given in Table 10.3 and pathological values in Table 10.4. Additional reference and pathological values have been reported by others [28, 34].

Quality Control

For each run, two quality control samples from non-IEM patients with a low and a high concentration of free urinary sialic acid and total sialic acid should be included and monitored (see Chap. 2).

Pitfalls and Limitations of the Screening Method

When increased free sialic acid is detected, bacterial contamination of the sample or bacterial infection, such as HUS should be considered as a possible cause, as well as diabetes mellitus type II, renal failure, or bladder tumors [28].

Finally, similar to the previously mentioned screening methods, detection of sialic acid by LC-MS/MS is a screening approach and also requires follow-up confirmatory testing for sialic acid metabolism disorders by genetic and enzyme activity testing.

10.5 Conclusion

In this chapter various screening approaches have been described to analyze urinary oligosaccharides and sialic acid excretions, primarily to detect oligosaccharidoses and disorders of sialic acid metabolism, but also sphingolipidoses, mucolipidoses, Pompe disease (GSD type II), MPS type IV B, and NGLY1 deficiency. Advantages and disadvantages for these methods exist but depending on specific questions, needs, and resources of individual (metabolic) laboratories one approach might be preferred over the other. It is expected, however, that MS-based approaches will replace the conventional TLC method in time.

Acknowledgments The authors would like to thank Jeroen van den Bosch and Michiel Bongaerts for their help in developing the UHPLC/HRAM-MS method and Frans Verheijen, David Vos, and Juditte de Ronde for suggestions and technical assistance with this chapter.

References

1. Bender DA (2023) Glycoproteins. In: Kennelly PJ, Botham KM, McGuinness OP, Rodwell VW, Weil PA (eds) Harper's illustrated biochemistry, 32e. McGraw Hill Education, New York
2. Lebrilla CB, Liu J, Widmalm G, Prestegard JH (2022) Oligosaccharides and polysaccharides. In: Varki A, Cummings RD, Esko JD, Stanley P, Hart GW, Aebi M et al (eds) Essentials of glycobiology [internet], 4th edn. Cold Spring Harbor Laboratory Press, New York
3. Conte F, van Buuringen N, Voermans NC, Lefeber DJ (2021) Galactose in human metabolism, glycosylation and congenital metabolic diseases: Time for a closer look. Biochim Biophys Acta Gen Subj 1865(8):129898. https://doi.org/10.1016/j.bbagen.2021.129898. S0304-4165(21)00056-8 [pii]
4. Varki A (2017) Biological roles of glycans. Glycobiology 27(1):3–49. https://doi.org/10.1093/glycob/cww086. cww086 [pii]
5. van Karnebeek CD, Bonafe L, Wen XY, Tarailo-Graovac M, Balzano S, Royer-Bertrand B et al (2016) NANS-mediated synthesis of sialic acid is required for brain and skeletal development. Nat Genet 48(7):777–784. https://doi.org/10.1038/ng.3578. ng.3578 [pii]
6. Verheijen FW (2008) Sialic acid. In: Blau N, Duran M, Gibson KM (eds) Laboratory guide to the methods in biochemical genetics. Springer, Berlin, pp 335–349
7. Futerman AH, van Meer G (2004) The cell biology of lysosomal storage disorders. Nat Rev Mol Cell Biol 5(7):554–565. https://doi.org/10.1038/nrm1423. nrm1423 [pii]
8. Beck M, Lukacs Z (2022) Oligosaccharidoses and sialic acid disorders. In: Blau N, Dionisi Vici C, Ferreira CR, Vianey-Saban C, van Karnebeek CDM (eds) Physician's guide to the diagnosis, treatment, and follow-up of inherited metabolic diseases. Springer, Cham, pp 1249–1265
9. Jones S, Wijburg FA (2022) Glycosaminoglycans and oligosaccharides disorders: glycosaminoglycans synthesis defects, mucopolysaccharidoses, oligosaccharidoses and sialic acid disorders. In: Saudubray J-M, Baumgartner MR, García-Cazorla Á, Walter J (eds) Inborn metabolic diseases: diagnosis and treatment. Springer, Berlin, pp 765–783
10. Winchester B (2005) Lysosomal metabolism of glycoproteins. Glycobiology 15(6):1R–15R. https://doi.org/10.1093/glycob/cwi041. cwi041 [pii]
11. Verheijen FW, Verbeek E, Aula N, Beerens CE, Havelaar AC, Joosse M et al (1999) A new gene, encoding an anion transporter, is mutated in sialic acid storage diseases. Nat Genet 23(4):462–465. https://doi.org/10.1038/70585
12. Miyaji T, Echigo N, Hiasa M, Senoh S, Omote H, Moriyama Y (2008) Identification of a vesicular aspartate transporter. Proc Natl Acad Sci U S A 105(33):11720–11724. https://doi.org/10.1073/pnas.0804015105. 0804015105 [pii], 4461 [pii]
13. Weiss P, Tietze F, Gahl WA, Seppala R, Ashwell G (1989) Identification of the metabolic defect in sialuria. J Biol Chem 264(30):17635–17636. S0021-9258(19)84615-X [pii]
14. Kayashima T, Matsuo H, Satoh A, Ohta T, Yoshiura K, Matsumoto N et al (2002) Nonaka myopathy is caused by mutations in the UDP-N-acetylglucosamine-2-epimerase/N-acetylmannosamine kinase gene (GNE). J Hum Genet 47(2):77–79. https://doi.org/10.1007/s100380200004
15. Wen XY, Tarailo-Graovac M, Brand-Arzamendi K, Willems A, Rakic B, Huijben K et al (2018) Sialic

acid catabolism by N-acetylneuraminate pyruvate lyase is essential for muscle function. JCI Insight 3(24). https://doi.org/10.1172/jci.insight.122373. 122373 [pii]
16. Ramsay SL, Meikle PJ, Hopwood JJ, Clements PR (2005) Profiling oligosaccharidurias by electrospray tandem mass spectrometry: quantifying reducing oligosaccharides. Anal Biochem 345(1):30–46. https://doi.org/10.1016/j.ab.2005.06.042. S0003-2697(05)00492-6 [pii]
17. Tsay GC, Dawson G, Sung SS (1976) Structure of the accumulating oligosaccharide in fucosidosis. J Biol Chem 251(19):5852–5859. S0021-9258(17)33030-2 [pii]
18. van Pelt J, Kamerling JP, Vliegenthart JF, Hoogeveen AT, Galjaard H (1988) A comparative study of the accumulated sialic acid-containing oligosaccharides from cultured human galactosialidosis and sialidosis fibroblasts. Clin Chim Acta 174(3):325–335. https://doi.org/10.1016/0009-8981(88)90059-9. 0009-8981(88)90059-9 [pii]
19. Xia B, Asif G, Arthur L, Pervaiz MA, Li X, Liu R et al (2013) Oligosaccharide analysis in urine by maldi-tof mass spectrometry for the diagnosis of lysosomal storage diseases. Clin Chem 59(9):1357–1368. https://doi.org/10.1373/clinchem.2012.201053. clinchem.2012.201053 [pii]
20. Piraud M, Pettazzoni M, Menegaut L, Caillaud C, Nadjar Y, Vianey-Saban C et al (2017) Development of a new tandem mass spectrometry method for urine and amniotic fluid screening of oligosaccharidoses. Rapid Commun Mass Spectrom 31(11):951–963. https://doi.org/10.1002/rcm.7860
21. Schindler D, Kanzaki T, Desnick RJ (1990) A method for the rapid detection of urinary glycopeptides in alpha-N-acetylgalactosaminidase deficiency and other lysosomal storage diseases. Clin Chim Acta 190(1-2):81–91. https://doi.org/10.1016/0009-8981(90)90282-w. 0009-8981(90)90282-W [pii]
22. Schymanski EL, Jeon J, Gulde R, Fenner K, Ruff M, Singer HP et al (2021) Identifying small molecules via high resolution mass spectrometry: communicating confidence. eGalactose in human metabolism, glycosylation and congenital metabolic diseases: Time for a closer look. Biochim Biophys Acta Gen Subj 1865(8):129898. https://doi.org/10.1016/j.bbagen.2021.129898. S0304-4165(21)00056-8 [pii]
23. Humbel R, Collart M (1975) Oligosaccharides in urine of patients with glycoprotein storage diseases. I. Rapid detection by thin-layer chromatography. Clin Chim Acta 60(2):143–145. https://doi.org/10.1016/0009-8981(75)90119-9. 0009-8981(75)90119-9 [pii]
24. Mathis D, Croft J, Chrastina P, Fowler B, Vianey-Saban C, Ruijter GJG (2022) The role of ERNDIM diagnostic proficiency schemes in improving the quality of diagnostic testing for inherited metabolic diseases. J Inherit Metab Dis 45(5):926–936. https://doi.org/10.1002/jimd.12523. JIMD12523 [pii]
25. Raymond K, Rinaldo P (2013) From art to science: oligosaccharide analysis by maldi-tof mass spectrometry finally replaces 1-dimensional thin-layer chromatography. Clin Chem 59(9):1297–1298. https://doi.org/10.1373/clinchem.2013.208793. clinchem.2013.208793 [pii]
26. Sewell AC (2008) Oligosaccharides. In: Blau N, Duran M, Gibson KM (eds) Laboratory guide to the methods in biochemical genetics. Springer, Berlin, pp 325–333
27. van den Bosch J, Oemardien LF, Srebniak MI, Piraud M, Huijmans JG, Verheijen FW et al (2011) Prenatal screening of sialic acid storage disease and confirmation in cultured fibroblasts by LC-MS/MS. J Inherit Metab Dis 34(5):1069–1073. https://doi.org/10.1007/s10545-011-9351-3. 9351 [pii]
28. van der Ham M, Prinsen BH, Huijmans JG, Abeling NG, Dorland B, Berger R et al (2007) Quantification of free and total sialic acid excretion by LC-MS/MS. J Chromatogr B Analyt Technol Biomed Life Sci 848(2):251–257. https://doi.org/10.1016/j.jchromb.2006.10.066. S1570-0232(06)00861-0 [pii]
29. Hagemeijer MC, van den Bosch JC, Bongaerts M, Jacobs EH, van den Hout JMP, Oussoren E et al (2023) Analysis of urinary oligosaccharide excretion patterns by UHPLC/HRAM mass spectrometry for screening of lysosomal storage disorders. J Inherit Metab Dis 46(2):206–219. https://doi.org/10.1002/jimd.12597
30. Bonesso L, Piraud M, Caruba C, Van Obberghen E, Mengual R, Hinault C (2014) Fast urinary screening of oligosaccharidoses by MALDI-TOF/TOF mass spectrometry. Orphanet J Rare Dis 9:19. https://doi.org/10.1186/1750-1172-9-19. 1750-1172-9-19 [pii]
31. Mak J, Cowan TM (2021) Detecting lysosomal storage disorders by glycomic profiling using liquid chromatography mass spectrometry. Mol Genet Metab 134(1-2):43–52. https://doi.org/10.1016/j.ymgme.2021.08.006. S1096-7192(21)00770-8 [pii]
32. Semeraro M, Sacchetti E, Deodato F, Coskun T, Lay I, Catesini G et al (2021) A new UHPLC-MS/MS method for the screening of urinary oligosaccharides expands the detection of storage disorders. Orphanet J Rare Dis 16(1):24. https://doi.org/10.1186/s13023-020-01662-8. 1662 [pii]
33. Bongaerts M, Bonte R, Demirdas S, Jacobs EH, Oussoren E, van der Ploeg AT et al (2020) Using out-of-batch reference populations to improve untargeted metabolomics for screening inborn errors of metabolism. Metabolites 11(1). https://doi.org/10.3390/metabo11010008. metabolites-11-00008 [pii]
34. Valianpour F, Abeling NG, Duran M, Huijmans JG, Kulik W (2004) Quantification of free sialic acid in urine by HPLC-electrospray tandem mass spectrometry: a tool for the diagnosis of sialic acid storage disease. Clin Chem 50(2):403–409. https://doi.org/10.1373/clinchem.2003.027169. clinchem.2003.027169 [pii]

35. Sluiter W, van den Bosch JC, Goudriaan DA, van Gelder CM, de Vries JM, Huijmans JG et al (2012) Rapid ultraperformance liquid chromatography-tandem mass spectrometry assay for a characteristic glycogen-derived tetrasaccharide in Pompe disease and other glycogen storage diseases. Clin Chem 58(7):1139–1147. https://doi.org/10.1373/clinchem.2011.178319. clinchem.2011.178319 [pii]

36. Jean-Claude, Michalski André, Klein (1999) Glycoprotein lysosomal storage disorders: α- and β-mannosidosis fucosidosis and α-N-acetylgalactosaminidase deficiency Biochimica et Biophysica Acta (BBA) - Molecular Basis of Disease 1455(2-3) 69-84. https://doi.org/10.1016/S0925-4439(99)00077-0

Metabolites of the Pentose Phosphate Pathway

11

Mirjam M. C. Wamelink and Desirée E. C. Smith

Contents

Abbreviations

C7-sugars	Heptuloses
CSF	Cerebrospinal fluid
EDTA	Ethylenediaminetetraacetic acid
GALE	Galactose-4-epimerase
GALK	Galactokinase
GALT	Galactose-1-phosphate uridyltransferase
GC-FID	Gas chromatograph with flame ionization detection
HCl	Hydrochloric acid
IS	Internal standard
LC-MS/MS	Liquid chromatography–tandem mass spectrometry
Min	Minutes
MRM	Multiple reaction monitoring
RPI	Ribose-5-phosphate isomerase
SHPK	Sedoheptulokinase
SORD	Sorbitol dehydrogenase
TALDO	Transaldolase
TKT	Transketolase

M. M. C. Wamelink (✉) · D. E. C. Smith
Laboratory Genetic Metabolic Diseases (F0-132), Department of Clinical Chemistry, Amsterdam UMC Location University of Amsterdam, Amsterdam, The Netherlands
e-mail: m.wamelink@amsterdamumc.nl; d.smith@amsterdamumc.nl

11.1 Introduction

For the diagnosis of defects in the non-oxidative part of the pentose phosphate pathway (PPP) and other pathways involving sugar metabolism, analysis of polyols and (seven-carbon) sugars is important. Polyols, or sugar alcohols, are formed by the reduction of simple sugars. Knowledge of the metabolism and functions of most of the polyols is very limited. Galactitol is a metabolic end product, derived from galactose. Elevated concentrations are found in urine and plasma from patients affected with galactose-1-phosphate uridyltransferase (GALT) deficiency, uridine diphosphate galactose-4-epimerase (GALE) defi-

N. Blau, F. M. Vaz (eds.), *Laboratory Guide to the Methods in Biochemical Genetics*,
https://doi.org/10.1007/978-3-031-58819-8_11

ciency, galactokinase (GALK) deficiency, and galactose mutarotase deficiency [1, 2]. Sorbitol is derived from glucose and can be converted into fructose. Elevated concentrations of sorbitol (glucitol) have been described in diabetes mellitus and together with xylitol in sorbitol dehydrogenase deficiency (OMIM 182500) and have been implicated in the pathophysiology of these metabolic disorders [3]. The pentitol xylitol is known to be an intermediate in the conversion of L-xylulose into D-xylulose. L-arabitol is derived from L-xylulose and is elevated in essential pentosuria (OMIM 260800), a benign inborn error of xylitol metabolism, which results from a partial deficiency of L-xylulose reductase [4] and in L-arabinosuria, which has been described in one patient with elevated excretion of L-arabinose and L-arabitol. L-arabinosuria is a presumed deficiency of L-arabitol dehydrogenase [5]. The polyols erythritol, D-arabitol, and ribitol are derived from metabolism of sugar-phosphates from the PPP and can accumulate in case of a defect in the non-oxidative part of the PPP [6]. The seven-carbon sugars (sedoheptulose and mannoheptulose) and polyols (sedoheptitol and perseitol) are thought to derive from sedoheptulose-7-phosphate, an intermediate of the PPP. See Fig. 11.1 for the different pathways and defects.

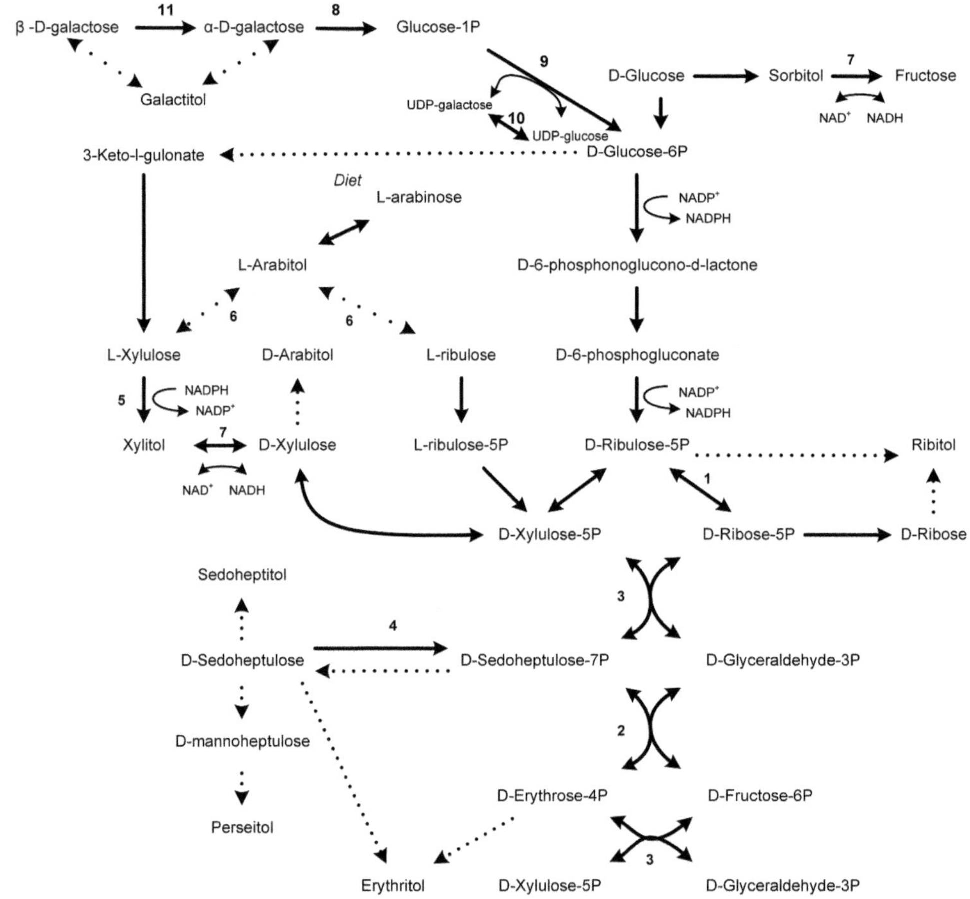

Fig. 11.1 The pentose phosphate pathway, the glucuronic acid pathway, and the pathway of galactose metabolism combined. The conversions displayed with the dotted arrow have not been proven in humans. (1) Ribose-5-phosphate isomerase; (2) transaldolase; (3) transketolase; (4) sedoheptulokinase; (5) L-xylulose reductase; (6) presumed L-arabitol dehydrogenase with conversion to either L-xylulose or L-ribulose; (7) sorbitol dehydrogenase; (8) galactokinase; (9) galactose-1-phosphate uridyltransferase; (10) galactose epimerase; (11) galactose mutarotase, P phosphate

More than 30 years ago, a gas chromatographic separation of urinary, plasma, and erythrocyte sugars and polyols was published [7] using trimethylsilyl (TMS) derivatives for enhanced detection. To date this method is still relevant and used for the diagnosis of the above-described defects.

For a more specific measurement of only the polyols or the heptuloses in urine a specific liquid chromatography–tandem mass spectrometry (LC-MS/MS) method is available [8, 9].

11.2 Clinical Significance

The measurement of polyols can help with diagnosing the following defects:

To date *deficiency of ribose-5-phosphate isomerase* (RPI, OMIM 608611) has been described in four cases that presented with psychomotor developmental delay, visual impairment, spasticity, and leukoencephalopathy. Arabitol and ribitol concentrations are strongly elevated in urine, plasma, and CSF [10].

Transaldolase (TALDO) deficiency (OMIM 606003) presents with hepato- and splenomegaly, cardiac abnormalities, liver function problems, anemia, thrombocytopenia, and abnormal skin (e.g., cutis laxa) with variable phenotype. In urine, erythritol, arabitol, ribitol, and C7-sugars and C7-polyols are elevated [11, 12]. Furthermore, sedoheptulose-7-phosphate was found to be elevated in blood spots, fibroblasts, and lymphoblasts from TALDO patients [13].

Transketolase (TKT) deficiency (OMIM 617044) was described in five patients of three families with common clinical findings: short stature, developmental delay, and congenital heart defect. Elevated excretion of erythritol, arabitol, ribitol, and pent(ul)ose-5P, as well as elevated plasma levels of erythritol, arabitol, and ribitol is found in affected patients [14].

Defects in galactose metabolism, e.g., GALT deficiency (OMIM 230400), GALE deficiency (OMIM 230350), and GALK deficiency (OMIM 230200) can present asymptomatic, only cataract or can be life threatening with multiorgan involvement and long-term complications. Galactitol and galactose are elevated in urine and plasma in all three defects. Galactose-1-phosphate in erythrocytes is only elevated in GALT and GALE deficiency.

Sorbitol dehydrogenase deficiency with peripheral neuropathy (SORDD, OMIM 618912) is an autosomal recessive disorder characterized by onset of distal muscle weakness mainly affecting the lower limbs and resulting in difficulty walking. Onset of symptoms is usually in the first or second decades of life, although later adult onset has been reported; the disorder is slowly progressive [15]. Elevated sorbitol and xylitol concentrations are found in urine and plasma. Previously SORD deficiency was associated with cataract, although no direct relation has been confirmed of this enzyme deficiency and (congenital) cataract [16].

Three other inborn errors with polyol abnormalities without a clear suspected clinical phenotype are:

Sedoheptulokinase (SHPK) deficiency (OMIM 617213) is the result of mutations in the SHPK gene. A 57-kb-deletion, the most common cause of nephropathic cystinosis, also results in SHPK deficiency. Two unrelated isolated SHPK deficient patients have been described with completely different clinical pictures suggesting the clinical phenotype may not be related to SHPK deficiency. In urine erythritol and sedoheptulose are elevated [17].

l-xylulose reductase deficiency (OMIM 260800) is a defect of the glucuronic oxidation pathway, with the biochemical feature of pentosuria and is a benign condition [4]. Strong elevations of L-arabitol in urine are found.

l-arabinosuria has been documented in only one patient with elevated excretion of L-arabinose and L-arabitol [5]. The defect has not been fully elucidated in humans.

11.3 Metabolites

11.3.1 Properties of Metabolites

Polyols, polyhydric alcohols, or sugar alcohols are organic compounds containing multiple hydroxyl groups. They can be formed by the reduction of sugars and are classified based on the numbers of C-atoms: erythritol and threitol (C4-polyols, tetritols), ribitol, arabitol, and xylitol (C5-polyols, pentitols), galactitol, sorbitol, and mannitol (C6-polyols, hexitols), and sedoheptitol and perseitol (C7-polyols, heptitols). Polyols are neutral compounds which do not have functional groups enabling their direct selective isolation from biological materials. A polyol has the general formula $(CHOH)_nH_2$.

Sugars are soluble carbohydrates and can be classified in monosaccharides (e.g., ribose, glucose, fructose) or disaccharides (two bonded monosaccharides, e.g., lactose, sucrose). A monosaccharide has the general formula $C_nH_{2n}O_n$. Oligosaccharides are carbohydrates with more than 2 monosaccharides bonded together.

11.3.2 Determination of Sugar and Polyol Analysis in Urine, Plasma, and CSF with Gas Chromatography [7]

Pre-analytical

A urine sample, EDTA, or heparin plasma or CSF can be used for the analysis of sugars and polyols. Diet may influence the results, therefore a sample taken before a meal is best. Bacterial contamination can decrease urinary sugar concentrations. Samples should be stored frozen.

Reagents and Chemicals

Reagent/chemical	Producer/vendor	Product number
D-arabinose	Sigma-Aldrich	A3131
D-arabitol	Sigma-Aldrich	A3381
BSTFA (*N,O*-bis(trimethylsilyl)-trifluoroacetamide)	Pierce	38830
D-cellobiose	Sigma-Aldrich	C-7252
i-Erythritol	Sigma-Aldrich	E-7500
D-erythrose	Sigma-Aldrich	E-7625
D-fructose	Sigma-Aldrich	F-2543
D-fucose	Sigma-Aldrich	F8150
n-Hexaan	Merck	1043720500
Galactitol	Calbiochem	317492
D-galactose	Sigma-Aldrich	G-6404
D-glucose	Merck	8337
Hydrochloric acid (HCL) 37%	VWR	20252290
i-Inositol	Sigma-Aldrich	I-5125
Lactose	Merck	7660
Maltose	Merck	5912
D-mannitol	Merck	4125
D-mannoheptulose	Sigma-Aldrich	M-4375
D-mannose	Merck	5984
Methanol	VWR	20.847.320
Perseitol	Sigma-Aldrich	P-8295
Ribitol	Fluka	2240
D-ribose	Sigma-Aldrich	R-7500
D-sorbitol	Aldrich	24.085-0
Sucrose	Sigma-Aldrich	S-7903
DL-Threitol	Sigma-Aldrich	T-2396
Tri-Sil TBT	Pierce	49016
Xylitol	Sigma-Aldrich	X-3375
D-melibiose	Sigma-Aldrich	M-5500
D-xylose	Sigma-Aldrich	X-3877
D-xylulose	Sigma-Aldrich	X-4625

Standard 1 contains 0.4 mmol/L of erythrose, threitol, erythritol, xylulose, arabinose, ribose, fucose, xylose, xylitol, arabitol, ribitol, fructose, galactose, glucose, mannitol, sorbitol, galactitol, myo-inositol, lactose, sucrose, maltose, cellobiose, and melibiose. Standard 2 contains 0.4 mmol/L of mannose. The internal standard (IS) consists of 0.4 mmol/L of mannoheptulose, perseitol, and trehalose.

Instrumentation

The measurement is performed on a gas chromatograph with flame ionization detection (GC-FID) and a split/splitless injector. Mass spectrometry can be performed for additional confirmation. Metabolites are separated on a Chrompack CP SIL 5 CB column, 25 m × 0.25 mm (i.d.), df = 0.12 μm. Gas flow rate (helium): 0.5 mL/min, split flow 17 mL/min, detector temperature 300 °C and injector temperature 275 °C. The oven temperature starts at 110 °C, increases by 1.5 °C/min to 156 °C, subsequently increases with 4 °C/min to 265 °C, then increases with 10 °C/min to 295 °C and is kept at 295 °C for 5 min for urine samples, and 20 min for plasma and CSF samples.

Analytical Procedure/Manual

Standards: Add 100 μL standard to 100 μL of 0.4 mM IS in a glass 10 mL tube.

Urine: Pipet 0.5 mL urine in an Eppendorf cup. Check the pH and adjust to pH 5–7 when necessary, with 1 mol/L HCL. Centrifuge urine samples at 3345 g for 5 min at room temperature. Use the amount of 1/(x × creatinine concentration) mL, where $x = 2$ when creatinine concentration is <1 mmol/L, if age is 0–3 months or when high values are expected and $x = 1$ is used for all other urine samples. Add 100 μL of 0.4 mM IS to the urine in glass 10 mL tubes.

Plasma and CSF: Add 500 μL of plasma or CSF to 100 μL of 0.4 mM IS in glass 10 mL tubes, add 2 mL methanol and mix the samples. Store the samples for 30 min at –20 °C and afterwards centrifuge for 10 min at 1890 g. Collect the supernatant and extract with 4 mL hexane. Transfer the methanol layers after centrifugation for 2 min at 3345 g at room temperature to a clean tube.

Evaporate all samples (standards, urine, plasma, and CSF) to dryness at 60 °C under a stream of nitrogen. Add 300 μL Tri-Sil-TBT and heat for 30 min at 100 °C for derivatization. After cooling add 1 mL water and extract with 1 mL hexane. Collect the hexane layers after centrifugation, 2 min at 3345 g at room temperature, and wash with 1 mL of 0.1 mol/L HCl solution. Pipet the hexane layers in a glass vial and add one drop of bis(trimethylsilyl)trifluoroacetamide (BSTFA). Dry the samples at room temperature under a stream of nitrogen and dissolve in 40 μL of hexane.

Calculation

Quantification of sugars/polyols is performed by comparing the peak area ratio of each analyte and its corresponding internal standard with that of standard 1 or 2. As some analytes give rise to more than one peak, ratios of both peaks are calculated for confirmation of the analyte. In case the calculated ratio is different from the standards one of the peaks is disturbed by another compound. Only one peak is used for quantification. The relevant peaks and their internal standards are indicated in Table 11.1

Pitfalls

When the samples are dried under nitrogen at 60 °C, stop immediately when the samples are dry otherwise polyols will evaporate.

The assessment of the concentration of glucose in plasma and CSF is inaccurate due to the high concentrations of this metabolite.

Post-analytical (e.g., Chromatograms of Pathological Samples)

See Fig. 11.2 for chromatograms of standards, urine, plasma, and CSF controls and urine samples of RPI, TALDO, TKT GALT deficient patients

Reference Values

Reference values in urine, plasma, and CSF are summarized in Table 11.2.

Table 11.1 Calculation information for sugar and polyol analysis GC-FID

Carbohydrate	Peak used for calculation[a]	Internal standard used[b]
Erythrose	1	IS-1
Threitol		IS-2
Erythritol		IS-2
Xylulose	1	IS-1
Arabinose	2	IS-1
Ribose	2	IS-1
Fucose	1	IS-2
Xylitol		IS-2
Arabitol		IS-2
Ribitol		IS-2
Xylose	2	IS-1
Mannose	2	IS-2
Fructose		IS-1
Galactose	1	IS-2
Glucose	1	IS-2
Mannitol		IS-2
Sorbitol		IS-2
Galactitol		IS-2
Myo-inositol		IS-2
Lactose	2	IS-3
Sucrose		IS-3
Maltose	1	IS-3
Cellobiose	1	IS-3
Melibiose	2	IS-3

[a]Based on retention time
[b]IS-1: Mannoheptulose, IS-2: Perseitol, IS-3: Trehalose

Pathological Values

Table 11.3 gives an overview of pathological values found in patients.

Secondary causes for elevated sugars and polyols are:

Glucose in urine may derive from glucose infusion, kidney dysfunction, diet, or untreated diabetes mellitus.

Fructose, sucrose, and lactose may derive from the diet.

Erythritol, xylitol, and sorbitol are used in food, medication, and supplements as sweeteners. Liver dysfunction can give elevated erythritol, arabitol, ribitol, galactose, and galactitol.

Quality Control

For quality control, pooled urine, plasma, and/or CSF (to match the type of sample included in the series) are included in each series.

11.3.3 Determination of Polyol Analysis in Urine by LC-MS/MS [8, 19]

Pre-analytical

A random urine sample can be used for the analysis of polyols. Store samples frozen. Repeated freeze-thawing does not influence the concentrations of polyols.

Reagents and Chemicals

Arabitol, erythritol, perseitol, sedoheptitol, sorbitol, threitol, and xylitol are purchased from Sigma-Aldrich. D-[U-$^{13}C_4$]-erythritol and D-$^{13}C_2$ [4,5]-arabitol are purchased from Omicron Biochemicals. Ribitol came from Fluka, galactitol from Calbiochem and mannitol and acetonitrile from Merck. Amberlite IRN 150 L monobed mixed resin came from Boehringer Ingelheim GmbH. [5,6,6-d3]-sorbitol is prepared by reduction of [5,6,6-d3]-glucose following a described procedure [20].

Instrumentation

Detection of the polyols is carried out on an API 5000 tandem mass spectrometer (Sciex) or a similar instrument equipped with an electrospray ion source operating in the negative mode. The mass spectrometer parameters are as follows: source temperature, 500 °C; ionspray voltage, 4500 volt and collision energy, 30 eV. Detection of the polyols is performed by multiple reaction monitoring (MRM). The MRM transitions (Q1/Q3) settings for the different polyols are listed in Table 11.4:

Data is acquired and processed using LC-MS/MS.

Analytical Procedure

Different amounts of urine are used depending on the creatinine concentration. Use 200 μL urine when creatinine <1.0 mmol/L; 100 μL urine when creatinine >1.0 and <10.0 mmol/L; 50 μL urine when creatinine >10.0 mmol/L. Distilled water is added to obtain a total volume of 200 μL. Then 2.5 nmol $^{13}C_4$-erythritol, 50 nmol $^{13}C_2$-arabitol, and 10 nmol D_3-sorbitol are added as IS. To the mix 20 μg of Amberlite IRN 150 L

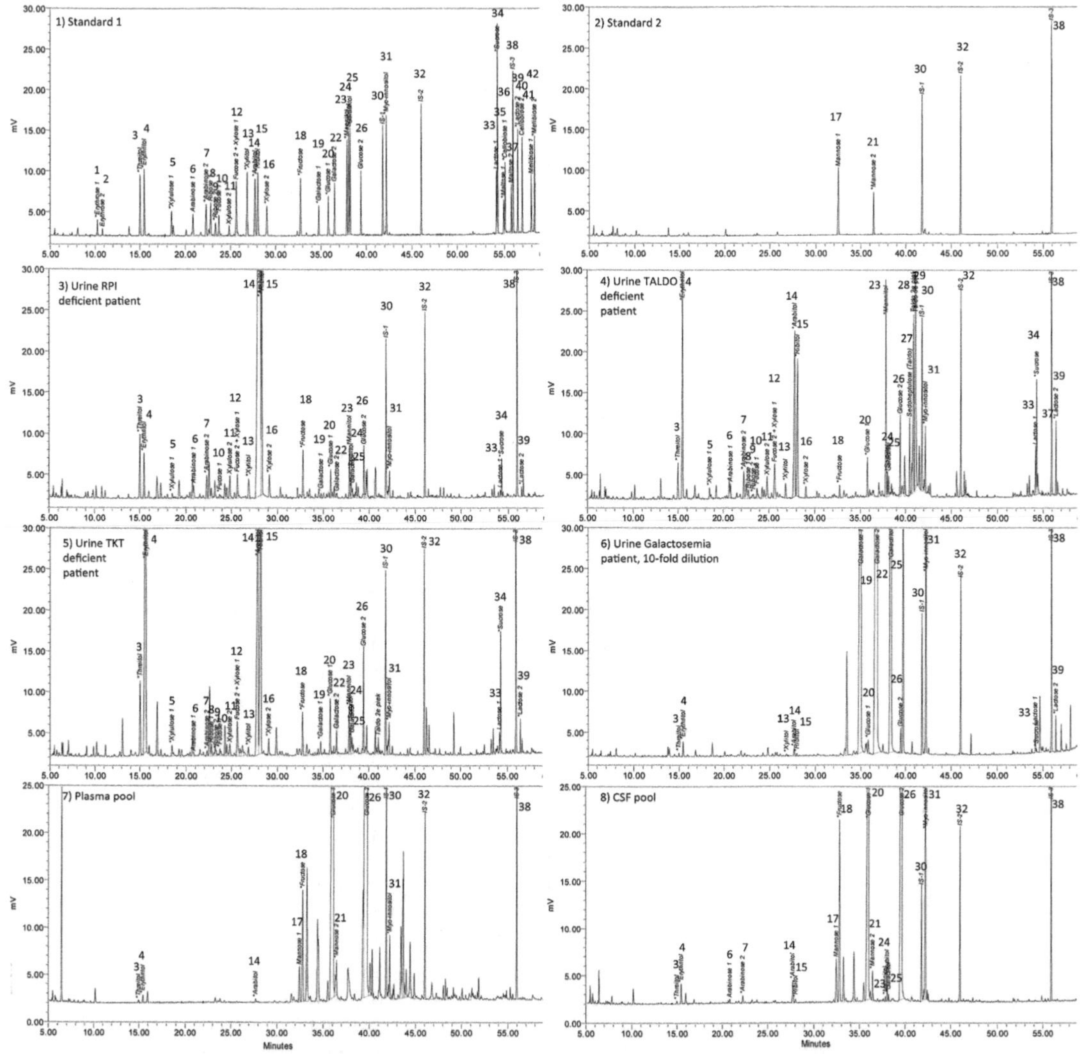

Fig. 11.2 Chromatograms of sugar and polyol analysis with GC-FID. (**1**) Standard 1; (**2**) standard 2; (**3**) urine RPI deficient patient; (**4**) urine TALDO deficient patient; (**5**) urine TKT deficient patient; (**6**) urine galactosemia patient 10 times diluted; (**7**) pooled plasma; (**8**) pooled CSF. Sugars and polyols are numbered in order of their retention times: (1) erythrose_1; (2) erythrose_2; (3) threitol; (4) erythritol; (5) xylulose_1; (6) arabinose_1; (7) arabinose_2; (8) ribose_1; (9) ribose_2; (10) fucose_1; (11) xylulose_2; (12) fucose_2 + xylose_1; (13) xylitol; (14) arabitol; (15) ribitol; (16) xylose_2; (17) mannose_1; (18) galactose_1; (20) glucose_1; (21) mannose_2; (22) galactose_2; (23) mannitol; (24) sorbitol; (25) galactitol; (26) glucose_2; (27) sedoheptulose_1; (28) sedoheptulose_2; (29) sedoheptulose_3; (30) IS1; (31) myo-inositol; (32) IS2; (33) lactose_1; (34) sucrose; (35) maltose_1; (36) cellobiose_1; (37) maltose_2; (38) IS3; 39 lactose_2; (40) cellobiose _2; (41) melibiose_1; (42) melibeiose_2

monobed mixed resin is added, mixed, and centrifuged for 5 min at 410 g. The supernatant is stored at –20 °C until injection.

Calibration

Calibration curves for all polyols are prepared from a standard solution and are included in all series. The concentration ranges of the calibration curves are (adapted in line with the concentrations found in urines): 0–4 nmol for ribitol, sedoheptitol, and perseitol, 0–8 nmol for xylitol and sorbitol, 0–12 nmol for galactitol, 0–20 nmol for threitol, 0–32 nmol for mannitol, 0–40 nmol for erythritol, and 0–48 nmol for arabitol.

Table 11.2 Reference values of relevant sugars and polyols in urine, plasma, and CSF determined by GC or LC-MS/MS

Urine carbohydrate	Polyolen and C7-sugars LC-MS/MS [8, 9] Age	Reference values mmol/mol creatinine	Sugars and polyols GC [7] Age	Reference values mmol/mol creatinine
Erythritol	0–4 mth	89–209	0–3 mth	58–162
	4 mth–5 yr	38–148	3–12 mth	89–158
	5–80 yr	14–88	1–2 yr	76–192
			2–6 yr	55–105
			6–18 yr	35–179
			18–77 yr	19–76
Ribitol	0–4 mth	0–25	0–3 mth	7–16
	4 mth–5 yr	1–14	3–12 mth	10–17
	5–80 yr	0–7	1–2 yr	9–24
			2–6 yr	8–11
			6–18 yr	4–11
			18–77 yr	<5
Arabitol	0–4 mth	30–151	0–3 mth	27–97
	4 mth–5 yr	22–103	3–12 mth	51–99
	5–80 yr	6–60	1–2 yr	52–88
			2–6 yr	32–78
			6–18 yr	16–89
			18–77 yr	10–44
Xylitol	0–5 yr	3–13	0–3 mth	4–22
	5–80 yr	0–12	3–12 mth	5–11
			1–2 yr	7–29
			2–6 yr	5–49
			6–18 yr	3–7
			18–77 yr	<20
Xylulose	–	–	0–80 yr	<5
Arabinose	–	–	0–3 mth	28–97
			3–12 mth	8–70
			1–2 yr	25–81
			2–6 yr	16–73
			6–18 yr	8–46
			18–77 yr	8–37
Sorbitol	0–4 mnd	2–12	0–3 mth	2–17
	4 mnd–5 yr	0–28	3–12 mth	2–13
	5–80 yr	0–15	1–2 yr	7–56
			2–6 yr	4–12
			6–18 yr	3–17
			18–77 yr	2–26
Sedoheptitol	0–80 yr	<1	–	–
Perseitol	0–80 yr	<1	–	–
Sedoheptulose	0–1 mth	<40	–	–
	1 mth–80 yr	<9		
Mannoheptulose	0–80 yr	<3	–	–

Plasma carbohydrate	Reference values (μmol/L) [18]
Erythritol, ribitol, arabitol, arabinose, xylitol, xylulose, and sorbitol	<5

CSF carbohydrate	Reference values (μmol/L)
Erythritol	12–33
Ribitol	<5
Arabitol	9–39
Arabinose	6–23

Table 11.3 Pathological findings of sugars and polyols in patients with metabolic disorders

Disorder	Abnormal metabolite(s)
Ribose-phosphate isomerase deficiency	Arabitol (u,p,c) ↑↑-↑↑↑ Ribitol (u,p,c) ↑↑-↑↑↑ Xylitol (u) n-↑
Transaldolase deficiency	Erythritol (u,p) n-↑↑ Arabitol (u,p) ↑-↑↑ Ribitol (u,p,c) ↑-↑↑ Sedoheptulose (u) ↑↑ Mannoheptulose (u) ↑↑ Sedoheptitol (u) ↑ Perseitol (u) ↑
Transketolase deficiency	Erythritol (u,p) ↑↑ Arabitol (u,p) ↑↑ Ribitol (u,p) ↑↑
Sedoheptulokinase deficiency	Sedoheptulose (u) ↑↑ Erythritol (u) ↑↑
Sorbitol dehydrogenase deficiency	Sorbitol (u,p) ↑↑ Xylitol (u,p) ↑↑
L-xylulose reductase deficiency	Arabitol (u) ↑ Xylulose (u,p) ↑-↑↑
L-arabinosuria[a]	Arabinose (u,p,c) ↑↑ Arabitol (u,p,c) ↑↑
Galactokinase, GALT and GALE deficiency	Galactose (u,p) ↑-↑↑↑ Galactitol (u,p) ↑-↑↑↑

u, urine; p, plasma; c, CSF
[a]Diet with fruit

Table 11.4 MRM transitions for LC-MS/MS methods

Metabolite	MRM transition (m/z)
Erythritol and threitol	121/89
13C4-Erythritol	125/92
Ribitol, arabitol, and xylitol	151/89
13C2-Arabitol	153/91
Galactitol, sorbitol, and mannitol	181/89
D3-Sorbitol	184/89
Sedoheptitol and perseitol	211/89
Sedoheptulose and mannoheptulose	209/119
13C2-xylulose	151/91

Liquid chromatography (Waters, Alliance 2695 pump, Milford, MA, USA) is performed using an Aminex HPX-87C column (250 × 4.0 mm, bead size 9 μm, Bio-Rad Laboratories, Veenendaal, The Netherlands). An isocratic eluent of 40% ACN/water is used, with the column kept at 65 °C. The flow rate is set to 0.4 mL/min and is split after the analytical column in a ratio of 1:2, resulting in an inlet flow into the tandem mass spectrometer of 200 μL/min; 20 μL of sample is injected onto the column and the total run time is 24 min.

Calculation

Quantification of urinary polyols is performed using the corresponding calibration curves. $^{13}C_4$-erythritol is used as an IS for the tetritols, $^{13}C_2$-arabitol for the pentitols and D_3-sorbitol for the hexitols and heptitols. Concentrations are expressed in mmol/mol creatinine.

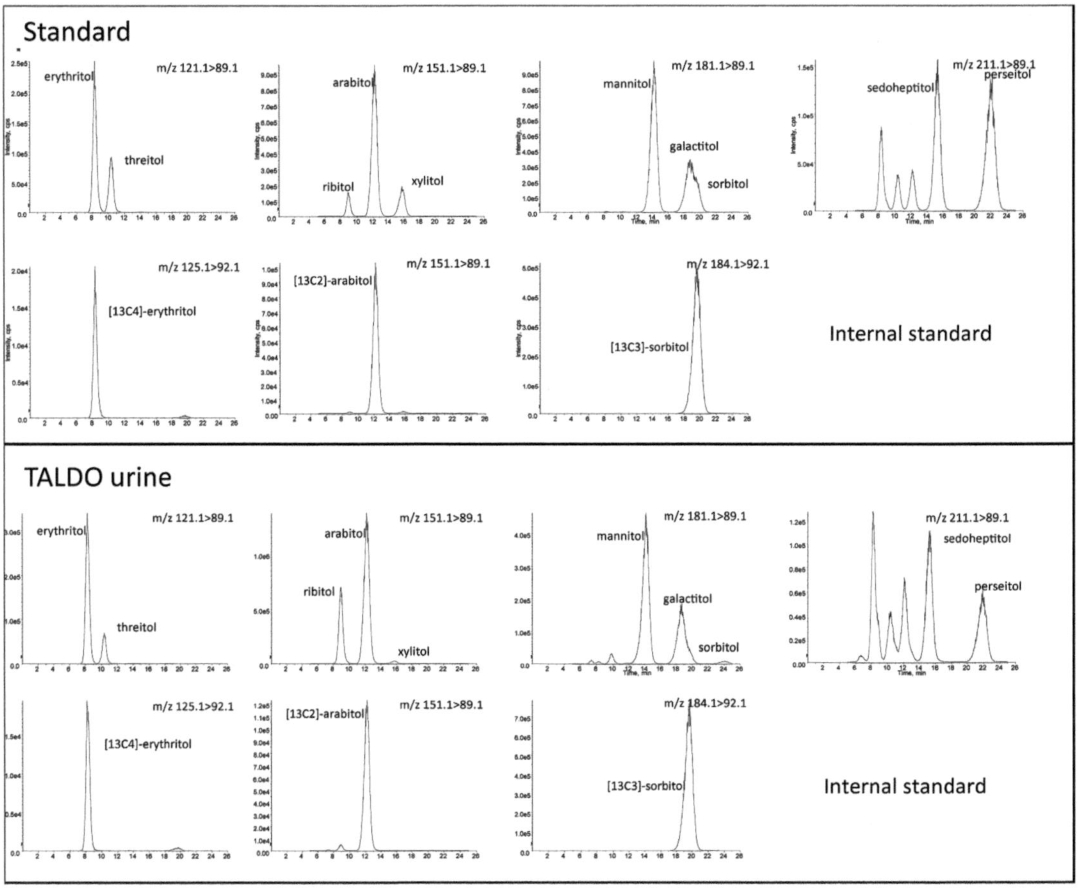

Fig. 11.3 MRM chromatograms of polyols of a standard and a urine from TALDO deficient patient measured by LC-MS/MS. MRM transitions are given for each mass transition

Post-analytical

See Fig. 11.3 for chromatograms of the polyol standard and from a patient urine with TALDO deficiency.

Reference Values

Urinary concentrations of polyols are age-dependent, as summarized in Table 11.2.

Pathological Values

Table 11.3 gives an overview of pathological values of urinary polyols found in patients.

Secondary causes for elevated polyols are:

Erythritol, xylitol, and sorbitol are used in food, medication, and supplements as sweeteners. Liver dysfunction can give elevated erythritol, arabitol, ribitol, and galactitol.

Quality Control

For quality control, 100 μL pooled urine spiked with 8 μmol/L sedoheptitol and 8 μmol/L perseitol is included in each series. In addition, four times dilution of this sample and a pooled urine spiked with 180 μmol/L erythritol, 90 μmol/L threitol, 218 μmol/L arabitol, 36 μmol/L xylitol, 18 μmol/L ribitol, 36 μmol/L sorbitol, 145 μmol/L mannitol, 55 μmol/L galactitol, 16 μmol/L sedoheptitol, and 16 μmol/L perseitol are included in each series. These three urine samples were chosen to obtain concentrations in the low, middle, and high part of the calibration curves.

11.3.4 Determination of C7-Sugars in Urine with LC-MS/MS [9]

Pre-analytical

A random urine sample can be used for the analysis of C7-sugars (heptuloses). Samples are stored frozen. Repeated freeze-thawing does not influence the concentrations.

Reagents and Chemicals

Mannoheptulose and sedoheptulosan were purchased from Sigma-Aldrich. $^{13}C_2$-Xylulose was purchased from Omicron Biochemicals, acetonitrile from Merck, HCl from VWR, and NH_2-extraction tubes from Varian Inc.

Sedoheptulose was prepared by dissolving sedoheptulosan (anhydro-sedoheptulose) in 1% HCl to a final concentration of 20 mmol/L and heating it in boiling water for 30 min [9].

Instrumentation

Detection of sedoheptulose and mannoheptulose was achieved using an API 5000 tandem mass spectrometer (Sciex) operating in the negative mode, by multiple reaction monitoring (MRM). The MRM transitions are listed in Table 11.4. Collision energy was 16 eV for sedoheptulose and 13 eV for mannoheptulose. Other settings were as described for the polyol method.

Analytical Procedure/Manual

Depending on creatinine concentration different quantities of urine are employed; 450 μL urine when creatinine is <1.0 mmol/L; 200 μL urine when creatinine is >1.0 and <10.0 mmol/L; and 100 μL urine when creatinine is >10.0 mmol/L. Calibration curves are prepared from standard solutions and range from 0 to 14 nmol for sedoheptulose and from 0 to 10 nmol for mannoheptulose. [$^{13}C_2$]-Xylulose (5 nmol) is added as IS. Add distilled water to obtain a minimum volume of 250 mL. Mix the samples with acetonitrile (1:1 v/v). Pretreat NH_2-bond extraction tubes with 0.5 mL acetonitrile, after which the samples are applied and the heptuloses are eluted with 0.5 mL distilled water and ready to be measured.

Liquid chromatography (Waters, Alliance 2695 pump, Milford, MA, USA) is performed as above for polyol measurements

Calculation

Quantification of urinary heptuloses is performed using the corresponding calibration curves. Concentrations are expressed in mmol/mol creatinine.

Post-analytical

See Fig. 11.4 for chromatograms of the standard, a control urine, and urine from a patient with TALDO deficiency.

Reference Values

Urinary concentrations of heptuloses are age-dependent, as summarized in Table 11.2.

Pathological Values

Elevated excretion of sedoheptulose and mannoheptulose is suggestive for TALDO deficiency. Elevated excretion of only sedoheptulose is suggestive of SHPK deficiency, see Table 11.3.

Quality Control

For quality control, a pooled urine (control low) together with a pooled urine spiked with heptuloses (control high) should be included in each series.

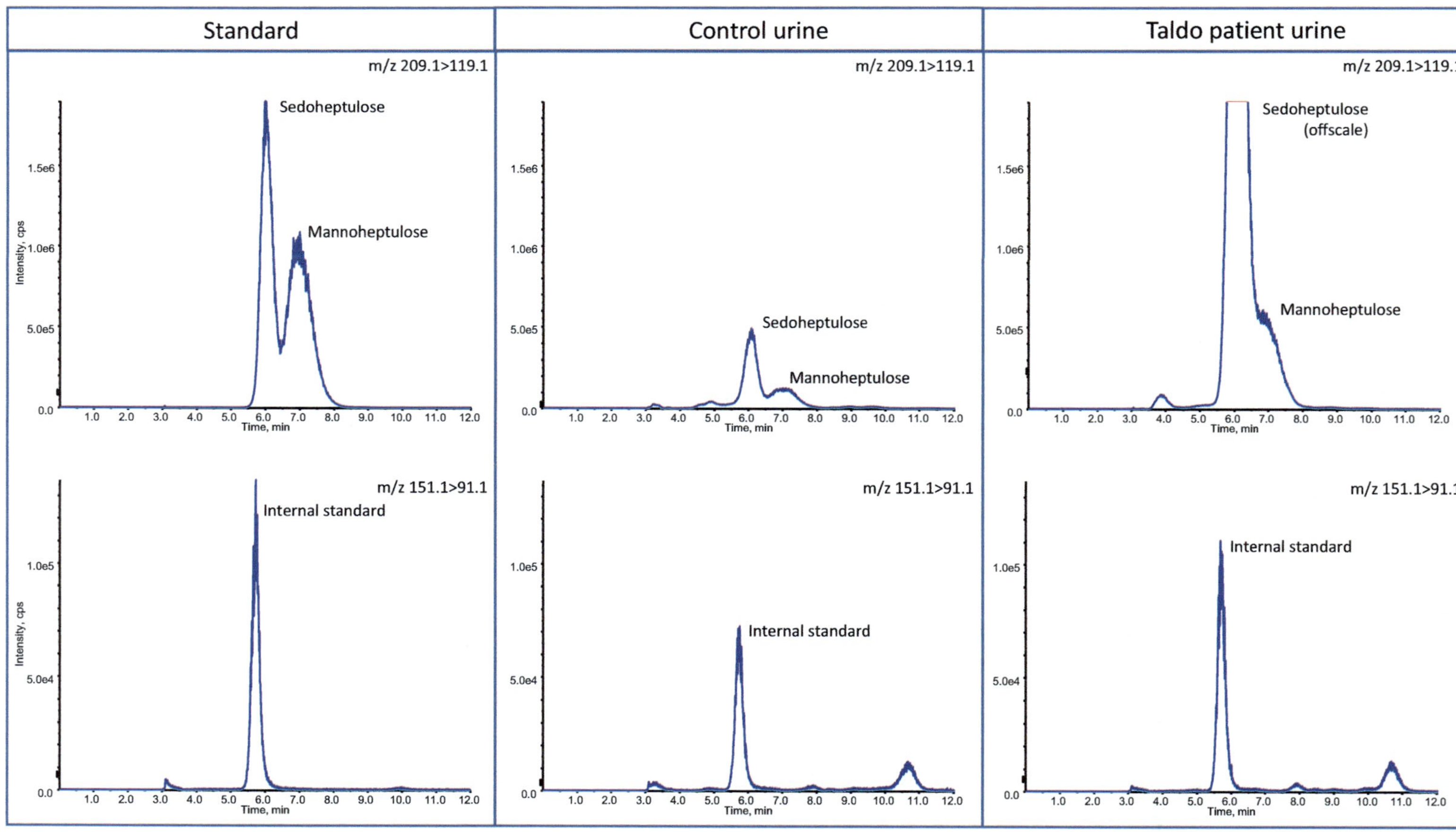

Fig. 11.4 MRM chromatogram of C7-sugars of a standard, a control urine, and a urine from TALDO deficient patient measured by LC-MS/MS. MRM transitions are given for each mass transition

11.4 Follow-Up Assays

To further distinguish between SHPK or TALDO deficiency sedoheptulose-7P can be measured in urine [9], in TALDO deficiency it is elevated and in SHPK deficiency it is not.

TALDO and TKT deficiency can be confirmed enzymatically in lymphoblasts, erythrocytes, or fibroblasts, RPI deficiency in fibroblasts or lymphoblasts. The cells are incubated with substrate, after which formation of sugar-phosphates is analyzed by LC-MS/MS [10, 11, 13]. Enzyme assays for TALDO, TKT, and RPI following the decrease of NADH at 340 nm with a spectrophotometer can also be used [21, 22].

References

1. Endres W, Shin YS (1985) Cataract and metabolic disease. J Inherit Metab Dis 13:509–516
2. Wada Y, Arai-Ichinoi N, Kikuchi A, Kure S (2022) β-Galactosidase therapy can mitigate blood galactose elevation after an oral lactose load in galactose mutarotase deficiency. J Inherit Metab Dis 45:334–339
3. Lee AY, Chung SS (1999) Contributions of polyol pathway to oxidative stress in diabetic cataract. FASEB J 13:23–30
4. Lane AB (1985) On the nature of L-xylulose reductase deficiency in essential pentosuria. Biochem Genet 23:61–72
5. Onkenhout W, Groener JE, Verhoeven NM, Yin C, Laan LA (2002) L-Arabinosuria: a new defect in human pentose metabolism. Mol Genet Metab 77:80–85
6. Wamelink MM, Struys EA, Jakobs C (2008) The biochemistry, metabolism and inherited defects of the pentose phosphate pathway: a review. J Inherit Metab Dis 31(6):703–717
7. Jansen G, Muskiet FA, Schierbeek H, Berger R, van der Slik SW (1986) Capillary gas chromatographic profiling of urinary, plasma and erythrocyte sugars and polyols as their trimethylsilyl derivatives, preceded by a simple and rapid prepurification method. Clin Chim Acta 157:277–293
8. Wamelink MM, Smith DE, Jakobs C, Verhoeven NM (2005) Analysis of polyols in urine by liquid chromatography-tandem mass spectrometry: A useful tool for recognition of inborn errors affecting polyol metabolism. J Inherit Metab Dis 28:951–963
9. Wamelink MM, Smith DE, Jansen EE, Verhoeven NM, Struys EA, Jakobs C (2007) Detection of transaldolase deficiency by quantification of novel seven-carbon chain carbohydrate biomarkers in urine. J Inherit Metab Dis 30(5):735–742
10. Huck HJ, Verhoeven NM, Struys EA, Salomons GS, Jakobs C, van der Knaap MS (2004) Ribose-5-phosphate isomerase deficiency: new inborn error in the pentose phosphate pathway associated with a slowly progressive leukoencephalopathy. Am J Hum Genet 74:745–751
11. Verhoeven NM, Huck JH, Roos B, Struys EA, Salomons GS, Douwes AC, van der Knaap MS, Jakobs C (2001) Transaldolase deficiency: liver cirrhosis associated with a new inborn error in the pentose phosphate pathway. Am J Hum Genet 68:1086–1092
12. Verhoeven NM, Wallot M, Huck JH, Dirsch O, Ballauf A, Neudorf U, Salomons GS, van der Knaap MS, Voit T, Jakobs C (2005) A newborn with severe liver failure, cardiomyopathy and transaldolase deficiency. J Inherit Metab Dis 2:169–179
13. Wamelink MM, Struys EA, Huck JH, Roos B, van der Knaap MS, Jakobs C, Verhoeven NM (2005) Quantification of sugar phosphate intermediates of the pentose phosphate pathway by LC-MS/MS: application to two new inherited defects of metabolism. J Chromatogr B Analyt Technol Biomed Life Sci 823:18–25
14. Boyle L, Wamelink MMC, Salomons GS, Roos B, Pop A, Dauber A, Hwa V, Andrew M, Douglas J, Feingold M, Kramer N, Saitta S, Retterer K, Cho MT, Begtrup A, Monaghan KG, Wynn J, Chung WK (2016) Mutations in TKT are the cause of a syndrome including short stature, developmental delay, and congenital heart defects. Am J Hum Genet 98(6):1235–1242
15. Cortese A, Zhu Y, Rebelo AP, Negri S, Courel S, Abreu L, Bacon CJ, Bai Y, Bis-Brewer DM, Bugiardini E, Buglo E, Danzi MC, Feely SME, Athanasiou-Fragkouli A, Haridy NA, Inherited Neuropathy Consortium, Isasi R, Khan A, Laurà M, Magri S, Pipis M, Pisciotta C, Powell E, Rossor AM, Saveri P, Sowden JE, Tozza S, Vandrovcova J, Dallman J, Grignani E, Marchioni E, Scherer SS, Tang B, Lin Z, Al-Ajmi A, Schüle R, Synofzik M, Maisonobe T, Stojkovic T, Auer-Grumbach M, Abdelhamed MA, Hamed SA, Zhang R, Manganelli F, Santoro L, Taroni F, Pareyson D, Houlden H, Herrmann DN, Reilly MM, Shy ME, Zhai RG, Zuchner S (2020) Biallelic mutations in SORD cause a common and potentially treatable hereditary neuropathy with implications for diabetes. Nat Genet 52(5):473–481
16. Shin YS, Rieth M, Endres W, Haas P (1984) Sorbitol dehydrogenase deficiency in a family with congenital cataracts. J Inherit Metab Dis 7(Suppl 2):151–152
17. Wamelink MM, Ramos RJ, van den Elzen AP, Ruijter GJ, Bonte R, Diogo L, Garcia P, Neves N, Nota B, Haschemi A, Tavares de Almeida I, Salomons GS (2015) First two unrelated cases of isolated sedoheptulokinase deficiency: A benign disorder? J Inherit Metab Dis 38(5):889–894
18. Kusmierz J, DeGeorge JJ, Sweeney D, May C, Rapoport SI (1989) Quantitative analysis of polyols in human plasma and cerebrospinal fluid. J Chromatogr 497:39–48

19. Verhoeven NM, Wamelink MMC, Jakobs C (2008) Polyols. In: Blau N, Duran M, Gibson KM (eds) Laboratory guide to the methods in biochemical genetics. Springer, Berlin
20. Karabinos JV, Ballun AT (1953) Direct reduction of aldoses and ketoses by Raney nickel. J Am Chem Soc 75:4501–4502
21. Banki K, Hutter E, Colombo E, Gonchoroff NJ, Perl A (1996) Glutathione levels and sensitivity to apoptosis are regulated by changes in transaldolase expression. J Biol Chem 271(51):32994–33001
22. Novello F, McLean P (1968) The pentose phosphate pathway of glucose metabolism. Measurement of the non-oxidative reactions of the cycle. Biochem J 107(6):775–791

Sphingolipids

12

André B. P. van Kuilenburg, Ana Pop,
Ben J. H. M. Poorthuis, and Susanna M. I. Goorden

Contents

Ben J. H. M. Poorthuis (Deceased)

A. B. P. van Kuilenburg (✉)
Laboratory Genetic Metabolic Diseases, Department of Clinical Chemistry, Amsterdam UMC Location University of Amsterdam,
Amsterdam, The Netherlands

Amsterdam Gastroenterology Endocrinology Metabolism, Amsterdam, The Netherlands
e-mail: a.b.vankuilenburg@amsterdamumc.nl

A. Pop
Laboratory Genetic Metabolic Diseases, Department of Clinical Chemistry, Amsterdam UMC Location University of Amsterdam,
Amsterdam, The Netherlands

S. M. I. Goorden
Center for Lysosomal and Metabolic Diseases, Department of Clinical Genetics, Erasmus University Medical Center, Rotterdam, The Netherlands

N. Blau, F. M. Vaz (eds.), *Laboratory Guide to the Methods in Biochemical Genetics*,
https://doi.org/10.1007/978-3-031-58819-8_12

List of Abbreviations

CCL18	Chemokine (C-C motif) ligand 18
C-triol	Cholestane-3-beta, 5-alpha, 6-beta-triol
ER	Endoplasmic reticulum
GM	Monosialoganglioside
IS	Internal standard
Lyso-Galcer	Lysogalactosylceramide (C18:1 sphingosine backbone)
Lyso-Gb3	Lyso-globotriaosylceramide (C18:1 sphingosine backbone)
Lyso-GlcCer	Lyso-glucosylceramide (C18:1 sphingosine backbone)
Lyso-HexCer	Lyso-hexosylceramide (C18:1 sphingosine backbone)
Lyso-SL	Lyso-sphingolipid
Lyso-SM	Lysosphingomyelin (C18:1 sphingosine backbone)
Lyso-SM-509	Lysosphingomyelin-509
PPCS	*N*-palmitoyl-*O*-phosphocholine-serine (previously known as lysosphingomyelin-509)
SPC	Sphingosylphosphorylcholine

12.1 Introduction

Sphingolipids are complex lipids which are formed from a sphingosine backbone, a fatty acid and are further differentiated by addition of various head groups. Concentrated in the so-called lipid rafts in the cell membrane, they are involved in cell signaling and transport functions. Sphingolipidoses are a group of genetic disorders caused by a deficiency in the lysosomal degradation or transport of these sphingolipids. The enzymes involved in the stepwise degradation of sphingolipids are exohydrolases with optimal activity at acidic pH. A deficiency of one of these enzymes gives rise to a particular sphingolipid storage disorder. Since most of the enzymes involved in the degradation of sphingolipids require an activator protein for activity in vivo, a deficiency of an activator protein gives rise to a storage disease with clinical features similar to the disease caused by a deficiency of the enzyme itself.

12.2 Properties of Metabolites

Chemically, sphingolipids comprise a family of molecules consisting of a sphingoid base (usually C18-sphingosine), a variable hydrophilic head group linked to the C1-hydroxyl group and a fatty acid chain (of variable composition) linked to the amino group (see Fig. 12.1). They are synthesized from ceramides (essentially sphingolipids without the head group), starting in the endoplasmic reticulum with subsequent modifications in the Golgi system. Based on the structure of the head group, two main classes can be identified: sphingolipids which contain sugar-based head groups, referred to as glycosphingolipids, and phosphosphingolipids which contain a phosphorylcholine- (in case of sphingomyelin), a phosphorylethanolamine-, or a phosphate head group.

Glycosphingolipids differ in their complexity of the sugar chain, including single to as much as six sugar residues. Glycosphingolipids containing sialic acid (*N*-acetyl-neuraminic acid in humans) are named gangliosides. Glycosphingolipids with galactose as a head group (galactosylceramide; also named cerebroside) can be sulfated giving rise to sulfatides. Both, galactosylceramide and sulfatides, are particularly abundant in myelin.

Degradation of sphingolipids occurs in the lysosome by stepwise degradation of sphingolipids by various acid hydrolases. Some of these hydrolases rely on activator proteins, so-called saposins, for full activity. These saposins are proteolytically cleaved in the lysosome from the common precursor protein prosaposin. They include saposin A (activator of galactocerebrosidase), saposin B (activator of arylsulfatase A), saposin C (activator of glucocerebrosidase and saposin D (activator of acid ceramidase). Deficiencies have been described for prosaposin and the individual saposins with exception of saposin D. In addition, hydrolysis of monosialoganglioside 2 (GM2) by β-Hexosaminidase A requires the GM2-activator protein (GM2-AP). Deficiency of GM2-AP leads to the AB variant, which is rarely encountered and indistinguishable from classical Tay-Sachs disease.

In case of a deficiency of acid hydrolases or activator proteins, specific sphingolipids accumulate inside the lysosome. These spingolipids are subsequently deacylated by promiscuous activity of acid ceramidase to the corresponding lyso-sphingolipids [1] (See Fig. 12.1). These hydrophilic sphingoid bases are easily released from cells to the plasma. Therefore, they can be used as plasma biomarkers informative of intralysosomal pathology [1].

In 2015 it was shown that a compound with a similar structure to lyso-SM, but with a m/z of 509 (therefore named lyso-SM-509), accumulates in plasma of Niemann-Pick type C patients [2]. Subsequent studies validated the use of this biomarker for Niemann-Pick type C and Niemann-Pick type A and B (alongside lyso-SM) [3–6]. The exact structure of lyso-SM-509 has recently been determined and reveals that it is the most abundant member of a novel class of lipids, called *N*-acyl-*O*-phosphocholineserines (APCSs) [7, 8]. Therefore, the metabolite has been renamed to *N*-palmitoyl-*O*-phosphocholineserine (PPCS).

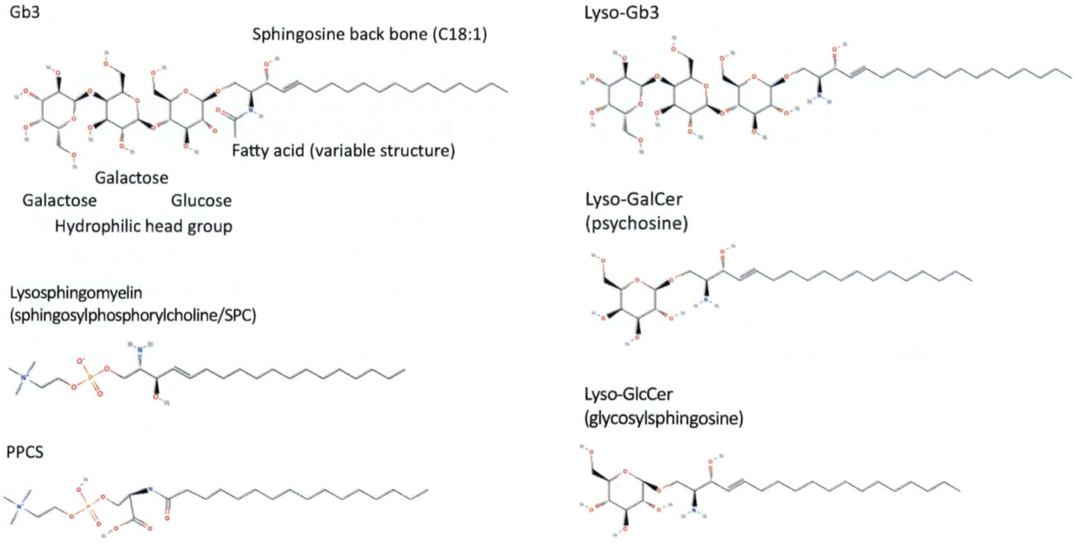

Fig. 12.1 Structure of selected (lyso)-sphingolipids

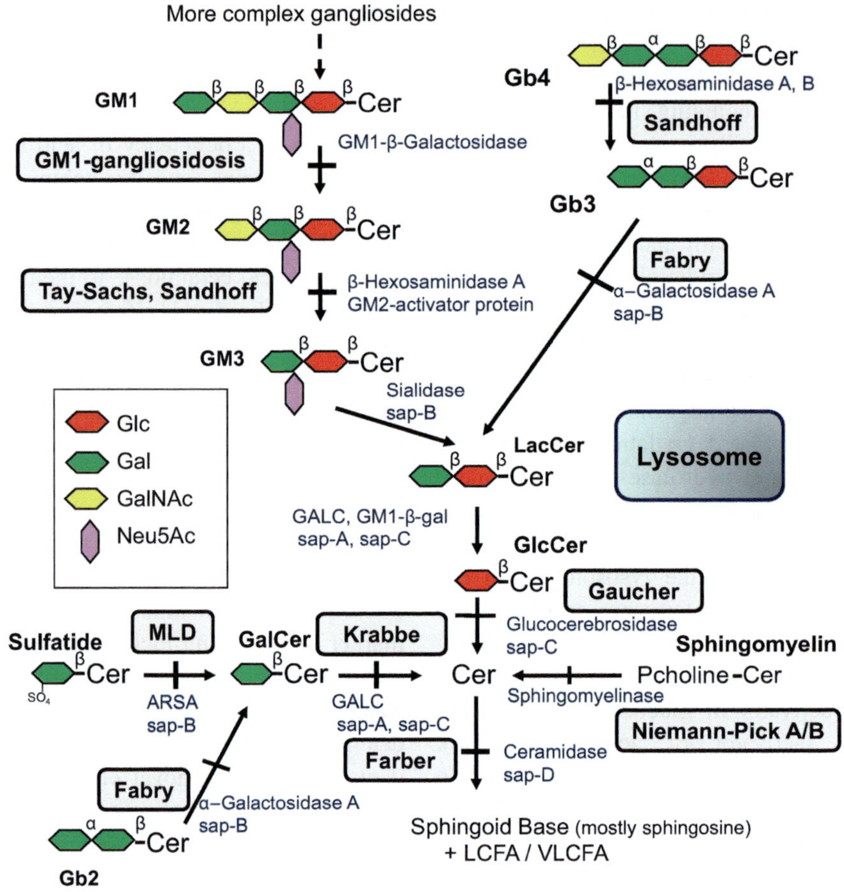

Fig. 12.2 Lysosomal degradation of sphingolipids. *ARSA* arylsulfatase A, *GALC* galactocerebrosidase, *GalCer* galactosylceramide (or galactocerebroside), *GalNAc* *N*-acetyl-galactosamine, *Gb2* galactobiosylceramide, *Gb3* globotriaosylceramide, *Gb4* globotetraosylceramide (globoside), *GlcCer* glucosylceramide (or glucocerebroside), *GM1* GM1 ganglioside. Adapted with permission from: Vanier, M.T., Caillaud, C., Levade, T. (2022). Disorders of Sphingolipid Synthesis, Sphingolipidoses, Niemann-Pick Disease Type C and Neuronal Ceroid Lipofuscinoses. In: Saudubray, JM., Baumgartner, M.R., García-Cazorla, Á., Walter, J. (eds) Inborn Metabolic Diseases. Springer, Berlin, Heidelberg. https://doi.org/10.1007/978-3-662-63123-2_40

The biosynthetic pathway and catabolism of this compound are still poorly understood.

12.3 Clinical Significance

Sphingolipidoses constitute a group of mainly autosomal recessive disorders (with exception of the X-linked Fabry disease), which are biochemically characterized by a defect in the lysosomal breakdown of sphingolipids (See Fig. 12.2). This group of disorders includes deficiencies of sphingolipid degrading enzymes and their activator proteins. Adjacent to these classical sphingolipidoses is Niemann-Pick disease type C, in which sphingolipids accumulate due to faulty lysosomal cholesterol trafficking.

While the sphingolipidoses share some clinical symptoms (organomegaly and neurological involvement), each of these disorders has a distinct and recognizable clinical phenotype related to specific organ involvement. The clinical onset of sphingolipidoses may present at any age, varying from a very severe presentation with hydrops fetalis to late-onset diseases with mainly psychiatric symptoms. In general, a

Table 12.1 Overview of sphingolipidoses and their main clinical symptoms

Sphingolipidosis	Gene	Enzyme/protein	Characteristic clinical symptoms
Fabry disease	*GLA*	Alpha-galactosidase	Acroparesthesias, cornea verticillata, angiokeratomas, kidney disease, left ventricular hypertrophy
Gaucher disease	*GBA* *PSAP*	Glucocerebrosidase Saposin C	Hepatosplenomegaly, cytopenia, bone pathology. In neuronopathic forms additional neurological symptoms (i.e., brain stem dysfunction, pyramidal tract syndrome, horizontal gaze palsy)
GM1-gangliosidosis	*GLB1*	Beta-galactosidase	Hepatosplenomegaly, coarse features, hypotonia, pyramidal tract syndrome, extrapyramidal symptoms
Krabbe disease	*GALC* *PSAP*	Galactocerebrosidase Saposin A	Pyramidal tract syndrome, hypertonia, ataxia, loss of vision, rapid psychomotor decline
Metachromatic leukodystrophy	*ARSA* *PSAP*	Arylsulfatase A Saposin B	Walking problems, pyramidal tract syndrome, peripheral neuropathy, ataxia, psychiatric symptoms
Niemann-Pick disease type A and B	*SMPD1*	Acid sphingomyelinase	Hepatosplenomegaly, interstitial lung disease, neurological symptoms (i.e., hypotonia, loss of motor skills) only in type A
Niemann-Pick disease type C	*NPC1*, *NPC2*	NPC1 and 2 protein	Neonatal cholestasis, hepatosplenomegaly, supranuclear vertical saccades palsy, ataxia, movement disorders, psychiatric symptoms
Prosaposin deficiency	*PSAP*	Prosaposin	Hepatosplenomegaly, hypotonia, massive myoclonic bursts, abnormal ocular movements, dystonia, seizures
Sandhoff disease	*HEXB*	Beta-subunit of hexosaminidase	Hypotonia, hyperacusis, loss of acquired milestones, ataxia, psychosis, pyramidal tract syndrome, cherry red-spot
Tay-Sachs disease	*HEXA*	Alpha-subunit of hexosaminidase	Hypotonia, hyperacusis, loss of acquired milestones, ataxia, psychosis, pyramidal tract syndrome, cherry red-spot

good correlation exists between the levels of residual enzyme activity and the clinical severity of the disease. The characteristic clinical symptoms of the sphingolipidoses for which diagnostic assays are described in this chapter are summarized in Table 12.1. For a more elaborate clinical description, including a detailed distinction of early and later onset phenotypes, readers are referred to excellent clinical textbooks [9, 10].

12.4 Methods

12.4.1 Metabolite Assays

While traditionally only enzyme activity measurements were available to establish a diagnosis of a sphingolipid storage disorder, recent years have seen the emergence of various LC-MS/MS assays measuring primary or secondary storage products [11]. These assays, mainly multiplex in origin, allow for a broader and faster screening, especially when the clinical phenotype (e.g., isolated hepatosplenomegaly) is not specific for one particular sphingolipidosis [6, 11].

Metabolite assays applicable to sphingolipidoses include analysis of oxysterols, mainly 7-ketocholesterol and cholestane-3-beta,5-alpha,6-beta-triol (C-triol), analysis of C-triol derived bile acids (3-beta,5-alpha,6-beta-trihydroxycholanic acid (TCA) and its glycine conjugate (TCG)), and (lyso-)sphingolipid assays, often combined with PPCS (reviewed in [11]). Oxysterols, produced by auto-oxidation of cholesterol and its derivatives TCA and TCG are all found increased in patients with Niemann-Pick type C [12–16]. Oxysterol measurements are already used in diagnostics for some time, while TCA and TCG are recently identified, promising biomarkers [13].

Sphingolipids are the primary storage products in sphingolipidoses but are only very mildly increased in plasma and urine of patients, often overlapping with controls, with the exception of urine sulfatides in MLD [17] and urine

Gb3 in classical male Fabry patients [18]. In contrast, measurement of lyso-sphingolipids, the deacylated forms of sphingolipids, allows for clear distinction between different patient groups and controls [11]. While their diagnostic value was initially validated in single lyso-sphingolipid assays [2, 3, 19–32], the current generation multiplex assays enable a comprehensive screening for sphingolipidoses, including Fabry disease, Gaucher disease, Krabbe disease, Niemann-Pick type A and B, and Niemann-Pick type C [4–6]. Moreover, some of the published methods also quantify lyso-GM1 and lyso-GM2 levels allowing for additional screening for GM1-gangliosidosis, Sandhoff disease, and Tay-Sachs disease [4]. Besides diagnosis, lyso-sphingolipid measurements have also been proven very suitable for therapy control [19, 21], phenotyping [21, 30], and risk stratification [33].

Lyso-GlcCer (also named glucosylsphingosine), which accumulates in Gaucher disease, and lyso-GalCer (also named psychosine), a main accumulation product in Krabbe disease, are isomers. Depending on the column and chromatographic conditions chosen, they can be separated [23, 27, 29] or measured as a sum. Because of the different clinical presentations of the two disorders and availability of specific enzyme assays, separation of the isomers is not considered mandatory for screening purposes.

PPCS was first supposed to be a lyso-sphingolipid (called lyso-SM-509) and therefore included in most of the published multiplex assays [4–6]. Thanks to its structural determination, authentic (internal) standards could be synthesized in order to generate more robust quantification methods for this biomarker [8, 34, 35].

12.4.2 Enzyme Assays

The determination of the enzyme activity in leukocytes and fibroblasts is the method of choice for confirmative diagnosis of new patients suspected of suffering from a sphingolipidosis. With respect to newborn screening for lysosomal storage diseases, the analysis of lysosomal enzymes in dried bloodspots, using mass spectrometry, is increasingly being used [36, 37]. Confirmation of an enzyme deficiency, by the demonstration of pathogenic variants in the relevant genes and/or by the demonstration of accumulating metabolites, may be necessary in some cases to rule out the possibility of pseudodeficiency [38, 39]. A pseudodeficiency is a phenomenon in which a deficient or decreased enzyme activity toward a synthetic substrate is found whereas the activity toward the natural substrate is sufficient to prevent clinical disease. Pseudodeficiencies have been reported for β-hexosaminidase A, β-hexosaminidaseAandB,β-galactocerebrosidase and arylsulfatase A [40, 41].

For diagnosis and monitoring of Gaucher disease and Niemann-Pick type A/B, the activity of the enzyme chitotriosidase in plasma and the plasma levels of chemokine CCL18 can be determined [21, 42, 43]. Chitotriosidase and CCl18 are both specifically produced and secreted by the pathological storage macrophages (Gaucher cells) [44].

12.5 Description of Specific Metabolite Assays

12.5.1 Lyso-Sphingolipid Panel

In this section a multiplex assay for measurement of lyso-Gb, lyso-HexCer, lyso-SM, and PPCS is described. Lyso-HexCer refers to the sum of lyso-GlcCer and lyso-GalCer, which are not chromatographically separated in this assay. Lyso-GlcCer (internal) standards are used for measurement and quantification of lyso-HexCer.

When this assay was developed the exact structure of lyso-SM-509 was not known and it was assumed that its structure resembled lyso-SM. Therefore lyso-SM (internal) standards are used in this assay for calculation of PPCS concentrations. Currently, (internal) standards for PPCS are commercially available and it is recommended to use these when implementing this assay in your own laboratory.

12.5.1.1 Principle

The lyso-sphingolipids are separated using reversed-phase UPLC combined with electrospray ionization tandem mass spectrometry and detection is performed using multiple reaction monitoring. Stable-isotope-labeled reference compounds are used as internal standards for all metabolites, except for lyso-Gb for which a glycinated analogue is used.

12.5.1.2 Pre-analytical

Lyso-sphingolipids can be measured in lithium-heparin plasma and EDTA-plasma yielding comparable results [4]. Store plasma samples at – 20 °C prior to measurement. No further specific prerequisites apply regarding collection or storage.

12.5.1.3 Reagents and Chemicals

Name	Type	Supplier
Ammonium formate	Chemical	Sigma-Aldrich
Chloroform	Chemical	Biosolve
Butanol	Chemical	Merck
Methanol	Chemical	Biosolve
Formic acid	Chemical	Merck
Lyso-Gb3	Standard	Sigma-Aldrich
Gly-Lyso-Gb3	Internal Standard	Matreya
Lyso-GlcCer	Standard	Avanti Polar Lipids
Lyso-GlcCer-d5	Internal Standard	Avanti Polar Lipids
Lyso-SM	Standard	Avanti Polar Lipids
Lyso-SM-d7	Internal Standard	Avanti Polar Lipids

12.5.1.4 Instrumentation

- Waters Acquity H-Class—XEVO TQ-XS UPLC-mass spectrometer system
- Waters BEH C18 guard column
- Waters Acquity BEH C18 column (200 × 2.1 mm; particle size, 1.7 μm)

12.5.1.5 Analytical Procedure/ Manual

Preparation of Reagents and Standards

Reagents:

- *Ammonium formate buffer, 100 mM, pH 3.2*: dissolve 630 mg ammonium formate in 100 ml 20% formic acid
- *Eluent A: MilliQ water: formic acid: 99.5 : 0.5 (v/v) with 1 mM ammonium formate* (make fresh):
  - Dissolve 32 mg ammonium formate in 500 ml MilliQ water
  - Add 2.5 ml formic acid and mix by gently swirling
- Eluent B: *methanol: formic acid: 99.5 : 0.5 (v/v) with 1 mM ammonium formate:*
  - Dissolve 32 mg ammonium formate in 500 ml methanol
  - Add 2.5 ml formic acid and mix by gently swirling
  - Degas for 10 min in an ultrasonic bath

Standards:

- *Stock solution lyso-Gb3 (1 mM):* dissolve 1 mg lyso-Gb3 in 1272 μl methanol:chloroform:MilliQ water: 1:2:0.1 (v/v/v)
- *Working standards lyso-Gb3 (20 and 2 μM):*
  - Pipette 80 μl stock solution (1 mM) in a 10 ml pyrex tube
  - Evaporate to dryness at 40 °C under a constant stream of nitrogen and dissolve in 4 ml methanol
  - Prepare the 2 μM standard by diluting the 20 μM working standard in methanol (pyrex tube)
- *Calibration solutions lyso-Gb3 (0.01–0.02–0.1–0.2 μM):*
  - Prepare 0.01–0.02–0.1–0.2 μM solutions by diluting the 2 μM standard in methanol (pyrex tubes)
- *Stock solution gly-lyso-Gb3 (1 mM), IS:* dissolve 1 mg lyso-Gb3 in 1186 μl methanol : chloroform : MilliQ water: 1: 2 : 0.1 (v/v/v)
- *Working standard gly-Lyso-Gb3 (20 μM), IS:*
  - Pipette 80 μl stock solution (1 mM) in a 10 ml pyrex tube
  - Evaporate to dryness at 40 °C under a constant stream of nitrogen and dissolve in 4 ml methanol
- *Stock solution lyso-GlcCer (5 mM):* dissolve 5 mg Lyso-GlcCer in 2166 μl methanol : chloroform 2 : 1 (v/v)
- *Working standard lyso-GlcCer (20 μM):*
  - Pipette 32 μl stock solution (5 mM) in a 10 ml pyrex tube

   - Evaporate to dryness at 40 °C under a constant stream of nitrogen and dissolve in 8 ml methanol
- *Calibration solutions lyso-GlcCer (0.1–0.2–1–2 μM):*
   - Prepare 0.1–0.2–1–2 μM solutions by diluting the 20 μM standard in methanol (pyrex tubes)
- *Stock solution lyso-GlcCer-d5 (1 mM), IS*: dissolve 1 mg lyso-GlcCer-d5 in 2143 μl methanol : chloroform 2 : 1 (v/v)
- *Working standard lyso-GlcCer-d5 (20 μM), IS:*
   - Pipette 80 μl stock solution (1 mM) in a 10 ml pyrex tube
   - Evaporate to dryness at 40 °C under a constant stream of nitrogen and dissolve in 4 ml methanol
- *Stock solution lyso-SM (5 mM):* dissolve 5 mg lyso-SM in 2152 μl methanol : chloroform 2 : 1 (v/v)
- *Working standard lyso-SM (20 μM):*
   - Pipette 32 μl stock solution (5 mM) in a 10 ml pyrex tube
   - Evaporate to dryness at 40 °C under a constant stream of nitrogen and dissolve in 8 ml methanol
- *Calibration solutions lyso-SM (0.1–0.2–1–2 μM):*
   - Prepare 0.1–0.2–1–2 μM solutions by diluting the 20 μM standard in methanol (pyrex tubes)
- *Stock solution lyso-SM-d7 (1 mM), IS*: dissolve 1 mg lyso-SM-d7 in 2120 μl methanol : chloroform 2 : 1 (v/v)
- *Working standard lyso-SM-d7 (20 μM), IS:*
   - Pipette 80 μl stock solution (1 mM) in a 10 ml pyrex tube
   - Evaporate to dryness at 40 °C under a constant stream of nitrogen and dissolve in 4 ml methanol
- *IS mix gly-lysoGb3, lyso-GlcCer-d5, and lyso-SM-d7 (0.1 μM):*
   - Pipette 40 μl of each of the 20 μM working standards in a 10 ml pyrex tube
   - Evaporate to dryness at 40 °C under a constant stream of nitrogen and dissolve in 8 ml methanol

12.5.1.6 Sample and Calibration Standard Preparation

Preparation of Calibration Standards

Calibration standards are prepared in a plasma pool obtained from anonymized plasma samples by spiking the lyso-Gb3, lyso-GlcCer, and lyso-SM calibration solutions as explained in section "standard." See final concentrations in table below which cover the expected measuring ranges. These calibration standards are further prepared in the same way as the patient samples and controls.

Lyso-Gb3 (nM)	Lyso-GlcCer (nM)	Lyso-SM (nM)
0	0	0
2	20	20
5	50	50
10	100	100
20	200	200
50	500	500
100	1000	1000
200	2000	2000

Preparation of Samples and Controls (and Calibration Standards)

- Pipette 50 μl plasma or control in a 2 ml tube
- Add 25 μl IS mix (0.1 μM)
- Add 25 μl MilliQ water
- Add 300 μl methanol and 150 μl chloroform and vortex for 30 s
- Put the sample at room temperature for 30 min and vortex once in a while
- Centrifuge 10 min at 16,060 g at 4 °C
- Transfer the supernatant to a 2 ml tube
- Add 150 μl methanol and 225 μl ammonium formate buffer (100 mM, pH 3.2) and vortex for 30 s
- Centrifuge 2 min at 16,060 g at 4 °C
- Pipette 400 μl of the upper phase and transfer it to a 2 ml tube
- Evaporate to dryness at 40 °C under a constant stream of nitrogen
- Add 700 μl MilliQ water and 700 μl butanol and vortex
- Centrifuge 2 min at 16,060 g at 4 °C

- Pipette 600 µl of the upper phase and transfer it to a 1.5 ml tube with screw cap
- Evaporate to dryness at 40 °C under a constant stream of nitrogen
- Dissolve the dried extract in 50 µl methanol assisted by vortexing for 30 s
- Add 50 µl MilliQ water and vortex shortly (5 s)
- Transfer 100 µl to a Gilson vial and cap

Chromatographic Conditions

A mobile phase gradient is used during a 5.5 min run according to the settings below. Flow rate is 0.25 ml/min.

Time (min)	% Eluent A	% Eluent B
Initial	100	0
2.5	0	100
4.0	0	100
5.0	100	0
5.5	100	0

MS Settings

Mass spectrometry detection is carried out by electron spray ionization in positive mode (ESI+) using a XEVO TQ-XS UPLC-mass spectrometer system in multiple reaction monitoring (MRM) mode according to the settings as outlined in the table.

Compound	Parent ion (m/z)	Daughter ion (m/z)	Cone Voltage (V)	Collision energy (eV)
Lyso-Gb3	786.5	282.3	45	35
Gly-Lyso-Gb3 (IS)	843.5	264.3	35	50
Lyso-HexCer	462.4	282.3	25	20
Lyso-GlcCer-d5 (IS)	467.4	287.4	25	20
Lyso-SM	465.4	184.1	30	25
Lyso-SM-d7 (IS)	472.4	184.1	30	25
PPCS	509.4	184.1	30	25

12.5.1.7 Calculation

Levels of lyso-SLs are calculated using an eight-point calibration curve in control plasma within the appropriate concentration range (0–200 nM for lyso-Gb3 and 0–2000 nM for Lyso-SM and lyso-HexCer). PPCS concentrations are determined using lyso-SM calibration curve and lyso-SM-d7 internal standard.

12.5.1.8 Reference Values and Pathological Values

For all sphingolipidoses a characteristic pattern of elevations can be detected as outlined in the table below [6].

	Lyso-Gb3	Lyso-HexCer	Lyso-SM	PPCS
Reference interval (2.5–97.5 percentile)	0.3–0.5	0.2–3.5	0.2–3.0	1.3–32.3
Fabry disease male, classical phenotype	↑↑	N/↑	N/↑	N/↑
Fabry disease male, non-classical phenotype	↑	N	N	N
Fabry disease female	↑	N	N	N
Gaucher disease	N/↑	↑↑	N/↑	N/↑
Krabbe disease	N	↑	N	N
Niemann-Pick disease type A and B	N	N	↑/↑↑	↑↑
Niemann-Pick disease type C	N	N	N/↑	↑↑

All values are in nmol/l; N, within reference range; ↑, elevated; ↑↑, highly elevated

Due to the rarity of the disorder, no samples of prosaposin deficient patients were tested with this method, but based on similar methods published it could be expected that plasma levels of lyso-Gb3 and lyso-HexCer are increased in this disorder [5, 45]. Lyso-SM and PPCS may be slightly increased or normal [5, 45].

12.5.1.9 Pitfalls

PPCS is a sensitive marker but not entirely specific. It can also be elevated in lysosomal acid lipase deficiency (LALD) patients [4, 6]. Also PPCS values may be elevated in NPC1 carriers [13] and in congenital disorders of glycosylation [46].

Furthermore, lyso-HexCer concentrations may be normal in rare, especially adult-onset cases of Krabbe disease [11].

12.5.1.10 Quality Control

Internal QC

In each run 2 control plasma samples (high and low pool) are measured. The low pool is prepared by mixing leftover (anonymized) plasma samples. To obtain the high pool, the low pool is spiked with lyso-Gb3, lyso-GlcCer, lyso-SM, and PPCS. Results of these internal control samples are analyzed using Westgard multiQC-rules.

External QC

Lyso-Gb3, lyso-GlcCer, and lyso-SM are included in the Erndim SAS scheme.

12.5.1.11 Conclusion

This multiplex lyso-sphingolipid assay is suitable for screening of Fabry disease, Gaucher disease, Krabbe disease, Niemann-Pick type A and B, and Niemann-Pick type C.

12.6 Description of Specific Enzyme Assays

12.6.1 Principle

The activity of the lysosomal enzymes involved in glycosphingolipid degradation is measured with homogenates from leukocytes or fibroblasts. Leukocytes isolated from EDTA-blood are the preferred material for first-line screening. Fibroblast cultures are started from skin biopsy samples using standard tissue-culture techniques. Fibroblasts are grown in Ham's F-10 medium until 1 week past confluence. Cells are harvested and cell pellets are stored at –80 °C. Leukocyte and fibroblast pellets are taken up in 0.2 ml ice-cold distilled water and sonicated briefly (3–5 s) at low power. Homogenates are either used fresh or stored at –20 °C until use.

All enzymes are measured at acidic pH with artificial fluorogenic substrates, except for arylsulfatase A, which is measured with a chromogenic substrate. The fluorogenic substrates are derivatives of 4-methylumbelliferone or 6-hexadecanoylamino-4-methylumbelliferone. The reactions with fluorogenic substrates are stopped with a basic buffer and the strong fluorescence of the liberated 4-methylumbelliferone is measured using a fluorometer.

12.6.2 Stock Solutions for the Preparation of Reaction Buffers

Assays are performed at acidic pH either in 0.1 M citrate buffer/0.2 M phosphate buffer or in acetate buffers.

1. Citric acid stock solution (0.1 M): dissolve 21.0 g citric acid monohydrate (MW 210.14) in 1 l water.
2. Phosphate stock solution (0.2 M): dissolve 35.6 g $Na_2HPO_4 \cdot 2H_2O$ (MW 177.99) in 1 l water.
3. Sodium acetate trihydrate stock solution (0.1 M): dissolve 13.6 g (MW 136.08) in 1 l water.
4. Acetic acid (0.1 M): dilute 5.7 ml glacial acetic acid (MW 60.05) to 1 l with water.

12.6.3 Stop Buffers

The stop buffer comprises 0.2 M $NaCO_3$/glycine, pH 10.5. Dissolve 21.2 g Na_2CO_3 (MW 105.98) in 400 ml water. Dissolve 3.75 g glycine (MW 75.07) in 100 ml water. Titrate 400 ml of the Na_2CO_3

solution with the glycine solution to pH 10.5. Add water to 1 l. This stop buffer is used in all assays unless indicated otherwise. When assays are adapted to microtiter plates, detergent (0.025% w/v Triton X-100) should be present in the stop buffers in order to minimize meniscus effects.

12.6.4 Standard 4-Methylumbelliferone Solution

Dissolve 17.6 mg 4-methylumbelliferone (MW 176) in 10 ml stop buffer to a final concentration of 10 mM. Dilute stepwise to 10 μM.

12.7 Acid β-Galactosidase

This enzyme is deficient in GM1-gangliosidosis (Fig. 12.2). The assay is based on the method described by Ho and O'Brien [47, 48].

12.7.1 Reagents and Chemicals

1. Reaction buffer: mix 57 ml 0.1 M citric acid with 43 ml 0.2 M phosphate solution. Adjust the pH to 4.3 with either citric acid or phosphate solution.
2. Substrate/buffer solution: dissolve 1.70 mg 4-methylumbelliferyl-β-D-galactopyranoside (MW 338.3) and 118.7 mg NaCl (MW 58.44) in 10 ml reaction buffer to a final concentration of 0.5 mM 4-methylumbelliferyl-β-D-galactopyranoside and 0.2 M NaCl. Warm up to a maximum of 60 °C, until all components are dissolved.

12.7.2 Assay Conditions

1. Incubate 10 μl homogenate (5 μg of leukocyte protein or 1–2 μg of fibroblast protein) with 100 μl substrate/buffer for 30 min at 37 °C.
2. Blank incubation: use water instead of homogenate.
3. Stop the reaction with 2.89 ml stop buffer.
4. Read the fluorescence at excitation wavelength 360 nm and emission wavelength 450 nm. Include a three-point calibration curve of 500, 1000, and 2000 pmol 4-methylumbelliferone; this curve should be linear.

12.7.3 Reference Values/Pathological Values

1. Reference range for leukocytes: 80–240 nml/mg h.
2. Pathological range for leukocytes: < 10 nmol/mg h.
3. Reference range for fibroblasts: 539–2233 nmol/mg h.
4. Pathological value for fibroblasts: <10 nmol/mg h.

12.8 β-Hexosaminidase A

This enzyme is deficient in GM2-gangliosidosis, B-variant (Tay-Sachs disease; (Fig. 12.2)). The assay is based on the method described by Inui and Wenger [49].

12.8.1 Reagents and Chemicals

1. Reaction buffer: mix 58.6 ml 0.1 M citric acid with 41.4 ml 0.2 M phosphate solution. Adjust the pH to 4.2 with either citric acid or phosphate solution.
2. Substrate/buffer solution: dissolve 12.5 mg 4-methylumbelliferyl-6-sulfo-2-acet-amido-2-deoxy-β-D-glucopyranoside (sodium salt, MW 483.5) in 2.58 ml of reaction buffer to a final concentration of 10 mM.

12.8.2 Assay Conditions

1. Incubate 20 μl homogenate (10 μg of leukocyte protein or 10 μg of fibroblast protein) with 20 μl reaction buffer and 10 μl substrate/buffer for 20 min at 37 °C.
2. Incubate 10 μl plasma with 30 μl reaction buffer and 10 μl substrate/buffer for 20 min at 37 °C.
3. Blank incubation: incubate with water instead of protein.

4. Stop the reaction with 2.95 ml stop buffer.
5. Read the fluorescence at excitation wavelength 360 nm and emission wavelength 450 nm. Include a three-point calibration curve of 500, 1000, and 2000 pmol 4-methylumbelliferone; this curve should be linear.

12.8.3 Reference Values/Pathological Values

1. Reference range for leukocytes: 243–562 nmol/mg h.
2. Pathological value for leukocytes: <15 nmol/mg h.
3. Reference range for fibroblasts: 1700–3900 nmol/mg h.
4. Pathological value for fibroblasts: <60 nmol/mg h.
5. Reference range for plasma: 85–260 nmol/ml h.
6. Pathological value for plasma: <10 nmol/mg h

12.8.4 Special Remarks/Pitfalls

β-Hexosaminidase A is also deficient in GM2-gangliosidosis, O-variant (Sandhoff disease), when measured with this assay, but to a lesser degree than in Tay-Sachs disease. When a normal activity of β-hexosaminidase A is found in a patient with characteristic symptoms of Tay-Sachs disease, the possibility of a GM2-activator protein deficiency should be considered (GM2-gangliosidosis AB variant; Fig. 12.2). This activator protein deficiency can be diagnosed by analysis of GM2-ganglioside in plasma and fibroblasts or by mutation analysis of the GM2-activator protein gene [50, 51].

12.9 β-Hexosaminidase A and B

These enzymes are deficient in GM2-gangliosidosis, O-variant (Sandhoff disease; (Fig. 12.2). The assay is based on the method described by O'Brien et al. [52].

12.9.1 Reagents and Chemicals

1. Reaction buffer: mix 55.9 ml 0.1 M citric acid with 44.1 ml 0.2 M phosphate solution. Adjust the pH to 4.4 with either citric acid or phosphate solution.
2. Substrate/buffer solution: dissolve 20 mg 4-methylumbelliferyl-2-acetamido-2-deoxy-β-D-glucopyranoside (MW 379.4) in 10.5 ml of reaction buffer to a final concentration of 5 mM.
3. 1% (w/v) bovine serum albumin (BSA) solution: dissolve 10 mg BSA (heat shock fraction) in 1 ml of water.

12.9.2 Assay Conditions

1. Incubate 45 µl homogenate (2 µg leukocyte protein or 0.5 µg of fibroblast protein) with 5 µl 1% BSA and 50 µl substrate/buffer for 30 min at 37 °C.
2. Incubate 5 µl plasma and 45 µl water with 100 µl substrate/buffer for 30 min at 37 °C.
3. Blank incubation: incubate with water instead of protein.
4. Stop the reaction with either 2.90 ml (for leukocytes and fibroblasts) or 2.85 ml (for plasma) stop buffer.
5. Read the fluorescence at excitation wavelength 360 nm and emission wavelength 450 nm. Include a three-point calibration curve of 500, 1000, and 2000 pmol 4-methylumbelliferone; this curve should be linear.

12.9.3 Reference Values/Pathological Values

1. Reference range for leukocytes: 600–3500 nml/mg h.
2. Pathological value for leukocytes: <125 nmol/mg h.
3. Reference range for fibroblasts: 7000–28000 nmol/mg h.
4. Pathological value fibroblasts: <250 nmol/mg h.
5. Reference range for plasma: 650–3000 nmol/ml h.

6. Pathological value for plasma: <100 nmol/mg h

12.10 β-Glucocerebrosidase

This enzyme is deficient in Gaucher disease (Fig. 12.2). The assay is based on the method described by Peters et al. [53] and Daniels et al. [54]

12.10.1 Reagents and Chemicals

1. Reaction buffer: mix 43 ml 0.1 M citric acid with 57 ml 0.2 M phosphate solution. Adjust the pH to 5.5 with either citric acid or phosphate solution.
2. 1.5% (w/v) Taurocholate: dissolve 3 mg taurocholate in 2 ml reaction buffer.
3. Substrate/buffer solution: dissolve 6.8 mg 4-methylumbelliferyl-β-D-glucopyranoside (MW 338.3) in 3.6 ml of reaction buffer to a final concentration of 5.6 mM. Add 0.4 ml of taurocholate in reaction buffer and mix. The final substrate concentration is 5 mM.

12.10.2 Assay Conditions

1. Incubate 50 μl homogenate (40 μg of leukocyte protein or 2 μg of fibroblast protein; preferably freshly prepared) with 100 μl substrate/buffer for 1 h at 37 °C.
2. Blank incubation: incubate with water instead of protein.
3. Stop the reaction with 2.85 ml stop buffer.
4. Read the fluorescence at excitation wavelength 360 nm and emission wavelength 450 nm. Include a three-point calibration curve of 500, 1000, and 2000 pmol 4-methylumbelliferone; this curve should be linear.

12.10.3 Reference Values/ Pathological Values

1. Reference range for leukocytes: 9–24 nmol/mg h.
2. Pathological value for leukocytes: <3 nmol/mg h.
3. Reference range for fibroblasts: 520–910 nmol/mg h.
4. Pathological value for fibroblasts: < 30 nmol/mg h.

12.10.4 Special Remarks/Pitfalls

A decreased β-glucocerebrosidase activity in leukocytes has also been observed in a patient with Niemann-Pick type C, while a normal β-glucosidase activity was present in fibroblasts [55].

12.11 Chitotriosidase

The activity of this enzyme is increased in Gaucher disease and may also be increased, but to a lesser extent, in other sphingolipidoses such as Niemann-Pick disease type A and B, Niemann-Pick disease type C, Krabbe disease, and GM1-gangliosidosis. The assay, using 4MU-deoxychitobiose, is based on the method described before [56, 57].

12.11.1 Reagents and Chemicals

1. Reaction buffer: mix 46.4 ml 0.1 M citric acid with 53.6 ml 0.2 M phosphate solution. Adjust the pH to 5.2 with either citric acid or phosphate solution. For reaction buffer with BSA 0.1% (w/v), dissolve 10 mg BSA in 10 ml reaction buffer.
2. Substrate/buffer solution: dissolve 5 mg 4-methylumbelliferyl-deoxy-chitobiose (MW 566.2) in 250 μl DMSO and add 1 ml reaction buffer with BSA. Vortex to dissolve, and warm up if needed. Further dilute to a final concentration of 0.25 mM 4-methylumbelliferyl-deoxy-chitobiose.

12.11.2 Assay Conditions

1. Incubate 10 μl plasma (undiluted, 1:25 and 1:50 diluted with PBS (0.1% BSA) for the

first time diagnostic test; or 1:X diluted for follow-up patients) with 40 µl substrate/buffer (with or without BSA) for 20 min at 37 °C.
2. Stop the reaction with 1.45 ml stop buffer.
3. Read the fluorescence at excitation wavelength 360 nm and emission wavelength 450 nm. Include a three-point calibration curve of 500, 1000, and 2000 pmol 4-methylumbelliferone; this curve should be linear.

12.11.3 Reference Values/Pathological Values

1. Reference range for plasma 17–235 nmol/(ml h)
2. Pathological value for plasma >200 µmol/l h.

12.11.4 Special Remarks/Pitfalls

When a strongly increased value of chitotriosidase activity is obtained in plasma, such as in Gaucher disease, the assay should be repeated with a diluted plasma sample, e.g., 1:50. Chitotriosidase is deficient in about 6% of the general population. This is an autosomal recessive trait and the high carrier frequency has to be taken into account when interpreting patient values [58]. Chitotriosidase measurements are also extremely useful for monitoring the effect of enzyme therapy or substrate reduction therapy in patients with Gaucher disease [57]. In case of a deficiency of chitotriosidase, the determination of the chemokine CCL 18/PARC is a good alternative [21, 59, 60]. Chitotriosidase (and lyso-HexCer) should also be measured when a normal β-glucocerebrosidase activity is found in a patient with symptoms characteristic of Gaucher disease. Increased chitotriosidase (and lyso-HexCer) is indicative of an activator protein/saposin C deficiency in such cases. As stated above, chitotriosidase may be moderately increased in other sphingolipidoses, but the sensitivity to detect patients is less than 100% [6, 43, 61–63].

12.12 α-Galactosidase A

This enzyme is deficient in Fabry disease (Fig. 12.2). The assay is based on the methods described by Desnick et al. [64] and Mayes et al. [65].

12.12.1 Reagents and Chemicals

1. Reaction buffer 0.1 M acetate buffer pH 4.5: mix 43 ml 0.1 M sodium acetate stock solution with 57 ml 0.1 M acetic acid. Adjust the pH to 4.5 with either acetate or acetic acid.
2. Substrate/buffer: dissolve 8.5 mg of 4-methylumbelliferyl-α-D-galactopyranoside (MW 338.3) in 5 ml reaction buffer to a final concentration of 5 mM.
3. 1.0 M *N*-acetylgalactosamine (MW 221): dissolve 2.21 g *N*-acetylgalactosamine in water and bring the final volume to 10 ml.

12.12.2 Assay Conditions

1. Incubate 35 µl homogenate (10 µg of leukocyte protein or 15 µg of fibroblast protein) with 15 µl 1.0 M *N*-acetylgalactosamine and 100 µl substrate/buffer for 30 min at 37 °C.
2. Incubate 25 µl plasma with 25 µl H2O and 100 µl substrate/buffer for 30 min at 37 °C.
3. Blank incubation: incubate with water instead of protein.
4. Stop the reaction with 2.85 ml stop buffer.
5. Read the fluorescence at excitation wavelength 360 nm and emission wavelength 450 nm. Include a three-point calibration curve of 500, 1000, and 2000 pmol 4-methylumbelliferone; this curve should be linear.

12.12.3 Reference Values/Pathological Values

1. Reference range for leukocytes: 32–70 nmol/mg h.

2. Pathological value for leukocytes hemizygotes: <3 nmol/mg h.
3. Pathological value for leukocytes heterozygotes: <30 nmol/mg h
4. Reference range for fibroblasts: 41–167 nmol/mg h.
5. Pathological value for fibroblasts: < 3 nmol/mg h.
6. Reference range for plasma: 3.2–14.3 nmol/ml h.
7. Pathological value for plasma: <1.5 nmol/ml h.

12.12.4 Special Remarks/Pitfalls

Fabry disease is an X-linked disorder and accurate carrier detection is important for genetic counseling and because enzyme therapy is available for clinically affected hemizygote and heterozygote patients. Because of random X-inactivation, not all heterozygotes can be reliably identified by enzyme assay, therefore mutation analysis and determination of lyso-Gb3 levels in plasma are strongly recommended.

12.13 β-Galactocerebrosidase

This enzyme is deficient in Krabbe disease (Fig. 12.2). The assay is essentially based on the method described by Wiederschain et al. [66].

12.13.1 Reagents and Chemicals

1. Oleic acid stock solution: dissolve 6 mg oleic acid (MW 282) in 1 ml of hexane (prepare fresh).
2. Taurocholate stock solution: dissolve 30 mg sodium taurocholate (MW 538) in 1 ml chloroform/methanol (2/1 v/v).
3. Stock solution 6-hexadecanoylamino-4-methylumbelliferyl-β-D-galactopyranoside (MW 591): dissolve 2.66 mg in 5 ml chloroform/methanol (2/1 v/v) to a final concentration of 0.9 mM.
4. Reaction buffer: mix 46.4 ml citric acid with 53.6 ml phosphate solution. Adjust the pH to 5.2 with either citric acid or phosphate solution.
5. Substrate/buffer solution (prepare fresh): Mix 75 μl oleic acid, 50 μl taurocholate, and 250 μl 6-hexadecanoylamino-4-methylumbelliferyl-β-D-galactopyranoside stock solution in a 1.5-ml vial and take the mixture to dryness under a gentle stream of nitrogen. Add 500 μl of reaction buffer at room temperature, mix vigorously on a Vortex mixer. Sonicate the solution. The final concentrations of oleic acid, taurocholate, and 6-hexadecanoylamino-4-methylumbelliferyl-β-D-galactopyranoside are 0.9 mg/ml, 3 mg/ml, and 0.45 mM, respectively.
6. Assay specific stop buffer: 0.1 M Na_2CO_3/$NaHCO_3$ pH 10.7, 0.2% (w/v) Triton X-100.

12.13.2 Assay Conditions

1. Incubate 10 μl homogenate (20 μg of leukocyte protein or 15 μg of fibroblast protein) with 20 μl substrate/buffer for 17 h at 37 °C.
2. Blank incubation: incubate with water instead of protein.
3. Stop the reaction with 1.47 ml assay specific stop buffer.
4. Read the fluorescence at excitation wavelength 360 nm and emission wavelength 450 nm. Include a three-point calibration curve of 500, 1000, and 2000 pmol 4-methylumbelliferone; this curve should be linear.

12.13.3 Reference Values/Pathological Values

1. Reference range for leukocytes: 20–102 nmol/mg·17 h.
2. Pathological value for leukocytes: < 4 nmol/mg·17 h.
3. Reference range for fibroblasts: 18–136 nmol/mg·17 h.

4. Pathological value for fibroblasts: < 2.5 nmol/mg·17 h.

12.13.4 Special Remarks

A low-activity allele for galactocerebrosidase due to polymorphic variants in the GALC (galactocerebrosidase) gene may be present. When polymorphic variants are present on both GALC alleles or when a low-activity GALC allele is present in the compound heterozygous state with a Krabbe allele, this may pose diagnostic difficulties, especially in patients with nonspecific neurological problems. Therefore, it is very important to measure Lyso-HexCer or specifically Lyso-GalCer (psychosine) in plasma of a suspected patient and to perform mutation analysis of the GALC gene [6].

12.14 Arylsulfatase A

This enzyme is deficient in metachromatic leukodystrophy (Fig. 12.2). The assay is based on the method described by Baum et al. [67] and modified by Lee-Vaupel et al. [68].

12.14.1 Reagents and Chemicals

1. Reaction buffer 0.5 M acetate buffer pH 5: mix 43 ml 0.5 M sodium acetate stock solution with 57 ml 0.5 M acetic acid. Adjust the pH to 5 with either acetate or acetic acid. Dissolve 9.93 g NaCl (MW 58.4) and 13.3 mg Na4P2O7 (MW 265.9) in the acetate buffer and bring the volume to 100 ml. The final concentrations of NaCl and Na4P2O7 are 1.7 M and 0.5 mM, respectively.
2. Substrate/buffer: dissolve 31.1 mg *p*-nitrocatechol sulfate (MW 311) in 10 ml of reaction buffer to a final concentration of 10 mM.
3. 1 M NaOH: dissolve 4 g NaOH in 100 ml water.

12.14.2 Assay Conditions

1. Incubate 50 μl homogenate (100 μg of leukocyte protein or 30 μg of fibroblast protein) with 200 μl substrate/buffer for 17 h at 0 °C in vials on melting ice in the cold room or refrigerator.
2. Blank incubation: incubate with water instead of protein.
3. Protein blank: incubate protein with water instead of substrate/buffer.
4. Stop the reaction with 250 μl 1 M NaOH. Read the extinction in a spectrophotometer in microcuvettes at 515 nm. The molar extinction coefficient of *p*-nitrocatechol at 515 nm is 12,400/Mol/cm.

12.14.3 Reference Values/Pathological Values

1. Reference range for leukocytes: 45–260 nmol/mg·17 h.
2. Pathological value for leukocytes: <15 nmol/mg·17 h.
3. Reference range for fibroblasts: 950–2580 nmol/mg·17 h.
4. Pathological value for fibroblasts: <60 nmol/mg·17 h.

12.14.4 Special Remarks/Pitfalls

In case of a deficiency of arylsulfatase A, at least one other sulfatase should be measured to exclude multiple sulfatase deficiency. Sulfatide excretion in urine should be measured and/or mutation analysis should be performed to confirm the diagnosis and to exclude a pseudodeficiency of arylsulfatase A, especially if the clinical symptoms are atypical [17, 69]. Sulfatide excretion should also be measured if a normal arylsulfatase A activity is found in a patient with symptoms characteristic of (juvenile) metachromatic leukodystrophy. In that case, increased urinary sulfatide excretion is indicative of an activator protein/saposin B deficiency.

12.15 Acid Sphingomyelinase

This enzyme is deficient in Niemann-Pick disease type A and B (Fig. 12.2). The assay is based on the method described by van Diggelen et al. [70].

12.15.1 Reagents and Chemicals

1. Reaction buffer 0.1 M sodium acetate buffer pH 5.2: mix 79 ml 0.1 M acetate with 21 ml 0.1 M acetic acid and bring the pH to 5.2 with either acetate or acetic acid solution. Dissolve 20 mg sodium taurocholate in 10 ml buffer to a final concentration of 0.2% (w/v).
2. Substrate/buffer: make a solution of 4 mg 6-hexadecanoylamino-4-methylumbelliferyl-phosphorylcholine (MW 610) in 5 ml reaction buffer to a final concentration of 1.32 mM. Briefly heat at 60 °C until the solution becomes clear.
3. Lysosphingomyelin (competitive inhibitor of acid sphingomyelinase): 12 mmol/l (MW 464.6): dissolve 5.6 mg lysosphingomyelin/ml reaction buffer at room temperature.
4. Assay specific stop buffer: 0.1 M Na_2CO_3/$NaHCO_3$ pH 10.7, 0.2% (w/v) Triton X-100.

12.15.2 Assay Conditions

1. Incubate 10 µl homogenate (30 µg of leukocyte protein or 5 µg of fibroblast protein) with 10 µl substrate/buffer and 10 µl reaction buffer or 10 µl lysosphingomyelin in reaction buffer for 17 h at 37 °C in the case of leukocytes, and 30 min at 37 °C in the case of fibroblasts.
2. Blank incubation: incubate with water instead of protein.
3. Stop the reaction by adding 1.47 ml stop buffer.
4. Read the fluorescence at excitation wavelength 360 nm and emission wavelength 450 nm. Include a three-point calibration curve of 500, 1000, and 2000 pmol 4-methylumbelliferone; this curve should be linear.

12.15.3 Reference Values/ Pathological Values

1. Reference range for leukocytes: 10–53 nmol/mg·17 h.
2. Pathological value for leukocytes: <4 nmol/mg·17 h.
3. Reference range for fibroblasts: 86–232 nmol/mg·h.
4. Pathological value for fibroblasts: <18 nmol/mg h.

12.15.4 Special Remarks/Pitfalls

A serious drawback of the assay is that it does not detect a deficiency of acid sphingomyelinase in patients who are homozygous or compound heterozygous for the Q292K variant [71]. However, the presence of this variant gives rise to an apparent lack of inhibition of acid sphingomyelinase by lysosphingomyelin in those patients as compared to controls [70]. The activity of acid sphingomyelinase can be decreased in fibroblasts of patients with Niemann-Pick type C [72, 73].

12.16 Filipin Staining for Cholesterol Storage

Cholesterol storage in lysosomes/endosomes is a hallmark of Niemann-Pick type C disease. The assay is based on the method described by Vanier et al. [74, 75].

12.16.1 Reagents and Chemicals

1. Hanks balanced salt solution (HBSS).
2. Phosphate buffered saline (PBS).
3. Formalin (acid free); filter before use.
4. Fixative (make fresh): mix 1 part of filtered formalin with nine parts PBS and add sucrose

to a final concentration of 0.1 M. Adjust the pH to 7.4.
5. Filipin stock solution: dissolve 25 mg filipin in 5 ml dimethylformamide; store at 4 °C.
6. Filipin assay solution: dilute 10 µl filipin stock solution in 1 ml PBS; make fresh.
7. 1,4-Diazabicyclo(2,2,2)octane (DABCO) stock solution: dissolve 1.25 g DABCO in 50 ml glycerol; store at 4 °C.
8. DABCO mounting solution: mix 3 ml PBS pH 7.4 with 27 ml DABCO stock solution; store at 4 °C.
9. Triton X-100 solution: dissolve 50 mg Triton X-100 in 100 ml PBS.

12.16.2 Assay Conditions

Grow the fibroblasts in Ham F-10 medium with 10% fetal bovine serum (FBS). Collect the cells and seed 30,000 cells on cover slips, in six-well plates, in Ham F-10 medium with 10% FBS; grow to near confluence. Remove the medium and wash the cells three times with 2 ml HBSS. Fix the cells with 1.5 ml fixative for minimum 1 h at room temperature. Optional, the plates can be sealed with tape and stored in the refrigerator at 4 °C until staining.

Let the plates come to room temperature and rinse the cover slips twice with PBS. Incubate the cover slips for 5 min with 2 ml Triton X-100 solution, remove, and rinse twice with 2 ml PBS. Remove the PBS and incubate the cover slips for 30 min with 2 ml of filipin assay solution, in the dark. Remove the filipin assay solution and rinse the cover slips twice with PBS. Remove the PBS and let the cover slips dry at ambient temperature. Mount the cover slips on a microscope slide with one drop of DABCO solution and seal with nail polish. Store at 4 °C. View the cells with a fluorescence microscope with a G365 band-pass excitation filter and either a 470-nm long-path emission filter with a 460 nm dichroic mirror or a 515-nm long-pass filter with a 500 nm dichroic mirror.

12.16.3 Reference Values/ Pathological Values

In the so-called "classical" biochemical phenotype, which represents more than 80% of the cases of Niemann-Pick type C disease, more than 80% of the cells show a characteristic perinuclear punctate fluorescence.

12.16.4 Special Remarks/Pitfalls

The assay described above has a limited sensitivity (80–85%) to detect patients, especially patients with the so-called "variant" biochemical phenotype [76, 77]. For increased sensitivity, the cells should be cultured in medium with 10% lipoprotein depleted FBS for 4–6 days and subsequently, for 1 extra day, in medium supplemented with 10% FBS as described above [75]. Of note, measurement of plasma biomarkers, in particular PPCS, has an estimated sensitivity approaching 100% and is considered to be a superior initial screening test for Niemann-Pick type C disease [13]. Since the specificity of these plasma biomarkers is suboptimal, filipin staining may have additional value when results of *NPC1/NPC2* mutation analysis and plasma biomarkers don't match.

12.17 Chemokine (C-C motif) Ligand 18 (CCL18)/PARC

Plasma CCL18 levels can be used as valuable alternative marker for monitoring the effect of enzyme replacement therapy in patients with Gaucher disease, especially in patients who are deficient in chitotriosidase activity [59, 60]. The assay is based on the method described by Boot et al. [59].

12.17.1 Reagents and Chemicals

1. Capture antibodies: monoclonal mouse anti-human CCL18/PARC antibody. Working

dilution (in PBS): 2.2 μg/ml; make fresh every time.
2. Detection antibodies: human CCL18/PARC Biotinylated Antibody. Working dilution (in DELFIA reaction buffer): 10.7 μg/ml; make fresh every time.
3. Washing buffer: PBS—0.05% Tween-20, make fresh every time.
4. Blocking buffer: PBS—1% BSA; make fresh every time.
5. DELFIA reaction buffer (5X). Dilute to 1X with water, and add Tween-20 to a final concentration of 0.01%. Store at 4 °C.
6. Streptavidin-HRP (streptavidin conjugated to horseradish peroxidase). Working dilution according to manufacturer's instructions; make fresh every time.
7. Stop solution: 2N H_2SO_4.
8. Standard for calibration curve: recombinant human CCL18. Working dilution: 500 ng/ml in reaction buffer; make fresh every time. Range: 0–8 ng/ml.
9. 3,3′,5,5′ tetramethylbenzidine slow kinetic (TMB) liquid substrate.

12.17.2 Assay Conditions

1. Coat wells (ELISA plate) by incubating with capture antibodies overnight, at room temperature.
2. Wash wells three times with washing buffer.
3. Block with blocking buffer for 1 h at room temperature.
4. Wash wells three times with washing buffer.
5. Add plasma samples (50, 2000, 4000, and 8000 times diluted)/standard curve and incubate for 2 h at room temperature.
6. Wash wells three times with washing buffer.
7. Add detection antibodies and incubate for 2 h at room temperature.
8. Wash wells three times with washing buffer.
9. Add streptavidin-HRP and incubate for 20 min at room temperature, in the dark.
10. Wash wells three times with washing buffer.
11. Add TMB, incubate for 15 min at room temperature, in the dark.
12. Stop the reaction with stop solution.
13. Read absorption at 450 and 570 nm.
14. Calculate the amount of CCL18/PARC based on the absorption (450–570 nm absorption) and the CCL18 standard calibration curve.

12.17.3 Reference Values/ Pathological Values

1. Reference range for plasma: 15–74 ng/ml
2. Pathological value for plasma: normal to 1860 ng/ml

12.18 Conclusion

Sphingolipidoses are a group of genetic disorders caused by a deficiency in the lysosomal degradation or transport of sphingolipids, a specialized lipid class. Screening for these disorders can be performed using multiplex lysosphingolipid assays and measurement of the macrophage activation markers chitotriosidase and CCL-18 and diagnosis can be confirmed by assessing specific enzyme activities.

Acknowledgments We thank the technicians of the Laboratory Genetic Metabolic Diseases, Amsterdam UMC, for setting up these diagnostic assays and for implementing and running them in the day-to-day patient diagnostics. We are grateful to Johannes M.F.G. Aerts for the permission to reuse part of the text of the previous version of this chapter. We would like to thank Marie Vanier for the permission to reuse Fig. 12.2 (originally published in Inborn Metabolic Diseases by Saudubray et al. 7th edition, 2022) for this chapter.

References

1. Ferraz MJ, Marques AR, Appelman MD, Verhoek M, Strijland A, Mirzaian M et al (2016) Lysosomal glycosphingolipid catabolism by acid ceramidase: formation of glycosphingoid bases during deficiency of glycosidases. FEBS Lett 590(6):716–725. https://doi.org/10.1002/1873-3468.12104
2. Giese AK, Mascher H, Grittner U, Eichler S, Kramp G, Lukas J et al (2015) A novel, highly sensitive and specific biomarker for Niemann-Pick type C1 disease.

Orphanet J Rare Dis 10:78. https://doi.org/10.1186/s13023-015-0274-1. [pii], 274 [pii]

3. Kuchar L, Sikora J, Gulinello ME, Poupetova H, Lugowska A, Malinova V et al (2017) Quantitation of plasmatic lysosphingomyelin and lysosphingomyelin-509 for differential screening of Niemann-Pick A/B and C diseases. Anal Biochem 525:73–77. https://doi.org/10.1016/j.ab.2017.02.019. S0003-2697(17)30099-4 [pii]
4. Pettazzoni M, Froissart R, Pagan C, Vanier MT, Ruet S, Latour P et al (2017) LC-MS/MS multiplex analysis of lysosphingolipids in plasma and amniotic fluid: A novel tool for the screening of sphingolipidoses and Niemann-Pick type C disease. PLoS One 12(7):e0181700. https://doi.org/10.1371/journal.pone.0181700. PONE-D-17-11267 [pii]
5. Polo G, Burlina AP, Kolamunnage TB, Zampieri M, Dionisi-Vici C, Strisciuglio P et al (2017) Diagnosis of sphingolipidoses: a new simultaneous measurement of lysosphingolipids by LC-MS/MS. Clin Chem Lab Med 55(3):403–414. https://doi.org/10.1515/cclm-2016-0340
6. Voorink-Moret M, Goorden SMI, van Kuilenburg ABP, Wijburg FA, Ghauharali-van der Vlugt JMM, Beers-Stet FS et al (2018) Rapid screening for lipid storage disorders using biochemical markers. Expert center data and review of the literature. Mol Genet Metab 123(2):76–84. https://doi.org/10.1016/j.ymgme.2017.12.431. S1096-7192(17)30678-9 [pii]
7. Sidhu R, Mondjinou Y, Qian M, Song H, Kumar AB, Hong X et al (2019) N-acyl-O-phosphocholineserines: structures of a novel class of lipids that are biomarkers for Niemann-Pick C1 disease. J Lipid Res 60(8):1410–1424. https://doi.org/10.1194/jlr.RA119000157. S0022-2275(20)32235-5 [pii], RA119000157 [pii]
8. Maekawa M, Jinnoh I, Matsumoto Y, Narita A, Mashima R, Takahashi H et al (2019) Structural determination of lysosphingomyelin-509 and discovery of novel class lipids from patients with Niemann-Pick disease type C. Int J Mol Sci 20(20). https://doi.org/10.3390/ijms20205018. ijms-20-05018 [pii]
9. Saudubray J-M, Berghe G, Walter JH (2022) Inborn metabolic diseases. Springer, Berlin
10. Blau N, Vici CD, Ferreira CR, Vianey-Saban C, van Karnebeek CDM (2022) Physician's guide to the diagnosis, treatment, and follow-up of inherited metabolic diseases. Springer, Cham
11. Piraud M, Pettazzoni M, Lavoie P, Ruet S, Pagan C, Cheillan D et al (2018) Contribution of tandem mass spectrometry to the diagnosis of lysosomal storage disorders. J Inherit Metab Dis 41(3):457–477. https://doi.org/10.1007/s10545-017-0126-3
12. Porter FD, Scherrer DE, Lanier MH, Langmade SJ, Molugu V, Gale SE et al (2010) Cholesterol oxidation products are sensitive and specific blood-based biomarkers for Niemann-Pick C1 disease. Sci Transl Med 2(56):56ra81. 2/56/56ra81 [pii]. https://doi.org/10.1126/scitranslmed.3001417
13. Jiang X, Ory DS (2021) Advancing diagnosis and treatment of Niemann-Pick C disease through biomarker discovery. Explor Neuroprot Ther 1(3):146–158. https://doi.org/10.37349/ent.2021.00012
14. Jiang X, Sidhu R, Porter FD, Yanjanin NM, Speak AO, te Vruchte DT et al (2011) A sensitive and specific LC-MS/MS method for rapid diagnosis of Niemann-Pick C1 disease from human plasma. J Lipid Res 52(7):1435–1445. https://doi.org/10.1194/jlr.D015735. S0022-2275(20)37038-3 [pii], d015735 [pii]
15. Jiang X, Sidhu R, Mydock-McGrane L, Hsu FF, Covey DF, Scherrer DE et al (2016) Development of a bile acid-based newborn screen for Niemann-Pick disease type C. Sci Transl Med 8(337):337ra63. https://doi.org/10.1126/scitranslmed.aaf2326. 8/337/337ra63 [pii]
16. Mazzacuva F, Mills P, Mills K, Camuzeaux S, Gissen P, Nicoli ER et al (2016) Identification of novel bile acids as biomarkers for the early diagnosis of Niemann-Pick C disease. FEBS Lett 590(11):1651–1662. https://doi.org/10.1002/1873-3468.12196. FEB212196 [pii]
17. Mirzaian M, Kramer G, Poorthuis BJ (2015) Quantification of sulfatides and lysosulfatides in tissues and body fluids by liquid chromatography-tandem mass spectrometry. J Lipid Res 56(4):936–943. https://doi.org/10.1194/jlr.M057232. S0022-2275(20)35574-7 [pii], m057232 [pii]
18. Vedder AC, Linthorst GE, van Breemen MJ, Groener JE, Bemelman FJ, Strijland A et al (2007) The Dutch Fabry cohort: diversity of clinical manifestations and Gb3 levels. J Inherit Metab Dis 30(1):68–78. https://doi.org/10.1007/s10545-006-0484-8
19. Mirzaian M, Wisse P, Ferraz MJ, Gold H, Donker-Koopman WE, Verhoek M et al (2015) Mass spectrometric quantification of glucosylsphingosine in plasma and urine of type 1 Gaucher patients using an isotope standard. Blood Cells Mol Dis 54(4):307–314. https://doi.org/10.1016/j.bcmd.2015.01.006. S1079-9796(15)00026-1 [pii]
20. Aerts JM, Groener JE, Kuiper S, Donker-Koopman WE, Strijland A, Ottenhoff R et al (2008) Elevated globotriaosylsphingosine is a hallmark of Fabry disease. Proc Natl Acad Sci U S A 105(8):2812–2817. https://doi.org/10.1073/pnas.0712309105. 0712309105 [pii], 9440 [pii]
21. Dekker N, van Dussen L, Hollak CE, Overkleeft H, Scheij S, Ghauharali K et al (2011) Elevated plasma glucosylsphingosine in Gaucher disease: relation to phenotype, storage cell markers, and therapeutic response. Blood 118(16):e118–e127. https://doi.org/10.1182/blood-2011-05-352971. S0006-4971(20)41034-1 [pii], 2011/352971 [pii]
22. Boutin M, Gagnon R, Lavoie P, Auray-Blais C (2012) LC-MS/MS analysis of plasma lyso-Gb3 in Fabry disease. Clin Chim Acta 414:273–280. https://doi.org/10.1016/j.cca.2012.09.026. S0009-8981(12)00461-5 [pii]
23. Chuang WL, Pacheco J, Zhang XK, Martin MM, Biski CK, Keutzer JM et al (2013) Determination of psychosine concentration in dried blood spots from newborns that were identified via newborn screening to be

at risk for Krabbe disease. Clin Chim Acta 419:73–76. https://doi.org/10.1016/j.cca.2013.01.017. S0009-8981(13)00039-9 [pii]
24. Gold H, Mirzaian M, Dekker N, Joao Ferraz M, Lugtenburg J, Codee JD et al (2013) Quantification of globotriaosylsphingosine in plasma and urine of fabry patients by stable isotope ultraperformance liquid chromatography-tandem mass spectrometry. Clin Chem 59(3):547–556. https://doi.org/10.1373/clinchem.2012.192138. clinchem.2012.192138 [pii]
25. Johnson B, Mascher H, Mascher D, Legnini E, Hung CY, Dajnoki A et al (2013) Analysis of lyso-globotriaosylsphingosine in dried blood spots. Ann Lab Med 33(4):274–278. https://doi.org/10.3343/alm.2013.33.4.274
26. Rolfs A, Giese AK, Grittner U, Mascher D, Elstein D, Zimran A et al (2013) Glucosylsphingosine is a highly sensitive and specific biomarker for primary diagnostic and follow-up monitoring in Gaucher disease in a non-Jewish, Caucasian cohort of Gaucher disease patients. PLoS One 8(11):e79732. https://doi.org/10.1371/journal.pone.0079732. PONE-D-13-14225 [pii]
27. Welford RW, Garzotti M, Marques Lourenco C, Mengel E, Marquardt T, Reunert J et al (2014) Plasma lysosphingomyelin demonstrates great potential as a diagnostic biomarker for Niemann-Pick disease type C in a retrospective study. PLoS One 9(12):e114669. https://doi.org/10.1371/journal.pone.0114669. PONE-D-14-29750 [pii]
28. Fuller M, Szer J, Stark S, Fletcher JM (2015) Rapid, single-phase extraction of glucosylsphingosine from plasma: A universal screening and monitoring tool. Clin Chim Acta 450:6–10. https://doi.org/10.1016/j.cca.2015.07.026. S0009-8981(15)00355-1 [pii]
29. Turgeon CT, Orsini JJ, Sanders KA, Magera MJ, Langan TJ, Escolar ML et al (2015) Measurement of psychosine in dried blood spots—a possible improvement to newborn screening programs for Krabbe disease. J Inherit Metab Dis 38(5):923–929. https://doi.org/10.1007/s10545-015-9822-z
30. Smid BE, van der Tol L, Biegstraaten M, Linthorst GE, Hollak CE, Poorthuis BJ (2015) Plasma globotriaosylsphingosine in relation to phenotypes of Fabry disease. J Med Genet 52(4):262–268. https://doi.org/10.1136/jmedgenet-2014-102872. jmedgenet-2014-102872 [pii]
31. Murugesan V, Chuang WL, Liu J, Lischuk A, Kacena K, Lin H et al (2016) Glucosylsphingosine is a key biomarker of Gaucher disease. Am J Hematol 91(11):1082–1089. https://doi.org/10.1002/ajh.24491
32. Nowak A, Mechtler TP, Desnick RJ, Kasper DC (2017) Plasma LysoGb3: A useful biomarker for the diagnosis and treatment of Fabry disease heterozygotes. Mol Genet Metab 120(1–2):57–61. https://doi.org/10.1016/j.ymgme.2016.10.006. S1096-7192(16)30325-0 [pii]
33. van der Veen SJ, Sayed ME, Hollak CEM, Brands MM, Snelder CKS, Boekholdt SM et al (2023) Early risk stratification for natural disease course in fabry patients using plasma globotriaosylsphingosine levels. Clin J Am Soc Nephrol. https://doi.org/10.2215/CJN.0000000000000239. 01277230-990000000-00201 [pii]
34. Sidhu R, Kell P, Dietzen DJ, Farhat NY, Do AND, Porter FD et al (2020) Application of N-palmitoyl-O-phosphocholineserine for diagnosis and assessment of response to treatment in Niemann-Pick type C disease. Mol Genet Metab 129(4):292–302. https://doi.org/10.1016/j.ymgme.2020.01.007. S1096-7192(19)30810-8 [pii]
35. Iwahori A, Maekawa M, Narita A, Kato A, Sato T, Ogura J et al (2020) Development of a diagnostic screening strategy for Niemann-Pick diseases based on simultaneous liquid chromatography-tandem mass spectrometry analyses of N-palmitoyl-O-phosphocholine-serine and Sphingosylphosphorylcholine. Biol Pharm Bull 43(9):1398–1406. https://doi.org/10.1248/bpb.b20-00400
36. Gelb MH, Scott CR, Turecek F (2015) Newborn screening for lysosomal storage diseases. Clin Chem 61(2):335–346. https://doi.org/10.1373/clinchem.2014.225771. clinchem.2014.225771 [pii]
37. Hong X, Daiker J, Sadilek M, Ruiz-Schultz N, Kumar AB, Norcross S et al (2021) Toward newborn screening of metachromatic leukodystrophy: results from analysis of over 27,000 newborn dried blood spots. Genet Med 23(3):555–561. https://doi.org/10.1038/s41436-020-01017-5. S1098-3600(21)04957-1 [pii]
38. Hong X, Sadilek M, Gelb MH (2020) A highly multiplexed biochemical assay for analytes in dried blood spots: application to newborn screening and diagnosis of lysosomal storage disorders and other inborn errors of metabolism. Genet Med 22(7):1262–1268. https://doi.org/10.1038/s41436-020-0790-9. S1098-3600(21)01185-0 [pii]
39. Kwon JM, Matern D, Kurtzberg J, Wrabetz L, Gelb MH, Wenger DA et al (2018) Consensus guidelines for newborn screening, diagnosis and treatment of infantile Krabbe disease. Orphanet J Rare Dis 13(1):30. https://doi.org/10.1186/s13023-018-0766-x. 766 [pii]
40. Thomas GH (1994) "Pseudodeficiencies" of lysosomal hydrolases. Am J Hum Genet 54(6):934–940
41. Filocamo M, Morrone A (2011) Lysosomal storage disorders: molecular basis and laboratory testing. Hum Genomics 5(3):156–169. https://doi.org/10.1186/1479-7364-5-3-156. D10117470183423K [pii], 1479-7364-5-3-156 [pii]
42. Boot RG, Verhoek M, Langeveld M, Renkema GH, Hollak CE, Weening JJ et al (2006) CCL18: a urinary marker of Gaucher cell burden in Gaucher patients. J Inherit Metab Dis 29(4):564–571. https://doi.org/10.1007/s10545-006-0318-8
43. Brinkman J, Wijburg FA, Hollak CE, Groener JE, Verhoek M, Scheij S et al (2005) Plasma chitotriosidase and CCL18: early biochemical surrogate markers in type B Niemann-Pick disease. J Inherit Metab Dis 28(1):13–20. https://doi.org/10.1007/s10545-005-4416-9

44. Aerts JM, Hollak CE, van Breemen M, Maas M, Groener JE, Boot RG (2005) Identification and use of biomarkers in Gaucher disease and other lysosomal storage diseases. Acta Paediatr Suppl 94(447):43–46. Discussion 37–8. https://doi.org/10.1111/j.1651-2227.2005.tb02110.x
45. Motta M, Tatti M, Furlan F, Celato A, Di Fruscio G, Polo G et al (2016) Clinical, biochemical and molecular characterization of prosaposin deficiency. Clin Genet 90(3):220–229. https://doi.org/10.1111/cge.12753
46. Dang Do AN, Chang IJ, Jiang X, Wolfe LA, Ng BG, Lam C et al (2023) Elevated oxysterol and N-palmitoyl-O-phosphocholineserine levels in congenital disorders of glycosylation. J Inherit Metab Dis 46(2):326–334. https://doi.org/10.1002/jimd.12595
47. Ho MW, O'Brien JS (1970) Stimulation of acid beta-galactosidase activity by chloride ions. Clin Chim Acta 30(2):531–534. https://doi.org/10.1016/0009--8981(70)90147-6. 0009-8981(70)90147-6 [pii]
48. Ho MW, O'Brien JS (1971) Differential effect of chloride ions on -galactosidase isoenzymes: a method for separate assay. Clin Chim Acta 32(3):443–450. https://doi.org/10.1016/0009-8981(71)90446-3. 0009-8981(71)90446-3 [pii]
49. Inui K, Wenger DA (1984) Usefulness of 4-methylum belliferyl-6-sulfo-2-acetamido-2-deoxy-beta-D-gluco pyranoside for the diagnosis of GM2 gangliosidoses in leukocytes. Clin Genet 26(4):318–321. https://doi.org/10.1111/j.1399-0004.1984.tb01066.x
50. Ganne B, Dauriat B, Richard L, Lamari F, Ghorab K, Magy L et al (2022) GM2 gangliosidosis AB variant: first case of late onset and review of the literature. Neurol Sci 43(11):6517–6527. https://doi.org/10.1007/s10072-022-06270-x
51. Blondel A, Kraoua I, Marcelino C, Khrouf W, Schlemmer D, Ganne B et al (2023) Plasma G(M2) ganglioside potential biomarker for diagnosis, prognosis and disease monitoring of GM2-Gangliosidosis. Mol Genet Metab 138(2):106983. https://doi.org/10.1016/j.ymgme.2022.106983. S1096-7192(22)00461-9 [pii]
52. O'Brien JS, Okada S, Chen A, Fillerup DL (1970) Tay-sachs disease. Detection of heterozygotes and homozygotes by serum hexosaminidase assay. N Engl J Med 283(1):15–20. https://doi.org/10.1056/NEJM197007022830104
53. Peters SP, Lee RE, Glew RH (1975) A microassay for Gaucher's disease. Clin Chim Acta 60(3):391–396. https://doi.org/10.1016/0009-8981(75)90083-2. 0009-8981(75)90083-2 [pii]
54. Daniels LB, Glew RH, Diven WF, Lee RE, Radin NS (1981) An improved fluorometric leukocyte beta-glucosidase assay for Gaucher's disease. Clin Chim Acta 115(3):369–375. https://doi.org/10.1016/0009--8981(81)90251-5. 0009-8981(81)90251-5 [pii]
55. Lo SM, McNamara J, Seashore MR, Mistry PK (2010) Misdiagnosis of Niemann-Pick disease type C as Gaucher disease. J Inherit Metab Dis 33 Suppl 3(03):S429–S433. https://doi.org/10.1007/s10545-010-9214-3
56. Aguilera B, Ghauharali-van der Vlugt K, Helmond MT, Out JM, Donker-Koopman WE, Groener JE et al (2003) Transglycosidase activity of chitotriosidase: improved enzymatic assay for the human macrophage chitinase. J Biol Chem 278(42):40911–40916. https://doi.org/10.1074/jbc.M301804200. S0021-9258(20)82885-3 [pii]
57. Schoonhoven A, Rudensky B, Elstein D, Zimran A, Hollak CE, Groener JE et al (2007) Monitoring of Gaucher patients with a novel chitotriosidase assay. Clin Chim Acta 381(2):136–139. https://doi.org/10.1016/j.cca.2007.02.042. S0009-8981(07)00137-4 [pii]
58. Boot RG, Renkema GH, Verhoek M, Strijland A, Bliek J, de Meulemeester TM et al (1998) The human chitotriosidase gene. Nature of inherited enzyme deficiency. J Biol Chem 273(40):25680–25685. https://doi.org/10.1074/jbc.273.40.25680. S0021-9258(19)59873-8 [pii]
59. Boot RG, Verhoek M, de Fost M, Hollak CE, Maas M, Bleijlevens B et al (2004) Marked elevation of the chemokine CCL18/PARC in Gaucher disease: a novel surrogate marker for assessing therapeutic intervention. Blood 103(1):33–39. https://doi.org/10.1182/blood-2003-05-1612. S0006-4971(20)50281-4 [pii]
60. Deegan PB, Moran MT, McFarlane I, Schofield JP, Boot RG, Aerts JM et al (2005) Clinical evaluation of chemokine and enzymatic biomarkers of Gaucher disease. Blood Cells Mol Dis 35(2):259–267. https://doi.org/10.1016/j.bcmd.2005.05.005. S1079-9796(05)00067-7 [pii]
61. Guo Y, He W, Boer AM, Wevers RA, de Bruijn AM, Groener JE et al (1995) Elevated plasma chitotriosidase activity in various lysosomal storage disorders. J Inherit Metab Dis 18(6):717–722. https://doi.org/10.1007/BF02436762
62. Ries M, Schaefer E, Luhrs T, Mani L, Kuhn J, Vanier MT et al (2006) Critical assessment of chitotriosidase analysis in the rational laboratory diagnosis of children with Gaucher disease and Niemann-Pick disease type A/B and C. J Inherit Metab Dis 29(5):647–652. https://doi.org/10.1007/s10545-006-0363-3
63. Vedder AC, Cox-Brinkman J, Hollak CE, Linthorst GE, Groener JE, Helmond MT et al (2006) Plasma chitotriosidase in male Fabry patients: a marker for monitoring lipid-laden macrophages and their correction by enzyme replacement therapy. Mol Genet Metab 89(3):239–244. https://doi.org/10.1016/j.ymgme.2006.04.013. S1096-7192(06)00170-3 [pii]
64. Desnick RJ, Allen KY, Desnick SJ, Raman MK, Bernlohr RW, Krivit W (1973) Fabry's disease: enzymatic diagnosis of hemizygotes and heterozygotes. Alpha-galactosidase activities in plasma, serum, urine, and leukocytes. J Lab Clin Med 81(2):157–171. 0022-2143(73)90283-7 [pii]
65. Mayes JS, Scheerer JB, Sifers RN, Donaldson ML (1981) Differential assay for lysosomal alpha-

galactosidases in human tissues and its application to Fabry's disease. Clin Chim Acta 112(2):247–251. https://doi.org/10.1016/0009-8981(81)90384-3. 0009-8981(81)90384-3 [pii]
66. Wiederschain G, Raghavan S, Kolodny E (1992) Characterization of 6-hexadecanoylamino-4-methylumbelliferyl-beta-D-galactopyranoside as fluorogenic substrate of galactocerebrosidase for the diagnosis of Krabbe disease. Clin Chim Acta 205(1–2):87–96. https://doi.org/10.1016/s0009--8981(05)80003-8. S0009-8981(05)80003-8 [pii]
67. Baum H, Dodgson KS, Spencer B (1959) The assay of arylsulphatases A and B in human urine. Clin Chim Acta 4(3):453–455. https://doi.org/10.1016/0009--8981(59)90119-6. 0009-8981(59)90119-6 [pii]
68. Lee-Vaupel M, Conzelmann E (1987) A simple chromogenic assay for arylsulfatase A. Clin Chim Acta 164(2):171–180. https://doi.org/10.1016/0009--8981(87)90068-4. 0009-8981(87)90068-4 [pii]
69. Beerepoot S, van Dooren SJM, Salomons GS, Boelens JJ, Jacobs EH, van der Knaap MS et al (2020) Metachromatic leukodystrophy genotypes in The Netherlands reveal novel pathogenic ARSA variants in non-Caucasian patients. Neurogenetics 21(4):289–299. https://doi.org/10.1007/s10048-020-00621-6. [pii]. 621 [pii]
70. van Diggelen OP, Voznyi YV, Keulemans JL, Schoonderwoerd K, Ledvinova J, Mengel E et al (2005) A new fluorimetric enzyme assay for the diagnosis of Niemann-Pick A/B, with specificity of natural sphingomyelinase substrate. J Inherit Metab Dis 28(5):733–741. https://doi.org/10.1007/s10545-005-0105-y
71. Harzer K, Rolfs A, Bauer P, Zschiesche M, Mengel E, Backes J et al (2003) Niemann-Pick disease type A and B are clinically but also enzymatically heterogeneous: pitfall in the laboratory diagnosis of sphingomyelinase deficiency associated with the mutation Q292 K. Neuropediatrics 34(6):301–306. https://doi.org/10.1055/s-2003-44668
72. Tamura H, Takahashi T, Ban N, Torisu H, Ninomiya H, Takada G et al (2006) Niemann-Pick type C disease: novel NPC1 mutations and characterization of the concomitant acid sphingomyelinase deficiency. Mol Genet Metab 87(2):113–121. https://doi.org/10.1016/j.ymgme.2005.07.025. S1096-7192(05)00226-X [pii]
73. Reagan JW Jr, Hubbert ML, Shelness GS (2000) Posttranslational regulation of acid sphingomyelinase in Niemann-Pick type C1 fibroblasts and free cholesterol-enriched Chinese hamster ovary cells. J Biol Chem 275(48):38104–38110. https://doi.org/10.1074/jbc.M005296200. S0021-9258(20)88530-5 [pii]
74. Vanier MT, Rodriguez-Lafrasse C, Rousson R, Gazzah N, Juge MC, Pentchev PG et al (1991) Type C Niemann-Pick disease: spectrum of phenotypic variation in disruption of intracellular LDL-derived cholesterol processing. Biochim Biophys Acta 1096(4):328–337. https://doi.org/10.1016/0925-4439(91)90069-l. 0925-4439(91)90069-L [pii]
75. Vanier MT, Latour P (2015) Laboratory diagnosis of Niemann-Pick disease type C: the filipin staining test. Methods Cell Biol 126:357–375. https://doi.org/10.1016/bs.mcb.2014.10.028. S0091-679X(14)00029-6 [pii]
76. Sun X, Marks DL, Park WD, Wheatley CL, Puri V, O'Brien JF et al (2001) Niemann-Pick C variant detection by altered sphingolipid trafficking and correlation with mutations within a specific domain of NPC1. Am J Hum Genet 68(6):1361–1372. https://doi.org/10.1086/320599. S0002-9297(07)61047-7 [pii], 012682 [pii]
77. Vanier MT (1997) Phenotypic and genetic heterogeneity in Niemann-Pick disease type C: current knowledge and practical implications. Wien Klin Wochenschr 109(3):68–73

13 Mitochondrial Oxidative Phosphorylation Enzymes

Richard J. Rodenburg

Contents

List of Abbreviations

ATP	Adenosine triphosphate
CAS	Chemical Abstracts Service
coQ1	Coenzyme Q1
coQ10	Coenzyme Q10
DCIP	Dichloroindophenol
DNA	Deoxyribonucleic acid
DQ	Decylubiquinone
GC-MS	Gas chromatography–mass spectrometry
LC-MS	Liquid chromatography–mass spectrometry
mtDNA	Mitochondrial DNA
NAD	Nicotinamide adenine dinucleotide
NADH	NAD + hydrogen
OXPHOS	Oxidative phosphorylation
QCC	Ubiquinol:cytochrome c oxidoreductase
RNA	Ribonucleic acid
SCC	Succinate:cytochrome c oxidoreductase
SDH	Succinate dehydrogenase
TCA	Tricarboxylic acid cycle
VUS	Variant of unknown clinical significance
WES	Whole exome sequencing
WGS	Whole genome sequencing

R. J. Rodenburg (✉)
Radboud Center for Mitochondrial Medicine (RCMM), Department of Pediatrics, Department of Genetics, 830 Translational Metabolic Laboratory (TML), Radboud University Medical Center, Nijmegen, The Netherlands
e-mail: richard.rodenburg@radboudumc.nl

N. Blau, F. M. Vaz (eds.), *Laboratory Guide to the Methods in Biochemical Genetics*,
https://doi.org/10.1007/978-3-031-58819-8_13

13.1 Introduction

In recent years, genetic tests are increasingly being used as screening diagnostic tests, and the investigations of the enzyme activities of the respiratory chain enzymes have shifted from screening tests to a confirmatory diagnostic test for the purpose of validating genetic findings. The advantages of the "genetics first" approach include the fact that genetic tests usually require a simple blood sample (or a urine sample, in the case of a suspected mtDNA-defect), have relatively short turn-around times, are less invasive than the tissue biopsy in the case of biochemical diagnostic tests, and can detect even the rarest pathogenic defects with generic genetic tests such as exome and genome sequencing. There are disadvantages as well, like for example the fact that a genetic screening test is not suitable to exclude a mitochondrial disease, this in contrast to biochemical testing of biopsies of clinically affected tissue. These down-sides do not weigh up to the advantages of genetic screening. Unfortunately, genetic tests are not a universal solution to diagnose all mitochondrial patients. From different patient cohort studies, the diagnostic yield varies between 40 and 60%. The remaining undiagnosed patients may instead suffer from other diseases with clinical overlap to mitochondrial disorders, may have a genetic defect that is not detected, or not recognized in the vast amount of data produced by WES/WGS. Moreover, in a subset of patients, genetic variants are detected of which it is unclear whether it is pathogenic or not. This latter group of patients may require subsequent biochemical testing in order to prove that the genetic variant identified in these patients is pathogenic or not.

13.2 Clinical Significance

The mitochondrial respiratory chain plays a central role in cellular energy metabolism. The respiratory chain consists of 4 multi-subunit enzymes (complex I, III, III, and IV). The respiratory chain, together with another multi-subunit enzyme, complex V, forms the oxidative phosphorylation system (Table 13.1). Defects in one, or a combination of, OXPHOS enzymes cause mitochondrial disease, a inheritable disorder with a very broad clinical spectrum in many aspects (age of onset, clinical symptoms, affected tissues and organs, and disease course). Some clinical features are shared by most mitochondrial disorders, as it often affects involves multiple tissues/organs that convert a lot of energy, such as skeletal muscle and brain tissue, but other organs can be affected as well. The disease course often is progressive. The patients with a severe form of mitochondrial disease often have a disease presentation at a young age, and in those patients the disease is often lethal. By contrast, adult patients often show relatively mild symptoms, such as exercise intolerance, muscle pain, and neurological symptoms. Please note that these are very general remarks and there are many other types of mitochondrial disorders. This broad clinical spectrum is one of the main reasons that the disease is not always recognized as a mitochondrial disease and establishing a diagnosis can be a time-consuming process, requiring multiple complex diagnostic tests. An important diagnostic test is the measurement of enzyme activity of the OXPHOS enzymes. Before holistic genetic tests were available, the enzyme activity tests were performed early on in the diagnostic process. This usually included measurements of OXPHOS enzymes in muscle biopsy specimens. Muscle tissue contains large amounts of mitochondria and often is clinically affected in mitochondrial disorders, and is accessible via a needle biopsy or surgical procedure. In cases with normal biochemical test results in muscle, a mitochondrial disorder is unlikely, in particular in cases with clear clinical symptoms in muscle. In cases with a biochemical defect in muscle, e.g.,

Table 13.1 The OXPHOS system

	Subunits		
Enzyme	mtDNA	nDNA	EC
Complex I	7	37	7.1.1.2
Complex II	0	4	1.3.5.1
Complex III	1	10	7.1.1.8
Complex IV	3	11	7.1.1.9
Complex V	2	18	7.1.2.2

an OXPHOS enzyme deficiency, the next diagnostic step used to be a genetic test of a small number of candidate genes. Nowadays, in the majority of cases, the genetic screening of candidate genes is done first. This involves exome or genome sequencing of all nuclear genes and of the mtDNA. As these genetic tests are performed without a priori knowledge of the biochemical phenotype, the interpretation of detected rare genetic variants can be challenging, and may result in the identification of genetic variants of unknown clinical significance (VUS). In that case, a biochemical test may be needed to confirm or exclude pathogenicity of such variants [1]. For this purpose, a targeted biochemical test is usually sufficient and in many cases it can be performed on cultured skin fibroblasts instead of a muscle biopsy.

13.3 Methods

The activity of the OXPHOS enzymes can be determined by spectrophotometry. For each of the five enzymes sensitive spectrophotometric assays have been developed. Details of these tests are given below. Depending on the circumstances, like the desired turn-around times, or the number of diagnostic samples to be tested, the assays can be performed either manually or by automated platforms. In our lab at the Radboudumc in Nijmegen, the Netherlands, these enzyme tests are being performed by an automated platform that allows the use of in-house developed enzyme tests [2]. The assays performed using this platform are based on existing manual assays developed by us or by others. In the case OXPHOS enzymes are being measured on a regular basis (such as in diagnostic laboratories) it is recommended to prepare the reaction mixtures described below in larger batches that can be stored in aliquots in the freezer. It is important to check newly prepared reaction mixtures by performing measurements on QC samples and compare the results with those of previous reaction mixture batches. Mixtures stored frozen remain of good quality for several months, nevertheless it is recommended to regularly test QC samples to check the quality of the mixture and compare results with previous runs.

Please note that the measuring conditions and assay reaction mixtures shown below may need minor tweaking depending on the spectrophotometric platform that is being used. For example, optimizing total volumes of the reaction mixture in the measuring cuvette. Also, the way in which inhibitors are added to reaction mixtures may require small adaptation in the way samples and reagents are handled in order to minimize unwanted carry-over problems, which may be a risk when using automated spectrophotometric platforms.

13.3.1 Samples

Given the heterogeneity of the clinical signs and symptoms caused by a pathological mitochondrial defect, an optimally informative diagnostic test result requires the investigation of mitochondrial activity of clinically affected tissues. However, an important factor that should be taken into consideration is the invasiveness of the sample collection procedure. For this reason, a brain biopsy is almost never considered. Very rarely, heart or liver tissue is collected for enzymatic testing, often in in the case that other surgical procedures create an opportunity to combine it with tissue sample collection for enzyme and mtDNA diagnostics. Biopsies of skeletal muscle are frequently tested, because of the fact that muscle tissue has a relatively high mitochondrial content, combined with the fact that many mitochondrial disease patients have skeletal muscle symptoms, and this tissue is accessible by a needle biopsy procedure on adult patients. In pediatric cases, a muscle biopsy using a surgical procedure is preferred over a needle biopsy. However, in recent years, investigations of tissue biopsy material for diagnostic screening purposes have been replaced by molecular genetic testing, in particular exome or genome sequencing, often combined with a high coverage test of the mtDNA. In the case the genetic screening has not resulted in a diagnosis because of negative test results, or in the case of the identification of novel genetic variants requiring follow-up investigation, tissue biopsy procedures are considered. Importantly, most genetic tests are not suitable to exclude the possibility of a mitochondrial disorder, and in situations where a mitochondrial disease needs to be excluded with more certainty than achievable by molecular genetic tests, a muscle biopsy for biochemical testing can be considered.

13.3.2 Sample Preparation Methods

In addition to the patient-derived tissue samples, the methods described below can be applied to various other types of samples, e.g., for research purposes. The enzyme assays are quite robust and can be used on a wide variety of tissue- and cultured cell types, such as tissue samples from laboratory animals, whole zebrafish embryos, Drosophila (whole flies and segments), tumor biopsy material, and a large number of cultured cell types like HeLa and HEK293, primary myoblasts, various induced pluripotent stem cells (IPSC)-derived samples and many more.

Samples from cultured cells are prepared as follows: Around ten million cultured cells are harvested using standard cell culturing procedures after which the cell pellets are usually stored at −20 °C until used for further sample preparation steps. Cell pellets are resuspended in 1.25 mL 10 mM Tris–HCl, pH = 7.6. The cell suspension is transferred to a Potter-Elvehjem tissue homogenizer and is homogenized by using a pestle 8 times, after which 250 μL 1.5 M sucrose is added to the homogenate and this is carefully but thoroughly mixed. A small amount of this homogenate (100 μL) is collected and stored frozen (<−70 °C). The remaining suspension is centrifuged for 10 min at 600 g, and the supernatant is centrifuged for 10 min at 14,000 g. The resulting pellet containing the mitochondria is resuspended in 450 μL 10 mM Tris–HCl, pH = 7,6, and is stored in aliquots (3 × 150 μL) frozen at <−70 °C until the measurements..

The first step in the preparation of muscle samples (either frozen or fresh) is to remove any visible fat or connective tissue, as well as any transportation buffer, after which the muscle tissue is weighed. The amount of required tissue is at least 50 mg for OXPHOS enzyme measurements. We advise to collect at least 100 mg in order to have enough tissue to repeat measurements if necessary, as well as to have residual tissue for other tests, like for example blue native-polyacrylamide gel electrophoresis (BN-PAGE), Western blots, metabolic/mitochondrial flux analysis, or proteomics. The muscle tissue is minced, e.g., using a sharp razor blade, and transferred to a Potter-Elvehjem tube to which SETH- buffer (0.25 M sucrose, 2 mM EDTA, 5.10^4 U/L heparin, 10 mM Tris–Cl pH 7.4) is added to end up with a 10% (w/V) suspension. The sample is homogenized using a pestle until the suspension is homogeneous. A small part of this homogenate is stored, e.g., for DNA extraction. The remainder of the homogenate is centrifuged for 10 min. at 600 g after which the supernatant is collected and stored at <−70 °C. This material is suitable for measurements of the OXPHOS enzymes except complex V, for which a more pure mitochondrial preparation is needed. This is prepared from 300 μL of the 600 g supernatant by centrifugation at 14,000 g for 10 min. The supernatant is removed and the brownish pellet is washed by resuspending it in 700 μL SETH buffer followed by another centrifugation step at 14,000 g for 10 min. The pellet is resuspended in 200 μL SETH, and stored frozen (<−70 °C) until being used for complex V measurements.

13.4 Enzyme Activity Measurements of Individual Complexes

The activity of the OXPHOS enzymes can be measured in different ways. Enzyme activity testing using spectrophotometry is widely used to diagnose OXPHOS enzyme deficiencies. These are discussed below. There are alternative approaches to investigate OXPHOS enzymes by spectrophotometry. For example, in-gel activity measurements can be performed. For this purpose, intact protein complexes in mitochondrial protein samples are separated on the basis of their molecular weight on a blue native polyacrylamide gel, after which the activities of the individual OXPHOS enzymes are determined by incubating the gel with enzyme reaction mixtures resulting in a staining that can be judged by eye for possible deficiencies. This is not a very accurate approach and it is not suitable for diagnostic applications. A much more straightforward test is a immunocapture assay in dipstick format that has been shown to detect complex I and complex IV deficiencies [3]. Such a fast and cheap test is suitable for a quick first screening, but not for an accurate investigation of all OXPHOS enzymes. At the other end of the spectrum is the proteomics-based holistic examination of the mitochondrial proteome, which provides information of protein expression for thousands of proteins in one analytical run. This approach detects protein expression levels, it does not detect enzyme activities, although information about a loss of function by a defect in the protein of interest can be reflected by aberrant expression levels of proteins acting in the same protein complex as the protein of interest. For this purpose, the spectrophotometric approach still is the method of choice for accurate diagnostic OXPHOS measurements. It should be noted that there are no universal reference values for OXPHOS enzyme activities, and each diagnostic center uses its own reference values, although we have shown that the ratio's of the measured OXPHOS enzyme activities were quite similar when compared between different diagnostic centers [4].

Below an overview is presented of the methods used to examine OXPHOS enzyme activity measurements in our center. All enzyme tests can be performed on a wide variety of spectrophotometric devices, and independent of the spectrophotometer being used the assays can be performed in the same manner. The first step is to add an optimal amount of reaction mixture to the cuvette. The optimal amount is mainly determined by the spectrophotometric device that is being used. In our lab, we use spectrophotometers with cuvettes with a reaction volume of 100 μL. The reactions, all performed at 37 °C, are started by the addition of a few microliters of sample. The change in absorbance is measured and from the rate of absorbance change the enzyme activity can be calculated using the principle of Lambert–Beer. It is strongly advised to test the optimal amount of each different sample type to be added to the reaction mixture. The observed activities should be linear for a few minutes, long enough to be able to calculate the enzyme activity reliably.

13.4.1 Complex I

Complex I is the largest enzyme of the OXPHOS enzyme complexes. It consists of 44 different subunits, including 7 proteins encoded by the mtDNA and 37 proteins encoded by genes located in the nuclear genome. In a large number of these genes, disease-causing genetic variants have been identified [5]. In addition, the assembly of complex I depends on several so-called assembly factors. These proteins play an essential role in the assembly of the building blocks of complex I [6]. The N-module of complex I that is oriented toward the mitochondrial matrix, oxidizes NADH to NAD+. The electrons coming from NADH are guided through several iron–sulfur clusters present in the N-domain of complex I toward coenzyme Q10 that resides in the Q-module of complex I nearest to the N-module. The oxidation of Q10 is linked with the translocation of protons from the mitochondrial intermembrane space to the mitochondrial matrix.

The activity of complex I (EC 1.6.5.3) can be measured in different ways. For diagnostic purposes, the measurement of NADH oxidation by spectrophotometry at a wavelength of 340 nm is widely used [7]. As an alternative approach, we have developed a spectrophotometric assay that makes use of DCIP as the final electron acceptor that can be measured at 600 nm [8]. One of the benefits of this assay is that it is more sensitive than the assay based on NADH measurements because of a two-fold higher molar extinction coefficient than that of NADH. As a blank, the activity of complex I is measured in the same reaction mixture to which the complex I inhibitor rotenone has been added. Measurement of complex I activity requires the addition of Q10 analogue to the reaction mixture, as the endogenously available coenzyme Q10 does not result in an optimal activity of complex I [7]. Often compounds similar to Q10, like decylubiquinone (DQ) or coQ1 are used for this purpose. It should be noted that although the enzyme reaction mixture with either DQ or coQ1 gives very similar results, the assay with DQ is less robust than with coQ1 due to the fact that the kinetics of the enzyme reaction with DQ is less favorable compared to the assay with coQ1 [9].

13.4.1.1 Complex I Method

Method: spectrophotometry, wavelength: 600 nm

Preparation of complex I incubation mix:

- 25 mM phosphate buffer (K_2HPO_4/KH_2PO_4, pH 7.6)
- 0.35% bovine serum albumin, fraction V (CAS 90604-29-8)
- 60 μmol/L DCIP (CAS 55486-00-5)
- 0.2 mmol/L NADH (CAS 108321-31-9)
- 70 μmol/L coenzyme Q_1 (CAS 727-81-1)

Execution of enzyme activity test: The reaction is performed at 37 °C and is started by the addition of the mitochondrial preparation to the incubation mix. A blank reaction that is identical to the reaction of the test sample and that contains complex I inhibitor (1 μmol/L rotenone), is run in parallel.

13.4.2 Complex II

Complex II (succinate dehydrogenase; SDH) consists of four subunits that are encoded by genes present in the nuclear DNA, and is the only OXPHOS enzyme complex that does not contain subunits encoded by the mtDNA. The enzyme catalyzes the conversion of succinate to fumarate in the citric acid cycle, and the reduction of FAD to FADH2 combined with the reduction of coenzyme Q10. Defects in this enzyme complex are known causes for mitochondrial disorders [10], but also can cause rare tumors including pheochromocytomas and paraganglioma's [11, 12]. Although both tumor development and mitochondrial diseases require a bi-allelic defect in one of the SDH genes, there have not been any reports yet of mitochondrial disease patients that also develop tumors, although there are a few reports about genetic variants that have been associated with mitochondrial disease and tumors, but not in the same families [13, 14]. Possibly, patients with a mitochondrial disease caused by SDH mutations do not survive until the young-adult age at which SDH-associated tumors develop.

Enzyme activity measurements by spectrophotometry are performed using a reaction mixture containing succinate, DQ, and the final electron acceptor DCIP with a read-out at 600 nm. In order to prevent complex I to interfere with the complex II assay, rotenone is added to the mixture. Blank measurements are performed by adding malonate, a competitive inhibitor of complex II.

13.4.2.1 Complex II Method

Method: spectrophotometry, wavelength: 600 nm

Preparation of complex II incubation mix:

- 80 mM phosphate buffer (K_2HPO_4/KH_2PO_4, pH 7.8)
- 0.2% bovine serum albumin, fraction V (CAS 90604-29-8)
- 80 μmol/L DCIP (CAS 55486-00-5)
- 0.2 mmol/L ATP (56-65-5)
- 2.0 mmol/L EDTA
- 80 μmol/L decylubiquinone (CAS 55486-00-5)
- 10 mmol/L succinate (CAS 110-15-6)

Execution of enzyme activity test: The reaction is performed at 37 °C and is started by the addition of the mitochondrial preparation. A blank reaction that is identical to the reaction of the test sample and that contains complex II inhibitor (5 mmol/L malonate), is run in parallel.

13.4.3 Complex III

Complex III deficiencies are probably more rare than other respiratory chain enzyme deficiencies, together with complex II (not taking oncogenetics into account). Although one could argue that this might be in part due to technical issues resulting in false positive results, the fact that genetic screening by exome sequencing + mtDNA sequencing do not result in large numbers of (putative) pathogenic variants (data not shown) indicates that by complex III enzyme measurements no significant numbers of complex III deficiencies have been missed (data not shown). Complex III is usually measured as antimycin A-sensitive ubiquinol:cytochrome c oxidoreductase (QCC) activity with the reduction of cytochrome c detected spectrophotometrically at 550 nm as a read-out [15]. Alternatively (or in addition) complex III can be measured as a complex I + III (NADH:cytochrome c oxidoreductase and as a complex II + complex III (succinate:cytochrome c oxidoreductase), although it should be mentioned that these combined enzyme reactions each have their own technical issues and are not equally sensitive to complex III deficiencies compared to complex I or II, and therefore measurements of isolated complex III are preferred [16]. We recommend to make use of cytochrome c of the highest purity, as inferior preparations will result in suboptimal results in the enzyme assay. In addition, we recommended to test the purity of each batch of cytochrome c, in order to verify the purity of batches. The cytochrome c is purified from muscle tissue of pig or horse. Suppliers normally determine the purity of cytochrome c batches but batches can deteriorate during storage and during transport. It has been reported that certain MT-CYB (encoding cytochrome b, a subunit of complex III encoded by the mtDNA) that could result in a catalytic defect may be missed by the QCC complex III test, which usually is measured against a blank reaction with the inhibitor antimycin A. Such variants may result in an abnormally diminished sensitivity toward the complex III inhibitor myxothiazol in comparison to wild type complex III in healthy control samples [17, 18].

13.4.3.1 Complex III Method

Method: spectrophotometry, wavelength: 550 nm

Preparation of complex III incubation mix:

- 50 mM phosphate buffer (K_2HPO_4/KH_2PO_4, pH 7.8)
- 1.0 mmol/L EDTA
- 3.0 mmol/L sodium azide (CAS 26628-22-8)
- 0.04% Tween20 (CAS 9005-64-5)
- 300 μmol/L decylubiquinol (CAS 55486-00-5)
- 50 μmol/L cytochrome c (CAS 9007-45-6)

Execution of enzyme activity test: The reaction is performed at 37 °C and is started by the addition of the mitochondrial preparation. It may be necessary to tweak the added amount of sample to obtain a linear measurement.

13.4.4 Complex IV

The complex IV assay is based on the method described by Cooperstein et al. [19]. Similar to complex III, the activity of complex IV (also known as cytochrome c oxidase) is investigating cytochrome c by spectrophotometry at 550 nm. For complex IV measurements, the substrate is reduced cytochrome c, which usually is a purified fraction from horse or pig heart protein preparations that is reduced by using sodium dithionite. Although the enzyme activity of complex IV will follow a first-order reaction constant, in practice the activity can be calculated by assuming a linear rate of the initial enzyme reaction.

13.4.4.1 Complex IV Method

Method: spectrophotometry, wavelength: 550 nm

Preparation of complex IV incubation mix:

- 30 mM phosphate buffer (K_2HPO_4/KH_2PO_4, pH 7.4)
- 70 μmol/L reduced cytochrome c (CAS 9007-45-6)

Execution of enzyme activity test: The reaction is performed at 37 °C and is started by the addition of the mitochondrial preparation.

13.4.5 Complex V

Complex V (F1Fo ATP synthase) is the terminal enzyme in the mitochondrial ATP production system. It uses the mitochondrial membrane potential that is maintained by the respiratory chain to produce ATP from ADP and pyrophosphate. The enzyme can also operate "in reverse" and can function as an ATPase and in this way contribute to the membrane potential. Most diagnostic tests measure complex V activity as an oligomycin-sensitive ATPase activity. Since cells usually contain several different ATPases that can interfere with the complex V ATPase assay, the use of oligomycin alone is not sufficient to obtain reliable diagnostic results. For an optimal diagnostic measurements, it is necessary to prepare a mitochondria-enriched fraction from tissue biopsies and cultured cells. Unfortunately, we and others have seen that reliable complex V measurements are not possible when mitochondrial preparations are made from frozen muscle biopsy material. It is not fully understood what is causing this. This phenomenon is not seen in mitochondrial preparations from "fresh" tissue biopsies, even if the mitochondrial preparations have been stored frozen after preparation. Pathogenic variants in nuclear DNA-encoded complex V subunits and assembly factors are quite rare. Genetic variants in the mtDNA-encoded subunits are more frequently observed, in particular in *MT-ATP6*. Establishing the pathogenicity of *MT-ATP6* genetic variants can be challenging. It is not uncommon that the classical enzyme test described above produces a normal enzyme activity in cases with pathogenic *MT-ATP6* variants. This is likely due to the fact that the enzyme test does not fully depend on a proper functioning *MT-ATP6* subunit. Unfortunately other biochemical tests that are expected to be disturbed do not universally show abnormalities [20]. Therefore, interpreting *MT-ATP6* genetic variants often relies on combining genotype–phenotype data from multiple families with the same variants [21].

13.4.5.1 Complex V Method

Method: spectrophotometry, wavelength: 550 nm

Preparation of complex V incubation mix:

- 25 mM phosphate buffer (K_2HPO_4/KH_2PO_4, pH 8.0)
- 0.2 mmol/L EGTA (CAS 67-42-5
- 5 mg/L Ap5A (CAS 102783-61-9)
- 0.3% bovine serum albumin, fraction V (CAS 90604-29-8)
- 250 mmol/L sucrose (CAS 57-50-1)
- 7.5 mmol/L $MgCl_2$.6 H_2O (CAS 7791-18-6)
- 50 mmol/L KCl (CAS 7447-40-7)
- 0.2 mmol/L NADH (CAS 108321-31-9)
- 0.1 mmol/L phosphoenolpyruvate (CAS 4265-07-0)
- 2.5 U/mL lactate dehydrogenase (CAS 232-617-8)
- 1.5 U/mL pyruvate kinase (CAS 9001-59-6)

Execution of enzyme activity test: The reaction is performed at 37 °C and is started by the addition of the mitochondrial preparation. A blank measurement is performed, consisting of the test sample incubation with the addition of oligomycin (CAS 1404-19-9).

13.4.5.2 Coenzyme Q10

Coenzyme Q10 (coQ10) is an essential component of the respiratory chain. Its molecular structure consists of a quinone ring structure that is derived from tyrosine and an isoprenoid tail derived from the mevalonate pathway that is shared among others by dolichol and cholesterol, and is synthesized in mitochondria. The biosynthesis of coQ10 requires at least 15 different proteins, and mutations in the corresponding genes can result in coQ10 deficiency, which leads to an impaired electron transport in the respiratory chain, causing mitochondrial disorders. Within the OXPHOS coQ10 is reduced by complex I and complex II, and oxidized by complex III. The quinone ring structure is the active part of coQ10 that can carry electrons between respiratory chain enzymes. A deficiency of this molecule results in an impaired respiratory chain resulting in a mitochondrial disorder. Several different analytical tests have been developed to measure coQ10 levels, including LC-electrochemical detection and LC-mass spectrometry [22, 23]. Since coQ10 deficiencies are quite rare, measurements of coQ10 levels in each diagnostic sample is not very efficient. An option could be to instead measure succinate:cytochrome c oxidoreductase

(SCC; complex II-coQ10-complex III) as part of an OXPHOS enzyme panel, in which the SCC-assay has a dual function, namely as an extra test in addition to the single enzyme tests to confirm complex II (and III) deficiencies. In coQ10 deficient cases, the SCC activity will be reduced, while isolated complex II activity will be normal. In such cases, addition of a coQ10 analogue (e.g., decylubiquinone) will result in normalization of the SCC enzyme activity [24]. Nowadays, genetic testing of patients often precedes functional/biochemical diagnostic testing and in many cases coQ10 measurements are performed to complement genetic findings, in particular in the case of the finding of a genetic variant of unknown clinical significance (VUS) in genes of the coQ10 biosynthetic pathway. For this purpose, measurements of coQ10 in blood would be an ideal solution to confirm a diagnosis without invasive biopsies. Unfortunately, results of blood testing for coQ10 do not always provide diagnostically accurate data, neither in plasma/serum nor in blood cells. Instead, measurements of coQ10 in cultured skin fibroblasts provide a good alternative as it is minimally invasive and in our experience is much more informative than blood tests, as cultured fibroblasts fully depend on the coQ10 biosynthetic pathway.

13.5 Assays Investigating the Overall Mitochondrial Energy Metabolism

In addition to investigations of the enzyme activity of individual OXPHOS enzymes, the activity of the intact OXPHOS in the context of the complete mitochondrial ATP-generating system can be investigated. This includes mitochondrial substrate transporters, the TCA cycle, and the OXPHOS system. In addition, there are many other proteins that directly or indirectly influence the rate of respiration, such as the mitochondrial translation (ribosomes, tRNAs, translation factors), the integrity of the mitochondrial inner membrane, and many more. In our experience, mitochondrial respiration tests are in general less sensitive and less accurate compared to single OXPHOS enzyme tests. On the other hand, information on the health status of the entire mitochondrial energy metabolism system can only be obtained by investigating parameters that depend on the intact mitochondrial energy metabolism [25]. Below, three different types of readout are discussed: respiration/oxygen consumption, ATP production, and substrate conversion. The reliability of these three methods of energy metabolism investigations in intact mitochondria requires that special care has to be taken to obtain intact mitochondria from test samples. The samples (tissue biopsies or cultured cells) may not be frozen prior to the measurements, as freezing can disrupt the integrity of the mitochondrial inner membrane, which can affect the mitochondrial membrane potential. This may interfere with mitochondrial membrane potential maintenance that is one of the driving factors in mitochondrial ATP production, respiration, and substrate oxidation. Mitochondrial integrity can be tested by investigating the stimulation of these parameters by the addition of ADP to the reaction mixture, as this should lead to an immediate response by these three types of parameters. In most of the mentioned examples, some method of sample pre-preparation is necessary. For tissue samples this can be a mitochondrial purification by homogenization of tissue material followed by differential centrifugation. In the case of testing cultured cells, the addition of digitonin results in a partial permeabilization when used diligently at low concentrations in order to prevent interference with the mitochondrial membrane potential [26, 27]. Importantly, it is necessary to objectively normalize the results, as the (partial) purifications are a potential cause for poor reproducibility of test results. This can be overcome by normalization of the data for mitochondrial content of the samples. Often, citrate synthase is measured for this purpose. Citrate synthase is a mitochondrial enzyme and is one of the enzymes of the TCA cycle. Deficiencies of citrate synthase have not been reported in literature to date, and citrate synthase activities are known to be relatively stable in cells with an OXPHOS deficiency. Mitochondrial citrate synthase is measurable by a simple spectrophotometric assay [28].

13.5.1 Respirometry/Oxygen Consumption

As a measure for the integral OXPHOS activity as central unit in mitochondrial energy metabolism, mitochondrial respiration can be measured. The oxygen consumption rate is correlated with the mitochondrial ATP production rates and with mitochondrial substrate oxidation rates, and although oxygen is exclusively consumed by complex IV, mitochondrial respiration is fully dependent on all components of the OXPHOS. The classical method to detect oxygen in solution makes use of the Clark electrode [29]. Nowadays, the platforms that are being used most widely to investigate mitochondrial respiration are the Agilent Seahorse XF analyzer and the Oroboros NextGen O2k respirometer. Each platform has its own benefits and drawbacks.

In our experience, one of the benefits of the Seahorse platform is the 96-wells plate format in which multiple samples can be examined simultaneously. Another advantage is that all consumables (plates, assay mixtures) are commercially available. A drawback of the Seahorse platform is the suboptimal reproducibility of the test results, which can be managed by performing multiple replicate measurements (4–6 wells per sample), and by combining control samples with test samples on the same plate. In addition to measuring respiration, the Seahorse platform can also measure the acidification of reaction mixtures caused by lactate production, which makes it possible to investigate shifts from oxidative energy metabolism in which the OXPHOS system plays a central role, to anaerobic energy metabolism in which cytosolic pyruvate is converted to lactate, instead of being imported by mitochondria and oxidized leading to intramitochondrial ATP generation.

The O2k respirometer has only two chambers that can be used simultaneously and therefore the throughput is relatively low. A major benefit of this platform is that it is very flexible with respect to assay conditions, and that it allows for the addition of substances while measuring, e.g., to add substrates, inhibitors, or drug, and to do titrations in a single experiment. Another advantage of this platform is that when oxygen levels in the reaction mixture become too low to do meaningful measurements, the chambers can be opened shortly to allow ambient oxygen to replenished oxygen levels in the reaction mixture, making it possible to continue measurements.

13.5.2 Substrate Oxidations

The measurement of the consumption of mitochondrial substrates like pyruvate and malate is a relatively sensitive approach to investigate energy metabolism of intact mitochondria. For this purpose, we have implemented assays that make use of ^{14}C-radiolabeled substrates of the TCA-cycle. During several steps of the TCA-cycle, carbons atoms coming from these substrates are released in the form of CO_2. The incubations are performed at exact time-points in order to be able to calculate substrate consumption rates. Reactions are stopped by the addition of perchloric acid, resulting in the release of the 14C-radiolabeled CO_2 produced by the mitochondria during the incubation with mitochondrial substrates. The released $14CO_2$ is trapped in hyamine after which the amount of $14CO_2$ is quantified using a liquid scintillation counter. By selectively using substrates with the radiolabels at specific positions within the substrate molecules the release of CO_2 can be used as a measure of the intact mitochondrial energy coverting activity. Additional diagnostic information can be obtained by the clever use of (non-labeled) co-substrates, in order to create assays condition in which there is an excess of substrates and co-substrates and in which the limiting factor is the mitochondrial ATP-generating system [30].

As an alternative to test substrate oxidations by radiochemical methods, substrates can be labeled by stable isotopes such as ^{13}C, and detection is done by means of mass spectrometry (LC-MS or GC-MS) [31]. This approach has been used widely in cancer research, but only scarcely in mitochondrial disease diagnostics.

13.5.3 ATP Production

As the end-point of mitochondrial energy metabolism, the production of ATP can be measured. In order to be able to detect defects in the energy metabolism pathway, measuring the rate of ATP production is far more informative than measuring steady state ATP levels, as the latter is only partly depending on the mitochondrial ATP production pathway, which in part is due to the fact that steady state ATP levels are in part depending on the ATP consumption rates, which may obscure defects in ATP synthesis. There are different methods to measure ATP: including luminometric measurements of ATP-dependent luminescence using luciferase [32], or by spectrophotometry using an ATP-dependent enzymatic activity [33].

13.5.3.1 Molecular Genetics

The implementation of whole exome sequencing as a molecular genetic diagnostic test has revolutionized the diagnostics of mitochondrial disorders in the past 15 years. Initially, both the costs and the data quality of exome sequencing were not optimal, but nowadays the costs have declined tremendously while the data quality has improved up to a level similar to what is achieved with targeted gene sequence analysis and has even led to the replacement of single genes tests altogether in our center. To date the number of genes in which variants have been associated with mitochondrial disorders is around 400 (depending on the definition of mitochondrial disorders), the genotype-phenotype correlation is variable, and there is considerable clinical overlap with other (non-mitochondrial) disorders. These factors can make it difficult to predict a certain gene defect on the basis of clinical signs and symptoms of individual patients. Hence, exome sequencing is very suitable to investigate all potential genes of interests in suspected mitochondrial disease patients. The introduction of exome sequencing in the genetic screening of suspected mitochondrial disease patients has resulted in a large number of "novel" gene defects and broadening of the clinical spectrum associated with mitochondrial disease genes. The implementation of exome (and genome) sequencing as a diagnostic test for mitochondrial disorders has also created new challenges, in particular in the interpretation of test results. These include the identification of genetic variants of unknown clinical significance, multiple pathogenic variants in a single patient, pathogenic variants with limited penetrance. Partly, these can be overcome by performing a trio exome sequencing test in which the parents are tested together with the proband. In the case of variants of unknown clinical significance, a targeted functional/biochemical tests can be performed in order to obtain specific diagnostic information about the pathogenicity of the VUS in the gene of interest [1, 34]. In addition, it is expected that artificial intelligence is going to play a role in the interpretation of genetic data and in combining genetic data with clinical signs and symptoms [35, 36]. Measurements of OXPHOS enzyme activities can provide crucial data to interpret such a VUS, especially those in genes that encode subunits of OXPHOS enzymes, but also of genes encoding proteins involved in the biosynthesis of OXPHOS enzymes, and many other proteins that can affect OXPHOS activity.

In addition to a defect in the nuclear genome, a significant number of patients have a primary defect in the mtDNA. The mtDNA contains a total of 13 genes that encode subunits of the OXPHOS. In addition, it contains genes encoding mitochondrial ribosomal RNAs and the genes encoding all of the mitochondrial tRNAs. In most of these genes, pathogenic variants have been identified in mitochondrial disease patients. As mutations in the mtDNA often are heteroplasmic, meaning that the variant is present in only a subset of the mtDNA copies in the cell, and the fact that heteroplasmy usually is tissue-specific complicates the identification of pathogenic mtDNA variants. Blood samples often have heteroplasmy levels that can be much lower than in clinically affected tissues like muscle or brain. For this reason, validating genetic findings may include a combination of segregation analysis in families, including the testing of multiple samples like blood, saliva, and urine sediment. In addition, functional tests can provide highly informative data. The creation of transmitochondrial cybrids with different heteroplasmy levels makes it possible to obtain information about genotype–phenotype correlation (i.e., heteroplasmy levels combined with OXPHOS enzyme activities in vitro).

It is important to note that the holistic genetic tests that are currently available, such as exome and genome sequencing, will not provide a diagnosis in all patients. The diagnostic yield of exome sequencing of mitochondrial disease patients is around 50% [25]. There are a number of possible additional tests to increase this rate, such as RNA sequencing [37], but still a significant percentage of patients will not be diagnosed by exome/genome sequencing. Unfortunately, a negative exome/genome result does not exclude the possibility of a mitochondrial disease, and in those cases a biochemical screening of the OXPHOS system by the methods described in this overview will be important to confirm or reject a mitochondrial disease diagnosis.

13.6 Conclusion

While OXPHOS enzyme activity tests have been used as diagnostic screening tools to investigate muscle and skin biopsies from patients suspected to have a mitochondrial disorder, the recent developments in genetic testing that make it possible to investigate nearly all candidate genes in a single test, such as exome sequencing, have led to a shift in the use of OXPHOS enzyme test as a tool to obtain functional/biochemical information to validate genetic findings. However, the methodology of these enzyme tests remains the same. Importantly, as the diagnostic yield of exome sequencing is around 40–60%, a relatively large subgroup of patients in whom a genetic diagnosis has not been found remains dependent on diagnostic enzyme measurements in order to obtain a biochemical diagnosis. The enzyme tests to examine individual OXPHOS enzymes are all spectrophotometric enzyme activities that are similar with respect to the overall assay design (though very different with respect to substrates, inhibitors, and detection wavelength). The assays can be executed both as manual assays and as an OXPHOS enzyme panel using automated spectrophotometric platforms. There is little doubt that the diagnostic yield of exome/genome sequencing will continue to increase due to various additional holistic generically applicable methods, such as RNA sequencing, protein expression profiling, and sophisticated data evaluation with the help of artificial intelligence methods. At this moment, we know that RNA sequencing can result in a modest increase of the diagnostic yield. Until new developments result in drastic improvements of the diagnostic yield of exome/genome sequencing, there is a need to have diagnostic biochemical OXPHOS tests available for both screening purposes and as methods to validate genetic findings in order to be able to reliably diagnose mitochondrial disease patients.

Acknowledgments On the basis of the principles of the OXPHOS enzyme assays as described by others before, Liesbeth Wintjes and Berendien Stoltenberg, both technician in our lab, have optimized the reaction conditions in order to increase the robustness of the assays, to minimize the use of inhibitors in these protocols, and have implemented the assays on automated spectrophotometric platforms in our laboratory. In addition to acknowledging their efforts, we are grateful for the technical support as well as stimulating discussions with the other technicians in the mitochondrial diagnostics team.

References

1. Rodenburg RJ (2018) The functional genomics laboratory: functional validation of genetic variants. J Inherit Metab Dis 41(3):297–307
2. Rodenburg RJ (2011) Biochemical diagnosis of mitochondrial disorders. J Inherit Metab Dis 34(2):283–292
3. Willis JH, Capaldi RA, Huigsloot M, Rodenburg RJ, Smeitink J, Marusich MF (2009) Isolated deficiencies of OXPHOS complexes I and IV are identified accurately and quickly by simple enzyme activity immunocapture assays. Biochim Biophys Acta 1787(5):533–538
4. Rodenburg RJ, Schoonderwoerd GC, Tiranti V, Taylor RW, Rotig A, Valente L et al (2013) A multicenter comparison of diagnostic methods for the biochemical evaluation of suspected mitochondrial disorders. Mitochondrion 13(1):36–43
5. Rodenburg RJ (2016) Mitochondrial complex I-linked disease. Biochim Biophys Acta 1857(7):938–945
6. Guerrero-Castillo S, Baertling F, Kownatzki D, Wessels HJ, Arnold S, Brandt U et al (2017) The assembly pathway of mitochondrial respiratory chain complex I. Cell Metab 25(1):128–139
7. Fischer JC, Ruitenbeek W, Trijbels JM, Veerkamp JH, Stadhouders AM, Sengers RC et al (1986) Estimation of NADH oxidation in human skeletal muscle mitochondria. Clin Chim Acta 155(3):263–273

8. Janssen AJ, Trijbels FJ, Sengers RC, Smeitink JA, van den Heuvel LP, Wintjes LT et al (2007) Spectrophotometric assay for complex I of the respiratory chain in tissue samples and cultured fibroblasts. Clin Chem 53(4):729–734
9. Janssen RJ, Nijtmans LG, van den Heuvel LP, Smeitink JA (2006) Mitochondrial complex I: structure, function and pathology. J Inherit Metab Dis 29(4):499–515
10. Fullerton M, McFarland R, Taylor RW, Alston CL (2020) The genetic basis of isolated mitochondrial complex II deficiency. Mol Genet Metab 131(1–2):53–65
11. Ben Aim L, Pigny P, Castro-Vega LJ, Buffet A, Amar L, Bertherat J et al (2019) Targeted next-generation sequencing detects rare genetic events in pheochromocytoma and paraganglioma. J Med Genet 56(8):513–520
12. MacFarlane J, Seong KC, Bisambar C, Madhu B, Allinson K, Marker A et al (2020) A review of the tumour spectrum of germline succinate dehydrogenase gene mutations: beyond phaeochromocytoma and paraganglioma. Clin Endocrinol (Oxf) 93(5):528–538
13. Renkema GH, Wortmann SB, Smeets RJ, Venselaar H, Antoine M, Visser G et al (2015) SDHA mutations causing a multisystem mitochondrial disease: novel mutations and genetic overlap with hereditary tumors. Eur J Hum Genet 23(2):202–209
14. Alston CL, Ceccatelli Berti C, Blakely EL, Olahova M, He L, McMahon CJ et al (2015) A recessive homozygous p.Asp92Gly SDHD mutation causes prenatal cardiomyopathy and a severe mitochondrial complex II deficiency. Hum Genet 134(8):869–879
15. Bentlage H, de Coo R, ter Laak H, Sengers R, Trijbels F, Ruitenbeek W et al (1995) Human diseases with defects in oxidative phosphorylation. 1. Decreased amounts of assembled oxidative phosphorylation complexes in mitochondrial encephalomyopathies. Eur J Biochem 227(3):909–915
16. Zheng XX, Shoffner JM, Voljavec AS, Wallace DC (1990) Evaluation of procedures for assaying oxidative phosphorylation enzyme activities in mitochondrial myopathy muscle biopsies. Biochim Biophys Acta 1019(1):1–10
17. Geier BM, Schagger H, Brandt U, Colson AM, Von Jagow G (1992) Point mutation in cytochrome b of yeast ubihydroquinone:cytochrome-c oxidoreductase causing myxothiazol resistance and facilitated dissociation of the iron-sulfur subunit. Eur J Biochem 208(2):375–380
18. Legros F, Chatzoglou E, Frachon P, Ogier De Baulny H, Laforet P, Jardel C et al (2001) Functional characterization of novel mutations in the human cytochrome b gene. Eur J Hum Genet 9(7):510–518
19. Cooperstein SJ, Lazarow A (1951) A microspectrophotometric method for the determination of cytochrome oxidase. J Biol Chem 189(2):665–670
20. Ganetzky RD, Stendel C, McCormick EM, Zolkipli-Cunningham Z, Goldstein AC, Klopstock T et al (2019) MT-ATP6 mitochondrial disease variants: phenotypic and biochemical features analysis in 218 published cases and cohort of 14 new cases. Hum Mutat 40(5):499–515
21. Stendel C, Neuhofer C, Floride E, Yuqing S, Ganetzky RD, Park J et al (2020) Delineating MT-ATP6-associated disease: from isolated neuropathy to early onset neurodegeneration. Neurol Genet 6(1):e393
22. Tang PH, Miles MV, DeGrauw A, Hershey A, Pesce A (2001) HPLC analysis of reduced and oxidized coenzyme Q(10) in human plasma. Clin Chem 47(2):256–265
23. Tang Z, Li S, Guan X, Schmitt-Kopplin P, Lin S, Cai Z (2014) Rapid assessment of the coenzyme Q10 redox state using ultrahigh performance liquid chromatography tandem mass spectrometry. Analyst 139(21):5600–5604
24. Jakobs BS, van den Heuvel LP, Smeets RJ, de Vries MC, Hien S, Schaible T et al (2013) A novel mutation in COQ2 leading to fatal infantile multisystem disease. J Neurol Sci 326(1–2):24–28
25. Wortmann SB, Oud MM, Alders M, Coene KLM, van der Crabben SN, Feichtinger RG et al (2022) How to proceed after "negative" exome: a review on genetic diagnostics, limitations, challenges, and emerging new multiomics techniques. J Inherit Metab Dis 45(4):663–681
26. Wanders RJ, Ruiter JP, Wijburg FA (1993) Studies on mitochondrial oxidative phosphorylation in permeabilized human skin fibroblasts: application to mitochondrial encephalomyopathies. Biochim Biophys Acta 1181(3):219–222
27. Vercesi AE, Bernardes CF, Hoffmann ME, Gadelha FR, Docampo R (1991) Digitonin permeabilization does not affect mitochondrial function and allows the determination of the mitochondrial membrane potential of Trypanosoma cruzi in situ. J Biol Chem 266(22):14431–14434
28. Srere PA (1969) Citrate synthase, EC 4.1.3.7, citrate oxaloacetate lyase (CoA-acetylating). Methods Enzymol 13:3–11
29. Clark LC Jr, Benington F, Morin RD (1964) The enzymatic oxidative deamination and effect on cat behavior mescaline and structurally-related beta-phenethylamines. Ala J Med Sci 1:417–429
30. Janssen AJ, Trijbels FJ, Sengers RC, Wintjes LT, Ruitenbeek W, Smeitink JA et al (2006) Measurement of the energy-generating capacity of human muscle mitochondria: diagnostic procedure and application to human pathology. Clin Chem 52(5):860–871
31. Gravel SP, Andrzejewski S, Avizonis D, St-Pierre J (2014) Stable isotope tracer analysis in isolated mitochondria from mammalian systems. Metabolites 4(2):166–183
32. Morciano G, Sarti AC, Marchi S, Missiroli S, Falzoni S, Raffaghello L et al (2017) Use of luciferase probes to measure ATP in living cells and animals. Nat Protoc 12(8):1542–1562
33. Lamprecht W, Stein P, Heinz F, Weisser H (1974) Phosphocreatine: determination with creatine kinase,

hexokinase, and glucose-6-phosphate dehydrogenase. In: Bergmeyer HU (ed) Methods of enzymatic analysis, 2nd edn. Academic Press, New York, pp 1777–1781

34. Ferreira EA, Veenvliet ARJ, Engelke UFH, Kluijtmans LAJ, Huigen M, Hoegen B et al (2023) Diagnosing, discarding, or de-VUSsing: a practical guide to (un) targeted metabolomics as variant-transcending functional tests. Genet Med 25(1):125–134
35. Aradhya S, Facio FM, Metz H, Manders T, Colavin A, Kobayashi Y et al (2023) Applications of artificial intelligence in clinical laboratory genomics. Am J Med Genet C Semin Med Genet 193(3):e32057
36. Dingemans AJM, Hinne M, Truijen KMG, Goltstein L, van Reeuwijk J, de Leeuw N et al (2023) PhenoScore quantifies phenotypic variation for rare genetic diseases by combining facial analysis with other clinical features using a machine-learning framework. Nat Genet 55(9):1598–1607
37. Kremer LS, Bader DM, Mertes C, Kopajtich R, Pichler G, Iuso A et al (2017) Genetic diagnosis of Mendelian disorders via RNA sequencing. Nat Commun 8:15824

Bile Acids

14

Sacha Ferdinandusse and Frédéric M. Vaz

Contents

List of Abbreviations

3β-HSOR	3β-Hydroxy-Δ^5-C_{27}-steroid oxidoreductase
5β-reductase	Δ^4-3-Oxosteroid 5β-reductase
ACOX2	Acyl-CoA oxidase 2
AMACR	α-Methylacyl-CoA racemase
BAAT	Bile acid CoA:amino acid N-acyltransferase
BACS	Bile acid-CoA ligase
CA	Cholic acid = 3α,7α,12α-trihydroxy-5β-cholanoic acid
CDCA	Chenodeoxycholic acid = 3α, 7α-dihydroxy-5β-cholanoic acid
CTX	Cerebrotendinous xanthomatosis
DBP	D-bifunctional protein
DCA	Deoxycholic acid = 3α,12α-dihydroxy-5β-cholanoic acid
DHCA	Dihydroxycholestanoic acid = 3α,7α-dihydroxy-5β-cholestanoic acid
EDTA	Ethylenediaminetetraacetic acid
ESI	Electrospray ionization
FIA	Flow injection analysis
HRMS	High resolution mass spectrometry
LC	Liquid chromatography
LCA	Lithocholic acid
m/z	Mass over charge
MRM	Multi-reaction monitoring
MS	Mass spectrometry
Q-TOF	Quadrupole time-of-flight
SCPx	Sterol carrier protein X
THCA	Trihydroxycholestanoic acid = 3α,7α,12α-trihydroxy-5β-cholestanoic acid
UDCA	Ursodeoxycholic acid = 3α, 7β-dihydroxycholanoic acid
UPLC	Ultra performance liquid chromatography
γ-GT	γ-Glutamyltranspeptidase

S. Ferdinandusse · F. M. Vaz (✉)
Laboratory Genetic Metabolic Diseases, Department of Clinical Chemistry and Pediatrics, Emma Children's Hospital, Amsterdam UMC Location University of Amsterdam,
Amsterdam, The Netherlands
e-mail: s.ferdinandusse@amsterdamumc.nl; f.m.vaz@amsterdamumc.nl

N. Blau, F. M. Vaz (eds.), *Laboratory Guide to the Methods in Biochemical Genetics*,
https://doi.org/10.1007/978-3-031-58819-8_14

14.1 Introduction

Bile acids are acidic sterols; they contain a steroidal ring structure and a side-chain with a terminal carboxylic acid function. For their general structure, see Fig. 14.1. They are synthesized in the liver and are the major end products of cholesterol metabolism. Because of their amphipathic structure, bile acids are detergents which is crucial for their role in facilitating excretion of bile lipids and cholesterol, and enabling the intestinal absorption of dietary lipids and lipid soluble compounds including vitamins. Bile acids are important for cholesterol homeostasis, are needed to induce bile flow and excretion of toxic substances, but they are also ligands to nuclear receptors with different regulatory functions. The two primary bile acids synthesized from cholesterol in man are cholic acid (CA) and chenodeoxycholic acid (CDCA). After synthesis they are converted into bile salts (glycine or taurine conjugates), which are excreted together with other bile components into the canaliculi and the resulting bile is transported to the gallbladder. After a meal, the bile is secreted into the small intestine. During intestinal passage, gut bacteria modify part of the bile salts resulting in secondary bile acids (deoxycholic acid (DCA), lithocholic acid (LCA), ursodeoxycholic acid (UDCA), hyodeoxycholic acid, and hyocholic acid). Ninety five percent of the bile acids/salts are recycled in the enterohepatic circulation and only 5% of the pool is lost in stool. This loss is compensated by hepatic biliary acid synthesis, thereby maintaining bile acid homeostasis and the bile acid pool [1]. Bile acid biosynthesis defects often result in a deficiency of the primary bile acids and the accumulation of potentially toxic bile acid intermediates. Patients can have cholestasis, jaundice, liver disease, fat malabsorption, fat-soluble vitamin deficiency, but also neurological disease. As most bile acid synthesis defects have a characteristic bile acid profile, bile acid analysis by tandem mass spectrometry in plasma and urine should be performed in patients with the listed symptoms above.

In the previous version of this book, a broad overview of different analytical methods to analyze bile acids and bile acid conjugates was presented [2]. As many of these methods are not used anymore in current metabolic laboratories, we defer the reader to this earlier version for these methods (and the following review and references therein [3]), and, in this chapter, concentrate on the bile acid analysis using electrospray ionization (ESI) mass spectrometry (MS).

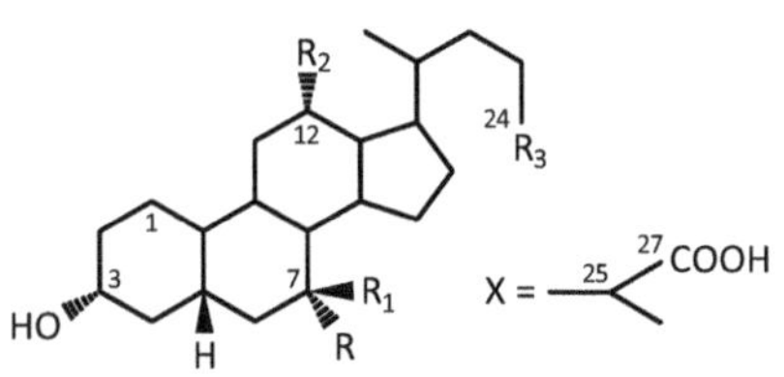

	R	R_1	R_2	R_3
CA	OH	H	OH	COOH
CDCA	OH	H	H	COOH
UDCA	H	OH	H	COOH
DCA	H	H	OH	COOH
LCA	H	H	H	COOH
DHCA	OH	H	H	X
THCA	OH	H	OH	X

Fig. 14.1 General structure of C_{24} and C_{27} bile acids. Abbreviations: cholic acid (CA), chenodeoxycholic acid (CDCA), ursodeoxycholic acid (UDCA), deoxycholic acid (DCA), lithocholic acid (LCA), dihydroxycholestanoic acid (DHCA), and trihydroxycholestanoic acid (THCA). The carboxylic acid function at C24 or C27 can remain free or make an amide bond with glycine or taurine

14.1.1 Clinical Significance

Bile acid biosynthesis from cholesterol constitutes three main processes: (1) modifications to the steroid nucleus, (2) modifications to the aliphatic side-chain, and (3) amidation. Different bile acid biosynthetic pathways have been delineated over the years, of which the classic or neutral pathway and the alternative or acidic pathway are the most commonly described [4, 5]. These pathways share the majority of the enzymes but the order of steps differ (Fig. 14.2). In the acidic/alternative pathway, first the acid function is introduced, followed by the nuclear modification, whereas in the classic/neutral pathway the order is reversed (Fig. 14.2). Both routes produce the C_{27}-bile acids, dihydroxycholestanoic acid (DHCA) and trihydroxycholestanoic acid

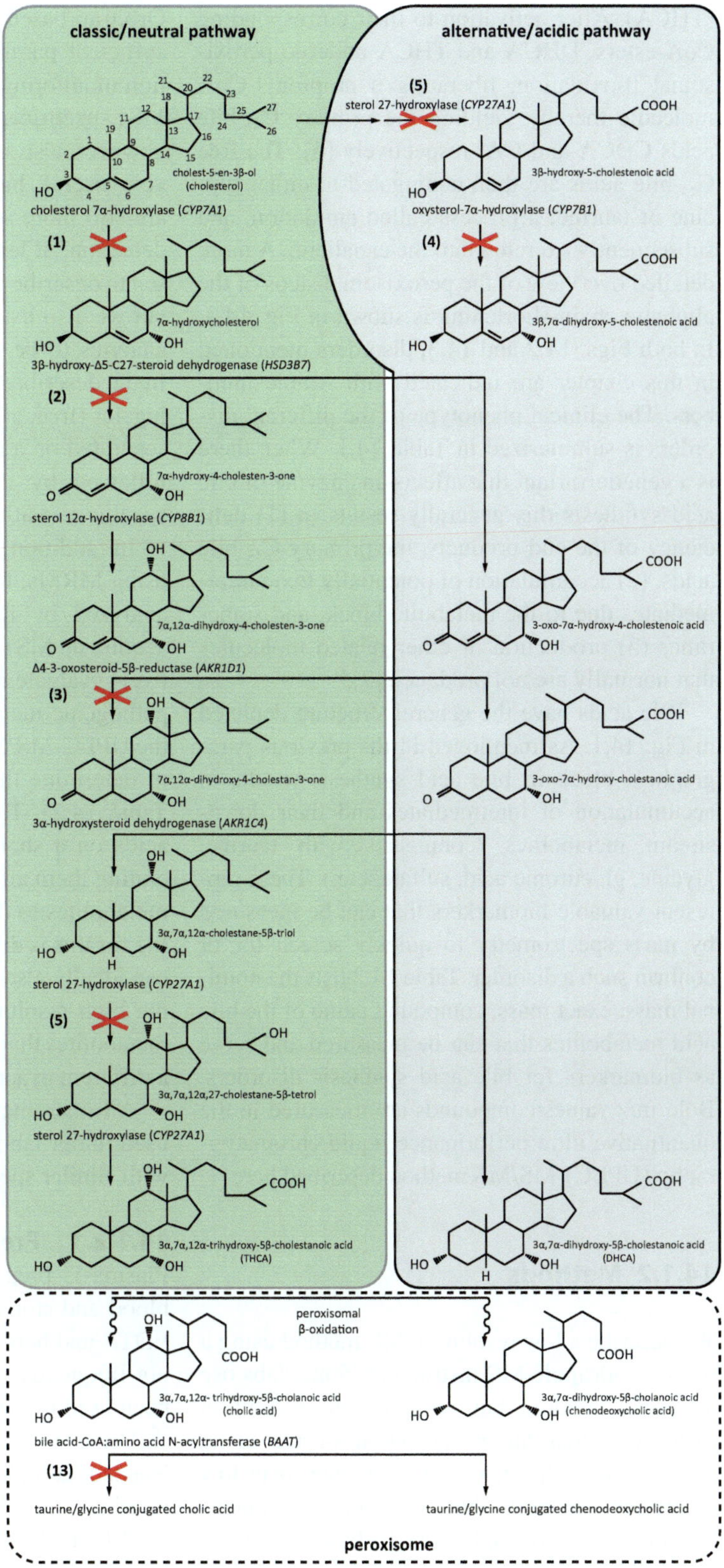

Fig. 14.2 Schematic representation of the intermediates, enzymes and corresponding genes of the classic/neutral pathway and the acid/alternative pathway of bile acid synthesis. Inborn errors mentioned in this chapter are indicated by red crosses and disease number between parenthesis. The disease numbers correspond to the disease numbers in the Tables. In the neutral pathway, modification of the sterol ring structure precedes the conversion to an acid, whereas in the acidic pathway an acid (= carboxylic group) first is formed followed by steroid nucleus modifications. Peroxisomal shortening of the aliphatic side-chain of dihydroxy- and trihydroxycholestanoic acid (DHCA and THCA) produces the primary bile acids chenodeoxycholic and cholic acid, respectively. These bile acids can be amidated to form taurine or glycine conjugates, i.e. the bile salts. Reproduced with permission from [8]

(THCA). After activation to their corresponding CoA-esters, DHCA and THCA undergo peroxisomal β-oxidation, liberating a propionyl-CoA molecule thereby yielding the primary C_{24}-bile acids CDCA and CA, respectively [6]. The free C_{24}-bile acids are then conjugated to either glycine or taurine, a process called amidation, and subsequently excreted into the canaliculi. A more detailed overview of the peroxisomal steps of the aliphatic chain-shortening is shown in Fig. 14.3. In both Figs. 14.2 and 14.3, disorders mentioned in this chapter are indicated with Arabic numbers. The clinical phenotype of the different disorders is summarized in Table 14.1. When there is a genetic variant that affects an enzyme of bile acid synthesis this generally results in (1) deficiency of the end products, the primary C_{24}-bile acids, (2) accumulation of potentially toxic intermediates due to the metabolic block, and sometimes (3) production of other related molecules that normally are not produced [7].

Bile acids have the general structure depicted in Fig. 14.1. As mentioned in the previous paragraph, a block in bile acid synthesis results in accumulation of intermediates and their downstream metabolites (conjugates with taurine, glycine, glucuronic acid, sulfate, etc.). These represent valuable biomarkers that can be measured by mass spectrometry to quickly screen for or confirm such a disorder. Table 14.2 lists the nominal mass, exact mass, compound name of the bile acid metabolites that can be measured and serve as biomarkers for bile acid synthesis disorders. Bold m/z values/compounds are measured in the quantitative ultra performance liquid chromatography (UPLC)-MS/MS method described here.

14.1.2 Methods

We describe a low resolution MS method using a triple-quadrupole MS instrument. Some labs use a similar method but with a high resolution instrument that allows detection of the exact mass of ions. This can have advantages over low resolution methods but is very instrument-specific. Time-of-flight (TOF)-based systems generally have higher limits of detection when compared to triple-quad instruments and Orbitrap-based systems generally cannot acquire sufficient parallel transitions, i.e. multiple reaction monitoring (MRM). The measurement of an MS1 spectrum on these high resolution machines, however, also is an option and these profiles are very useful, but the triple-quad based methods are still more selective and have lower limits of detection, at least for now. Hence, we have chosen to describe a triple quadrupole-based method, but we also list the exact masses of the different analytes to be comprehensive and future-proof. In the described method, bile acids and their conjugates (free, glycine, and taurine conjugates) are separated on a reversed-phase analytical column followed by the measurement of metabolite-specific transitions, MRMs, on an ESI-(HR)MS.

In addition to the UPLC-MS/MS analysis using MRMs, for each sample the extract is also analyzed by flow injection analysis (FIA) to acquire an MS1 spectrum. This is used to qualitatively evaluate the bile acid profile and search for pathogenic metabolites that are not measured in the UPLC-MS/MS analysis but are needed/useful to determine the disorder in question (also see Table 14.2). This is done by trapping the bile acids on a short column (a guard column) and eluting them after the diversion of salts and polar metabolites to the waste. A total ion current profile for this is displayed in Fig. 14.4. This method can ideally also be performed on a HRMS where the high resolution will allow better distinction metabolites that have a similar unit mass but have a different exact mass.

All instrumentation that is mentioned here is used in our laboratory, but alternative equipment with similar specifications can be used.

14.1.2.1 Pre-Analytical

Plasma is prepared by centrifugation of whole blood and stored at −20 °C until analysis. Both EDTA and heparin plasma can be used. Serum is not frequently used but is also possible and has been used in the past. Needed: 50 μl.

Urine is stored at −20 °C until analysis. Needed: 200 μl. Only qualitative analysis is done in urine.

Bile acid metabolites are generally very stable, also at room temperature, but storage of samples at −20 °C until analysis is advised.

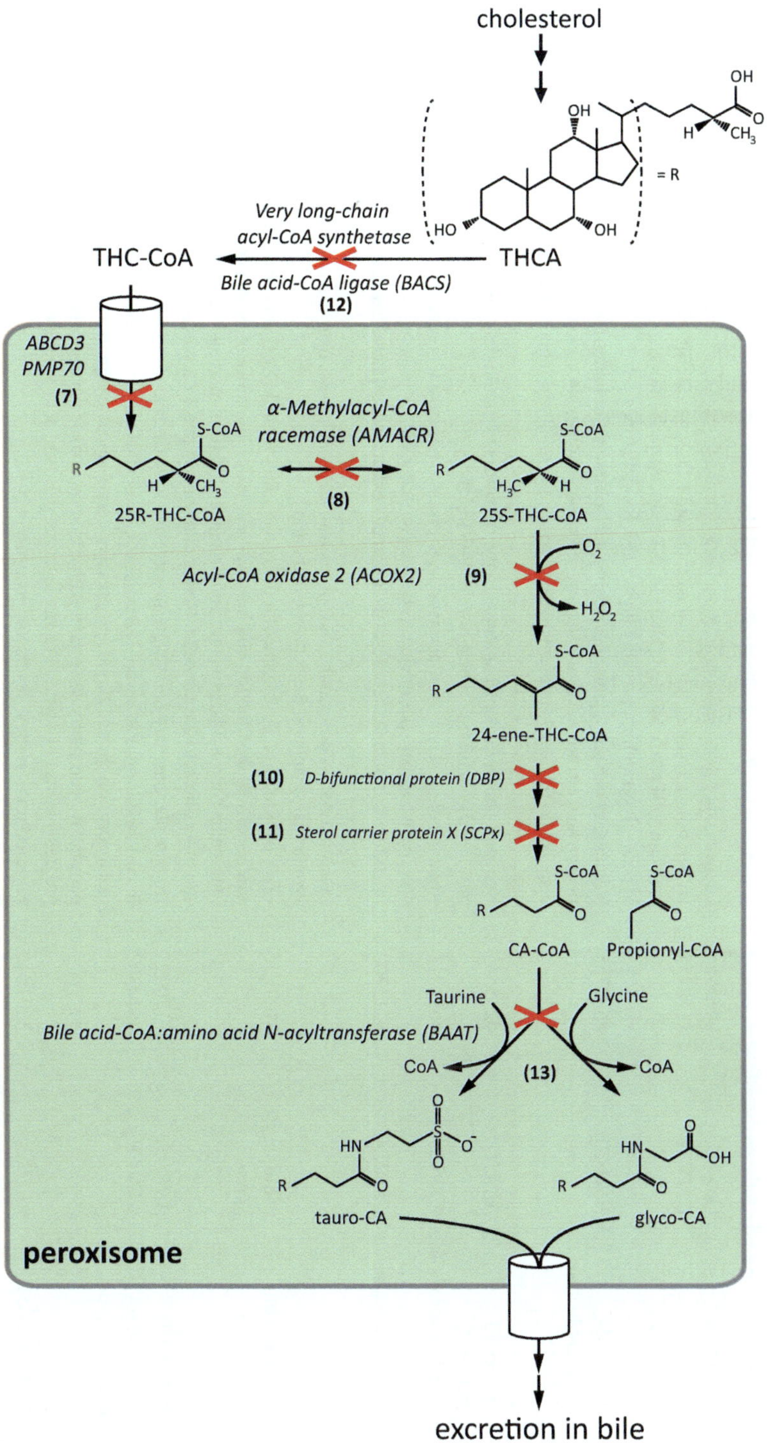

Fig. 14.3 Schematic representation of the enzymes and bile acid intermediates of the peroxisomal steps in bile acid biosynthesis. Known inborn errors mentioned in this chapter are indicated by red crosses and disease number between parenthesis. The disease numbers correspond to the disease numbers in the Tables. In short, the C_{27}-bile acid intermediate trihydroxycholestanoic acid (THCA) is formed via several enzymatic steps from cholesterol and then activated to CoA-ester at the endoplasmic reticulum. Subsequently, THC-CoA is transported into the peroxisome via the transporter ABCD3 after which the side-chain is cleaved via peroxisomal beta-oxidation generating choloyl-CoA (CA-CoA) followed by amidation via bile acid CoA:amino acid N-acyltransferase (BAAT). Reproduced with permission from [8]

14.1.2.2 Reagents and Chemicals

All reagents are minimally of analytical grade and preferably ultrapure LC-MS grade, especially for the HRMS method:

- Acetonitrile
- Ethanol
- Methanol
- Ammoniumformiate

Table 14.1 Clinical aspects of bile acid synthesis disorders[a]

#	Enzyme/protein	Gene	OMIM	Clinical/laboratory features	References
1	Cholesterol 7α-hydroxylase	*CYP7A1*	*118455	Hypercholesterolemia with elevated total and LDL-cholesterol which is resistant to Hydroxymethylglutaryl-CoA reductase inhibitors, hypertriglyceridemia, possibly higher risk for gallstones	[9, 10]
2	3β-hydroxy-Δ^5-C_{27}-steroid oxidoreductase	*HSD3B7*	#607765	Neonatal cholestasis, jaundice, itching, fat-soluble vitamin deficiency, steatorrhea, cirrhosis, seizures/tetany, hepatosplenomegaly, rickets, favorable response to bile acid supplementation, hypocalcemia, prolonged coagulation times, normal γ-GT activity	[10–19]
3	Δ^4-3-oxosteroid 5β-reductase	*AKR1D1*	#235555	Neonatal jaundice, hepatosplenomegaly, fat-soluble vitamin deficiency, prolonged coagulation times, normal γ-GT activity	[10, 12, 20–24]
4	Oxysterol 7α-hydroxylase	*CYP7B1*	#613812	Neonatal cholestasis, fatal neonatal liver disease, hepatosplenomegaly, fibrosis, progressive spastic paraplegia, cataract. Normal γ-GT	[10, 25–30]
5	Sterol 27-hydroxylase	*CYP27A1*	#213700	In childhood: neonatal cholestasis, intractable diarrhea, bilateral cataract, developmental delay In adulthood: tendon xanthomas, premature atherosclerosis, neuropsychiatric symptoms, peripheral neuropathy, dementia	[10, 31–34]
6	Zellweger spectrum disorders	Multiple *PEX* genes	Multiple	Liver dysfunction, developmental delay and other neurological abnormalities, adrenocortical dysfunction and hearing- and vision impairment. In plasma very long-chain fatty acids, phytanic and pristanic acid (diet- and age-dependent) and pipecolic acid are elevated. Reduced plasmalogens in erythrocytes	[35]
7	ABCD3/PMP70	*ABCD3*	*170995	Hepatosplenomegaly and severe progressive liver disease	[36]
8	α-Methylacyl-CoA racemase	*AMACR*	#614307	In childhood: mild to severe liver disease, cholestasis, and fat-soluble vitamin malabsorption. In adulthood: progressive neurological symptoms including peripheral neuropathy, encephalopathy, tremor, spastic paraparesis, and progressive loss of vision. Plasma levels of pristanic acid are elevated	[37–46]
9	Acyl-CoA oxidase 2	*ACOX2*	*601641	Elevated transaminase levels and liver fibrosis	[47, 48]
10	D-bifunctional protein	*DBP/HSD17B4*	#261515	Liver dysfunction, developmental delay and other neurological abnormalities, and hearing- and vision impairment. Very long-chain fatty acids, phytanic and pristanic acid (latter two diet- and age-dependent) are elevated	[49]
11	Sterol carrier protein X	SCPx/SCP2	#613724	Leukoencephalopathy with dystonia and motor neuropathy. Elevated plasma pristanic acid	[50, 51]
12	Bile acid-CoA ligase	*BACS/SLC27A5*	*603314	Jaundice, with extensive fibrosis and architectural distortion at liver biopsy, failure to thrive, rickets, fat-soluble vitamin deficiency	[52]
13	Bile acid CoA:amino acid N-acyltransferase	*BAAT*	*602938	Fat malabsorption, fat-soluble vitamin deficiency, coagulopathy, rickets, growth retardation, transient neonatal cholestatic hepatitis	[53–55]

[a] Table was adapted from [7] Properties of metabolites

Table 14.2 Overview of (exact) masses of bile acids and intermediates in disorders of bile acid synthesis

Nominal mass	Exact mass	Compound name	Disorder
375	375.2905	3β-hydroxy-5-cholanoic acid (lithocholic acid)	
389	389.2697	Dihydroxy-cholenoic acid	3β-HSOR (#2)
391	391.2885	Dihydroxy-cholanoic acids (e.g., chenodeoxycholic acid)	BAAT (#13), BACS (#12)
405	405.2646	Trihydroxy-cholenoic acids, 3β,7α,12α-trihydroxy-5-cholenoic acid	3β-HSOR (#2)
407	407.2803	Trihydroxy-cholanoic acids (e.g., cholic acid)	BAAT (#13), BACS (#12)
431	431.3169	3β,7α-dihydroxy-cholestenoic acid	3β-HSOR (#2)
432	432.3119	3β-hydroxy-5-cholanoic acid (lithocholic acid) glycine	
433	433.3323	Dihydroxy-cholestanoic acid (e.g., DHCA)	Zellweger (#6), AMACR (#8), DBP (#10)
444	444.2755	7α-hydroxy-3-oxo-4-cholenoic acid glycine	5β-reductase (#3)
446	446.2912	3β,7α-dihydroxy-5-cholenoic acid glycine	3β-HSOR (#2)
447	447.3116	3β,7α,12α-trihydroxy-cholestenoic acid	3β-HSOR (#2)
448	448.3069	Dihydroxy-cholanoic acids (e.g., chenodeoxycholic acid, deoxycholic acid) glycine	
449	449.3273	Trihydroxy-cholestanoic acid (e.g., THCA)	Zellweger (#6), AMACR (#8), DBP (#10)
453	453.2316	3β-hydroxy-5-cholenoic acid sulfate	Oxysterol 7α-hydroxylase (#4)
460	460.2705	7α,12α-dihydroxy-3-oxo-4-cholenoic acid glycine	5β-reductase (#3)
462	462.2861	3β,7α,12α-trihydroxy-5-cholenoic acid glycine	3β-HSOR (#2)
464	464.3018	Trihydroxy-cholanoic acids glycine (e.g., cholic acid)	
465	465.3222	Tetrahydroxy-cholestanoic acid (e.g., OH-THCA, varanic acid)	Zellweger (#6), AMACR (#8), DBP (#10)
465	465.3044	Cholesterol sulfate	
469	469.2265	3β,7α-dihydroxy-5-cholenoic acid sulfate	3β-HSOR (#2)
471	471.2422	Dihydroxy-cholanoic acid sulfate	BAAT (#13), BACS (#12)
472	472.3432	3β-hydroxy-5-cholestanoic acid glycine	Oxysterol 7α-hydroxylase (#4)
480	480.2789	3β-hydroxy-5-cholenoic acid taurine	Oxysterol 7α-hydroxylase (#4)
480	480.2967	Tetrahydroxy-cholanoic acids glycine	
482	482.2946	3β-hydroxy-5-cholanoic acid (lithocholic acid) taurine	
485	485.2215	3β,7α,12α-trihydroxy-5-cholenoic acid sulfate	3β-HSOR (#2)
487	487.2371	Trihydroxy-cholanoic acid sulfate	BAAT (#13), BACS (#12)
488	488.3381	3β,7α-dihydroxy-cholestenoic acid glycine	
490	490.3538	Dihydroxy-cholestanoic acid (e.g., DHCA) glycine	Zellweger (#6), AMACR (#8), DBP (#10)
494	494.2582	7α-hydroxy-3-oxo-4-cholenoic acid taurine	5β-reductase (#3)
498	498.2895	Dihydroxy-cholanoic acids (chenodeoxycholic acid) taurine	
506	506.3487	Trihydroxy-cholestanoic acid (e.g., THCA) glycine	Zellweger (#6), AMACR (#8), DBP (#10)
507	507.3327	C_{29}-dicarboxylic acid	Zellweger (#6)
510	510.2531	7α,12α-dihydroxy-3-oxo-4-cholenoic acid taurine	5β-reductase (#3)
510	510.2531	3β-hydroxy-5-cholenoic acid glycine and sulfate	Oxysterol 7α-hydroxylase (#4)
514	514.2844	Trihydroxy-cholanoic acids (cholic acid) taurine	
522	522.3259	3β-hydroxy-5-cholestanoic acid taurine	Oxysterol 7α-hydroxylase (#4)
526	526.2480	3β,7α-dihydroxy-5-cholenoic acid glycine and sulfate	3β-HSOR (#2)

(continued)

Table 14.2 (continued)

Nominal mass	Exact mass	Compound name	Disorder
528	528.2637	Dihydroxy-cholanoic acids (e.g., chenodeoxycholic acid) glycine and sulfate	
530	530.2793	Tetrahydroxy-cholanoic acids taurine	
540	540.3364	Dihydroxy-cholestanoic acid (e.g., DHCA) taurine	Zellweger (#6), AMACR (#8), DBP (#10)
542	542.2429	3β,7α,12α-trihydroxy-5-cholenoic acid glycine and sulfate	3β-HSOR (#2)
552	552.3000	7α,12α-dihydroxy-3-oxo-4-cholestenoic acid taurine	5β-reductase (#3)
554	554.3157	Trihydroxy-cholestenoic acid (e.g., THCA:1) taurine	DBP (#10)
556	556.3313	Trihydroxy-cholestanoic acid (e.g., THCA) taurine	Zellweger (#6), AMACR (#8), DBP (#10)
567	567.3175	Dihydroxy-cholanoic acids glucuronide	BAAT (#13), BACS (#12)
570	570.3106	Tetrahydroxy-cholestenoic acids (e.g., OH-THCA:1) taurine	DBP (#10)
572	572.3263	Tetrahydroxy-cholestanoic acids (e.g., OH-THCA) taurine	Zellweger, AMACR,DBP
583	583.3124	Trihydroxy-cholanoic acids glucuronide	BAAT (#13), BACS (#12)
586	586.3055	Pentahydroxy-cholestenoic acids (e.g., diOH-THCA:1) taurine	DBP (#10)
588	588.3212	Pentahydroxycholestanoic acid taurine	Zellweger (#6)
609	609.3644	5β-cholestanetriol-?-one glucuronide, 5β-cholestenetetrol glucuronide	CTX (#5)
611	611.3801	5β-cholestanetetrol glucuronide	CTX (#5)
611	611.3437	Tetrahydroxy-27-nor-5β-cholestane-24-one glucuronide	SCPx
613	613.3593	27-nor-5β-cholestanepentol glucuronide	SCPx
625	625.3593	Tetrahydroxy-5β-cholestane-?-one glucuronide, 5β-cholestenepentol glucuronide	CTX (#5)
627	627.3750	5β-cholestanepentol glucuronide	CTX (#5)
627	627.3386	Pentahydroxy-27-nor-5β-cholestane-24-one glucuronide	SCPx (#11)
629	629.3543	27-nor-5β-cholestanehexol glucuronide	SCPx (#11)
641	641.3543	5β-cholestanepentol-?-one glucuronide, 5β-cholestenehexol glucuronide	CTX (#5)
643	643.3699	5β-cholestanehexol glucuronide	CTX (#5)
643	643.3335	Hexahydroxy-27-nor-5β-cholestane-24-one glucuronide	SCPx (#11)
645	645.3492	27-nor-5β-cholestaneheptol glucuronide	SCPx (#11)
651	651.3862	Dihydroxy-cholanoic acid glycine and N-acetylglucosamine	
659	659.3648	5β-cholestaneheptol glucuronide	CTX (#5)
701	701.3689	Dihydroxy-cholanoic acid taurine and N-acetylglucosamine	

3β-HSOR (#2) = 3β-hydroxy-Δ^5-C27-steroid oxidoreductase deficiency; 5β-reductase (#3) = Δ^4–3-oxosteroid 5β-reductase deficiency; CTX (#5) = cerebrotendinous xanthomatosis; Zellweger (#6) = Zellweger spectrum disorders, AMACR (#8) = α-Methylacyl-CoA racemase deficiency; DBP (#10) = D-bifunctional protein deficiency; SCPx (#11) = Sterol Carrier protein X; BACS (#12) = Bile acid-CoA ligase; BAAT (#13) = Bile acid CoA:amino acid N-acyltransferase. The numbers indicated behind the different disorders (#X) correspond to the numbers in Figs. 14.2 and 14.3. Bold m/z values/compounds are measured in the quantitative UPLC-MS/MS method described here

- Ammonium hydroxide 25% solution
- Water
- Standards of bile acids and conjugates: [free, glycine, taurine, preferably all also stable-isotope labeled versions to be used as internal standards, if available]:
  - CA
  - CDCA

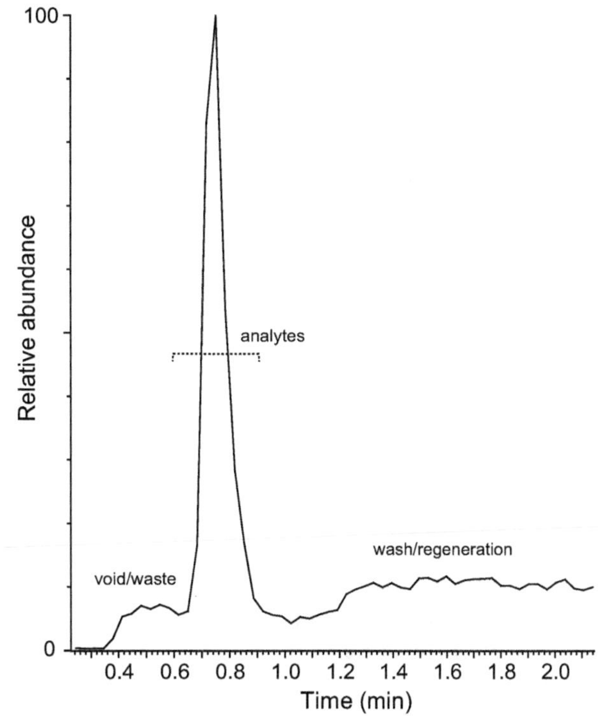

Fig. 14.4 Flow profile of FIA-MS/MS method showing time vs total ion current demonstrating the retention of the analytes on the (guard) column to separate them from salts/ interfering substances

- LCA
- UDCA
- DCA
- DHCA
- THCA

14.1.2.3 Instrumentation

- 4 ml glass vials with Teflon cap (Alltech, nr. 98110)
- Sample concentrator with heating (N_2 as drying gas) (Techne, FCS400D, Techne, DB-3)
- Ultrasonic bath for Eluant degassing (Branson, 3510)
- Guard column (of UPLC): C18-guard, Waters BEH pre-column 2.1 × 5 mm (Waters, 186003975)
- UPLC-MS/MS analytical column: Waters BEH C18 100 × 2.1 mm, 1.7 µm (Waters, 186002352)
- FIA-MS/MS column: Discovery C18 Supelguard Guard Cartridge (SKU 505188)
- UPLC with column oven
- Triple-quad MS: Waters TQ-XS, or equivalent
- HRMS: Q-TOF or Orbitrap-based MS equipped with ESI-source

14.1.3 UPLC-MS/MS (Quantitative)

- Eluant A: 3.75 mM ammoniumformiate, pH 8.1:methanol (1:3, v/v), incubate 10 min in ultrasonic bath to degas
- Eluant B: Methanol:water (1:3, v/v)
- Stock solution of bile acids: 2.5 mM in water
- Work solution of bile acids: 10 µM of bile acids in water:methanol (2:1, v/v)
- Standard mix 2.5: 1 ml of work solution +2.2 ml water +0.8 ml ethanol → 2.5 µM of bile acids
- IS mix 2.5: Stable-isotope labeled internal standards in 20% ethanol (v/v) at concentration of about 2.5 µM

14.1.4 FIA-MS/MS (Qualitative)

- Eluant A: 0.025% ammonium hydroxide in ultra-pure water (v/v)
- Eluant B: Acetonitrile:water (9:1, v/v)

14.1.4.1 Analytical Procedure

- Urine: Centrifuge 200 µl for 15 min at 1000 × g at room temperature. Transfer 50 µl of the supernatant to an UPLC autosampler vial and add 50 µl internal standard mix.
- Plasma/serum: Pipet 50 µl of sample in an 1.5 ml Eppendorf cup and add 50 µl internal standard mix and mix by vortexing. While vortexing, add 250 µl acetonitrile. Add another 250 µl of acetonitrile (total 500 µl) and vortex again thoroughly. Centrifuge for 10 min at 16,000 × g at 4 °C. Transfer supernatant to 4 ml vial and take to dryness in the sample concentrator at 40–50 °C under nitrogen flow. Resuspend the sample in 100 µl methanol:water (1:3, v/v).

14.1.5 Liquid Chromatography for UPLC-MS/MS (Quantitative)

Five µl of the extract is injected into the UPLC system. Oven temperature is 50 °C. Below is the gradient table for the Waters Acquity used in our laboratory (Table 14.3). This should be fine-tuned to the system that is used.

14.1.6 Pump Settings for FIA-MS/MS (Qualitative)

Ten µl of the extract is injected into the system. Analysis is done at room temperature (column oven off). Below is the FIA table used in our laboratory (Table 14.4). This should be fine-tuned to the system that is used.

14.1.7 Mass Spectrometry for UPLC-MS/MS (Quantitative)

ESI in the negative ion mode is used. Specific MRM of internal standards are not given as these may differ depending on how they are labeled, which affects the MRM. These are settings for the Waters TQ-XS. For some compounds the daughter is the same as the parent. This is done as metabolites fragment poorly and the application of the listed collision parameters do result in fragmentation of interfering substances (Table 14.5).

14.1.8 Mass Spectrometry for FIA-MS/MS (Qualitative)

ESI ionization in the negative ion mode is used. MS1 scan (continuum) from m/z 300 to m/z 705 with a cone voltage of 70, scan time of 0.2 s during 5 min.

14.1.8.1 Calculation

Mass Spectrometry for UPLC-MS/MS (Quantitative)

In each analytical run, a standard with 2.5 µM of most analytes is included to which the internal standard has been added. To calculate the concentration in the sample, a 1-point calibration based on the 2.5 µM is used. The abundance of the analyte in the sample is divided by the abundance of the corresponding internal standard and this "response" is divided by the response of the

Table 14.3 Gradient program for UPLC-MS/MS

Time	%A	%B	Flow (ml/min)
Initial	100	0	0.450
0.10	100	0	0.450
6.50	47	53	0.600
6.60	0.2	99.8	0.600
8.45	0.2	99.8	0.600
8.50	100	0	0.500
12.00	100	0	0.450

Table 14.4 Pump settings for FIA-MS/MS

Time	%A	%B	Flow (ml/min)
Initial	100	0	0.300
1.00	100	0	0.300
2.00	0	100	0.300
5.00	0	100	0.300
5.10	100	0	0.300
8.00	100	0	0.300

Table 14.5 MRM transitions and other mass spectrometric parameters

Metabolite	Parent (m/z)	Daughter (m/z)	Dwell (s)	Collision energy (eV)	Cone voltage (V)
Free LCA	375.3	375.3	0.004	10	70
Free CDCA/UDCA/DCA	391.3	391.3	0.004	10	70
Free CA	407.3	407.3	0.004	10	70
Free DHCA	433.3	433.3	0.004	30	70
Free THCA	449.3	449.3	0.004	22	70
Free OH-THCA	465.3	465.3	0.004	10	70
g-CDCA/UDCA/DCA	448.3	73.9	0.004	30	70
g-CA	464.3	73.9	0.004	30	70
Free OH-THCA	482.3	80.0	0.004	70	70
t-CDCA/UCDA/DCA	498.3	80.0	0.004	70	70
t-CA	514.3	80.0	0.004	70	70
C_{29}-bile dicarboxylic acid	507.3	463.3	0.004	40	70
t-DHCA	540.3	80.0	0.004	70	70
t-THCA	556.3	80.0	0.004	70	70
OH-t-THCA	572.3	80.0	0.004	70	70

standard 2.5 μM (also analyte/IS), which is multiplied by the concentration of the standard, in this case 2.5 μM. CDCA and UCDA, like DCA and CA, are isobaric and are separated on the analytical column allowing separate quantification. Given that conjugation depends on both genetic background and availability of taurine/glycine no reference ranges can be calculated for specific forms. Hence, the free forms of the bile acids and the glycine- and taurine-conjugates are summed and the total CA, CDCA, DCA, UDCA, THCA, and DHCA are reported.

Mass Spectrometry for FIA-MS/MS (Qualitative)

Spectra are processed by subtracting the background and spectral smoothing (Savitzky Golay) where specific parameters are machine and software-dependent. Using Table 14.1 and reference spectra, a qualitative evaluation of the spectrum is performed by the trained laboratory specialist together with the technician.

14.1.8.2 Post-Analytical

Figure 14.5 shows some example spectra of plasma and urine of cholestasis, CTX, a Zellweger spectrum disorder, and DBP deficiency.

14.1.8.3 Reference Values

See Table 14.6.

14.1.8.4 Pathological Values

See Tables 14.1 and 14.7 for accumulating or deficient metabolites in the different bile acid synthesis disorders. It is important to realize that depending on the disorder, the profile can vary considerably and in some disorders may even be without obvious abnormalities.

14.1.8.5 Quality Control

In each analytical run, these samples are to be included for quality control:

- Standard 2.5 μM
- Poolplasma with normal levels of bile acids (low/normal control)
- Poolplasma of Zellweger patients and/or cholestatic patients (high control). Alternatively, a low/normal control plasma can be spiked with standards to create a high control.

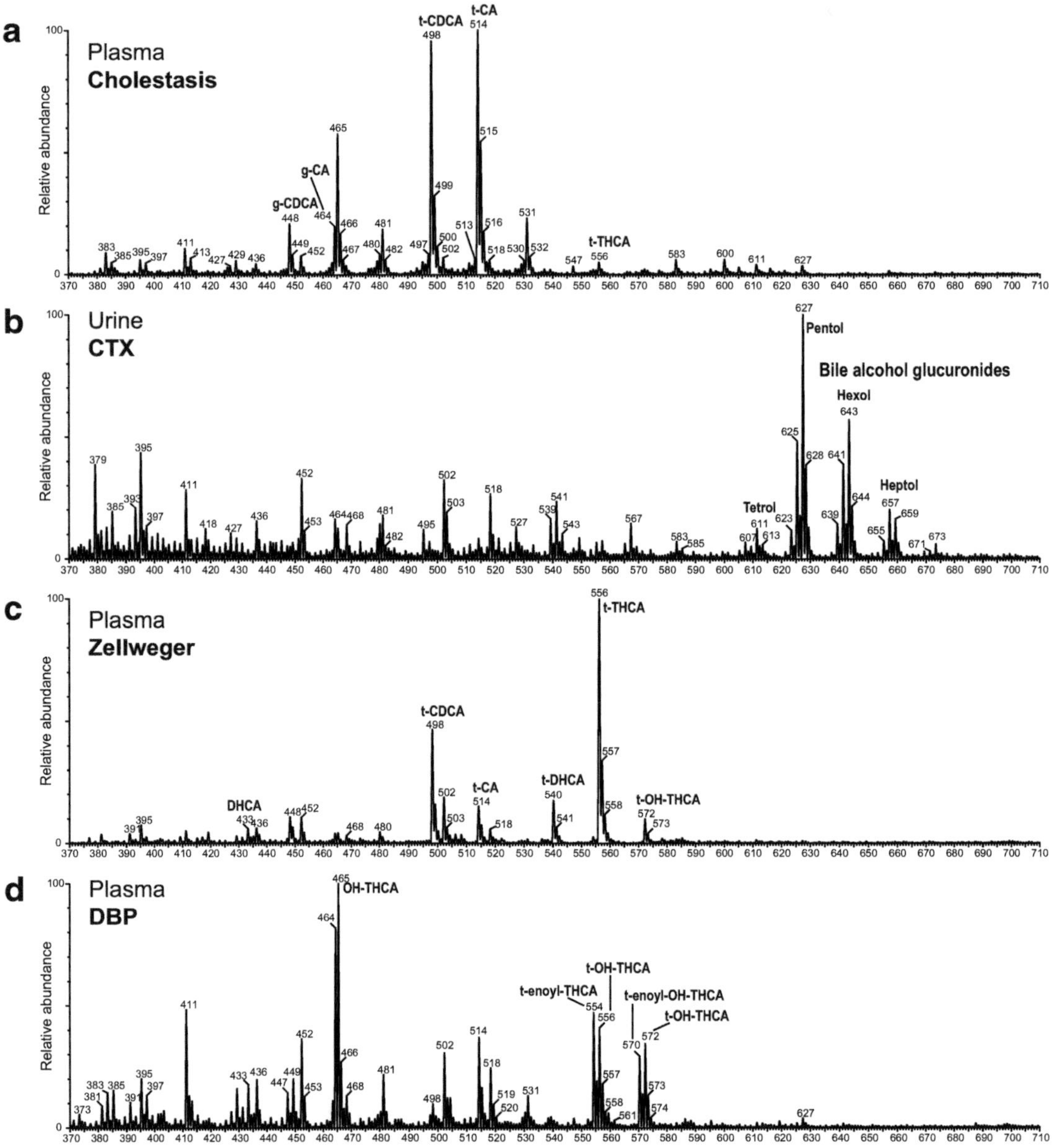

Fig. 14.5 Spectra acquired by FIA-MS/MS of (**a**) plasma from a patient with cholestasis showing elevated primary (mainly conjugated) bile acids and some elevation of tauro-trihydroxycholestanoic acid (t-THCA), which is normal in case of cholestasis, (**b**) urine from a patient with CTX showing accumulation of characteristic bile alcohol glucuronides, (**c**) plasma from a patient with a Zellweger spectrum disorder with accumulation of C_{27}-bile acid intermediates; free and taurine-conjugated dihydroxycholestanoic acid (t-DHCA) and tauro-hydroxy-trihydroxycholestanoic acid (t-OH-THCA), and (**d**) plasma of a DBP patient showing, similar to the Zellweger spectrum disorder patient, accumulation of C_{27}-bile acid intermediates, but in addition also the enoyl-forms of THCA (produced by the acyl-CoA oxidase), t-enoyl-(OH)-THCA

Table 14.6 Reference ranges of total bile acids

Metabolite	Reference range (μM)
CDCA	0.7–10
CA/DCA (age-dependent)	0.1–4.7
LCA	Not detectable, <0.02
DHCA	0–0.02
THCA	0–0.08
C_{29}-bile dicarboxylic acid	0–0.001 (should not be present)

Table 14.7 Biochemical aspects of bile acid synthesis disorders

Deficiency	Plasma bile acid profile	Urinary bile acid profile
Cholesterol 7α-hydroxylase (#1)	No abnormalities. Possibly more CDCA and corresponding secondary bile acids	No abnormalities
3β-hydroxy-Δ^5-C_{27}-steroid oxidoreductase (#2)	Low/absent primary bile acids/salts Accumulation of Δ^5-3β-bile acids: 3β,7α-dihydroxy-5-cholenoic acid: • m/z 389 (free) • m/z 449 (glycine) • m/z 469 (sulfate) • m/z 526 (glycine and sulfate) 3β,7α,12α-trihydroxy-5-cholenoic acid: • m/z 405 (free) • m/z 462 (glycine) • m/z 485 (sulfate) • m/z 542 (glycine and sulfate)	Like plasma
Δ^4-3-oxosteroid 5β-reductase (#3)	Like urine. Low/absent primary bile acids	7α-hydroxy-3-oxo-4-cholenoic acid: • m/z 444 (glycine) • m/z 494 (taurine) 7α,12α-dihydroxy-4-cholenoic acid: • m/z 460 (glycine) • m/z 510 (taurine) 7α,12α-dihydroxy-3-oxo-4-cholestenoic: • m/z 552 (taurine)
Oxysterol 7α-hydroxylase (#4)	27-Hydroxycholesterol (other analysis) 25-hydroxycholesterol (other analysis) 3β-hydroxy-5-cholenoic acid: • m/z 430 (glycine) • m/z 480 (taurine) 3β-hydroxy-5-cholestanoic acid: • m/z 472 (glycine) • m/z 522 (taurine)	3β-Hydroxy-5-cholenoic acids: • m/z 453 3β-hydroxy-5-cholenoic acid (sulfate) • m/z 480 3β-hydroxy-5-cholenoic acid (taurine) • m/z 510 3β-hydroxy-5-cholenoic acid (glycine and sulfate)
Sterol 27-hydroxylase (#5)	Bile alcohol glucuronides (in order of abundance): • m/z 611 cholestanetetrol • m/z 595 cholestanetriol • m/z 627 cholestanepentol • m/z 643 cholestanehexolCholestene or oxo-derivatives of bile alcohols (with 2 Da less: m/z 609, 593, 625, 641) are also observed	Bile alcohol glucuronides (in order of abundance): • m/z 627 cholestanepentol • m/z 611 cholestanetetrol • m/z 643 cholestanehexol • m/z 659 cholestaneheptol • m/z 675Cholestene or oxo-derivatives of bile alcohols (with 2 Da less: m/z 625, 609, 641, 657, 673) are also observed

(continued)

Table 14.7 (continued)

Deficiency	Plasma bile acid profile	Urinary bile acid profile
Zellweger spectrum disorders (#6)	C_{27}-bile acid intermediates: • m/z 433 DHCA (free) • m/z 449 THCA (free) • m/z 465 OH-THCA (free) • m/z 540 DHCA (taurine) • m/z 556 THCA (taurine) • m/z 572 OH-THCA (taurine) • m/z 588 diOH-THCA (taurine) • m/z 509 C_{29}-dicarboxylic acid C_{24}-bile acid levels can be normal or even elevated	C_{27}-bile acid intermediates: • m/z 556 THCA (taurine) • m/z 572 OH-THCA (taurine) • m/z 588 diOH-THCA (taurine) Bile alcohol glucuronides can be present
ABCD3 (#7)	Unconjugated and taurine conjugated DHCA and THCA, C_{29}-dicarboxylic acid	Not known
α-Methylacyl-CoA racemase (#8)	25R-isomers of DHCA and THCA: • m/z 433 DHCA (free) • m/z 449 THCA (free) • m/z 540 DHCA (taurine) • m/z 556 THCA (taurine)	Similar to Zellweger, C_{27}-bile acid intermediates: • m/z 556 THCA (taurine) • m/z 572 OH-THCA (taurine) • m/z 588 diOH-THCA (taurine)
Acyl-CoA oxidase 2 (#9)	• m/z 433 DHCA (free) • m/z 449 THCA (free) • m/z 540 DHCA (taurine) • m/z 556 THCA (taurine) Decreased levels of primary bile acids	Unconjugated or taurine conjugated (OH-)DHCA and (OH-)THCA
D-bifunctional protein (#10)	C_{27}-bile acid intermediates: • m/z 433 DHCA (free) • m/z 449 THCA (free) • m/z 465 OH-THCA (free) • m/z 540 DHCA (taurine) • m/z 554 THCA:1 (taurine) • m/z 556 THCA (taurine) • m/z 570 OH-THCA:1 (taurine) • m/z 572 OH-THCA (taurine) • m/z 588 diOH-THCA (taurine)	• m/z 570 OH-THCA:1 (taurine) • m/z 586 diOH-THCA:1 (taurine)
Sterol carrier protein X (#11)	No significant abnormalities	• m/z 611 tetrahydroxy-27-nor-5β-cholestane-24-one (glucuronide) • m/z 613 27-nor-5β-cholestanepentol (glucuronide) • m/z 627 pentahydroxy-27-nor-5β-cholestane-24-one (glucuronide) • m/z 629 27-nor-5β-cholestanehexol (glucuronide) • m/z 643 hexahydroxy-27-nor-5β-cholestane-24-one (glucuronide) • m/z 645 27-nor-5β-cholestaneheptol (glucuronide)
Bile acid-CoA ligase (#12)	>85% unamidated primary bile acids: • m/z 391 CDCA • m/z 407 CA (most abundant)	>85% unamidated primary bile acids: • m/z 391 CDCA • m/z 407 CA (most abundant) Can be present: • m/z 471 CDCA (sulfate) • m/z 487 CA (sulfate) • m/z 567 CDCA (glucuronide) • m/z 583 CA (glucuronide)

(continued)

Table 14.7 (continued)

Deficiency	Plasma bile acid profile	Urinary bile acid profile
Bile acid CoA:amino acid N-acyltransferase (#13)	Absence of amidated primary bile acids. Only unamidated primary bile acids present • m/z 391 CDCA • m/z 407 CA (most abundant) Can be present: • m/z 471 CDCA (sulfate) • m/z 487 CA (sulfate) • m/z 567 CDCA (glucuronide) • m/z 583 CA (glucuronide)	Same as plasma

The numbers indicated behind the different disorders (#X) correspond to the numbers in Figs. 14.2 and 14.3

14.2 Conclusion

In this chapter, we have described the current state-of-the-art mass spectrometric analysis of bile acids in plasma and urine, which is suitable for the identification/confirmation of bile acid biosynthesis defects.

Acknowledgments The authors acknowledge Albert Bootsma for expert technical assistance.

References

1. Heubi J, Setchell K, Bove K (2007) Inborn errors of bile acid metabolism. Semin Liver Dis 27:282–294
2. Blau N, Duran M, Gibson KM (2008) Editors. Laboratory guide to the methods in biochemical genetics. Springer, Berlin
3. Griffiths WJ, Sjovall J (2010) Bile acids: analysis in biological fluids and tissues. J Lipid Res 51:23–41
4. Russell DW (2003) The enzymes, regulation, and genetics of bile acid synthesis. Annu Rev Biochem 72:137–174
5. Russell DW (2009) Fifty years of advances in bile acid synthesis and metabolism. J Lipid Res 50:S120–S125
6. Ferdinandusse S, Denis S, Faust PL, Wanders RJA (2009) Bile acids: the role of peroxisomes. J Lipid Res 50:2139–2147
7. Vaz FM, Ferdinandusse S (2017) Bile acid analysis in human disorders of bile acid biosynthesis. Mol Aspects Med 56:10–24
8. Blau N, Dionisi Vici C, Ferreira CR, Vianey-Saban C, van Karnebeek CDM (eds) (2022) Physician's guide to the diagnosis, treatment, and follow-up of inherited metabolic diseases. Springer, Cham
9. Pullinger CR, Eng C, Salen G, Shefer S, Batta AK, Erickson SK et al (2002) Human cholesterol 7alpha-hydroxylase (CYP7A1) deficiency has a hypercholesterolemic phenotype. J Clin Invest 110:109–117
10. Baumgartner MR, Fowler B, Thöny B, Duran M, Gibson KM, Dionisi-Vici C (2014) Physician's guide to the diagnosis, treatment, and follow-up of inherited metabolic diseases. Blau N, Duran M, Gibson KM, Dionisi Vici C, editors. Springer, Berlin
11. Clayton PT, Leonard JV, Lawson AM, Setchell KD, Andersson S, Egestad B et al (1987) Familial giant cell hepatitis associated with synthesis of 3 beta, 7 alpha-dihydroxy-and 3 beta, 7 alpha, 12 alpha-trihydroxy-5-cholenoic acids. J Clin Invest 79:1031–1038
12. Akobeng AK, Clayton PT, Miller V, Super M, Thomas AG (1999) An inborn error of bile acid synthesis (3beta-hydroxy-delta5-C27-steroid dehydrogenase deficiency) presenting as malabsorption leading to rickets. Arch Dis Child 80:463–465
13. Buchmann MS, Kvittingen EA, Nazer H, Gunasekaran T, Clayton PT, Sjövall J et al (1990) Lack of 3 beta-hydroxy-delta 5-C27-steroid dehydrogenase/isomerase in fibroblasts from a child with urinary excretion of 3 beta-hydroxy-delta 5-bile acids. A new inborn error of metabolism. J Clin Invest 86:2034–2037
14. Schwarz M, Wright AC, Davis DL, Nazer H, Björkhem I, Russell DW (2000) The bile acid synthetic gene 3β-hydroxy-Δ5-C27-steroid oxidoreductase is mutated in progressive intrahepatic cholestasis. J Clin Invest 106:1175–1184
15. Ichimiya H, Egestad B, Nazer H, Baginski ES, Clayton PT, Sjovall J (1991) Bile acids and bile alcohols in a child with hepatic 3 beta-hydroxy-delta 5-C27-steroid dehydrogenase deficiency: effects of chenodeoxycholic acid treatment. J Lipid Res 32:829–841
16. Yamato Y, Kimura A, Murai T, Yoshimura T, Kurosawa T, Terazawa S et al (2001) 3beta-hydroxy-delta5 -C27-steroid dehydrogenase deficiency: diagnosis and treatment. J Paediatr Child Health 37:516–519
17. Jacquemin E, Setchell KD, O'Connell NC, Estrada A, Maggiore G, Schmitz J et al (1994) A new cause of progressive intrahepatic cholestasis: 3 beta-hydroxy-C27-steroid dehydrogenase/isomerase deficiency. J Pediatr 125:379–384
18. Cheng JB, Jacquemin E, Gerhardt M, Nazer H, Cresteil D, Heubi JE et al (2003) Molecular genetics of 3beta-hydroxy-Delta5-C27-steroid oxidoreduc-

tase deficiency in 16 patients with loss of bile acid synthesis and liver disease. J Clin Endocrinol Metab 88:1833–1841
19. Zhang W, Jha P, Wolfe B, Gioiello A, Pellicciari R, Wang J et al (2015) Tandem mass spectrometric determination of atypical 3β-hydroxy-Δ5-bile acids in patients with 3β-hydroxy-Δ5-C27-steroid oxidoreductase deficiency: application to diagnosis and monitoring of bile acid therapeutic response. Clin Chem 61:955–963
20. Setchell KDR, Suchy FJ, Welsh MB, Zimmer-Nechemias L, Heubi J, Balistreri WF (1988) Δ4-3-oxosteroid 5β-reductase deficiency described in identical twins with neonatal hepatitis. A new inborn error in bile acid synthesis. J Clin Invest 82:2148–2157
21. Shneider BL, Setchell KD, Whitington PF, Neilson KA, Suchy FJ (1994) Delta 4-3-oxosteroid 5 beta-reductase deficiency causing neonatal liver failure and hemochromatosis. J Pediatr 124:234–238
22. Clayton PT, Mills KA, Johnson AW, Barabino A, Marazzi MG (1996) Delta 4-3-oxosteroid 5 beta-reductase deficiency: failure of ursodeoxycholic acid treatment and response to chenodeoxycholic acid plus cholic acid. Gut 38:623–628
23. Lemonde HA, Custard EJ, Bouquet J, Duran M, Overmars H, Scambler PJ et al (2003) Mutations in SRD5B1 (AKR1D1), the gene encoding delta(4)-3-oxosteroid 5beta-reductase, in hepatitis and liver failure in infancy. Gut 52:1494–1499
24. Gonzales E, Cresteil D, Baussan C, Dabadie A, Gerhardt M-F, Jacquemin E (2004) SRD5B1 (AKR1D1) gene analysis in delta(4)-3-oxosteroid 5beta-reductase deficiency: evidence for primary genetic defect. J Hepatol 40:716–718
25. Dai D, Mills PB, Footitt E, Gissen P, McClean P, Stahlschmidt J et al (2014) Liver disease in infancy caused by oxysterol 7α-hydroxylase deficiency: successful treatment with chenodeoxycholic acid. J Inherit Metab Dis 37:851–861
26. Setchell KDR, Schwarz M, O'Connell NC, Lund EG, Davis DL, Lathe R et al (1998) Identification of a new inborn error in bile acid synthesis: mutation of the oxysterol 7alpha-hydroxylase gene causes severe neonatal liver disease. J Clin Invest 102:1690–1703
27. Kimura A, Nittono H, Mizuochi T, Ueki I, Kurosawa T, Muto A et al (2012) Neonatal cholestasis with increased 3β-monohydroxy-δ5bile acids in serum and urine: not necessarily primary oxysterol 7α hydroxylase deficiency. Clin Chim Acta 413:1700–1704
28. Ueki I, Kimura A, Nishiyori A, Chen H-L, Takei H, Nittono H et al (2008) Neonatal cholestatic liver disease in an Asian patient with a homozygous mutation in the oxysterol 7alpha-hydroxylase gene. J Pediatr Gastroenterol Nutr 46:465–469
29. Arnoldi A, Crimella C, Tenderini E, Martinuzzi A, D'Angelo MG, Musumeci O et al (2012) Clinical phenotype variability in patients with hereditary spastic paraplegia type 5 associated with CYP7B1 mutations. Clin Genet 81:150–157
30. Biancheri R, Ciccolella M, Rossi A, Tessa A, Cassandrini D, Minetti C et al (2009) White matter lesions in spastic paraplegia with mutations in SPG5/CYP7B1. Neuromuscul Disord 19:62–65
31. Inoue K, Kubota S, Seyama Y (1999) Cholestanol induces apoptosis of cerebellar neuronal cells. Biochem Biophys Res Commun 256:198–203
32. Seyama Y (2003) Cholestanol metabolism, molecular pathology, and nutritional implications. J Med Food 6:217–224
33. Pierre G, Setchell K, Blyth J, Preece MA, Chakrapani A, McKiernan P (2008) Prospective treatment of cerebrotendinous xanthomatosis with cholic acid therapy. J Inherit Metab Dis 31
34. Verrips A, Hoefsloot LH, Steenbergen GC, Theelen JP, Wevers RA, Gabreëls FJ et al (2000) Clinical and molecular genetic characteristics of patients with cerebrotendinous xanthomatosis. Brain 123(Pt 5):908–919
35. Klouwer FCC, Berendse K, Ferdinandusse S, Wanders RJA, Engelen M, Poll-The BT (2015) Zellweger spectrum disorders: clinical overview and management approach. Orphanet J Rare Dis 10:151
36. Ferdinandusse S, Jimenez-Sanchez G, Koster J, Denis S, Van Roermund CW, Silva-Zolezzi I et al (2015) A novel bile acid biosynthesis defect due to a deficiency of peroxisomal ABCD3. Hum Mol Genet 24:361–370
37. Smith EH, Gavrilov DK, Oglesbee D, Freeman WD, Vavra MW, Matern D et al (2010) An adult onset case of alpha-methyl-acyl-CoA racemase deficiency. J Inherit Metab Dis 33:S349–S353
38. Clarke CE, Alger S, Preece MA, Burdon MA, Chavda S, Denis S et al (2004) Tremor and deep white matter changes in alpha-methylacyl-CoA racemase deficiency. Neurology 63:188–189
39. Thompson SA, Calvin J, Hogg S, Ferdinandusse S, Wanders RJA, Barker RA (2008) Relapsing encephalopathy in a patient with alpha-methylacyl-CoA racemase deficiency. J Neurol Neurosurg Psychiatry 79:448–450
40. Van Veldhoven PP, Meyhi E, Squires RH, Fransen M, Fournier B, Brys V et al (2001) Fibroblast studies documenting a case of peroxisomal 2-methylacyl-CoA racemase deficiency: possible link between racemase deficiency and malabsorption and vitamin K deficiency. Eur J Clin Invest 31:714–722
41. Verhagen JMA, Huijmans JG, Williams M, van Ruyven RLJ, Bergen AAB, Wouters CH et al (2012) Incidental finding of alpha-methylacyl-CoA racemase deficiency in a patient with oculocutaneous albinism type 4. Am J Med Genet A 158A:2931–2934
42. Haugarvoll K, Johansson S, Tzoulis C, Haukanes BI, Bredrup C, Neckelmann G et al (2013) MRI characterisation of adult onset alpha-methylacyl-coA racemase deficiency diagnosed by exome sequencing. Orphanet J Rare Dis 8:1
43. Kapina V, Sedel F, Truffert A, Horvath J, Wanders RJA, Waterham HR et al (2010) Relapsing rhabdomyolysis due to peroxisomal alpha-methylacyl-coa racemase deficiency. Neurology 75:1300–1302

44. Dick D, Horvath R, Chinnery PF (2011) AMACR mutations cause late-onset autosomal recessive cerebellar ataxia. Neurology 76:1768–1770
45. Ferdinandusse S, Denis S, Clayton PT, Graham A, Rees JE, Allen JT et al (2000) Mutations in the gene encoding peroxisomal alpha-methylacyl-CoA racemase cause adult-onset sensory motor neuropathy. Nat Genet 24:188–191
46. Setchell KDR, Heubi JE, Bove KE, O'Connell NC, Brewsaugh T, Steinberg SJ et al (2003) Liver disease caused by failure to racemize trihydroxycholestanoic acid: gene mutation and effect of bile acid therapy. Gastroenterology 124:217–232
47. Vilarinho S, Sari S, Mazzacuva F, Bilgüvar K, Esendagli-Yilmaz G, Jain D et al (2016) *ACOX2* deficiency: a disorder of bile acid synthesis with transaminase elevation, liver fibrosis, ataxia, and cognitive impairment. Proc Natl Acad Sci 113:11289–11293
48. Monte MJ, Alonso-Peña M, Briz O, Herraez E, Berasain C, Argemi J et al (2017) ACOX2 deficiency: an inborn error of bile acid synthesis identified in an adolescent with persistent hypertransaminasemia. J Hepatol. European Association for the Study of the. Liver 66:581–588
49. Ferdinandusse S, Denis S, Mooyer PAW, Dekker C, Duran M, Soorani-Lunsing RJ et al (2006) Clinical and biochemical spectrum of D-bifunctional protein deficiency. Ann Neurol 59:92–104
50. Ferdinandusse S, Kostopoulos P, Denis S, Rusch H, Overmars H, Dillmann U et al (2006) Mutations in the gene encoding peroxisomal sterol carrier protein X (SCPx) cause leukencephalopathy with dystonia and motor neuropathy. Am J Hum Genet 78:1046–1052
51. Horvath R, Lewis-Smith D, Douroudis K, Duff J, Keogh M, Pyle A et al (2015) SCP2 mutations and neurodegeneration with brain iron accumulation. Neurology 85:1909–1911
52. Chong CPK, Mills PB, McClean P, Gissen P, Bruce C, Stahlschmidt J et al (2012) Bile acid-CoA ligase deficiency-a new inborn error of bile acid metabolism. J Inherit Metab Dis 35:521–530
53. Setchell KDRR, Heubi JE, Shah S, Lavine JE, Suskind D, Al-Edreesi M et al (2013) Genetic defects in bile acid conjugation cause fat-soluble vitamin deficiency. Gastroenterology 144:945–955.e6. quiz e14–5
54. Carlton VEH, Harris BZ, Puffenberger EG, Batta AK, Knisely AS, Robinson DL et al (2003) Complex inheritance of familial hypercholanemia with associated mutations in TJP2 and BAAT. Nat Genet 34:91–96
55. Hadžić N, Bull LN, Clayton PT, Knisely AS (2012) Diagnosis in bile acid-CoA: amino acid N-acyltransferase deficiency. World J Gastroenterol 18:3322–3326

Cholesterol Biosynthesis Metabolites

15

Hans R. Waterham and Frédéric M. Vaz

Contents

List of Abbreviations

BSTFA	N,N-(Bis-trimethylsilyl)-trifluoroacetamide
CDPX	2X-linked dominant chondrodysplasia punctata 2
CHILD	Congenital hemidysplasia with ichthyosiform erythroderma and limb defects
CTX	Cerebrotendinous xanthomatosis
FID	Flame ionization detector
HEM	Hydrops–ectopic calcification–Moth-eaten
HIDS	Hyper IgD and periodic fever syndrome
HMG	3-Hydroxy-3-methylglutaryl
LDL	Low density lipoprotein
MA	Mevalonic aciduria
MKD	Mevalonate kinase deficiency
P	Phosphate
PMKD	Phosphomevalonate kinase deficiency
PP	Pyrophosphate
SLOS	Smith-Lemli-Opitz syndrome
SMO	C4-Methyl sterol oxidase
TMCS	Trimethylchlorosilanel

H. R. Waterham (✉) · F. M. Vaz
Laboratory Genetic Metabolic Diseases, Department of Laboratory Medicine, Amsterdam UMC, University of Amsterdam, Amsterdam, The Netherlands
e-mail: h.r.waterham@amsterdamumc.nl; f.m.vaz@amsterdamumc.nl

N. Blau, F. M. Vaz (eds.), *Laboratory Guide to the Methods in Biochemical Genetics*, https://doi.org/10.1007/978-3-031-58819-8_15

15.1 Isoprenoid/Cholesterol Biosynthesis

Mammalian cells acquire cholesterol either by de novo synthesis from acetyl-CoA or through LDL receptor-mediated uptake of low density lipoprotein (LDL) particles that contain cholesterol esterified with long-chain fatty acids. These LDL-cholesterol esters are subsequently hydrolyzed in lysosomes, after which free cholesterol molecules become available for synthesis of membranes, steroid hormones, bile acids, or oxysterols [1, 2].

The de novo synthesis of cholesterol is performed by the isoprenoid biosynthesis pathway. In addition to cholesterol, this pathway produces a variety of sterol and non-sterol isoprenoids that have important cellular functions in such diverse processes as cell growth and differentiation, protein glycosylation, protein isoprenylation, signal transduction pathways, and mitochondrial electron transport (Fig. 15.1) [1–3].

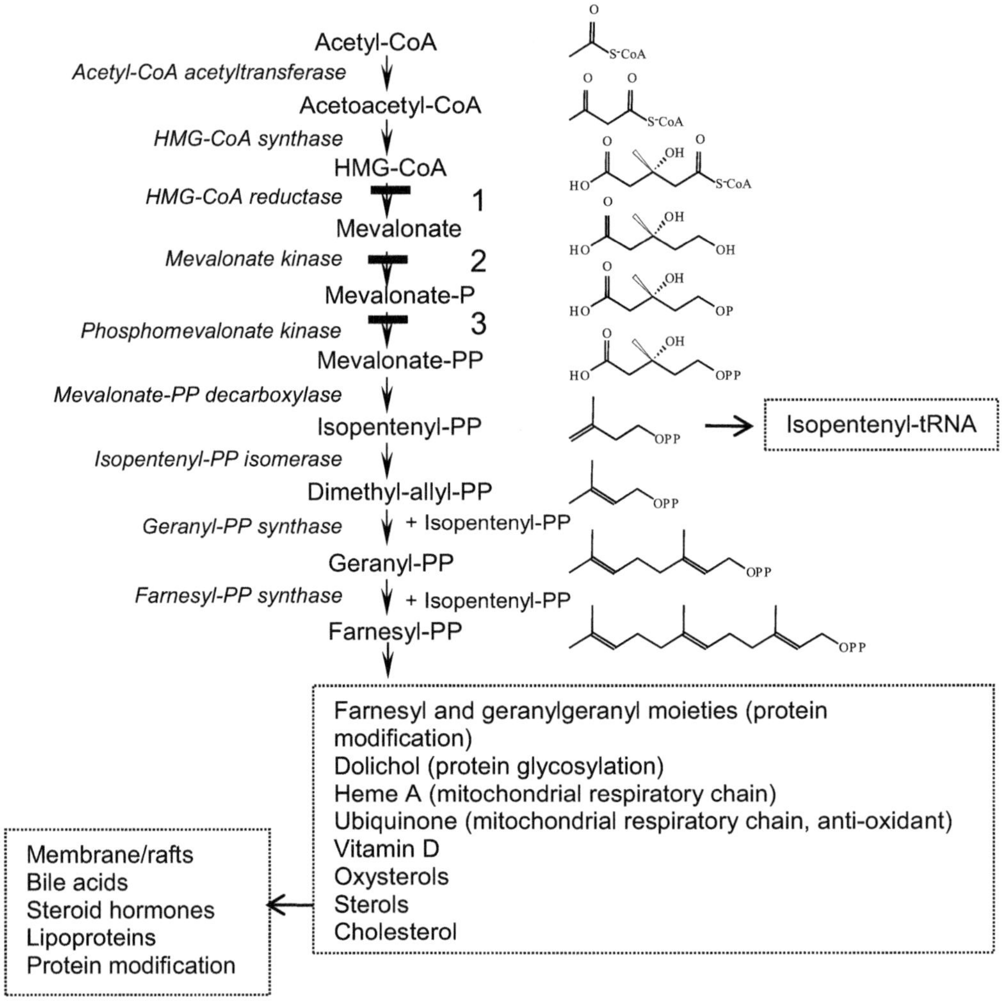

Fig. 15.1 The pre-squalene part of the isoprenoid biosynthetic pathway showing the enzymes and intermediate non-sterol isoprenoids involved. Currently known disorders in this part of the pathway are indicated by *solid bars* across the arrows. (1) HMG-CoA reductase deficiency; (2) Mevalonate kinase deficiency; (3) Phosphomevalonate kinase deficiency. *CoA*, coenzyme A; *HMG*, 3-hydroxy-3-methylglutaryl; *P*, phosphate; *PP*, pyrophosphate

The synthesis of all isoprenoids starts with acetyl-CoA, which in a series of 6 different enzyme reactions is converted into isopentenyl-PP, the basic C-5 isoprene unit that is used for the synthesis of all subsequent isoprenoids (Fig. 15.1) [3]. At the level of farnesyl-PP the pathway divides into several branches involved in the production of the various isoprenoid end products. One of the major branches involves the cholesterol biosynthetic part of the pathway, of which squalene is the first committed intermediate in the production of sterols. Following cyclization of squalene, lanosterol is produced. To eventually produce cholesterol from lanosterol, a series of at least 8 different enzyme reactions is required, including one demethylation at C-14, two demethylations at C-4, one isomerization of $\Delta^{8(9)}$ to Δ^{7}, three reductions of the Δ^{24}, Δ^{14}, and Δ^{7} double bonds, and one desaturation between C-5 and C-6 (Fig. 15.2). Because the sequence of enzyme reactions required for the conversion of lanosterol to cholesterol may vary, two major routes involving the same enzymes have been proposed which, depending on the timing of reduction of the Δ^{24} double bond, postulate either 7-dehydrocholesterol or desmosterol as the ultimate precursor of cholesterol [3].

15.2 Defects of Isoprenoid/ Cholesterol Biosynthesis

At present, 12 different inherited disorders have been linked to specific enzyme defects in the isoprenoid/cholesterol biosynthetic pathway [3–6]. Most of the underlying biochemical and genetic causes of the disorders have been resolved after the finding of abnormally increased levels of intermediate metabolites in tissues and/or body fluids of patients (Table 15.1). These characteristic metabolites can still be used as a first indication for diagnosis of the disorders, but this needs to be followed by genetic testing to establish the defect (Table 15.1). Metabolite analysis may also assist in confirming a putative genetic diagnosis obtained by next generation sequencing (genetics first approach).

15.3 Pre-squalene Disorders

Currently, four different disorders that affect enzymes in the pre-squalene part have been identified (Fig. 15.1). Most common of these are the previously described **Hyper IgD and periodic fever syndrome** (HIDS) and **Mevalonic aciduria** (MA), which represent the mild and severe clinical and biochemical ends of the **mevalonate kinase deficiency** (MKD) spectrum [7, 8]. Both defects are due to biallelic pathogenic variants in the *MVK* gene causing a deficiency of the enzyme mevalonate kinase, which is less severe in HIDS than in MA as also reflected in the clinical presentation. Patients present with an early-onset autoinflammatory disorder characterized by lifelong recurring episodes of fever and inflammation, which last 3–7 days, recur on average every 4–6 weeks and are associated with abdominal pain, vomiting and diarrhea, cervical lymphadenopathy, hepatosplenomegaly, arthralgia, and skin rash. Disease onset occurs mostly in the first year of life; episodes often occur without clear cause but may be provoked by vaccinations, physical and emotional stress, and minor trauma. Severe MKD patients may additionally present with developmental delay, ataxia, cerebellar atrophy, hypotonia, severe failure to thrive and dysmorphic features, and often die in early infancy. Biochemical hallmark of MKD is the accumulation of mevalonic acid in body fluids and tissues, which can be detected by organic acid analysis or, preferably, by stable isotope dilution gas chromatography-mass spectrometry (GC-MS) [9]. Cholesterol levels are usually normal to low-normal. Confirmative diagnostic options include direct measurement of mevalonate kinase activities in white blood cells or primary skin fibroblasts from patients and/or genetic analysis of the *MVK* gene.

Recently, two patients with **phosphomevalonate kinase deficiency** (PMKD) due to biallelic

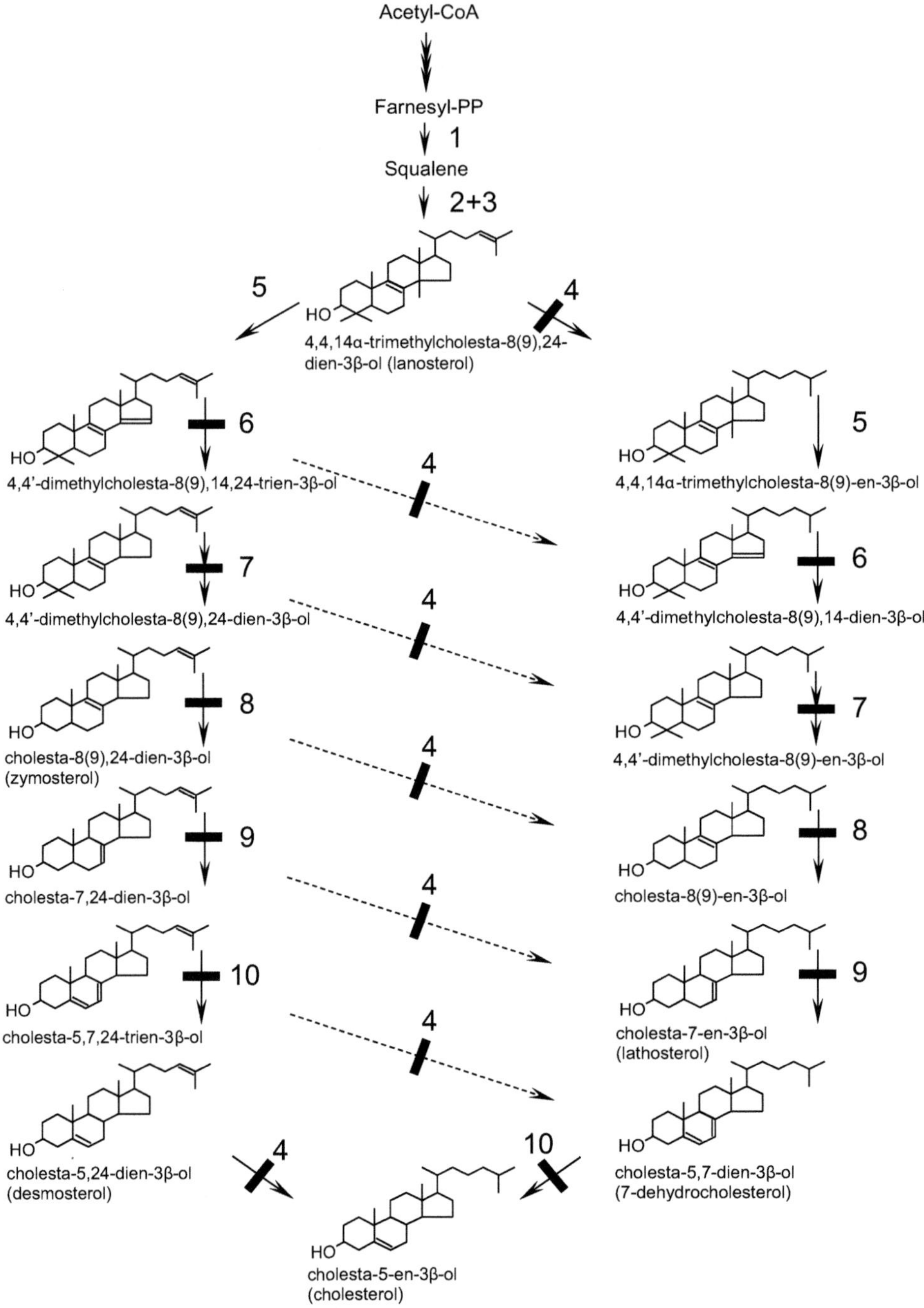

Fig. 15.2 The post-squalene or cholesterol synthesis part of the isoprenoid biosynthetic pathway showing the enzymes and intermediate sterol isoprenoids involved. Enzymes are numbered as follows: (1) squalene synthase (squalene synthase deficiency); (2) squalene epoxidase; (3) 2,3-oxidosqualene sterol cyclase; (4) 3β-hydroxysterol Δ^{24}-reductase (desmosterolosis); (5) lanosterol C-14 demethylase (lanosterol C14-demethylase deficiency); (6) 3β-hydroxysterol Δ^{14}-reductase (HEM dysplasia); (7) sterol C-4 demethylase complex (including a 3β-hydroxysteroid dehydrogenase defective in CHILD syndrome and C4-methylsterol monooxygenase 1 in SMO deficiency); (8) 3β-hydroxysterol Δ^{8}-Δ^{7} isomerase (CDPX2); (9) 3β-hydroxysterol Δ^{5}-desaturase (lathosterolosis); (10) 3β-hydroxysterol Δ^{7}-reductase (SLOS). Currently known disorders in this part of the pathway are indicated by *solid bars* across the arrows

Table 15.1 Inherited disorders of isoprenoid/cholesterol biosynthesis and disorders affecting cholesterol biosynthesis

Disorder	Defective enzyme	Increased metabolites	Gene	Chromosome
Involved in isoprenoid/ cholesterol biosynthesis				
HMG-CoA reductase deficiency	HMG-CoA reductase	Unknown	*HMGCR*	5q13.3
Mevalonate kinase deficiency (HIDS/MA)	Mevalonate kinase	Mevalonic acid	*MVK*	12q24
Phosphomevalonate kinase deficiency	Phosphomevalonate kinase	Mevalonic acid	*PMVK*	1q21.3
Squalene synthase deficiency	Squalene synthase	Farnesol Saturated and unsaturated branched-chain dicarboxylic acids Farnesol-derived glucuronides	*FDFT1*	8p23.1
Lanosterol C14-demethylase deficiency	Lanosterol C14-demethylase	Lanosterol 7-dehydrocholesterol	*CYP51A1*	7q21.2
C4-methyl sterol oxidase deficiency	C4-Methylsterol monooxygenase 1	C4-methyl sterols	*MSMO1*	4q32.3
CHILD syndrome	3β-hydroxysteroid dehydrogenase	Not detectable	*NSDHL*	Xq28
Conradi-Hunermann syndrome (CDPX2)	3β-hydroxysterol Δ^8-Δ^7-isomerase	Cholesta-8(9)-en-3β-ol 8(9)-dehydrocholesterol	*EBP*	Xp11.22-23
HEM skeletal dysplasia	3β-hydroxysterol Δ^{14}-reductase	Cholesta-8,14-dien-3β-ol	*LBR*	1q42
Lathosterolosis	3β-hydroxysterol Δ^5-desaturase	Lathosterol	*SC5D*	11q23.3
Desmosterolosis	3β-hydroxysterol Δ^{24}-reductase	Desmosterol	*DHCR24*	1p31.1-p33
Smith–Lemli–Opitz syndrome	3β-hydroxysterol Δ^7-reductase	7-dehydrocholesterol 8(9)-dehydrocholesterol	*DHCR7*	11q13
Affecting cholesterol biosynthesis				
POR deficiency	NADPH-cytochrome P450 reductase	Lanosterol (abnormal urinary steroid pattern)	*POR*	7q11.23
CTX	Sterol 27-hydroxylase	Cholesterol Cholestanol	*CYP27A1*	2q33-qter
Sitosterolemia	Sterolin-1 or sterolin-2	Cholesterol Phytosterols (ß-sitosterol, campesterol, stigmasterol) and 5α-saturated stanols	*ABCG5* *ABCG8*	2p21 2p21

pathogenic variants in the *PMVK* gene have been identified [10, 11]. Similar as MKD patients, the PMK-deficient patients present with inflammatory symptoms and increased mevalonic acid in body fluids and tissues.

Finally, several individuals from one family were recently diagnosed with **HMG-CoA reductase deficiency** due to a pathogenic homozygous loss-of-function missense variant in the *HMGCR* gene encoding HMG CoA-reductase, which resulted in a late-onset progressive form of limb girdle muscular dystrophy with similar features as statin-induced myopathy [12]. Statins are HMG-CoA reductase inhibitors that are widely used drugs to lower LDL cholesterol in individuals at risk of cardiovascular disease, including hypercholesterolemia. Cholesterol levels were normal in the patients; levels of other metabolites, such as HMG, have not been mentioned.

15.4 Post-squalene Disorders

Currently, eight inherited disorders due to defects of enzymes specifically involved in the synthesis of cholesterol have been described (Fig. 15.2)

[3–6]. Patients afflicted with these disorders often present with multiple morphogenic and congenital anomalies including internal organ, skeletal and/or skin abnormalities, underlining the pivotal role of cholesterol in human embryogenesis and development.

For most disorders (increased levels of) characteristic intermediate sterol species can be detected in cells, tissues, and/or body fluids reflecting the specific enzyme defect (Table 15.1), while cholesterol levels are often normal or low-normal. Conclusive diagnosis is usually reached by genetic testing.

Squalene synthase deficiency is caused by biallelic pathogenic variants in the *FDFT1* gene causing a deficiency of squalene synthase, which catalyzes the condensation of 2 farnesyl-PP moieties to form squalene, which is the first isoprenoid committed to sterol synthesis [13]. The deficiency results in low total and LDL-cholesterol and elevated farnesol in plasma and a unique urine metabolic profile with increased saturated and unsaturated branched-chain dicarboxylic acids and glucuronides derived from farnesol. Clinical symptoms may include profound global developmental delay, structural brain malformations, 2/3 syndactyly of the toes, and facial dysmorphisms. Laboratory diagnosis may involve GC-MS analysis of plasma sterols and organic acid analysis of urine.

Desmosterolosis is caused by biallelic pathogenic variants in the *DHCR24* gene encoding 3β-hydroxysterol Δ^{24}-reductase, which catalyzes the reduction of the Δ^{24} double bond of sterol intermediates (including desmosterol) [14, 15]. The deficiency results in elevated levels of desmosterol in plasma, tissue, and cultured patient cells. Clinical symptoms may include major CNS involvement, agenesis of the corpus callosum, dilated ventricles, facial dysmorphic features, arthrogryposis, severe failure to thrive, learning difficulties, epilepsy.

Lanosterol C14-Demethylase deficiency is due to biallelic pathogenic variants in the *CYP51A1* gene encoding lanosterol C14-Demethylase, which converts lanosterol into 4,4-dimethylcholesta-8(9),14,24-trien-3β-ol. Sterol profiling may reveal elevated lanosterol and 7-dehydrocholesterol levels in plasma, tissue, and cultured patient cells. Clinical symptoms include in particular cataracts and may include liver disease, developmental delay, and spastic diplegia [16, 17].

Hydrops-Ectopic Calcification-Moth-eaten (HEM) Skeletal Dysplasia or **Greenberg Skeletal Dysplasia** is caused by biallelic pathogenic variants in the *LBR* gene affecting the 3β-hydroxysterol Δ^{14}-reductase region of the encoded lamin B receptor [18]. This causes a deficiency of 3β-hydroxysterol Δ^{14}-reductase activity, required for the reduction of the Δ^{14} double bond in early sterol intermediates resulting in elevated levels of Δ^{14}-sterol intermediates that can be detected in tissues and cells of fetuses with HEM skeletal dysplasia. HEM—or Greenberg skeletal dysplasia is characterized by early in utero lethality. Affected fetuses typically present with severe fetal hydrops, short-limb dwarfism, an unusual "moth-eaten" appearance of the markedly shortened long bones, bizarre ectopic ossification centers and a marked disorganization of chondro-osseous histology, and may present with polydactyly and additional non-skeletal malformations [19].

The sterol C4-demethylase complex comprises 3 enzymes which jointly catalyze the sequential removal of the two methyl groups at the C4 position of early sterol precursors. Deficiencies of two of the enzymes have been identified as underlying cause of two different disorders. **C4-Methyl Sterol Oxidase Deficiency** (SMO Deficiency) is caused by biallelic pathogenic variants in the *MSMO1* gene encoding C4-methyl sterol oxidase. Sterol profiling may reveal elevated C4-methyl sterols in plasma and cultured skin fibroblasts. Clinical symptoms may include severe psoriasiform dermatitis, arthralgias, immune dysfunction, congenital cataracts, failure to thrive, short stature, microcephaly, and developmental delay [20, 21].

CHILD (Congenital hemidysplasia with ichthyosiform erythroderma and limb defects) syndrome is inherited as an X-linked dominant trait and due to heterozygous pathogenic variants in the *NSDHL* gene causing a deficiency of

3β-hydroxysteroid dehydrogenase [22]. As a consequence, one would expect elevated 4-methyl sterol precursors; however, these precursors are hardly or not detectable in plasma of patients. Cholesterol levels are normal. CHILD syndrome mostly affects females and is often lethal in hemizygous males, although few males with hypomorphic variants have been reported. Clinical symptoms show a striking unilateral distribution affecting the right side of the body more often than the left and include ichthyosiform skin lesions, alopecia, nail involvement and limb reduction defects with calcific stippling of the epiphysis.

X-linked dominant chondrodysplasia punctata 2 (CDPX2), also known as Conradi-Hünermann syndrome is also inherited as an X-linked dominant trait and due to heterozygous pathogenic variants in the *EBP* gene encoding 3β-hydroxysterol Δ^8-Δ^7 isomerase, which results in elevated plasma and cellular levels of cholesta-8(9)-en-3β-ol and 8-dehydrocholesterol, while plasma cholesterol levels are often (low) normal [23, 24]. The defect mostly affect females, and patients show similar clinical symptoms as CHILD syndrome patients, but then often bilateral and asymmetric. Additional symptoms include cataracts and skeletal abnormalities including short stature, asymmetric rhizomelic shortening of limbs, calcific stippling of the epiphyseal regions, and craniofacial defects.

Lathosterolosis is caused by biallelic pathogenic variants in the *SC5D* gene, causing a deficiency of 3β-hydroxysterol Δ^5-desaturase, which introduces the C5-C6 double bond in lathosterol to produce 7-dehydrocholesterol. Consequently, elevated levels of lathosterol can be detected in plasma, tissue, and cultured cells of patients. Cholesterol levels can be normal. Clinical symptoms may include growth failure, microcephaly, ptosis, cataracts, short nose, micrognathia, prominent alveolar ridges, ambiguous genitalia, bilateral syndactyly of the second and third toes, and bilateral postaxial hexadactyly of the feet [25].

Smith–Lemli–Opitz Syndrome (SLOS) is the most common defect of cholesterol biosynthesis and caused by biallelic pathogenic variants in the *DHCR7* gene, causing a deficiency of 3β-hydroxysterol Δ^7-reductase, which catalyzes the predominant final step in cholesterol biosynthesis, i.e. the reduction of the C7-C8 double bond of 7-dehydrocholesterol to produce cholesterol [26, 27]. Consequently, low cholesterol and increased 7-dehydrocholesterol levels are detected in plasma, cells, and tissues of most SLOS patients. In addition, increased 8-dehydrocholesterol plasma levels can be detected. Clinical symptoms may include a large variety of morphogenic and congenital anomalies constituting a clinical continuum ranging from hardly recognizable through mild to very severe. Frequent symptoms are a characteristic craniofacial appearance with microcephaly, a short nose with broad nasal bridge and anteverted nares, a long philtrum, micro-/retrognathia and often blepharoptosis, low-set, posteriorly rotated ears, cleft or high-arched palate, pale hair, and broad or irregular alveolar ridges. Common limb abnormalities include cutaneous syndactyly of the second and third toes, short proximally placed thumbs and, in more severe cases, postaxial polydactyly. Also common are genital abnormalities, congenital heart defects, renal, adrenal, lung, and gastrointestinal anomalies, profound prenatal and postnatal growth retardation, neonatal ascites, cholestatic jaundice, intellectual disability, feeding difficulties, behavioral problems, sleeping disorder, and sunlight sensitivity [28].

15.5 Disorders Affecting Isoprenoid/Cholesterol Biosynthesis

A few disorders have been described that are not due to a primary defect of any of the enzymes involved in isoprenoid biosynthesis, but to a secondary effect on one or more of the enzymes, the entire flux of the pathway and/or the levels of cholesterol or derivatives thereof.

Cerebrotendinous xanthomatosis (CTX) is a lipid-storage disease due to biallelic pathogenic variants in the *CYP27A1* gene encoding the bile acid biosynthetic enzyme sterol 27-hydroxylase.

Although a defect of bile acid synthesis, CTX patients typically have large deposits of cholesterol and cholestanol (the 5-alpha-dihydro derivative of cholesterol) in virtually every tissue, particularly the Achilles tendons, brain, and lungs [29]. Plasma cholesterol concentrations can be low normal in CTX patients and in untreated patients 7- and 8-dehydrocholesterol and 7- and 8-lathosterol can be highly elevated [30].

Sitosterolemia can be caused by biallelic pathogenic variants in the *ABCG5* or the *ABCG8* gene, which encode two ABC transporter proteins involved in limiting intestinal absorption and promoting biliary excretion of sterols. The defect is characterized by unrestricted intestinal absorption of both cholesterol and plant-derived cholesterol-like molecules, such as β-sitosterol, campesterol, and stigmasterol. HMG-CoA reductase levels are reduced causing a severely reduced overall hepatic cholesterol biosynthesis but an enhanced total and receptor-mediated uptake of low density lipoprotein. Patients with this disorder have high levels of plant sterols (and cholesterol) in plasma and develop tendon and tuberous xanthomas, accelerated atherosclerosis, and premature coronary artery disease.

POR deficiency is caused by biallelic pathogenic variants in the *POR* gene encoding NADPH-cytochrome P450 oxidoreductase (POR), which may result in a deficiency of lanosterol C14-demethylase associated with elevated plasma lanosterol, similar as found in patients with pathogenic variants in the *CYP51A1* gene encoding lanosterol C14-demethylase. POR is required for the activity of all known human microsomal (type II) P450 enzymes, which play important roles in steroid and sterol synthesis, including lanosterol C14-demethylase, as well as in hepatic drug metabolism [31]. Accordingly, POR deficiency is characterized by an abnormal urinary steroid pattern, including significantly elevated metabolites associated with deficiency of 17α-hydroxylase and 21-hydroxylase. Increased plasma lanosterol can be an indication for lanosterol C14-demethylase deficiency as well as POR deficiency. To discriminate between these two possibilities, genetic testing of the respective genes is indicated.

15.6 Properties of Metabolites

See Table 15.2.

15.7 Methods

15.7.1 Principle

Following addition of internal standards to samples, sterols are saponified with KOH in ethanol, extracted with hexane, dried under nitrogen, and derivatized with BSTFA and 1% TMCS. The trimethylsilylated derivatives of the sterols are separated by gas chromatography (GC) and detected using a flame ionization detector (FID). Sterol quantification is based on the detected concentration of the internal standard. The identity of the detected sterols can be verified using gas chromatography/mass spectrometry (GC/MS) of the trimethylsilylated derivatives of the sterols.

15.7.2 Pre-analytical

15.7.2.1 Sample/Matrix Requirements

Sterol analysis and quantification in plasma, serum, and cerebral spinal fluid (CSF) samples can be performed without pre-extraction. Analysis of sterols from tissues and cells requires a pre-extraction work-up.

Requirements for the different matrices are as follows:

Plasma: Isolated from total blood samples obtained by venipuncture and collected in standard heparin or EDTA glass tubes. Separated from cells and harvested as supernatant after centrifugation for 10 min at 1600 × *g* at room temperature. Storage at –18 °C or lower (stable for >1 year). Minimum amount required for analysis: 150 μL.

Serum: Isolated from total blood samples obtained by venapuncture and collected in glass tubes. Separated from cells and harvested as supernatant after centrifugation for 10 min at 1600 × *g* at room temperature following a clotting period of at least 1 h at room temperature. Storage at –18 °C or lower (stable for >1 year). Minimum amount required for analysis: 150 μL.

CSF: No pre-treatment necessary. Storage at –18 °C or lower (stable for >1 year). Minimum amount required for analysis: 150 μL.

Tissues: Small pieces of tissue, stored at –80 °C. Minimum amount required for analysis: 10 mg.

Cells: Lymphoblasts, primary skin fibroblasts or other cells cultured for 4–7 days in lipoprotein-depleted medium, harvested by centrifugation and used fresh or after storage as pellets at –18 °C or lower. Minimum amount required for analysis: ~1 × 10^6 cells.

Table 15.2 Properties of metabolites used as internal standards and/or for the diagnosis of the different disorders (see also Table 15.1)

Trivial name	Official name	Molecular formula	Molecular weight	Chemical structure
Mevalonic acid	Mevalonic acid	$C_6H_{12}O_4$	132.14	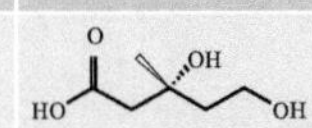
Farnesol	Farnesol	$C_{15}H_{26}O$	222.37	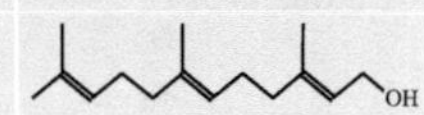
Epicoprostanol	5ß-cholestan-3α-ol	$C_{27}H_{48}O$	388.67	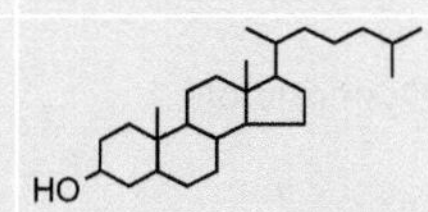
Campesterol	24α-methyl-5-cholestene-3ß-ol	$C_{28}H_{48}O$	400.68	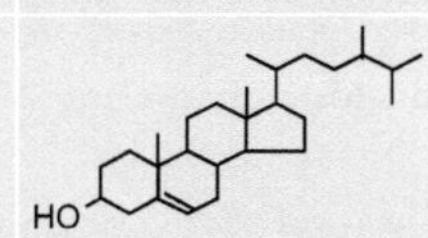
Campestanol	24α-methyl-5-cholestan-3ß-ol	$C_{28}H_{50}O$	402.70	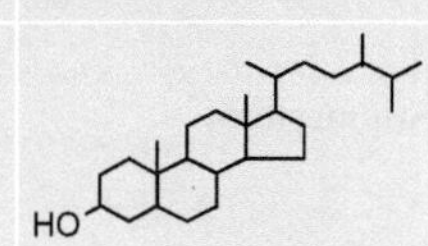
Coprostanol	5ß-cholestan-3ß-ol	$C_{27}H_{48}O$	388.67	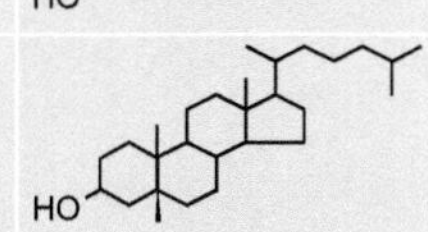
Cholestanol	5α-cholestan-3ß-ol	$C_{27}H_{48}O$	388.67	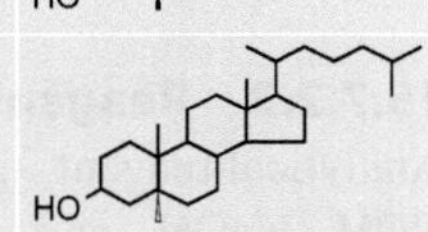
Cholesterol	5α-cholesten-3ß-ol	$C_{27}H_{46}O$	386.65	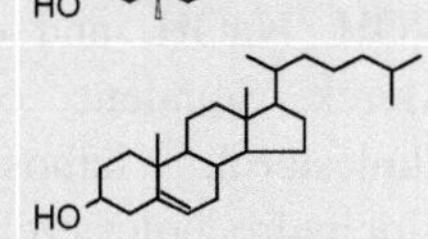
C4-methyl sterols	4,4′-dimethylcholesta-8(9)-en-3β-ol	$C_{29}H_{50}O$	414.71	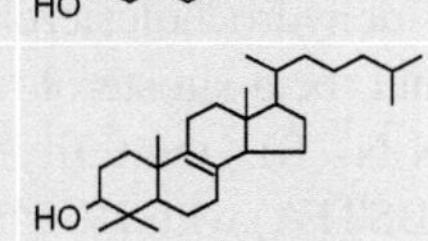
Lanosterol	4,4,14α-trimethylcholesta-8(9),24-dien-3β-ol	$C_{30}H_{50}O$	426.72	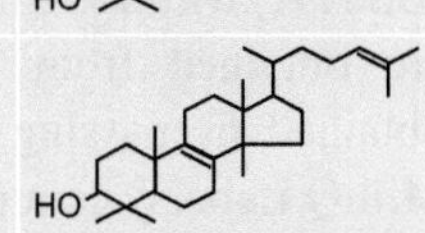
7- Dehydrocholesterol	5,7-cholestdien-3ß-ol	$C_{27}H_{44}O$	384.64	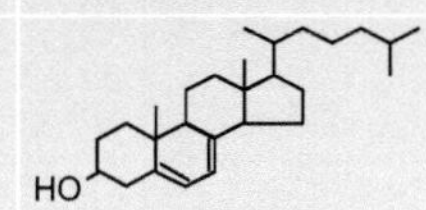

(continued)

Table 15.2 (continued)

Trivial name	Official name	Molecular formula	Molecular weight	Chemical structure
8(9)- dehydrocholesterol	5,8-cholestdien-3ß-ol	$C_{27}H_{44}O$	384.64	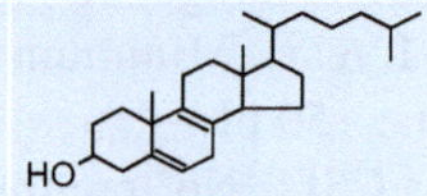
Cholesta-8(9)-en-3β-ol	Cholesta-8(9)-en-3β-ol	$C_{27}H_{46}O$	386.65	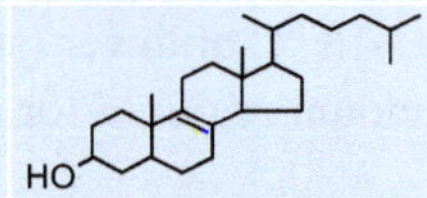
Cholesta-8,14-dien-3β-ol	Cholesta-8,14-dien-3β-ol	$C_{27}H_{44}O$	384.64	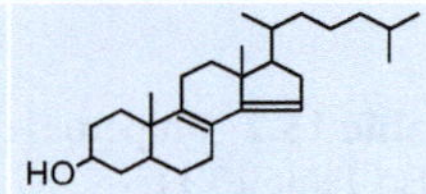
7-Lathosterol	5α-cholest-7-en-3ß-ol	$C_{27}H_{46}O$	386.65	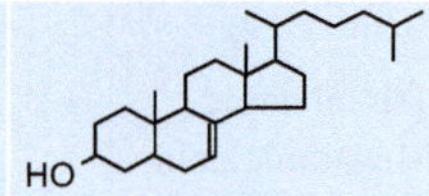
8-Lathosterol	5α-cholest-8-en-3ß-ol	$C_{27}H_{46}O$	386.65	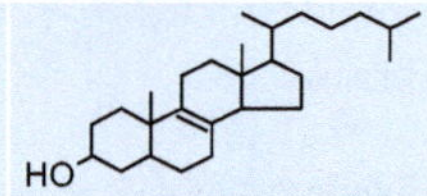
Desmosterol	5,24-cholestdien-3ß-ol	$C_{27}H_{44}O$	384.64	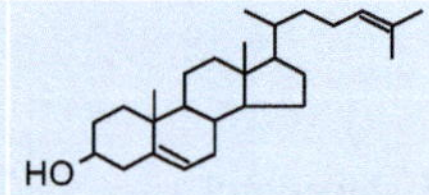
ß-sitosterol	24α-ethyl-5-cholestene-3ß-ol	$C_{29}H_{50}O$	414.71	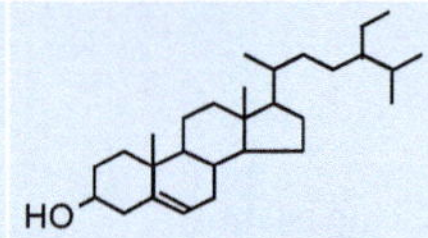
Stigmasterol	24α-ethyl-5,22-cholestdien-3ß-ol	$C_{29}H_{48}O$	412.69	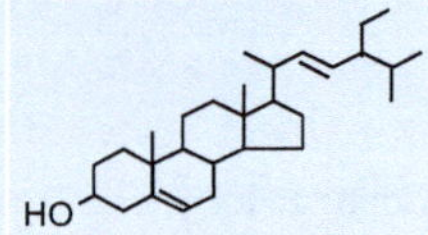

15.7.2.2 Reagents and Chemicals

Analytical/reagent grade n-hexane, heptane, KOH, NaOH, and ethanol are obtained from Merck. Squalene, epicoprostanol, cholestanol, cholesterol, lathosterol, desmosterol and 7-dehydrocholesterol, stigmasterol, campesterol and beta-sitosterol are obtained from Sigma. N,N-(Bis-trimethylsilyl)-trifluoroacetamide (BSTFA) with 1% trimethylchlorosilane (TMCS) are obtained from Pierce. MilliQ water was obtained by passing deionized water through a MilliQ Labo system (Millipore).

15.7.2.3 Preparation of Reagents

- Internal standard: Prepare a solution of ~0.8 mM epicoprostanol in ethanol (store at 4 °C for up to 2 years)
- Saponification reagent: Dissolve 1 g KOH in 5 mL MilliQ water and add 45 mL ethanol to make a 0.3 N KOH/EtOH solution (store at RT for up to 2 years)
- Sterol mixture: Dissolve 1 mg squalene, epicoprostanol, cholestanol, cholesterol, 7-lathosterol, desmosterol, campesterol, stigmasterol, stigmastanol, 7-dehydrocholesterol,

and β-sitosterol in 20 mL heptane (store at 4 °C for at least 2 years)
- BSTFA with 1% TMCS is used as obtained from the supplier.

15.7.2.4 Instrumentation

Gas Chromatography (GC)
- Agilent Gas Chromatograph type 7890B-2 with flame ionization detector (FID) plus injector (or similar instrument) connected to PC with Chromeleon 7.2 integration software (Thermo) (or similar software)
- Capillary columns:
  GC: CPsil 5; 25 m (l) × 0.25 mm (ID) × 0.25 μm(f)) (Varian CP7441)
  GC: CPsil 19; 25 m (l) × 0.25 mm (ID) × 0.20 μm(f)) (Varian CP7712) (or similar columns)

Gas Chromatography/Mass Spectrometry (GC/MS)
- Agilent Gas Chromatograph type 7890B-2 coupled to an Agilent MSD5977A Mass Selective Detector (MSD), with ChemStation/Masshunter software (Agilent) (or similar instrument)
- Capillary column:
  GCMSD: CPsil5 Low bleed MS; 25 m (l) × 0.25 mm (ID) × 0.25 μm(f)) (Varian CP7842) (or similar column)

Other
- Hydrogen generator
- Temperature-controlled incubator
- Sample concentrator
- Heating block
- Sonicator

15.7.2.5 GC and GC/MS Settings

	GC settings	GC/MS settings
Column	CP SIL5-CB	CP SIL5-CB
Injection	Splitless	Splitless
Carrier	Helium (1.5 mL/min)	Helium (1.5 mL/min)
Pressure	Constant flow	Constant flow
Detection	FID	MSD
H_2	Flow: 30 ± 5 mL/min	–
N_2	Flow: 25 ± 5 mL/min	–
Air	Flow: 400 ± 50 mL/min	–
Detector temp.	280 °C	290 °C
Injector temp.	290 °C	290 °C
Injection volume	1 μL	2 μL
Purge	2 min	2 min

15.7.3 Temperature Program for Both GC and GC/ MS

Retention time (min)	°C/min	Target	Hold time (min)
0	0	100 °C	–
2	0	100 °C	2
30	30	280 °C	16

15.7.4 Analytical Procedures

15.7.4.1 Extraction of Sterols from Plasma and Serum

- Thaw plasma, serum and Internal Standard on ice
- Set temperature-controlled incubator at 80 °C
- Collect and label clean 4 mL glass vials including screw caps and Teflon liner (Alltech)
- Transfer 50 μL plasma or serum into a 4 mL glass vial
- Add 50 μL Internal Standard using a glass pipette
- Add 0.5 mL saponification reagent and close vial tightly with screw cap and insert
- Mix contents carefully and incubate for at least 2 h at 80 °C
- Cool sample to room temperature
- Add 250 μL milliQ water
- Add 1 mL n-hexane and mix for 10 s by vortex
- Transfer 300 μL from upper hexane layer into clean 4 mL glass vial
- Evaporate samples to dryness under N_2 at 40 °C
- Dissolve the dried contents in 120 μL BSTFA +1% TMCS
- Incubate the sample for 30 min at 80 °C

- Cool sample to room temperature
- Transfer 60 μL of the sample into a crimp top vial and close vial with cap and insert (Alltech). This sample is ready for GC analysis
- Transfer 60 μL of the sample into a crimp top vial and close vial with cap and insert (Alltech). This sample is ready for GC/MS analysis

15.7.4.2 Extraction of Sterols from Cerebral Spinal Fluid

- Same procedure as for plasma and serum

15.7.4.3 Extraction of Sterols from Tissues

- Thaw small piece of tissue on ice
- Set temperature-controlled incubator at 80 °C
- Collect and label clean 4 mL glass vials including screw caps and Teflon liner (Alltech)
- Homogenize tissue in 1 mL saponification reagent containing 10 μM Internal Standard, transfer suspension into a 4 mL glass vial and close vial tightly with screw cap and insert
- Mix contents carefully and incubate for at least 2 h at 80 °C
- Cool sample to room temperature
- Add 500 μL milliQ water
- Add 2 mL n-hexane and mix for 10 s by vortex
- Transfer in 500 μL from upper hexane layer into a clean 4 mL glass vial
- Evaporate samples to dryness under N_2 at 40 °C
- Dissolve the dried contents of the vial in 100 μL BSTFA +1% TMCS
- Incubate the sample for 30 min at 80 °C
- Transfer the sample after cooling down into a crimp top vial and close vial with cap and insert (Alltech). This sample is ready for GC/MS analysis

15.7.4.4 Extraction of Sterols from Cells

- Thaw cell pellets on ice
- Set temperature-controlled incubator at 80 °C
- Collect and label clean 4 mL glass vials including screw caps and Teflon liner (Alltech)
- Resuspend cells in 1 mL saponification reagent containing 10 μM Internal Standard, transfer suspension into a 4 mL glass vial and close vial tightly with screw cap and insert
- Mix contents carefully and incubate for at least 2 h at 80 °C
- Cool sample to room temperature
- Add 500 μL milliQ water
- Add 2 mL n-hexane and mix for 10 s by vortex
- Transfer in 500 μL from upper hexane layer into 1 mL glass vial
- Evaporate samples to dryness under N_2 at 40 °C
- Dissolve the dried contents of the vial in 100 μL BSTFA +1% TMCS
- Incubate the sample for 30 min at 75 °C
- Transfer the sample after cooling down into a crimp top vial and close vial with cap and insert (Alltech). This sample is ready for GC/MS analysis

15.7.4.5 GC Analysis of Extracted Sterols

- Use GC settings described above
- Bring H_2, N_2 and Air to pressure
- Ignite the FID
- Re-inject 1 μL heptane without sterols until baseline in chromatogram is stable
- Inject standard sterol mixture to confirm the retention times of the various sterols
- Inject samples to be analyzed

15.7.4.6 GC/MS Analysis of Extracted and Derivatized Sterols

- Use GC/MS settings described above
- Inject standard sterol mixture
- Inject samples to be analyzed

15.7.4.7 Calculation

The concentration of each sterol in plasma, serum, or liquor is calculated in μmol/L through integration of the corresponding peak area multiplied with the amount of Internal Standard added (in nmol) multiplied with the response factor of the metabolite, divided by the peak area of the internal standard multiplied by the number of milliliters of the sample.

15.7.5 Post-analytical

15.7.5.1 Interpretation

The detection of specific and often isolated intermediate sterols that are not observed in control samples or in higher concentrations than the reference range may be indicative for a specific enzyme defect in cholesterol biosynthesis (Table 15.1). Such elevated levels of intermediate sterols often but not necessarily coincide with decreased levels of cholesterol. The finding of elevated levels of intermediate sterols warrants further diagnostic investigations, including sterol profiling in cultured cells and genetic testing.

15.7.5.2 Chromatograms

See Fig. 15.3.

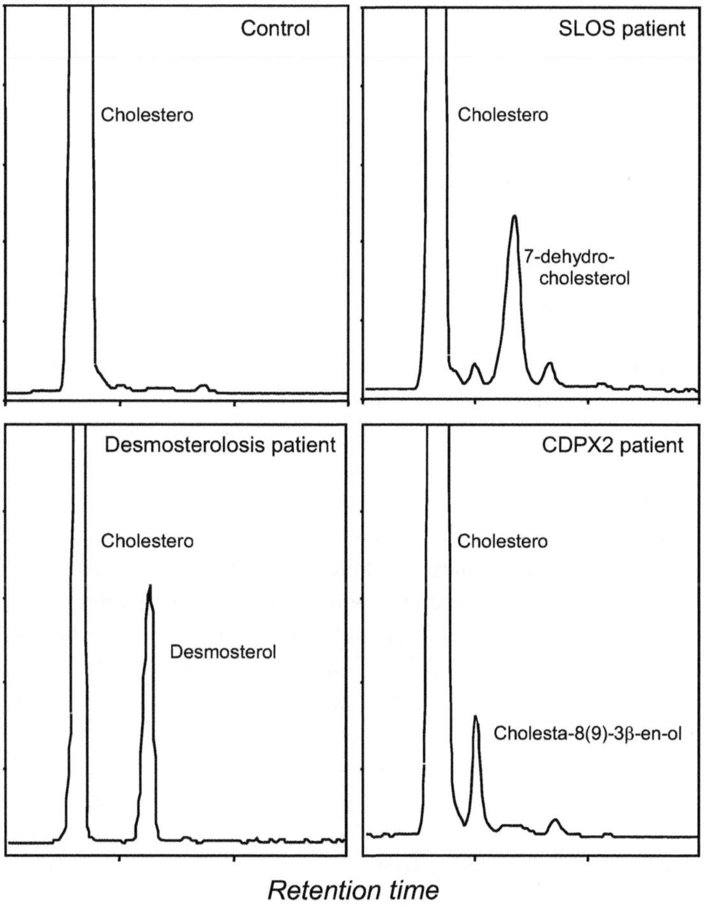

Fig. 15.3 Examples of sterol analysis in patients with defective cholesterol biosynthesis. GC-MS analysis of trimethylsilyl derivatives of sterols extracted from primary skin fibroblasts of a control subject, a Smith–Lemli–Opitz syndrome (SLOS) patient and a Conradi–Hunermann syndrome (CDPX2) patient, and lymphoblasts of a desmosterolosis patient cultured in lipoprotein-deficient medium reveals the accumulation of sterol intermediates indicative of a defect in cholesterol biosynthesis. Similar spectra can be obtained by sterol analysis of plasma of such patients

15.7.5.3 Reference Values

Sterol	Age	Plasma/serum (μmol/L)
%Cholestanol/cholesterol	0–100 days	0.12–0.72
%Cholestanol/cholesterol	>100 days	0.1–0.28
7-dehydrocholesterol	>0 days	0
7-lathosterol	>0 days	0–12
8,24-cholestadien-3β-ol	>0 days	0–0.2
8-dehydrocholesterol	>0 days	0
8-Lathosterol	>0 days	0–2
Campesterol	>0 days	0–23
Cholestanol	0–100 days	2.8–19
Cholestanol	>100 days	3.5–10
Cholesterol	0–365 days	1881–4887
Cholesterol	366–6570 days	2468–4827
Cholesterol	>6571 days	2889–5985
Desmosterol	>0 days	0–7
Lanosterol	>0 days	0–0.2
Sitostanol	>0 days	0–2
Sitosterol	>0 days	0–16
Stigmasterol	>0 days	0–10

15.7.5.4 Typical Pathological Values

Intermediate sterol levels higher than observed (if observed at all) in control samples in conjunction with specific clinical symptoms may point to a defect in cholesterol biosynthesis and should be followed up by additional studies as outlined below. Although cholesterol levels are often reduced they may be (low) normal.

Note: Some medications may exert an effect on cholesterol biosynthesis and can result in elevated concentrations of intermediate sterol levels. Known examples include haloperidol, aripiprazole, and trazodone, which inhibit the enzyme 3β-hydroxysterol Δ^7-reductase causing elevated levels of 7-dehydrocholesterol, the metabolite that also accumulates in Smith–Lemli–Opitz syndrome, due to deficiency of this enzyme [28, 32], and amiodarone, which results in accumulation of desmosterol [33].

15.7.5.5 Quality Control

As internal control for sterol analysis in plasma, serum, and liquor, each series of samples includes an 150 μL aliquot of a large batch of control plasma samples in which cholestanol and cholesterol concentrations are determined and compared with previously determined concentrations. The values for these should fall within ±2 standard deviations. The control plasma sample is composed of pooled plasma samples of different individuals stored as 150 μL aliquots at −18 °C.

15.7.5.6 Follow-up Assays

Assays to determine the activities of enzymes defective in the 8 known disorders of post-squalene cholesterol biosynthesis have only been described for 3β-hydroxysterol Δ7-reductase (defective in Smith–Lemli–Opitz syndrome) and 3β-hydroxysterol Δ24-reductase, but are not useful for diagnostic purposes.

As an alternative, primary skin fibroblasts or lymphoblasts of patients suspected to be affected with a post-squalene cholesterol biosynthesis defect can be cultured for 3–7 days in medium supplemented with fetal calf serum depleted from lipoproteins to induce cholesterol biosynthesis whereupon the specific defect can be determined by sterol analysis using GC-MS as described above. This procedure will readily identify patients affected with most sterol biosynthesis disorders except patients with CHILD syndrome.

For conclusive diagnosis of all defects, molecular testing through sequencing of the respective genes is essential. Molecular testing is also the primary choice for prenatal testing.

15.8 Conclusion

In this chapter, we have described an analytical method that can be used to measure different sterol intermediates in different matrices and how these measurements can be used for the (differential) diagnosis of patients with an inherited disorder of cholesterol biosynthesis.

References

1. Goldstein JL, Brown MS (1990) Regulation of the mevalonate pathway. Nature 343(6257):425–430
2. Luo J, Yang H, Song BL (2020) Mechanisms and regulation of cholesterol homeostasis. Nat Rev Mol Cell Biol 21(4):225–245. https://doi.org/10.1038/s41580-019-0190-7
3. Waterham HR (2006) Defects of cholesterol biosynthesis. FEBS Lett 580(23):5442–5449. https://doi.org/10.1016/j.febslet.2006.07.027
4. Herman GE, Kratz L (2012) Disorders of sterol synthesis: beyond Smith-Lemli-Opitz syndrome. Am J Med Genet C Semin Med Genet 160C(4):301–321. https://doi.org/10.1002/ajmg.c.31340
5. Waterham HRC, PT (2022) Disorders of isoprenoid/cholesterol synthesis. In: Jean-Marie Saudubray MRB, Garcia-Cazorla A, Walter JH (eds) Inborn metabolic diseases: diagnosis and treatment. 7th edn Springer-Verlag GmbH, Heidelberg, Germany, pp 693–704
6. Kratz LE, Kelley RI (2022) Disorders of cholesterol biosynthesis. In: Nenad Blau CDV, Ferreira CR, Vianey-Saban C, van Karnebeek CDM (eds) Physician's guide to the diagnosis, treatment, and follow-up of inherited metabolic diseases, vol II, 2nd edn. Springer Nature, Cham, Switzerland, pp 1057–1076
7. van der Hilst JCH, Bodar EJ, Barron KS, Frenkel J, Drenth JPH, van der Meer JWM, Simon A, International HSG (2008) Long-term follow-up, clinical features, and quality of life in a series of 103 patients with hyperimmunoglobulinemia D syndrome. Medicine (Baltimore) 87(6):301–310. https://doi.org/10.1097/MD.0b013e318190cfb7
8. Politiek FA, Waterham HR (2021) Compromised protein prenylation as pathogenic mechanism in mevalonate kinase deficiency. Front Immunol 12:724991. https://doi.org/10.3389/fimmu.2021.724991
9. Hoffmann GF, Sweetman L, Bremer HJ, Hunneman DH, Hyanek J, Kozich V, Lehnert W, Nyhan WL, Speidel I, Trefz FK et al (1991) Facts and artefacts in mevalonic aciduria: development of a stable isotope dilution GCMS assay for mevalonic acid and its application to physiological fluids, tissue samples, prenatal diagnosis and carrier detection. Clin Chim Acta 198(3):209–227. https://doi.org/10.1016/0009-8981(91)90355-g
10. Yildiz C, Gezgin Yildirim D, Inci A, Tumer L, Cengiz Ergin FB, Sunar Yayla ENS, Esmeray Senol P, Karacayir N, Egritas Gurkan O, Okur I, Ezgu FS, Bakkaloglu SA (2023) A possibly new autoinflammatory disease due to compound heterozygous phosphomevalonate kinase gene mutation. Joint Bone Spine 90(1):105490. https://doi.org/10.1016/j.jbspin.2022.105490
11. Berner J, van de Wetering C, Jimenez Heredia R, Rashkova C, Ferdinandusse S, Koster J, Weiss JG, Frohne A, Giuliani S, Waterham HR, Castanon I, Brunner J, Boztug K (2023) Phosphomevalonate kinase deficiency expands the genetic spectrum of systemic autoinflammatory diseases. J Allergy Clin Immunol 152(4):1025–1031.e2. https://doi.org/10.1016/j.jaci.2023.06.013
12. Yogev Y, Shorer Z, Koifman A, Wormser O, Drabkin M, Halperin D, Dolgin V, Proskorovski-Ohayon R, Hadar N, Davidov G, Nudelman H, Zarivach R, Shelef I, Perez Y, Birk OS (2023) Limb girdle muscular disease caused by HMGCR mutation and statin myopathy treatable with mevalonolactone. Proc Natl Acad Sci U S A 120(7):e2217831120. https://doi.org/10.1073/pnas.2217831120
13. Coman D, Vissers L, Riley LG, Kwint MP, Hauck R, Koster J, Geuer S, Hopkins S, Hallinan B, Sweetman L, Engelke UFH, Burrow TA, Cardinal J, McGill J, Inwood A, Gurnsey C, Waterham HR, Christodoulou J, Wevers RA, Pitt J (2018) Squalene synthase deficiency: clinical, biochemical, and molecular characterization of a defect in cholesterol biosynthesis. Am J Hum Genet 103(1):125–130. https://doi.org/10.1016/j.ajhg.2018.05.004
14. Waterham HR, Koster J, Romeijn GJ, Hennekam RC, Vreken P, Andersson HC, FitzPatrick DR, Kelley RI, Wanders RJ (2001) Mutations in the 3beta-hydroxysterol Delta24-reductase gene cause desmosterolosis, an autosomal recessive disorder of cholesterol biosynthesis. Am J Hum Genet 69(4):685–694
15. Schaaf CP, Koster J, Katsonis P, Kratz L, Shchelochkov OA, Scaglia F, Kelley RI, Lichtarge O, Waterham HR, Shinawi M (2011) Desmosterolosis-phenotypic and molecular characterization of a third case and review of the literature. Am J Med Genet A 155A(7):1597–1604. https://doi.org/10.1002/ajmg.a.34040
16. Gillespie RL, O'Sullivan J, Ashworth J, Bhaskar S, Williams S, Biswas S, Kehdi E, Ramsden SC, Clayton-

Smith J, Black GC, Lloyd IC (2014) Personalized diagnosis and management of congenital cataract by next-generation sequencing. Ophthalmology 121 (11):2124-2137. e2121–2122. doi:https://doi.org/10.1016/j.ophtha.2014.06.006
17. Patel N, Anand D, Monies D, Maddirevula S, Khan AO, Algoufi T, Alowain M, Faqeih E, Alshammari M, Qudair A, Alsharif H, Aljubran F, Alsaif HS, Ibrahim N, Abdulwahab FM, Hashem M, Alsedairy H, Aldahmesh MA, Lachke SA, Alkuraya FS (2017) Novel phenotypes and loci identified through clinical genomics approaches to pediatric cataract. Hum Genet 136(2):205–225. https://doi.org/10.1007/s00439-016-1747-6
18. Waterham HR, Koster J, Mooyer P, Noort GG, Kelley RI, Wilcox WR, Wanders RJ, Hennekam RC, Oosterwijk JC (2003) Autosomal recessive HEM/Greenberg skeletal dysplasia is caused by 3 beta-hydroxysterol delta 14-reductase deficiency due to mutations in the lamin B receptor gene. Am J Hum Genet 72(4):1013–1017
19. Porter FD, Herman GE (2011) Malformation syndromes caused by disorders of cholesterol synthesis. J Lipid Res 52(1):6–34. https://doi.org/10.1194/jlr.R009548
20. He M, Kratz LE, Michel JJ, Vallejo AN, Ferris L, Kelley RI, Hoover JJ, Jukic D, Gibson KM, Wolfe LA, Ramachandran D, Zwick ME, Vockley J (2011) Mutations in the human SC4MOL gene encoding a methyl sterol oxidase cause psoriasiform dermatitis, microcephaly, and developmental delay. J Clin Invest 121(3):976–984. https://doi.org/10.1172/JCI42650
21. Kalay Yildizhan I, Gokpinar Ili E, Onoufriadis A, Kocyigit P, Kesidou E, Simpson MA, McGrath JA, Kutlay NY, Kundakci N (2020) New homozygous missense MSMO1 mutation in two siblings with SC4MOL deficiency presenting with psoriasiform dermatitis. Cytogenet Genome Res 160(9):523–530. https://doi.org/10.1159/000511126
22. Liu XY, Dangel AW, Kelley RI, Zhao W, Denny P, Botcherby M, Cattanach B, Peters J, Hunsicker PR, Mallon AM, Strivens MA, Bate R, Miller W, Rhodes M, Brown SD, Herman GE (1999) The gene mutated in bare patches and striated mice encodes a novel 3beta-hydroxysteroid dehydrogenase. Nat Genet 22(2):182–187. https://doi.org/10.1038/9700
23. Braverman N, Lin P, Moebius FF, Obie C, Moser A, Glossmann H, Wilcox WR, Rimoin DL, Smith M, Kratz L, Kelley RI, Valle D (1999) Mutations in the gene encoding 3 beta-hydroxysteroid-delta 8, delta 7-isomerase cause X-linked dominant Conradi-Hunermann syndrome. Nat Genet 22(3):291–294. https://doi.org/10.1038/10357
24. Derry JM, Gormally E, Means GD, Zhao W, Meindl A, Kelley RI, Boyd Y, Herman GE (1999) Mutations in a delta 8-delta 7 sterol isomerase in the tattered mouse and X-linked dominant chondrodysplasia punctata. Nat Genet 22(3):286–290. https://doi.org/10.1038/10350. jderry@immunex.com
25. Brunetti-Pierri N, Corso G, Rossi M, Ferrari P, Balli F, Rivasi F, Annunziata I, Ballabio A, Russo AD, Andria G, Parenti G (2002) Lathosterolosis, a novel multiple-malformation/mental retardation syndrome due to deficiency of 3beta-hydroxysteroid-delta5-desaturase. Am J Hum Genet 71(4):952–958. https://doi.org/10.1086/342668
26. Fitzky BU, Witsch-Baumgartner M, Erdel M, Lee JN, Paik YK, Glossmann H, Utermann G, Moebius FF (1998) Mutations in the Delta7-sterol reductase gene in patients with the Smith-Lemli-Opitz syndrome. Proc Natl Acad Sci U S A 95(14):8181–8186. https://doi.org/10.1073/pnas.95.14.8181
27. Waterham HR, Wijburg FA, Hennekam RC, Vreken P, Poll-The BT, Dorland L, Duran M, Jira PE, Smeitink JA, Wevers RA, Wanders RJ (1998) Smith-Lemli-Opitz syndrome is caused by mutations in the 7-dehydrocholesterol reductase gene. Am J Hum Genet 63(2):329–338
28. Kelley RI, Hennekam RC (2000) The Smith-Lemli-Opitz syndrome. J Med Genet 37(5):321–335. https://doi.org/10.1136/jmg.37.5.321
29. Nobrega PR, Bernardes AM, Ribeiro RM, Vasconcelos SC, Araujo D, Gama VCV, Fussiger H, Santos CF, Dias DA, Pessoa ALS, Pinto W, Saute JAM, de Souza PVS, Braga-Neto P (2022) Cerebrotendinous Xanthomatosis: a practice review of pathophysiology, diagnosis, and treatment. Front Neurol 13:1049850. https://doi.org/10.3389/fneur.2022.1049850
30. Wolthers BG, Walrecht HT, van der Molen JC, Nagel GT, Van Doormaal JJ, Wijnandts PN (1991) Use of determinations of 7-lathosterol (5 alpha-cholest-7-en-3 beta-ol) and other cholesterol precursors in serum in the study and treatment of disturbances of sterol metabolism, particularly cerebrotendinous xanthomatosis. J Lipid Res 32(4):603–612
31. Idkowiak J, Cragun D, Hopkin RJ, Arlt W (1993) Cytochrome P450 oxidoreductase deficiency. In: Adam MP, Mirzaa GM, Pagon RA et al (eds) GeneReviews((R))
32. Hall P, Michels V, Gavrilov D, Matern D, Oglesbee D, Raymond K, Rinaldo P, Tortorelli S (2013) Aripiprazole and trazodone cause elevations of 7-dehydrocholesterol in the absence of Smith-Lemli-Opitz syndrome. Mol Genet Metab 110(1–2):176–178. https://doi.org/10.1016/j.ymgme.2013.04.004
33. Simonen P, Lehtonen J, Lampi AM, Piironen V, Stenman UH, Kupari M, Gylling H (2018) Desmosterol accumulation in users of amiodarone. J Intern Med 283(1):93–101. https://doi.org/10.1111/joim.12682

Porphyrins, Porphobilinogen, and δ-Aminolevulinic Acid

16

Matthew J. Schultz, Patricia L. Hall, and Silvia Tortorelli

Contents

List of Abbreviations

AHPs	Acute hepatic porphyrias
AIP	Acute intermittent porphyria
ALA	δ-Aminolevulinic acid
ALAD	Aminolevulinic acid dehydratase
CEP	Congenital erythropoietic porphyria
EPP	Erythropoietic protoporphyria
HCP	Hereditary coproporphyria
HEP	Hepatoerythropoietic porphyria
PBG	Porphobilinogen
PBGD	Porphobilinogen deaminase
PCT	Porphyria cutanea tarda
UPGC	Uroporphyrinogen III carboxylase
UROD	Uroporphyrinogen III decarboxylase
VP	Variegate porphyria
XLP	X-linked protoporphyria

M. J. Schultz · P. L. Hall · S. Tortorelli (✉)
Department of Laboratory Medicine and Pathology, Mayo Clinic, Rochester, MN, USA
e-mail: tortorelli.silvia@mayo.edu

16.1 Introduction

Porphyrias are disorders arising from inherited or acquired enzymatic defects in the heme biosynthetic pathway. Currently eight porphyria disorders are described which are often grouped into acute hepatic porphyrias and cutaneous porphyrias. As a group, genetic porphyrias demonstrate various modes of inheritance including autosomal dominant, autosomal recessive, and X-linked transmission. Some porphyrias exhibit reduced penetrance demonstrating the importance of modifier genes and environmental influences on clinical expressivity. Conversely, most individuals with porphyria cutanea tarda have an acquired enzyme deficiency which can be induced by similar environmental and genetic factors. These complexities combined with the expanding availability of treatments highlight the importance of biochemical testing for this pathway.

Acute hepatic porphyrias (AHPs) are clinically characterized by episodic neurovisceral attacks

N. Blau, F. M. Vaz (eds.), *Laboratory Guide to the Methods in Biochemical Genetics*,
https://doi.org/10.1007/978-3-031-58819-8_16

presenting as severe abdominal pain. Acute porphyria attacks or crises are also associated with nausea, vomiting, tachycardia, and hypertension and can be further complicated by mental changes, convulsions, peripheral neuropathy that may progress to respiratory paralysis, and hyponatremia. Attacks can be life threatening. Certain drugs, alcohol, endocrine factors, calorie restriction, stress, and infections are precipitating factors that must be avoided by these patients. Due to reduced penetrance, many individuals with a pathogenic variant in one of the involved genes do not experience an acute episode, and few individuals will have more than one in their life.

Cutaneous porphyrias present with sun sensitivity in form of blisters in the case of porphyria cutanea tarda (PCT) and congenital erythropoietic porphyria (CEP) and as painful edematous reaction in the case of protoporphyrias. Hereditary coproporphyria (HCP) and variegate porphyria (VP), two AHPs, can also have cutaneous reaction like PCT. PCT is mainly acquired and several factors, causing hepatic iron overload, as hemochromatosis, alcohol, estrogen use, infections (HCV, in particular), smoking, contribute to the clinical expressivity.

Porphyrin biochemical analysis has multiple preanalytic and analytic challenges and caveats. Proper consideration of preanalytic factors and assay design are critical to obtaining the correct diagnosis. Testing the correct specimen is essential since diagnosis often requires testing a specimen where the accumulating porphyrin intermediate is detected. Specimen collection timing, especially for acute porphyrias, and specimen handing requirements can also increase the likelihood of an informative analysis. Following preanalytical consideration, accurate biochemical analysis and interpretation is of paramount importance for a correct diagnosis.

16.2 Part Background

16.2.1 Porphyrin Pathways Biochemistry

Heme is synthesized in all nucleated cells, but most prominently in the bone marrow for hemoglobin synthesis and in the liver for vital hemoproteins, i.e. cytochrome P450 enzymes involved in drug metabolism. The heme biosynthetic pathway begins in the mitochondria where the initial substrates glycine and succinyl-coenzyme A form δ-aminolevulinic acid (ALA) in a reaction catalyzed by X-linked ALA synthase. Eight enzymes, encoded by nine genes, are required for complete synthesis. The first and last three enzymatic steps are located within the mitochondria while the remaining steps occur in the cytosol (Fig. 16.1). Deficiency in each enzymatic step of the pathway leads to a described porphyria condition (Table 16.1). Porphyrins and/or porphyrin precursors accumulate in different organs or tissues depending on which enzyme is affected resulting in different symptomatology.

Delta-Aminolevulinic acid (ALA) and porphobilinogen (PBG) are porphyrin precursors. The first is a non-proteinogenic amino acid, while the latter is a cyclic pyrrole, formed by condensation of two ALA molecules catalyzed by ALA dehydrogenase. The basic porphyrin structure consists of four mono pyrrole rings connected by methene bridges to form a tetrapyrrole ring (Fig. 16.2). The substituted aromatic ring strongly absorbs electromagnetic radiation leading to the photosensitivity of porphyrin metabolites. Uroporphyrinogen III metabolism proceeds enzymatically as well as non-enzymatically with isomers differing in their carboxylic acid position. Both isomer series can yield diagnostic information; however, only the III series isomers continue through the biosynthetic pathway to heme.

As the pathway proceeds porphyrins have a decreasing number of carboxylic acids. Their solubility, and consequently the best body fluid to test, depends on the number of the carboxylic groups. Uroporphyrin has eight carboxylic acid groups and is the most soluble in water. Protoporphyrin has two carboxylic groups and is hydrophobic, hence it is not excreted in urine, but in feces. Coproporphyrin which has four carboxylic acid groups and intermediate solubility, is excreted in both urine and feces. Knowledge of the solubility of different porphyrin metabolites allows the proper choice of specimen to analyze for suspected defects.

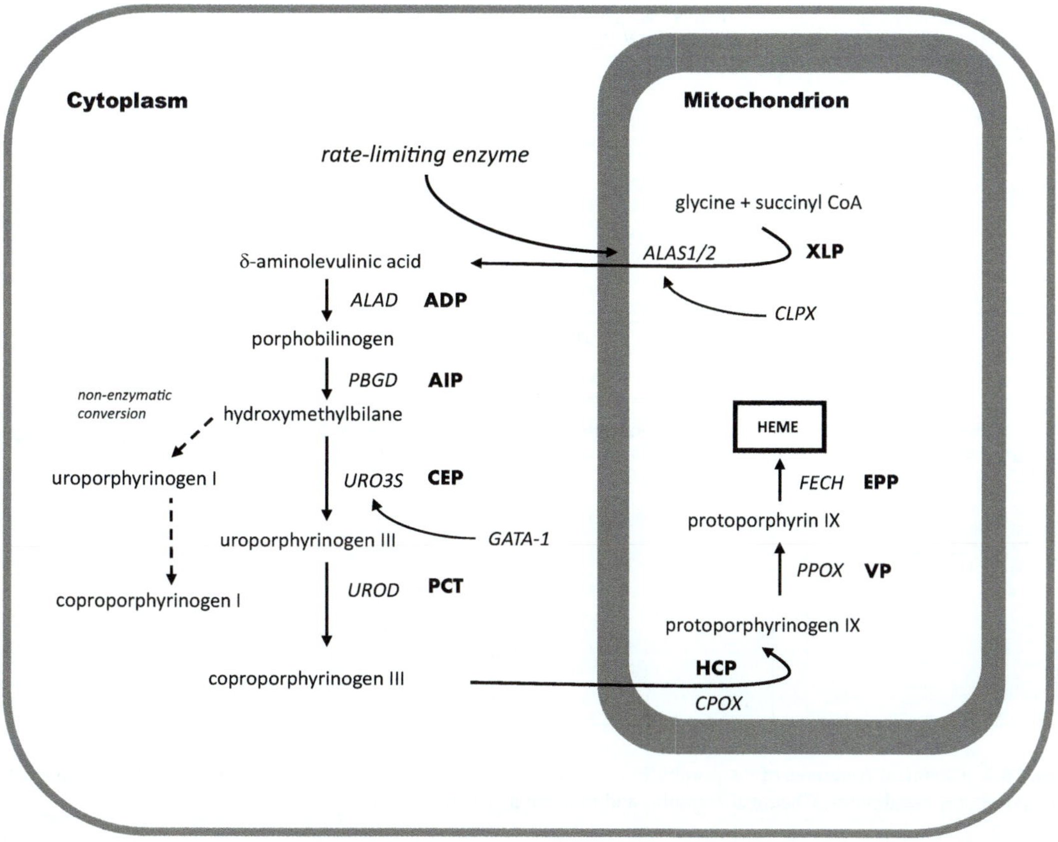

Fig. 16.1 Heme biosynthetic pathway. Porphyria disorder abbreviation shown in bold with associated gene in italics. Solid arrows represent enzymatic conversation while dashed arrows represent non-enzymatic conversation. Disease abbreviations: XLP: X-linked protoporphyria, ALAD: aminolevulinic acid dehydratase, AIP: acute intermittent porphyria, CEP: congenital erythropoietic porphyria, PCT: porphyria cutanea tarda, HCP: hereditary coproporphyria, VP: variegate porphyria, EPP: erythropoietic protoporphyria

Table 16.1 Genes, transmission modality and enzymes involved in the porphyrias

Disease	Gene	Inheritance	Enzyme/protein involved
Acute porphyrias			
ALAD deficiency porphyria	*ALAD*	AR	ALA dehydratase
Acute Intermittent Porphyria	*HMBS*	AD	PBG deaminase aka hydroxymethylbilane synthase
Mixed porphyrias			
Hereditary Coproporphyria	*CPOX*	AD	coproporphyrinogen III oxidase
Variegate Porphyria	*PPOX*	AD	protoporphyrinogen oxidase
Cutaneous porphyrias			
X-linked Protoporphyria	*ALAS2*	XL	ALA synthase
Congenital Erythropoietic Porphyria	*UROS*	AR	uroporphyrinogen III co-synthase
	GATA1	XL	GATA-1 protein
Porphyria Cutanea Tarda	*UROD*	AD[a]	uroporphyrinogen III decarboxylase
Hepatoerythropoietic Porphyria	*UROD*	AR	uroporphyrinogen III decarboxylase
Erythropoietic Protoporphyria	*FECH*	AR	Ferrochelatase
	CLPX	AD	AAA+ protein unfoldase

[a] Only the familial cases. Two thirds of cases are acquired

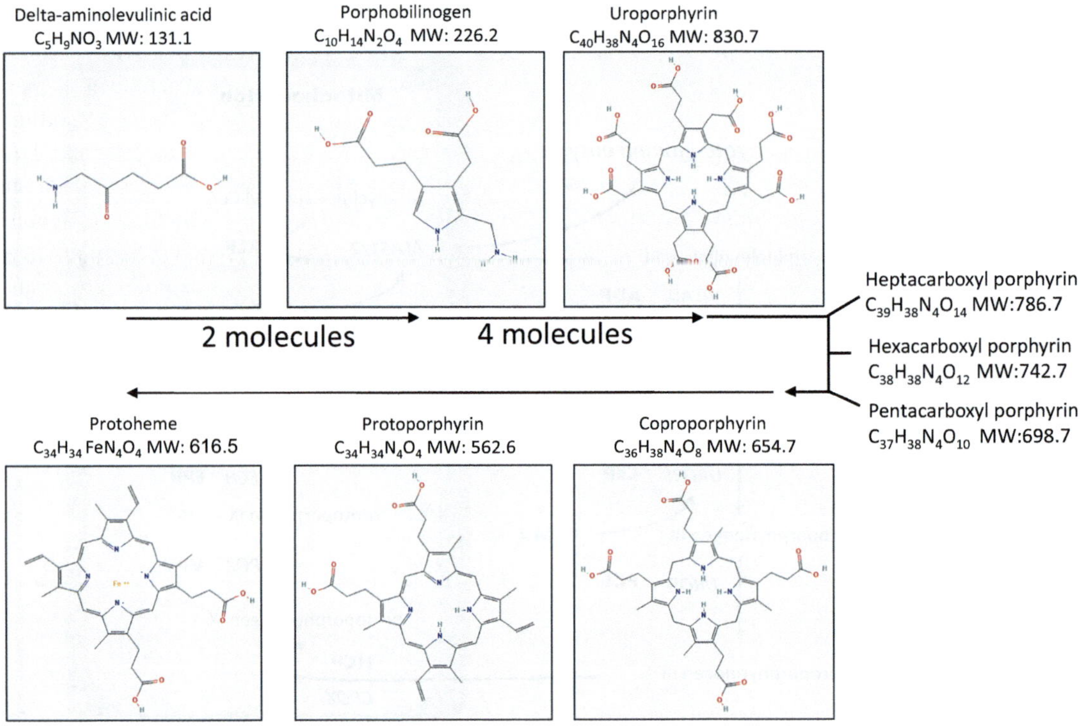

Fig. 16.2 Chemical structures of the porphyrin precursors, δ-aminolevulinic acid (ALA) and porphobilinogen (PBG), and porphyrin metabolites. Chemical formulas and molecular weights are labeled

16.2.2 Acute Porphyrias

When an enzymatic block, like a defect of porphobilinogen deaminase (PBGD), causes a direct or indirect ALA synthase induction and an accumulation of porphyrin precursors, such as ALA and porphobilinogen (PBG), the resulting phenotype will be characterized by episodic crisis of severe abdominal pain, nausea, vomiting, electrolyte imbalance (e.g., Na^+ and Mg^+), hypertension and tachycardia, the typical clinical picture of the acute porphyrias. Hemin treatment induces negative feedback to reduce aminolevulinic acid synthase (ALAS) activity. Peripheral neuropathy mimicking Guillain Barré syndrome, behavioral abnormalities, and seizures are usually present during an acute attack [1] and, in a number of patients, they become chronic [2]. A minority of affected individuals have multiple attacks, most often women. Due to nonspecific symptoms patients can endure a considerable delay to diagnosis if the condition is not suspected clinically.

Acute porphyrias include acute intermittent porphyria, ALAD deficiency porphyria, hereditary coproporphyria, and variegate porphyria although the latter two can exhibit skin manifestations. Measurement of PBG and elevation of PBG is a specific indicator of an acute porphyria, with the exception of ALAD deficiency where only ALA is elevated. Analysis of porphyrin metabolites during an acute attack is useful to making a specific diagnosis as excretion of PBG and porphyrins can be normal between attacks. First tier PBG and urine screening can demonstrate PBG elevation and abnormal accumulation of porphyrin metabolites which is a characteristic of acute porphyrias.

16.2.3 Cutaneous Porphyrias

Accumulation of porphyrins in the upper dermis and their interaction with visible light are responsible for sun sensitivity, the main symptom of the cutaneous porphyrias. Because of their chemical structure, porphyrins react to the sun by absorbing light at 400 to 410 nm (Soret peak) wavelengths that can penetrate the deep layers of the dermis. According to the type and location of porphyrin accumulation, patients can present with one of two dermatologic reactions: a severe response to the ultraviolet rays in sunlight characterized by blisters or a more immediate and painful photosensitivity with no blisters [3].

Approximately two third of cases of porphyria cutanea tarda (PCT) have a sporadic form where only an UROD inhibitor reduces the hepatic enzymatic uroporphyrinogen III decarboxylase (UROD) activity of approximately 20% of normal. The interaction of several genetic and environmental factors has been linked to the production of uroporphomethene, the UROD inhibitor, from uroporphyrinogen through an iron-dependent oxidation [4]. Normally enzymatic deficiencies responsible for the different porphyrias are caused by pathogenic variants in the gene encoding for the respective enzyme, but in rare instances, other genes are involved in the pathogenesis of protoporphyria (*CLPX*) and congenital erythropoietic porphyria (*GATA1*), respectively [5].

16.2.4 Clinical Biochemical Testing

Because of both acute and cutaneous porphyrias can present with nonspecific symptoms, variable presentations, minor abnormalities in routine lab tests, and lack of symptoms or findings in between attacks diagnosis can be difficult and delayed for years [6, 7]. Selecting the correct test in the right matrix at the right time is paramount for a prompt and correct diagnosis. Specific diagnosis of porphyria can require testing in multiple specimen types, see testing algorithm proposed in Fig. 16.3.

In the majority of porphyrias, excess metabolites are detected in urine. Urinary porphobilinogen during acute attacks and porphyrin fractionation in urine when patients present with blistering are the first line tests. For the differential diagnosis of AHPs, fecal porphyrin fractionation is indicated since patterns of excretion are characteristic for each disease. When a protoporphyrias is suspected, analysis of erythrocyte protoporphyrins will help in diagnosing EPP, XLP or excluding porphyria as cause of sun sensitivity.

For a correct diagnosis of a specific type of porphyria, several tests are required and biochemical evidence of a derangement of the path-

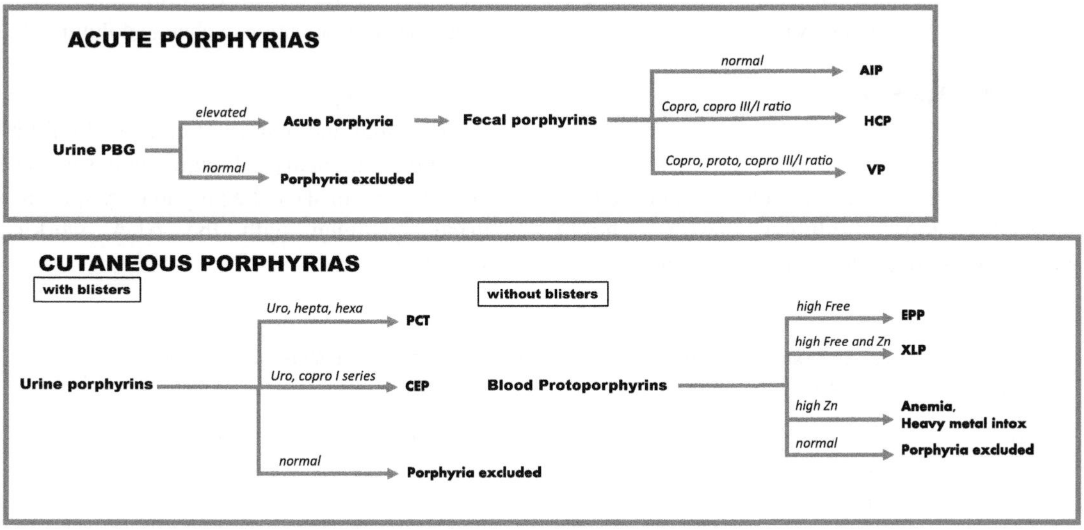

Fig. 16.3 Algorithms for diagnosis of acute (on top) and cutaneous (below) porphyrias

way is a necessary requirement. In fact, although it has been shown some genotype–phenotype correlation the presence of pathogenic variants is not sufficient to determining the phenotype but precipitating factors and likely other modifying gene(s) also play a major role in the pathogenesis of acute attacks [8].

16.3 Porphyrin Metabolite Testing

Porphyrias are characterized by the accumulation of heme precursors. Metabolite testing can provide a specific diagnosis and is relevant for the therapeutic monitoring of acute episodes. Due to differing porphyrin solubilities testing the correct specimen type is critical to obtaining clinically relevant results. Porphyrin levels can also be elevated due to environmental exposure to lead and other heavy metals.

16.3.1 Quantitation of Aminolevulinic Acid in Urine

Aminolevulinic acid (ALA) is increased due to an inherited acute porphyria or secondary inhibition of ALA dehydratase (can be caused by acute or chronic lead intoxication, tyrosinemia type I, alcoholism, and pregnancy).

16.3.1.1 Method

Principle
The excretion of ALA in urine is quantified by stable isotope dilution with liquid chromatography-tandem mass spectrometry.

Preanalytical
1. Specimens

2.0 mL random urine with no preservatives. Abstain from alcohol for 24 h prior to collection. Samples should be stored either frozen or refrigerated.

2. Reagents and Chemicals

- Mobile Phase A (Water +0.1% formic acid)
- Mobile Phase B (Methanol +0.1% formic acid)
- Aminolevulinic acid standards (dilutions should be done with 0.4% acetic acid)
- ALA Internal standard: 5-Aminolevulinic Acid13C5, 15N (dilutions should be done with 0.4% acetic acid)

3. Instrumentation & Supplies

- LC-MS/MS system
- Vortex mixer
- 0.2 μm Nylon filter vial
- C18 HPLC column (4.6 × 150 mm; 5 μm)
- Pre-packed ALA/PBG columns (BioRad)
- Glass wool

4. Calibration

ALA working standard should be prepared at a concentration of 100 nmol/mL. Strip urine of ALA using the BioRad cation exchange column.

5. Quality Control
A normal and abnormal control should be prepared and analyzed with each run. The abnormal control can be prepared by spiking a urine specimen with the ALA working standard.

#	ALA working IS (μL)	ALA working Std (μL)	Stripped urine (μL)	0.4% Acetic acid (μL)	Theoretical Conc. (nmol/mL)	IS Conc (nmol/mL)	Conc. Ratio
1	200	0	100	100	0.00	25	0
2	200	10	100	90	5.0	25	0.2
3	200	25	100	75	12.5	25	0.5
4	200	50	100	50	25	25	1.0
5	200	100	100	0	50	25	2.0

Analytical Procedure

Specimen Preparation

1. Label a filter vial for each patient sample, control, and a blank.
2. Add 200 μL of patient or control urine to each vial. Use 200 μL of water for the blank.
3. Add 200 μL of ALA internal standard (25 nmol/mL) to each filter vial.
4. Attach the filter plunger, but do not press down. Vortex each filter unit.
5. Press filter plunger into the bottom of the filter vial. Verify no bubbles are present. Samples are ready for analysis.
6. Data Acquisition—Samples should be analyzed and quantified using available software on the LC-MS/MS instrument in use. The software on the instrument should be able to calculate the amount of ALA in each patient sample.

Reference Ranges

All reference ranges should be either obtained in one's own laboratory according to all appropriate regulations. Comparison to published reference ranges may be helpful. Our reference ranges are below.

Age range	ALA concentration
<1 year	≤10 nmol/mL
1–17 years	≤20 nmol/mL
≥18 years	≤15 nmol/mL

16.3.2 Quantitation of Porphobilinogen in Urine

Porphobilinogen (PBG) excretion is elevated during the acute phase of porphyrias with neurovisceral symptoms (AIP, HCP, VP). PBG excretion may be normal outside of an acute attack in an affected individual.

16.3.2.1 Method

Principle

The excretion of PBG in random urine samples is quantified by stable isotope dilution with liquid chromatography-tandem mass spectrometry.

Preanalytical

1. Specimens

2.0 mL random urine with no preservatives. Abstain from alcohol for 24 h prior to collection. Samples should be stored either frozen or refrigerated.

2. Reagents and Chemicals

- Mobile Phase (600 mL acetonitrile, 1400 mL water, 2 mL formic acid)
- PBG standards:
- Stock standard (100 mg/L); volume made up with 0.5% sodium carbonate
- Working standard (4.0 mg/L); volume made up 0.5 N HCl
- $^{13}C2$-PBG (Internal Standard)
- Stock internal standard (100 mg/L); volume made up with 0.5% sodium carbonate
- Working internal standard (2.75 mg/L); volume made up with 0.5% sodium carbonate
- Artificial urine

3. Instrumentation and Supplies

- LC-MS/MS system
- Vortex mixer
- 0.2 μm Nylon filter vial
- C18 HPLC column (3.3 cm × 4.6 mm, 3 μm)
- Waters Oasis HLB SPE column

4. Calibration

PBG working standard should be prepared to a concentration of 4.0 mg/L and made up to volume with artificial urine and 0.5 N HCl as shown in the table in the calculations paragraph. Prior to use the artificial urine should be brought to a pH of 1–3 using dropwise addition of 12 N HCl.

5. Quality Control

A normal and abnormal control should be prepared and analyzed with each run. The abnormal control can be prepared by spiking an aliquot of artificial urine with the PBG working standard.

Analytical Procedure

Specimen Preparation

1. Prepare acidified water by adding 1 drop of 12 N HCl to 5 mL water.
2. Label 16 × 100 mm disposable glass test tubes for all samples that need to be analyzed, including QCs.
3. Aliquot 5.0 mL of patient sample or QC into the corresponding test tube. pH each sample and record the pH. Specimens with an initial pH below 5.0 should not be analyzed.
4. Adjust the pH of each sample to between 1 and 3 by drop wise addition of 12 N HCl. Record the number of drops. Keep samples cold until analysis can be completed.
5. Prep the appropriate number of SPE cartridges and wash with 1 mL of acetic acid, followed by 1 mL of water.
6. Add 100 µL of PBG working internal standard to each labeled 12 × 75 mm tube. Transfer 1 mL of acidified urine to the respective 12 × 75 mm tubes and mix.
7. Transfer 1 mL of the sample + internal standard solution to the SPE column. Elute the sample under positive pressure and discard the eluate from this step. Wash with 0.5 mL water and discard the water.
8. Depending on sample volumes, prepare either a 96 well deep well plate or appropriate sample vials and place under respective SPE cartridges. Add 1 mL acetic acid and elute under positive pressure.
9. Data acquisition Samples should be analyzed and quantified using available software on the LC-MS/MS instrument in use. The software on the instrument should be able to calculate the amount of PBG in each patient sample.
10. Calculations

(Raw PBG)(4.42)(PBGU Multiplication Factor) = micromoles PBG/L

Raw PBG = Calculated concentration from LC-MS/MS software

4.42 = Conversion from mg to µmol for PBG

PBGU Multiplication Factor—See below table

Number of drops	pH volume (mL)	PBG multiplication factor
1	5.0	1.00
2–5	5.1	1.02
6–8	5.2	1.04
9–12	5.3	1.06
13–15	5.4	1.08
16–18	5.5	1.10
19–21	5.6	1.12
22–30	(# of drops * 0.029 + 5)	pH volume/5

Reference Ranges

Reference ranges for PBG do not vary with age. Published reference ranges should be verified in one's own laboratory according to appropriate regulations.

Random Urine PBG ≤ 1.3 µmol/L

16.3.3 Quantitation of Aminolevulinic Acid and Porphobilinogen in Plasma

Porphobilinogen (PBG) and aminolevulinic acid (ALA) are elevated during the acute phase of neurologic porphyrias (AIP, HCP, VP).

16.3.3.1 Method

Principle

The concentrations of PBG and ALA in plasma are simultaneously quantified using stable isotope dilution LC-MS/MS.

Preanalytical

1. Specimens

1 mL Sodium heparin plasma, stored in amber tubes. LiHep, EDTA, and ACD A/B are also acceptable specimen types.

2. Reagents and Chemicals

- Porphobilinogen standards
- PBG stock standard (5.0 mM) made up to volume with 0.5% sodium carbonate. Prepare fresh day of use.

- PBG intermediate standard (100 μM) made up to volume with 0.5% sodium carbonate
- PBG working standards (make up to 10 mL total volume with sodium carbonate)
- $^{13}C_2$ PBG (Internal Standard)
  - PBG stock internal standard (5 mM); make up to volume with 0.5% sodium carbonate
  - PBG working internal standard (40 μM); make up to volume with 0.5% sodium carbonate
- ALA Standards
  - ALA stock standard (5.0 mM); make up to volume with 0.4% acetic acid
  - ALA intermediate standard (100 μM); make up to volume with 0.4% acetic acid
  - ALA working standards (make up to 50 mL total volume with 0.4% acetic acid)

Concentration (μM)	Volume of PBG intermediate std (μL)
1.0	100
3.0	300
5.0	500
7.0	700
10.0	1000

Concentration (μM)	Volume of ALA intermediate std (μL)
0.5	250
1.0	500
2.0	1000
3.0	1500
5.0	2500

- Mobile Phases
- Mobile Phase A (500 mL water, 5 mL formic acid, 315.3 mg ammonium formate)
- Mobile Phase B (1000 mL acetonitrile, 2.5 mL formic acid)

3. Instrumentation and Supplies

- LC-MS/MS system
- Vortex mixer
- Plate dryer
- Waters Amide HILIC HPLC column 5 μm, 2.1 × 50 mm
- Strata-X-C 33 u polymeric strong cation SPE tubes

4. Calibration

ALA and PBG curves are generated separately. ALA curves should be run with each analysis. PBG curves should be reverified with each new lot of working internal standard.

5. Quality Control

A normal and abnormal control should be prepared and analyzed with each run. The abnormal control can be prepared by spiking a urine specimen with the ALA working standard.

Analytical Procedure

Specimen Preparation

1. Bring all patient samples, controls, and standards to room temperature.
2. Prepare the appropriate number of SPE cartridges into the manifold and add a waste container.
3. Label microcentrifuge tubes for all standards, QC and patient samples. Add 500 μL 5% FA in water and 25 μL PBG working IS.
4. Add 200 μL of pooled plasma to the zero calibrator. For other calibrators, add 200 μL of pooled plasma and 200 μL of each working standard.
5. Add 200 μL of water to the blank and 200 μL of sample/QC to each control and patient tube. Vortex all tubes for 5 s. Centrifuge all tubes for 2 min at 13,000 RPM.
6. Condition SPE tubes by adding 1 mL of 5% FA in methanol to each column. Using positive pressure, draw the conditioning solution through the SPE column into the waste container. Flow rate should be approximately 1 mL/min. Discard the waste.
7. Equilibrate the SPE column by adding 1 mL of 5% FA in water to each column. Draw through using positive pressure. Discard this liquid.

8. Remove sample tubes from the centrifuge and load SPE columns with 600 μL of supernatant from each tube.
9. Draw the sample through the SPE tube using positive pressure at a flow rate of 1 mL/min. Discard this filtrate.
10. Wash the SPE column with 1 mL of 5% FA in water. Use positive pressure for a flow rate of approximately 1 mL/min. Discard this filtrate.
11. Wash the SPE column with 1 mL of 5% FA in methanol. Use positive pressure for a flow rate of approximately 1 mL/min. Discard this filtrate.
12. Remove the waste collection containers and replace with either labeled autosampler vials or a deep well plate, depending on patient volumes.
13. Elute the analytes into the labeled collection device using 0.5 mL of 5% NH_4OH in methanol. This step should be performed in a fume hood. Draw through using positive pressure.
14. Dry these specimens under room temperature nitrogen.
15. Add 100 μL of water to the residue and agitate for 5 s. Add 400 μL of acetonitrile to each well and mix three times by aspiration.
16. Seal with plate.
17. Data acquisition samples should be analyzed and quantified using available software on the LC-MS/MS instrument in use. The software on the instrument should be able to calculate the amount of ALA in each patient sample. Curves should properly account for the endogenous levels of the analytes.

Reference Ranges

All reference ranges should be either obtained in one's own laboratory according to all appropriate regulations. Comparison to published reference ranges may be helpful. Our reference ranges are below. There are no age-based reference ranges for these analytes.

Analyte	Reference range
Porphobilinogen	≤0.5 μmol/L
Aminolevulinic acid	≤0.5 μmol/L

16.3.4 Urine Porphyrin Fractionation

16.3.4.1 Method

Principle

Urine porphyrins are extracted and separated by HPLC and the eluted porphyrins are quantified by comparison of their fluorescence intensity to that of known porphyrin standards. This method quantitates urinary porphyrins uroporphyrin (octacarboxyl porphyrin), heptacarboxyl porphyrin, hexacarboxyl porphyrin, pentacarboxyl porphyrin, and coproporphyrin (tetracarboxyl porphyrin). The III and I isomer series are not resolved in favor of a faster runtime.

Preanalytical

1. Specimens

Random or 24-h urine sample. For 24 h collection, add 5 g sodium carbonate at start of collection. Random collection or aliquot of 24-h collection is stored in an amber bottle protected from light and stored and shipped at −20 °C.

2. Reagents and Chemicals

- Methanol
- Phosphate buffer, pH 3.5
- Hydrochloric acid, concentrated (12 N)
- Mesoporphyrin IX stock standard [2.0 μg/dL]
- Porphyrin calibration standard

  – [Uroporphyrin III, Heptacarboxylporphyrin I, Hexacarboxylporphyrin I, Pentacarboxylporphyrin I, Coproporphyrin III]. From 10 mg/dL stock solutions of each porphyrin analyte above, add 200 μL to a 1 L amber volumetric flask. Dilute to volume with 1.5 N HCl.

3. Supplies and Instrumentation

- Analytical C18 column 3.9 mm × 300 mm, 10 μm particle size
- pH meter
- HPLC system

- Fluorescence spectrophotometer detector (excitation: 408 nm, emission: 625 nm)
- Centrifuge

4. Calibration

Calibration is performed daily using a 1:1 mix of Porphyrin Calibration Standard and Working Injection Marker Solution on each in-use analytical system. Porphyrin retention time and concentration is monitored.

5. Quality Control

Low and high control specimens are included on each batch with individual porphyrins monitored longitudinally. Blank samples are monitored for potential carryover.

Analytical Procedure

Specimen Preparation

1. Prepare blank solution (1 drop of 12 N HCl to 5 mL water) and pipette porphyrin calibration standard (PCS) with mesoporphyrin XI injection marker solution.
2. Review PCS and blank chromatograms on HPLC system for system suitability.
3. Aliquot 5.0 mL of patient sample and controls into labeled glass test tube.
4. Take initial pH of each 5.0 mL aliquot using the pH meter.
5. Acidify urine to a pH of 1.0 to 3.0 by titrating dropwise with 12 N HCl, recording number of drops used. Place each tube place on ice until batch is processed. * For difficult to titrate samples, specimen can be centrifuged to remove debris.
6. Centrifuge the specimen at 2000 RPM for 5 min at 0 °C.
7. Pipette 500 μL of urine supernatant and 500 μL injection marker into an autosampler vial.
8. Swirl to mix the contents in the autosampler vial and analyze by HPLC equipped with fluorescence detector.

Note: HPLC method to be determined by laboratory

Our laboratory uses a 12-min runtime. I series and III series isomers are not resolved.

9. Calculation

Random urine collections are calculated and reported as nmol of porphyrin per liter utilizing the following formula. Please note, in many methods these values are normalized to creatinine.

$$\frac{C\text{std}(\mu g/\text{dL})\times(Ax)\times\left[V\text{pH}(\text{mL})\right]\times 10\,\text{dL}/\text{L}\times\text{CF}}{(A\text{std})\times 5.0\,\text{mL}}$$

C_{std} = Standard concentration (μg/dL)

A_x = Area count of analyte

V_{24h} = Volume of 24 h collection (mL/24 h)

V_{pH} = Volume of pH'd urine (mL) ***see pH Dilution Chart below***

CF = Porphyrin conversion factor (μg to nmol)

A_{std} = Area count of standard

Note: Random sample volume should account for dropwise pHing in sample preparation

Porphyrin Molar Conversation Factor table

Urine porphyrin	Molar conversion calculation	Conversion factor (CF)
Uroporphyrin	(1 μg) × (1 μmol/830.76 μg) × (1000 nmol/μmol)	1.20
Heptacarboxyl porphyrin	(1 μg) × (1 μmol/786.75 μg) × (1000 nmol/μmol)	1.27
Hexacarboxyl porphyrin	(1 μg) × (1 μmol/742.74 μg) × (1000 nmol/μmol)	1.35
Pentacarboxyl porphyrin	(1 μg) × (1 μmol/698.73 μg) × (1000 nmol/μmol)	1.43
Coproporphyrin	(1 μg) × (1 μmol/654.72 μg) × (1000 nmol/μmol)	1.53

Reference Intervals

Laboratories should establish their own reference ranges prior to implementation

Random Collections (<20 h) or Results Normalized per Liter (<100 mL)

Uroporphyrin (nmol/L)	≤30
Heptacarboxylporphyrin (nmol/L)	≤7
Hexacarboxylporphyrin (nmol/L)	≤2
Pentacarboxylporphyrin (nmol/L)	≤5
Coproporphyrin (nmol/L)	≤110

Pathological Values

Abnormal porphyrin excretion is observed in a recognizable profile (Table 16.2) and quantified values. Example profile of a patient with PCT is shown in Fig. 16.4. In our laboratory, porphobilinogen is determined for each urine porphyrin profile analyzed aiding in the identification of acute porphyrias. PCT is reliably identified by this method both by pattern of excretion and the presence of a unique detected species. Acute porphyrias are recognizable by

Table 16.2 Prominent abnormalities in different matrices according to the different clinical entities

	Urine		Fecal	Blood	Plasma	
Disease	ALA/PBG	Porphyrin	Porphyrin	Porphyrin	nm	Porphyrin
XLP			(Proto)	Zn and free Proto	606	Proto
ALAD porphyria	ALA			Zn Proto		
AIP	ALA,PBG					
CEP		Uro, Copro (I series)	Uro, copro (I series)	Zn Proto		Uro, Copro
PCT		Uro, Penta	Isocorpro		599	Uro, Penta
HCP	ALA,PBG	Copro	Copro, CI/III ratio			
VP	ALAPBG	Copro	Copro, Proto, CI/III ratio		602[a]	
EPP				Free Proto	606	Proto

Uro: uroporphyrin, Copro: coproporphyrin, Penta: pentacarboxylporphyrin, Proto: protoporphyrin (Zinc complexed or free protoporphyrin for blood)

[a] Alternative methods use emission at 626nm. Non-informative test/condition combinations are left blank

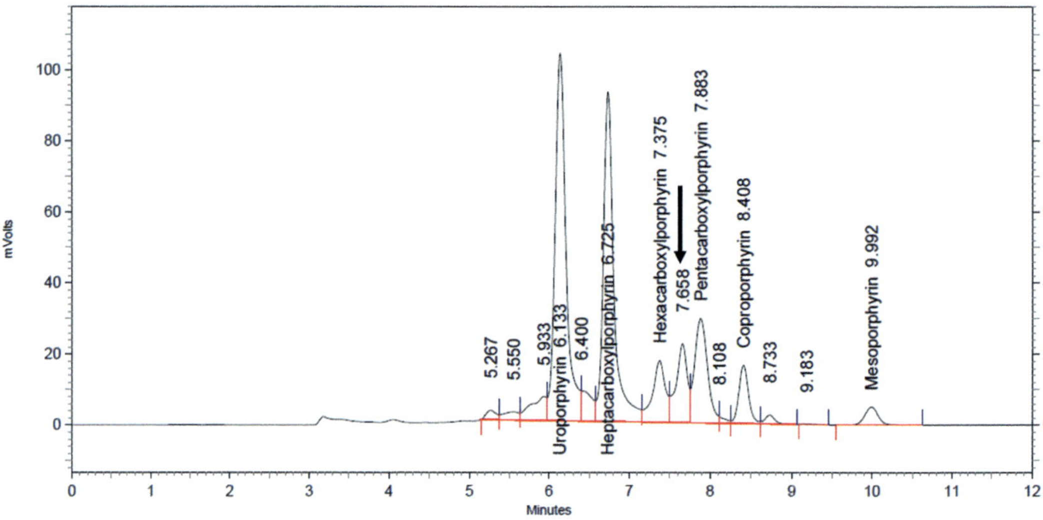

Fig. 16.4 Typical HPLC urine profile of a PCT patient, with elevations of uroporphyrin, heptacarboxylporphyrin and, in lesser degree, of hexacarboxylporphyrin and pentacarboxylporphyrin. Arrow indicates isohexacarboxylporphyrin

a common, similar profile which can be discriminated by further testing in additional specimen types. In our experience, random urine collections yield similar diagnostic information to 24-h collections, and we strongly discourage 24-h collection.

Limitations

- Thymol interferes in the hexacarboxyl porphyrin region of the chromatography.
- Individuals should abstain from alcohol 24 h before and during collection. Alcohol ingestion prior to collection may lead to elevated excretion of coproporphyrin.
- Samples should be protected from light.
- Samples requiring greater than 30 drops of 12 N HCl for titration are canceled for being super saturated.
- Samples collected when patient is asymptomatic are most likely to be uninformative.
- This method does not normalize to creatinine which is the preferred method.

16.3.5 Plasma Total and Fractionation

16.3.5.1 Method

Principle

The total plasma porphyrins concentration is determined by extracting the porphyrins from plasma with a mixture of ethyl acetate and acetic acid. The porphyrins are then back extracted into dilute hydrochloric acid. Total porphyrins are quantitated on this extract via scanning spectrofluorometry. If the total plasma porphyrin concentration is elevated, the extract is submitted for HPLC analysis, whereby the individual porphyrin analytes are separated by differential partitioning between a stationary aliphatic surface and a moving aqueous buffer solution. As the porphyrins emerge add separately from the column, the quantity of each is monitored with a fluorescence detector. *This method is a departure from the common emission peak detection from excitation at 405 nm in plasma (See Fig. 16.5).

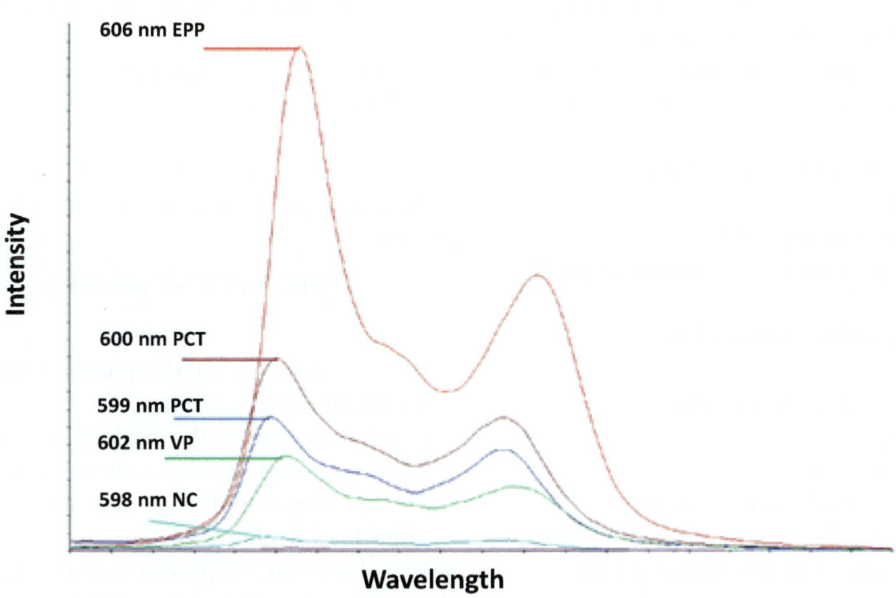

Fig. 16.5 Fluorescence emission scans of plasma from patients with different porphyrias. Different wavelengths correspond to different porphyrin elevations and therefore are suggestive of specific porphyrias. EPP: erythropoietic protoporphyria; PCT: porphyria cutanea tarda; VP: variegate porphyria; NC: normal control

Preanalytical

1. Specimen

Green top (heparin or lithium) Lavender top (EDTA)

Note: Centrifuge specimen and store at −20 protected from light. Abstinence from alcohol is recommended 24 h prior to testing.

2. Reagents and Chemicals

- Ethyl acetate
- Glacial acetic acid
- Hydrochloric acid
- Methanol
- Ethanol
- Sodium phosphate, monobasic
- Phosphoric acid
- Saline, 0.9%
- Uroporphyrin III Standard [1.0 μg/dL] in 1 N HCl
- Porphyrin Calibration Standard (for fractionation if performed)
  - [Uroporphyrin III, Heptacarboxylporphyrin I, Hexacarboxylporphyrin I, Pentacarboxylporphyrin I, Coproporphyrin III]. From 10 mg/dL stock solutions of each porphyrin analyte above, add 200 μL to a 1 L amber volumetric flask. Dilute to volume with 1.5 N HCl.
- RBC extraction reagent 4:1 ethyl acetate: glacial acetic
- Phosphate buffer, pH 3.5
- Protoporphyrin working standard [3 μg/dL]

3. Supplies and Instrumentation:

- Fluorescence spectrophotometer
- Centrifuge
- Multi-tube vortex
- HPLC system (for fractionation)
- Analytical column, C18, 3.9 mm × 300 mm, 10 μm particle size (for fractionation)

4. Calibration

Calibration is performed daily on a Uroporphyrin III standard. Retention time and concentration is monitored.

5. Quality Control

Pooled low and high control specimens are included on each batch with individual porphyrins monitored longitudinally. Blank samples are monitored for potential carryover.

Analytical Procedure

Specimen Preparation

1. Transfer 500 μL of blank (0.9% saline), Uroporphyrin III standard solution, or patient sample into a 16 × 125 mm glass test tubes.
2. Add 3.0 mL RBC Extraction Mixture.
3. Vortex for a minimum of 1 min using the pulse mode on the multi-tube vortex with the speed set to 2000 RPM.
4. Centrifuge tubes for 5 min at 2000 RPM.
5. Transfer the supernatant fluid into its corresponding labeled 16 × 125 mm glass test tube and discard cell pellet waste.
6. Add 3.0 mL 1.5 N HCl to the supernatant.
7. Vortex for a minimum of 1 min using the pulse mode on the multi-tube vortex with the speed set at 2000 rpm.
8. Allow the aqueous layer to visibly separate from the organic layer and remove organic layer.
9. Analyze by scanning spectrophotometer.
10. Elevated total porphyrin samples are reflexed to fractionation analysis by HPLC.
11. Samples are analyzed along with

Scanning spectrometer: μg total porphyrin/dL plasma

$$\left(FPE^{*}VPE^{*}CS\right)/\left(FS^{*}VP\right)$$

FPE = Fluorescence of the plasma extract, in relative fluorescence units

FS = Fluorescence of the working low uroporphyrin standard, in relative fluorescence units

VPE = Volume (mL) of the total plasma extract = 3.8 mL

VP = Volume (mL) of plasma assayed = 0.5 mL

CS = Concentration of the working low uroporphyrin standard, in μg/dL

12. Calculations

Plasma porphyrin fractions, μg porphyrin/dL plasma.

$$\frac{\text{Fractionated individual porphyrin concentration}}{\text{Total porphyrin concentration}} \times (\text{Total porphyrin concentration}) = \mu g \text{ porphyrin / dL plasma}$$

Reference Values

- Total plasma porphyrin analysis: reference <1 μg/dL
- Fractionated analyses are qualitatively reviewed. Expected quantitative values <1.0 μg/dL for each porphyrin species.

Pathological Values

In total plasma porphyrin analysis, samples with a concentration greater than a predetermined value (in our laboratory 1 μg/dL) are subjected to HPLC analysis for individual porphyrin determination.

Plasma porphyrin profiles are recognizable for several types of cutaneous and acute porphyria (see Table 16.2). Although analysis in plasma is not preferred for diagnosis, abnormal profiles are observed in the cutaneous porphyrias and traditionally has been used for the diagnosis of VP during acute episodes.

Limitations

- Hemolysis results in an elevated protoporphyrin concentration

16.3.6 Erythrocytes or Whole Blood Porphyrin Analysis

16.3.6.1 Method

Principle

Following cell lysis, the porphyrins are extracted from washed erythrocytes into an organic solvent. A back-extraction is performed using acid to extract the porphyrins into aqueous phase. Total porphyrins are quantified by measurement of the fluorescence intensity of the acid extract in a fluorescence spectrophotometer.

Preanalytical

1. Specimen

Washed erythrocytes.

2. Chemicals and Reagents

- Saline, 0.9%
- Celite (diatomaceous earth)
- Ethyl acetate
- Glacial acetic acid
- Hydrochloric acid (HCl)(12 N)
- 1.5N HCl
- Methanol
- Protoporphyrin working standard [3 μg/dL]
- Celite solution reagent
- Combine 10 g of Celite and combine with 190 mL of 0.9% Saline
- RBC extraction mixture: 4:1 ethyl acetate: glacial acetic acid

3. Supplies and Instrumentation

- Fluorescence spectrophotometer
- Centrifuge
- Multi-tube vortex
- HPLC system (for fractionation)

4. Calibration

Calibration using the Protoporphyrin Working Standard [3 μg/dL] is performed each day the instrument is used for analysis monitoring concentration and wavelength between 602 and 610 nm.

5. Quality control

High and low controls are validated and run with each batch. The performance of these specimens is monitored overtime.

Analytical Procedure

1. Transfer 500 µL of well-mixed Celite Solution into labeled test tube for samples, blank, and controls.
2. Vortex to mix all tubes lightly for ~10 s to ensure cell lysis.
3. Add 3 mL of RBC extraction solvent to each tube.
4. Vortex tubes 2 min using the pulse mode on the multi-tube vortex-mixer with the speed set at 2000 RPM.
5. Centrifuge tubes for 5 min at 2000 RPM.
6. Decant entire extract to its corresponding clean, labeled test tube, and discard cell pellet waste.
7. Add 3 mL of 1.5 N HCl to extract supernatant.
8. Vortex at least 45 s using the pulse mode on the multi-tube vortex mixer with the speed set at 2000 RPM.
9. Allow the aqueous layer to visibly separate from the organic layer, this takes ~5 min.
10. Under ventilation, remove the organic layer and discard.
11. Evaluate porphyrin emission at 605 nm (excitation at 408 nm) in the fluorescence spectrophotometer.
12. Calculations:

$$\left(\frac{\text{Sample fluorescence}}{\text{Protoporphyrin std fluorescence}}\right)\left(\frac{\text{Volume of cell suspension}}{\text{Volumen of packed cells}}\right)\left(\frac{4.5}{0.1}\right)(3.0)(1.04)$$

Assay constants in calculation:

4.5 = Final volume (mL) of aqueous extract
0.1 = Specimen volume (mL) extracted
µg/dL = Protoporphyrin working standard concentration
1.04 = Extraction efficiency factor

Reference Values

Total erythrocyte porphyrins <80 µg/dL

Pathological Values

Total protoporphyrin levels are determined and those samples exceeding an established cutoff (80 µg/dL in our laboratory) are fractionated for Zn and free protoporphyrin determination.

Erythrocyte Zn and free protoporphyrin levels and the ratio can distinguish multiple porphyria types (See Table 16.2). The most frequent cause of a mild to moderate elevation of Zn-complexed protoporphyrin is iron deficient anemia. Chronic exposure to lead and other heavy metals results in elevation in Zn-complexed protoporphyrin.

Zn-complexed protoporphyrin is found elevated in autosomal recessive porphyrias such as ALAD porphyria and CEP.

Limitations

- Hemolyzed samples will yield an artificially higher protoporphyrin value.

16.3.7 Fecal Porphyrin Determination

16.3.7.1 Method

Principle

The porphyrins are extracted from a sample of homogenized feces with acidified acetone-water and analyzed by HPLC. A gradient of aqueous ammonium phosphate and methanol separates the porphyrins by their polarity and isomer status. The fluorescence characteristics of the porphyrins comprise a basis for a high level of analytic specificity. The individual porphyrins are identified by comparison of retention times with those of standards.

Preanalytical

1. Specimen

24-h fecal collection, minimum 100 g. 10 g of homogenized sample used for analysis.

2. Chemicals and Reagents

- Methanol
- Ammonium hydroxide
- Phosphoric acid
- Ammonium phosphate buffer [100 mM], pH 6.0
- Hydrochloric acid
- Acetone, HPLC grade
- Fecal protoporphyrin IX intermediate standard [10.0 μg/dL]
- Porphyrin isomer calibration standard [5.0 μg/dL]
- Saline, 0.9%

3. Supplies and Instrumentation

- 50 mL polypropylene plug seal tube, Par Ex #: 263541
- Autosampler clear vials, Par Ex #: 294012
- Autosampler vial caps, Par Ex #: 294013
- Guard cartridge—C18, 4 × 3.0 mm
- Analytical column—C18, 4 micron, 3.9 × 150 mm
- Centrifuge
- Balance
- Multi-tube vortex
- pH meter
- HPLC system

4. Calibration

Blank samples (0.5 N HCl) and porphyrin standard solution are run with each batch to assess potential carryover and system performance.

5. Quality Control

Calibration peaks are reviewed for peak symmetry, baseline resolution, retention time consistency, and concentration. Coproporphyrin III retention time and concentration are recorded and monitored for each batch.

Analytical Procedure

1. Weigh 1.00 g ± 0.05 g homogenized patient sample into a tared, 50 mL tube.
2. Dispense 10 mL of 0.5 N HCl into each tube.
3. Dispense 10 mL of acetone into each tube.
4. Cap and vortex.
5. Centrifuge for 5 min at 2000 RPM.
6. Pipette 0.5 mL of the clarified fluid into a labeled autosampler vial.
7. Add 0.5 mL of 0.5 N HCl into each autosampler vial, cap, and analyze by HPLC.
8. Excitation 408 nm emission 630 nm.
9. Calculations:

$$\frac{\text{porphyrin ug}}{24\,\text{h}} = \frac{2(\text{TF})(20\,\text{ml}+1)\left(\text{measured concentration}\,\frac{ug}{dl}\right)}{1\left(\frac{100\,\text{ml}}{dl}\right)} = 0.42(\text{TF})(\text{measured concentration})$$

TFv = Total grams of fecal homogenate/24 h excretion
Assay constants in calculation:

2 = Accounts for the times 2 dilution step
20 = Volume of acetone: 0.5 N HCl used
1 = Grams of fecal homogenate assayed

Reference Values

Uroporphyrin I	<120	Pentacarboxylporphyrin I	<20
Uroporphyrin III	<50	Pentacarboxylporphyrin III	<20
Heptacarboxylporphyrin I	<40	Isopentacarboxylporphyrins	<80
Heptacarboxylporphyrin III	<40	Coproporphyrin I	<500
Isoheptacarboxylporphyrins	<30	Coproporphyrin III	<400
Hexacarboxylporphyrin I	<10	Isocoproporphyrins	<200
Hexacarboxylporphyrin III	<10	Protoporphyrins	<1500
Isohexacarboxylporphyrins	<10	Coproporphyrin III/I ratio	<1.20 (no units)

The units are micrograms porphyrin excreted per 24 h (μg/24 h). The preferred method is analyte quantitation over dry sample mass.

Pathological Values

Increased fecal porphyrin excretions and recognizable profiles are observed in multiple porphyria types (see Table 16.2). An example profile from a patient with variegate porphyria is shown in Fig. 16.6. This analysis is useful in biochemically distinguishing acute porphyria types. In quiescent phases, as well as prior to puberty, fecal porphyrin excretion may be within normal limits. EPP and XLDPP patients may have elevated protoporphyrin levels, however, these disorders cannot be diagnosed by fecal analysis alone.

Limitations

- Abstain from red meat or aspirin-containing medications for 3 days prior to collection.
- No barium or laxatives may be used 24 h prior to or during collection.

16.3.8 Porphyria Enzyme Analysis

Porphyrin enzyme analysis should not be used as a first tier analysis, but rather as confirmation when other testing suggests a specific enzyme deficiency. Particularly in the case of PCT, where many cases can be acquired differentiating between inherited and acquired porphyrias can be important for prognosis, as well as for recurrence risk and family planning. Identification of

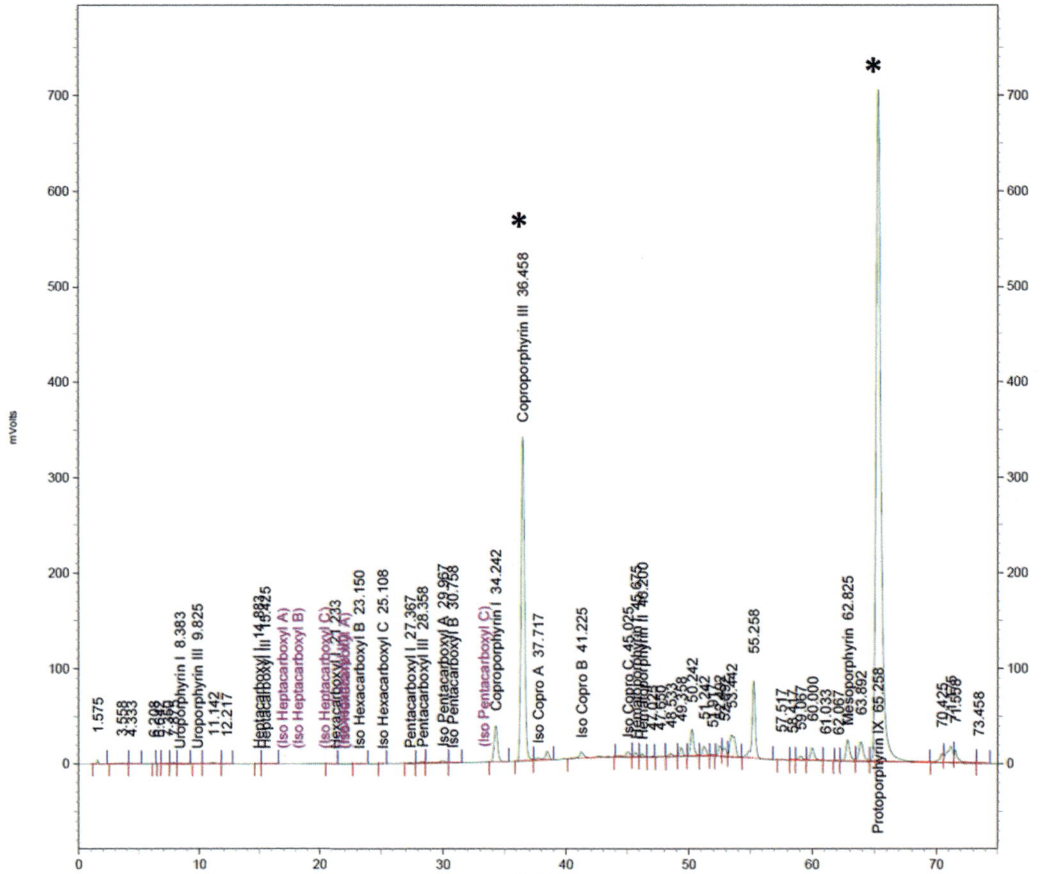

Fig. 16.6 HPLC fecal profile of a Variegate porphyria patient. *denotes characteristic elevations of coproporphyrin of the III series and protoporphyrin in the disease profile. The copro III/copro I ratio is also elevated

decreased enzyme activity in vitro can be an important piece of information in making this determination.

All porphyrin enzyme analyses are performed on washed erythrocytes. Each is based on the same principle of incubating a targeted substrate with the patient specimen, and then measuring the amount of product generated using a fluorescence spectrophotometer. The enzyme, substrate, and product combinations are shown in the table below.

Porphyria enzyme testing substrates and products

Enzyme	Condition	Substrate	Product
Porphobilinogen deaminase	Acute intermittent porphyria	Porphobilinogen	Uroporphyrin
Aminolevulinic acid dehydratase	ALAD porphyria	Aminolevulinic acid	Uroporphyrin
Uroporphyrinogen III synthase	Congenital erythropoietic porphyria	Aminolevulinic acid	Porphyrin I & III isomers (proportion of III isomers)
Uroporphyrinogen decarboxylase	Porphyria cutanea tarda and hepatoerythropoietic porphyria	Aminolevulinic acid	Ratio of coproporphyrin to uroporphyrin

Analytic procedures for porphobilinogen deaminase and uroporphyrinogen decarboxylase are outlined below.

16.3.8.1 Porphobilinogen Deaminase Activity

Method

Principle

Porphobilinogen deaminase activity is measured by calculating the production of uroporphyrin from porphobilinogen in incubated, lysed erythrocytes. Porphobilinogen deaminase (PBGD) activity is decreased in acute intermittent porphyria.

Preanalytical

1. Specimens

Washed erythrocytes (1.0 mL) or whole blood (3–5 mL; EDTA, Lithium Heparin, Sodium Heparin)

2. Reagents and Chemicals

- Sodium phosphate dibasic
- Citric acid monohydrate
- Sodium hydroxide
- Phosphoric acid
- Porphobilinogen
- Trichloroacetic acid
- Triton X-100
- Hydrochloric acid
- Saline (0.9%)
- Methanol
- Uroporphyrin standard (660 μg/dL)

3. Instrumentation

- Fluorescence spectrophotometer
- Water bath
- Centrifuge
- Vortex mixer

4. Calibration

Calibration is performed with each analytical batch using the Uroporphyrin standard. A single point calibration is used, and the λ_{max} should be between 596 and 604 nm.

5. Quality Control

Pooled blood samples can be used to create normal and abnormal controls. Incubation at 50 °C for 1 h in a water bath should produce deficient enzyme activity to function as an abnormal control.

Analytical Procedure

Specimen Preparation

1. Prepare PBG substrate to a final concentration of 1.4 mM. Adjust the pH to 7.5 using

concentrated phosphoric acid or 10% NaOH. Make up to volume with Phosphate-Citrate buffer. Prepared PBG substrate is stable for 1 year when stored at −70 °C. New lots of substrate should be verified prior to use. Substrate should be thawed at ambient temperature only, prior to use. Warming during thawing can cause increased levels of uroporphyrin.

2. On the day of testing, thaw the appropriate amount of PBG substrate for the number of samples to be analyzed (0.5 mL/sample).
3. Thaw patient samples and controls at ambient temperature.
4. Add 1.0 mL Triton-X solution (2 mL Triton X-100 in 1 L, made up to volume using water) to a labeled 16 × 125 mm disposable glass test tube for each blank, control, standard, and patient sample on the run.
5. Pipet 100 μL of each patient sample, control, and standard into the corresponding well. Use 0.9% saline for the blank. Patient samples should be well mixed, but do not vortex.
6. Shake each tube gently by hand to mix.
7. Verify the water bath temperature is 37.0 °C +/− 0.2 °C. Bring all tubes up to 37 °C by placing them in the water batch for 3 min.
8. Add 0.5 mL of PBG substrate to each reaction tube. Shake briefly by hand to mix. Cap all tubes.
9. Incubate all tubes for 30 min at 37 °C in a covered water bath w/shaker set @ 130 RPM.
10. After incubation is complete, remove tubes from water bath and remove all caps. Add 5.0 mL of 10% trichloroacetic acid solution to each tube to quench the reaction. Quench tubes in the same order substrate were added.
11. Cover all tubes with parafilm. Vortex using multi-tube vortex mixer with speed = 2000 on pulse mode.
12. After vortexing is complete, centrifuge tubes at 2000 RPM for 15 min to clarify the supernatant. If any patient samples remain cloudy after centrifugation, filter prior to reading the fluorescence.
13. Data acquisition

Exact steps for data acquisition will depend on the fluorometer model being used.

Measure the fluorescence of each standard, blank, control, and sample according to instrument instructions. The excitation wavelength should be 408.00 nm. The emission wavelength should be set to the peak maxima determined by analysis of the uroporphyrin standard. It should be between 596 and 604 nm.

14. Calculations

PBGD Activity

$$\left(\frac{6.60}{0.100}\right)\left(\frac{120.4}{1800}\right)\left(\frac{Vts}{Ve}\right)\left(\frac{Fx}{Fstd}\right)=(4.41)\left(\frac{Vts}{Ve}\right)\left(\frac{Fx}{Fstd}\right)$$

6.60 = Final volume of the assay solution in mL

0.100 = Specimen volume in mL

120.4 = Concentration of the uroporphyrin calibration standard in nmol/L

1800 = Time of incubation with PBG substrate in sec

V_{ts} = Total volume of washed, resuspended erythrocytes in mL

V_e = Volume of packed erythrocytes which were washed

F_x = Fluorescence intensity of the test sample

F_{std} = Fluorescence intensity of the calibration standard

Reference Ranges and Pathological Values

PBGD reference ranges	(nmol/L/s)
Normal	>7.0
Indeterminate	6.0—6.9
Reduced	<6.0

Limitations

- Some patients (~5–10%) with AIP have normal RBC PBGD activity.
- RBC PBGD activity measured during an acute episode can be in the normal range [9].

16.3.8.2 UROD Enzyme Analysis

Methods

Principle

Washed red blood cells are incubated for 3 h after addition of the substrate aminolevulinic acid (ALA). The porphyrins formed are measured by HPLC. Uroporphyrinogen decarboxylase (UROD) activity in the patient's red blood cells is determined by the ratio of coproporphyrin vs. uroporphyrin formed. UROD activity of approximately 50% of normal is suggestive of PCT type II. Individuals affected with HEP have activity closer to 10% of normal.

Preanalytical

1. Specimens

Washed erythrocytes (1.0 mL) or whole blood (3–5 mL; EDTA, lithium heparin, sodium heparin)

2. Reagents and Chemicals

- Sodium phosphate dibasic
- Citric acid monohydrate
- Phosphate-citrate buffer combine 33.5 g of $NaPO_4$ dibasic, 2.0 g citric acid to ~400 mL water, pH to 7.5.
- 5-aminolevulinic acid
- PENZ substrate, 22 nM
- ALA diluted with phosphate-citrate buffer, pH 6.8
- Phosphoric acid monobasic
- Phosphoric acid
- Phosphate buffer
- Add 55.6 g $NaPO_4$ monobasic to 750 mL water in a 2 L beaker. Titrate to pH 3.5 with phosphoric acid. Filter and store in brown bottle.
- Ethyl acetate
- Acetic acid, glacial
- RBC extraction mixture
- Combine 1600 mL ethyl acetate and 400 mL glacial acetic acid in a 2 L bottle.
- Methanol
- Hydrochloric acid, (12 N)
- Saline (0.9%)
- Methanol
- Porphyrin calibration standard
  - [Uroporphyrin III, Heptacarboxylporphyrin I, Hexacarboxylporphyrin I, Pentacarboxylporphyrin I, Coproporphyrin III]. From 10 mg/dL stock solutions of each porphyrin analyte above, add 200 μL to a 1 L amber volumetric flask. Dilute to volume with 1.5 N HCl

3. Instrumentation

- Fluorescence spectrophotometer
- Water bath
- Centrifuge
- Vortex mixer
- HPLC instrument

4. Calibration

Symmetric peak shape, retention time is within 0.2 min of the retention time from the previous run, the concentration of the porphyrin peaks corresponds to 1.80 to 2.20 μg/dL

5. Quality Control

Pooled blood samples can be used to create normal and abnormal controls as in PBGD activity section.

Analytical Procedure

1. Pipette 1 mL of water to one of the 20 × 125 tubes for each reagent blank, control, and patient sample.
2. Pipet 100 μL of 0.9% saline (blank), control, and well-mixed patient sample (not vortexed) into its corresponding labeled UPGD tube. Swirl gently to mix.
3. Prior to incubation (or blank extraction), pipette 500 μL of PENZ substrate into each tube.

4. Cap all tubes and gently swirl to mix.
5. Place control and patient tubes in 37.0 °C water bath. Place on shaker at 130 RPM for 3 h.
6. At the end of the 3 h incubation, remove control and patient tubes from the water bath.
7. Remove caps and dispense 10 mL RBC extraction mixture into each of the tubes. Recap the tubes.
8. Vortex for 1 min using the pulse mode on the multi-tube vortex mixer with the speed set at 2000.
9. Centrifuge for 5 min at 2000 RPM.
10. Decant supernatant into its corresponding labeled 20 × 125 tube.
11. To the tubes containing supernatant, dispense 3.0 mL of 1.5 N HCl. 18. Cap and vortex for 1 min using the pulse mode on the multi-tube vortex mixer. Set in rack, uncap tubes, and allow enough time for the bottom aqueous layer to separate from the upper organic layer (~1 min).
12. Using a disposable glass pipette, transfer bottom layer extract into the third corresponding, labeled 20 × 125 tube.
13. Repeat the 1.5 N HCl extraction twice more, combining extracts from each extraction into the corresponding tube.
14. Cap and gently invert tube containing the acid extracts to mix.
15. Transfer extract into its corresponding autosampler vial, cap, and place in its corresponding position in the autosampler for HPLC analysis. This method and subsequent calculations below injects 50 μL of the sample and develop a chromatogram.
16. Calculation

UPGD Activity

$$\frac{\text{nmol porphyrin formed}}{\text{L} \times \text{s}} = \frac{\text{pmol ng } Vf\, 1000\,\text{ml/L}}{\text{ng } Vi\, Ve\, 10{,}800\,\text{s}} \frac{1\,\text{nmol}}{1000\,\text{pmol}}$$

Where:

ng = Amount of porphyrin in the injected volume

V_i = Volume (mL) of fluid injected 0.050

V_f = Total volume (mL) of pooled extracts 12.00

V_e = Volume (mL) of suspension which was extracted 0.100

Each porphyrin product should be calculated using the corresponding formula:

$$\frac{\text{nmol uroporphyrin formed}}{\text{L} \times \text{s}} = (0.2674)(\text{ng uroporphyrin})$$

$$\frac{\text{nmol coproporphyrin formed}}{\text{L} \times \text{s}} = (0.3394)(\text{ng coproporphyrin})$$

Record and report

$$\text{UPGD activity} = \frac{\text{nmol coproporphyin formed}}{\text{nmol uroporphyrin formed}}$$

Reference Ranges and Pathological Values

or = 1.00 RU (normal)

0.80–399 RU (indeterminate)

<0.80 RU (indicative of PCT type II)

Limitations

Sample handling is a cause of falsely decreased enzymes, particularly for UPGD.

RBC UPGD activity is normal in patients with the acquired form of PCT (PCT type II)

16.4 Conclusion

Biochemical analyses for the diagnosis of porphyrias are of paramount importance. The presence of a pathogenic variant in one of the genes of the heme biosynthesis pathway is no sufficient to ascribe the symptomatology to the disease. Given the low penetrance, the high impact of environmental factors, and the elevated number of acquired cases of PCT, a biochemical diagnosis is essential. The right specimen at the right time is the key for the correct diagnosis of these group of diseases.

References

1. Gerischer LM et al (2021) Acute porphyrias—a neurological perspective. Brain Behav 11(11):e2389
2. Gouya L et al (2020) EXPLORE: a prospective, multinational, natural history study of patients with acute hepatic porphyria with recurrent attacks. Hepatology 71(5):1546–1558
3. Poblete-Gutiérrez P et al (2006) The porphyrias: clinical presentation, diagnosis and treatment. Eur J Dermatol 16(3):230–240
4. Phillips JD et al (2007) A porphomethene inhibitor of uroporphyrinogen decarboxylase causes porphyria cutanea tarda. Proc Natl Acad Sci U S A 104(12):5079–5084
5. Yasuda M, Chen B, Desnick RJ (2019) Recent advances on porphyria genetics: inheritance, penetrance & molecular heterogeneity, including new modifying/causative genes. Mol Genet Metab 128(3):320–331
6. Bonkovsky HL et al (2014) Acute porphyrias in the USA: features of 108 subjects from porphyrias consortium. Am J Med 127(12):1233–1241
7. Lala SM, Naik H, Balwani M (2018) Diagnostic delay in erythropoietic protoporphyria. J Pediatr 202:320–323.e2
8. von und zu Fraunberg M et al (2002) Clinical and biochemical characteristics and genotype-phenotype correlation in Finnish variegate porphyria patients. Eur J Hum Genet 10(10):649–657
9. Kostrzewska E, Gregor A (1986) Increased activity of porphobilinogen deaminase in erythrocytes during attacks of acute intermittent porphyria. Ann Clin Res 18(4):195–198

Congenital Disorders of Glycosylation

17

Nastassja Himmelreich, Jürgen Lübbehusen, Christian Körner, Dirk J. Lefeber, and Christian Thiel

Contents

List of Abbreviations

A	Ampère
ALG	Asparagine-linked glycosylation
aqua bidest.	Double-distilled water
CDG	Congenital disorders of glycosylation
Dev	Development
ECL	Enhanced chemi-luminescence
EDTA	Ethylenediaminetetraacetic acid
ER	Endoplasmic reticulum
GDP	Guanosine diphosphate
GlcNAc	N-acetylglucosamine
HPLC	High performance liquid chromatography
IEF	Isoelectric focusing
Man	Mannose
MPI	Phosphomannose isomerase
NADP	Nicotinamide adenine dinucleotide phosphate
NADPH	Reduced nicotinamide adenine dinucleotide phosphate
p.a.	Pro analysis
PBS	Phosphate-buffered saline
PGM1	Phosphoglucomutase 1
PMM2	Phosphomannomutase 2
PNGase F	N-glycosidase F
rpm	Rounds per minute
Sep.	Separation
TFA	Trifluoroacetic acid
Tris	Tris(hydroxymethyl)-aminomethane

Christian Körner (Deceased)

N. Himmelreich · J. Lübbehusen · C. Thiel (✉)
Department 1, Center for Pediatric and Adolescent Medicine, University Heidelberg, Heidelberg, Baden-Wuerttemberg, Germany
e-mail: Christian.Thiel@med.uni-heidelberg.de

D. J. Lefeber (✉)
Department of Neurology, Radboud University Medical Center, Nijmegen, The Netherlands

Translational Metabolic Laboratory, Department of Human Genetics, Radboud University Medical Center, Nijmegen, The Netherlands
e-mail: Dirk.Lefeber@radboudumc.nl

N. Blau, F. M. Vaz (eds.), *Laboratory Guide to the Methods in Biochemical Genetics*,
https://doi.org/10.1007/978-3-031-58819-8_17

UDP	Uridine diphosphate
V	Volt
v	Volume
W	Watt
w	Weight
× g	x-fold gravity

17.1 Introduction

17.1.1 Glycosylation in General

The term "Glycosylation" describes the co- and posttranslational linkage of sugar moieties onto newly synthesized proteins and lipids. This complex metabolic process, which has been found in all forms of life from bacteria to man, comprises one of the most widespread and variable forms of protein modifications. The sugar moieties, which are also termed "oligosaccharides" or "glycans," are necessary for protein quality control, directed transport, and biological activity [1, 2]. Glycosylation affects a variety of physicochemical properties of proteins on behalf of their stability, solubility, and polarity. Glycoproteins play an important role in a lot of complex biological processes such as growth, differentiation, organ development, signal transduction, immunologic defense, but are also concerned in pathologic processes like tumor progression [3]. In man, the complex glycosylation apparatus comprises a huge variety of proteins such as glycosyltransferases, glycosidases, sugar-nucleotide transporters as well as proteins in sugar metabolism needed for the allocation of donor substrates (e.g., the polyisoprenoid and hexosamine pathways), needed for the intracellular transport of glycoproteins (e.g., the Conserved Oligomeric Golgi (COG) complex) or for the maintenance of pH or ion gradients to ensure the functioning of the different glycosylation machineries for N-glycosylation, O-glycosylation (O-mannose, O-glucose, O-fucose, O-GlcNAc, O-GalNAc), C-mannosylation, the glycosaminoglycans (GAG), the glycophosphatidylinositol (GPI) anchored proteins and glycolipids [4, 5].

17.1.2 Congenital Disorders of Glycosylation

Defects in one of the glycosylation pathways lead to "Congenital Disorders of Glycosylation" (CDG), a still rapidly expanding group of mostly autosomal recessive inherited metabolic diseases (~190 defects by the middle of 2024) [6–8]. Due to the importance of glycosylation, patients normally present with a multisystemic disorder which is mostly combined with severe neurological impairment [9]. Notably, CDG appear with a highly variable phenotype even within the same CDG subtype and same genetic variants. Typical clinical characteristics comprise growth retardation either due to disturbed nutrition uptake and recurrent vomiting or hormone disturbance. Microcephaly often leads to ataxia and nerve conduction velocity is reduced. Furthermore, inverted nipples and fat pads at the bottom and the upper arms. Moreover, strabismus, retinitis pigmentosa, optic atrophy, coloboma of the iris, and cataracts have been described. Some CDG patients suffer from hypogonadism and hyperinsulinism. Orthopedical problems present in some cases with osteopenia, exostosis, and contractures of the joints. Coagulation abnormalities appear quite frequently with thrombosis, hemophilia, and phlebitis. During the first months after birth, life-threatening cardiac problems may appear which are caused by cardiomyopathy, pericarditis, and pericardial effusion. Gastrointestinal defects in CDG often appear with chronical diarrhea, hepatomegaly, and protein-losing enteropathy. Renal complications present with proteinuria, microcysts, and proximal tubulopathy. Most of the CDG patients already present with clinically striking symptoms within the first year of life. However, in some cases, the symptoms can also develop later. The mortality rate is about 20% in the first 2–3 years of life [10, 11].

Nomenclature of CDG is based on the abbreviation of the affected protein in combination with "-CDG," e.g. PMM2-CDG in case of a defect in the phosphomannomutase 2 (PMM2). Based on the glycosylation pattern of serum transferrin, which still represents the gold standard marker in CDG diagnostics, N-glycosylation defects can be subdivided into CDG-I and CDG-II defects. CDG-I comprises deficiencies which either affect the biosynthesis of lipid-linked oligosaccharides (LLO) or the transfer of oligosaccharides onto newly synthesized proteins by the oligosaccharyltransferase complex in the endoplasmic reticulum. CDG-II affects the subsequent trimming and elongation of N-glycans in the endoplasmic reticulum and the Golgi apparatus. Less technically, the CDG-I group represents the early glycosylation defects and the CDG-II group contains the late glycosylation defects [12–14].

17.1.3 Diagnostics and Markers for CDG

All patients suffering from unclear multi-organic diseases, especially in combination with mental and psychomotor retardation, strabismus, cerebellar atrophy, and blood clotting problems, should be investigated for CDG [15].

CDG diagnostics primarily starts with checking the very common N-glycosylation, in which the sugar chains are connected to the glycoprotein via asparagine residues. The 78 kDa transferrin loaded with two biantennary N-glycan chains of the complex type serves as a marker protein and still represents the gold standard in worldwide diagnostics for CDG defects in N-glycosylation. Transferrin is analyzed by using HPLC, capillary zone electrophoresis, isoelectric focusing (IEF), or mass spectrometry. IEF displays a very robust and inexpensive method that has established itself worldwide in analysis. However, more precise results of the sugar chains can be achieved with the help of mass spectrometry. In this chapter, we present both, the method of IEF and MS. Another important marker protein in the context of N-glycosylation is alpha-1-antitrypsin (A1AT) which can be used in parallel to transferrin in children younger than 12 weeks old as some of these children show significant hypoglycosylation when analyzing transferrin, which can be misinterpreted as CDG-II. (16; see also 2.1.4.2 and 2.4). Furthermore, Apolipoprotein CIII (ApoCIII) is available for the analysis of defects within the very common O-glycosylation of core 1 mucin types. This is particularly evident in the case of superordinate glycosylation defects (e.g., the various COG-CDG defects) that affect both N- and O-glycosylation, with a conspicuous band pattern in the analysis.

It is important to note that, regardless of the laboratory method chosen, not all CDG types can be routinely identified from a serum or plasma sample due to the lack of established marker proteins. Problematic here are, for example, O-mannosylation defects (as Walker–Warburg Syndrome, Muscle Eye Brain Disease) for which immunohistochemical staining in muscle could be performed, GPI anchor defects (e.g., PIG-CDGs) for some of which blood cell surface antigens can be analyzed, or O-fucosylation disorders (POFUT1- and POFUT2-CDG) [16]. In the last years, isofocusing of the bikunin protein in plasma has been suggested as a diagnostic option for the so-called linkeropathies, a subgroup of CDG due to defects in biosynthesis of the oligosaccharide that links glycosaminoglycans to their core protein. For biochemical diagnosis of these CDG, we refer to literature. In addition, in some patients, the defects are so mild that underglycosylation cannot always be recognized even with established glycosylation markers. A negative result in the analysis does not necessarily rule out a CDG defect. If the suspicion of CDG persists (due to clinical symptoms or genetic data), it may make sense to re-examine another sample. An analysis in an alternative material than blood like skin fibroblasts can also be useful. Those examinations in cells can be carried out in specialized laboratories.

17.2 Methods

17.2.1 Isoelectric Focusing (IEF) of Serum Transferrin

17.2.1.1 Principles

In most CDG cases, loss or shortening of oligosaccharide moieties linked to glycoproteins has been observed. Therefore, the iron-binding glycoprotein transferrin is used for initial CDG diagnostics by IEF since this protein is abundant in serum (about 5% of all plasma proteins) and the structure of its N-linked oligosaccharide moieties is very uniform. By IEF, the charge of transferrin is analyzed in an electric field. In healthy persons, transferrin is mainly found in the tetrasialo form due to the linkage of the protein to two oligosaccharide chains carrying four negatively charged terminal sialic acid residues (Fig. 17.1).

In case of CDG-I, it was observed that in addition to tetrasialo transferrin, di- and asialotransferrin bands appear due to the loss of sialic acid residues (Fig. 17.1, left, lanes 2–7). This loss is caused by the partial absence of complete oligosaccharide chains. Notably, the early glycosylation defect PGM1-CDG is a chameleon and can present with a combined CDG-I/CDG-II pattern (Fig. 17.1, left, lane 8). In contrast to CDG-I, changes in charge of serum transferrin of most known CDG-II types are due to shortening of the oligosaccharide moieties (Fig. 17.1, right, lanes 2–8).

17.2.1.2 Pre-analytical

Specimen

To investigate the glycosylation state of transferrin by IEF, 0.1 ml of serum is required.

Reagents and Chemicals

- 200 μM Fe(III)-citrate p.a. (Merck, Darmstadt, Germany)
- 150 mM NaCl p.a. (Roth, Karlsruhe, Germany)
- Formaldehyde (36.5–38%; Sigma-Aldrich, Taufkirchen, Germany)
- Glutardialdehyde 25% (Merck, Darmstadt, Germany)
- Acetic acid 100% p.a. (Roth, Karlsruhe, Germany)
- Ethanol 99.8% p.a. (Roth, Karlsruhe, Germany)
- Silver nitrate p.a. (Merck, Darmstadt, Germany)

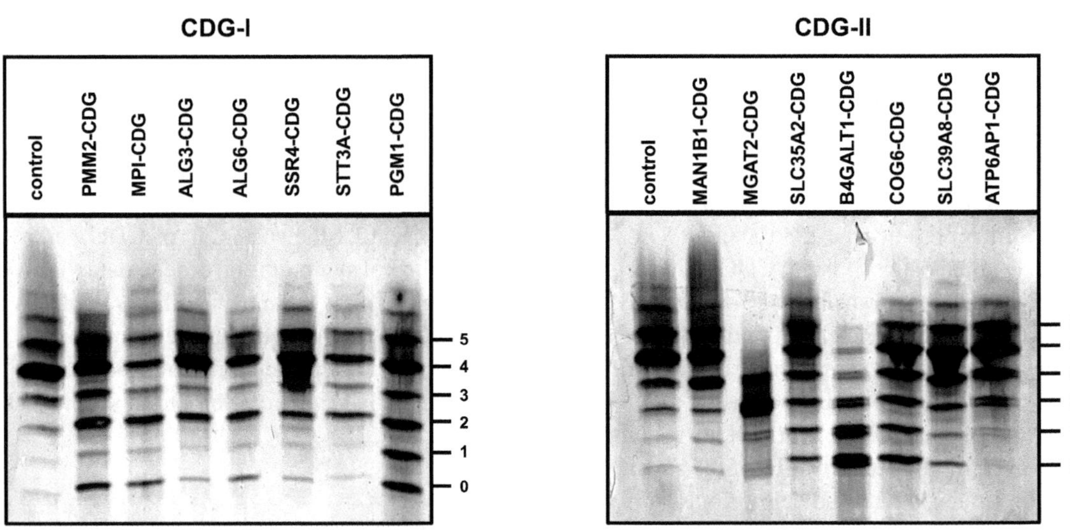

Fig. 17.1 IEF of serum transferrin of CDG patients. Sera from a healthy control and seven CDG-I patients (left) and of 7 CDG-II patients were analyzed by IEF. In case of the healthy person, the main form of the protein carries four negatively charged sialic acid residues (4), even though small amounts of penta (5) and trisialo transferrin (3) are detectable. Additional disialo (2) and asialo transferrin (0) forms are indicators for CDG-I. Please note that PGM1-CDG can mimic a CDG-II defect. (Right) In CDG-II, often additional trisialo- and monosialotransferrin forms occur which are due to the loss of either one or three sialic acid residues. MGAT2-CDG displays with a unique transferrin pattern characterized by a strong appearance of a band coeluting with disialo transferrin. Isoforms of transferrin which are independent from a pathological phenotype can cause double bands in IEF (lanes 3, 5, and 8). Numbers 0–5 indicate asialo to pentasialo transferrin

- Trichloroacetic acid Sigma Ultra 99% (Sigma-Aldrich, Taufkirchen, Germany)
- Sodium carbonate p.a. (Merck, Darmstadt, Germany)
- Antibody: Rabbit anti human transferrin (DakoCytomation, Hamburg, Germany)
  - Fixation: 20% trichloroacetic acid (Sigma-Aldrich, Taufkirchen, Germany)
  - Wash solution 1: 50% ethanol (Roth, Karlsruhe, Germany)/10% acetic acid (Roth, Karlsruhe, Germany)
  - Wash solution 2: 10% ethanol (Roth, Karlsruhe, Germany)/5% acetic acid (Roth, Karlsruhe, Germany)
  - Developer: 2.5% (w/v) sodium carbonate (Merck, Darmstadt, Germany)/0.015% formaldehyde (v/v) (Sigma-Aldrich, Taufkirchen, Germany)
  - Stop solution: 50 mM EDTA (Titriplex III, Roth, Karlsruhe, Germany)

Instrumentation

- PhastSystem™ (Amersham Biosciences, Freiburg, Germany)
- PhastGel pH 4–6.5 (Amersham Biosciences, Freiburg, Germany)
- PhastGel Sample Applicator 8/1 (Amersham Biosciences, Freiburg, Germany)

Calibration

Calibration of the IEF is performed by using serum of a healthy control person as well as serum of CDG patients suffering from different types of the CDG.

Running Conditions

IEF

Sep 1.1	2000 V	2.5 mA	3.5 W	15 °C	75 V h
Sep 1.2	200 V	2.5 mA	3.5 W	15 °C	15 V h
Sep 1.3	2000 V	5 mA	3.5 W	15 °C	185 V h

- Sample Applicator down at 1.2, 0 V h
- Sample Applicator up at 1.3, 0 V h
- Extra Alarm to sound at 1.1, 73 V h

Silver Staining

Dev 1.1	Fixation	5 min	20 °C
Dev 1.2	Wash solution 1	2 min	50 °C
Dev 1.3	Wash solution 2	2 min	50 °C
Dev 1.4	Wash solution 2	4 min	50 °C
Dev 1.5	8.3% glutardialdehyde	6 min	50 °C
Dev 1.6	Wash solution 2	3 min	50 °C
Dev 1.7	Wash solution 2	5 min	50 °C
Dev 1.8	Aqua bidest.	2 min	50 °C
Dev 1.9	Aqua bidest.	2 min	50 °C
Dev 1.10	0.5% silver nitrate	10 min	50 °C
Dev 1.11	Aqua bidest.	0.5 min	50 °C
Dev 1.12	Aqua bidest.	0.5 min	50 °C
Dev 1.13	Developer	0.5 min	50 °C
Dev 1.14	Developer	12 min	50 °C
Dev 1.15	Stop solution	5 min	50 °C
Dev 1.16	Aqua bidest.	5 min	50 °C

17.2.1.3 Analytical

Procedure

To saturate serum transferrin with iron, 10 µl of serum and 10 µl 200 µM Fe(III)-citrate are added to 30 µl aqua bidest. After incubation for 10 min at room temperature the mixture is diluted 1:50 with distilled water. 1 µl is loaded onto the PhastGel Sample Applicator 8/1 and the IEF is performed as described above. After separation, gels are incubated with rabbit anti human transferrin antibody (dilution 1:3 in 150 mM NaCl) for 40 min followed by washing in 150 mM NaCl overnight. Automated silver staining of gels is performed as described above using the PhastSystem™.

17.2.1.4 Post Analytical

Interpretation

Data analysis is carried out by comparing the patterns of controls and known CDG types with the patients. IEF of transferrin from controls shows predominantly the tetrasialo form of the protein whereas in case of CDG-I patients additional di- and asialo bands appear (Fig. 17.3).

Pitfalls

IEF of serum transferrin can detect a wide range of CDG cases. Nevertheless, some defects as NUS1-CDG, DDOST-CDG, TUSC3-CDG, TRAPC11-CDG, MOGS-CDG and defects within the fucosylation of N-glycans (GFUS-CDG, SLC35C1-CDG, FUT8-CDG, FUT9-CDG) are missed by transferrin analysis. In the known cases of CDG, biochemical prenatal diagnostics from fetal blood is not reliable [17, 18] and should be addressed by genetic analysis. Moreover, the use of transferrin shows an analytical problem in the time span from birth up to 12 weeks of life. In this developmental period oligosaccharide moieties N-linked to proteins are often incomplete, resembling an abnormal CDG pattern and leading to false-positive results [19]. Here, IEF of alpha-1-antitrypsin or analysis of N-glycans derived from whole serum glycoproteins by mass spectrometry displays effective diagnostic alternatives. Furthermore, heavy alcohol consumption, galactosemia, and fructosemia can result in abnormal IEF patterns. As well, variants in the protein backbone of transferrin, which lead either to loss or addition of charged amino acid residues, might resemble CDG-like IEF patterns. To avoid false-positive results, desialylation of transferrin by neuraminidase treatment (see Sect. 2.3, Fig. 17.2) or an IEF of an alternative glycoprotein like α-1-antitrypsin (see Sect. 2.4) should be performed.

17.2.2 IEF of Blood Samples from Guthrie Cards

17.2.2.1 Sample Preparation

A punched-out piece of the Guthrie card is incubated in 50 µl of distilled water for 10 min at room temperature. 10 µl of Fe(III)-citrate (200 µM) and 50 µl of distilled water are added to the sample and incubation is performed for

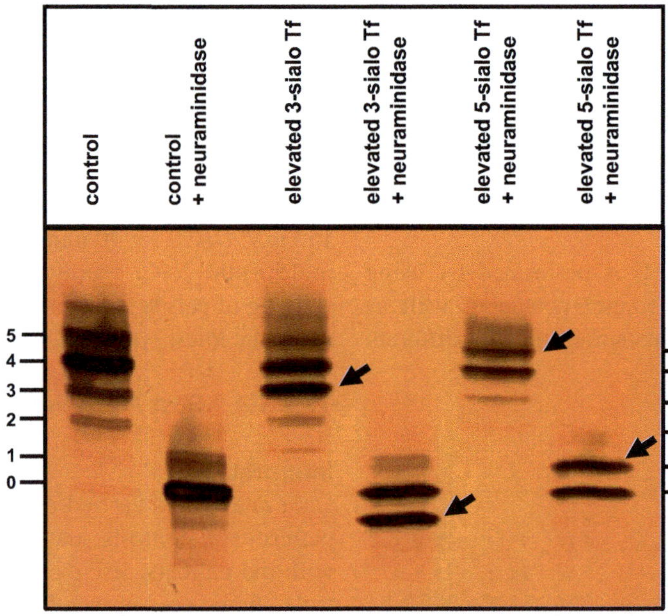

Fig. 17.2 Pitfall in CDG diagnostics due to variants in the protein backbone of transferrin. Striking abnormal IEF patterns are repeatedly detected in transferrin IEF analysis, characterized foremost by elevated trisialo (lane 3, arrow) and pentasialo (lane 5, arrow) transferrin bands, hereby mimicking a CDG-II defect. In these cases, it is advisable to perform a neuraminidase digestion of the transferrin's N-glycans to avoid false-positive results. Due to the loss of the terminal sialic acid residues on the sugar moieties, most of the transferrin's charge is lost which is accompanied by its shift to the asialo transferrin size (lane 2). Presence of two bands on the IEF gel after neuraminidase treatment (lanes 4 and 6, arrows) indicates a variant affecting the size of the transferrin, but no glycosylation defect. Numbers 0–5 indicate asialo to pentasialo transferrin

10 min at room temperature. 1 µl of dissolved material is subsequently investigated by PhastGel as described above.

17.2.2.2 Pitfalls

A lot of questionable CDG cases on behalf of IEF analysis belong to blood derived from Guthrie cards. In these cases, IEF investigations should be repeated with serum.

17.2.3 Desialylation of Transferrin by Neuraminidase Treatment

17.2.3.1 Principle

Due to mutations in the protein backbone of transferrin which lead to changes in charged functional residues of amino acids, shifts in IEF appear, which resemble CDG patterns. To avoid false-positive results, desialylation of transferrin by neuraminidase treatment (Fig. 17.2) or IEF of an alternative glycoprotein like α-1-antitrypsin (Sect. 2.4) should be conducted (Fig. 17.3).

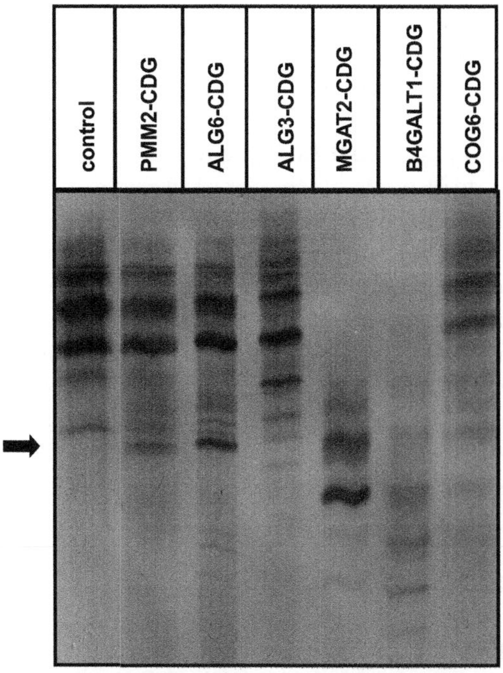

Fig. 17.3 IEF pattern of α-1-antitrypsin. Sera from a control (lane 1), three CDG-I (PMM2-, ALG6-, ALG3-CDG; lanes 2–4), and three CDG-II patients (MGAT2-, B4GALT1-, COG6-CDG; lanes 5–7) were analyzed by IEF. In contrast to the pattern of a healthy control, additional bands with a cathodal shift appear. In abnormal patterns (lanes 2–7), the position of the first additional abnormal cathodal band is indicated by an arrow. This band and also all bands below indicate a glycosylation deficiency

17.2.3.2 Pre-analytical

Specimen

For neuraminidase digestion, 6 µl of serum from a control and a patient is required.

Reagents and Chemicals

- Neuraminidase (1 U/100 µl) (Roche, Mannheim, Germany)
- 1 M sodium acetate (pH 5.0) (Merck, Darmstadt, Germany)
- 10 mM Tris (pH 7.0) (p.a., Roth, Karlsruhe, Germany)/0.9% NaCl (p.a., Roth, Karlsruhe, Germany)
- Dialysis membrane (cut-off: 12–14 kDa) (neolab, Heidelberg, Germany)

Instrumentation

- Heating block (Eppendorf, Hamburg, Germany)
- Speed-vac (Bachofer, Weilheim, Germany)
- Magnetic stirrer (Heidolph, Kelheim, Germany)

Calibration

In order to compare sera from patients suffering from unclear CDG types, sera of healthy persons and of patients with already defined CDG are used as controls for neuraminidase treatment studies.

17.2.3.3 Analytical

Procedure

Neuraminidase digestion of transferrin is performed from 6 µl of serum, 2.5 µl 1 M sodium acetate (pH 5.0), 15.5 µl aqua bidest., and 1 µl neuraminidase (0.01 U). Incubation is carried out over night at 37 °C followed by dialysis against aqua bidest. at 4 °C for 12 h and subsequent drying by lyophilization. Dried samples are resuspended in 1.5 ml 10 mM Tris (pH 7.0)/0.9% NaCl and analyzed by IEF followed by silver staining as described above.

17.2.3.4 Post Analytical

Interpretation

Data analysis is carried out by comparing the patterns of questionable patients with those of healthy controls and already defined CDG patients. After neuraminidase treatment, controls and CDG patients will predominantly show the asialo form of transferrin. In case of mutations, which affect the protein backbone of transferrin, additional bands will appear (Fig. 17.2).

17.2.4 IEF of α-1-Antitrypsin

17.2.4.1 Principle

In addition to the IEF of serum transferrin, IEF of alternative serum glycoproteins like α-1-antitrypsin can be used to confirm or to rule out CDG. Notably and in contrast to transferrin, the glycosylation status of α-1-antitrypsin is already fully existent shortly after birth allowing for the investigation of CDG in the first weeks of life [19]. Human plasma α-1-antitrypsin is normally fully glycosylated at amino acid residues 46, 83, and 247 [20]. Due to combinations of tri- and diantennary complex type oligosaccharides and shortened isoforms of the protein, a pattern of seven bands in IEF analysis of α-1-antitrypsin of a healthy control is detectable. In case of CDG, additional bands with a cathodal shift appear on the gel (Fig. 17.3).

17.2.4.2 Pre-analytical

Specimen

1–3 μl of serum of a control and the patient are used for the test.

Reagents and Chemicals

- Pharmalyte™ pH 4.2–4.9 (Amersham Biosciences, Freiburg, Germany)
- Antibody: Rabbit anti human α-1-antitrypsin (DakoCytomation, Hamburg, Germany)
- For reagents and chemicals, see Sect. 2.1.2.2

Instrumentation

- PhastSystem™ (Amersham Biosciences, Freiburg, Germany)
- PhastGel Dry Gel (Amersham Biosciences, Freiburg, Germany)
- PhastGel Sample Applicator 8/1 (Amersham Biosciences, Freiburg, Germany)
- PhastGel Cassette (Amersham Biosciences, Freiburg, Germany)

Calibration

Calibration of IEF is carried out with sera from healthy controls as well as sera derived from CDG patients with defined types of the CDG.

Running Conditions

IEF

Sep 1.1	2000 V	2.5 mA	3.5 W	15 °C	75 V h
Sep 1.2	200 V	2.5 mA	3.5 W	15 °C	15 V h
Sep 1.3	2000 V	2.5 mA	3.5 W	15 °C	100 V h
Sep 1.4	1000 V	2.5 mA	3.5 W	15 °C	280 V h

- Sample Applicator: Down at 1.2, 0 V h
- Sample Applicator: Up at 1.3, 100 V h
- Extra Alarm: To sound at 1.1, 73 V h

Silver Staining

Dev 1.1	Fixation	5 min	20 °C
Dev 1.2	Wash solution 1	2 min	50 °C
Dev 1.3	Wash solution 2	2 min	50 °C
Dev 1.4	Wash solution 2	4 min	50 °C
Dev 1.5	8.3% glutardialdehyde	6 min	50 °C
Dev 1.6	Wash solution 2	3 min	50 °C
Dev 1.7	Wash solution 2	5 min	50 °C
Dev 1.8	Aqua bidest.	2 min	50 °C
Dev 1.9	Aqua bidest.	2 min	50 °C
Dev 1.10	0.5% silver nitrate	10 min	50 °C
Dev 1.11	Aqua bidest.	0.5 min	50 °C
Dev 1.12	Aqua bidest.	0.5 min	50 °C
Dev 1.13	Developer	0.5 min	50 °C
Dev 1.14	Developer	12 min	50 °C
Dev 1.15	Stop solution	5 min	50 °C
Dev 1.16	Aqua bidest.	5 min	50 °C

17.2.4.3 Analytical

Procedure

IEF PhastGels are rehydrated for 30 min at room temperature in a solution containing aqua bidest. and 60 ml/l Pharmalyte (4.2–4.9) in a PhastGel Cassette. The sera are diluted 1:150 and 0.85 µl of the samples are loaded on the PhastGel Sample Applicator 8/1 for IEF. After separation, the gels are incubated with a rabbit anti human α-1-antitrypsin antibody for 40 min followed by washing in 150 mM NaCl overnight. Automated silver staining of the gels by the PhastSystem™ is performed as described above.

17.2.4.4 Post Analytical

Interpretation

Data are analyzed by comparing the α-1-antitrypsin IEF patterns of patients to healthy controls and patients with already defined CDG. A normal pattern of α-1-antitrypin in the patient indicates changes in the protein moiety of serum transferrin instead of a general glycosylation deficiency.

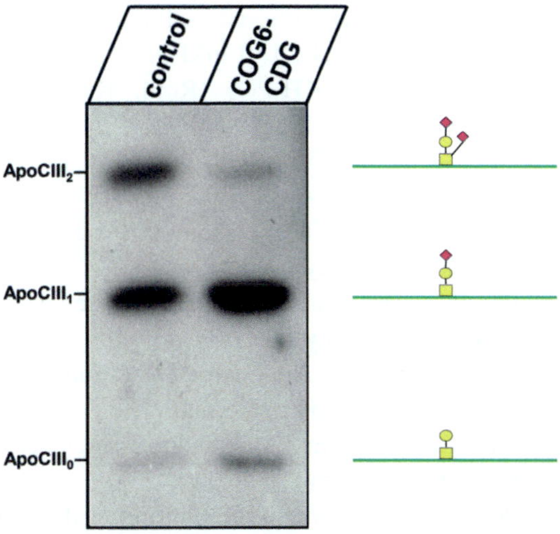

Fig. 17.4 Analysis of core-1 mucin type O-linked glycans from Apolipoprotein CIII (ApoCIII). Differentially charged oligosaccharide structures linked to the serum protein ApoCIII from a control (lane1) and a COG6-CDG patient (lane 2) derived serum were separated by isoelectric focusing and further investigated by staining with an ApoCIII antibody. $ApoCIII_2$, $ApoCIII_1$, and $ApoCIII_0$ indicate disialo, monosialo, and asialo apolipoprotein variants

17.2.5 IEF of Apolipoprotein CIII (ApoCIII)

17.2.5.1 Principles

IEF of ApoCIII is a simple and rapid technique that allows quantitative determination of the three ApoCIII isoforms [21]. ApoCIII isoforms from plasma may be used as markers for the biosynthesis of core 1 mucin type O-glycans. This glycan type is widely expressed throughout the human body and is abundant in the nervous system. It is considered as the most common type of O-glycan. Core 2 mucin type O-glycans are synthesized from unsubstituted core 1 mucin type O-glycans. Therefore, a defect in the biosynthesis of unsubstituted core 1 O-glycans will also affect the synthesis of core 2 mucin type O-glycans. Performing an IEF of transferrin in combination with ApoCIII allows screening of patients with combined N- and O-glycosylation deficiencies (e.g., the COG-CDGs, Fig. 17.4) or sole O-mucin type defects (e.g., GALNT2-CDG).

17.2.5.2 Pre-analytical

Specimen

1 to 3 µl of serum of a healthy person and a patient is needed for the test.

Reagents and Chemicals

- Urea p.a. (Roth, Karlsruhe, Germany)
- Pharmalyte™ (pH 4,2-4,9) (Amersham Biosciences, Freiburg, Germany)
- Servalyt pH 3–5 (Serva, Heidelberg, Germany)
- Nitrocellulose membrane ECL (Amersham Biosciences, Freiburg, Germany)
- Rabbit anti human apoC-III antibody (Biotrend, Cologne, Germany)
- Goat anti rabbit horseradish peroxidase-conjugated antibody (Dianova, Hamburg, Germany)
- Tween-20 (Roth, Karlsruhe, Germany)
- Super Signal West Pico Chemiluminescent Substrate (ECL) (Perbio, Bonn, Germany)

- PBS (phosphate-buffered saline): 10 mM Na_2HPO_4 (pH 7.4), 150 mM NaCl, 2.5 mM KCl, 1.5 mM KH_2PO_4 (Roth, Karlsruhe, Germany)

Instrumentation
- PhastSystem™ (Amersham Biosciences, Freiburg, Germany)
- PhastGel Dry Gel (Amersham Biosciences, Freiburg, Germany)
- PhastGel Sample Applicator 8/1 (Amersham Biosciences, Freiburg, Germany)
- PhastGel Cassette (Amersham Biosciences, Freiburg, Germany)
- Pierce™ ECL Western blotting substrate (Thermo Fisher, Darmstadt, Germany)
- Fusion SL4 detector (PeqLab, Erlangen, Germany)

Calibration
The IEF pattern from the control person is used as standard.

Running Conditions for IEF

Sep 1.1	2000 V	2.5 mA	3.5 W	15 °C	75 V h
Sep 1.2	200 V	2.5 mA	3.5 W	15 °C	15 V h
Sep 1.3	1000 V	2.5 mA	3.5 W	15 °C	100 V h
Sep 1.4	2000 V	2.5 mA	3.5 W	15 °C	495 V h

- Sample Applicator down at 1.2, 0 V h
- Sample Applicator up at 1.3, 100 V h
- Extra Alarm to sound at 1.1, 73 V h

17.2.5.3 Analytical

Procedure for IEF
Dry IEF gels are rehydrated for 1.5 h at room temperature in a solution of 8 mol/l urea and 60 ml/l ampholytes, which contains Pharmalyte (4.2–4.9) and Servalyt (3.0–5.0) in a ratio of 2:1. 0.5 µl of serum is applied to the gel and IEF is performed as described above.

Procedure for Western Blot
Proteins are blotted onto a nitrocellulose membrane by diffusion for 1 h at 60 °C. Nonspecific sites on the nitrocellulose membrane are blocked with 50 g/l non-fat dried milk dissolved in PBS/0.1% Tween-20. The membrane is washed in PBS/0.1% Tween-20. The primary rabbit anti human ApoC-III antibody is diluted 1000-fold in PBS/0.1% Tween-20. The secondary goat anti rabbit horseradish peroxidase-conjugated antibody is diluted 10,000-fold in the same solution. 0.1 ml ECL detection reagent per cm^2 of membrane is added. After adding Pierce™ ECL Western blotting substrate, the protein signal was detected by light emission on a Fusion SL4 detector (PeqLab). ImageJ was used for quantification (https://imagej.nih.gov/ij/).

17.2.5.4 Post Analytical

Interpretation
Interpretation of the data is performed by comparing the patterns of the healthy controls with the patients. The glycosylation pattern of ApoCIII is age-dependent. Kids under 1 year of age present with less $ApoCIII_2$ than $ApoCIII_1$, whereas persons with 1 year of age or older show balanced $ApoCIII_2$ and $ApoCIII_1$ isoforms. Abnormal ratios or even increased presence of $ApoCIII_0$ indicates hypoglycosylation.

17.2.6 CDG Screening and In-Depth Analysis by Intact Transferrin Mass Spectrometry

17.2.6.1 Principle
Transferrin isofocusing provides information on the number of sialic acids on transferrin. Analysis of transferrin by mass spectrometry provides detailed information on the glycan structures and on the presence or absence of glycans. While the latter is affected for ER defects and some cytosolic defects in CDG-I, abnormal glycan structures allow to diagnose specific CDG-II subtypes. In addition, recent evidence shows that some CDG-I defects also result in abnormal glycan structures on transferrin, such as ALG1-CDG, or total plasma glycans. To annotate individual glycan structures at the intact transferrin protein, high-resolution mass spectrometry is required. For this diagnostic test, transferrin is immunopurified from plasma or serum and analyzed by

mass spectrometry. Alternatively, glycans can be released from transferrin or total serum/plasma by N-glycosidase F digestion for subsequent mass spectrometric analyses [22].

17.2.6.2 Pre-analytical: Purification of Transferrin

Specimen

10 µl of plasma or serum of a control and the patient is needed. Samples should be stored at -20 °C until use. Shipment can be performed at room temperature for samples that arrive within 2–3 days, otherwise frozen.

Reagents and Chemicals

- Capture Select Transferrin Affinity Matrix beads (191306010, Life Technologies)
- Pierce Centrifuge Columns 0.8 ml Polyethylene filter, 30 µm pore size (nr.89868 Thermo Scientific)

Procedure

- Mix the beads (taken in 1.0 M NaCl, pH 7.4) and pipet 20 µl beads (cut the tip of the pipet) on the Pierce centrifuge column, which is placed in a 2 ml Eppendorf tube.
- Wash the beads by pipetting 700 µl 10 mM Tris-HCI, pH 7 on the column, spin the column in a table centrifuge and remove the filtrate. Repeat the washing 4×.
- Close the lower side of the column with a piece of parafilm. Pipet 100 µl of a diluted plasma/serum solution on the column (5 µl plasma/serum diluted with 95 µl 0.9% NaCl). Incubate for 20 min at room temperature on a shaking platform (mix at 800 rpm). Remove the parafilm and spin the column in a table centrifuge.
- Wash the beads by pipetting 700 µl 10 mM Tris-HCI, pH 7 on the column, spinning down and removing the filtrate. Repeat this step 5x.
- Place the column in a new Eppendorf tube of 1.5 ml for elution. Pipet 50 µl elution buffer on the column (0.1 M Glycine-HCI, pH 2.7). Incubate for ~1 min at room temperature, and spin down for 15 s. Neutralize (to pH 7) the eluate with 1.0 M Tris-HCI, pH 9. Check pH and directly measure the sample or store at -80 °C until analysis. Isolated transferrin should be stored at -80 °C or measured within 24 h.

17.2.6.3 Analytical: Measurement of Transferrin by QTOF Mass Spectrometry

Reagents and Chemicals

- Transferrin standard (glycosylated human protein from serum, Sigma-Aldrich, Taufkirchen, Germany)
- QC samples

Instrumentation

- Microfluidic 6540 HPLC-chip-QTOF instrument (Agilent Technologies)

Procedure

Centrifuge the eluted fractions briefly (1 min at 12,000 rpm) and transfer the supernatant to polypropylene vials (Agilent Technologies, cat. no. 9301-0978) in the autosampler. Two microliters of sample are injected on the microfluidic HPLC C8-chip. Separation is performed by using a short binary gradient of solution A (water, acetonitrile, formic acid 97/3/0.1 [vol/vol/vol]) and solution B (water, acetonitrile, formic acid 3/97/0.1 [vol/vol/vol]). The column is first equilibrated at a flow rate of 0.4 ml/min for the nanopump and 4 ml/min for the capillary pump. The 10-min gradient was programmed as follows: 0–1 min, 3%–30% B; 1–4 min, 30%–50% B, and continued with 50% B to 7 min; 7–7.5 min, B increased to 100%, and continued with 100% B to 8.5 min, and finally, 3% B from 9.5–10 min to equilibrate the chip column before the next sample injection. To minimize carryover, a blank run was performed in between each sample by injection of 2 ml MilliQ water. The instrument performed automatic tuning using a commercial calibrating solution (Agilent Technologies, cat. no. G1969-85000). An internal calibrating solution was introduced by a constant flow (100 ml/min) via a dual nebulizer electrospray source, containing the internal reference masses (m/z 118.0863 and 922.0098) in positive ion mode.

The nanoelectrospray voltage was set to 1800 V, with a drying gas of 4 l/min nitrogen at 300 °C. The fragmentor voltage was 250 V, and the skimmer voltage was 65 V. Data were acquired at 2 GHz, in extended dynamic range mode. First, test the system by injecting a commercial transferrin standard and test for correct peak shape and mass. The series of (patient) samples is run together with QC samples (a control and an abnormal CDG sample).

Data Analysis

For data analysis, use Agilent Mass Hunter Qualitative Analysis Software B.04.00. The Agilent BioConfirm Software is used to deconvolute the charge distribution raw data to reconstructed mass data. Multiple charged spectra are transformed with an S/N ratio of 30, an input data range of m/z 1000–3000, and an output data range of 73,000–83,000 Da. The software will generate a list of peak abundances. In case there was no aglycosylated transferrin or monoglycosylated transferrin present in the samples, abundances were estimated at the average reconstructed masses of 75,145 and 77,351 Da, respectively.

Interpretation/Pitfalls

Abundances of the mass peaks are used to calculate the ratios between abnormally glycosylated transferrin peaks and diglycosylated transferrin (see Fig. 17.5 for examples of control and CDG profiles). For detection of CDG-I, the ratio of aglycosylated transferrin (peak 1, not present in controls) and diglycosylated transferrin and of monoglycosylated transferrin (peak 4, reference range < 4%) and diglycosylated transferrin is calculated. For CDG-II diagnostics, abnormal glycan peaks are annotated on basis of the molecular

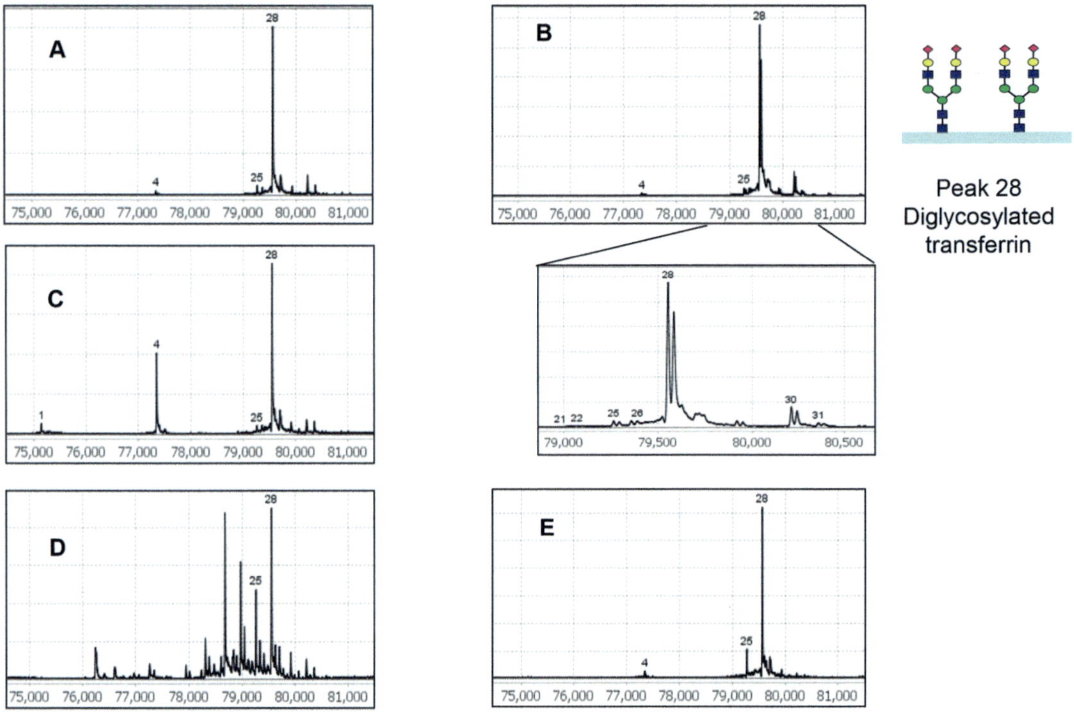

Fig. 17.5 Deconvoluted high-resolution mass spectra of immunopurified transferrin. (**a**) Control spectrum with peak 28 as main peak, representing diglycosylated transferrin and minor peaks 25 (missing one sialic acid) and 4 (missing one glycan). (**b**) Control spectrum with polymorphism showing double peaks in the zoomed spectrum. (**c**) Example of a CDG-I profile with accumulating peaks 4 and 1 (no glycosylation). (**d**) Example of a strongly abnormal CDG-II profile with a large range of peaks that represent truncated glycan structures. (**e**) Example of a mildly abnormal CDG-II profile with an increase of peak 25

mass difference with diglycosylated transferrin (peak 28). In controls, these are usually not present except for peak 25 which corresponds to the mass of diglycosylated transferrin that lacks one sialic acid.

A potential pitfall in transferrin diagnostics is the presence of polymorphisms. By mass spectrometry, polymorphisms can be easily observed as double peaks with a mass distance that is smaller than a single monosaccharide, which avoids misinterpretation.

17.3 Appendix

17.3.1 Analysis of Phosphomannomutase 2 (PMM2) Activity

17.3.1.1 Principle

PMM2 is the enzyme which is deficient in the most widespread form of CDG known so far, PMM2-CDG, affecting about 60–70% of all known CDG patients. It is caused by biallelic variants in the *PMM2* gene which result in a severe decrease of PMM2 activity, thereby disturbing the interconversion of mannose-1-phosphate and mannose-6-phosphate and subsequently the synthesis of GDP-mannose in the cytosol. Enzyme activity is determined photometrically by a NADPH-linked coupled enzymatic test [23].

17.3.1.2 Pre-analytical

Specimen

120 µg of protein isolated from fibroblasts or leukocytes are needed for the assay.

Reagents and Chemicals

- 1 M Hepes, (pH 7.1) (Roth, Karlsruhe, Germany)
- 1 M KCl (Roth, Karlsruhe, Germany)
- 0.2 M dithiothreitol (Merck, Taufkirchen, Germany)
- 10 mg/ml leupeptin (Merck, Taufkirchen, Germany)
- 10 mg/ml antipain (Merck, Taufkirchen, Germany)
- 5 mM $MgCl_2$ (Merck, Darmstadt, Germany)
- 0.7 µM glucose-1,6-bisphosphate (Merck, Taufkirchen, Germany)
- 0.25 mM NADP (Merck, Taufkirchen, Germany)
- 10 µg/ml phosphoglucose isomerase (2 mg/ml; Boehringer, Mannheim, Germany)
- 10 µg/ml glucose-6-phosphate dehydrogenase (5 mg/ml; Boehringer, Mannheim, Germany)
- 3.5 µg/ml phosphomannose isomerase (3 mg/ml; Merck, Taufkirchen, Germany)
- mM mannose-1-phosphate (Merck, Taufkirchen, Germany)
- µM mannose 1,6-bisphosphate
- Aqua bidest.

Instrumentation

- 96 well plate (quartz glass) (Hellma, Germany)
- Spectra Max plus photometer (Molecular Devices, Germany).
- Beckman Microfuge Tubes Polypropylene (Beckman Coulter 357448)

Calibration

PMM2 activity from fibroblasts or leukocytes of a healthy control is compared to the equivalent cellular material of the patient.

17.3.1.3 Analytical

Procedure

Samples are kept on ice throughout the procedure. Fibroblasts (derived by trypsinization not scraping) or leukocytes (derived from blood samples) pellets are resuspended in 200 µl of lysis buffer (20 mM Hepes [pH 7.1], 25 mM KCl, 1 mM dithiothreitol, 10 µg/ml leupeptin, 10 µg/ml antipain). Next, the volume is adjusted to 1 ml by adding HPLC water. After a freeze thaw cycle at −80 °C, cells are disrupted by passing 20 times through a 22 g syringe followed by centrifugation for 15 min at 100,000 × g. The supernatant is transferred to new vials and the protein concentration is determined by standard procedures. Four measurements per control and patient (each with and without substrate in a duplicate approach) are conducted, respectively. Phosphomannomutase activity is measured from 30 µg protein in 50 mM HEPES (pH 7.1), 5 mM $MgCl_2$, 0.25 mM NADP, 1 µM mannose 1,6-bisphosphate, 10 µg/ml glucose 6-phosphate dehydrogenase, 10 µg/ml phosphoglucose isomerase, and 3.5 µg/ml phosphomannose isomerase. All incubations are carried out at 37 °C for 2 h with continuous recording of absorbance at $OD_{340\,nm}$ and $OD_{400\,nm}$ in a two-channel spectrophotometer. Enzymatic reaction is started by the addition of 0.1 mM mannose-1-phosphate as substrate. A parallel assay is measured and recorded without substrate.

Calculation

Enzyme activity is investigated in a double beam spectrophotometer with or without substrate in 96 well plate (quartz glass), respectively. On behalf of the slope of the substrate containing sample compared to the substrate free sample over 120 min of continuous measurement, PMM2 activity is determined.

17.3.1.4 Post Analytical

Interpretation

PMM2 activity of less than 50% indicates PMM2-CDG and of 50-80% hints to carriership. Mutational analysis of the corresponding *PMM2* gene should be carried out whenever PMM2 activity is 80% or less.

17.3.2 Analysis of Phosphomannose Isomerase (MPI) Activity

17.3.2.1 Principle

MPI-CDG is caused by MPI deficiency. This enzyme catalyzes the conversion of Frc-6-P to Man-6-P in the cytosol. The clinical phenotype presents different from the wide majority of other CDG types with gastrointestinal bleedings and hepatopathy but without neurological impairment. Determination of MPI activity is performed by a coupled optical test [24].

17.3.2.2 Pre-analytical

Specimen

120 µg of protein isolated from fibroblasts or leukocytes are needed for the assay.

Reagents and Chemicals

- 1 M Hepes, (pH 7.1) (Roth, Karlsruhe, Germany)
- 1 M KCl (Roth, Karlsruhe, Germany)
- 0.2 M dithiothreitol (Merck, Taufkirchen, Germany)
- 10 mg/ml leupeptin (Merck, Taufkirchen, Germany)
- 10 mg/ml antipain (Merck, Taufkirchen, Germany)
- 1 M $MgCl_2$ (Merck, Darmstadt, Germany)
- mM NADP (Merck, Taufkirchen, Germany)
- Phosphoglucose isomerase (2 mg/ml; 350 U/mg) (Boehringer, Mannheim, Germany),
- Glucose-6-phosphate dehydrogenase (5 mg/ml) (Boehringer, Mannheim, Germany),
- M mannose-6-phosphate (Merck, Taufkirchen, Germany)
- Aqua bidest.

Instrumentation

- 96 well plate (quartz glass) (Hellma, Germany)
- Spectrophotometer (U-2000, Hitachi, Hamburg, Germany)
- Beckman Microfuge Tubes Polypropylene (Beckman Coulter 357448)

Calibration

MPI activity from fibroblasts or leukocytes of a healthy control is compared to the equivalent cellular material of the patient.

17.3.2.3 Analytical

Procedure

Samples are kept on ice throughout the procedure. Fibroblasts (derived by trypsinization not scraping) or leukocytes (derived from blood samples) pellets are resuspended in 200 µl of lysis buffer (20 mM Hepes [pH 7.1], 25 mM KCl, 1 mM dithiothreitol, 10 µg/ml leupeptin, 10 µg/ml antipain). Next, the volume is adjusted to 1 ml by adding HPLC water. After a freeze thaw cycle at -80 °C, cells are disrupted by passing 20 times through a 22 g syringe followed by centrifugation for 15 min at 100,000 × g. The supernatant is transferred to new vials, the protein concentration is determined and adjusted to a concentration of 600 µg/ml with lysis buffer. Four measurements per control and patient (each with and without substrate in a duplicate approach) are conducted, respectively. The MPI activity is determined from 30 µg protein (in 50 µl lysis buffer) by addition of 50 µl reaction buffer (100 mM Hepes [pH 7.1], 10 mM $MgCl_2$, 0.5 mM NADP, 1 µg phosphoglucose isomerase, and 1 µg glucose-6-phosphate dehydrogenase). Enzymatic reaction is started by the addition of mannose-6-phosphate (1 mM final concentration). A parallel assay is measured and recorded without substrate. Incubation of samples is carried out at room temperature and enzymatic activity is followed at $OD_{340\ nm}$ and $OD_{400\ nm}$ over 30 min.

Calculation

Enzyme activity is investigated in a double beam spectrophotometer with or without substrate in cuvettes, respectively. On behalf of the slope of the substrate containing sample compared to the substrate free sample over 30 min of continuous measurement, MPI activity is determined.

17.3.2.4 Post Analytical

Interpretation

MPI activity of less than 50% indicates MPI-CDG and of 50-80% hints to carriership. Mutational analysis of the corresponding *MPI* gene should be carried out whenever MPI activity is 80% or less.

17.3.3 Analysis of Phosphoglucomutase 1 (PGM1) Activity

17.3.3.1 Principle

PGM1-CDG is caused by PGM1 deficiency. This enzyme catalyzes the transfer of phosphate between the 1 and 6 positions of glucose in the cytosol. Reduced PGM1 activity affects the synthesis of UDP-glucose and UDP-galactose, thereby leading to a mixed CDG-I/CDG-II patterns in IEF of transferrin. Hepatopathy, Bifid uvula, growth retardation, myopathy, hypoglycemia, and dilated cardiomyopathy without neurological symptoms are common clinical characteristics. Determination of PGM1 activity is performed by a coupled optical test [25].

17.3.3.2 Pre-Analytical

Specimen

120 µg of protein isolated from fibroblasts or leukocytes are needed for the assay.

Reagents and Chemicals

- 1 M Hepes, (pH 7.1) (Roth, Karlsruhe, Germany)
- 1 M KCl (Roth, Karlsruhe, Germany)
- 0.2 M dithiothreitol (Merck, Taufkirchen, Germany)
- 10 mg/ml leupeptin (Merck, Taufkirchen, Germany)
- 10 mg/ml antipain (Merck, Taufkirchen, Germany)
- 5 M $MgCl_2$ (Merck, Darmstadt, Germany)
- mM NADP (Merck, Taufkirchen, Germany)
- 50 mM glucose-1,6-diphosphate (Merck, Taufkirchen, Germany)
- Glucose-6-phosphate dehydrogenase (5 mg/ml) (Boehringer, Mannheim, Germany)
- M glucose-1-phosphate (Merck, Taufkirchen, Germany)
- Aqua bidest.

Instrumentation

- 96 well plate (quartz glass) (Hellma, Germany)
- Spectrophotometer (U-2000, Hitachi, Hamburg, Germany)
- Beckman Microfuge Tubes Polypropylene (Beckman Coulter 357448)

Calibration

PGM1 activity from fibroblasts or leukocytes of a healthy control is compared to the equivalent cellular material of the patient.

17.3.3.3 Analytical

Procedure

Samples are kept on ice throughout the procedure. Fibroblasts (derived by trypsinization not scraping) or leukocytes (derived from blood samples) pellets are resuspended in 200 µl of lysis buffer (20 mM Hepes [pH 7.1], 25 mM KCl, 1 mM dithiothreitol, 10 µg/ml leupeptin, 10 µg/ml antipain). Next, the volume is adjusted to 1 ml by adding HPLC water. After a freeze thaw cycle at -80 °C, cells are disrupted by passing 20 times through a 22 g syringe followed by centrifugation for 15 min at 100,000 × g. The supernatant is transferred to new vials and the protein concentration is determined by standard procedures. Four measurements per control and patient (each with and without substrate in a duplicate approach) are conducted, respectively. For each measurement 30 μg protein in volume of 135 μl (50 mM Hepes, pH 7.1) are needed. Next, 30 μl chemical mix (0.5 mM NADP, 10 mM MgCl2, 700 μM Gluc-1,6-phosphate in 50 mM Hepes, pH 7.1) and 30 μl enzyme mix (1.6 μl Glucose-6-Phosphate-Dehydrogenase [5 mg/ml] in 998.4 μl 50 mM Hepes, pH 7.1) are added, mixed, and transferred to a 96 well quartz glass plate. Finally, the reaction is started by adding 15 μl 0.1 M Glucose-1-Phosphate as substrate into two wells and by adding 15 μl water into the other two wells, respectively. Mix well. Measurement is carried out at 37 °C and enzymatic activity is followed at $OD_{340\ nm}$ and $OD_{400\ nm}$ over 30 min.

Calculation

Enzyme activity is investigated in a double beam spectrophotometer with and without substrate in cuvettes, respectively. On behalf of the slope of the substrate containing sample compared to the substrate free sample over 30 min of continuous measurement, PGM1 activity is determined.

17.3.3.4 Post Analytical

Interpretation

PGM1 activity of less than 50% indicates PGM1-CDG and of 50-80% hints to carriership. Mutational analysis of the corresponding *PGM1* gene should be carried out whenever PGM1 activity is 80% or less.

17.3.4 Specimen Collection for CDG Diagnostics

See Table 17.1 (Specimen).

Table 17.1 Specimen collection for CDG diagnostics. Depending on the analysis, different patient materials are required in the laboratory, which also have to be prepared in different ways for shipping. Samples should reach the laboratory within 24 h

Test	Material needed	Transport	Annotation
IEF of transferrin	Serum	From autumn to spring: cold pack; in summer: frozen	Analysis from Guthrie cards also possible but not optimal
IEF of alpha-1-antitrypsin, apolipoprotein CIII	Serum or plasma	From autumn to spring: cold pack; in summer: frozen	Analysis from Guthrie cards also possible but not optimal
Mass spectrometry of transferrin or total plasma/serum N-glycans	Serum, plasma	From autumn to spring: cold pack; in summer: frozen	Analysis from Guthrie cards also possible but not optimal
PMM2-CDG enzyme activity	EDTA blood, leukocytes, fibroblasts	Blood: non-frozen; leukocytes: frozen; fibroblasts: in culture	Analysis should be announced in advance
MPI-CDG enzyme activity	EDTA blood, leukocytes, fibroblasts	Blood: non-frozen; leukocytes: frozen; fibroblasts: in culture	Analysis should be announced in advance
PGM1-CDG enzyme activity	EDTA blood, leukocytes, fibroblasts	Blood: non-frozen; leukocytes: frozen; fibroblasts: in culture	Analysis should be announced in advance
Other enzyme assays (needs expert experience)	EDTA blood, leukocytes, fibroblasts	Blood: non-frozen; leukocytes: frozen; fibroblasts: in culture	Analysis should be announced in advance

(continued)

Table 17.1 (continued)

Test	Material needed	Transport	Annotation
Mutational analysis	Genomic DNA, total RNA, leukocytes, fibroblasts	DNA: cold pack, not frozen; RNA: frozen; leukocytes: frozen; fibroblasts: in culture	Analysis from Guthrie cards also possible but not optimal
Prenatal analysis by sequencing	Genomic DNA derived from chorionic villi or amniotic fluid cells (exclusion of maternal contamination needed)	Cold pack, not frozen	Analysis should be announced in advance

17.4 Conclusion

CDG diagnostics is challenging. The vast majority of CDG patients present with a defect from the area of N-glycosylation. Although the methods for diagnostics presented here can only help to identify some of the 190 CDG types known to date, most patients can nevertheless be detected.

In addition, and in view of the increasing number of genetic studies and the associated detection of variants of unclear significance (VUS) in glycosylation genes, the procedures described here can help to avoid false-positive CDG findings which is of great importance for patients.

References

1. Varki (2017) Biological roles of glycans. Glycobiology 27(1):3–49
2. Schjoldager KT, Narimatsu Y, Joshi HJ, Clausen H (2020) Global view of human protein glycosylation pathways and functions. Nat Rev Mol Cell Biol 21(12):729–749
3. Reily C, Stewart TJ, Renfrow MB, Novak J (2019) Glycosylation in health and disease. Nat Rev Nephrol 15(6):346–366
4. D'Souza Z, Sumya FT, Khakurel A, Lupashin V (2021) Getting sugar coating right! The role of the Golgi trafficking machinery in glycosylation. Cells 10(12):3275
5. Bieberich E (2023) Synthesis, processing, and function of N-glycans in N-glycoproteins. Adv Neurobiol 29:65–93
6. Ondruskova N, Cechova A, Hansikova H, Honzik T, Jaeken J (2021) Congenital disorders of glycosylation: still "hot" in 2020. Biochim Biophys Acta Gen Subj 1865(1):129751
7. Sosicka P, Ng BG, Freeze HH (2022) Chemical therapies for congenital disorders of glycosylation. ACS Chem Biol 17(11):2962–2971
8. Ng BG, Freeze HH, Himmelreich N, Blau N, Ferreira CR (2024) Clinical and biochemical footprints of congenital disorders of glycosylation: proposed nosology Mol Genet Metab 142(1):108476 https://doi.org/10.1016/j.ymgme.2024.108476
9. Paprocka J, Jezela-Stanek A, Tylki-Szymańska A, Grunewald S (2021) Congenital disorders of glycosylation from a neurological perspective. Brain Sci 11(1):88
10. Altassan R, Péanne R, Jaeken J, Barone R, Bidet M, Borgel D, Brasil S, Cassiman D, Cechova A, Coman D, Corral J, Correia J, de la Morena-Barrio ME, de Lonlay P, Dos Reis V, Ferreira CR, Fiumara A, Francisco R, Freeze H, Funke S, Gardeitchik T, Gert M, Girad M, Giros M, Grünewald S, Hernández-Caselles T, Honzik T, Hutter M, Krasnewich D, Lam C, Lee J, Lefeber D, Marques-de-Silva D, Martinez AF, Moravej H, Õunap K, Pascoal C, Pascreau T, Patterson M, Quelhas D, Raymond K, Sarkhail P, Schiff M, Seroczyńska M, Serrano M, Seta N, Sykut-Cegielska J, Thiel C, Tort F, Vals MA, Videira P, Witters P, Zeevaert R, Morava E (2019) International clinical guidelines for the management of phosphomannomutase 2-congenital disorders of glycosylation: diagnosis, treatment and follow up. J Inherit Metab Dis 42(1):5–28
11. Lefeber DJ, Freeze HH, Steet R, Kinoshita T (2022) Congenital disorders of glycosylation. In: Varki A, Cummings RD, Esko JD, Stanley P, Hart GW, Aebi M, Mohnen D, Kinoshita T, Packer NH, Prestegard JH, Schnaar RL, Seeberger PH (eds) Essentials of glycobiology [internet], 4th edn. Cold Spring Harbor Laboratory Press, Cold Spring Harbor, NY. Chapter 45
12. Sparks SE, Krasnewich DM (2005) Congenital disorders of N-linked glycosylation and multiple pathway overview. Aug 15 [updated 2017 Jan 12]
13. Linders PTA, Peters E, Ter Beest M, Lefeber DJ, van den Bogaart G (2020) Sugary logistics gone wrong: membrane trafficking and congenital disorders of glycosylation. Int J Mol Sci 21(13):4654
14. Wilson MP, Matthijs G (2021) The evolving genetic landscape of congenital disorders of glycosylation. Biochim Biophys Acta Gen Subj 1865(11):12997

15. Lipiński P, Tylki-Szymańska A (2021) Congenital disorders of glycosylation: what clinicians need to know? Front Pediatr 9:715151
16. Francisco R, Marques-da-Silva D, Brasil S, Pascoal C, Dos Reis FV, Morava E, Jaeken J (2019) The challenge of CDG diagnosis. Mol Genet Metab 126(1):1–5
17. Clayton P, Winchester B, Di Tamaso E, Young E, Keir G, Rodeck C (1993) Carbohydrate-deficient glycoprotein syndrome: normal glycosylation in the fetus. Lancet 341:956
18. Stibler H, Skovby F (1994) Failure to diagnose carbohydrate-deficient glycoprotein syndrom prenatally. Pediatr Neurol 11:71
19. Thiel C, Meßner-Schmitt D, Hoffmann GF, Körner C (2013) Screening for congenital disorders of glycosylation in the first weeks of life. J Inherit Metab Dis 36(5):887–892
20. Clerc F, Reiding KR, Jansen BC, Kammeijer GS, Bondt A, Wuhrer M (2016) Human plasma protein N-glycosylation. Glycoconj J 33(3):309–343
21. Wopereis S, Grünewald S, Morava E, Penzien JM, Briones P, García-Silva MT, Demacker PMN, Huijben KMLC, Wevers RA (2003) Apolipoprotein C-III Isofocusing in the diagnosis of genetic defects in O-glycan biosynthesis. Clin Chem 49:1839–1845
22. Abu Bakar N, Ashikov A, Brum JM, Smeets R, Kersten M, Huijben K, Keng WT, Speck-Martins CE, de Carvalho DR, de Rizzo IMPO, de Mello WD, Heiner-Fokkema R, Gorman K, Grunewald S, Michelakakis H, Moraitou M, Martinelli D, van Scherpenzeel M, Janssen M, de Boer L, van den Heuvel LP, Thiel C, Lefeber DJ (2022) Synergistic use of glycomics and single-molecule molecular inversion probes for identification of congenital disorders of glycosylation type-1. J Inherit Metab Dis 45(4):769–781
23. Körner C, Lehle L, von Figura K (1998) Abnormal synthesis of mannose 1-phosphate derived carbohydrates in carbohydrate-deficient glycoprotein syndrome type I fibroblasts with phosphomannomutase deficiency. Glycobiology 8(2):165–171
24. Niehues R, Hasilik M, Alton G, Körner C, Schiebe-Sukumar M, Koch HG, Zimmer KP, Wu R, Harms E, Reiter K, von Figura K, Freeze HH, Harms HK, Marquardt T (1998) Carbohydrate-deficient glycoprotein syndrome type Ib. Phosphomannose isomerase deficiency and mannose therapy. J Clin Invest 101:1414–1420
25. Wong SY, Gadomski T, van Scherpenzeel M, Honzik T, Hansikova H, Holmefjord KSB, Mork M, Bowling F, Sykut-Cegielska J, Koch D, Hertecant J, Preston G, Jaeken J, Peeters N, Perez S, Nguyen DD, Crivelly K, Emmerzaal T, Gibson KM, Raymond K, Abu Bakar N, Foulquier F, Poschet G, Ackermann AM, He M, Lefeber DJ, Thiel C, Kozicz T, Morava E (2017) Oral D-galactose supplementation in PGM1-CDG. Genet Med 19(11):1226–1235

Homocysteine, S-Adenosylmethionine and S-Adenosylhomocysteine

18

Brian Fowler and Alessio Cremonesi

Contents

List of Abbreviations

ADK	Adenosine kinase
Hcy	Homocysteine
MAT	Methionine adenosyl transferase
MeCbl	Methylcobalamin
MRM	Multiple reaction monitoring
MTHFR	Methylenetetrahydrofolate reductase
SAM	S-adenosylmethionine

B. Fowler (✉)
Division of Metabolism and Molecular Pediatrics, University Children's Hospital, Zürich, Switzerland
e-mail: Brian.Fowler@unibas.ch

A. Cremonesi
Division of Clinical Chemistry and Biochemistry, University Children's Hospital Zurich, University of Zurich, Zurich, Switzerland

18.1 Introduction

Homocysteine (Hcy) metabolism is closely linked to that of the essential amino acid methionine and thus plays a central role in several vital biological processes. Methionine itself is needed for protein synthesis and donates methyl groups for the synthesis of a broad range of vital methylated compounds. It is also a main source of sulphur and acts as the precursor for several other sulphur-containing amino acids such as cystathionine, cysteine and taurine. In addition, it donates the carbon skeleton for polyamine synthesis [1, 2].

Hcy is also important in the metabolism of folate and in the breakdown of choline.

Hcy levels are determined by its synthesis from methionine involving several enzymes, its remethylation to methionine and its breakdown by trans-sulphuration.

The metabolisms of Hcy-related pathways in man are shown in Fig. 18.1. Methionine is converted to its active form S-adenosylmethionine

N. Blau, F. M. Vaz (eds.), *Laboratory Guide to the Methods in Biochemical Genetics*,
https://doi.org/10.1007/978-3-031-58819-8_18

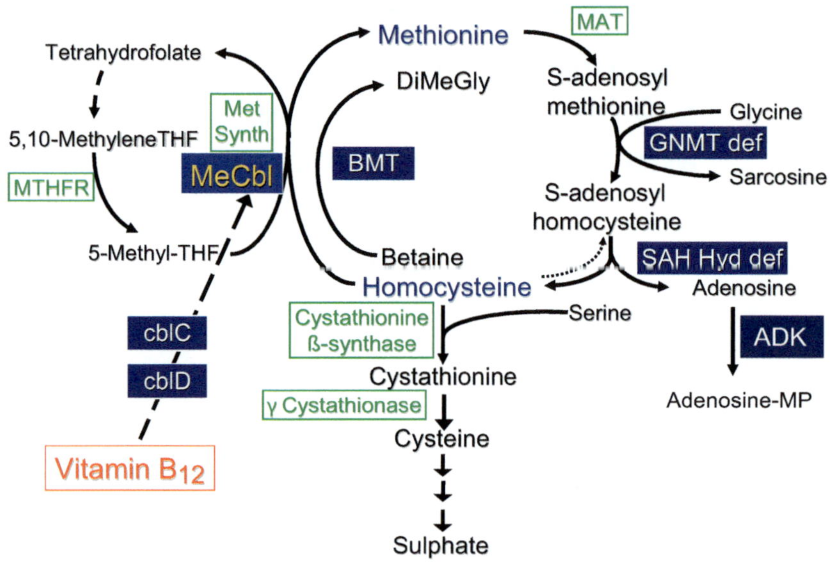

Fig. 18.1 Outline of homocysteine metabolism. *BMT* betaine methyltransferase; *cblC, cblD* the C and D types of cobalamin defects; *GNMT def* glycine N-methyltransferase deficiency; *MAT* methionine adenosyl transferase; *MeCbl* methylcobalamin; *Met Synth* methionine synthase; *MTHFR* methylenetetrahydrofolate reductase; *SAH Hyd def* S-adenosylhomocysteine hydrolase deficiency; *ADK* adenosine kinase; *Adenosine-MP*, adenosine monophosphate

(AdoMet) by methionine adenosyltransferase (MAT). AdoMet is the methyl group donating cosubstrate for a very wide range of methyltransferases involved in the formation of important methylated compounds, for example creatine, epinephrine, dopamine, phosphatidylcholine, methylated proteins and methylated DNA [3]. The highly abundant glycine methyltransferase, which transfers a methyl group to glycine-forming sarcosine (N-methylglycine) [4] is a particularly important methyltransferase that fulfils the need for a high-capacity utilisation of SAM and is considered to be an integral enzyme in the trans-sulphuration sequence.

The loss of a methyl group from AdoMet in each of these reactions yields S-adenosylhomocysteine (AdoHcy) and this is subsequently hydrolysed to adenosine and Hcy by AdoHcy-hydrolase. Hcy sits at a metabolic branch point and can be remethylated to methionine, by two reactions. One is the 5-methyltetrahydrofolate-dependent reaction catalysed by methionine synthase which itself is reductively methylated by cobalamin (vitamin B_{12}) and AdoMet requiring methionine synthase reductase. 5-Methyltetrahydrofolate is generated from 5,10 methylenetetrahydrofolate by methylenetetrahydrofolate reductase. The second remethylation reaction is catalysed by betaine methyltransferase which is restricted to liver, kidney and brain whilst methionine synthase is widely distributed.

As well as remethylation Hcy can be degraded in the trans-sulphuration pathway which first involves condensation of Hcy with serine forming cystathionine, then break down of this compound to cysteine and α-oxo-butyrate. These reactions are catalysed by cystathionine ß-synthase and γ-cystathionase respectively both requiring pyridoxal 5′ phosphate (vitamin B_6) as coenzyme. Cysteine is needed both for protein synthesis and is the precursor of the important antioxidant glutathione. A number of enzymatic reactions [5] lead to the oxidation of the sulphur atom and further breakdown of cysteine ultimately forming inorganic sulphate.

The metabolic control of methionine metabolism is complex and involves, for example changes in enzyme levels in particular tissues, mechanisms linked to the kinetic properties of the various enzymes and their interaction with metabolic effectors [6, 7].

A particularly important metabolic effector is AdoMet. This inhibits the low Km isoenzymes of MAT, and methyleneTHF reductase, inactivates betaine methyltransferase but activates MAT III (the high Km isoenzyme) and cystathionine ß-synthase. Therefore high methionine intake and therefore AdoMet levels favours trans-sulphuration and when levels are low methionine is conserved. AdoHcy potently inhibits AdoMet-dependent methyltransferases and both Hcy remethylating enzymes. Another important control mechanism is the export of Hcy from cells into extracellular space and into plasma which occurs as soon as intracellular levels increase [8].

Elevated Hcy occurs in a wide range of disease processes. This is not surprising bearing in mind the central role of this metabolite in a number of closely related pathways and because of its links to processes such as polyamine synthesis, methylation and redox potential.

A large elevation of Hcy in body fluids and tissues is found in several *genetic enzyme deficiencies*, the homocystinurias, including CBS deficiency [9], deficiency of MTHFR [methionine synthase and methionine synthase reductase], as well as defects of intracellular cobalamin metabolism namely the cblF, cblC, cblD and cblJ defects [10]. It is noteworthy that low levels of total Hcy have been described in sulphite oxidase deficiency [11].

Elevated Hcy levels can also occur in *nutritional deficiencies* of vitamin B_{12} (cobalamin) and folate. Folate or vitamin B_{12} deficiency can even lead to total Hcy levels as high as those seen in genetic homocystinurias [12]. Also, a reciprocal relationship between blood levels or intake of these vitamins and total Hcy concentrations [13] has been shown in many studies.

Also, a number of *drugs* such as the antifolate compound methotrexate, or the anaesthetic gas nitrous oxide, may interfere with methionine metabolism and lead to mild increases in Hcy [2]. Abnormal renal function has been shown to lead to increased plasma Hcy, for example in end-stage renal disease patients [14].

The association of moderately elevated Hcy and several diseases, particularly various forms of vascular disease [15] has been extensively studied and documented although a role in causation is controversial Less strong links have also been shown with neural tube defects, neuropsychiatric disorders, impaired cognitive function, dementia, cancer and osteoporosis, for example [16].

Because of these ever-widening interests, the measurement of plasma total Hcy is undertaken in many clinical chemistry and routine laboratories. Various methods are employed including HPLC assays, conventional amino acid analysis, capillary electrophoresis, gas chromatography with or without mass spectrometry, liquid chromatography with tandem mass spectrometry and in many routine clinical chemistry laboratories immunoassays. In this chapter, those methods that are often available in laboratories involved in investigation of inborn errors of metabolism are described, namely HPLC and tandem MS.

The laboratory measurement of AdoMet and AdoHcy has become key in studies of pathogenesis of elevated Hcy as well as the differential diagnosis of hypermethioninaemia due to deficiency of methionine-S-adenosyltransferase, glycine methyltransferase, S-adenosylhomocysteine hydrolase and adenosine kinase [17].

Until the middle 1990s, AdoMet and AdoHcy had been almost exclusively studied in tissues or CSF. Recent studies have shown that AdoMet and AdoHcy are readily detectable in isolated erythrocytes or whole blood and reduced levels of AdoMet have been reported in whole blood from patients with coronary artery disease [18]. Application of HPLC with fluorescence detection has allowed detection of these two compounds in plasma although levels are in the nanomolar range and clearly much lower than those in tissues [19, 20]. Clear increases were shown in normal subjects following administration of methionine [20] or AdoMet [21]. Subsequently, several studies have confirmed the ability to measure these two key compounds in plasma [22, 23]. The measurement of both compounds in CSF can also be useful in the diagnosis of remethylation defects as well as in monitoring their treatment.

18.2 Properties of Hcy, AdoMet and AdoHcy

Hcy is a sulphydryl amino acid of molecular weight 135.2. It oxides readily to its disulphide form homocysteine and can be converted to the thiolactone form in an acid solution.

In normal plasma Hcy exists in various forms, the sulphydryl form (ca. 1%), bound to the cysteine residues of proteins (ca. 70%) and bound to free cysteine as cysteine-Hcy mixed disulphide (ca. 30%). When levels are elevated the disulphide form homocysteine is formed. All these forms can be converted to Hcy by chemical reduction and then measured as "total Hcy".

AdoHcy comprises an adenosyl moiety attached to Hcy through the latter's S atom. Its molecular weight is 384.4 and it absorbs UV light strongly with a λ max of 260 nm at pH 7.

AdoMet comprises an adenosyl moiety attached to methionine at the sulphur atom. The presence of three substituent groups on the sulphur atom produces a positively charged sulphonium compound that is thermodynamically unstable rendering high reactivity to AdoMet. Its molecular weight is 399.4 and it absorbs UV light strongly with a λ max of 260 nm at pH 7. The compound is unstable under many conditions. Stable solid forms include the hydrogen sulphate and disulphate di-*p*-toluene sulphate salts. Four stereoisomers exist and the active form is L at the 2-carbon and—at the sulphonium centre.

18.3 Methods

18.3.1 Measurement of Total Hcy in Plasma Using Isocratic High-Performance Liquid Chromatography: Fluorescence Detection

18.3.1.1 Principle of Assay

Thiols exist in biological systems in various oxidised and reduced forms. In plasma, very little Hcy is found in the free sulphydryl form and it mainly exists as cysteine-Hcy mixed disulphide or bound to cysteine residues of plasma proteins or as its disulphide homocysteine but the latter only when levels are high. Routinely measured total Hcy constitutes all of these forms together and is produced by chemical reduction.

The method described here is based on two reactions, first the reduction of the disulphide bond between Hcy and other thiols or the cysteine residue of proteins by the reducing compound tri-n-butylphosphine followed by the reaction of Hcy and other thiols with the fluorogenic thiol-specific reagent ammonium 7-fluorobenzo-2-oxa-1,3-diazole-4-sulphonate (SBDF).

Reaction of tri-n-butyl phosphine as a reducing agent:

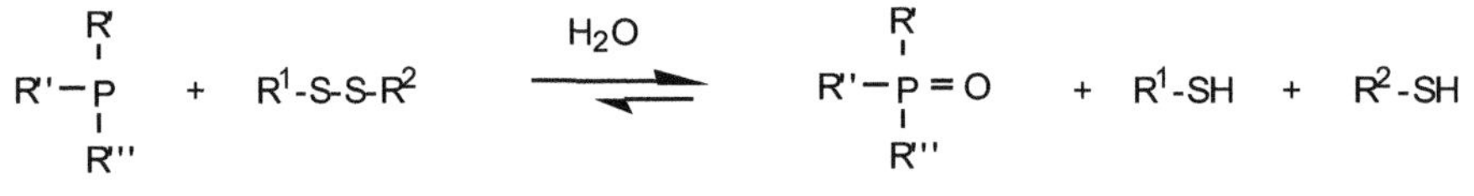

Reaction of the derivatisation of homocysteine with 7-fluorobenzo-2-oxa-1,3-diazole-4-sulfonate, which is a fluorescent derivatising reagent:

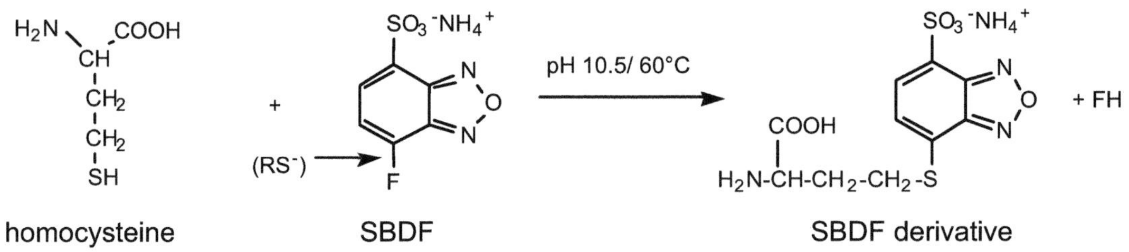

The derivatives are separated by reverse phase-HPLC using isocratic elution and detection and quantitation by fluorescence [24, 25].

18.3.1.2 Pre-analytical

Specimen:

A minimum volume of 400 µL plasma is required. EDTA or lithium heparin blood, collected after overnight fasting and separated as soon as possible by centrifugation at 2000 *g* for 10 min at 4 °C. If separation cannot be performed within 1 h blood can be kept on ice for a maximum of 1 h. Plasma is stored frozen at −20 °C until analysis or sent from outside on dry ice. Hcy is stable in plasma at room temperature for 24 h, at 4 °C for 1 week and at −20 °C for many years. Serum should not be used since allowing blood to coagulate leads to a delay in centrifugation.

Pooled plasma collected from healthy subjects is used for standard preparation to avoid matrix effects and for internal control. Pooled plasma is kept in 0.5 mL aliquots at −20 °C until required.

Reagents and chemicals:

All chemicals should be of the highest purity available or HPLC grade.

Solutions for reduction and derivatisation:

- Reduction buffer: 0.1 mol/L borate, pH 9.5.
- Dissolve 6.11 g $K_2B_4O_7 \times 4\ H_2O$ (Sigma, O-5754) in H_2O, adjust pH to 9.5 with 2 mol/L NaOH and makeup to 200 mL with dem. water.
- Internal standard: 0.25 mmol/L cysteamine in 0.1 mol/L borate buffer.
- Dissolve 3.78 mg cysteamine (Sigma, M-6500) and 93.55 mg NaEDTA (2 mmol/L, Merck 1.08418) in 0.1 mol/L borate buffer in a final volume of 125 mL.
- tri-n-butylphosphine, 10%:
- 1 volume of tri-n-butylphosphine (Sigma P-6918) is added to 9 volumes of N,N-dimethylformamide (Aldrich 31,993-7). This must always be freshly prepared.
- Perchloric acid 0.6 mol/L:
- 5 mL 70% $HClO_4$ (Merck, cat. no. 519) and 37.2 mg NaEDTA are made up to 100 mL with dem. water.
- Derivatising buffer: 0.2 mol/L borate, pH 10.5:
- Dissolve 6.11 g $K_2B_4O_7 \times 4\ H_2O$ and 186 mg NaEDTA in H_20, adjust the pH to 10.5 with 1 mol/L NaOH and make up to 100 mL with dem. water.
- 7-fluorobenzo-2-oxa-1,3-diazole-4-sulfonic acid, (SBDF):
- Dissolve 1 mg SBDF ammonium salt (Sigma F-4383) in 1 mL of borate buffer, 1 mol/L, pH 9.5 made up by dissolving 61.1 g $K_2B_4O_7 \times 4\ H_2O$, adjusting the pH to 9.5 with 2 mol/L NaOH and making up to 200 mL with dem. water.

Mobile phase for HPLC:

- Dissolve 13.61 g KH_2PO_4 in about 900 mL H_2O, add 45 mL acetonitrile (=45%), adjust pH to 2.1 with about 6 mL 85%, ortho-phosphoric acid, makeup to 1 L with dem. water and vacuum filter (0.45 µM filter, Schleisser and Schuell, reference 10410212).

Hcy standards:

- Hcy stock solution, 4 mmol/L:

  Dissolve 5.4 mg DL-homocysteine (Fluka 54501) in 10 mL 0.1 mol/L borate buffer solution containing internal standard.
- Hcy working standards:

  Standards of 6.3–400 µmol/L are prepared by serial dilution of the stock solution in an internal standard solution.

Both stock and working standards are stable for at least 1 year and stored in aliquots at −20 °C.

Instrumentation

General laboratory equipment, heating block at 60 °C for sample derivatisation, laboratory centrifuge, microcentrifuge, vortex mixer.

HPLC system consisting of a buffer pump preferably with degasser, autosampler with cooling system, fluorescence detector (excitation 384 nm, emission 516 nm) and an integration and data handling system.

HPLC column: Nucleosil 120 C18, 5 µm (4.6×250 mm) with a guard column (4.6×20 mm) containing the same column packing material.

Calibration

Calibration is performed using three standards selected to cover the range expected in the sample batch. (usually 4.2, 8.4 and 33.6 μmol/L), together with pooled plasma to allow for any matrix effect, plus a pooled plasma blank.

Quality Control

Quality assessment is performed by including pooled plasma as an internal control in every run. External QC is achieved using samples obtained from the ERNDIM Special Assays Scheme run according to the scheme schedule.

Commercially available control plasma samples level I + II obtained from Chromosystems GmbH München are included with each batch of samples.

Detection limit

Sample concentration of 1 μmol/L

Precision

Within-run CV (n = 10) was shown to be 4.5% and 2.5% for standards of 16 and 67 μmol/L, respectively.

Between-run CV (n = 20) was 6.1% and 1.9% for standards of 16 μmol/L and 67 μmol/L, respectively.

Recovery

98.9% ± 6.0 over a concentration range of 4–67 μmol/L.

18.3.1.3 Analytical

Procedure

The reduction reaction is carried out in 1.5 mL Eppendorf tubes. Frozen samples are thawed out, thoroughly mixed and if necessary centrifuged to remove particulate material.

For samples and pooled plasma blank, 150 μL plasma and 50 μL internal standard in 0.1 mol/L borate buffer are added to 20 μL of 10% tri-n-butylphosphine. For each Hcy standard 150 μL pooled plasma and 50 μL of Hcy standard (internal standard is included) are mixed with 20 μL of tri-n-butylphosphine. Tubes are left to stand on ice for 30 min. Samples are deproteinized by the addition of 125 μL 0.6 mol/L perchloric acid, thorough vortex mixing and centrifugation at 13,000 rpm for 5 min (Eppendorf centrifuge). The supernatant is reserved for further processing.

Derivatisation is performed in 5 mL glass tubes by mixing 100 μL of the supernatant with 200 μL 0.2 mol/L, pH 10.5, borate buffer and 100 μL freshly prepared SBDF solution. After thorough vortex mixing the tubes are sealed with plastic stoppers and incubated at 60 °C for 60 min in a heating block. The reaction is stopped by placing tubes on ice. Samples are filtered through a 0.45-μm membrane (4 mm, Titan PVDF filters, Infochroma ag, reference 8804-PV-4) using a 1-mL plastic syringe filter and used for HPLC analysis.

If the analysis is not performed immediately samples can be stored at −20 °C.

HPLC conditions

The HPLC analysis is performed isocratically with a flow rate of 2.0 mL/min and a run time of 10 min. Usual pressures are between 22 and 24 MPa (230–240 bar). The injection volume is 25 μL. The fluorimeter is set with an excitation wavelength of 384 nm and emission at 516 nm.

Calculation:

A four-point linear regression plot is constructed for calibration: *x*-axis: added Hcy concentration [μmol/L]; *y*-axis: ratio of areas of added Hcy/internal standard.

Concentrations of Hcy in samples are calculated by dividing the ratio of the area of Hcy/IS area by the slope of the calibration line.

Thus, tHcy concentration = (Ratio Hcy/IS)/slope (Fig. 18.2)

18.3.1.4 Post-analytical

Interpretation

The separation obtained with a normal plasma sample is shown in Fig. 18.3.

The order of elution is cysteine (3.5 min), cysteamine (4.0 min), cysteinyl-glycine (5.0 min) and Hcy (6.5 min) with a peak width of under 1 min and complete resolution of peaks.

Chromatogram

HPLC chromatogram of SBDF-derivatized thiols in human plasma:

a—cysteine, b—cysteamine, c—cysteinyl-glycine, d—Hcy

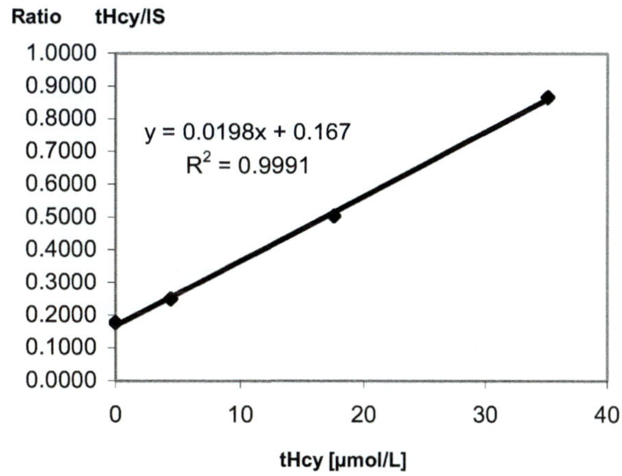

Fig. 18.2 Standard curve of total Hcy in human plasma

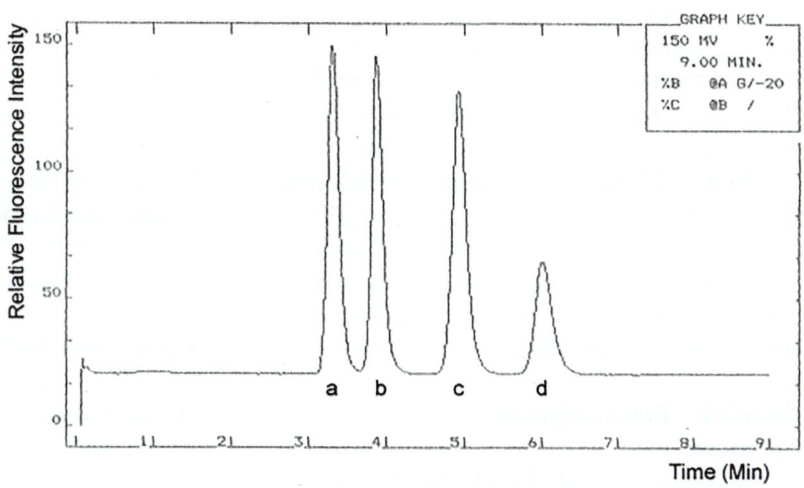

Fig. 18.3 Chromatogram of SBDF-derivatized thiols in human plasma

Table 18.1 Reference values for total homocysteine

		Total homocysteine µmol/L		
	n	Mean	SD	Range
Men and women aged 21–64 years	65	8.1	2.5	3.3–13.3
Men[a]				8–18
Pre-menopausal women[a]				6–15
Post-menopausal women[a]				6–19
Children, aged 2 months to 18 years[b]				3.3–8.3
Children aged 11–15 years[b]				4.7–10.3

[a]Values from Dr. Ries Duran
[b]Reference [26]

Reference values

Ideally, each laboratory should obtain its own reference values.

Those obtained in our own laboratory are summarised in Table 18.1.

Table 18.2 Typical pathological values of tHcy

Vitamin B12 deficiency:	40–100 µmol/L
Folate deficiency:	25–65 µmol/L
Renal failure:	23–30 µmol/L
Severe renal failure:	30–100 µmol/L
Cystathionine-beta-synthase defect:	100–500 µmol/L
Methyl THF reductase defect:	100–250 µmol/L
Methionine synthase deficiency (cblC, cblD, cblF, cblE, cblG defects)	100–250 µmol/L

Typical pathological values of total Hcy

Pitfalls

– There is a time and temperature-dependent increase of plasma total Hcy in the presence of erythrocytes, for example 5–15% per hour at room temperature [27]. Thus, delay in separation of plasma after blood drawing must be avoided to prevent a falsely elevated Hcy level.

- Low purity of N,N-dimethylformamide may result in the presence of a negative peak immediately preceding the cysteine peak although this should not interfere with integration.

18.3.2 Measurement of S-Adenosylmethionine and S-Adenosylhomocysteine in Plasma and CSF by High-Performance Liquid Chromatography with Fluorescence Detection

18.3.2.1 Principle of Assay

The reaction of AdoMet and AdoHcy with mono chloro-acetalaldehyde produces the etheno-derivatives which are strongly fluorescent.

Mechanism of the derivatisation of AdoMet and AdoHcy and chloroacetaldehyde (Fig. 18.4):

These derivatives are separated by isocratic HPLC and detected and quantitated by fluorescence. Our methods have been modified from those of Wagner et al. [28].

18.3.2.2 Pre-analytical

Specimen:

A minimum of 2 mL EDTA blood is placed immediately on ice and centrifuged within 30 min at 2000 *g* for 5 min at 4 °C.

Plasma is immediately deproteinised by adding 1 mL to 0.625 mL of 10% $HClO_4$ and mixing thoroughly by vortex. The mixture is frozen and stored at −70 °C until analysis. See Sect. 18.3.3.2 (specimen) regarding the need for immediate deproteinization.

1 mL aliquots of pooled plasma collected from healthy subjects are prepared and stored in the same way and are used to prepare AdoMet and AdoHcy calibration standards to allow for matrix effects and also as an internal quality control.

CSF is collected in plain tubes, transported to lab on dry ice and stored at −70 °C until analysis. For external requests samples must be sent on dry ice. Blood-containing samples must be immediately centrifuged at 4 °C and 2000 *g* for 3 min and the supernatant stored in a plain tube.

Reagents and chemicals:

All chemicals should be of the highest purity available or HPLC grade.

Solutions for derivatisation and deproteinization

- 55% Chloroacetaldehyde
- Mix 55 mL chloroacetaldehyde (Fluka, cat. no. 22760) with 45 mL dem. water
- 3 mol/L sodium acetate buffer
- Dissolve 40.82 g sodium acetate trihydrate (Merck cat. no. 1.06267) in 100 mL dem. water
- 10% perchloric acid
- Make 14.28 mL 70% $HClO_4$ (Merck, cat. no. 519) up to 100 mL with dem. water
- 0.4 mol/L perchloric acid
- Make 3.33 mL 70% $HClO_4$ (Merck, cat. no. 519) up to 100 mL with dem. water

Mobile phases for HPLC

- AdoMet. 0.1 mol/L sodium acetate/2.3% acetonitrile with 10 mmol/L heptanesulphonic acid:
- Dissolve 13.61 g sodium acetate trihydrate GR (Merck cat. no. 1.06267) in 900 mL H_2O and add 2 g 1-heptanesulphonic acid, sodium salt (Sigma, cat. no. H2766/H8109) and 23 mL acetonitrile (Fisher Scientific A/0630/PB15). Adjust the pH to 4.3 with 100% glacial acetic acid, makeup to 1000 mL with dem. water and vacuum filter as above.
- AdoHcy. 0.1 mol/L sodium acetate/2.5% acetonitrile:
- Dissolve 13.61 g sodium acetate trihydrate GR (Merck cat. no. 1.06267) in 900 mL dem. water and add 9 mL acetic acid glacial and 25 mL acetonitrile. Adjust the pH to 4.2 with 100% glacial acetic acid, makeup to 1000 mL with dem. water and vacuum filter as above.

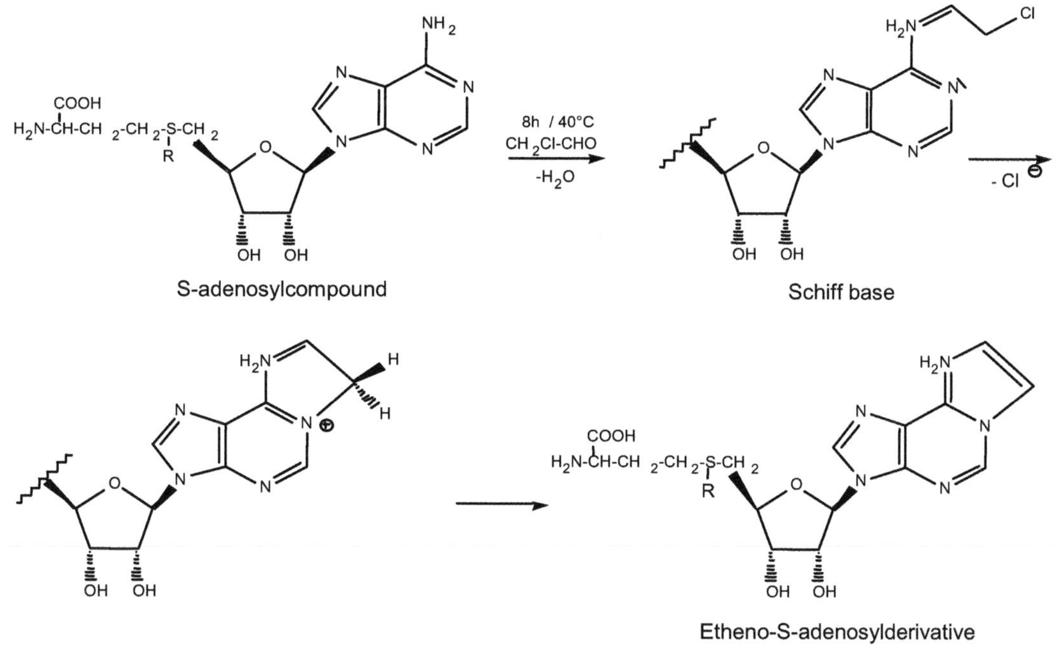

Fig. 18.4 Mechanism of the derivatisation of S-adenosylmethionine (AdoMet) and S-adenosylhomocysteine (AdoHcy) with chloroacetaldehyde

AdoMet and AdoHcy standards:

- AdoMet (34.2 µmol/L)/AdoHcy (68.6 µmol/L) stock solutions:

  Accurately weigh approximately 1.18 mg S-adenosylmethionine hydrogensulphonate (Boehringer Mannheim GmbH, cat. no. 102 407, MW: 496.2 g/mol) or 1.32 mg S-adenosylhomocysteine (Sigma cat. no. A 9384, MW 384.4 g/mol). These are dissolved in dem. water (usually ca. 5 mL) to give a concentration for AdoMet of 0.342 mmol/L (after correction for the purity declared for the particular batch) and for AdoHcy of 0.686 mmol/L. These solutions are immediately diluted tenfold with 0.4 mol/L perchloric acid and can be stored in aliquots at −20 °C for at least 6 months.
- Calibration standards are prepared freshly by serial dilution of the stock solutions in 0.4 mol/L perchloric acid (3.43 µmol/L to 0.0267 µmol/L).

Standards are stable for at least 6 months at −20 °C.

Instrumentation

General laboratory equipment, heating block at 60 °C for sample derivatisation laboratory centrifuge, microcentrifuge.

HPLC system consisting of two separate HPLC pumps or a single pump with gradient forming facility, preferably with degasser, autosampler with cooling system, fluorescence detector (excitation 384 nm, emission 516 nm) and an integration and data handling system.

HPLC columns

- AdoMet:
- Hydrosphere C-18, 3 µm (150 × 3 mm) with a guard column (10 × 3 mm) containing the same packing material.
- AdoHcy:
- Phenomenex, Gemini C-18, 3 µm (150 × 3 mm) with a guard column (4 × 2 mm) containing the same packing material.

Calibration

For both AdoMet and AdoHcy calibration is performed using three standards, e.g., 53, 106 and

210 nmol/L for AdoMet and 40, 80 and 160 nmol/L for AdoHcy, each added to pooled plasma SNF, plus a pooled plasma blank.

Quality Control

Neither AdoMet nor AdoHcy have so far been included in any external quality control scheme. Control samples are not commercially available.

A pooled plasma sample is included in each batch as internal quality control.

Detection limit:

This depends greatly on the sensitivity of the fluorimeter used but the following levels should be achievable.

AdoMet and AdoHcy, 2–5 nmol/L per sample.

Precision

1. For AdoMet standards of 27 nmol/L and 107 nmol/L, the within-run CV (n = 10) was shown to be 5.5% and 2.5%, respectively, with a between-run CV (n = 20) of 7.1% and 1.8%, respectively.
2. For AdoHcy standards of 20 nmol/L and 81 nmol/L the within-run CV (n = 10) was 3.6% and 2.0%, respectively, and the between-run CV (n = 20) was 8.2% and 1.7%, respectively.

Recovery

AdoMet: 103% ± 9.8 over a concentration range of 27–528 nmol/L.

AdoHcy: 92% ± 3.0 over a concentration range of 20–160 nmol/L.

18.3.2.3 Analytical

AdoMet and AdoHcy are separated and measured as their etheno-derivatives using reverse phase-HPLC with isocratic elution in separate runs using ion-pairing for AdoMet. In both runs the column is flushed with 100% acetonitrile followed by re-equilibration with the mobile phase between sample injections.

18.3.2.4 Procedure

Derivatization procedure for AdoMet and AdoHcy

Frozen deproteinized samples are thawed out, thoroughly mixed and centrifuged at 2000 *g* for 5 min at 4 °C. If the supernatant is turbid it should be clarified by filtration as above. Derivatisation is performed in 5 mL Pyrex tubes. For plasma samples 200 μL supernatant is added to 50 μL chloroacetaldehyde and thoroughly mixed followed by the addition of 45 μL 3 mol/L sodium acetate to adjust the pH to 3.5–4.0. Standards are prepared by adding 100 μL of pooled plasma supernatant, to 100 μL of each standard solution or 100 μL 0.4 mol/L perchloric acid for the blank. 50 μL of 55% chloroacetaldehyde is added to each tube with thorough mixing followed by the addition of 45 μL 3 mol/L sodium acetate to adjust the pH to 3.5–4.0.

Tubes are sealed with plastic stoppers and incubated at 39 °C for 8 h in a heating block.

The reaction is stopped by placing the tubes on ice. The samples can be immediately analysed by HPLC or kept at −20 °C until analysis.

HPLC conditions for AdoMet:

The HPLC analysis is performed by pumping 100% mobile phase for 54 min at 0.45 mL/min. then changing to 100% acetonitrile for 9 min. The column is re-equilibrated with mobile phase for 8 min prior to the subsequent injection. Usual pressures are between ~20 and 21 MPa (210 bar). 25 μL samples are injected.

The fluorimeter is set with an excitation wavelength of 270 nm and emission at 410 nm.

HPLC Conditions for AdoHcy:

The HPLC analysis is performed isocratically with a flow rate of 0.5 mL/min and a run time of 71 min. Usual pressures are between ~20 and 21 MPa (210 bar).

25 μL samples are injected.

The fluorimeter is set with an excitation wavelength of 270 nm and emission at 410 nm.

Calculation:

Concentrations of AdoMet or AdoHcy in plasma samples are determined by dividing the peak area in the unknown run by the slope of the calibration line and multiplying by 1.625 to allow for the dilution of plasma.

Thus AdoMet or AdoHcy concentration = Area/slope × 1.625 (see Fig. 18.5).

18.3.2.5 Post-analytical

Interpretation

Chromatogram

Peak identification

Chromatography of one or two samples spiked with AdoMet (210 nmol/L) and AdoHcy (160 nmol/L) is performed within each batch to aid in identification of peaks.

Typical elution times are 28–34 min for etheno-AdoMet and 8–10 min for etheno-AdoHcy.

The separations obtained for AdoMet (a) and AdoHcy(a) with a normal plasma sample are shown in Figs. 18.6 and 18.7.

Pitfalls

AdoMet and AdoHcy are unstable in neutral and alkaline solution must be kept acid.

Collected plasma has to be immediately deproteinised with 10% perchloric acid.

Delayed deproteinization may result in decomposition of AdoMet/AdoHcy.

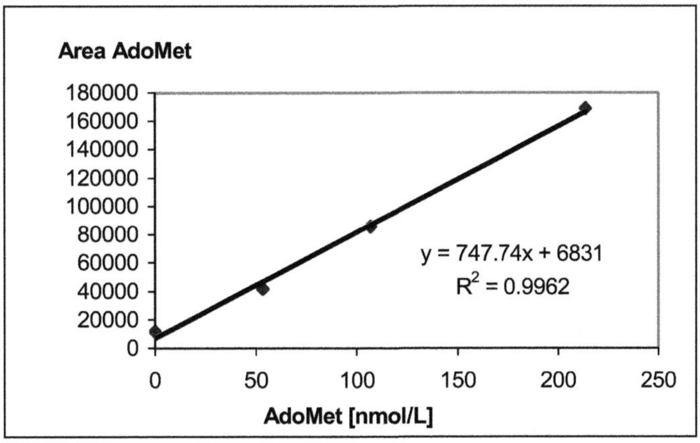

a

: Standard curve of AdoHcy in plasma

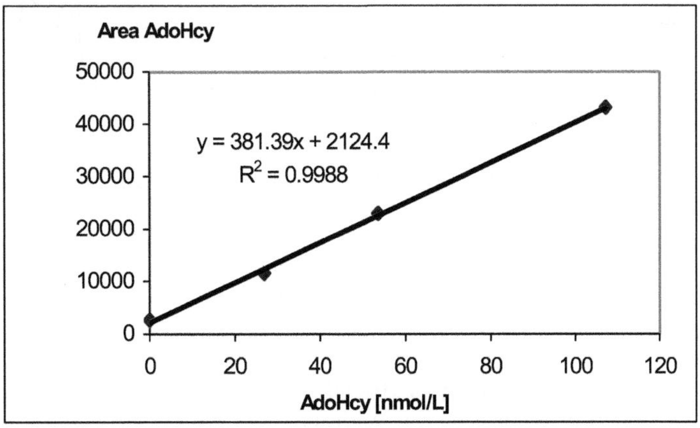

b

Fig. 18.5 (**a**) Standard curve of AdoMet in human plasma. (**b**) Standard curve of AdoHcy in plasma

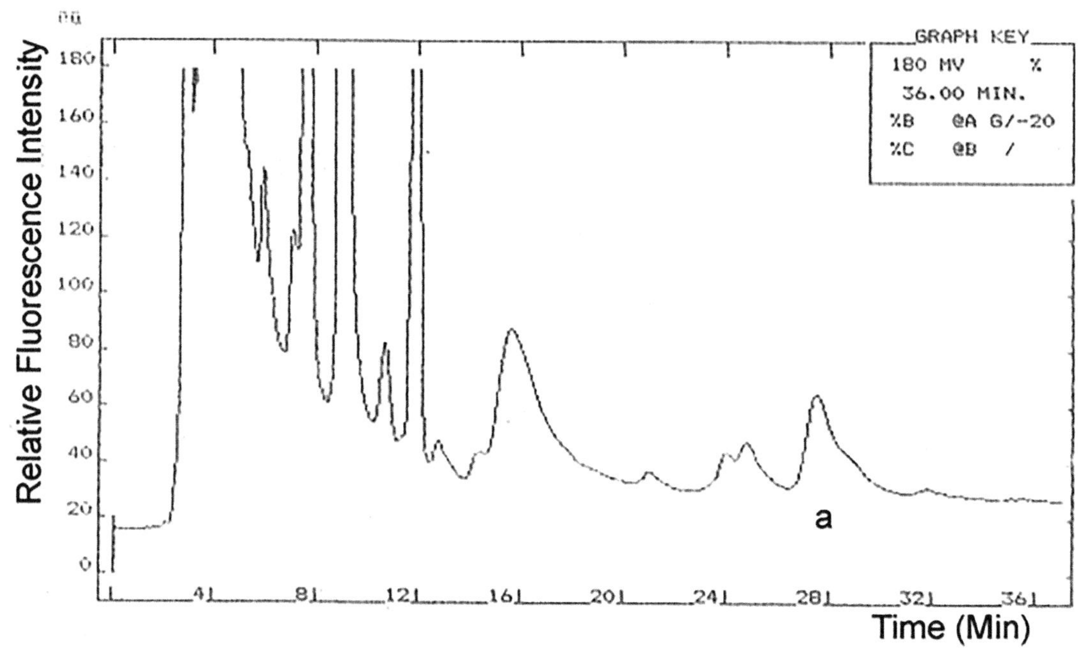

Fig. 18.6 Chromatogram of etheno-AdoMet in plasma

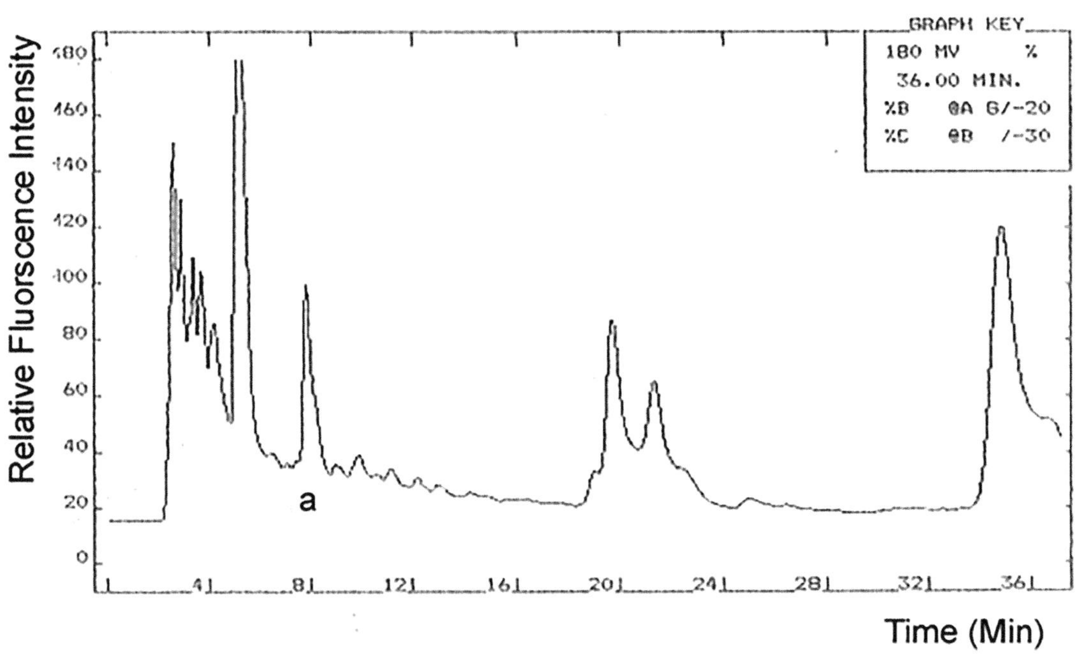

Fig. 18.7 Chromatogram of etheno-AdoHcy in plasma

To achieve optimal pH, crucial for the derivatisation step, control the pH with indicator paper after adding 3 mol/L sodium acetate.

Care needs to be taken in distinction of the true AdoMet and AdoHcy peaks from closely eluting substances.

18.3.3 Measurement of Methionine, Total Homocysteine, Ado Met and Ado Hcy in Plasma and CSF by HPLC and Tandem Mass-Spectrometry

18.3.3.1 Principle of Assay

Measurement of Met, Hcy, Ado Met and Ado Hcy in plasma and CSF is achieved using preliminary separation by liquid chromatography followed by mass spectrometric determination with selective multiple reaction monitoring (MRM). Ionisation is achieved by electrospray ionisation in the positive mode. An MRM spectrum of the appropriate masses is drawn.

18.3.3.2 Pre-analytical

Specimen

EDTA or lithium heparin blood, collected after overnight fasting and separated as soon as possible by centrifugation at 2000 *g* for 10 min at 4 °C. If separation cannot be performed within 1 h blood can be kept on ice slurry for a maximum of 1 h. Prolonged storage of blood at room temperature is not recommended because of the instability of AdoMet, AdoHcy and Hcy. Similarly, plasma should be processed and analysed immediately or stored frozen at −20 °C until analysis. Plasma samples from outside must be sent on dry ice.

Immediate deproteinization of blood is recommended in the section on HPLC determinations. This is recommended if the samples are analysed several months after collection. However, this step is not necessary if samples are measured right after collection or within the first 3 months (SAM: −13%; SAH: +11%).

CSF is collected in plain tubes, transported to lab on dry ice and stored at −70 °C until analysis. For external requests samples must be sent on dry ice. Blood-containing samples must be immediately centrifuged at 4 °C and 2000 *g* for 3 min and the supernatant stored in a plain tube.

Reagents and chemicals: All chemicals should be of the highest purity available or HPLC grade.

1. Water, HPLC quality.
2. 0.5 M DTT: 77 mg dissolved in 1 mL H_2O, must always be freshly prepared.
3. 5 M K_2CO_3: 138.2 g K_2CO_3 dissolved in 200 mL H_2O and kept at room temperature.
4. 1 mM HCl: 3 mL 37% HCL added to a 1 L measuring flask almost filled with H_2O then made up to the mark.
5. Mobile phase AdoMet/AdoHcy-A: Water/FA/HFBA 100:0.1:0.05 (v/v/v): Pipette 1 mL formic acid and 0.5 mL heptafluorobutyric acid into a 1 L measuring flask almost filled with water, makeup to the mark with water, mix well and store at RT.
6. Mobile phase AdoMet/AdoHcy-B: Methanol/FA/HFBA 100:0.1:0.05 (v/v/v): Pipette 1 mL formic acid and 0.5 mL heptafluorobutyric acid into a 1 L measuring flask almost filled with methanol, makeup to the mark with methanol, mix well and store at RT.
7. Mobile phase Met/tHcy-A: Water/FA 100:0.1 (v/v): Pipette 1 mL formic acid into a 1-L measuring flask almost filled with water, make-up to the mark with water, mix well and store at RT.
8. Mobile phase Met/tHcy-B: Methanol/Water/FA 80:20:0.1 (v/v/v): Add 1 mL formic acid to 800 mL methanol in a 1 L measuring flask, make up to the mark with water, mix well and store at RT.

Standard solutions:

- Stock standard solutions (can be stored at −70 °C for up to 5 years).
- Methionine 1 mM: 1.492 L-methionine (Sigma-Aldrich M9375) dissolved in 10 mL H_2O.
- Homocysteine 1 mM: 1.352 mg homocysteine (Sigma-Aldrich 69453) dissolved in 10 mL H_2O.
- AdoMet 1 mM: 3.844 mg S-adenosylmet-hionine (Sigma-Aldrich A2408) dissolved in 10 mL 1 mM HCl.
- AdoHcy 1 mM: 3.944 mg S-adenosylhomocysteine (Sigma-Aldrich A9384) dissolved in 10 mL 1 mM HCl.
- Prepare a standard mixture by adding 500 μL of each standard to 500 μL methanol to a final concentration of 200 μmol/L.

- Stable isotope internal standard stock solutions
- Methionine 1 mM: 1.532 mg L-methionine-d4 (CDN isotopes, D7086) dissolved in 10 mL H_2O.
- Homocysteine 1 mM: 1.392 mg homocysteine-d4 (CDN isotopes D6452) dissolved in 10 mL H_2O.
- AdoMet 1 mM: 4.015 mg S-adenosylmethionine-d3 (CDN isotopes D4093) dissolved in 10 mL H_2O.
- AdoHcy 1 mM: 3.943 mg S-adenosylhomocysteine-13C10 (Cambridge isotopes CLM 8906) dissolved in 10 mL H_2O.
- Internal standards working solution.
- Mix 50 µl of each stock internal standard and 50 µL H2O ~~methanol~~ to a final concentration of 200 µmol/L.
- Mix 100 µL of this solution with 400 µL H_2O to give a working solution of 40 µM.

Calibration solutions
Calibrators are prepared by serial dilution of the standard mixture in H_2O/Fa 100:0.1 (v/v) and addition of 20 µL of 40 µM internal standard working solution to obtain the following end concentrations: 5, 10, 25, 100, 500 nmol/L, 1, 5, 10, 50, 100 µmol/L. All calibrators are measured but two different calibration ranges are used for quantification: from 5 nmol/L to 5 µmol/L for AdoMet and AdoHcy and from 100 nmol/L to 100 µmol/L for Met and Hcy.

Instrumentation

- General laboratory equipment, laboratory centrifuge, vortex mixer.
- Tandem mass spectrometer, e.g., SCIEX Triplequad 5500 (AB Sciex) LC-MS/MS coupled with HPLC pump, e.g., UHPLC Ultimate 3000 XRS (Thermo Fisher).
- Analytical column: ACQUITY UPLC BEH C18 1.7 µm 2.1 × 100 mm.

QC samples
Plasma—pooled heparinised plasma is used for internal quality control. 220 µL are immediately frozen in liquid nitrogen and stored at −70 °C.

CSF—220 µL aliquots of pooled CSF are snap frozen in 1.5 mL tubes and stored at −70 °C.

Inter-day precision (CV):

AdoMet in plasma: 9.5%
AdoHcy in plasma: 9.1%
Met in plasma: 8.2%
Hcy in plasma: 3.2%
AdoMet in CSF: 5.1%
AdoHcy in CSF: 4.2%
Met in CSF: 12.6%
Hcy in CSF: 4.5%

Mean recovery (n = 4)

AdoMet in plasma: 94.2%
AdoHcy in plasma: 117.3%
Met in plasma: 106.8%
Hcy in plasma: 100.0%
AdoMet in CSF: 92.2%
AdoHcy in CSF: 116.0%
Met in CSF: 90.3%
Hcy in CSF: 113.3%

18.3.3.3 Analytical Procedure

Two sample preparations are used for either Met/Hcy or AdoMet/AdoHcy.

Plasma: Met/Hcy

100 µL plasma is mixed with 20 µL freshly prepared 0.5 M DTT, 20 µL of internal standard working solution is added. The mixture is vortex mixed and incubated at RT for 15 min. 260 µL MeOH is added, mixture is vortex mixed and centrifuged for 10 min at 14,000 *g* (RT). The supernatant is dried over nitrogen at 40 °C for 40–50 min and the residue is resuspended in 200 µL mobile phase A (water/for-

mic acid 100:0.1) and if necessary re-centrifuged for 3 min at 14,000 *g*. This solution is placed in LC vials for analysis with 2 μL injected.

Plasma: AdoMet/AdoHcy

160 μL plasma is mixed with 20 μL of internal standard working solution, vortex mixed and 12 μL 70% $HClO_4$ are added. The mixture is vortex mixed and centrifuged for 10 min at 14,000 *g* (4 °C). Supernatant is transferred to a fresh tube and 8 μL 5M K_2CO_3 is added, vortex mixed and centrifuged for 5 min at 14,000 *g* (4 °C). This supernatant is transferred to LC vials for analysis with 10 μL injected.

CSF: Met/Hcy

CSF is centrifuged at 14,000 *g* (4 °C) for 5 min, 160 μL is mixed with 20 μL freshly prepared 0.5 M DTT and 20 μL of internal standard working solution is added. The mixture is vortex mixed and incubated at RT for 15 min. The sample is centrifuged for 5 min at 14,000 *g* (4 °C) and the supernatant is transferred to LC vials for analysis with 2 μL injected.

CSF: AdoMet/AdoHcy

CSF is centrifuged at 14,000 *g* (4 °C) for 5 min, 160 μL is mixed with 20 μL of internal standard working solution and 20 μL mobile phase A (0.1% formic acid). The mixture is vortex mixed and centrifuged for 5 min at 14,000 *g* (4 °C). The supernatant is transferred to LC vials for analysis with 10 μL injected.

LC conditions

	AdoMet/ AdoHcy	Met/Hcy
Flow rate	260 μL/min	300 μL/min
Column temperature	30 °C	30 °C
Injection volume	10 μL	2 μL
Autosampler temperature	10 °C	10 °C
Gradient	10–95% B within 5.1 min	5–50% B within 2.5 min
Total run time	5.5 min	4 min

MS conditions

	Q1	Q3	DP (V)	CE (V)
Met (quant)	150	61	60	15
Met (qual)	150	104	60	8
Hcy (quant)	136	90	60	15
Hcy (qual)	136	56	60	25
AdoMet (quant)	399	250	60	23
AdoMet (qual)	399	136	60	40
AdoHcy (quant)	385	88	60	60
AdoMet (qual)	385	136	60	25
Met-IS	154	63	60	27
Hcy-IS	140	59	60	25
AdoMet-IS	402	250	60	21
AdoHcy-IS	396	89	60	65

18.3.3.4 Post-analytical and Interpretation (Figs. 18.8, 18.9 and 18.10)

For reference values, see Tables 18.1 and 18.3.

For typical pathological values, see Tables 18.2 and 18.4.

Pitfalls

Inadequate pre-analytical conditions may lead to falsely low concentrations of homocysteine and AdoMet and high values of AdoHCy. Patients receiving medication with AdoMet may show elevated Hcy levels. Treatment with a number of drugs including methotrexate, carbamazepine, phenytoin, various anticonvulsants and 6-azauridine can also lead to falsely elevated Hcy levels.

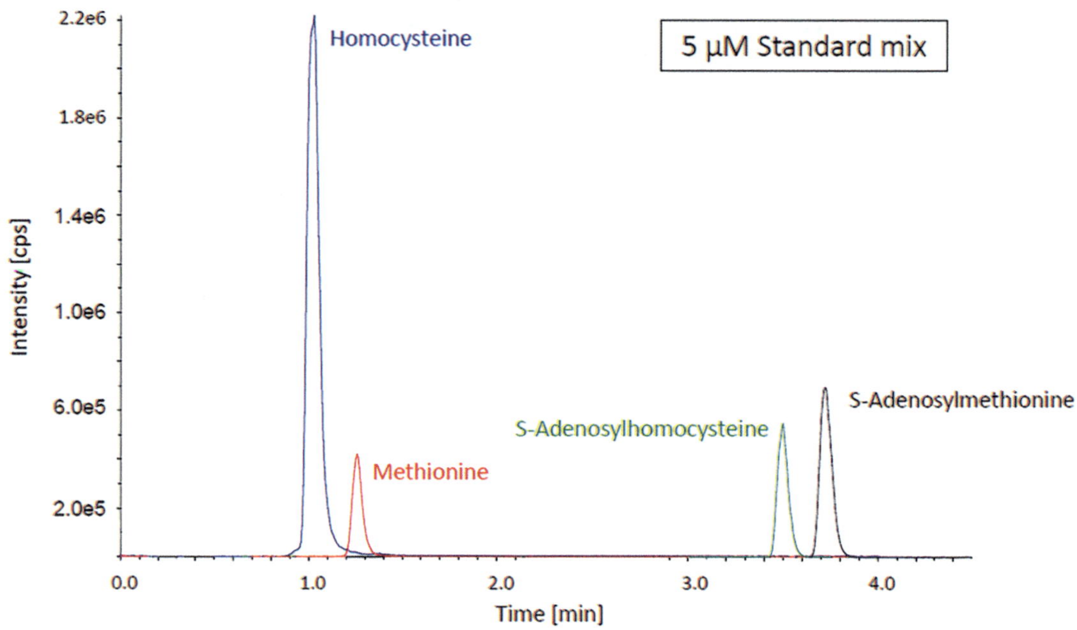

Fig. 18.8 High-performance liquid chromatography-tandem mass spectrometric analysis of Met, Hcy, AdoMet and AdoHcy in a standard mixture containing 5 µmol/L of each compound. Note that Met and Hcy are measured with a different LC method than AdoMet and AdoHcy but the four resulting chromatographic peaks are put together for simplicity. The total run time is 4.0 min for Met/Hcy and 5.5 min for AdoMet/AdoHcy. Also note that for each compound isotopically labelled internal standards as well as additional transitions for qualitative purposes are used, which are not depicted

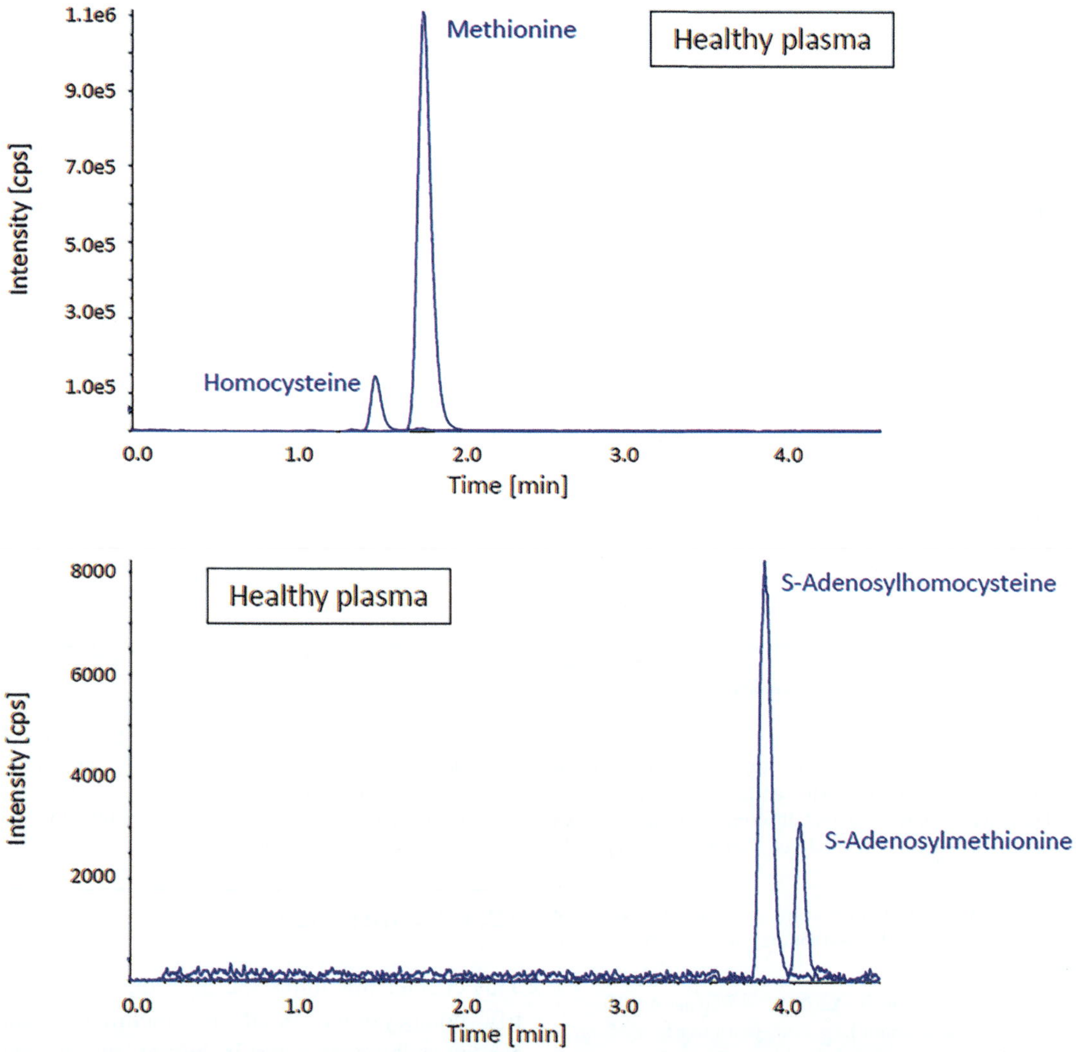

Fig. 18.9 Upper panel: Chromatogram of Met and Hcy in a healthy plasma sample. The analytes were evaluated by multiple reaction monitoring with the following quantitative transitions: m/z 150 → 61 for Met; m/z 136 → 90 for Hcy. For each analyte, the transition giving the highest signal intensity was used. Lower panel: Chromatogram of AdoMet and AdoHcy in a healthy plasma sample. The analytes were evaluated by multiple reaction monitoring with the following quantitative transitions: m/z 399 → 250 for AdoMet; m/z 385 → 88 for AdoHcy. For each analyte, the transition giving the highest signal intensity was used

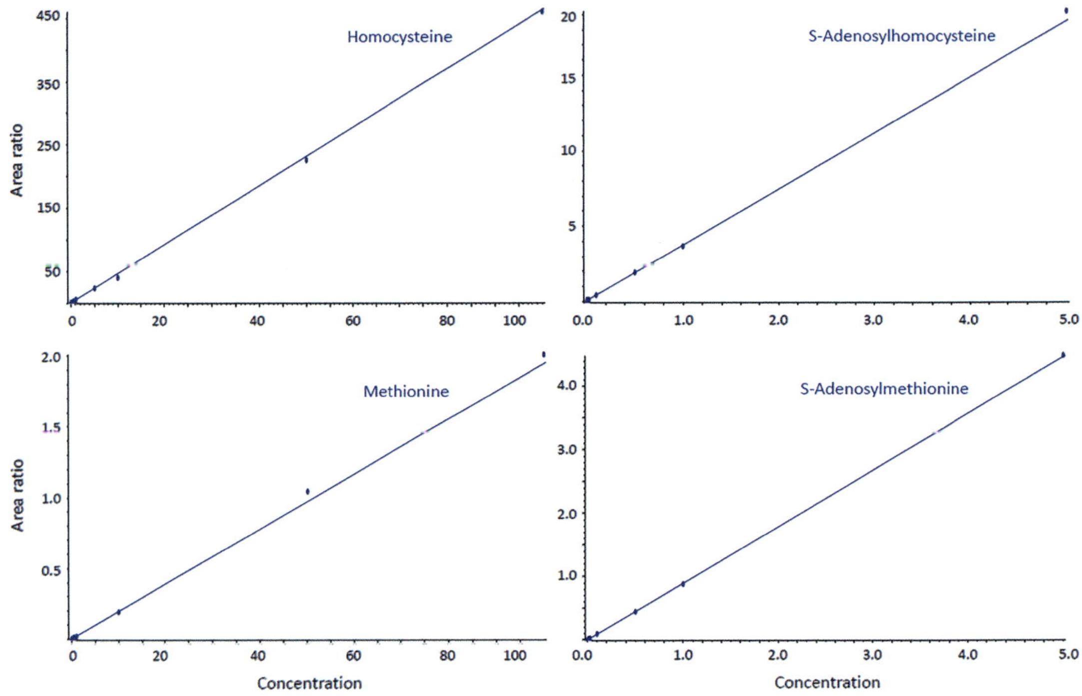

Fig. 18.10 Standard curves of Met, Hcy, AdoMet and AdoHcy diluted in aqueous solutions. Concentrations on the *x*-axis are in µmol/L. All standard curves are linear and have a $R^2 > 0.99$. Note that different calibration ranges are used for each compound according to the expected physiological concentrations: from 5 nmol/L to 5 µmol/L for AdoMet and AdoHcy and from 100 nmol/L to 100 µmol/L for Met and Hcy. If in a sample the concentration of one analyte is above the highest calibrator, this sample is diluted using mobile phase A

Table 18.3 Control values for S-adenosylmethionine (AdoMet) and S-adenosylhomocysteine (AdoHcy) in plasma

	Plasma AdoMet (nmol/L)	Plasma AdoHcy (nmol/L)
Range	29.6–118.2	9.6–48.7
Mean + SD	56.4 ± 17.4	25.5 ± 7.9
n	64	66

Table 18.4 Typical pathological values for AdoMet and AdoHcy

Pathological condition	AdoMet (nmol/l)	AdoHcy (nmol/l)
Renal disease:	220–840	530–1590
Methionine: S-adenosyltransferase deficiency:	48–120	10–15
Glycine n-methyltransferase deficiency:	1000–4000	Normal
S-adenosyl-homocysteine hydrolase deficiency:	800–3000	330–5000
Cystathionine β-synthase deficiency:	900–2000	150–1700

18.4 Conclusions

This chapter highlights the procedures to quantify homocysteine, S-adenosylmethionine and S-adenosylhomocysteine in human plasma and CSF to diagnose different inborn errors of metabolism and well as vitamin B_{12} and folate deficiencies. Previously HPLC coupled with fluorescence detection after derivatization has been successfully used to quantify these analytes. This approach has been slowly replaced in recent years by liquid chromatography coupled with mass spectrometry due to higher analytical specificity and sensitivity. The presented methods can be easily transferred to other laboratories after some optimization steps. However, each laboratory is encouraged to establish its own reference ranges. Particular care should be paid to the preanalytical procedure because of the instability of these analytes in whole blood and plasma at room temperature.

References

1. Finkelstein JD (1990) Methionine metabolism in mammals. J Nutr Biochem 1:228–237
2. Fowler B (2005) Homocysteine: overview of biochemistry, molecular biology, and role in disease processes. Semin Vasc Med 5(2):77–86
3. Clarke S, Banfield K (2001) S-adenosylmethionine-dependent methyltransferases. In: Carmel R, Jacobsen D (eds) Homocysteine in health and disease. Cambridge University Press, Cambridge, UK, pp 63–78
4. Baric I (2009) Inherited disorders in the conversion of methionine to homocysteine. J Inherit Metab Dis 32:459–471
5. Johnson JL, Duran M (2001) Molybdenum cofactor deficiency and isolated sulfite oxidase deficiency. In: Scriver CR, Beaudet AL, Sly WS, Valle D (eds) The metabolic and molecular bases of inherited disease, 8th edn. McGraw-Hill, New York, NY, pp 3163–3177
6. Finkelstein JD (2000) Pathways and regulation of homocysteine metabolism in mammals. Semin Thromb Hemost 26:219–225
7. Finkelstein JD (1998) The metabolism of homocysteine: pathways and regulation. Eur J Pediatr 157(Suppl 2):S40–S44
8. Fowler B (2001) Transport and tissue distribution of homocysteine and related S-adenosyl compounds. In: Carmel R, Jacobsen D (eds) Homocysteine in health and disease. Cambridge University Press, Cambridge, UK, pp 163–174
9. Fowler B, Blom H (2017) Disorders of transsulfuration. In: Sarafoglou K, Hoffmann GF, Roth KS (eds) Pediatric endocrinology and inborn errors of metabolism, 2e. McGraw Hill
10. Watkins D, Morel CF, Rosenblatt DS (2017) Inborn errors of folate and cobalamin transport and metabolism. In: Sarafoglou K, Hoffmann GF, Roth KS (eds) Pediatric endocrinology and inborn errors of metabolism, 2e. McGraw Hill
11. Sass JO, Nakanishi T, Sato T, Shimizu A (2004) New approaches towards laboratory diagnosis of isolated sulphite oxidase deficiency. Ann Clin Biochem 41:157–159
12. Stabler SP, Marcell PD, Podell ER, Allen RH, Savage DG, Lindenbaum J (1988) Elevation of total homocysteine in the serum of patients with cobalamin or folate deficiency detected by capillary gas chromatography –mass spectrometry. J Clin Invest 81:466–474
13. Selhub J, Jacques PF, Wilson PW, Rush D, Rosenberg IH (1993) Vitamin status and intake as primary determinants of homocysteinemia in an elderly population. JAMA 270:2693–2698
14. Löhrer RMT, Angst CP, Brunner FP, Haefeli WE, Fowler B (1998) Evidence for disturbed S-adenosylmethionine : AdoHcy ratio in patients with end-stage renal failure: a cause for disturbed methylation reactions? Nephrol Dial Transplant 13:656–661
15. Wang Y, Chen S, Yao T, Li D, Wang Y, Li Y, Wu S, Cai J (2014) Homocysteine as a risk factor for hypertension: a 2-year follow-up study. PLoS One 9(10):e108223
16. Smith AD, Refsum H, Bottiglieri T et al (2018) Homocysteine and dementia: an international consensus statement. J Alzheimers Dis 62:561–570
17. Barić I, Kožich V, Fowler B (2022) Disorders of sulfur amino acid and hydrogen sulfide metabolism. In: Blau N, Dionisi Vici C, Ferreira CR, Vianey-Saban C, van Karnebeek CDM (eds) Physician's guide to the diagnosis, treatment, and follow-up of inherited metabolic diseases. Springer, Cham
18. Loehrer FMT, Angst CP, Haefeli WE, Jordan PP, Ritz R, Fowler B (1996) Low whole-blood S-adenosylmethionine and correlation between 5-methyltetrahydrofolate and homocysteine in coronary artery disease. Arter Thromb Vasc Biol 16:727–733
19. Capdevila A, Wagner C (1998) Measurement of plasma S-adenosylmethionine and A-adenosylhomocysteine as their fluorescent isoindoles. Analyt Biochem 264:180–184
20. Loehrer FMT, Haefeli WE, Angst CP, Browne G, Frick G, Fowler B (1996) Effect of methionine loading on 5-methyltetrahydrofolate, S-adenosylmethionine and S-adenosylhomocysteine in plasma of healthy humans. Clin Sci 981:79–86
21. Loehrer FMT, Schwab R, Angst CP, Haefeli WE, Fowler B (1997) Influence of oral S-adenosylmethionine on plasma 5-methyltetrahydrofolate, S-adenosylhomocysteine, homocysteine and methionine in healthy humans. J Pharmacol Exp Therapeut 282:845–850
22. Castro R, Rivera I, Struys EA, Jansen EE, Ravasco P, Camilo ME, Blom HJ, Jakobs C, Tavares de Almeida I (2003) Increased homocysteine and S-adenosylhomocysteine concentrations and DNA hypomethylation in vascular disease. Clin Chem 49:1292–1296
23. Gellekink H, van Oppenraaij-Emmerzaal D, van Rooij A, Struys EA, den Heijer M, Blom HJ (2005) Stable-isotope dilution liquid chromatography-electrospray injection tandem mass spectrometry method for fast, selective measurement of S-adenosylmethionine and S-adenosylhomocysteine in plasma. Clin Chem 51:1487–1492
24. Araki A, Sako Y (1987) Determination of free and total homocysteine in human plasma by high-performance liquid chromatography with fluorescence detection. J Chromatogr 422:43–52
25. Ubbink JB, Hayward Vermaak WJ, Bissbort S (1991) Rapid high-performance liquid chromatographic assay for total homocysteine levels in human serum. J Chromatogr 565:441–446

26. Ueland PM, Refsum H, Stabler SP, Malinow MR, Andersson A, Allen RH (1993) Total homocysteine in plasma or serum: methods and clinical applications. Clin Chem 39:1764–1779
27. Vilaseca MA, Moyano D, Ferrer I, Artuch R (1997) Total homocysteine in pediatric patients. Clin Chem 43:690–692
28. Wagner J, Hirth Y, Claverie N, Danzin C (1986) A sensitive high-performance liquid chromatographic procedure with fluorimetric detection for the analysis of decarboxylated S-adenosylmethionine and analogues in urine samples. Anal Biochem 154:604–617

19 Lactate, Pyruvate, Acetoacetate, and 3-Hydroxybutyrate

Anne Vassault, Christelle Corne, and Jean François Benoist

Contents

List of Abbreviations

3OHB	3-Hydroxybutyrate
ACAC	Acetoacetate
ALT	Alanine aminotransferase
HBDH	Hydroxybutyrate dehydrogenase
KB	Ketone bodies
L	Lactate
LDH	Lactate dehydrogenase
NEFA	Non-esterified fatty acids
P	Pyruvate

A. Vassault (✉)
Laboratoire de Biochimie, Hopital Necker Enfants Malades, Paris-Cité University France, Paris, France
e-mail: anne.vassault@orange.fr

C. Corne (✉)
Laboratoire des maladies métaboliques et dépistage néonatal, Institut de Biologie et pathologie, Grenoble, France
e-mail: CCorne@chu-grenoble.fr

J. F. Benoist (✉)
Laboratoire de Biochimie, Hopital Necker Enfants Malades, Paris-Saclay Université France, Paris, France
e-mail: jean-francois.benoist@aphp.fr

19.1 Introduction

Lactate (L), pyruvate (P), acetoacetate (ACAC), and 3-hydroxybutyrate (3OHB) are intermediary metabolites that normally occur in blood and play an essential role in energy production. Their relative blood concentrations are an expression of nutritional balance, providing a view of the metabolic disturbances arising in a patient. Associated or not to non-esterified fatty acids (NEFA), and

N. Blau, F. M. Vaz (eds.), *Laboratory Guide to the Methods in Biochemical Genetics*,
https://doi.org/10.1007/978-3-031-58819-8_19

glucose (GLU) measurements, they are useful tools to investigate intermediary metabolism in health and disease, particularly in inherited metabolic diseases.

The determination of these metabolites in biological fluids is useful in the early detection, diagnosis, and treatment follow-up of abnormalities regarding enzymes defects of pyruvate “carrefour” (PDH or PC defects), of tricarboxylic acids cycle, of neo-glucogenesis, of liver glycogenolysis, of fatty acids oxidation, of ketogenesis, ketolysis or mitochondrial respiratory chain.

19.2 Clinical Significance

Under physiological conditions, the blood levels of lactate, pyruvate, acetoacetate, and 3-hydroxybutyrate are defined by carefully regulated production and utilization pathways, at different sites in the body. Many inherited metabolic diseases reflect their underlying enzymatic defect(s) in an abnormal blood pattern of these compounds. It should be noted that the metabolic profiles of the major organs, such as brain, muscle, and liver are very different.

Lactate is a metabolite whose plasma level reflects the equilibrium between its production and its consumption by different tissues. Lactate is the end-product of anaerobic glycolysis, and is an important energy source for some tissues such as heart, muscles, and kidney.

Lactate accumulation in blood higher than 5 mmol/L leads to lactic acidosis. Hyperlactatemia can be observed in ischemic conditions or in many inherited metabolic diseases (pyruvate dehydrogenase deficiency, pyruvate carboxylase deficiency, etc.).

During the fasting period, lactate is used by the liver and kidneys to provide glucose (gluconeogenesis). A decrease in blood lactate level is observed during the first 15 h of fast (about 30% regardless of age). In the postprandial period, lactate is oxidized in mitochondria, producing energy for the heart, muscles, and kidneys. Metabolic pathways are given in Fig. 19.1. The ratio of lactate to pyruvate remains unchanged independently of the fasting state.

Lactate and pyruvate are interconverted within the cytoplasm, depending upon the NAD/NADH2 ratio. In the cytoplasm, lactate to pyruvate ratio reflects the oxidoreduction state.

Pyruvate is an intermediate metabolite, a product of the metabolism of carbohydrates or protein metabolism. It is the end point of glycolysis. In mitochondria, pyruvate may be partially oxidized to acetyl-coenzyme A (acetyl-CoA) by PDH or fully oxidized to CO_2 and H_2O, or carboxylated by PC into oxaloacetate (Fig. 19.2).

Pyruvate is transformed to lactate by lactate dehydrogenase (LDH) under anaerobic conditions.

Ketone bodies (KBs) circulate in the blood as 3-hydroxybutyrate (3OHB) and acetoacetate (ACAC). The blood concentrations of these metabolites depend upon the equilibrium between their production by the liver (ketogenesis) and consumption at the peripheral level (ketolysis). Abnormalities of KBs metabolism manifest biologically as ketosis, hypoketotic hypoglycemia, and perturbation of the 3OHB/ACAC ratio [1, 2].

In the fed state, the KB concentrations do not usually exceed 0.2 mmol/L, except in the neonatal period when higher concentrations are observed. The blood concentration of KBs increases during the fasted state due to the PDH inhibition by mitochondrial acetyl-CoA and mitochondrial NADH, with an associated increase in the 3OHB/ACAC ratio. KBs are primarily synthesized in the liver from acetyl-CoA, the product of fatty acid mitochondrial oxidation.

Interconversion between ACAC and 3OHB depends upon the NAD/NADH2 ratio due to an oxidoreduction mechanism involving hydroxybutyrate dehydrogenase (HBDH) mainly localized in the mitochondria. Fatty acids are transported to the liver to undergo beta-oxidation during fasting.

Acyl-CoA is introduced into mitochondria following coenzyme A esterification in the cytoplasm. Mitochondrial entry depends upon a double membrane transport involving carnitine acyltransferases II and I. Excess acetyl-CoA is used for KBs synthesis. KBs are transported in the blood flow and metabolized in most tissues except the liver and the red blood cells via the

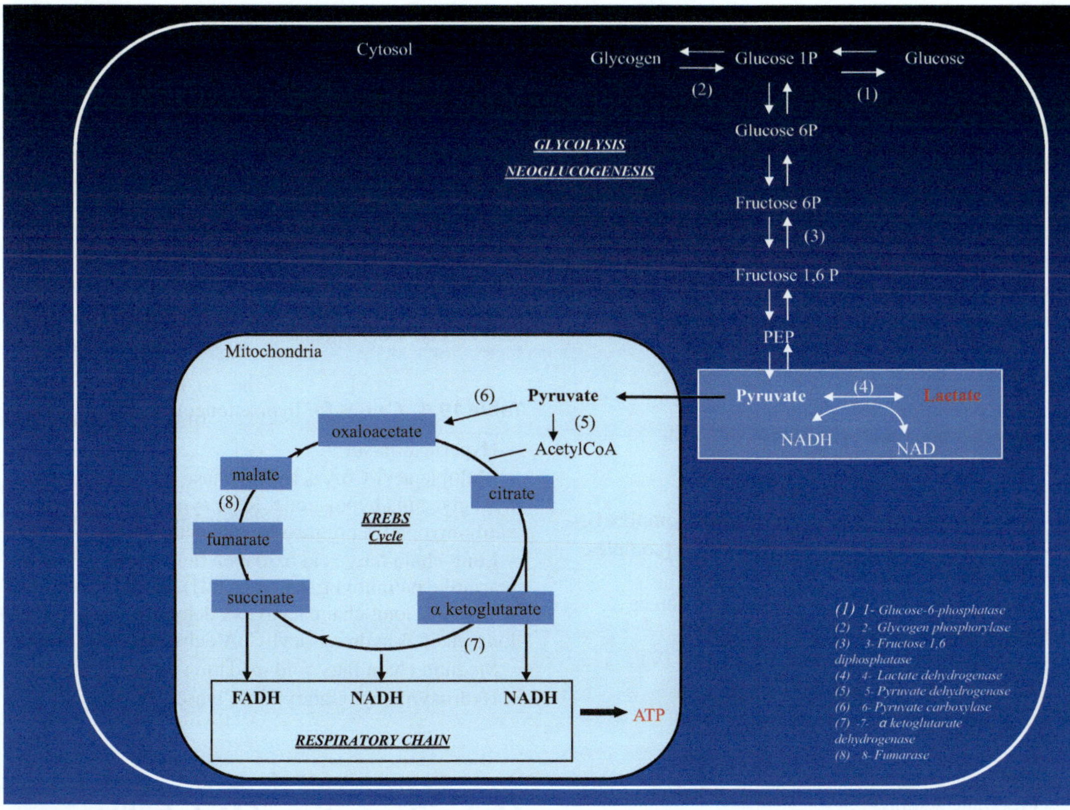

Fig. 19.1 Lactate and pyruvate metabolic pathways

Fig. 19.2 Pyruvate pathways

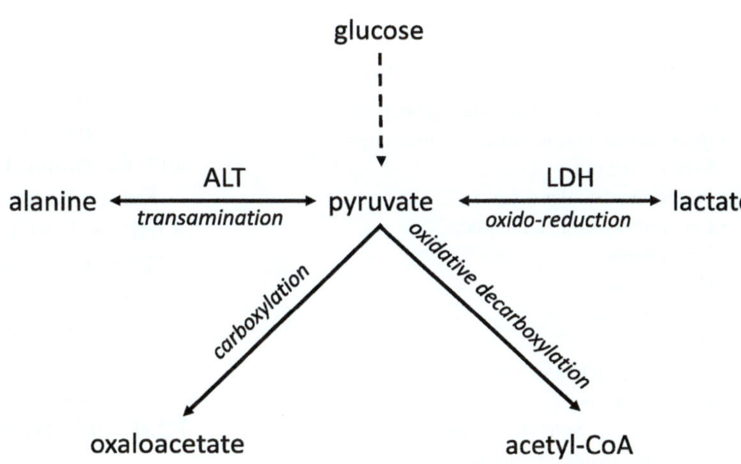

Krebs cycle. KBs are necessary to provide energy to the brain during fasting, a true alternative substrate to glucose.

The main causes of acquired hyperlactatemia and hereditary hyperlactatemia are given in Tables 19.1 and 19.2, respectively [3, 4].

Table 19.1 Main etiologies for hereditary hyperlactatemias

I Primary hyperlactatemias
• Glycogen metabolism disorders
– Amylo-1,6-glucosidase defect
– Liver phosphorylase defect
– Glycogen synthetase defect
• Gluconeogenesis defects
– Glucose 6-phosphatase defect
– Fructose 1,6 biphosphatase defect
– Phosphoenolpyruvate carboxykinase defect
• Pyruvate disorders
– Pyruvate carboxylase defect
– Pyruvate dehydrogenase defect
• Krebs cycle abnormalities
– Cetoglutarate dehydrogenase defect
– Fumarase defect
• Respiratory chain abnormalities
– NADH coenzyme Q reductase defect (complex I)
– Succinate coenzyme Q reductase defect (complex II)
– Coenzyme Q cytochrome C reductase defect (complex III)
– Cytochrome C oxidase defect (complex IV)
– ATPase defect (complex V)
II Secondary hyperlactatemias
• Organic acidemias
– Methylmalonic aciduria
– Propionic aciduria
– Isovaleric aciduria
• Congenital hyperammonemia: citrullinemia

Table 19.2 Main causes for acquired hyperlactatemia

With anoxia resulting in tissue hypoperfusion:
• Heart stroke, endotoxins, or hemorrhagic
• Serious anemia
• Intensive muscular disease
• Left ventricular insufficiency
• Convulsions
• Hypocapnia
Without anoxia:
• Diabetes
• Renal insufficiency
• Reye syndrome
• Intoxications: Salicylic, cyanide, alcohol, antiretroviral drugs, biguanides

The main causes of hyperketonemia and hypoketonemia are given in Tables 19.3 and 19.4 [3, 4].

Table 19.3 Causes for hyperketonemias

Over production
– Diabetes insulinodependant
– Methylmalonic, propionic, or isovaleric acidemias
– Pyruvate carboxylase and multiple carboxylase deficiency
– Neoglucogenesis enzymes deficiency: glucose-6-phosphatase, fructose-1,6-diphosphatase or abnormality of glycogen synthesis (glycogen synthase)
Decreasing ketone body utilization at peripheral level
– Ketolysis defect: succinyl-CoA transferase, acetoacetyl-CoA thiolase

Table 19.4 Causes for hypoketonemias

– Hyperinsulinism
– Multiple acyl-CoA dehydrogenase (MAD): hypoglycemia hypoketotic, Reye syndrome, cardiomyopathy, progressive myopathy
– Long-chain fatty acid oxidation defect: CPT I and II (carnitine palmitoyl transferase I, II), systemic carnitine deficiency, long-chain acyl-CoA dehydrogenase defect, long-chain 3-hydroxy-acyl-CoA dehydrogenase defect
– Medium chain fatty acid oxidation defect (MCAD)
– Hydroxymethyl glutaryl CoA lyase defect

19.3 Properties of Metabolites

Lactate (CH3-CHOH-COOH) is the anionic conjugate base of *lactic acid*, arising from deprotonation of the carboxy group. The isomer l(+) is the only one the human body can be metabolized.

Pyruvate (CH3-CO-COOH), designed 2-oxopropanoate is the conjugate base of pyruvic acid, the simplest alpha-ketoacid.

Ketone bodies (KBs) circulate in the blood as 3OHB (CH3-CHOH-CH2-COOH) and ACAC (CH3-CO-CH2-COOH). Acetone formed by non-enzymatic decarboxylation of acetoacetate is not involved in metabolism.

19.4 Methods

For simultaneous blood measurement of the four metabolites (lactate, pyruvate, acetoacetate, 3-hydroxybutyrate), deproteinized blood by perchloric acid is used as a sample. Spectrophotometric enzymatic methods according to reactions given in Fig. 19.3 were devel-

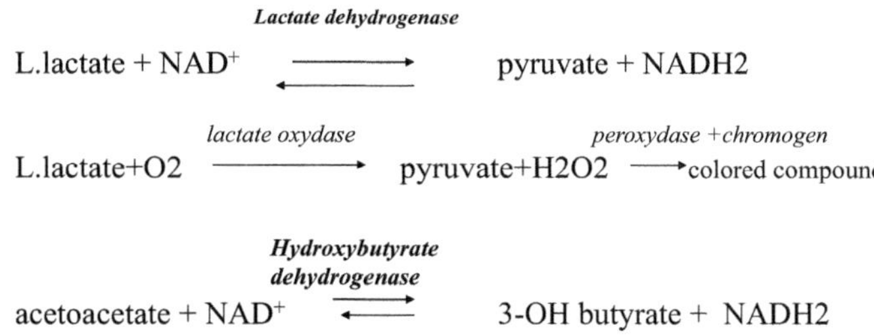

Fig. 19.3 Reactions involved for enzymatic measurement of 3-OH butyrate, acetoacetate, lactate, and pyruvate

oped for automated analyzers to minimize sample volume and improve precision [5–9].

19.4.1 Pre-analytical

Patient Preparation

Patient preparation includes the evaluation of clinical indications and nutritional conditions (i.e., fed state, fasted state, and fasting duration) must be precisely defined.

The patient must be fully informed of all procedures and at rest before blood sampling or invasive procedures. If fasting is required, the fast is scheduled for 24 h (according to age), but can be interrupted.

Beginning at night, sampling is performed at 2, 15, 20, and 24 h of the fast if the age of the child allows it.

Specimen:

LAC and 3OHB can be measured in blood, plasma, urine, and CSF. PYR and ACAC are highly unstable in blood, so, deproteinization has to be performed immediately after sampling.

LAC, PYR, 3OHB, and ACAC determination can be measured in the same sample after deproteinization using perchloric acid [9–11].

Blood collection:

Blood is collected as follows:

1. Blood sampling (venipuncture) must be performed without tourniquet, venestasis, and muscular hand activity. The use of a catheter is recommended. Once the catheter (Cathelon; Critikon, Tampa, Florida, USA) is installed, the first milliliter (1–2 volumes of the catheter) should be discarded before a blood sample is taken.
2. Blood volume: 0.6 ml–1 ml into a heparinized tube without gel.
3. The period of the blood collection is determined according to the diagnostic and therapeutic processes chosen, in accordance with the suspected disease (before meals, after meals, fasting state, fasting test, loading test, etc.). For example, when screening for mitochondrial cytopathy, redox has to be evaluated at the fed state or after a glucose loading test and after 10 h fast.
4. A report in the patient's charts and examination requests the exact time of sampling and delays after the last meals.
5. Avoid capillary blood
6. Choose venous or arterial blood and report which has been taken.

Sampling volume

Blood lactate, pyruvate, acetoacetate, 3-hydroxybutyrate, glucose, and REDOX ratio must be measured using the same sample for a reliable calculation of REDOX status.

Blood pre-treatment

Deproteinization using acids is necessary to avoid glycolysis but has to be practiced as soon as possible after sampling (within 5 min).

Different procedures are described. Perchloric acid is the most widely used agent: blood collected in a tube containing heparin is immediately deproteinized with a perchloric acid solution (1 mol/L) stored at + 2-8 °C (two volumes for one volume of blood).

Deproteinized samples may be frozen for up to 5 days.

Plasma lactate and 3OHB

Lactate and 3OHB can be measured in plasma if pyruvate to lactate ratio or 3OHB/ACAC ratio is not required. Blood is collected in a tube with an anticoagulant (heparin, EDTA, oxalate) associated with an anti-glycolytic agent (fluoride—in order to avoid an artefactual increase in lactate levels). The tube is transported in ice water and plasma is separated from blood within 15 min following sampling.

Lactate levels in plasma sampled according to the aforementioned sampling conditions are stable for up to 1 week at +2-8°C and for up to 1 month at −18-22 °C.

Lactate in CSF

CSF sampling is performed by lumbar puncture avoiding hemorrhagic puncture and hemolysis. No pre-treatment of the sample thus obtained is necessary [12, 13].

Specimen preparation procedure

Deproteinization

Deproteinization is carried out immediately after collection (<5 min) as follows:

1. Measure accurately 1 ml (or 0.5 ml) blood collected in a tube containing heparin, in a recipient containing 2 ml (or 1 ml) perchloric acid 1 M (which was measured accurately and refrigerated at +2-8 °C previously).
2. Agitate the blood/perchloric acid mixture thoroughly to obtain a brown coagulum.
3. Keep the tube frozen (−18-22 °C) for a period of less than 5 days.

Neutralization

1. After thawing, deproteinized samples are centrifuged (+2-8 °C) during 10 min at 3000 × *g*. Check and record the accuracy of the distributed volumes.
2. Mix 1 ml of the supernatant with 0.5 ml of phosphate buffer. If the volume of the supernatant available is <1 ml, keep the ratio: one volume of supernatant to half a volume of buffer.
3. Mix. Place the tubes for 10 min at −18-22 °C.
4. Check that the pH is 7 with pH paper (range 1–14).
5. If needed, adjust pH as follows:
   - If pH <7 add phosphate buffer M (50–100 µl)
   - If pH >7 add $HClO_4$ M (50–100 µl)

Record the added volume to take it into account for calculation.

6. Centrifuge for 5 min at 3500 × *g* at +2-8 °C.

The stability of lactate, pyruvate, 3OHB et ACAC supernatant is verified for 5 days at −18-22 °C.

19.4.2 Reagents and Chemicals

Lists of reagents and recipes of working solutions and reagents are provided in Tables 19.5 and 19.6, respectively. The lactate reagent is prepared according to the manufacturer's instructions. The

Table 19.5 List of reagents (*HBDH* hydroxybutyrate dehydrogenase, *LDH* lactate dehydrogenase, *3OHB* 3 hydroxybutyrate)

Name	Provider	Reference	Storage temperature
Perchloric acid 70–72%	Merck	100519	15–25 °C
Lactic acid	Sigma Supelco	L2250 PHR1215	2–8 °C 2–8 °C
ß NAD grade I	Sigma	N 1511	−18-22 °C
ß NADH grade II	Sigma	N 8129	−18-22 °C
Lactate PAP	Randox	LC 2389	2–8 °C
Phosphate tripotassic buffer	Sigma	04347	20–25 °C
Tris buffer	Merck	648314	15–30 °C
Paper pH 1–14	MN	902 04	4–30 °C
Acetoacetate powder	Sigma	A8509	-18-22 °C
D, L 3OHB powder	Sigma	H6501	2–8 °C
3 HBDH	Roche	3HBDB-RO	2–8 °C
LDH	Roche	LLDH-RO 11644793001	2–8 °C −18-220°C
NaHCO3	Merck Supelco	P8574 1.06329	2–8 °C 2–8 °C
Pyruvate powder	Sigma	P8574	2–8 °C

Table 19.6 Reagent preparation and working solutions. Lactate reagent was prepared according to the manufacturer's instructions

	ACAC	3OHB	Pyruvate
TRIS 0.1 mol/L buffer	*pH 6.98*: dissolve 3.0285 g of TRIS buffer into 200 ml distilled water. Adjust pH exactly to 6.98. Then, adjust the volume to 250 ml.	*pH 9.5*. dissolve 3.0285 g of TRIS buffer into 200 ml of distilled water. Adjust pH exactly to 9.5. Then, adjust the volume to 250 ml.	*pH 7.4*: dissolve 3.082 g of TRIS buffer into 200 ml distilled water. Adjust the pH with 1.2 M HCl. Then, adjust the volume to 250 ml.
Stability at +2-8 °C	1 month	1 month	1 month
β NADH reduced, stock solution, 0.6 mmol/L			Dissolve 0.85 mg of NADH into 2 ml $NaHCO_3$ 5% Stability 7 days at +2-8 °C
Working solutions: R1	*β NADH reduced solution at 0.4 mmol/L:* dissolve 2.8 mg of NADH into 10 ml of TRIS buffer (pH 6.98)	*Solution of NAD 12 mmol/L* dissolve 80 mg of NAD into 10 ml of TRIS buffer (pH 9.5)	*Solution of NAD 0.12 mmol/L:* dilute 1.2 ml of stock solution NADH into 4.8 ml TRIS buffer pH 7.4
R2	Mix: 750 µl of TRIS pH 6.98 + 250 µl HBDH	Mix: 750 µl TRIS buffer pH 9.5 + 250 µl HBDH	Mix: 80 µl of LDH solution (7500 U/ml) to 4720 µl of TRIS buffer pH 7.4.

reagent required for the deproteinization step is a solution of perchloric acid 1 mol/L (to dilute 8.6 ml of $HCLO_4$ 70% in 100 ml distilled water or 11 ml of $HCLO_4$ 60% in 100 ml distilled water) that required for the neutralization is phosphate tripotassic buffer 1 mol/L: add 5.3 g of phosphate tripotassic in 25 ml distilled water. This solution will be stable for 1 year at 25 °C.

19.4.3 Instrumentation and Calibration

An automated method can be run after validation using multiparametric open analyzers [6, 8, 9].

For example, KONELAB 30 (Thermo Fisher Clinical Scientific, Waltham, USA)

If there is no standard for pyruvate, ACAC, and 3OHB, the results can be calculated using the molecular extinction coefficient of NADH (6.3 $mmol^{-1}.cm^{-1}$).

A lactate standard is provided in the kit.

19.4.4 Analytical Procedure/Manual

Methods

The same deproteinized blood was used for lactate, pyruvate, 3OHB, and ACAC assays using enzymatic spectrophotometric measurement (Fig. 19.3) [7, 9].

Enzymatic methods using spectrophotometric measurement (Fig. 19.3). The enzymes involved are LDH for pyruvate, lactate oxidase for lactate, HBDH for acetoacetate, and 3-hydroxybutyrate.

Procedure

The operation procedure for this method is given in Table 19.7. Note that lactate is quantified according to the manufacturer's instructions.

19.4.5 Calculation

For the filtrate samples, results have to be recalculated according to the dilution ratio (e.g., × 4.5).

If acid-base or buffer was added to correct pH, an additional correction must be made according to the following equation: calculation has to be practiced

$$\frac{\text{observed result} \times \text{final volume}}{\text{initial volume}}$$

Table 19.7 Operating procedure

	Pyruvate	Acetoacetate	3-Hydroxybutyrate
Sample (neutralized supernatant)	60 µl	60 µl	40 µl
R1 (buffer + coenzyme)	100 µl	105 µl	120 µl
Mix thoroughly, incubate at 37 °C, measure absorbance at 340 nm A1			
R2 (enzyme) (µl)	20	12	10
Wait at 37 °C for (s)	300	450	600
Measure absorbance at 340 nm A2			
Calculation factor (µmol/L) taking into account the sample dilution (×4.5)	475 2142	466 2686	671 3019

Table 19.8 Preparation of quality control samples

	Lactate	Pyruvate	ACAC	3OHB
Stock solution	50 mmol/L	10 mmol/L	20 mmol/L	10 mmol/L
Dissolve into 10 ml albumin 50 g/l solution	0.048 g of lactate (sodium salt)	11 mg of Na pyruvate (Na salt)	25.5 mg acetoacetate	25.7 mg of D-L, 3-hydroxybutyrate
Stability	6 months at −18-22 °C	6 months at −18-22 °C	6 months at −18-22 °C	6 months at −18-22°C
Prepare three different level solutions				
Level 1	0,25 mmol/L	100 µmol/L	10 µmol/L	0.25 mmol/L
Level 2	0,5 mmol/L	200 µmol/L	50 µmol/L	0.5 mmol/L
Level 3	1 mmol/L	400 µmol/L	100 µmol/L	1 mmol/L
Stability	Do not store	Do not store	Do not store	Do not store
Target values and acceptable limits	0.25 mmol/L ± 0.030 0.5 mmol/L ± 0.075 1 mmol/L ± 0.15	100 µmol/L ± 15 200 µmol/L ± 30 400 µmol/L ± 60	10 µmol/L ± 1.5 50 µmol/L ± 7.5 100 µmol/L± 15	0.25 mmol/L ± 0.03 0.5 mmol/L ± 0.075 1 mmol/L ± 0.15

Table 19.9 Analytical performances of the methods [14]

	Lactate	Pyruvate	Acetoacetate	3-Hydroxybutyrate
Reproducibility within run (CV %) range (mmol/L)	<3 0.8–8	<3 0.1–0.5	<3 0.1–2	<3 0.4–5
Reproducibility run to run (CV %) range (mmolL)	<5 0.8–8	<5 0.1–0.5	<5 0.1–2	<5 0.4–5
Linearity limits (mmol/L)	0–15	0–15	0–15	0–9
Detection limit (mmol/L)	0.1	0.01	0.02	0.02

19.4.6 Quality Control

The Internal Quality Control samples can be prepared as described in Table 19.8. Target values are calculated according to the amount of weighed product and values obtained must be within ±15% of the target value. External Quality Assessment Scheme are achieved from ERNDIM (Quality assurance in laboratory testing for IEM (Inborn Error of Metabolism), The Netherland for Lactate and 3OHB.

19.4.7 Validation of the Methods

The results of method validation [14] are given in Table 19.9.

19.4.8 Post-analytical

Interpretation: Biological variation

Blood lactate values are usually lower than plasma level (−15%) depending upon hematocrit values.

Results are expressed as mmol/L.

Ketone body concentration depends on whether the patient is in a fed or fasted state and of its age. Very low concentrations are observed in the fed state, rising as the fasting period increases with variation according to age (e.g., KB elevations are higher in younger, more precocious children, see Fig. 19.4) [15]. In the fed state, KB levels are less than 0.2 mmol and the ratio 3-OHB/ACAC is less than 1. The blood concentration increases during fasting. The 3OHB/ACAC ratio increases with the concentration level during fasting (2.5–3.5).

Plasmatic values of 3OHB are 10–20% higher than blood concentration levels. No difference has been observed between venous and arterial blood in this regard.

19.5 Reference Values

Reference values are given according to age, fed state, fasting state, diet, and nutritional status (Table 19.10).

Figure 19.4 shows the evolution of KB level as a function of fasting time and age [9, 16]. The reference values for lactate/creatinine.

URINE: Lactate to creatinine ratio <0.2 mmol/mmol creatinine

Physiological value of ketonuria is less than 2 mmol/24 h.

CSF [12, 13]

Lactate: 1.1 to 2.2 mmol/L

Pyruvate: 0.05 to 0.15 mmol/L

L/P ratio: 11 to 15

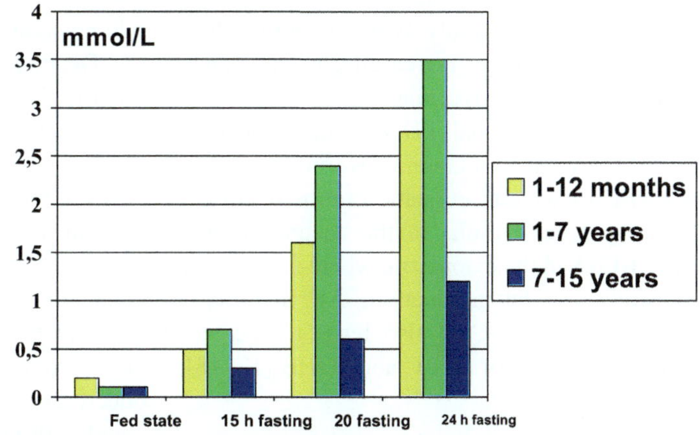

Fig. 19.4 Ketone bodies according to fasting time and age

Table 19.10 Reference values for blood lactate, pyruvate, acetoacetate, and 3-hydroxybutyrate according to age, fed state, fasting state

	Lactate (L) (mmol/L)	Pyruvate (P) (mmol/L)	Ratio L/P	3-OH-butyrate (mmol/L)	ACAC (mmol/L)	Ketone bodies (mmol/L)	Rapport 3-OHB/AA	Non-esterified free fatty acid (mmol/L)
Children (0–1 year)								
Fed state, 1 h after meals	0.6–2.2	0.04–0.14	6–14	0.10–0.20	0.10–0.25	0.10–0.30	<1	0.02–0.4
Children (1–7 years)								
Fasting 10 h	0.7–1.8	0.09–0.17	6–14	0.02–0.3	0.04–0.2	0.02–0.6	<2.5	0.2–1.2
Fed state	0.9–1.8	0.08–0.17	6–14	0.02–0.1	0.04–0.13	0.02–0.2	<1	0.02–0.4
Children (7–15 years) and adults								
Fasting 10 h	0.7–0.9	0.04–0.12	6–14	0.02–0.3	<0.2	0.1–0.4	0.4–2.3	0.2–1.2
Fed state	1.0–1.55	0.08–0.16	6–14	0.02–0.1	0.04–0.13	<0.20	<1	0.02–0.4

19.6 Pathological Values

Respiratory Chain Abnormalities

Data from a patient affected with mitochondrial respiratory chain abnormalities (multiple complexes I, III, IV defect) are given in Table 19.11. Permanent hyperlactatemia is associated with an increased lactate/pyruvate ratio and postprandial ketosis with an increased 3OHB/AA ratio. Pyruvate, the product of glycolysis, is metabolized in the mitochondria to acetyl-coenzyme A by pyruvate dehydrogenase (PDH). Pyruvate can be reduced to lactate by LDH or may be utilized for gluconeogenesis. Mitochondrial respiratory chain dysfunction results in decreased acetyl-CoA metabolism. Pyruvate metabolism shifts toward other metabolic routes, including reduction to lactate and gluconeogenesis. Lactate cannot be cleared as rapidly as it is being produced resulting in acidosis. Increased gluconeogenesis can result in hyperglycemia.

A fully functional PDH complex leads to acetyl-CoA accumulation. Overproduction of acetyl-CoA without utilization in the respiratory chain complex, results in the accumulation of acetyl-CoA in the cytoplasm, where it serves as a substrate for fat production. An inability to metabolize acetyl-CoA also leads to increased circulating levels of ACAC and 3OHB [17, 18].

Table 19.11 Data from a patient affected with a complex I, III, and IV deficiency

	Lunch		Dinner	
	Before meals	After meals	Before meals	After meals
Lactate (mmol/L)	6	8.5	5	8.9
Pyruvate (mmol/L)	0.24	0.31	0.18	0.28
L/P	24	27.5	28	32
Acetoacetate (mmol/L)	0.22	0.14	0.16	0.20
3-OH butyrate (mmol/L)	1.53	0.90	1.10	1.27
Ketone bodies (mmol/L)	1.75	1.04	1.26	1.45
3OHB/AA	7	6.4	6.9	6.4
Glycemia (mmol/L)	5	5.8	3.3	5.2

PC Defect

In the initial step of gluconeogenesis, pyruvate is carboxylated to oxaloacetate by PC (Fig. 19.2). Two forms of PC deficiency are observed [1]. Patients with type B disease develop symptoms in the neonatal period, including severe lactic acidemia, hypotonia, seizures, failure to thrive, psychomotor retardation, and hepatomegaly.

PC deficiency develops symptoms in the neonatal period (type B) with severe lactic academia, hypotonia, seizures, failure to thrive, psychomotor retardation, and hepatomegaly. The lactate: pyruvate ratio is increased due to a deficiency in aspartate, which is involved in maintaining the mitochondrial redox status. Hypoglycemia is mild and helps to distinguish this condition from type I glycogen storage disease (glucose-6-phosphatase deficiency), which also produces lactic acidosis. Most patients with type B disease do not synthesize any PC protein.

Death usually occurs within the first 6 months of life. Type A disease manifests as less severe lactic acidemia (3–6 mmol/L) with acute metabolic decompensation usually observed in association with an illness or fasting. Those patients demonstrate low muscle tone and developmental delays, mental retardation with cerebral atrophy and abnormalities of brain myelinization.

Metabolically, the acetyl-CoA generated is diverted to ketogenesis and urea cycle activity is decreased, leading to hyperammonemia associated with fasting hypoglycemia, increased lactatemia associated with increased lactate: pyruvate ratio, and increased ketonemia with a 3OHB/AA ratio <1. Data from a patient affected by a PC defect are presented in Table 19.12.

PDH Defect

Table 19.12 Fed and fasting state for a patient affected with PC defect

	Before meals	After meals
Lactate (mmol/L)	22	11
Pyruvate (mmol/L)	0.55	0.22
L/P	40	50
ACAC (mmol/L)	0.80	0.50
3OHB (mmol/L)	0.35	0.25
KB (mmol/L)	1.15	0.80
3OHB/AA	0.4	0.50
Glucose (mmol/L)	2.4	5.6

Table 19.13 Data from a patient affected with a PDH defect

	Breakfast		Lunch	
	Before meals	After meals	Before meals	After meals
Lactate (mmol/L)	9.3	13	8.8	10
Pyruvate (mmolLl)	1.07	1.5	0.66	0.9
L/P	8.7	8.7	13	11
Acetoacetate (mmol/L)	0.07	0.03	0.04	0.06
3-OH butyrate (mmol/L)	<0.02	<0.02	<0.02	<0.02
Ketone bodies (mmol/L)	<0.20	<0.20	<0.20	<0.20
3OHB/AA	<1	<1	<1	<1

Data from a patient affected with a PDH defect are presented in Table 19.13. In general, hyperlactatemia (increasing with meals) associated with a normal lactate: pyruvate ratio and normal ketonemia are observed [1]. The ketogenic diet is a rational treatment for PDH complex deficiency.

Fatty Acid Mitochondrial Oxidation Defect (Beta-Oxidation Defect)

Hypoketotic hypoglycemia was observed in a patient presenting with hypotrophy and neonatal hypoglycemia. He was hospitalized at 6 months of age for anorexia, hypotonia, progressive asthenia, and failure to thrive [1, 19, 20].

Associated findings included liver insufficiency, icterus, and hepatomegaly. Metabolic evaluation of lactate, pyruvate, ACAC, and 3OHB associated with non-esterified fatty acids (Table 19.14) resulted in an evaluation of a fatty acid oxidation defect. Characterization of plasma acylcarnitines and oxidation of long-chain fatty acids (myristic and palmitic acids) in cultured fibroblasts demonstrated a long-chain 3-hydroxyacyl-coenzyme A dehydrogenase defect (LCHAD).

Ketolysis disorder:

Succinyl-CoA:3-ketoacid CoA transferase (SCOT) defect:

SCOT catalyzes the transformation of ACAC into acetoacetyl-CoA in the mitochondria of extra-hepatic tissues. This enzyme defect may be suggested in cases of severe ketoacidosis often associated with neurologic dysfunction [20].

Table 19.14 Data from a patient affected with a long-chain 3-hydroxyacyl-coenzyme A dehydrogenase defect (LCHAD). *NEFA* non-esterified fatty acids

	Before meals	1 h after	3 h after	4 h after
Lactate (mmol/L)	1.5	2.2	1.5	2.3
Pyruvate (mmol/L)	0.15	0.18	0.15	0.18
L/P	10	12	10	13
Acetoacetate (mmol/L)	0.06	0.07	0.07	0.05
3-OH butyrate (mmol/L)	0.05	0.02	0.04	0.04
Ketone bodies (mmol/L)	<0.20	<0.20	**<0.20**	**<0.20**
3OHB/AA	<1	<1	<1	<1
UEFA (mmol/L)	1.0	0.07	**1.20**	**3.50**
Glucose (mmol/L)	4	5.8	**2.5**	**2.4**

Bold values are typical of the pathology

Table 19.15 Data from a patient affected with a succinyl Coenzyme A: oxo acid transferase deficiency

	Fed state	20 h fasting
Lactate (mmol/L)	1.5	1.3
Pyruvate (mmol/L)	0.15	0.11
L/P	10	12
Acetoacetate (mmol/L)	0.25	**4.2**
3-OH butyrate (mmol/L)	0.20	**6.0**
Ketone bodies (mmol/L)	0.45	**10**
3OHB/AA	<1	1.4
UEFA (mmol/L)	0.10	3.0
Glucose (mmol/L)	4.5	3.0

Bold values are typical of the pathology

The data presented in Table 19.15 are derived from a 2-year-old child in the fed and fasting states and document the results of a metabolic evaluation of this patient. Permanent hyperketonemia increased with fasting contrasting with non-esterified fatty acids which are lower in comparison to the KB concentration, suggesting a defect in utilizing KBs in the periphery, not overproduction of KB from NEFA.

Pitfalls

Pre-analytical pitfalls [21, 22]

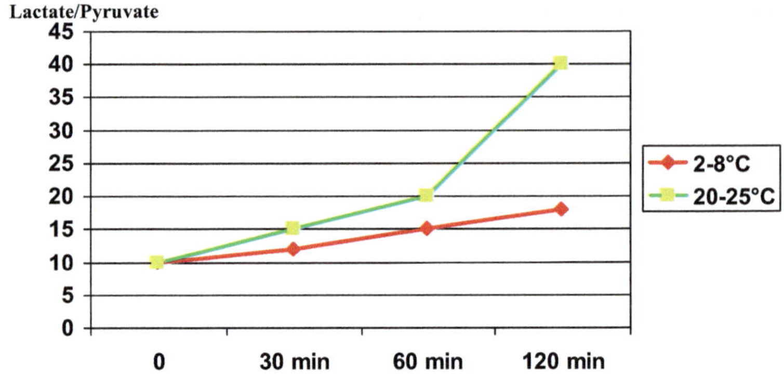

Fig. 19.5 Blood storage temperature and delay

The main sources of variation are linked to the time of collection, treatment, and sample transport and **storage**.

Deproteinization

1. An increased lactate: Pyruvate ratio is observed when the delay between the moment of blood collection and deproteinization exceeds 5 min (Fig. 19.5).
2. An inappropriate volume of the deproteinized sample will lead to errors in the measurement of L/P ratio (Fig. 19.5).
3. An inappropriate pH of the filtrate may reflect a non-uniform dilution ratio between the blood sample volume and $HClO_4$ volume or can be due to errors in $HClO_4$ molarity or instability due to inappropriate storage of the acid.
4. Lactate and pyruvate, 3OHB, and ACAC must be measured using the same deproteinized sample. If not, the lactate: pyruvate ratio may be incorrect.

Sample Stability

Pyruvate and acetoacetate in blood are not stable at 20–25 °C [10, 11].

19.7 Conclusion

Measurement of lactate, pyruvate, 3OHB, and ACAC in the blood is a very useful tool in order to identify metabolic profiles specific or highly suggestive of a given disorder for orientation of the diagnosis. Conditions for the preparation of the patient, collection of the samples, and technical specifications are essential for the reliability of the interpretation of the results.

References

1. Poggi-Travert F, Martin D, Billette de Villemeur T, Bonnefont JP, Vassault A, Rabier D, Charpentier C, Kamoun P, Munnich A, Saudubray JM (1996) Metabolic intermediates in lactic acidosis: compounds, samples and interpretation. J Inherit Metab Dis 19(4):478–488
2. Mitchell GA, Kassovska-Bratinova S, Boukaftane Y, Robert MF, Wang SP, Ashmarina L, Lambert M, Lapierre P, Potier E (1995) Medical aspects of ketone body metabolism. Clin Invest Med 18(3):193–216
3. Bonnefont JP, Saudubray JM, Vassault A (1991) Dosage des acides lactique, pyruvique et des corps cétoniques. Application au diagnostic des hyprerlactacidémies et des états d'acidocétose chez l'enfant. In Progrès en Pédiatrie, J.M. Saudubray, Maladies métaboliques. Doin editeurs, Paris, pp 29–40
4. Duke T (1999) Dysoxia and lactate. Arch Dis Child 81:343–350
5. Lu J, Pulsipher BS, Grenache DG (2015) Development of an enzymatic assay to measure lactate in perchloric acid-precipitated whole blood. Clin Chim Acta 444:208–211
6. Artuch R, Vilaseca MA, Farre C, Ramon F (1995) Determination of lactate, pyruvate, beta-hydroxybutyrate and acetoacetate with a centrifugal analyser. Eur J Clin Chem Clin Biochem 33:529–533
7. Kientsch-Engel RI, Siess EA (1981) D(-)-3-hydroxybutyrate and acetoacetate. Methods of enzymatic analysis Bergmeyer, vol VIII, 3rd edn, pp 60–69
8. Li PK, Lee JT, Mac Gillivray MH, Schaefer PA, Siegel JH (1980) Direct, fixed-time kinetic assays for beta-hydroxybutyrate and acetoacetate with a centrifugal analyser or a computer-backed spectrophotometer. Clin Chem 26:1713–1717

9. Vassault A, Bonnefont JP, Specola N, Saudubray JM (1991) Lactate, pyruvate and ketone bodies. In: Hommes FA (ed) Techniques in diagnostic human biochemical genetics, vol XIX. Wiley-Liss, pp 285–308
10. Laun RA, Rapsch B, Abel W, Schröder O, Röher H-D, Ekkernkamp A, Schulte K-M (2001) The determination of ketone bodies: preanalytical, analytical and physiological considerations. Clin Exp Med 1:201–209
11. Wolff F, El Khattabi C, Bourdon F, Willems D (2009) Comparison of high-performance liquid chromatography with fluorescence detection versus enzymatic assay to measure blood pyruvate in clinical practice. Clin Biochem 42:1099–1103
12. Hutchesson A, Preece MA, Gray G, Green A (1997) Measurement of lactate in cerebrospinal fluid in investigation of inherited metabolic disease. Clin Chem 43:158–161
13. Benoist JF, Alberti C, Leclercq S, Rigal O, Jean-Luois R, Ogier de Baulny H, Porquet D, Biou D (2003) Cerebrospinal fluid lactate and pyruvate concentration and their ratio in children: age-related reference intervals. Clin Chem 49:487–494
14. Boulet L, Van Noolen L, Bosson C, Faure P, Corne C (2014) Mise au point et validation en portée B selon les recommandations du Cofrac d'un dosage de pyruvate et de corps cétoniques sur un automate ouvert. Ann Biol Clin 72(3):312–320
15. Francois B, Colomb V, Bonnefont JP, Goulet O, Benhariz M, Vassault A, Rabier D, Ricour C (1997) Tolerance to starvation in children on long-term total parenteral nutrition. Clin Nutr 16:113–117
16. Bonnefont JP, Specola NB, Vassault A, Lombes A, Ogier H, de Klerk JB, Munnich A, Coude M, Paturneau-Jouas M, Saudubray JM (1990) The fasting test in paediatrics: application to the diagnosis of pathological hypo- and hyperketotic states. Eur J Pediatr 150(2):80–85
17. Garcia Silva MT, Bonnefont JP, Rotig A, Romero N, Vassault A, Colonna M, Coude M, Rabier D, Munnich A, Fardeau M et al (1989) Respiratory chain diseases in infancy. Clinical presentation and diagnosis. An Esp Pediatr 31(5):421–430
18. Munnich A, Rustin P, Rotig A, Chretien D, Bonnefont JP, Nuttin C, Cormier V, Vassault A, Parvy P, Bardet J et al (1992) Clinical aspects of mitochondrial disorders. J Inherit Metab Dis. 15(4):448–455
19. Saudubray JM, Mitchell G, Bonnefont JP, Schwartz G, Nuttin C, Munnich A, Brivet M, Vassault A, Demaugre F, Rabier D et al (1992) Approach to the patient with a fatty acid oxidation disorder. Prog Clin Biol Res 375:271–288
20. Saudubray JM, Specola N, Middleton B, Lombes A, Bonnefont JP, Jakobs C, Vassault A, Charpentier C, Day R (1987) Hyperketotic states due to inherited defects of ketolysis. Enzyme 38(1–4):80–90
21. Carragher FM, Bonham JR, Smith JM (2003) Pitfalls in the measurement of some intermediary metabolites. Ann Clin Biochem 40:313–320
22. Payne B (2004) Pitfalls in the measurement of some intermediary metabolites: stabilization of lactate and pyruvate. Ann Clin Biochem 41:83

GABA, GHB, and Homocarnosine

20

Erland Arning, Erwin E. W. Jansen, Cornelis Jakobs, and Brandi Wasek

Contents

List of Abbreviations

aCSF	Artificial cerebrospinal fluid
CSF	Cerebrospinal fluid
EDTA	Ethylenediaminetetraacetic acid
GABA	γ-Aminobutyric acid
GABA-T	GABA-transaminase
GAD	Glutamic acid decarboxylase
GC-MS	Gas chromatography-mass spectrometry
GHB	γ-Hydroxybutyric acid
HCl	Hydrochloric acid
LC-MS/MS	Liquid chromatography-tandem mass spectrometry
MRM	Multiple response monitoring
SSADH	Succinic semialdehyde dehydrogenase deficiency

E. Arning (✉) · B. Wasek
Institute of Metabolic Disease, Baylor Scott & White Research Institute, Dallas, TX, USA
e-mail: erland.arning@bswhealth.org

E. E. W. Jansen · C. Jakobs
Clinical Chemistry, Metabolic Unit, VU University Medical Center, Amsterdam, Netherlands

20.1 Introduction

γ-Aminobutyric acid (GABA) and homocarnosine (γ-aminobutyryl-L-histidine) are bioactive amines while γ-hydroxybutyric acid (GHB), a short-chain fatty acid, all have diverse roles in intermediary metabolism and central neurotransmission. GABA, a primary inhibitory neurotransmitter in the brain, is produced by the pyridoxine-dependent enzyme glutamic acid decarboxylase (GAD) from the excitatory amino acid glutamate. Catabolism of GABA requires two enzymes, GABA-transaminase (GABA-T) and succinate semialdehyde dehydrogenase (SSADH) [1]. GABA is also present in non-neuronal tissues and may exist in peripheral and central tissues as a dipeptide conjugated form as homocarnosine, homoanserine, GABA-lysine, and possibly other unknown GABA conjugates. GAD is the rate-limiting enzyme for the synthesis of GABA and is present throughout the central nervous system and select peripheral cells. Three inherited disorders of GABA metabolism have been identified; these are

N. Blau, F. M. Vaz (eds.), *Laboratory Guide to the Methods in Biochemical Genetics*,
https://doi.org/10.1007/978-3-031-58819-8_20

SSADH deficiency, GABA-T deficiency, and homocarnosinosis [1]. GABA-T deficiency and homocarnosinosis are extremely rare, however, these disorders may be underdiagnosed due to the requirement of CSF determination of free and total GABA and homocarnosine, respectively. Whereas SSADH deficiency, can be detected in routine urinary organic acids due to elevated GHB. The clinical presentation of these cases varies considerably and may include psychomotor retardation, convulsions, ataxia, hypotonia, hyperflexia, and delayed speech development.

GHB is a short-chain fatty acid which is derived from the catabolism of GABA. Accumulation of GHB is the hallmark of SSADH deficiency, which is a rare autosomal recessive disorder with pathological variants in the gene aldehyde dehydrogenase 5 family member A1 (*ALDH5A1)* variants [2]. Currently, there are approximately 450 cases of SSADH deficiency reported in the literature [3]. Normal neuronal disposal of GABA begins with transamination to succinic semialdehyde by GABA-T, which is subsequently converted to succinic by SSADH. Since GHB's discovery, several pharmaceutical uses of GHB have been approved in the USA and internationally for narcolepsy and alcohol withdrawal [4–6]. Despite the therapeutic uses of GHB, it has been dominated by a high prevalence of abuse. Illicit abuse of GHB is predominantly used for decreased inhibition, growth hormone release, and subject sedation [4]. However, these effects carry many adverse effects such as respiratory depression, hypothermia, coma, and death [4]. Unlike persistent accumulation of GHB in SSADH deficiency, detection of the presence of GHB due to illicit administration can be challenging due to its rapid metabolism and short half-life.

Homocarnosine is a brain-specific dipeptide composed of histidine and GABA, that is synthesized by carnosine synthetase in glial cells and hydrolyzed by carnosinase. In some brain regions, concentrations exceed 1 mM, and there is evidence that specific neuronal tracts employ homocarnosine as a storage form of GABA [1]. The specific function of homocarnosine in the brain has been widely explored, and the predominance of data suggests a role as an osmoregulatory or a neuroprotective antioxidant [7, 8]. Homocarnosinosis is a very rare disorder characterized by the accumulation of homocarnosine in the brain in three siblings of only one family [9]. The phenotypes related to homocarnosinosis can vary but can include intellectual disability, spastic paraplegia, and abnormal pigmentation of the skin and/or retina.

20.2 Analyte Properties and Formulae

Metabolites discussed in this chapter can be quantitated underivatized or derivatized due to the bifunctional nature of GABA and homocarnosine having both amine and carboxyl moieties. The physiological matrix of choice for GABA and homocarnosine is CSF, although these analytes can be determined in various other biological fluids and tissues. Determination of GHB is likely better suited for plasma or dried blood spots for screening purposes for SSADH deficiency. Free and total GABA are significantly elevated in both GABA-T and SSADH deficiencies, therefore making quantitation in CSF diagnostically relevant [10–12] (Fig. 20.1).

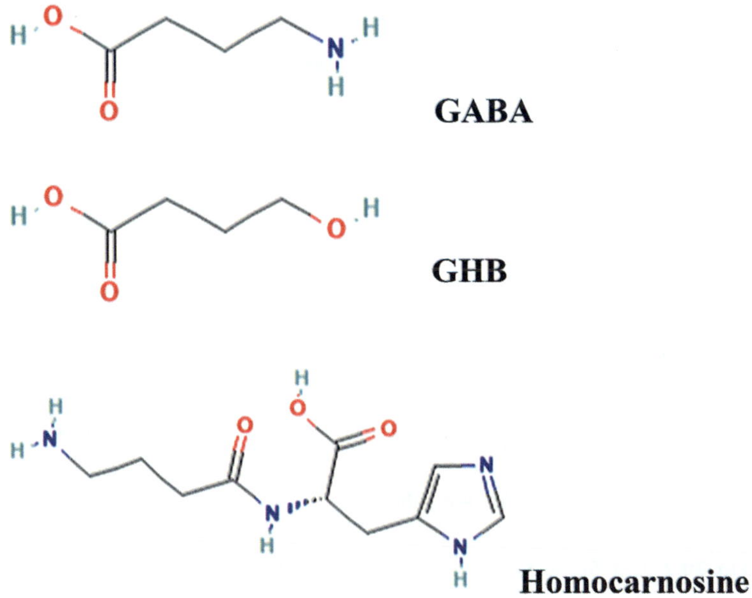

Fig. 20.1 Structures of the metabolites: GABA, GHB, and homocarnosine

20.3 Methods

20.3.1 Quantitation of GABA in CSF

Principle

GABA in CSF is present as a free and peptide-linked form. There are numerous free and total GABA methods that have been published utilizing various separation and detection techniques HPLC-fluorescence [13] HPLC-electrochemical detection [14], GC-MS [15], CE-MS [16], and LC-MS [17]. Analysis of free GABA is performed by simple stable isotope dilution. Quantitation of total GABA requires hydrolysis of peptide-linked forms of GABA to yield free amino acids. Both free and total GABA are separated by liquid chromatography. Quantitation utilizes multiple response monitoring (MRM) for GABA (qualifier and quantifier) and $^{2}H_{2}$-GABA (quantifier). This LC-MS/MS method for quantitation of free and total GABA described below was compared with the GC-MS method by Struys EA [15] through specimen exchange in 2009. Excellent correlation was observed for both free and total GABA (Fig. 20.2) between the methods.

20.3.1.1 Pre-analytical

Specimen

Lumbar CSF specimens drawn at any time during the day will be acceptable and no additional patient preparation is required. Optimal volume of 1.0 mL (minimum 0.25 mL) CSF should be collected in a regular sterile CSF collection tube. Due to the rostrocaudal gradient for CSF, the first 2 mL that emerges from the spinal tap is not suitable for GABA analysis. Collected CSF should be clear and immediately frozen at the bedside on dry ice. If CSF is blood contaminated, the specimen should be placed on wet ice and centrifuged within 5 min of collection. Transfer the clear CSF to another vial and freeze on dry ice. Store CSF at −80 °C until the time of testing.

Reagents and Chemicals

1. Hydrochloric acid (HCL) 36.5–38%—Sigma-Aldrich H1758
2. Formic acid ≥98%—Sigma-Aldrich 33015
3. Heptafluorobutyric acid ≥99.5%—Supelco 52411

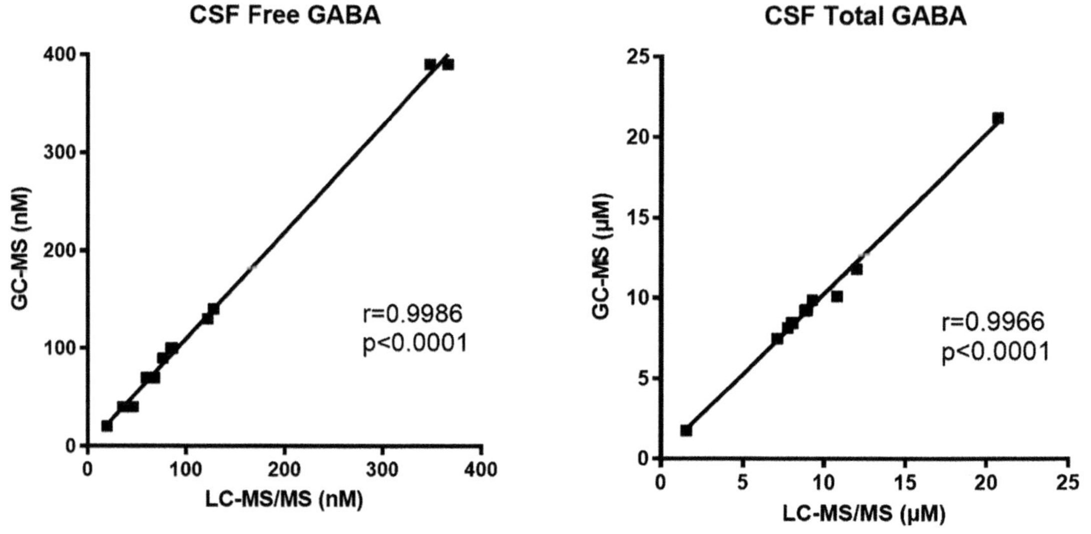

Fig. 20.2 CSF free and total GABA method correlation between LC-MS/MS and GC-MS (n = 12)

4. Water (Optima LC/MS grade)—Fisher Scientific W6-4
5. Acetonitrile (Optima LC-MS grade) >99.9%—Fisher Scientific A955-4
6. Methanol (HiPerSolv CHROMANORM) ≥99.5%—VWR BDH85800
7. Sodium chloride BioUltra ≥99.5%—Sigma-Aldrich 71376
8. Potassium chloride puriss ≥99.5%—Sigma-Aldrich 60130
9. Calcium chloride dihydrate BioUltra ≥99.5%—Sigma-Aldrich 21097
10. Magnesium chloride hexahydrate BioXtra ≥99.0%—Sigma-Aldrich M2670
11. Sodium phosphate dibasic BioUltra ≥99.5%—Sigma-Aldrich 71636
12. Sodium phosphate monobasic monohydrate BioUltra ≥99.0%—Sigma-Aldrich 71505
13. Mobile phases
    (a) Mobile phase A (1 L): 0.5% formic acid/0.25% heptafluorobutyric acid in LC-MS grade water
    (b) Mobile phase B (1 L): 0.5% formic acid/0.25% heptafluorobutyric acid in LC-MS grade acetonitrile
14. Autosampler wash solution (1 L): 10% methanol in LC-MS grade water
15. Deproteinizing solution (100 mL): 90% acetonitrile/0.1% formic acid/9.9% LC-MS grade in water
16. GABA BioXtra ≥99.0%—Sigma-Aldrich A5835 (Stock standard 1 mM in 0.1 M HCl)
17. $^{2}H_2$-GABA 98 atom% ^{2}H—CDN Isotopes D-1731 (Stock internal standard: 1 mM in 0.1 M HCl)
18. Working internal standard solutions
    (a) Free GABA $^{2}H_2$-GABA: 2 μM in LC-MS grade water
    (b) Total GABA $^{2}H_2$-GABA: 40 μM in LC-MS grade water
19. Artificial CSF (aCSF)
    (a) 2X Solution A: 296 mM Sodium chloride, 6 mM Potassium chloride, 2.8 mM Calcium chloride dihydrate, and 1.6 mM Magnesium chloride hexahydrate in LC-MS grade water.
    (b) 2X Solution B: 2.4 mM Sodium phosphate dibasic and 0.4 mM Sodium phosphate monobasic monohydrate in LC-MS grade water.
    (c) 1X aCSF: 1:1 (v:v) 2X solution A : 2X solution B.

Instrumentation

1. Mass spectrometer: Sciex 4000 QTRAP, operating in the negative chemical ionization mode
2. Liquid chromatography: Shimadzu Prominence liquid chromatograph system
3. Computer: Sciex Analyst software version 1.6.2
4. Analytical Column: Phenomenex EZfaast, 3 μm, 250 × 2 mm
5. Guard Column: Phenomenex Security Guard, 5 μm, 4 × 3 mm
6. Boiling Water Bath
7. Metal screw-top rack for 1.5 mL tubes

Calibration

Calibration curves are established and quantified for free and total GABA for each batch of samples. The calibration curve for free GABA is prepared in 1X aCSF: 25 nM, 50 nM, 100 nM, 200 nM, and 400 nM. The calibration curve for total GABA is prepared in 1X aCSF: 2.5 μM, 5 μM, 10 μM, 20 μM, and 40 μM. The lower limit of quantitation is 6 nM and 0.6 μM for free and total GABA, respectively. The upper limit of quantitation is 1000 nM and 80 μM free and total GABA, respectively.

Quality Control

CSF pools are used to prepare low- and high-quality controls (QC) for free and total GABA, which are analyzed at the beginning and end of every batch of samples. Free GABA target values for low control is 60–100 nM and high control target is 300–400 nM. Total GABA target values for low control is 4–8 μM and high control target is 30–40 μM. A blank (1X aCSF) is also included at the beginning of the assay and after the high QC of each batch of samples.

20.3.1.2 Analytical

Procedure

The procedure is fully described by Arning E et al. [18]. Bring standard, internal standard, and QC to room temperature. Low and High QC is placed at the beginning of the batch and end of the batch. Detailed sample preparation is described below:

- Sample Preparation-Free GABA (CSF)
  1. To labeled 1.5 mL microcentrifuge tubes, pipette 50 μL sample (calibrators, QC, patient CSF, or blank).
  2. Add 50 μL of Free $^2H_2$-GABA internal standard working solution.
  3. Cap and vigorously vortex for 3 s.
  4. Transfer 90 μL of each prepared sample into a 96-well microtiter plate and cover it with a silicone cover.
  5. Place the completed 96-well microtiter plate onto a refrigerated autosampler (4 °C).
  6. Inject 12 μL of sample onto LC-MS/MS. Representative LC-MS/MS ion chromatograms for free GABA and I.S. are shown in Fig. 20.3.

- Sample Preparation-Total GABA (CSF)
  1. To labeled 1.5 mL microcentrifuge tubes, pipette 50 μL sample (calibrators, QC, patient CSF, or blank).
  2. Add 50 μL of Total $^2H_2$-GABA internal standard working solution.
  3. Add 200 μL 6N HCl.
  4. Cap and vigorously vortex tubes for 3 s.
  5. Place tubes in metal screw-top rack and place in boiling water bath for 4 h.
  6. After boiling, remove samples and allow them to reach room temperature (18–24 °C).
  7. To new labeled 1.5 mL microcentrifuge tubes, add 180 μL deproteinizing solution.
  8. Transfer 20 μL of hydrolyzed (boiled) sample to a 1.5-mL microcentrifuge tube containing deproteinizing solution and mix well by vortex.
  9. Transfer 160 μL of the prepared sample into corresponding work list position in a 96-well microtiter plate and cover it with a silicone cover.
  10. Place completed 96-well microtiter plate onto a refrigerated autosampler (4 °C).
  11. Inject 10 μL of sample onto LC-MS/MS. Representative LC-MS/MS ion chromatograms for total GABA and I.S. are shown in Fig. 20.3.

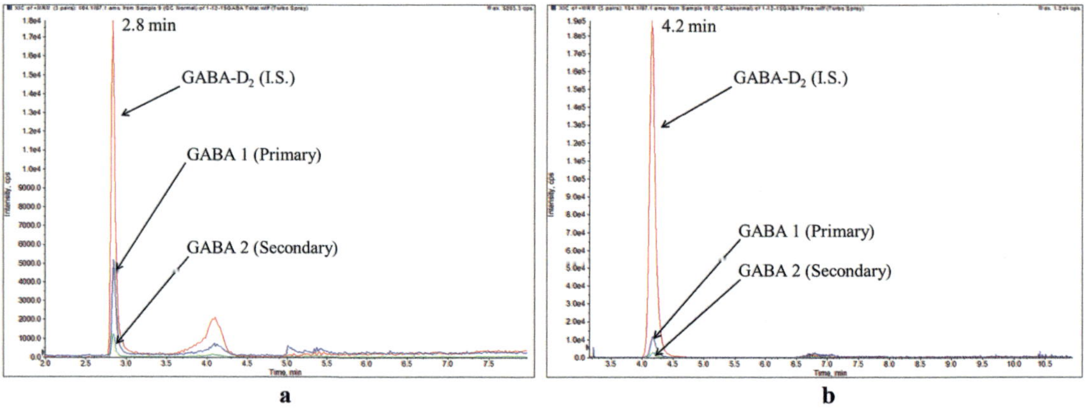

Fig. 20.3 Representative LC-MS/MS ion chromatogram (**a**) free GABA 1 (quant m/z 104.1 > 87.1), GABA 2 (qual m/z 104.1 > 69.1), $^{2}H_{2}$-GABA (m/z 106.0 > 89.1), free GABA concentration shown is 169 nM. (**b**) Total GABA concentration is shown as 8.8 μM

- LC-MS/MS Methodology

Quantitation of free and total GABA was performed on a Shimadzu Prominence liquid chromatograph system coupled to a Sciex 4000QTRAP® with Analyst software version 1.6.2. Chromatographic separation of GABA was achieved on a Phenomenex EZfaast, 3 μm, 250 × 2 mm. Specific LC method parameters are shown in Table 20.1. Mass spectrometry conditions were as follows: Entrance potential 10 V, curtain gas 10 psi, CAD gas-medium, Ion spray 5500 V, Source temp 600 °C, GS1 40 psi, GS2 40 psi, and Q1 and Q3 ran at Unit resolution.

Table 20.1 LC parameters for (a) free and (b) total GABA

(a) HPLC (Free GABA)[a]		
Column temp.	40°C	
Flow rate	0.230 mL/min	
Gradient	Time (min)	Mobile Phase A (%)
	0.0	90
	5.0	25
	5.1	0
	6.0	0
	6.1	90
	10.0	Stop
(b) HPLC (Total GABA)[a]		
Column temp.	40°C	
Flow rate	0.230 mL/min	
Gradient	Time (min)	Mobile Phase A (%)
	0.0	90
	4.0	40
	4.1	90
	8.0	Stop

[a]Optimized for Shimadzu Prominence liquid chromatography system equipped with Phenomenex Ezfasst, 3 μm, 250 × 2 mm analytical column; Mobile phase A: 0.5% formic acid-0.25% heptafluorobutyric acid in water; Mobile phase B: 0.5% formic acid-0.25% heptafluorobutyric acid in methanol

Calculation

Standard curves are generated based on linear regression of the analyte/internal standard peak-area ratio (y) versus analyte concentration (x) using the primary ions indicated in Table 20.2. The standard curves are weighted 1/x.

20.3.1.3 Post-analytical Quality Assurance

1. Acceptability of each run is confirmed if QC is within previously defined limits using Westgard rules.
2. Confirmatory analysis based on the ratio of two MRM transitions per assay; Free GABA 2 or Total GABA 2(qualifier)/Free GABA 1 or Total GABA 1 (quantifier) should be ±20% of the established value for the assay based upon the calibration curve.
3. Column performance is ensured by monitoring retention times of the beginning and end QCs, acceptance is <5% variation.

Table 20.2 Free and total GABA MRM settings

	MRM transition					
Compound	Q1 (m/z)	Q3 (m/z)	Dwell time (ms)	DP (V)	CE (V)	CXP (V)
GABA 1	104.1[a]	87.1[a,b]	300	36	15	6
GABA 2	104.1[a]	69.1[a,c]	300	36	23	4
GABA-D_2	106.0[a]	89.1[a,b]	300	26	15	6

[a]Optimized m/z may change based on tuning parameters and instrument used
[b]Primary ions for GABA quantification
[c]Secondary ion used for MRM ratio confirmation

4. Ion enhancement or suppression is monitored by comparing internal standard intensities of the QC and patient samples to the mean of the internal standard intensities of the calibration curves to a previously defined acceptable criteria range.

Interpretation
Calculated concentrations of free and total GABA are compared with the reference range values. Elevations of GABA, β-alanine, and homocarnosine may suggest a defect in GABA-T. Low CSF-free and total GABA may be attributed to febrile seizures or stiff person syndrome [19, 20]. CSF GABA results should be interpreted in context with patients' clinical symptoms and medications. Vigabatrin, topiramate, valproic acid and other anti-seizure medications are associated with increasing CSF GABA [21].

Reference Values
Age-specific reference ranges were adopted from previously established ranges published by the reference laboratory used for correlation validation studies. No significant gender difference is observed in CSF free and total GABA. See Table 20.3 for age-specific reference range [15].

Pitfalls
Approximately 98% of GABA in CSF is present in the conjugated form through peptide bonds with other amino acids. CSF specimen collection, storage, and thawing require rigorous efforts to prevent bound GABA from being converted into free GABA, which will yield false-positive values for free GABA. CSF should be immediately frozen after collection and stored frozen until time of analysis to obtain reliable free GABA results.

Table 20.3 Age-specific reference range for CSF GABA (total and free)

Age	Free GABA (nM)	Total GABA (μM)
<2 years	17–67	4.2–13.4
>2 years	32–170	3.3–12.2

20.3.2 Quantitation of GHB in Plasma

Principle
GHB is a short-chain fatty acid that is a metabolite of the degradation of the inhibitory neurometabolite GABA. The LC-MS/MS method presented here quantitates GHB in the free-underivatized form. The method utilizes simple stable isotope dilution followed by ultrafiltration for the removal of proteins. Chromatographic separation is achieved by UPLC on a T3 HSS 100 × 2.1 mm, 1.8 μm with a total analysis time of 4 min. GHB and $^2H_6$-GHB are quantified by MRM in negative-ion mode. The response of the unlabeled GHB m/z 103/85 and $^2H_6$-GHB m/z 109/90.

20.3.2.1 Pre-analytical

Specimen
Heparinized or EDTA plasma specimens drawn at any time during the day will be acceptable. No patient preparation is required. Specimen should be placed on wet ice following collection and separated by centrifugation within 30 min. Separated plasma should be transferred to a cryovial and stored at −80 °C until time of testing.

Reagents and Chemicals
1. Water (Optima LC/MS grade)—Fisher Scientific W6-4
2. Acetonitrile (Optima LC-MS grade) >99.9%—Fisher Scientific A955-4
3. Formic acid ≥98%—Sigma-Aldrich 33015
4. Mobile phases

(a) Mobile phase A (1 L): 0.1% formic acid in LC-MS grade water
(b) Mobile phase B (1 L): 0.1% formic acid in LC-MS grade acetonitrile

5. GHB sodium salt (certified reference material)—Cerilliant G-001-1ML 1.0 mg/mL in methanol
6. $^2H_6$-GHB sodium salt (certified reference material)—Cerilliant G-006-1ML 1.0 mg/mL in methanol
7. $^2H_6$-GHB sodium salt working internal standard: 2.5 μM in LC-MS grade water

Instrumentation

1. Mass spectrometer: Waters Xevo TQ MS, operating in the negative ionization mode
2. Liquid chromatography: Ultra-high-pressure liquid chromatography (Acquity UPLC, Waters)
3. Mass Spec Software: Waters MassLynx V4.1
4. Analytical Column: T3 HSS 100 × 2.1 mm, 1.8 μm
5. Guard Column: Acquity UPLC BEH C18 1.7 μm
6. Refrigerated centrifuge
7. Amicon Ultra-0.5 centrifugal filter unit (10 kDa cutoff)

Calibration

Calibration curves for GHB are established and quantified for each batch of samples. The calibration curve for GHB is prepared in water: 3.9 μM, 15.6 μM, 62.5 μM, 250 μM, and 1000 μM. Lower limit of quantitation is 1.0 μM and the upper limit of quantitation is 5000 μM.

Quality Control

Plasma pools are used to prepare low-, mid-, and high-QCs for GHB, which are analyzed at the beginning and end of every batch of samples. GHB target values for low control is 15.6 μM, mid-control is 398 μM and high control is 1589 μM. A blank (water) is also included at the beginning of the assay and after the high QC of each batch of samples.

20.3.2.2 Analytical Procedure

This procedure is modified for analysis of plasma from previously published methods for analysis of GHB in dried blood spots [22]. Bring standards, internal standards, and QC to room temperature. QC is placed at the beginning of the batch and end of the batch. Detailed sample preparation is described below:

- Sample Preparation GHB (plasma)
  1. To labeled 1.5 mL microcentrifugal filter units, pipette 200 μL of 2.5 μM of $^2H_6$-GHB working internal standard.
  2. Add 10 μL of sample (calibrators, QC, patient plasma, blank).
  3. Cap and vigorously vortex tubes for 3 s.
  4. Centrifuge for 20 min at 14,000 rpm at 4 °C.
  5. Transfer 150 μL filtrate into corresponding work list position in a 96-well microtiter plate and cover it with silicone cover.
  6. Place the completed 96-well microtiter plate onto a refrigerated autosampler (10 °C).
  7. Inject 5 μL of sample onto LC-MS/MS. Representative LC-MS/MS ion chromatograms for GHB and internal standard are shown in Fig. 20.4.

- LC-MS/MS Methodology

Quantitation of plasma GHB was performed on a Waters Acquity UPLC chromatography system coupled to a Xevo TQ MS (Waters) with MassLynx software version 4.1. Chromatographic separation of GHB was achieved on a Waters T3 HSS 100 × 2.1 mm, 1.8 μm. LC was operated with a flow rate of 0.25 mL/min with a step gradient from 3% to 98% acetonitrile utilizing a divert valve for the first minute. The gradient was held for 2.8 min at 3% and then quickly to 98% for 0.5 min followed by a 0.7 min equilibration, which delivers a 4 min total run time. Mass spectrometry conditions were as follows: capillary

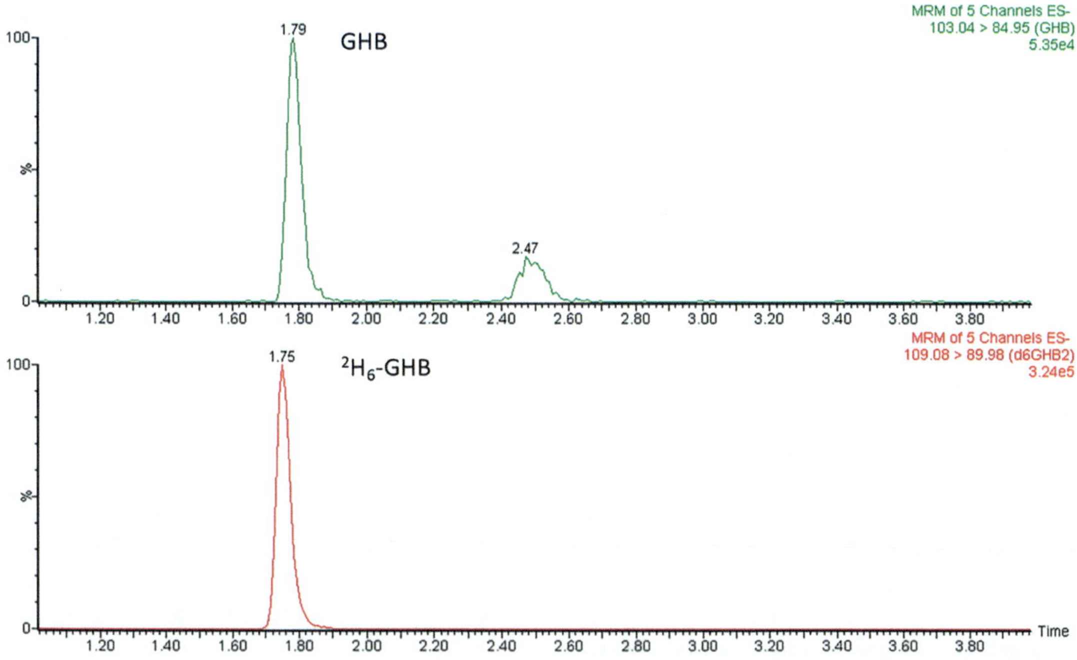

Fig. 20.4 Representative LC-MS/MS ion chromatogram of GHB (m/z 103 > 85) and $^{2}H_{6}$-GHB (m/z 109 > 90) with GHB concentration shown as 16 μM

2.00 kV, cone 28.00 V, Extractor 3.00 V, source temperature 150 °C, desolvation temperature 450 °C, cone gas flow 20 L/h, desolvation gas flow 1000 L/h, collision gas flow 0.15 mL/min, and collision energy 9 V.

Calculation

A standard curve is generated based on linear regression of the analyte/internal standard peak-area ratio (*y*) versus analyte concentration (*x*). The standard curve is weighted 1/*x*.

20.3.2.3 Post-analytical

Quality assurance

Acceptability of each run is confirmed if QC is within ±3SD of previously defined limits.

Interpretation

SSADH deficiency leads to accumulation of GABA, GHB, homocarnosine, and several other metabolites [23]. There is significant debate on the concentration cutoff of plasma GHB to minimize both false positives and false negatives in subjects with SSADH deficiency. In plasma, the recommended cutoff for GHB is approximately 38.4 μM, which has been demonstrated to minimize the number of false negatives [24]. No significant gender difference is observed in plasma GHB.

Pitfalls

GHB shares MRM transitions (m/z 109/57, and m/z 103/73) with its isomers α-hydroxybutyric acid, α-hydroxyisobutyric acid, β-hydroxybutyric acid, and β-hydroxyisobutyric acid [25, 26]. These transitions are not used for quantitation, but they are acquired and monitored to ensure proper separation of GHB from GHB isomers.

20.3.3 Quantitation of Homocarnosine in CSF

Principle

Homocarnosine is a dipeptide of GABA and L-histidine. After deproteinizing the sample with ethanol, the mixtures are centrifuged. The clear supernatant is evaporated to dryness and derivatized with butanol. The derivatized sample is then evaporated to dryness and redissolved in the mobile phase. The homocarnosine-butyl derivatives (Fig. 20.5) are quantified using LC-MS/MS operating in the positive mode. With MRM, the transitions of m/z 297.0/212.0 for homocarnosine and m/z 299.0/212.0 for $^{2}H_2$-L-homocarnosine are quantified.

20.3.3.1 Pre-analytical

Specimen

Lumbar CSF specimens drawn at any time during the day will be acceptable and no additional patient preparation is required. Optimal volume of 0.5 mL (minimum 0.25 mL) CSF should be collected in a regular sterile CSF collection tube. Due to the rostrocaudal gradient for CSF, the first 2 mL that emerges from the spinal tap is not suitable for homocarnosine analysis. Collected CSF should be clear and immediately frozen at the bedside on dry ice. If CSF is blood contaminated, the specimen should be placed on wet ice and centrifuged within 5 min of collection. Transfer the clear CSF to another vial and freeze on dry ice. Store CSF at −20 °C until the time of testing.

Reagents and Chemicals

1. Water (Optima LC/MS grade)—Fisher Scientific W6-4
2. Ethanol (LiChrosolv) ≥99.9%—Supelco 111727
3. Methanol (HiPerSolv CHROMANORM) ≥99.5%—VWR BDH85800
4. Ammonium acetate (ACS reagent) ≥97%—Sigma-Aldrich 238074
5. 10% 6 M HCl in butanol solution
6. L-Homocarnosine >99%—Sigma-Aldrich H4885
7. $^{2}H_2$-L-Homocarnosine 93% synthesized in-house [27]

Instrumentation

1. Mass spectrometer: Sciex API 3000 triple quadrupole tandem mass spectrometer operating in the positive chemical ionization mode.
2. Liquid chromatography: Perkin-Elmer Series 200 system.
3. Computer: Analyst for Windows NT (version 1.3.1).
4. Analytical Column: Symmetry C18 analytical column (3.9 × 150 mm; 5 μm bead size, Waters).
5. Cooled centrifuge.

Calibration

A calibration curve with aqueous calibrators containing 0, 0.1, 0.5, 1.0, and 5.0 μM homocarnosine is quantified for each batch of samples. The

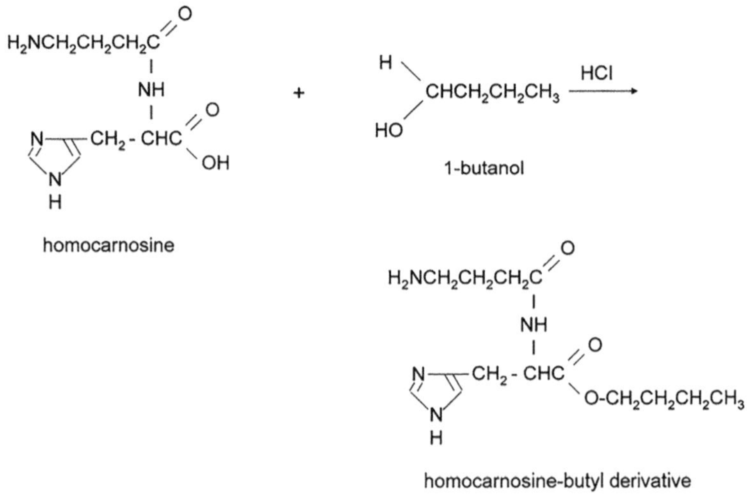

Fig. 20.5 Formation of the homocarnosine-butyl derivative from homocarnosine and 1-butanol

curve is plotted and results determined. The calibration curve is linear to at least 5.0 μM.

Quality Control
A CSF pool is measured for internal control in every batch of samples.

20.3.3.2 Analytical

Procedure
This procedure has been previously described in detail by Jansen et al. [27]. Bring standard, internal standard, QC, and unknown CSF samples to room temperature. Detailed sample preparation is described below:

- Sample Preparation Homocarnosine (CSF)
  1. To labeled 1.5 mL microcentrifuge tubes, pipette 100 μL of sample (calibrators, QC, patient CSF, or blank).
  2. Add 10 μL of 0.1 mM $^2H_2$-L-homocarnosine in water.
  3. Deproteinize the samples by adding 300 μL ethanol (99%).
  4. Mix the mixtures thoroughly and centrifuge at 20,000 × *g* for 5 min at 4 °C.
  5. Transfer the clear supernatant to a vial and evaporate to dryness under a gentle steam of nitrogen at 40 °C.
  6. Add 100 μL of 10% 6 M HCl in butanol solution and derivatize the samples at 60 °C for 15 min.
  7. After cooling, evaporate the samples to dryness at 40 °C under a stream of nitrogen.
  8. Redissolve the residues in a 100-μL mobile phase consisting of water-methanol (50:50, by volume) containing 25 mM ammonium acetate.
  9. Transfer the solutions to new tubes and centrifuge for 5 min at 10,000 × *g* at 4 °C.
  10. Approximately 1–10 μL of the clear supernatant is injected into the LC-MS/MS system. Representative ion chromatograms for homocarnosine-butyl derivatives and internal standard are shown in Fig. 20.6.

- LC-MS/MS Methodology

Separation of homocarnosine is achieved employing a Symmetry C18 analytical column (3.9 × 150 mm; 5-μm bead size; Waters) using water-methanol (50:50, by volume) containing 25 mM ammonium acetate as the mobile phase. The flow rate is set to 0.9 mL/min and is split into a ratio of 1:4, producing an inlet flow of 180 μL/min into the mass spectrometer. Detection of the homocarnosine-butyl derivatives is performed on a triple quadrupole tandem mass spectrometer with the turbo ion electrospray source operating in the positive mode. The temperature of the source is set to 400 °C; turbo ion gas (nitrogen) is used at a flow rate of 8 L/min and the ion spray voltage is set at 4200 V; the declustering potential is set at 30 V and the collision energy is 20 V. Data is acquired and processed using Analyst for Windows NT (version 1.3.1). For the MRM, the following transitions are performed: m/z 297.0 to m/z 212.0 for homocarnosine and m/z 299.0 to m/z 212.0 for $^2H_2$-L-homocarnosine.

Calculation
The CSF homocarnosine concentration is calculated by interpolating the observed analyte/internal standard peak-area ratio into the linear regression line for the calibration curve, which is obtained by plotting the peak-area ratios against analyte concentration.

20.3.3.3 Post-analytical

Quality assurance
Acceptability of each run is confirmed if QC is within ±3SD of previously defined limits.

Interpretation
Calculated homocarnosine concentrations are compared with the reference range values. Large quantities of GABA and GABA conjugates, including homocarnosine, are found in CSF derived from patients with GABA-T and SSADH deficiencies. CSF homocarnosine can also be elevated in patients with homocarnosinosis.

Reference Values
No significant gender difference is observed in CSF homocarnosine concentrations.

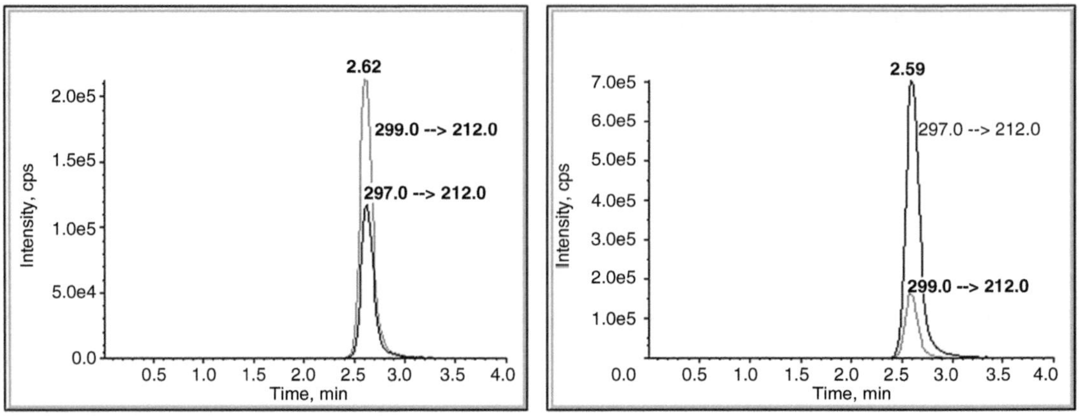

Fig. 20.6 Chromatograms of a pooled CSF sample (**a**) and from a succinate-semialdehyde-dehydrogenase-deficient patient aged 4 years (**b**). Transition m/z 297.0 → m/z 212.0 for homocarnosine; transition m/z 299.0 → m/z 212.0 for $^2H_2$-L-homocarnosine

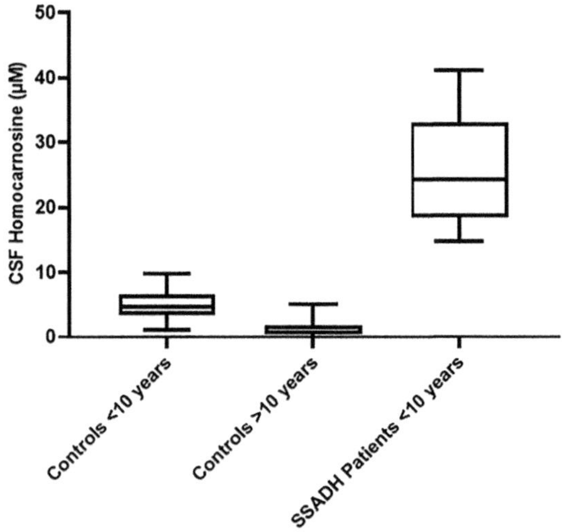

Fig. 20.7 Reference values for homocarnosine: at <10 years = 1.12–9.84 µmol/l (n = 21); at >10 years = <0.02–5.06 µmol/l (n = 52)

Homocarnosine concentration is highly age dependent. A comparison between two control groups should be made, individuals <10 years and individuals >10 years of age (Fig. 20.7).

Pitfalls

The same pitfalls for homocarnosine determination may be applied to the case of total GABA. Avoid excessive freeze-thaw cycles for CSF; thaw only prior to aliquoting for measurement, and freeze CSF immediately upon isolation.

20.3.4 Alternative Analytical Methodologies

GABA and homocarnosine can also be quantitated by the amino acid analyzer, ultra-high-pressure liquid chromatography coupled with UV/Vis and LC-MS/MS [28, 29]. However, sensitivity is significantly lower (µM vs nM) compared to methods described in this chapter. Also note that quantitation of GABA by amino acid profile will only include free GABA, as this methodology does not include a hydrolysis step to cleave peptide-bound GABA. Additionally, elevated GHB may be detected by GC-MS following derivatization to form volatile trimethylsilyl derivatives in routine urinary organic acids [18].

20.4 Conclusion

Three analytical methods for quantitating metabolites within the metabolic pathway of GABA have been described. Levels of these metabolites should be viewed in the context of patient's clinical symptoms. In addition to enzyme deficiencies in the GABA metabolic pathway discussed in this chapter are other disorders associated with GABA. One example is the discovery of antibodies against GAD. GAD exists as two distinct isoforms, GAD67 and GAD65. Several neurological autoimmunities have been shown to be linked to subjects with high titers of autoantibodies against GAD. Significant increases in anti-GAD antibodies are associated with Stiff Person Syndrome, autoimmune epilepsy, limbic encephalitis, and Type-1 diabetes mellitus [30]. Subjects with these disorders have been shown to have low CSF GABA compared to controls. Additionally, there are patients with pyridoxine-dependent seizures. This is an autosomal recessive disorder which is presumed to be due to reduced binding of pyridoxal-5-phosphate to GAD [1]. Recently, further molecular investigation in this original family with homocarnosinosis revealed mutations in the SPG11 gene, which is among a group of genes responsible for hereditary spastic paraplegias [31]. This new insight into the cause of the original patients' phenotype, suggests that the elevation of homocarnosine remains a biochemical anomaly with unknown significance. In conclusion, inherited metabolic disorders in GABA metabolism may be underdiagnosed due to the limited availability of laboratories performing these tests.

References

1. Gibson KM, Jakobs C (2001) Disorders of beta- and gamma-amino acids in free and peptide-linked forms. In: Scriver CR, Beaudet AL, Sly WS, Valle D (eds) The metabolic and molecular bases of inherited disease, 8th edn. McGraw-Hill, New York, pp 2079–2105
2. Gibson KM, Sweetman L, Nyhan WL, Jakobs C, Rating D, Siemes H, Hanefeld F (1983) Succinic semialdehyde dehydrogenase deficiency: an inborn error of gamma-aminobutyric acid metabolism. Clin Chim Acta 133(1):33–42
3. Didiasova M, Banning A, Brennenstuhl H, Jung-Klawitter S, Cinquemani C, Opladen T, Tikkanen R (2020) Succinic semialdehyde dehydrogenase deficiency: an update. Cells 9(2):477
4. White CM (2017) Pharmacological, pharmacokinetic, and clinical assessment of illicitly used gamma-hydroxybutyrate. J Clin Pharmacol 57(1):33–39
5. Carter LP, Pardi D, Gorsline J, Griffiths RR (2009) Illicit gammas-hydroxybutyrate (GHB) and pharmaceutical sodium oxybate (Xyrem): differences in characteristics and misuse. Drug Alcohol Depend 104(1–2):1–10
6. Van den Brink W, Addolorato G, Aubin HJ, Benyamina A, Caputo F, Damatteir M et al (2018) Efficacy and safety of sodium oxybate in alcohol-dependent patients with a very high drinking risk level. Addict Biol 23(4):969–986
7. Vaughan-Jones RD, Spitzer KW, Swietach P (2006) Spatial aspects of intracellular pH regulation in heart muscle. Prog Biophys Mol Biol 90(1–3):207–224
8. Boldyrev AA, Stvolinsky SL, Tyulina OV, Koshelev VB, Hori N, Carpenter DO (1997) Biochemical and physiological evidence that carnosine is an endogenous neuroprotector against free radicals. 17(2):259–271
9. Gjessing LR, Sjaastad O (1974) Letter: Homocarnosinosis: a new metabolic disorder associated with spasticity and mental retardation. Lancet 2(7887):1028
10. National Center for Biotechnology Information (2022) PubChem Compound Summary for CID 119, gamma-Aminobutyric acid. https://pubchem.ncbi.nlm.nih.gov/compound/gamma-Aminobutyric-acid. Retrieved 9 Dec 2022
11. National Center for Biotechnology Information (2022) PubChem Compound Summary for CID 10413, 4-Hydroxybutanoic acid. https://pubchem.ncbi.nlm.nih.gov/compound/4-Hydroxybutanoic-acid. Retrieved 9 Dec 2022
12. National Center for Biotechnology Information (2022) PubChem Compound Summary for CID 10243361, Homocarnosine. https://pubchem.ncbi.nlm.nih.gov/compound/Homocarnosine. Retrieved 9 Dec 2022
13. Goldsmith RF, Earl JW, Cunningham AM (1987) Determination of γ-aminobutyric acid and other amino acids in cerebrospinal fluid by reversed-phase liquid chromatography. Clin Chem 33(10):1736–1740
14. Naini AB, Vontzalidou E, Cote LJ (1993) Isocratic HPLC assay with electrochemical detection of free γ-aminobutyric acid in cerebrospinal fluid. Clin Chem 39(2):247–250
15. Struys EA, Guérand WS, ten Brink HJ, Jakobs C (1999) Combined method for the determination of gamma-aminobutyric and beta-alanine in cerebrospi-

nal fluid by stable isotope dilution mass spectrometry. J Chromatogr B 732:245–249
16. Song Y, Shenwu M, Dhossche DM, Liu Y (2005) A capillary liquid chromatographic/tandem mass spectrometric method for the quantification of γ-aminobutyric acid in human plasma and cerebrospinal fluid. J Chromatogr B 814:295–302
17. Eckstein JA, Ammerman GM, Reveles JM, Ackermann BL (2008) Analysis of glutamine, pyroglutamate, and GABA in cerebrospinal fluid using ion pairing HPLC with positive electrospray LC/MS/MS. J Neurosci Methods 171:190–196
18. Arning E, Wasek B, Bottiglieri T (2022) Quantitation of γ-aminobutyric acid in cerebrospinal fluid using liquid chromatography-electrospray-tandem mass spectrometry. Methods Mol Biol 2546:35–43
19. Rating D, Siemes H, Loscher W (1983) Low CSF GABA concentration in children with febrile convulsions, untreated epilepsy, and meningitis. J Neurol 230:217–225
20. Dalakas MC, Li M, Fujii M, Jacobowitz DM (2001) Stiff person syndrome: quantification, specificity, and intrathecal synthesis of GAD65 antibodies. Neurol 57(5):780–784
21. Porter RJ, Dhir A, Macdonald RL, Rogawski MA (2012) Mechanisms of action of antiseizure drugs. Handb Clin Neurol 108:663–681
22. Forni S, Pearl PL, Gibson KM, Yu Y, Sweetman L (2013) Quantitation of gamma-hydroxybutyric acid in dried blood spots: feasibility assessment for newborn screening of succinic semialdehyde dehydrogenase (SSADH) deficiency. Mol Genet Metab 109(3):255–259
23. Malaspina P, Roullet JB, Pearl PL, Ainslie GR, Vogel KR, Gibson KM (2016) Succinic semialdehyde dehydrogenase deficiency (SSADHD): pathophysiological complexity and multifactorial trait associations in a rare monogenic disorder of GABA metabolism. Neurochem Int 99:7284
24. Wernli C, Finochiaro S, Volken C, Andresen-Streichert H, Buettler A, Gygax D, Salomons GS, Jansen EE, Ainslie GR, Vogel KR, Gibson KM (2017) Targeted screening of succinic semialdehyde dehydrogenase deficiency (SSADHD) employing an enzymatic assay for γ-hydroxybutyric acid (GHB) in biofluids. Mol Genet Metab Rep 11:81–89
25. Johansen SS, Windberg CN (2011) Simultaneous determination of γ-hydroxybutyrate (GHB) and its analogues (GBL, 1,4-BD, GVL) in whole blood and urine by liquid chromatography coupled to tandem mass spectrometry. J Anal Toxicol 35:8–14
26. Stout PA, Simons KD, Kerrigan S (2010) Quantitative analysis of gamma-hydroxybutyrate at endogenous concentrations in hair using liquid chromatography tandem mass spectrometry. J Forensic Sci 55(2):531–537
27. Jansen EEW, Gibson KM, Shigematsu Y, Jakobs C, Verhoeven N (2006) A novel, quantitative assay for homocarnosine in cerebrospinal fluid using stable-isotope dilution liquid-chromatography mass spectrometry. J Chromatogr B Analyt Technol Biomed Life Sci 18:196–200
28. Peake RWA, Law T, Hoover PN, Gaewsky L, Shkreta A, Kellogg MD (2013) Improved separation and analysis of plasma amino acids by modification of the MassTrak™ AAA Solution Ultraperformance® liquid chromatography method. Clin Chim Acta 423:75–82
29. Roper SM, Weindel AL, Dietzen DJ (2022) Comprehensive determination of amino acids for diagnosis of inborn errors of metabolism. Methods Mol Biol 2546:55–64
30. Herard K, Khanni JL, Alusma Hibbert K, Samules CR, Espinosa PS (2019) Neurological disorders associated with glutamic acid decarboxylase antibodies. Cureus 11(5):e4738
31. Sjaastad O, Blau N, Rydning SL, Peters V, Rødningen O, Stray-Pederson A, Krossnes B, Tallaksen C, Koht J (2018) Homocarnisinosis: A historical update and findings in the SPG11 gene. Acta Neurol Scand 138(3):245–250

Oxalate, Glycolate, and Glycerate

21

Martin Poms

Contents

List of Abbreviations

HCl	Hydrochloric acid
HPIC	High-pressure ion chromatography
HPLC	High-performance liquid chromatography
MW	Molecular weight
NaOH	Sodium hydroxide
OMIM®	Online Mendelian Inheritance in Man®
PH-I	Primary oxaluria type one
PH-II	Primary oxaluria type two
QC	Quality control

M. Poms (✉)
Division of Clinical Chemistry and Biochemistry, University Children's Hospital, Zurich, Switzerland
e-mail: martin.poms@kispi.uzh.ch

21.1 Introduction

Measurement of oxalate in urine is important for the diagnosis of primary hyperoxaluria (OMIM® 259900 and 260000) and the secondary forms resulting from excessive intake or abnormal intestinal absorption of oxalate [1, 2]. Determination of glycolic acid is essential for the diagnosis of primary hyperoxaluria type I (PH-I), caused by low or absent activity of the liver-specific peroxisomal alanine:glyoxylate aminotransferase. In addition, to estimate the risk of stone formation in calcium oxalate urolithiasis and nephrocalcinosis, simultaneous determination not only of calcium, but also of citrate (a potent inhibitor of calcium oxalate and calcium phosphate crystallization) and other constituents (electrolytes, phosphate, and sulfate) are required to calculate the urinary calcium saturation [3]. Primary hyperoxaluria type II (PH-II) is caused by low or absent activity of D-glycerate dehydrogenase and hydroxypyruvate reductase activity, leading to elevated urinary excretion of both oxalate and L-glycerate [1]. High-pressure ion chro-

N. Blau, F. M. Vaz (eds.), *Laboratory Guide to the Methods in Biochemical Genetics*,
https://doi.org/10.1007/978-3-031-58819-8_21

matography (HPIC) is the method of choice for the quantification of oxalic acid, glycolic acid and glyceric acid, often also referred to by their conjugated base names oxalate, glycolate, and glycerate [4, 5]. In addition, preservation and storage of the liquid samples may influence the stability of oxalate and glycolate. The use of urinary filter spots is a practical alternative approach for the collection and safe transport of samples to be analyzed for many metabolic disorders [6].

21.2 Properties of Oxalate, Glycolate, and Glycerate

Oxalic acid (or ethanedioic acid, $H_2C_2O_4$; molecular weight, MW = 90.03 g/mol) is a dicarboxylic acid with the structure HOOC-COOH (Fig. 21.1). The anions of oxalic acid, as well as its salts, and esters, are known as oxalates. Due to the presence of two carboxyl groups, oxalate is a strong organic acid and possesses reducing properties. The root and leaves of rhubarb contain dangerously high concentrations of oxalic acid, posing a potential hazard. Foods that are edible, but still contain significant concentrations of oxalic acid include in decreasing order buckwheat, star fruit (carambola), black pepper, parsley, poppy seed, rhubarb stalks, amaranth, spinach, chard, beets, cocoa, chocolate, most nuts, most berries, and beans. Bodily oxalic acid may also be synthesized via the metabolism of glyoxylic acid or spontaneous chemical degradation of unused ascorbic acid (vitamin C) [7].

Glycolic acid (or hydroxyacetic acid; $C_2H_4O_3$; MW = 76.05 g/mol) is the smallest α-hydroxy acid (Fig. 21.1). In its pure form, glycolic acid is a colorless, crystalline solid. It is very soluble in water (0.1 g/mL), alcohols, acetone, and ethyl acetate. Glycolic acid is isolated from sugarcane, sugar beets, and unripe grapes.

Glyceric acid (or D-2,3-dihydroxypropanoic acid; $C_3H_6O_4$, MW = 125.10 g/mol) is a hydroxy acid derived from the oxidation of C^1 of glycerol or glyceraldehyde to a carboxyl group; various phosphorylated derivatives of L-glycerate are important intermediates in glucose metabolism (Fig. 21.1).

21.3 Methods

All three analytes oxalate, glycolate, and glycerate are analyzed using HPIC with conductivity detection. Oxalate, sulfate, and citrate are measured in the same run and glycolate and glycerate are measured on the same system, but under different conditions with a different method. The background conductivity is minimized using an anion suppressor unit.

21.3.1 Pre-analytical

Ascorbic acid is the main endogenous precursor of oxalic acid and in healthy persons up to 30% of urinary oxalate can originate from ascorbate. Ascorbate is extremely unstable in neutral and alkaline solutions and degrades to oxalate nonenzymatically [8]. Acidification to pH 1–2 is required to prevent deposition and to dissolve insoluble calcium oxalate.

21.3.1.1 Urine

Twenty-four hour or spot urine should be collected in the presence of HCl. Add 10 mL of 20% (v/v) HCl per liter of urine. For 24-h urine, it is mixed in its entirety, volume is determined in a graduated cylinder and acidified to pH = 2 before aliquots are frozen. Keep samples frozen at −20 °C. Thaw and thoroughly vortex the sample. To ensure complete dissolution of oxalate, the sample may be heated at 60 °C for 20 min. Centrifuge for 3 min at 3000 × *g*. For sulfate,

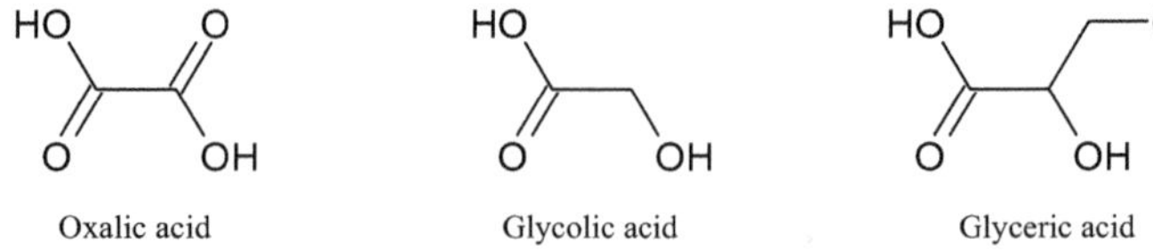

Fig. 21.1 Chemical structures of relevant metabolites in primary hyperoxaluria

oxalate, and citrate analysis, dilute supernatant with 30 mM boric acid to a final creatinine concentration of 0.5–1 mM. For glycolate and glycerate analysis, dilute 200 μL supernatant with 800 μL water (dilution factor 5).

Alternatively, acidified urine can be collected on a filter paper for easier shipment and transport. The urine is preserved with 20% HCl as described above. Filter paper strips (3 × 5 cm filter paper backing, cat. No. 165-0921, Bio-Rad) are dipped into urine to 1 cm below the upper edge. Excess urine is wiped off, and the filter is left to dry at room temperature. Urinary filter paper spots are then cut into small pieces, shaken in 3 mL of water for 15 min at room temperature and sonicated for 5 min at room temperature. The extracts are filtered through an Ultrafree-MC filter with 10-kDa cutoff (Millipore, USA) at 2000 g for 10 min at room temperature. The clear supernatant is analyzed by HPIC.

21.3.1.2 Plasma

Place Li-heparinized blood (3–5 mL) directly on ice and immediately centrifuge at 1550 × *g* for 5 min at 4 °C. This procedure should be completed within 10 min of vein puncture. Place approximately 2.5 mL plasma in the outer chamber of a Centrisart I ultrafiltration vial (Sartorius, NY, USA). Centrifuge at 4 °C by increasing speed from 0 × *g* to 1550 × *g* over 5 min and keep at 1550 × *g* for 30 min. Determine volume of the ultrafiltrate and add 30 μL 1 M HCl per mL ultrafiltrate. Ultrafiltrate should be analyzed immediately or frozen at −20 °C.

Cave: Recovery of the ultrafiltration of samples may vary significantly between different laboratories. It is therefore advisable that longitudinal monitoring of primary hyperoxaluria patients with deteriorating kidney function is performed by a single consistent laboratory using a defined methodology [9].

21.3.2 Reagents and Chemicals

21.3.2.1 Oxalate, Sulfate, and Citrate

The chemicals required are:

1. Water, HPLC quality.
2. HCl, 37% (MW: 36.46).
3. NaOH, 50–52% (MW: 40.00).
4. Boric acid (H_3BO_3, MG: 61.83; Merck no. 1.00165).
5. Sodium sulfate, water free (Na_2SO_4, MW: 142.04; Merck no. 1.06649).
6. Citric acid monohydrate ($C_6H_8O_7.H_2O$, MW: 210.14; Merck no. 1.00243).
7. Sodium oxalate ($C_2O_4Na_2$, MW: 134.0; Sigma no. 71800).
8. Sodium phosphate (NaH_2PO_4, MW: 119.98; Sigma no. S8282).

Stock solutions:

1. 20% (v/v) HCl: Add 300 mL of water to a 1-L graduated cylinder. Add 541 mL of 37% HCl and dilute with water to 1000 mL. Store at room temperature.
2. 1 M HCl: dilute 8.26 mL of 37% HCl with water to 100 mL. Store at room temperature.
3. 20 mM sulfate stock: dissolve 710 mg Na_2SO_4 in 250 mL of water. Store at −20 °C.
4. 2 mM oxalate stock: dissolve 67 mg sodium oxalate in 10 mL of Store at −20 °C.
5. 2 mM citrate stock: dissolve 105 mg citric acid monohydrate in 250 mL of water. Store at −20 °C.
6. 20 mM phosphate solution: dissolve 24 mg sodium phosphate in 100 mL water. Store at −20 °C.
7. 30 mM boric acid: dissolve 1.856 g of boric acid in 1000 mL of water. Store at 4 °C.

Standard solutions:

Standard solution should be freshly prepared before each analysis in the following order (Table 21.1):

1. 5 mL of the 30 mM boric acid solution
2. Add matrix (20% HCl solution and 10 mM phosphate solution)
3. Add stock solutions
4. Add 30 mM boric acid solution to final volume of 10 mL

Quality control

It is recommended to use two standard mixtures as internal quality controls covering the low and high range of the calibration:

QC 1 Low: Urine from a healthy donor (50–100 mL) with preferably low sulfate concentration (low meat diet) is diluted with 30 mM boric acid solution to a creatinine concentration of 3–10 mM. For measurement, appropriately dilute the sample so that the resulting final concentrations correspond to approximately 100 μM sulfate, 7 μM oxalate, and 40 μM citrate.

QC 2 High: Add 2.2 mL sulfate, 1.2 mL oxalate, 2 mL citrate, and 0.3 mL phosphate stock solutions to 40 mL of QC 1 Low sample. Acidify with 0.05 mL 20% HCl and dilute to 100 mL with 30 mM boric acid. Dilute as needed to achieve final concentrations of approximately 500 μM sulfate, 40 μM oxalate, and 150 μM citrate.

21.3.2.2 Glycolate and Glycerate

The following chemicals are required:

1. Water, HPLC quality.
2. HCl, 37% (MW 36.46).
3. NaOH, 50% (MW 40.00).
4. Disodium tetraborate decahydrate ($Na_2B4O_7.10\ H_2O$; MW 381.37; Merck no. 1.06308).
5. Sodium carbonate, water free (Na_2CO_3; MW 105.99; Merck no. 1.06395).
6. Glycolic acid ($C_2H_4O_3$; MW: 76.05; Sigma no. 124737).
7. Glyceric acid hemicalcium salt ($C_3H_5O_4.1/2\ Ca$; MW: 143.12; Sigma no. 367494).

Stock solutions:

1. Water, twice distilled, 18.2 mΩ quality, degassed.
2. 20% (v/v) HCl: Add 300 mL of water to a 1 L graduated cylinder. Add 541 mL of 37% HCl and dilute with water to 1000 mL. Store at room temperature.
3. 0.1 M NaOH: dilute 5.2 mL of 50% NaOH with water to a final volume of 1000 mL. Store at room temperature.
4. 1 M HCl: dilute 8.26 mL of 37% HCl with water to 100 mL. Store at room temperature.
5. 50 mM Sodium tetraborate solution: dissolve 19.1 g of sodium tetraborate in 1000 mL of water. Store at room temperature.
6. 50 mM Sodium carbonate solution: dissolve 5.3 g sodium carbonate in 1000 mL of water. Store at room temperature.
7. 10 mM Glycolate stock solution: dissolve 76.1 mg glycolic acid in 100 mL of water. Store at −20 °C.
8. 10 mM Glycerate standard: dissolve 143 mg glyceric acid in 100 mL of water. Store at −20 °C.

Standard solution should be freshly prepared before each analysis according to Table 21.2.

Store at aliquots at −20 °C.

Control: Urine from a healthy donor is spiked with glycolate and glycerate stock solutions to a

Table 21.1 Standard solution table for oxalate, sulfate, and citrate

	Stock solution [μL]			Matrix [μL]		Final concentration [μM]		
Standard	Sulfate	Oxalate	Citrate	20% HCl	Phosphate	Sulfate	Oxalate	Citrate
1	50	25	50	5	250	100	5	10
2	100	50	100	5	250	200	10	20
3	250	125	250	5	250	500	25	50
4	500	250	500	5	250	1000	50	100
5	1000	500	1000	5	250	2000	100	200

Table 21.2 Standard solution table for glycolate and glycerate

	Stock solution [μL]		Working solution [mL]	Final concentration [μM]
Standard	Glycolate	Glycerate	H_2O	Glycolate/glycerate
1	40	40	20	20
2	100	100	20	50
3	200	200	20	100
4	500	500	20	250
5	1000	1000	20	500

final concentration of approximately 30 μM and 400 μM, respectively. Store at aliquots at −20 °C.

21.3.3 Instrumentation

Thermo Liquid Chromatograph ICS-AS-AP-5000 with a conductivity detector, suppressor ADRS 600 (2 mm), and AS-AP Autosampler.

21.3.3.1 Oxalate, Sulfate, and Citrate

1. IonPac NG1 guard column (2 × 50 mm, Thermo).
2. IonPac AG 11-HC guard column (2 × 50 mm, Thermo).
3. IonPac AS 11 analytical column (2 × 250 mm, Thermo).
4. Eluent A: H_2O.
5. Eluent B: 100 mM NaOH.
6. See Table 21.3

21.3.3.2 Glycolate and Glycerate

1. CarboPac PA1 guard column (2 × 50 mm, Dionex).
2. CarboPac PA1 analytical column (2 × 250 mm, Dionex).
3. Eluent A: H_2O.
4. Eluent B: 100 mM NaOH.
5. Eluent C: 50 mM sodium tetraborate.
6. Eluent D: 50 mM sodium carbonate.
7. Flow: 0.5 mL/min.
8. See Table 21.4

21.4 Analytical

Fill the autosampler vials with the corresponding standard mixtures, quality control sample, and patient samples. For plasma ultrafiltrates use low-volume inserts. Inject 25 μL of the sample.

Calculation For the external standard method, measure creatinine in the urine samples and calculate the results per mol of creatinine (oxalate, citrate, glycolate, and glycerate = mmol/mol creatinine; sulfate = mol/mol creatinine).

Performance The linearity and limit of detection for oxalate, citrate, sulfate as well as glycerate and glycolate is shown in Table 21.5.

Table 21.3 Gradient table for oxalate, sulfate, and citrate

Time [min]	Flow rate	A (H_2O) [%]	B (NaOH) [%]
0.0	0.260	93	7
2.5	0.260	93	7
5.0	0.260	90	10
12.0	0.260	88	12
12.1	0.260	85	15
18.9	0.300	79	21
23.0	0.300	55	45
23.1	0.300	40	60
23.2	0.300	40	60
23.3	0.300	93	7
33.0	0.300	93	7
40.0	0.260	93	7

Table 21.4 Gradient table for glycolate and glycerate

Time [min]	A (H_2O) [%]	B (NaOH) [%]	C (Na_2CO_3) [%]	D ($Na_2B_4C_7$) [%]
0.0	68	30	0	2
15.0	67	31	0	2
18.0	48	50	0	2
18.9	48	50	0	2
19.0	0	0	100	0
19.9	0	0	100	0
20.0	0	0	0	100
21.0	0	0	0	100
21.1	8	90	0	2
23.0	8	90	0	2
23.3	68	30	0	2
43.0	68	30	0	2

Table 21.5 Performance parameters for oxalate, sulfate, citrate, glycolate, and glycerate

	Oxalate	Sulfate	Citrate	Glycolate	Glycerate
Linearity [μmol/L]	5–2000	10–2000	5–2000	5–5000	5–5000
LoD [μmol/L]	1	1	1	4	4

Table 21.6 Reference values for urinary oxalate, glycolate, citrate, glycerate (given in mmol/mol creatinine), and sulfate (given in mol/mol creatinine), and plasma oxalate, citrate, and sulfate (given in µmol/L)

Urine	Oxalate [mmol/mol]	Sulfate [mol/mol]	Citrate [mmol/mol]	Glycolate [mmol/mol]	Glycerate [mmol/mol]
<6 months	<325–360	0.96–5.08	>120	11–109	<10
7 months to 2 years	<132–174	1.22–8.03	>120	22–139	<10
2–5 years	<98–101	1.45–3.86	>120	17–103	<10
5–15 years	<70–82	0.7–4.97	>80	18–92	<10
>16 years	<40	0.63–4.78	>80	6–80	<10
Plasma	**Oxalate [µmol]**	**Sulfate [µmol]**	**Citrate [µmol]**	**Glycolate [µmol]**	**Glycerate [µmol]**
All ages	4.31–8.55	64.4–406	24.5–134	–	–

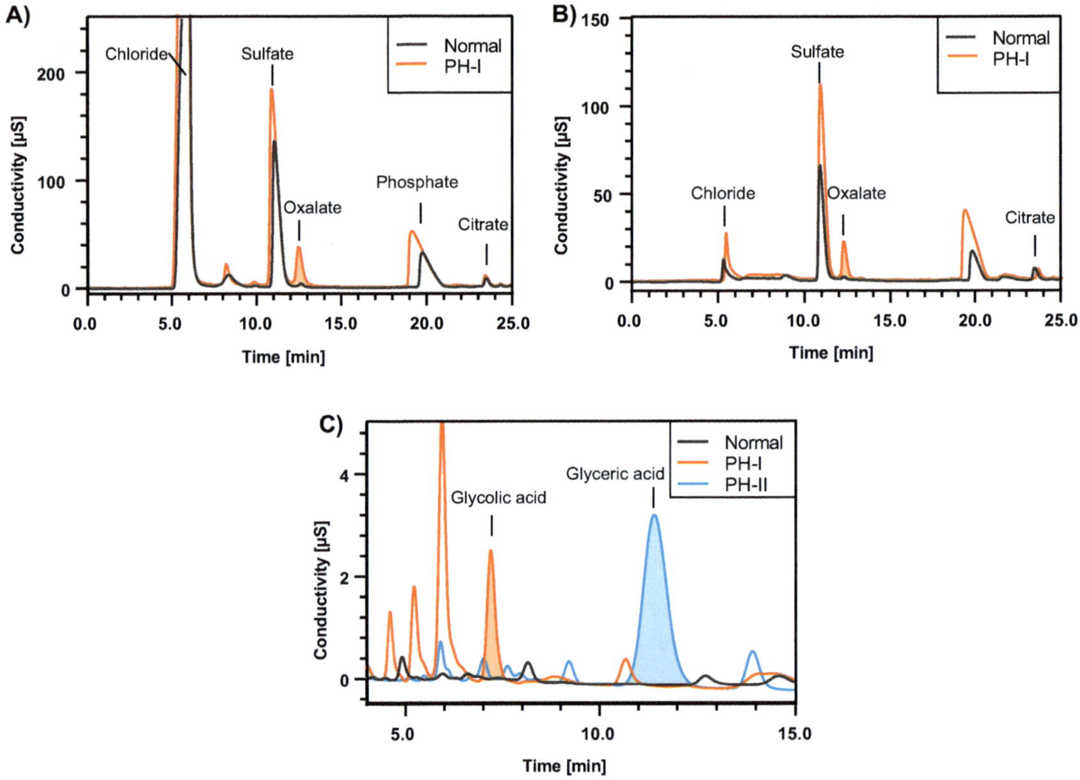

Fig. 21.2 Chromatographic separation of chloride, sulfate, oxalate, phosphate, and citrate in (**a**) urine and (**b**) plasma. Chromatographic separation of glycolate and glycerate in (**c**) urine. Samples from patients with primary hyperoxaluria type I and type II are shown in orange and blue, respectively

21.5 Post-analytical

21.5.1 Reference Values

Reference values for urine and plasma are given in Table 21.6.

21.5.2 Pathological Values

Figure 21.2 shows chromatograms of healthy controls, and patients with PH-I and PH-II. Tables 21.7 and 21.8 summarize the laboratory findings for patients with PH-I and PH-II, respectively.

Table 21.7 Laboratory findings in patients with primary hyperoxaluria type I (modified from [1])

	Renal failure (all ages)	Neonate	Infant	Children	Adolescent	Adult
Creatinine	++	N-↑	N-↑	N-↑	N-↑	N-↑
Urea	++	N-↑	N-↑	N-↑	N-↑	N-↑
Oxalate (U)	N-↑	↑-↑↑	↑-↑↑	↑-↑↑	↑-↑↑	↑-↑↑
Glycolate (U)	N-↑	N-↑	N-↑	N-↑	N-↑	N-↑
Oxalate (P)	↑↑↑	↑-↑↑	↑-↑↑	↑-↑↑	↑-↑↑	↑-↑↑

Table 21.8 Laboratory findings in patients with primary hyperoxaluria type II (modified from [1])

	Renal failure (all ages)	Neonate	Infant	Children	Adolescent	Adult
Creatinine	++	N-↑	N-↑	N-↑	N-↑	N-↑
Urea	++	N-↑	N-↑	N-↑	N-↑	N-↑
Oxalate (U)	N-↑	↑-↑↑	↑-↑↑	↑-↑↑	↑-↑↑	↑-↑↑
Glycerate (U)	?	↑-↑↑	↑-↑↑	↑-↑↑	↑-↑↑	↑-↑↑
Oxalate (P)	↑↑↑	↑-↑↑	↑-↑↑	↑-↑↑	↑-↑↑	↑-↑↑

Measurement of urinary oxalate, glycolate, and glycerate is pivotal for diagnosing primary hyperoxaluria types I and II, while assessing calcium, citrate, and various constituents via ion chromatography aids in estimating stone formation risk in urolithiasis and nephrocalcinosis. Additionally, preserving liquid samples and utilizing urinary filter spots offer practical solutions for stable sample storage and transport in metabolic disorder analysis.

References

1. Von Schnakenburg C, Latta K (2003) Hyperoxaluria. In: Blau N, Duran M, Blaskovics ME, Gibson KM (eds) Physician's guide to the laboratory diagnosis of metabolic diseases. Springer Berlin Heidelberg, Berlin, pp 509–518
2. Williams HE (1978) Oxalic acid and the hyperoxaluric syndromes. Kidney Int 13(5):410–417
3. Leumann E, Hoppe B, Neuhaus T, Blau N (1995) Efficacy of oral citrate administration in primary hyperoxaluria. Nephrol Dial Transplant 10(Suppl 8):14–16
4. Hoppe B, Kemper MJ, Hvizd MG, Sailer DE, Langman CB (1998) Simultaneous determination of oxalate, citrate and sulfate in children's plasma with ion chromatography. Kidney Int 53(5):1348–1352
5. Politi L, Chiaraluce R, Consalvi V, Cerulli N, Scandurra R (1989) Oxalate, phosphate and sulphate determination in serum and urine by ion chromatography. Clin Chim Acta 184(2):155–165
6. Blau N, Matasovic A, Lukasiewicz-Wedlechowicz A, Heizmann CW, Leumann E (1998) Simultaneous determination of oxalate, glycolate, citrate, and sulfate from dried urine filter paper spots in a pediatric population. Clin Chem 44(7):1554–1556
7. Knight J, Madduma-Liyanage K, Mobley JA, Assimos DG, Holmes RP (2016) Ascorbic acid intake and oxalate synthesis. Urolithiasis 44(4):289–297
8. Mazzachi BC, Teubner JK, Ryall RL (1984) Factors affecting measurement of urinary oxalate. Clin Chem 30(8):1339–1343
9. Stokes F, Acquaviva-Bourdain C, Hoppe B, Lieske JC, Lindner E, Toulson G et al (2020) Plasma oxalate: comparison of methodologies. Urolithiasis 48(6):473–480

Glycerol and Glycerol Phosphates 22

Katrina M. Dipple

Contents

List of Abbreviations

cGKD	Complex form of glycerol kinase deficiency
G-3-P	Glycerol-3-phosphate
GC-MS	Gas chromatography-mass spectrometry
GIS	Glycerol intolerance syndrome
GK	Glycerol kinase
GKD	Glycerol kinase deficiency
iGKD	Isolated form of glycerol kinase deficiency

K. M. Dipple (✉)
Division of Genetic Medicine, Department of Pediatrics, University of Washington, Seattle, WA, USA

Genetic Medicine, Seattle Children's Hospital, Seattle, WA, USA

Center for Translational and Clinical Research, Seattle Children's Research Institute, Seattle, WA, USA
e-mail: Katrina.dipple@seattlechildrens.org

22.1 Introduction

Glycerol and glycerol phosphates (glycerol-3-phosphate G-3-P) are measured using standard technologies including gas chromatography-mass spectrometry GC-MS and routine clinical chemistry. They are useful in the diagnosis of disorders of glycerol metabolism such as glycerol kinase deficiency and Glycerol Intolerance Syndrome (GIS). DNA sequencing and enzymatic analysis can also be done to help in diagnosis.

22.2 Clinical Significance

Disorders of glycerol metabolism include Glycerol Kinase (GK), Deficiency (GKD), and GIS. Of these, only GKD is well characterized and has a defined biochemical defect [1].

GKD is due to mutations, insertions, or deletions within the *GK* gene on Xp21. There are two

N. Blau, F. M. Vaz (eds.), *Laboratory Guide to the Methods in Biochemical Genetics*,
https://doi.org/10.1007/978-3-031-58819-8_22

distinct forms of GKD [1, 2]. The complex form (cGKD) involves large deletions and other genes in this region including the adrenal hypoplasia congenita, and Duchenne muscular dystrophy loci. The isolated form (iGKD) can be either symptomatic or asymptomatic and is due only to mutations, insertions, or deletions within the *GK* gene on Xp21.

Patients with both forms of GKD have hyperglyceroluria and hyperglycerolemia [1, 2]. The glyceroluria is detected on routine gas chromatography-mass spectrometry (GC-MS) of urine done as part of the metabolic work-up (see Chap. 5 in this book). The hyperglycerolemia is detected by routine clinical laboratory testing either as pseudo-hypteriglyceridemia or elevated glycerol. In addition, glycerol can be quantified using isotope dilution methods or commercially available kits.

Glycerol kinase activity can be determined from leukocytes, fibroblasts, transformed lymphoblastoid cell lines, or liver tissue [3]. Individuals with GKD also have decreased GK activity in other tissues including kidney, small intestines, and adrenal glands [2]. The kinase activity can be assessed directly by measuring glycerol phosphorylation, or indirectly by incorporation or oxidation.

22.2.1 Properties of Analyte

Glycerol is a neutral, hygroscopic, colorless, clear liquid [2]. Its chemical name is 1,2,3-propanetriol and its chemical formula is $C_3H_8O_3$ (Fig. 22.1a). It has a molecular mass of 92.0 g/mol. It is a neutral compound; however, since it is often detected by GC-MS of urine done during the workup of a metabolic crisis, GKD is often considered together with the organic acidemias [2].

Glycerol is phosphorylated to glycerol-3-phosphate (G-3-P) by glycerol kinase (ATP:glycerol-3-phosphotransferase, EC 2.7.1.30) [2]. G-3-P is also known as sn-glycerol 3-phosphate, glycerophosphoric acid, phosphatidyl glycerol, and sn-Gro-1-P (Fig. 22.1b). It has a chemical formula of $C_3H_9O_6P$ and a molecular mass of 172.07 g/mol (http://pubchem.ncbi.nlm.nih.gov).

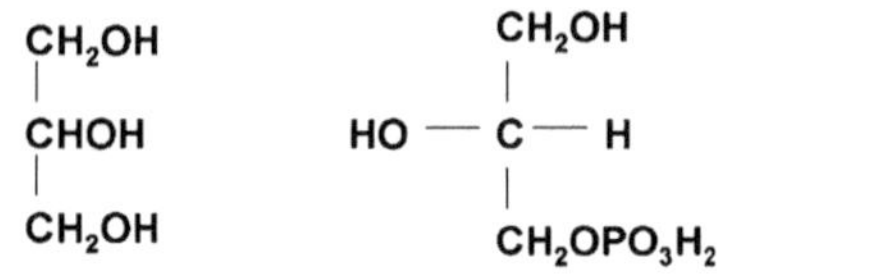

Fig. 22.1 Structure of glycerol (**a**) and sn-glycerol-3-phosphate (**b**)

22.3 Methods

22.3.1 Glyceroluria

Principle

Because glycerol is a neutral compound, it can be detected by GC-MS of urine performed by the usual protocols (see Chap. 5 in this book) if a solvent extraction procedure is used [2, 4, 5]. If, however, the sample for urine organic acid analysis is treated by an ion-exchange procedure, then glycerol will be in the column wash fraction with other neutral compounds and not in the organic acid fraction. Glycerol is not well extracted by organic solvents, but it is so highly abundant in patients with GKD that it is therefore easily detected by GC-MS [4]. In addition, glycerol concentration can also be determined from commercially available kits such as the Millipore Sigma serum triglyceride determination kit (TR0100, Sigma, St. Louis MO, USA), Millipore Sigma Free glycerol determination kit (FG0100, Sigma, St. Louis MO, USA), or an isotope dilution protocol as detailed below for glycerolemia [6].

Glycerol is present in many medications, ointments, suppositories, and emollient solutions. Therefore, many cases of glyceroluria are due to contamination of the sample by exogenous sources [2].

22.3.2 Pre-analytical

Specimen

Urine is obtained using standard techniques including sterile midstream catch or catheterization. In infants, urine specimens obtained by placing a bag in the genitourinary region are undesirable due to contamination of glycerol from ointment and glycerin suppositories, unless the perineum is washed and rinsed well and no suppositories have been used recently [2, 4].

Reagents and Chemicals

These must be glycerol free. A solvent extraction procedure must be used prior to GC-MS [2, 4, 5]. If the sample is prepared with ion-exchange chromatography, the glycerol will remain in the neutral fraction and not be detected by GC-MS [2, 4]. Hydrochloric acid (5 N), pH paper, sodium sulfate (anhydrous), ethyl acetate (free of contaminants), diethyl ether (anhydrous, peroxide-free, and free of contaminants except for 2,6-ditertbutylcresol, which is an antioxidant found in all ether), and malonic acid (26.25 mg/50 ml methanol).

Instrumentation

Standard GC-MS is used for organic acid analysis (see Chap. 5 in this book). Data regarding calibration and quality control in this procedure can be found in Chap. 5 of this book.

22.3.3 Analytical

Procedure: The Solvent Extraction Method

This method is based on the protocol described by Goodman and Markey [4]. Determine urine creatinine using standard methods. Add malonic acid (26.25 mg/50 ml) to 1 ml urine in a 12–13 ml round-bottom, glass-stoppered test tube as indicated in Table 22.1. Add 2–3 drops of 5 N HCl until the pH is 1, as determined by pH paper. Add 3 ml ethyl acetate, stopper, shake vigorously (1 min), and centrifuge briefly. Transfer the upper (ethyl acetate) phase to a clean tube. Re-extract the lower (urine, aqueous) phase with 3 ml diethyl ether, stopper, and shake (1 min). Take the upper (ether) phase and combine it with the previous ethyl acetate extract. Repeat the diethyl ether extraction. Take the upper ether phase and pool the extracts. Discard the lower (aqueous) phase. To the extracts, add 2 g sodium sulfate, stopper, shake vigorously (1 min), then centrifuge (5 min). Transfer the anhydrous extract into a 12–13 ml conical, glass-stoppered tube, and evaporate to dryness at 37 °C under nitrogen. The sample is now ready for derivatization and injection into the GS-MS as described in Chap. X of this book.

Table 22.1 Malonic acid needed for solvent extraction of urine

Creatine[a] (mg/100 ml) urinary	Malonic acid (26.25 mg/50 ml methanol) (μl)
1–5	25
6–10	50
11–20	100
21–40	200
41–60	300
61–80	400
81–100	500

[a]If >100 mg/100 ml, use 0.5 ml of urine for solvent extraction and use the volume of malonic acid corresponding to one-half of the creatine concentration. Adapted from [4]

22.3.4 Post-analytical

Interpretation/Chromatograms

The interpretation of results and chromatograms are described in Chap. 5 of this book.

Reference Values

In urine, the normal values of glycerol are ≤0.2 mM; not detectable as mmol/mmol creatinine; or ≤1 mmol/24 h [2].

Typical Pathological Values

Typical pathological values in urine are 41–345 mM; 90–193 mmol/mmol creatinine; or 11–360 mmol/24 h [2].

Quality Control

Glycerol contamination from exogenous sources such as lotions, medications, and suppositories is quite common [2]. A solvent extraction procedure must be used to detect the glycerol [2, 4, 5]. If, however, the sample for urine organic acid analysis is treated by an ion-exchange procedure, then glycerol will be in the column wash fraction and therefore not detected. Over evaporation during solvent extraction will lead to loss of organic acids [4].

22.4 Glycerolemia

Principle
In the blood, hyperglycerolemia is detected by routine clinical laboratory work up and is often misdiagnosed as pseudo-hypertriglyceridemia [2, 7] primarily because the clinical laboratory work-ups do not perform a blank to account for the high levels of glycerol in the patient's blood secondary to the GKD. If the clinical laboratory measures triglycerides as the glycerol released after lipolysis, the individual with GKD will be identified as secondary to hypertriglyceridemia [2]. This is pseudo-hypertriglyceridemia [7]. However, if the clinical laboratory uses a solvent extraction and colorimetry method, there is no interference from the free glycerol secondary to GKD [2, 8]. Glycerol can be detected by many routine methods including oxidation, esterification, dehydration reactions, ether formation, enzymatic determination, and chromatographic methods [9]. Alternatives include the determination of glycerol concentration using an enzymatic-based isotope-dilution method, as described here [6], or commercially available kits such as the Millipore Sigma serum triglyceride determination kit (TR0100, Sigma, St. Louis MO, USA), Millipore Sigma Free glycerol determination kit (FG0100, Sigma, St. Louis MO, USA).

22.4.1 Pre-analytical

Specimen
Blood is drawn using standard techniques in a non-glycerol-containing collection tube.

Reagents and Chemicals
These must be glycerol free. One potential source is the lubricant used between the stopper and tube of some Vacutainer collection devices. Glycerol standards (3–10 mM), ethanol (100% chilled), GK solution (0.2 ng/μl; made from dry *Escherichia coli* GK, G4509, Sigma-Aldrich, in 1 M Tris/10 mM ethylenediaminetetraacetic acid—EDTA—pH 7.4), ATP (100 mM), magnesium sulfate (40 mM), sodium fluoride (62.5 mM), 1 M Tris/10 mM EDTA pH 7.4, β-mercaptoethanol (200 mM made from 14.3 M standard solution), ^{14}C glycerol, diethylaminoethyl (DEAE) cellulose, water, glycerol (1 M), and scintillation vials.

Instrumentation
Standard scintillation counter and water baths.

Calibration
Standard curve using glycerol standards.

Analytical Procedure
Prepare a cocktail containing 6 μl of 100 mM ATP, 10 μl of 40 mM $MgSO_4$, 10 μl of 62.5 mM NaF, 10 μl of 1 M Tris/10 mM EDTA pH 7.4, and 10 μl of 200 mM β-mercaptoethanol per sample needed (total volume of cocktail is 46 μl per sample). Prepare a 1.5-ml microfuge tube for each sample and four standards (0, 3, 8, and 10 mM glycerol) and put 4–6 μl of cocktail into each tube. Add ^{14}C glycerol (to final concentration of 25 μM in a final total volume of 100 μl). Add water to make the volume of ^{14}C-glycerol 24 μl. Add 10 μl of sample or standard. Add 20 μl of GK solution, mix, and incubate at 37 °C in a water bath for 20 min. Stop the reaction by adding 100 μl of chilled 100% ethanol and place on ice for at least 5 min. Centrifuge on maximum speed for 5 min. Spot 100 μl of sample onto DEAE cellulose. Allow the discs to dry fully then wash them in water for 5 min using very gentle agitation. Discard the water in radioactive waste and the wash filters with 1 M glycerol. Repeat and the discard waste in the radioactive waste. Repeat the water wash twice. Place the discs briefly on a paper towel and then put them into scintillation vials containing scintillation fluid. Vortex vigorously until the paper is dispersed. Quantify ^{14}C using a scintillation counter.

Calculation
Prepare a standard curve using the glycerol concentrations and the radioactivity on the filter paper.

22.4.2 Post-analytical

Interpretation
Use the standard curve to determine the concentration of the samples.

Reference Values

In serum and plasma, normal glycerol levels are 0.02–0.27 mM [2] or less than 1.6 mg/dl [4].

Typical Pathological Values

Typical pathological values range from 1.8 to 8.3 mM [2].

Quality Control

If ^{3}H-labeled glycerol-3-phosphate is used, the radioactivity is self-absorbed by the DEAE cellulose, and must be acid-eluted for scintillation counting. DEAE cellulose breaks very easily so washes must be very gentle.

22.5 Follow-Up Enzyme Assays

Glycerol kinase activity can be measured directly or indirectly; however, these assays are not available as a clinical test. The direct method is as has been described previously [5, 10] and is used in the isotope dilution method indicated above. The amount of protein and the incubation time vary between cell types and it is important to be within the linear range of the assay for the given cell type. The indirect methods involve incorporation of ^{14}C from glycerol into macromolecules and its subsequent oxidation to $^{14}CO_2$ [3, 11].

22.6 DNA Testing

Deletions causing cGKD can be detected using fluorescence in situ hybridization analysis or by chromosomal microarrays [12], which are available as clinical tests. For iGKD, the causative mutation can be determined by direct DNA sequencing [13] of the *GK* gene or through exome or genome sequencing.

Conclusion Glycerol and glycerol phosphates are measured with standard methods and are important in diagnosing disorders of glycerol metabolism including glycerol kinase deficiency. This can be augmented by DNA sequencing and SNP arrays. Enzymatic testing for glycerol kinase deficiency is possible but not available clinically.

References

1. Dipple KM, McCabe ERB (2003) Disorders of glycerol metabolism. In: Blau N, Duran M, Blaskovics ME, Gibson KM (eds) Physician's guide to the laboratory diagnosis of metabolic diseases. Springer, Berlin, pp 369–376
2. McCabe E (2001) Disorders of glycerol metabolism. In: Scriver C, Beaudet A, Sly W, Valle D, Childs B, Vogelstein B (eds) Metabolic and molecular bases of inherited diseases. McGraw-Hill, New York, pp 2217–2237
3. Sjarif DR, Hellerud C, van Amstel JK, Kleijer WJ, Sperl W, Lacombe D, Sass JO, Beemer FA, Duran M, Poll-The BT (2004) Glycerol kinase deficiency: residual activity explained by reduced transcription and enzyme conformation. Eur J Hum Genet 12:424–432
4. Goodman SI, Markey SP (1981) Diagnosis of organic acidemias by gas chromatography-mass spectrometry. Alan R. Liss, New York
5. McCabe ER, Fennessey PV, Guggenheim MA, Miles BS, Bullen WW, Sceats DJ, Goodman SI (1977) Human glycerol kinase deficiency with hyperglycerolemia and glyceroluria. Biochem Biophys Res Commun 78:1327–1333
6. Seltzer WK, McCabe ER (1984) Human and rat adrenal glycerol kinase: subcellular distribution and bisubstrate kinetics. Mol Cell Biochem 62:43–50
7. Goussault Y, Turpin E, Neel D, Dreux C, Chanu B, Bakir R, Rouffy J (1982) "Pseudohypertriglyceridemia" caused by hyperglycerolemia due to congenital enzyme deficiency. Clin Chim Acta 123:269–274
8. Rose CI, Haines DS (1978) Familial hyperglycerolemia. J Clin Invest 61:163–170
9. Wu LC, Cheng CM (2005) Flow-injection enzymatic analysis for glycerol and triacylglycerol. Anal Biochem 346:234–240
10. Guggenheim MA, McCabe ER, Roig M, Goodman SI, Lum GM, Bullen WW, Ringel SP (1980) Glycerol kinase deficiency with neuromuscular, skeletal, and adrenal abnormalities. Ann Neurol 7:441–449
11. McCabe ER, Sadava D, Bullen WW, McKelvey HA, Seltzer WK, Rose CI (1982) Human glycerol kinase deficiency: enzyme kinetics and fibroblast hybridization. J Inherit Metab Dis 5:177–182
12. Stanczak CM, Chen Z, Zhang YH, Nelson SF, McCabe ERB (2007) Deletion mapping in Xp21 for patients with complex glycerol kinase deficiency using SNP mapping arrays. Hum Mutat 28:235–240
13. Zhang Y, Dipple KM, Vilain E, Huang BL, Finlayson G, Therrell BL, Worley K, Deininger P, McCabe ER (2000) AluY insertion (ivs4-52ins316alu) in the glycerol kinase gene from an individual with benign glycerol kinase deficiency

Riboflavin (B2) and FAD/FMN Metabolites

23

Annet M. Bosch, Marie van Dijk, and Susanna M. I. Goorden

Contents

Abbreviations

FAD	Flavin adenine dinucleotide
FMN	Flavin mononucleotide
RTD	Riboflavin Transporter Deficiency
RFVT	Riboflavin Transporter
MFT	Mitochondrial FAD transporter
MADD	Multiple Acyl-CoA Dehydrogenase Deficiency
DBS	Dried blood spot

23.1 Introduction

Riboflavin (vitamin B2) is a water-soluble vitamin and is the precursor in the synthesis of flavin mononucleotide (FMN) and flavin adenine dinucleotide (FAD), which function as active cofactors for enzymes called flavoproteins. These flavoproteins play an essential role in numerous redox reactions, involving fat, carbohydrate, and protein metabolism.

A. M. Bosch
Emma Children's Hospital, Department of Pediatrics, Division of Metabolic Diseases, Amsterdam Gastroenterology Endocrinology Metabolism Institute, Amsterdam UMC location University of Amsterdam, Amsterdam, The Netherlands

M. van Dijk
Department of Clinical Chemistry, Laboratory Genetic Metabolic Diseases, Amsterdam UMC location University of Amsterdam, Amsterdam, The Netherlands

S. M. I. Goorden (✉)
Department of Clinical Genetics, Center for Lysosomal and Metabolic Diseases, Erasmus University Medical Center, Rotterdam, The Netherlands
e-mail: s.goorden@erasmusmc.nl

N. Blau, F. M. Vaz (eds.), *Laboratory Guide to the Methods in Biochemical Genetics*, https://doi.org/10.1007/978-3-031-58819-8_23

Riboflavin-related inborn errors comprise genetic defects in riboflavin transport and intracellular riboflavin metabolism. These inborn errors have only recently been identified, as the association of Brown-Vialetto-Van Laere syndrome and Fazio Londe syndrome (now renamed Riboflavin Transporter Deficiency (RTD)) with RFVT2 and RFVT3 deficiency was first reported in 2010 [1, 2], while deficiency of FAD synthase and of the Mitochondrial FAD Transporter (MFT) were first reported in 2016 [3, 4].

More frequent than genetic defects, restricted diets, especially diets without dairy, may cause a nutritional riboflavin deficiency. In the literature, skin problems were reported in severe combined vitamin B deficiencies, in which the role of riboflavin is unclear. There is some evidence that poor riboflavin status interferes with iron handling and contributes to the development of anemia at the time of low iron intake [5]. A maternal riboflavin deficiency is associated with false-positive newborn screening for fatty acid oxidation disorders in the neonate [6, 7] but may also give rise to a life-threatening clinical and biochemical neonatal picture resembling neonatal Multiple Acyl-CoA Dehydrogenase Deficiency (MADD) [8–10].

Flavin levels (riboflavin, FAD, and FMN) are generally measured by HPLC fluorescence-based methods. Decreased flavin levels may indicate genetic defects of riboflavin transport, a nutritional vitamin B2 deficiency, or in rare cases a combination.

23.2 Properties of Metabolites

Riboflavin is made up of an isoalloxazine ring bound to a ribitol side chain (Fig. 23.1). FMN is its monophosphorylated form, while FAD contains an adenosine diphosphate moiety (Fig. 23.1). Riboflavin, FMN and FAD are collectively referred to as flavins, named after the Latin word "flavus" (yellow), due to their yellow-coloring when oxidized.

Flavins are water-soluble vitamins which are present in a wide variety of food products, including dairy products, green vegetables, and meat. In a Western diet milk and other dairy products constitute the major source [5]. Dietary sources contain riboflavin but also protein-bound FMN and FAD [5]. Prior to uptake in the intestine, these protein-bound forms are hydrolyzed to riboflavin by hydrolases present at the enterocyte brush border. The uptake of riboflavin in the enterocyte is mediated by the apically expressed RFVT3 (*SLC52A3*) transporter and the basally expressed RFVT2 (*SLC52A2*) and RFVT1 (*SLC52A1*) transporters. RFVT2 is also important for transport into neuronal cells, while RFVT1 is essential for transplacental transport.

Intracellularly, FMN and FAD, which are formed from riboflavin by the action of respectively riboflavin kinase and FAD synthase (Fig. 23.1), play a role in numerous redox reactions. More specifically, there are around 90 different flavin-dependent proteins, referred to as the flavoproteome, that play a role in lipid and protein metabolism, the mitochondrial electron transport chain, and cellular stress responses and apoptosis [11]. In general, flavoproteins are central in aerobic metabolism as they can catalyze both one- and two-electron transfer reactions. Typically, flavoproteins tightly bind either FAD or FMN non-covalently, although a subset of flavoproteins do bind the flavins covalently to the polypeptide chain [12].

To emphasize the importance of flavins, FAD is a cofactor for the Electron Transfer Flavoprotein (ETF) and ETF Ubiquinone Oxidoreductase (ETF:QO), which shuttle electrons from a number of mitochondrial flavoprotein dehydrogenases (predominantly Acyl-CoA Dehydrogenases (ACADs)) involved in fatty acid and amino acid metabolism to ubiquinone in the mitochondrial inner membrane, which in turn transfers electrons to complex III) [11]. The FMN prosthetic group of NADH:ubiquinone oxidoreductase (complex I) is essential for shuttling electrons to complex II. To function inside the mitochondrion both FAD and FMN are transported by the mitochondrial FAD transporter (MFT; *SLC25A32*) [13].

It appears that excess consumption of flavins is not harmful; riboflavin exceeding tissue requirements is excreted in urine as riboflavin or

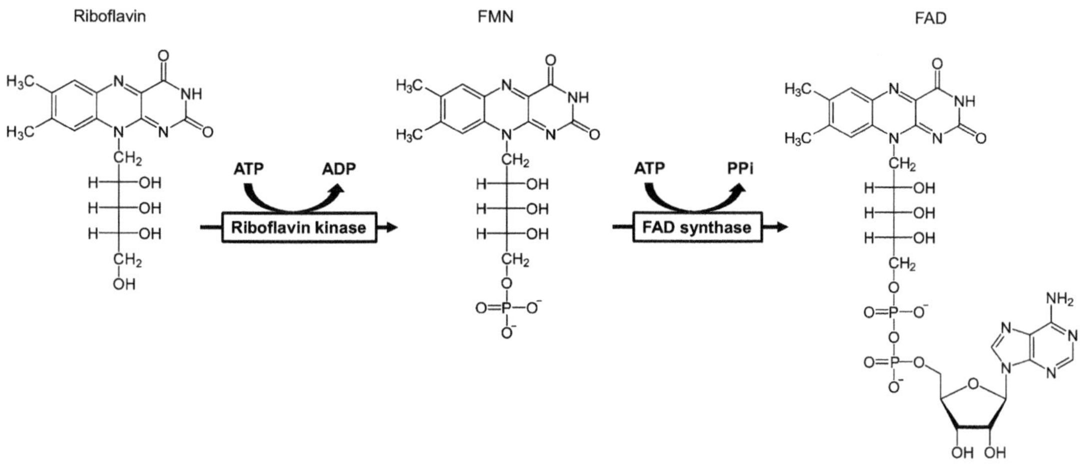

Fig. 23.1 Conversion of riboflavin via FMN to FAD

as its metabolites, 7-hydroxymethylriboflavin and lumiflavin [5]. Therefore, in contrast to many other B vitamins, there is no storage site from which flavins can be mobilized in times of low dietary intake, and vitamin B2 deficiency needs to be treated by supplementation or diet change.

23.3 Clinical Significance

23.3.1 Riboflavin Transporter Deficiency (RTD)

This is a treatable disease of riboflavin transport. Patients may present at all ages. The youngest patients present within the first year of life, usually with a rapidly progressive life-threatening hypotonia, muscular weakness, bulbar palsy, and respiratory insufficiency resulting from muscle weakness and diaphragmatic paralysis. At older ages, symptoms of sensorimotor axial neuropathy and hearing loss are more prominent. Approximately 50% of patients demonstrate an abnormal plasma or Dried Blood Spot (DBS) acylcarnitine profile when untreated, with an increase in short-, medium-, and long-chain acylcarnitines that is reminiscent of profiles of multiple acyl-CoA dehydrogenase deficiency. Flavin concentrations in plasma may be normal or low. Untreated patients all rapidly deteriorate with a limited life expectancy. Oral supplementation of riboflavin is life-saving and most patients demonstrate a striking improvement, while a minority only show stabilization of symptoms. Patients with riboflavin transporter deficiency type 2 (RFVT2) and type 3 (RFVT3) are reported to have similar phenotypes [14, 15].

23.3.2 Disorders of Intracellular Riboflavin Metabolism

23.3.2.1 FAD Synthase Deficiency

Patients with FAD synthase deficiency present either with an early onset form with severe progressive generalized hypotonia in the neonatal period or early infancy, sometimes associated with cardiomyopathy, leading to respiratory failure or cardiac arrest and early death, or with a late onset form with exercise intolerance, and muscle weakness in childhood or adolescence [3, 16–19]. In contrast to transporter defects, no cranial and no peripheral neuropathy is present. Patients present with abnormal plasma or DBS acylcarnitine profiles, exhibiting an increase in short-, medium-, and long-chain acylcarnitines, and urinary organic acid profiles demonstrate an increase in Ethylmalonic Acid (EMA), lactic acid, and, less constantly, glutaric acid, 2-hydroxyglutaric acid, and acylglycine derivatives. Mutation analysis of the *FLAD1* gene (exome, gene panels, or Sanger sequencing) revealed that mutations in the FADS domain (FAD synthesis) are associated with the late-onset phenotype, while mutations in the MPTb domain (FAD cleavage) are associated with the severe early phenotype. No clinical improvement was reported with riboflavin treatment in the early onset form of the disease [19]. The prognosis is severe with death occurring before the first year of life. Conversely, the late-onset form is associated with a clear clinical improvement with low doses of riboflavin [3, 16]. Flavin levels prior to treatment were only reported in two patients [3]. In plasma, flavin levels in these patients were found to be in the reference range, while erythrocyte FMN levels were elevated [3]. It remains to be established whether this is a constant finding with diagnostic potential for this disorder.

23.3.2.2 Mitochondrial FAD Transporter (MFT) Deficiency

The first two published patients with MFT deficiency [4, 20] presented with exercise intolerance, muscle weakness, and vomiting during exercise, starting in infancy or adolescence, leading to a more severe neuromuscular phenotype with progressive dysarthria, swallowing difficulties, myoclonic jerks and need of wheelchair around 50 years in one individual [20]. The acylcarnitine and organic acid profiles were suggestive of Multiple Acyl-CoA Dehydrogenase Deficiency (MADD) in both. Treatment with low doses of riboflavin resulted in a dramatic improvement in symptoms and a normalization of biochemical biomarkers. Symptoms reappeared when vitamin B2 was stopped in one patient [4]. In 2021, three more patients with FAD transporter deficiency were reported, presenting with recurrent episodes of hypoketotic hypoglycemia, without a neuromuscular phenotype [21]. All had acylcarnitine profiles suggestive of MADD and were homozygous for the c.272 G>T variant of the *SLC25A32* gene. After starting low-dose riboflavin supplementation (besides Q10 and carnitine in individual patients) there was no recurrence of the hypoglycemia [21]. Flavin levels prior to treatment were not reported in any of these MFT-deficient patients.

23.3.3 Dietary Riboflavin Deficiency

The recommended intake of riboflavin ranges from 0.3 mg/day in infancy to 1.6 mg/day in lactating women [22]. Riboflavin deficiency usually occurs with other vitamin B deficiencies. It is, therefore, unclear if the symptoms historically reported in the literature, such as skin disorders, are truly caused by riboflavin deficiency in patients without an underlying genetic disease. Severe riboflavin deficiency is very rare in the Western World, but its incidence may increase with the rise of plant-based diets in case of insufficient supplementation.

Maternal riboflavin deficiency secondary to maternal vegan, vegetarian, or lactose-free diets in pregnancy was reported to cause false-positive neonatal screening results for fatty acid oxidation disorders in asymptomatic infants [6, 7]. Life-threatening symptoms of riboflavin deficiency (an MADD-like illness) have been reported in newborns of mothers who were carriers of a riboflavin transporter deficiency or had a severe dietary riboflavin deficiency [6, 8–10].

23.4 Methods

Currently, mostly HPLC fluorescence-based methods are used to measure flavins. Multiplex methods to measure riboflavin using LC-MS/MS were recently developed for clinical use but are mainly aimed at detecting alimentary vitamin deficiencies (for example, see Chen et al. [23, 24]). Therefore, HPLC fluorescence-based method to measure riboflavin, FMN, and FAD in plasma is described here.

In this method, flavins are released from their binding proteins and the sample is deproteinized by adding perchloric acid. Subsequently, flavins are separated by reversed-phase UHPLC using a buffer gradient. Detection is performed by means of their native fluorescence. Individual flavins are identified based on respective retention time and peak shape.

23.4.1 Pre-analytical

Flavins can be measured in plasma, erythrocytes, or whole blood lysates. EDTA is preferred as anticoagulant since it prevents degradation of FMN and FAD into riboflavin by (Ca^{2+}/Mg^{2+}-dependent) pyrophosphatases. When measuring in plasma of heparinized blood local reference values should be established (verification/use of EDTA-based reference values does not suffice). Please note that in erythrocytes only trace amounts of riboflavin are present while FMN and FAD are enriched in erythrocytes compared to plasma [25]. Therefore, hemolysis may lead to falsely elevated FAD and FMN concentrations in plasma, with no effect on riboflavin levels. As flavins are light sensitive [26], it is recommended to protect samples (and standards) from light. Store plasma samples shielded from light at −20 °C prior to measurement.

23.4.2 Reagents and Chemicals

FAD and FMN standards are purchased from Sigma; riboflavin standard from Merck; galactoflavin (internal standard) from MSD. Potassium dihydrogen phosphate (KH_2PO_4), perchloric acid (PCA), disodium hydrogen phosphate dihydrate ($Na_2HPO_4.2H_2O$), and ethylenedinitrilotetraacetic acid (EDTA) are purchased from Merck; methanol (99.8%) and acetonitrile (ACN) from Biosolve.

23.4.3 Instrumentation

UHPLC system with fluorescence detection:

- Dionex UltiMate 3000 RS UHPLC with quaternary pump, autosampler, and a thermostatted column compartment.
- Thermo Scientific RSLC (rapid separation liquid chromatography) Acclaim Polar Advantage (PA) column (100 × 2.2 mm, 2 μm).
- Shimadzu RF-20A xs Fluorescence detector.

23.4.4 Analytical Procedure/Manual

23.4.4.1 Preparation of Reagents and Standards

Reagents:

- *Stock solution 1 M KH_2PO_4 with 1 mM EDTA*: dissolve 136 grams KH_2PO_4 and 372 mg EDTA in 1000 ml distilled water.
- *Eluens D = 0.05 M KH_2PO_4 ± 13% (v/v) acetonitril (pH 4.8):* dilute 25 ml 1 M KH_2PO_4 with distilled water until total volume of 500 ml is reached. Add 75 ml acetonitril.
- *Eluens B = methanol +9.1% (v/v) distilled water*: Add 100 ml distilled water to 1-liter methanol.
- *Primary 0.67 M KH_2PO_4 buffer (pH 4.65):* dissolve 11,876 grams KH_2PO_4 in 100 ml distilled water.
- *Secondary 0.67 M $Na_2HPO_4.2H_2O$ buffer (pH 9.25):* dissolve 9078 grams $Na_2HPO_4.2H_2O$ in 100 ml distilled water.
- *Na/K-buffer, pH 8.0:* dilute 0.5 ml primary KH_2PO_4 buffer with 9.5 ml secondary $Na_2HPO_4.2H_2O$ buffer. Beware of crystallization of the buffers. Heat to 35 °C shortly before use.

Standards:

The calibration standard contains riboflavin, FAD, and FMN; galactoflavin is added to samples and calibration standard as internal standard:

- *Stock standard FAD (656 μM):* dissolve 27.2 mg FAD in 50 ml distilled water.
- *Stock standard FMN (792 μM):* dissolve 24.9 mg FMN in 50 ml distilled water.
- *Stock standard riboflavin (120 μM):* dissolve 22.5 mg riboflavin in 500 ml distilled water.
- For the *mixed stock calibration standard* dilute FAD, FMN, and riboflavin standards in a volumetric flask with distilled water to obtain the following concentrations:

	Concentration (nM)
FAD	13,116
FMN	3959
RF	5984

Aliquot in tubes containing 500 µl and store in dark at −20 °C.

- The *calibration standard* is freshly made on the day of measurement (stability 4–6 h) and is further processed similarly to plasma samples. To obtain the calibration standard dilute the mixed stock calibration standard 1:100 with distilled water to obtain the following concentrations:

- *Stock standard galactoflavin (1.2 µM):* dissolve 50 mg galactoflavin in 100 ml 0.01 N HCl.
- *Working standard galactoflavin (1.2 nM):* dilute the stock standard 1000× with 0.01 M HCL.

	Dilution	Concentration (nM)
FAD	1:100	131
FMN	1:100	40
RF	1:100	60

Sample and calibration standard preparation

Pipette samples according to the scheme below:

	Amount (µL)	Amount (µL)
Calibration standard or blank	100	
Plasma sample		100
IS = GF 0.5 mg/l	20	20
0.6 N PCA	50	
1 N PCA		50

Place samples at 4 °C for 30 min. Centrifuge samples 10 min at 13,680 g.

Continue pipetting in a Gilson vial according to the scheme below (same volumes for standard and plasma samples):

	Amount (µL)	Amount (µL)
Supernatant	170	170
0.7 M Na/K-buffer pH 8.0	100	100

Mix the Gilson vials on a vortex mixer for at least 5 min (pH should be: 7.0 ± 0.2).

Place the vials in the autosampler.

Chromatographic conditions

Flavins are separated on a reverse phase RSLC Acclaim PA column at 18 °C. Injection volume is 10 µL.

The gradient profile is according to the following table:

Step	Time (min)	Flow (ml/min)	% B	% D
0	Equilibration	0.2	0	100
1	2.0	0.4	50	50
2	2.2	0.4	90	10
3	2.6	0.4	90	10
4	2.8	0.4	0	100
5	7.0	0.4	0	100
6	8.0	0.2	0	100

Detection is performed by the fluorescence detector: excitation: $\lambda = 445$ nm; emission: $\lambda = 530$ nm.

23.4.5 Calculation

After calibration with the calibration standard, concentrations of flavins in control and patient samples are calculated using the internal standard method by the Chromeleon software.

23.4.6 Post-analytical (e.g., Chromatograms of Pathological Samples)

A typical chromatogram of flavins and the internal standard galactoflavin in EDTA-plasma is shown in Fig. 23.2.

23.4.7 Reference Values

Metabolite	Specimen	Age group	Sex	Reference values (nmol/l)
Riboflavin	EDTA-plasma	All	All	3.9–49
FAD	EDTA-plasma	All	All	46–114
FMN	EDTA-plasma	All	All	2.8–21.4

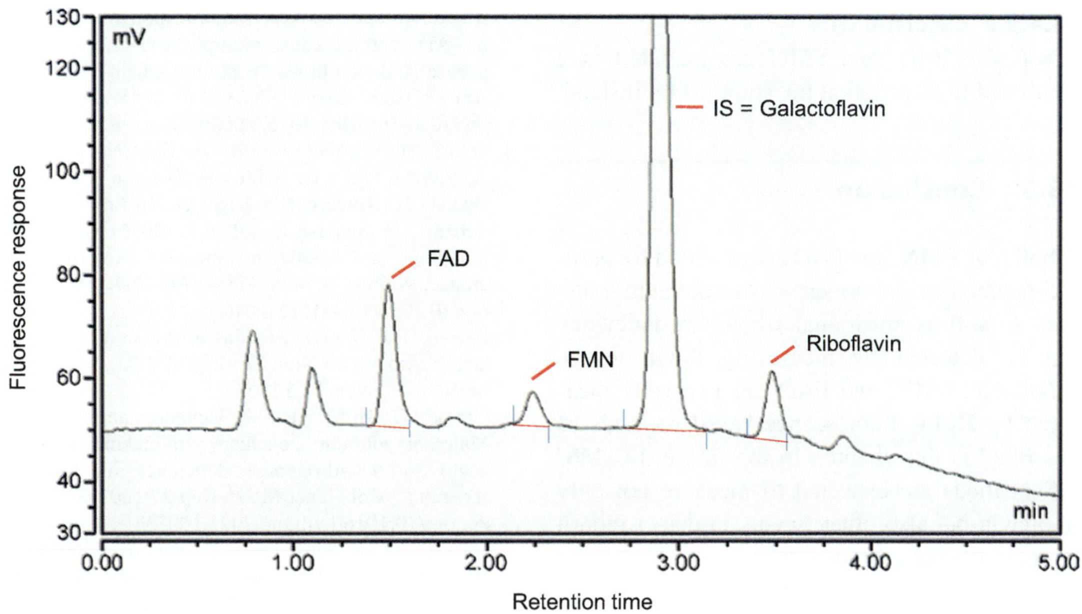

Fig. 23.2 A chromatogram of flavins and the internal standard galactoflavin in human EDTA-plasma

23.4.8 Pathological Values

In general, plasma flavin levels, in particular, riboflavin and FMN, are good indicators of vitamin B2 status and are typically decreased in vitamin B2 deficiency and respond to riboflavin suppletion [25]. In some cases of vitamin B2 deficiency riboflavin levels may be decreased, while FMN and especially FAD levels can (still) be normal [25]. This might be due to differential regulation of riboflavin-metabolizing enzymes which lead to conservation of FAD at the expense of riboflavin [27]. In addition, unbound flavins are relatively labile and quickly converted to riboflavin and excreted, while FMN and FAD bound to flavoproteins are less available for hydrolysis, resulting in so-called metabolic trapping [5].

In riboflavin transporter deficiency (RFVT2 or 3) decreased flavin levels are found in some patients, but not in all (exact distribution unknown due to limited availability and use of this test) [15]. Therefore, a suspicion of riboflavin transporter deficiency should not be rejected based on normal flavin levels [15]. It is unknown whether flavin concentrations are altered in riboflavin metabolism defects (MFT and FAD synthase deficiency) [28].

Besides flavins, acylcarnitine profiles and urine organic acids are often abnormal in patients with functional vitamin B2 deficiencies (nutritional or genetic) and show MADD-like patterns. In these patterns varying elevations of acylcarnitines can be found (in range of C4-C18-carnitines), while in organic acids analysis excretion of ethylmalonic acid, glutaric acid, 2-hydroxyglutaric acid, dicarboxylic acids and glycine conjugates may be increased, all reflecting impaired activity of flavin-dependent dehydrogenases [29].

23.4.9 Quality Control

23.4.9.1 Internal QC

In each run, two control plasma samples (high and low pool) are measured. The low pool is prepared by mixing left over (anonymized) plasma samples. To obtain the high pool the low pool is spiked with FMN, FAD, and riboflavin standards. Results of these internal control samples are analyzed using Westgard multiQC rules.

23.4.9.2 External QC

Riboflavin, FAD, and FMN are included in a combined EQC program for vitamins by Instand.

23.5 Conclusion

Riboflavin, FMN, and FAD are essential for aerobic metabolism. Inborn errors in riboflavin transport as well as nutritional riboflavin deficiency can be detected by measuring flavin levels. Riboflavin, FMN, and FAD are generally measured by HPLC fluorescence-based methods as described in this chapter. In the future, LC-MS/MS methods are expected to measure not only riboflavin but also other flavins to detect inborn errors in riboflavin transport.

Acknowledgments We thank the technicians of Laboratory Genetic Metabolic Diseases, Amsterdam UMC, for setting up this HPLC method for measurement of flavins and implementing and running this in day-to-day patient diagnostics.

References

1. Bosch AM, Abeling NG, Ijlst L, Knoester H, van der Pol WL, Stroomer AE, Wanders RJ, Visser G, Wijburg FA, Duran M, Waterham HR (2011) Brown-Vialetto-Van Laere and Fazio Londe syndrome is associated with a riboflavin transporter defect mimicking mild MADD: a new inborn error of metabolism with potential treatment. J Inherit Metab Dis 34(1):159–164. https://doi.org/10.1007/s10545-010-9242-z
2. Green P, Wiseman M, Crow YJ, Houlden H, Riphagen S, Lin JP, Raymond FL, Childs AM, Sheridan E, Edwards S, Josifova DJ (2010) Brown-Vialetto-Van Laere syndrome, a ponto-bulbar palsy with deafness, is caused by mutations in c20orf54. Am J Hum Genet 86(3):485–489. https://doi.org/10.1016/j.ajhg.2010.02.006
3. Olsen RKJ, Konarikova E, Giancaspero TA, Mosegaard S, Boczonadi V, Matakovic L, Veauville-Merllie A, Terrile C, Schwarzmayr T, Haack TB, Auranen M, Leone P, Galluccio M, Imbard A, Gutierrez-Rios P, Palmfeldt J, Graf E, Vianey-Saban C, Oppenheim M, Schiff M, Pichard S, Rigal O, Pyle A, Chinnery PF, Konstantopoulou V, Moslinger D, Feichtinger RG, Talim B, Topaloglu H, Coskun T, Gucer S, Botta A, Pegoraro E, Malena A, Vergani L, Mazza D, Zollino M, Ghezzi D, Acquaviva C, Tyni T, Boneh A, Meitinger T, Strom TM, Gregersen N, Mayr JA, Horvath R, Barile M, Prokisch H (2016) Riboflavin-responsive and -non-responsive mutations in FAD synthase cause multiple acyl-CoA dehydrogenase and combined respiratory-chain deficiency. Am J Hum Genet 98(6):1130–1145. https://doi.org/10.1016/j.ajhg.2016.04.006
4. Schiff M, Veauville-Merllie A, Su CH, Tzagoloff A, Rak M, Ogier de Baulny H, Boutron A, Smedts-Walters H, Romero NB, Rigal O, Rustin P, Vianey-Saban C, Acquaviva-Bourdain C (2016) SLC25A32 mutations and riboflavin-responsive exercise intolerance. N Engl J Med 374(8):795–797. https://doi.org/10.1056/NEJMc1513610
5. Powers HJ (2003) Riboflavin (vitamin B-2) and health. Am J Clin Nutr 77(6):1352–1360. https://doi.org/10.1093/ajcn/77.6.1352
6. Gramer G, Hoffmann GF, Hennermann JB (2021) Maternal vitamin deficiency mimicking multiple acyl-CoA dehydrogenase deficiency on newborn screening. Mol Genet Metab Rep 27:100738. https://doi.org/10.1016/j.ymgmr.2021.100738
7. Hagemeijer MC, Oussoren E, Ruijter GJG, Onkenhout W, Huidekoper HH, Ebberink MS, Waterham HR, Ferdinandusse S, de Vries MC, Huigen M, Kluijtmans LAJ, Coene KLM, Blom HJ (2021) Abnormal VLCADD newborn screening resembling MADD in four neonates with decreased riboflavin levels and VLCAD activity. JIMD Rep 61(1):12–18. https://doi.org/10.1002/jmd2.12223
8. Ho G, Yonezawa A, Masuda S, Inui K, Sim KG, Carpenter K, Olsen RK, Mitchell JJ, Rhead WJ, Peters G, Christodoulou J (2011) Maternal riboflavin deficiency, resulting in transient neonatal-onset glutaric aciduria Type 2, is caused by a microdeletion in the riboflavin transporter gene GPR172B. Hum Mutat 32(1):E1976–E1984. https://doi.org/10.1002/humu.21399
9. Jaeger B, Corpeleijn W, Dijsselhof M, Goorden S, Haverkamp J, Langeveld M, Waterham H, Westerbeek E, Bosch AM (2022) Mind the B2: life-threatening neonatal complications of a strict vegan diet during pregnancy. Neonatology 119(6):777–780. https://doi.org/10.1159/000526334
10. Mosegaard S, Bruun GH, Flyvbjerg KF, Bliksrud YT, Gregersen N, Dembic M, Annexstad E, Tangeraas T, Olsen RKJ, Andresen BS (2017) An intronic variation in SLC52A1 causes exon skipping and transient riboflavin-responsive multiple acyl-CoA dehydrogenation deficiency. Mol Genet Metab 122(4):182–188. https://doi.org/10.1016/j.ymgme.2017.10.014
11. Mosegaard S, Dipace G, Bross P, Carlsen J, Gregersen N, Olsen RKJ (2020) Riboflavin deficiency-implications for general human health and inborn errors of metabolism. Int J Mol Sci 21(11). https://doi.org/10.3390/ijms21113847
12. Joosten V, van Berkel WJ (2007) Flavoenzymes. Curr Opin Chem Biol 11(2):195–202. https://doi.org/10.1016/j.cbpa.2007.01.010
13. Spaan AN, Ijlst L, van Roermund CW, Wijburg FA, Wanders RJ, Waterham HR (2005) Identification of the human mitochondrial FAD transporter and its

potential role in multiple acyl-CoA dehydrogenase deficiency. Mol Genet Metab 86(4):441–447. https://doi.org/10.1016/j.ymgme.2005.07.014
14. Jaeger B, Bosch AM (2016) Clinical presentation and outcome of riboflavin transporter deficiency: mini review after five years of experience. J Inherit Metab Dis 39(4):559–564. https://doi.org/10.1007/s10545-016-9924-2
15. O'Callaghan B, Bosch AM, Houlden H (2019) An update on the genetics, clinical presentation, and pathomechanisms of human riboflavin transporter deficiency. J Inherit Metab Dis 42(4):598–607. https://doi.org/10.1002/jimd.12053
16. Auranen M, Paetau A, Piirila P, Pohju A, Salmi T, Lamminen A, Lofberg M, Mosegaard S, Olsen RK, Tyni T (2017) Patient with multiple acyl-CoA dehydrogenation deficiency disease and FLAD1 mutations benefits from riboflavin therapy. Neuromuscul Disord 27(6):581–584. https://doi.org/10.1016/j.nmd.2017.03.003
17. Garcia-Villoria J, De Azua B, Tort F, Mosegaard S, Ugarteburu O, Texido L, Morales-Romero B, Olsen RKJ, Ribes A (2018) FLAD1, encoding FAD synthase, is mutated in a patient with myopathy, scoliosis and cataracts. Clin Genet 94(6):592–593. https://doi.org/10.1111/cge.13452
18. Muru K, Reinson K, Kunnapas K, Lillevali H, Nochi Z, Mosegaard S, Pajusalu S, Olsen RKJ, Ounap K (2019) FLAD1-associated multiple acyl-CoA dehydrogenase deficiency identified by newborn screening. Mol Genet Genomic Med 7(9):e915. https://doi.org/10.1002/mgg3.915
19. Yildiz Y, Olsen RKJ, Sivri HS, Akcoren Z, Nygaard HH, Tokatli A (2018) Post-mortem detection of FLAD1 mutations in 2 Turkish siblings with hypotonia in early infancy. Neuromuscul Disord 28(9):787–790. https://doi.org/10.1016/j.nmd.2018.05.009
20. Hellebrekers D, Sallevelt S, Theunissen TEJ, Hendrickx ATM, Gottschalk RW, Hoeijmakers JGJ, Habets DD, Bierau J, Schoonderwoerd KG, Smeets HJM (2017) Novel SLC25A32 mutation in a patient with a severe neuromuscular phenotype. Eur J Hum Genet 25(7):886–888. https://doi.org/10.1038/ejhg.2017.62
21. Al Shamsi B, Al Murshedi F, Al Habsi A, Al-Thihli K (2022) Hypoketotic hypoglycemia without neuromuscular complications in patients with SLC25A32 deficiency. Eur J Hum Genet 30(8):976–979. https://doi.org/10.1038/s41431-021-00995-7
22. WHO (2004) Vitamin and mineral requirements in human nutrition, 2nd edn
23. Chen P, Tang Y, He Q, Liu L, Zhou Z, Song Y, Zhang N, Wang B, Zhou H, Shi H, Jiang J (2022) A sensitive UPLC-MS/MS method for simultaneous quantification of one-carbon metabolites & co-factors in human plasma. J Pharm Biomed Anal 219:114944. https://doi.org/10.1016/j.jpba.2022.114944
24. Diniz M, Dias N, Andrade F, Paulo B, Ferreira A (2019) Isotope dilution method for determination of vitamin B2 in human plasma using liquid chromatography-tandem mass spectrometry. J Chromatogr B Analyt Technol Biomed Life Sci 1113:14–19. https://doi.org/10.1016/j.jchromb.2019.03.001
25. Hustad S, McKinley MC, McNulty H, Schneede J, Strain JJ, Scott JM, Ueland PM (2002) Riboflavin, flavin mononucleotide, and flavin adenine dinucleotide in human plasma and erythrocytes at baseline and after low-dose riboflavin supplementation. Clin Chem 48(9):1571–1577
26. Treadwell GE, Cairns WL, Metzler DE (1968) Photochemical degradation of flavins. V. Chromatographic studies of the products of photolysis of riboflavin. J Chromatogr 35(3):376–388. https://doi.org/10.1016/s0021-9673(01)82399-2
27. Fass S, Rivlin RS (1969) Regulation of riboflavin-metabolizing enzymes in riboflavin deficiency. Am J Physiol 217(4):988–991. https://doi.org/10.1152/ajplegacy.1969.217.4.988
28. Balasubramaniam S, Christodoulou J, Rahman S (2019) Disorders of riboflavin metabolism. J Inherit Metab Dis 42(4):608–619. https://doi.org/10.1002/jimd.12058
29. Blau N (Ed CDVE), Ferreira CR (Editor), Vianey-Saban C (Editor), van Karnebeek CDM (Editor) (2022) Physician's guide to the diagnosis, treatment, and follow-up of inherited metabolic diseases. In: Karnebeek CDMv (ed) Physician's guide to the diagnosis, treatment, and follow-up of inherited metabolic diseases. Springer

B_6 Vitamers

24

Déborah Mathis

Contents

Abbreviations

AA	Acetic acid
ACN	Acetonitrile
Alb	Albumin
ALP	Alkaline phosphatase
CRP	C-reactive protein
CSF	Cerebrospinal fluid
DD	Differential diagnosis
ERT	Enzyme replacement therapy
FA	Formic acid
GPI	Glycophosphatidylinositol anchor
Hb	Hemoglobin
5-HIAA	5-Hydroxyindoleacetic acid
HFBA	Heptafluorobutyric acid
HPLC	High-performance liquid chromatography
HVA	Homovanillic acid
IEM	Inborn error of metabolism
IG	Intestinal glycosidase
IP	Intestinal phosphatase
ISOD	Isolated sulfite oxidase deficiency
LC-MS/MS	Liquid chromatography-tandem mass spectrometry
MOCOD	Molybdenum cofactor deficiency
3-MT	3-Methoxytyrosine
PA	Pyridoxic acid
PFP	Pentafluorophenylpropyl
PL	Pyridoxal
PLP	Pyridoxal 5′-phosphate
PLPBP	PLP-binding protein
PM	Pyridoxamine
PMP	Pyridoxamine 5′-phosphate
PN	Pyridoxine
PNP	Pyridoxine 5′-phosphate

D. Mathis (✉)
University Institute of Clinical Chemistry, Inselspital, Bern University Hospital, University of Bern, Bern, Switzerland
e-mail: deborah.mathis@insel.ch

N. Blau, F. M. Vaz (eds.), *Laboratory Guide to the Methods in Biochemical Genetics*,
https://doi.org/10.1007/978-3-031-58819-8_24

PNG	Pyridoxine 5′-β-glucoside
PNPO	Pyridox(am)ine phosphate oxidase
PROSC	See PLPBP
QC	Quality control
TCA	Trichloroacetic acid
TNSALP	Tissue non-specific alkaline phosphatase
VLA	Vanillactic acid

24.1 Introduction

Chemical Structure of Vitamin B_6 Vitamin B_6 (B_6) is a water-soluble vitamin present in food with various chemical structures, which are called the B_6 vitamers. All B_6 vitamers contain a pyridine ring with different functional groups at the 4 and 5′ positions [1]. The presence or absence of a phosphate group at the 5′-position and an aldehyde, amine, or alcohol functional group at the 4-position result in pyridoxal 5′-phosphate (PLP), pyridoxal (PL), pyridoxamine 5′-phosphate (PMP), and pyridoxamine (PM), pyridoxine 5′-phosphate (PNP), pyridoxine (PN), respectively. Pyridoxine 5′-β-glucoside (PNG) is also present in food (Fig. 24.1).

Metabolism of Vitamin B_6 B_6 vitamers need to be taken up from the diet or produced from the gut microbiota [2]. Good sources are meat, fish, and eggs, in which vitamin B_6 is found mainly as PLP and PL. In legumes, nuts, bananas, and potatoes, PN, PNP, and PNG are the main vitamers present [2, 3]. Once in the intestine, B_6 vitamers are absorbed in non-phosphorylated forms. Thus, the phosphorylated B_6 vitamers and glucoside of PN are first hydrolyzed by Intestinal Phosphatases (IP) and Intestinal Glycosidase (IG), respectively (Fig. 24.2) [2]. Rephosphorylation and conversion of PMP and PNP into the active cofactor PLP occur either already in the enterocytes or mainly in the liver by Pyridoxal Kinase (PK) and Pyridox(am)ine Phosphate Oxidase (PNPO), respectively. PNPO is a cytosolic oxygen- and FMN-dependent oxidase, expressed in all cell types [4]. Plasma transport of PLP occurs bound to the lysine 190 residue of albumin [2]. PLP reaches the brain and periphery and is again dephosphorylated prior entering the cells by Tissue Non-Specific Alkaline Phosphatase (TNSALP). TNSALP is an ectoenzyme tethered to cell membranes by a Glycophosphatidylinositol (GPI) anchor. Once in the brain and other target cells, PL is "trapped" as PLP and exerts its function as a cofactor in enzyme-catalyzed reactions.

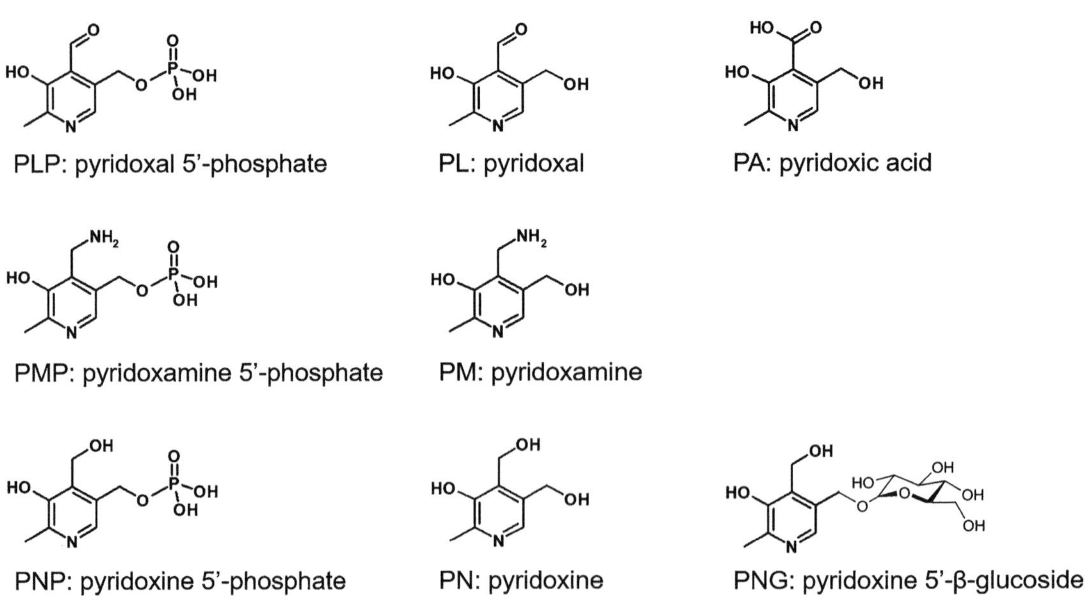

Fig. 24.1 Chemical structure of the B_6 vitamers and the degradation product pyridoxic acid

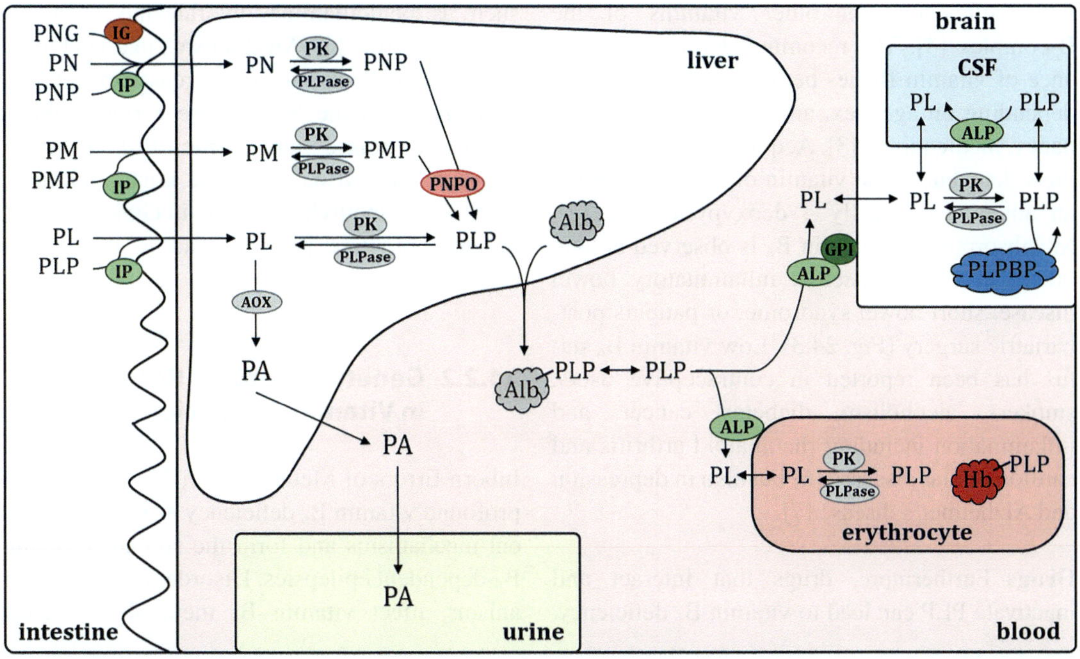

Fig. 24.2 Scheme of vitamin B_6 metabolism in the different body compartments. Alb: albumin; ALP: alkaline phosphatase; CSF: cerebrospinal fluid; GPI: glycophosphatidylinositol anchor; Hb: hemoglobin; IG: intestinal glycosidase; IP: intestinal phosphatase; PA: pyridoxic acid; PK: pyridoxal kinase; PL: pyridoxal; PLP: pyridoxal 5′-phosphate; PLPBP: PLP-binding protein; PM: pyridoxamine; PMP: pyridoxamine 5′-phosphate; PN: pyridoxine; PNG: pyridoxine 5′-β-glucoside; PNP: pyridoxine 5′-phosphate; PNPO: pyridox(am)ine phosphate oxidase

At this stage, PMP can be produced due to incomplete catalysis, which is converted back to PLP by PNPO. Cellular PLP homeostasis is of utmost importance to prevent inappropriate stress reactions and the intracellular-free PLP concentrations are maintained at approximately 1 μM [2]. Mechanisms that regulate PLP homeostasis involve negative feedback to PNPO and PK by PLP itself and binding of PLP to PLP-binding protein (PLPBP, previously PROSC). Excess of vitamin B_6 is oxidized in the liver to Pyridoxic Acid (PA) prior to excretion in the urine (Fig. 24.2).

Function of Vitamin B_6 All B_6 vitamers need to be converted into the active form PLP to exert the physiological function of vitamin B_6 involving its reactive aldehyde as cofactor in a variety of enzyme-catalyzed reactions. The aldehyde of PLP reacts with amino groups to produce a Schiff base that further reacts to complete the reaction and recover PLP [5]. PLP is involved in more than 140 reactions playing central roles in the metabolism of amino acids, neurotransmitters, folate, methylation, carbohydrates, lipids, and in proteins, polyamines, and heme synthesis [2].

This chapter describes the diverse pathologies associated with vitamin B_6 deficiency, the functional markers of vitamin B_6 deficiency, as well as the indications and analytical methodologies for the measurement of B_6 vitamers.

24.2 Clinical Significance of Vitamin B_6 Deficiency

24.2.1 Acquired Causes of Vitamin B_6 Deficiency

Acquired vitamin B_6 deficiency alone is rare and the condition is mostly occurring in combination with deficiencies of other vitamins of the B-complex [4]. The recommended daily allowance of vitamin B_6 lies between 1.3 and 1.7 mg depending on age, sex, and other factors (pregnancy, medications) [3]. Acquired causes include diets deficient in the vitamin or diets containing an antagonist, usually 4′-deoxypyridoxine [4]. Malabsorption of vitamin B_6 is observed among others in coeliac disease, inflammatory bowel disease, short bowel syndrome, or patients post-bariatric surgery (Fig. 24.3). Low vitamin B_6 status has been reported in contraceptive users, smokers, alcoholism, diabetes, cancer, and inflammation including rheumatoid arthritis and cardiovascular diseases [6] but also in depression and Alzheimer's disease [7].

Drugs Furthermore, drugs that interact and inactivate PLP can lead to vitamin B_6 deficiency, such as penicillamine, hydralazine, isoniazid, and cycloserine [2]. Mechanisms that lead to the deficiency not only include increased excretion in urine but also the Schiff base formed inhibits pyridoxal kinase and some PLP-dependent enzymes [4]. Also, pyridoxal kinase inhibitors such as 4-O-methylpyridoxine in Ginkgo nuts [8] or aminophylline [9] can lead to deficiencies (Fig. 24.3).

24.2.2 Genetic Disorders Resulting in Vitamin B₆ Deficiency

Inborn Errors of Metabolisms (IEMs) leading to profound vitamin B_6 deficiency arise from different mechanisms and form the so-called vitamin B_6-dependent epilepsies. Disorders, which mechanisms affect vitamin B_6 metabolism include

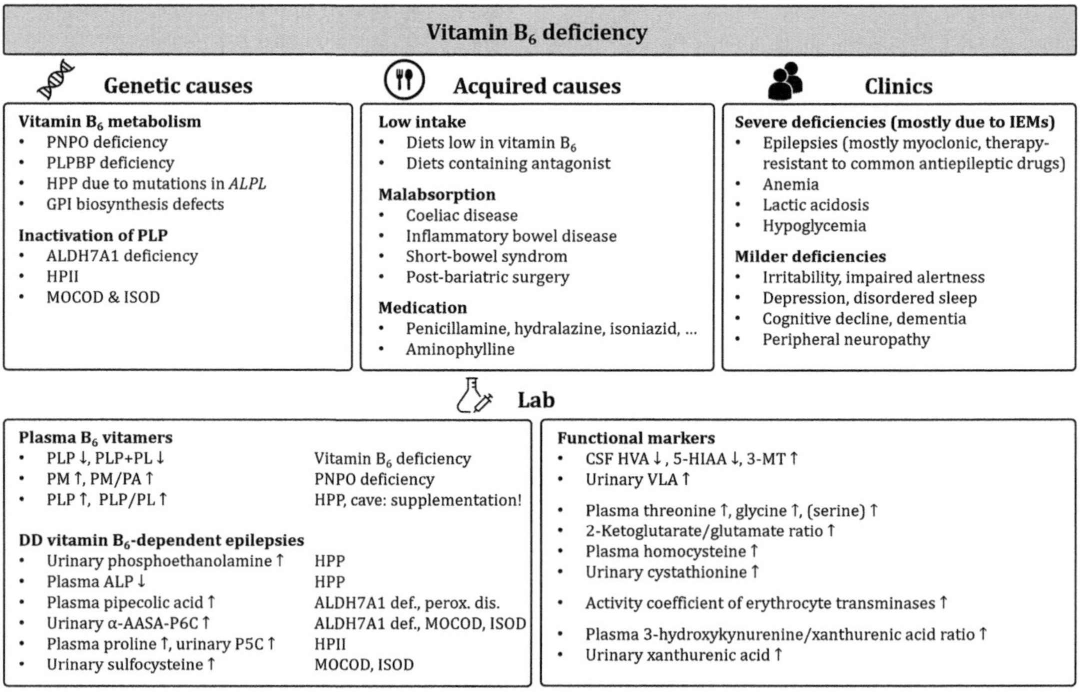

Fig. 24.3 Vitamin B_6 deficiency: genetic and acquired causes, clinical consequences, and laboratory work-up. ALP: alkaline phosphatase; CSF: cerebrospinal fluid; DD: differential diagnosis; GPI: glycophosphatidylinositol anchor; 5-HIAA: 5-hydroxyindoleacetic acid; HVA: homovanillic acid; IEM: inborn error of metabolism; ISOD: isolated sulfite oxidase deficiency; MOCOD: molybdenum cofactor deficiency; 3-MT: 3-methoxytyrosine; PA: pyridoxic acid; perox. dis.: peroxisomal disorders; PL: pyridoxal; PLP: pyridoxal 5′-phosphate; PM: pyridoxamine; PN: pyridoxine; PNPO: pyridox(am)ine phosphate oxidase; VLA: vanillactic acid

PNPO deficiency, PLPBP deficiency, and hypophosphatasia (HPP) due to mutations in *ALPL*. Some IEMs causing deficiency in proteins involved in GPI anchor process can also occasionally cause vitamin B_6-responsive seizures [10]. Other disorders arising from mechanisms that inactivate PLP include ALDH7A1 (antiquitin) deficiency, hyperprolinemia type II (HPII), molybdenum cofactor deficiency (MOCOD), and isolated sulfite oxidase deficiency (ISOD) (Fig. 24.2).

PNPO Deficiency (OMIM #610090) Pyridox(am)ine Oxidase (PNPO) is the enzyme that converts PMP and PNP into PLP (synthesis and recycling). PNPO deficiency can be treated with pharmacological doses of PLP or in some cases PN. Elevated PM and the PM/PA ratio are diagnostic for PNPO deficiency [11].

PLPBP Deficiency (OMIM #617290) PLP-binding protein (PLPBP, formerly PROSC) is involved in the intracellular homeostasis of PLP [12, 13]. There is no biomarker known to date.

Hypophosphatasia (HPP) due to Mutations in *ALPL* (OMIM #241500 #241510 #146300) HPP leads to deficiency in tissue non-specific alkaline phosphatase (TNSALP) and to an impaired cellular uptake of PLP [14]. The disease is characterized by elevated plasma PLP and PLP/PL ratio, low activity of ALP, and increased excretion of phosphoethanolamine in urine [15, 16].

ALDH7A1 Deficiency (OMIM #266100) Also called Antiquitin deficiency or α-aminoadipic semialdehyde dehydrogenase deficiency. ALDH7A1 is an enzyme in the degradation pathway of lysine and its deficiency is the most common cause of vitamin B_6-dependent epilepsy. The disorder leads to the accumulation of α-aminoadipic semialdehyde (AASA) and Δ^1-piperideine-6-carboxylate (P6C) that react and inactivate PLP to form a Knoevenagel condensation product [17, 18]. Specific biomarkers are elevated plasma pipecolic acid and urinary AASA-P6C [16].

Hyperprolinemia type II (HPII, OMIM #239510) Deficiency of the enzyme Δ^1-pyrroline-5-carboxylate (P5C) dehydrogenase involved in the degradation of proline and leading to accumulation of proline and P5C that react and inactivate PLP, similarly to ALDH7A1 deficiency [19].

Molybdenum Cofactor Deficiency (MOCOD, OMIM #252150 #252160 #615501) and Isolated Sulfite Oxidase Deficiency (ISOD, OMIM #272300) MOCOD and ISOD lead to the accumulation of sulfite, a nucleophile capable of reacting with the aldehyde of PLP to produce an inactive sulfonate [20].

24.2.3 Manifestation of Vitamin B_6 Deficiency

Because of the activity of PLP in numerous enzymatic reactions playing important roles in the brain, vitamin B_6 deficiency mostly presents with serious neurological symptoms. In the case of the IEMs that lead to severe vitamin B_6 deficiency, therapy-resistant seizures (often myoclonic) usually of early onset is the leading symptom [2]. This can be accompanied by other consequences including anemia, lactic acidosis, or hypoglycemia. Other brain-specific symptoms include irritability, impaired alertness, depression, disordered sleep, cognitive decline, dementia, and autonomic dysfunction [3]. In later-onset and in adults, vitamin B_6 deficiency manifests mainly as peripheral neuropathy [2].

Life-long treatment with pharmacological doses of PN or in some cases, PLP in vitamin B_6-dependent epilepsies mostly leads to seizure control. However, a high number of patients still develop intellectual disabilities [21]. Furthermore, high oral supplemental doses of pyridoxine can have neurotoxic and photosensitive effects [4] and affect the redox status by increasing oxidative stress [21].

24.2.4 Indications for the Assessment of Vitamin B_6 Status and the Analysis of the B_6 Vitamers

Assessment of vitamin B_6 status (mostly in combination with vitamin B_1 status) may be indicated in the various acquired conditions stated above (malabsorption syndromes, alcoholism, depression, etc.). Lately, it was also recommended for the differential diagnosis (DD) of dementia [22]. In most cases, vitamin B_6 status is assessed by the sole measurement of plasma PLP, although this has been shown to be inaccurate, especially in a diseased population [6]. Determination of the B_6 vitamers, PLP in combination with PL and PA would offer a better approach (see point 2.2).

In the DD of vitamin B_6-dependent epilepsies, the determination of the B_6 vitamers in plasma, including PM and the PM/PA ratio is necessary to detect PNPO deficiency [11, 16]. In addition, plasma PLP in combination with the PLP/PL ratio is useful for the diagnosis of HPP [23].

24.3 Laboratory Assessment of Vitamin B_6 Status

As with other B vitamins that act as cofactors in enzymatic reactions, biochemical assessment can be made directly by analyzing the vitamin/vitamers or by analyzing functional markers, arising from the impaired enzymatic reactions [3]. Since those functional markers (see below) are not specific for vitamin B_6 deficiency, a combination of several functional markers with direct measurement of PLP and the B_6 vitamers offers usually the best approach [4].

24.3.1 Functional Biomarkers of Vitamin B_6 Deficiency

Neurotransmitters and Metabolites PLP plays an essential role in amino acid metabolism and is a rate-limiting cofactor in the synthesis of neurotransmitters (dopamine, serotonin, γ-aminobutyric acid (GABA), and noradrenaline). Even mild deficiencies of vitamin B_6 downregulate their biosynthesis [3]. As a consequence, decreased levels of CSF homovanillic acid (HVA) and 5-hydroxyindoleacetic acid (5-HIAA) with increased CSF 3-methoxytyrosine (3-MT) are observed with concomitant increased urinary excretion of vanillactic acid (VLA) [1].

Plasma Threonine, Glycine (and Serine) As consequences of the role of PLP-dependent enzymes in the one-carbon metabolism (glycine cleavage system, serine hydroxymethyltransferase) and in threonine dehydratase [3], raised plasma threonine, glycine, and sometimes serine are observed, although those changes are not present in all patients [2]. Serine is not always raised concomitant to glycine because the accumulation of glycine can be converted into serine.

2-Ketoglutarate/Glutamate Ratio Increased ratio reflects a functional alteration of the PLP-dependent aminotransferases.

Plasma Homocysteine and Urinary Cystathionine As a consequence of the role of PLP as a cofactor in homocysteine catabolism reactions (transsulfuration pathway), low vitamin B_6 status (together with low vitamin B_{12} and folate) has been linked to increased levels of homocysteine in blood [4]. Also, increased excretion of cystathionine has been linked to inadequate vitamin B_6 status [3].

Erythrocyte Transaminase Activities The activity coefficient of erythrocyte aspartate (or alanine) aminotransferases (EAST, EALT) with addition of in vitro PLP over basal condition reflects the vitamin B_6 status (higher coefficient means a lower status) [3].

Tryptophan Metabolites PLP acts as a cofactor in enzymes involved in the kynurenine pathway. The substrate/product ratio 3-hydroxykynurenine/xanthurenic acid in plasma is a marker of vitamin B_6 status and is better than 3-hydroxykynurenine concentration alone [3]. Increased urinary xanthurenic acid is observed in vitamin B_6 deficiency [1]. Also, measurement of urinary xanthurenic acid following an oral load of L-tryptophan (2–5 g) is a test used in studies of vitamin B_6 nutriture [4].

24.3.2 B_6 Vitamers as Direct Biomarkers: PLP, PL, PM, PN, and PA

24.3.2.1 B_6 Vitamers Distribution in Body Fluids

With normal vitamin B_6 intake (no supplementation), the B_6 vitamers PLP and PL and the metabolite PA are present in plasma (PLP > PA > PL) with no significant amounts of PN and PM [11, 23]. The most abundant B_6 vitamer in erythrocytes is PLP, containing about 60% of whole blood PLP [22]. In CSF, the relative B_6 vitamer composition differs from that of plasma with PL > PLP > PA > PM [23]. In urine, PA represents >90% of vitamin B_6 and the excretion is highly dependent on recent vitamin B_6 intake [6].

24.3.2.2 Factors Affecting PLP Concentration

Limitations of Plasma PLP as Sole Marker of Vitamin B_6 Status In many cases, vitamin B_6 status is assessed by direct measurement of PLP in the circulation, although it is known that various factors affect plasma PLP concentration that are not related to vitamin B_6 status. Those factors include inflammation, albumin concentration, alkaline phosphatase activity (ALP), and inorganic phosphate [6].

Relation of ALP Activity with Plasma PLP Plasma PLP is inversely associated with ALP activity, best illustrated by markedly raised plasma PLP in HPP [6]. Single nucleotide polymorphisms (SNPs) in *ALPL* have also been associated with increased PLP and PLP/PL ratio in plasma [24]. Inorganic phosphate inhibits ALP, which indirectly affects plasma PLP levels, as illustrated by increased levels of PLP in postmenopausal women (elevated serum inorganic phosphate due to bone loss) [6]. Also, low plasma PLP is observed in conditions with elevated ALP such as in liver and bone diseases, cancer, pregnancy, and diabetes [6]. In conclusion, because many common pathological factors affect the activity of ALP, which is tightly linked to plasma PLP, plasma PLP alone may not be a valid biomarker to assess vitamin B_6 status in a diseased population [6].

PLP in Erythrocytes Plasma PLP does not necessarily correlate with erythrocyte PLP. As ALP is an ectoenzyme with very low activity inside the cells, erythrocyte PLP does not inversely correlate with ALP activity but is rather associated with increased uptake of PL by the cell. Thus, erythrocyte PLP is not increased in HPP but is increased in critically ill patients with systemic inflammation and hypoalbuminemia (concurrent to the marked reduction in plasma PLP, see above). Erythrocyte PLP may thus be a better indicator of vitamin B_6 status than plasma PLP [6].

PLP in CSF CSF PLP has been shown to be age dependent [25] and to correlate with plasma PLP [23]. It has been shown to be low in PNPO deficiency, ALDH7A1 deficiency, MOCOD and in patients with epilepsy of unclear etiology, reflecting most likely vitamin B_6 deficiency [11, 20, 25].

24.3.2.3 Interpretation of Plasma B_6 Vitamers

Plasma PLP is a good marker for vitamin B_6 status in a healthy population. In a diseased population, it may be affected by factors not related to vitamin B_6 (see point 2.2.2).

Plasma PL shows a stronger correlation than plasma PLP with erythrocyte PLP, thus plasma PL could be used as a surrogate marker of intracellular PLP [26].

Plasma PLP + PL (total B_6 aldehyde) has been suggested as being a better marker for vitamin B_6 status, as this may correct for partial degradation of PLP to PL during non-optimal sample handling [27]. Furthermore, PLP + PL may be recommended as a direct marker of vitamin B_6 status in subjects with altered ALP activity (see point 2.2.2) [6].

Plasma PLP/PL ratio is useful to assess the activity of ALP and is strongly increased in HPP with no vitamin B_6 supplementation. Plasma PLP/PL ratio is reduced by more than 50% in critically ill patients with elevated CRP, elevated ALP, and low serum albumin as compared with healthy controls [6]. The PLP/PL ratio is also of importance in the follow-up of HPP under enzyme replacement therapy (ERT) [28].

Plasma PM and PM/PA ratio: Markedly elevated plasma PM and PM/PA ratio are measured in patients with PNPO deficiency irrespective of supplementation with vitamin B_6. The PM/PA ratio delineates even more clearly than PM when the patient is supplemented with vitamin B_6. The PM/PA ratio is thus a specific biomarker for PNPO deficiency [11]. In control subjects, PM and PN are mostly below the detection limits but may be measured shortly after pyridoxine dosage.

Plasma PA is not protein bounded in plasma and shows a strong correlation with plasma PL and PLP. After supplementation, PA and PL correlate strongly but not PA and PLP, probably due to the high regulation of PLP [11]. The time of blood sampling after vitamin B_6 administration is a critical aspect affecting strongly the concentrations of the B_6 vitamers, especially PN, PL, and PA (see point 4.1). Plasma PA increases with renal dysfunction, which decreases the specificity of plasma PA as a biomarker of vitamin B_6 status [6].

24.4 Properties of the B_6 Vitamers

The B_6 vitamers are small molecules, which are easily ionizable as cations to be detected by mass spectrometry. The degradation product PA has a natural fluorescence that enables its detection by high-performance liquid chromatography (HPLC) with fluorescence detection without derivatization, whereas PLP and PL need precolumn derivatization. The chemical properties including structure and transitions for detection by mass spectrometry are summarized in Table 24.1.

Table 24.1 Abbreviations, names, chemical formula, structure, and transitions of the B_6 vitamers and the degradation product pyridoxic acid

Abb.	Name	Chemical formula	Structure	Transitions (positive ionization mode) [11]
PLP	Pyridoxal 5′-phosphate	$C_8H_{10}NO_6P$	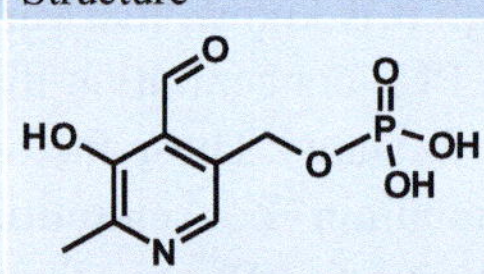	248.1 → 150.1/121.9 PLP-d3: 251.1 → 153.1/124.9
PL	Pyridoxal	$C_8H_9NO_3$	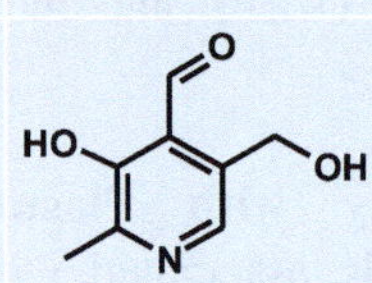	168.1 → 150.1/94.4 PL-d3 171.1 → 153.1/97.4
PMP	Pyridoxamine 5′-phosphate	$C_8H_{13}N_2O_5P$	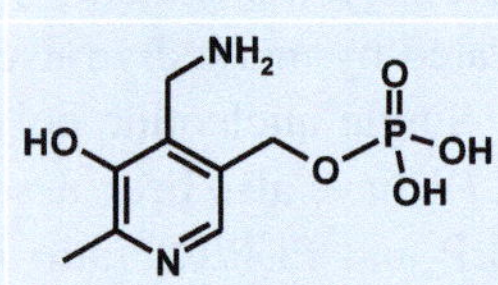	249.2 → 232.2 [29]
PM	Pyridoxamine	$C_8H_{12}N_2O_2$	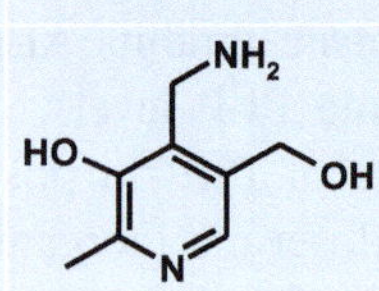	169.1 → 134.1/152.1 PM-d3 172.1 → 137.1/155.1

(continued)

Table 24.1 (continued)

Abb.	Name	Chemical formula	Structure	Transitions (positive ionization mode) [11]
PNP	Pyridoxine 5′-phosphate	$C_8H_{12}NO_6P$	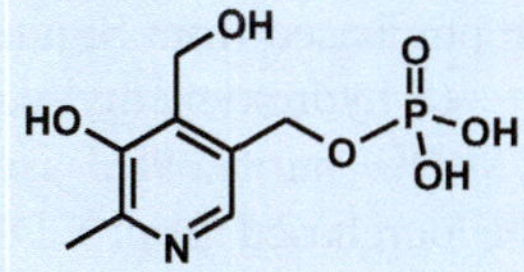	250.1 → 134.1 [30]
PN	Pyridoxine	$C_8H_{11}NO_3$	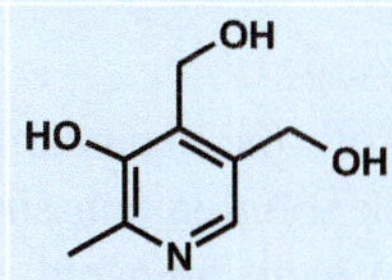	170.4 → 134.17/152.1 PN-d2 172.4 → 136.1/154.1
PA	Pyridoxic acid	$C_8H_9NO_4$	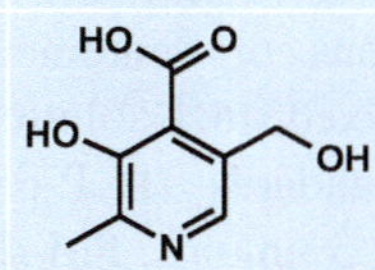	184.1 → 148.1/166 PA-d2 186.4 → 150.1/168

24.5 Methods

24.5.1 Pre-analytical

Matrix B_6 vitamers are usually analyzed in plasma. As B_6 vitamers in CSF and plasma correlate well [23], there is no evident advantage for CSF measurement. B_6 vitamers in urine, especially pyridoxic acid, are a measure of recent vitamin B_6 intake and are good to assess vitamin B_6 requirements but not to assess vitamin B_6 status [6].

Stability B_6 vitamers and especially PLP are sensitive to light but relatively stable at room temperature for at least 48 h. PA is stable at light and room temperature [16]. This pre-analytic property allows time for normal specimen transport (light protected) [31]. For longer storage periods, it is recommended to freeze the samples.

Time of Blood Sampling After Vitamin B_6 Supplementation The variability in the concentrations of B_6 vitamers, especially PN, PL, and PA is highly dependent on the time of blood sampling after vitamin B_6 administration. Specifically, PN concentrations rapidly raised after intake of a single oral dose of 200 mg PN hydrochloride in three healthy probands and returned to baseline after approximately 4 h. PL and PA concentrations raised also rapidly exceeding PN concentration and were close to baseline after 24 h [11]. Thus, it is recommended to draw blood samples immediately before vitamin B_6 administration.

24.5.2 Reagents and Chemicals

Reagents Heptafluorobutyric acid (HFBA, ≥95.5%), trichloroacetic acid (TCA, ≥99%), acetic acid (AA, ≥99.7%), formic acid (FA, ≥98%), and acetonitrile (ACN, ≥99.9%, HPLC-grade) can be purchased from Sigma-Aldrich.

Standards Pyridoxal 5′-phosphate monohydrate (PLP, ≥97%), pyridoxal hydrochloride (PL, ≥99%), pyridoxamine dihydrochloride (PM, ≥98%), pyridoxine hydrochloride (PN, ≥98%), and 4-pyridoxic acid (PA, ≥98%) can be purchased from Sigma-Aldrich.

Deuterated Internal Standards Pyridoxal 5′-phosphate (methyl-d_3) (PLP-d_3, 70% purity, 97% isotope enrichment) and 4-pyridoxic acid (5-hydroxymethyl-d_2) (PA-d_2, 100% purity, >98% isotope enrichment) can be purchased from Buchem BV. Pyridoxal (methyl-d_3) hydrochloride (PL-d_3, 98% purity, 98% isotope enrichment) and pyridoxamine (methyl-d_3) dihydrochloride (PM-d_3, 98% purity, 98% isotope enrichment) can be purchased from Sigma-Aldrich. Pyridoxine (5-hydroxymethyl-d2) hydrochloride (PN-d2, 98.9% purity, 98% isotope enrichment) can be purchased from CDN isotopes.

Solutions Stock solutions of the standards (5 mmol/L) and a mixed stock solution containing all standards (10 μmol/L) were prepared in Milli-Q Water. Stock solutions of deuterated standards (1 mmol/L) and a mixed stock solution containing all deuterated standards (PLP-d3: 10 μmol/L, PL-d3 and PA-d2: 5 μmol/L, PM-d3 and PN-d2: 2 μmol/L) were prepared in 0.3 mol/L TCA. Solutions of PA and PA-d2 were prepared in 2 mmol/L NaOH. All solutions were stored at −80 °C.

24.5.3 Instrumentation

B_6 vitamers can be analyzed by LC-MS/MS or HPLC with fluorescence detection (although PM and PN determination have not been reported with the latter method). The advantages of an LC-MS/MS method are clearly the high specificity and the simple work-up. Only protein precipitation under acidic conditions is required before injection, which is in sharp contrast to HPLC methods. There, precolumn derivatization either as semicarbazone or pyridoxic acid phosphate is required because of the lack of a fluorophore in their chemical structure [4, 22].

LC-MS/MS Instrument Settings The LC-MS/MS methods can include the B_6 vitamers and the degradation product PA only, [11, 30, 32] or can be part of a more extended panel, for example, for the DD of treatable early-onset inherited metabolic epilepsies [16]. All methods described below operate in positive ionization mode. The transitions are listed in Table 24.1.

AB Sciex 4000/5500 QTRAP with Thermo Scientific DionexUltiMate 3000 [11, 16].

LC Parameters Pentafluorophenylpropyl (PFP)-packing Discovery HS F5 3 μm 10 cm x 2.1 mm HPLC column (Supelco) operating at a flow rate of 250 μL/min. Analysis run time was 16 min, starting with 100% of solvent A (3.7% AA, 0.01% HFBA in Milli-Q water) for 0.5 min, followed by a 5.5 min gradient increasing the proportion of solvent B (0.1% FA in ACN) to 30%, then in 4 min to 95% and kept at 95% for 0.5 min. The system was re-equilibrated for 5 min at 100% of solvent A.

MS parameters-Source Curtain Gas: 30, Collision Gas: High, Ion Spray Voltage: 5500, Temperature: 550, Ion Source Gas 1: 40, Ion Source Gas 2: 60, Interface Heater: on. *Transitions*: Entrance Potential EP: 10 Volts, Declusturing potential DP: 30–90 Volts, Collision Cell Exit Potential (CXP): 4–25 Volts, Collision Energy (CE) 19–33 Volts.

Waters XEVO-TQ MS triple quadrupole mass spectrometer with Acquity UPLC [30]

LC parameters Acquity HSS-T3 analytical column, at 25 °C, at a flow rate of 400 μL/min, with a binary gradient of solvent A (650 mM AA with 0.01% HFBA) and B (acetonitrile). Gradient: 0–0.5 min (0% B); 2 min (20% B); 2.1–2.3 min (100% B); 2.4–3.6 min (0% B).

MS parameters Capillary voltage 0.5 kV, cone voltage 18 kV. Source and desolvation temperatures were 150 °C and 550 °C, respectively.

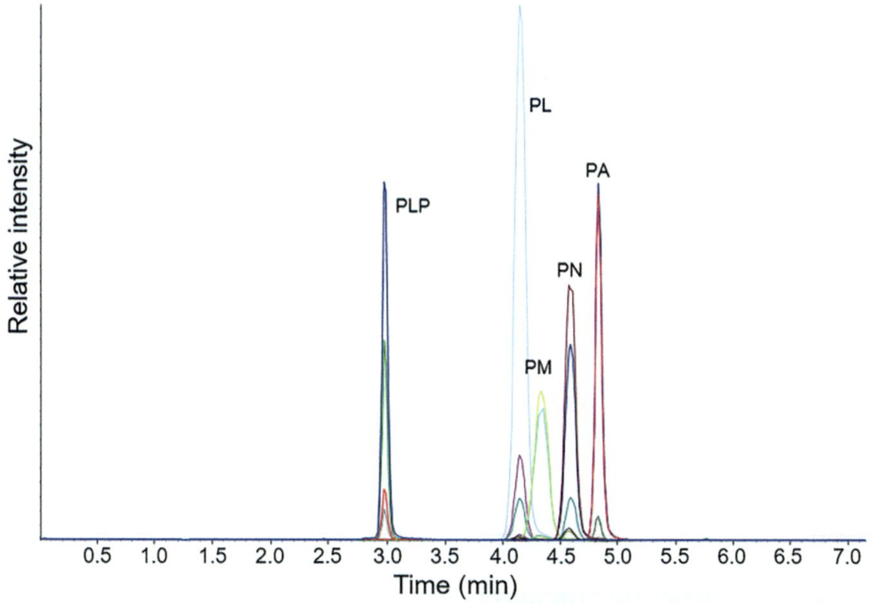

Fig. 24.4 MRM traces of the B_6 vitamers. Chromatogram illustrating the separation of PLP, PL, PM, PN, and PA in a standard mixture on a PFP-packing column (Supelco Discovery HS F5 3 μm 10 cm × 2.1 mm) [11]

24.5.4 Analytical Procedure

80 μL of a 0.6-M TCA solution containing the internal standards (PLP-d_3: 200 nmol/L, PL-d_3 and PA-d_2: 100 nmol/L, PM-d_3 and PN-d_2: 40 nmol/L) was added to 80 μL of plasma in an Eppendorf tube. The sample was vortexed and put on ice for 20 min in the dark. The tubes were centrifuged at 4 °C with 16,000 g for 10 min and the supernatant was transferred into an HPLC vial and analyzed immediately. The vitamin B_6 vitamers were analyzed by multiple reaction monitoring in positive ion mode (Fig. 24.4). Calibration curves were prepared in Milli-Q water with 7 individual points (1, 3, 10, 30, 100, 300, 1000 nmol/L) [11, 30, 32].

Dilution of Samples Due to the high dynamic range in concentrations of the B_6 vitamers due to supplementation, it is recommended to analyze the samples undiluted but if needed also diluted (typically at dilution factor 20) to be able to measure in the calibration range.

24.5.5 Post-analytical: Interpretation of B_6 Vitamer Profile

- Decreased PLP and PLP + PL are signs of low vitamin B_6 status.
- Increased PLP and PLP/PL ratio (no supplementation) is observed in HPP due to *ALPL* mutations. On supplementation with vitamin B_6, PLP and sometimes the PLP/PL ratio as well are increased, however, independently of the genetic background.
- Increased PM and PM/PA ratio is diagnostic for PNPO deficiency.
- Increased PA can be due to impaired renal function.
- Consider the time of blood sampling after vitamin B_6 supplementation: PN, PL, and PA rise shortly after intake of vitamin B_6.

24.5.6 Reference Values

24.5.6.1 Plasma Concentrations [nmol/L]

B_6 Vitamers	Children (n = 50, 0–18 years), range [11]	Children (n = 24, 4.3–16 years), range [32]	Children (n = 42, 1–18 years), range [33]	Adults (n = 523, 18–63 years), 2.5p-97.5p [23]	Adults (n = 120, 20–67 years), 2.5p-97.5p [31]
PLP	10–289	46–321	20.5–151	19.8–200	23–223
PL	4–85	4.6–18.1	8.8–58.7	4.2–24.5	n/a
PM	<1	n.d.	n/a	n/a	n/a
PMP	n/a	n.d.-0.62	n/a	n/a	n/a
PN	<1	n.d.	n/a	n/a	n/a
PNP	n/a	n.d.	n/a	n/a	n/a
PA	4–564	16.4–139	8.8–104	6.1–107	n/a
PM/PA	<0.01	n/a	n/a	n/a	n/a
PLP/PL	n/a	5.2–18.6	1.0–4.2	2.3–13.7	n/a

n/a: not available
n.d.: not detected

24.5.6.2 CSF Concentrations* [nmol/L]

B_6 Vitamers	Children (n = 20, 0.7–16 years), range [30]	Children (n = 41, 1–18 years), range [33]	Adults (n = 523, 18–63 years), 2.5p-97.5p [23]	Children (n = 7, 0–30 days), range [20]	Children (n = 37, 1–12 min), range [20]	Children (n = 28, 1–2 years), range [20]	Children (n = 49, >3 years), range [20]
PLP	8.8–42.0	11.0–33.7	6.9–34.0	26–69	10–136	8–86	8–41
PL	14.8–42.5	16.1–55.7	17.4–54.5	n/a	n/a	n/a	n/a
PM	0.1–0.5	0.3–0.9	0.06–0.9	n/a	n/a	n/a	n/a
PN	<0.03	n/a	n/a	n/a	n/a	n/a	n/a
PA	0.09–4.1	<0.09–3.1	<0.09–3.8	n/a	n/a	n/a	n/a
PLP/PL	n/a	0.4–1.4	0.2–1.1	n/a	n/a	n/a	n/a

* PLP concentration in CSF is age dependent [25]

24.5.6.3 Urine

Daily excretion of PA is 0.8 mg/day [4].

24.5.7 Pathological Values

24.5.7.1 Plasma Concentrations [nmol/L]

B_6 Vitamers	PNPO def. no therapy (n = 1) [11]	PNPO def. vitB_6 suppl. (n = 5) [11]	PNPO def. PLP suppl. (n = 2) [32]	HPP no therapy (n = 1) [11]	HPP vitB_6 suppl. (n = 2) [11]
PLP	7	412–1080	580–633	1885	1590–2030
PL	3	389–6800	427–5798	69	1110–5570
PM	26	206–2050	193–2731	<2.7	<1–21
PN	<1	1100–11,100	575–599	<0.3	2–2330
PA	4	583–4940	793–7926	54	1920–5950
PM/PA	6.5	0.12–0.62	0.24–0.35	<0.05	<0.01
PLP/PL	2.33	0.05–1.39	0.1–1.36	27.3	0.36–1.43

24.5.7.2 CSF Concentrations

PLP concentrations in CSF of four newborns (7–16 days) with PNPO deficiency were reported to be 3.6, 12, 14, and 18 nmol/L [25].

24.5.8 Quality Control

Internal QC Three internal Quality Control (QC) samples prepared from plasma pools (native and spiked with about 100–200 and 400–500 nM of each B_6 vitamers) should be analyzed with each run.

External QC Apart from PLP, which is commercially available as external QC, the other B_6 vitamers and PA have no external QC available to date. So far, PNPO deficiency has not been sent as diagnostic proficiency testing in a qualitative ERNDIM scheme, as the disease is not readily detected in urine. Hypophosphatasia samples have already been sent several times, because phosphoethanolamine is easily detected in urine [34].

24.6 Conclusion

The determination of B_6 vitamers is valuable for the assessment of vitamin B_6 status and for the differential diagnosis of vitamin B_6-dependent epilepsies.

Plasma PLP is influenced by factors not related to vitamin B_6 status, such as inflammation (reduces PLP), high alkaline phosphatase activity (reduces PLP), low serum albumin (reduces PLP), or elevated inorganic phosphate (increases PLP) [6]. Thus, determining the vitamin B_6 status with the sole determination of plasma PLP in a disease population may not be accurate and the use of other matrixes (e.g., erythrocytes) or the concomitant analysis of other B_6 vitamers (PL, PLP + PL, PA) may be more appropriate.

Analyzing plasma B_6 vitamers PLP, PL, PM, PN, and PA is useful for the differential diagnosis of vitamin B_6-dependent epilepsies:

- Plasma PM concentration and even more strikingly the PM/PA ratio are increased in PNPO deficiency, independent of vitamin B_6 supplementation, and represent thus a specific biomarker for this IEM.
- Plasma PLP and the PLP/PL ratio are increased in HPP due to mutations in *ALPL* due to a deficient activity of TNSALP. In this condition, vitamin B_6 supplementation may cover the specificity of the biomarker, as PLP and sometimes the PLP/PL ratio are also highly elevated in vitamin B_6-supplemented subjects. However, in this case, the other biomarkers of HPP, namely low ALP and high urinary phosphoethanolamine are still evident. The PLP/PL ratio is also of importance in the follow-up of HPP under ERT.

The following aspects are important to consider when analyzing and interpreting plasma B_6 vitamers:

- B_6 vitamers are light sensitive but are stable for at least 24 h at room temperature, allowing normal shipment.
- The time of blood sampling after vitamin B_6 administration is a critical aspect strongly affecting the concentrations of the B_6 vitamers, especially PN, PL, and PA. Thus, it is recommended to draw blood samples immediately before vitamin B_6 administration.
- PA increases with impaired renal function.

References

1. Clayton PT (2006) B6-responsive disorders: a model of vitamin dependency. J Inherit Metab Dis 29(2–3):317–326. https://doi.org/10.1007/s10545-005-0243-2
2. Wilson MP, Plecko B, Mills PB, Clayton PT (2019) Disorders affecting vitamin B(6) metabolism. J Inherit Metab Dis 42(4):629–646. https://doi.org/10.1002/jimd.12060
3. Kennedy DO (2016) B vitamins and the brain: mechanisms, dose and efficacy – a review. Nutrients 8(2):68. https://doi.org/10.3390/nu8020068
4. Tietz (2005) Textbook of clinical chemistry and molecular diagnostics. 4th edn, Elsevier Saunders
5. Eliot AC, Kirsch JF (2004) Pyridoxal phosphate enzymes: mechanistic, structural, and evolutionary considerations. Annu Rev Biochem

73:383–415. https://doi.org/10.1146/annurev.biochem.73.011303.074021
6. Ueland PM, Ulvik A, Rios-Avila L, Midttun O, Gregory JF (2015) Direct and functional biomarkers of vitamin B6 status. Annu Rev Nutr 35:33–70. https://doi.org/10.1146/annurev-nutr-071714-034330
7. Albersen M, Groenendaal F, van der Ham M, de Koning TJ, Bosma M, Visser WF, Visser G, de Sain-van der Velden MG, Verhoeven-Duif NM (2012) Vitamin B6 vitamer concentrations in cerebrospinal fluid differ between preterm and term newborn infants. Pediatrics 130(1):e191–e198. https://doi.org/10.1542/peds.2011-3751
8. Hasegawa S, Oda Y, Ichiyama T, Hori Y, Furukawa S (2006) Ginkgo nut intoxication in a 2-year-old male. Pediatr Neurol 35(4):275–276. https://doi.org/10.1016/j.pediatrneurol.2006.05.008
9. Glenn GM, Krober MS, Kelly P, McCarty J, Weir M (1995) Pyridoxine as therapy in theophylline-induced seizures. Vet Hum Toxicol 37(4):342–345
10. Kuki I, Takahashi Y, Okazaki S, Kawawaki H, Ehara E, Inoue N, Kinoshita T, Murakami Y (2013) Vitamin B6-responsive epilepsy due to inherited GPI deficiency. Neurology 81(16):1467–1469. https://doi.org/10.1212/WNL.0b013e3182a8411a
11. Mathis D, Abela L, Albersen M, Burer C, Crowther L, Beese K, Hartmann H, Bok LA, Struys E, Papuc SM, Rauch A, Hersberger M, Verhoeven-Duif NM, Plecko B (2016) The value of plasma vitamin B6 profiles in early onset epileptic encephalopathies. J Inherit Metab Dis 39(5):733–741. https://doi.org/10.1007/s10545-016-9955-8
12. Darin N, Reid E, Prunetti L, Samuelsson L, Husain RA, Wilson M, El Yacoubi B, Footitt E, Chong WK, Wilson LC, Prunty H, Pope S, Heales S, Lascelles K, Champion M, Wassmer E, Veggiotti P, de Crecy-Lagard V, Mills PB, Clayton PT (2016) Mutations in PROSC disrupt cellular pyridoxal phosphate homeostasis and cause vitamin-B(6)-dependent epilepsy. Am J Hum Genet 99(6):1325–1337. https://doi.org/10.1016/j.ajhg.2016.10.011
13. Plecko B, Zweier M, Begemann A, Mathis D, Schmitt B, Striano P, Baethmann M, Vari MS, Beccaria F, Zara F, Crowther LM, Joset P, Sticht H, Papuc SM, Rauch A (2017) Confirmation of mutations in PROSC as a novel cause of vitamin B ((6)) -dependent epilepsy. J Med Genet 54(12):809–814. https://doi.org/10.1136/jmedgenet-2017-104521
14. Whyte MP (2010) Physiological role of alkaline phosphatase explored in hypophosphatasia. Ann N Y Acad Sci 1192:190–200. https://doi.org/10.1111/j.1749-6632.2010.05387.x
15. Campistol J, Plecko B (2015) Treatable newborn and infant seizures due to inborn errors of metabolism. Epileptic Disord 17(3):229–242. https://doi.org/10.1684/epd.2015.0754
16. Mathis D, Beese K, Ruegg C, Plecko B, Hersberger M (2020) LC-MS/MS method for the differential diagnosis of treatable early onset inherited metabolic epilepsies. J Inherit Metab Dis 43(5):1102–1111. https://doi.org/10.1002/jimd.12244
17. Mills PB, Struys E, Jakobs C, Plecko B, Baxter P, Baumgartner M, Willemsen MA, Omran H, Tacke U, Uhlenberg B, Weschke B, Clayton PT (2006) Mutations in antiquitin in individuals with pyridoxine-dependent seizures. Nat Med 12(3):307–309. https://doi.org/10.1038/nm1366
18. Crowther LM, Mathis D, Poms M, Plecko B (2019) New insights into human lysine degradation pathways with relevance to pyridoxine-dependent epilepsy due to antiquitin deficiency. J Inherit Metab Dis 42(4):620–628. https://doi.org/10.1002/jimd.12076
19. Farrant RD, Walker V, Mills GA, Mellor JM, Langley GJ (2001) Pyridoxal phosphate de-activation by pyrroline-5-carboxylic acid. Increased risk of vitamin B6 deficiency and seizures in hyperprolinemia type II. J Biol Chem 276(18):15107–15116. https://doi.org/10.1074/jbc.M010860200
20. Footitt EJ, Heales SJ, Mills PB, Allen GF, Oppenheim M, Clayton PT (2011) Pyridoxal 5′-phosphate in cerebrospinal fluid; factors affecting concentration. J Inherit Metab Dis 34(2):529–538. https://doi.org/10.1007/s10545-011-9279-7
21. Crowther LM, Poms M, Zandl-Lang M, Abela L, Hartmann H, Seiler M, Mathis D, Plecko B (2022) Metabolomics analysis of antiquitin deficiency in cultured human cells and plasma: relevance to pyridoxine-dependent epilepsy. J Inherit Metab Dis 46:129. https://doi.org/10.1002/jimd.12569
22. Roelofsen-de Beer R, van Zelst BD, Wardle R, Kooij PG, de Rijke YB (2017) Simultaneous measurement of whole blood vitamin B1 and vitamin B6 using LC-ESI-MS/MS. J Chromatogr B Analyt Technol Biomed Life Sci 1063:67–73. https://doi.org/10.1016/j.jchromb.2017.08.011
23. Albersen M, Bosma M, Luykx JJ, Jans JJ, Bakker SC, Strengman E, Borgdorff PJ, Keijzers PJ, van Dongen EP, Bruins P, de Sain-van der Velden MG, Visser G, Knoers NV, Ophoff RA, Verhoeven-Duif NM (2014) Vitamin B-6 vitamers in human plasma and cerebrospinal fluid. Am J Clin Nutr 100(2):587–592. https://doi.org/10.3945/ajcn.113.082008
24. Loohuis LM, Albersen M, de Jong S, Wu T, Luykx JJ, Jans JJM, Verhoeven-Duif NM, Ophoff RA (2018) The alkaline phosphatase (ALPL) locus is associated with B6 Vitamer levels in CSF and plasma. Genes (Basel) 10(1). https://doi.org/10.3390/genes10010008
25. Ormazabal A, Oppenheim M, Serrano M, Garcia-Cazorla A, Campistol J, Ribes A, Ruiz A, Moreno J, Hyland K, Clayton P, Heales S, Artuch R (2008) Pyridoxal 5′-phosphate values in cerebrospinal fluid: reference values and diagnosis of PNPO deficiency in paediatric patients. Mol Genet Metab 94(2):173–177. https://doi.org/10.1016/j.ymgme.2008.01.004
26. Vasilaki AT, McMillan DC, Kinsella J, Duncan A, O'Reilly DS, Talwar D (2008) Relation between pyridoxal and pyridoxal phosphate concentrations in plasma, red cells, and white cells in patients with criti-

cal illness. Am J Clin Nutr 88(1):140–146. https://doi.org/10.1093/ajcn/88.1.140

27. Hustad S, Eussen S, Midttun O, Ulvik A, van de Kant PM, Morkrid L, Gislefoss R, Ueland PM (2012) Kinetic modeling of storage effects on biomarkers related to B vitamin status and one-carbon metabolism. Clin Chem 58(2):402–410. https://doi.org/10.1373/clinchem.2011.174490
28. Akiyama T, Kubota T, Ozono K, Michigami T, Kobayashi D, Takeyari S, Sugiyama Y, Noda M, Harada D, Namba N, Suzuki A, Utoyama M, Kitanaka S, Uematsu M, Mitani Y, Matsunami K, Takishima S, Ogawa E, Kobayashi K (2018) Pyridoxal 5′-phosphate and related metabolites in hypophosphatasia: effects of enzyme replacement therapy. Mol Genet Metab 125(1–2):174–180. https://doi.org/10.1016/j.ymgme.2018.07.006
29. Van den Eynde MDG, Scheijen J, Stehouwer CDA, Miyata T, Schalkwijk CG (2021) Quantification of the B6 vitamers in human plasma and urine in a study with pyridoxamine as an oral supplement; pyridoxamine as an alternative for pyridoxine. Clin Nutr 40(7):4624–4632. https://doi.org/10.1016/j.clnu.2021.05.028
30. van der Ham M, Albersen M, de Koning TJ, Visser G, Middendorp A, Bosma M, Verhoeven-Duif NM, de Sain-van der Velden MG (2012) Quantification of vitamin B6 vitamers in human cerebrospinal fluid by ultra performance liquid chromatography-tandem mass spectrometry. Anal Chim Acta 712:108–114. https://doi.org/10.1016/j.aca.2011.11.018
31. Panton KK, Farup PG, Sagen E, Sirum UF, Asberg A (2013) Vitamin B6 in plasma - sample stability and the reference limits. Scand J Clin Lab Invest 73(6):476–479. https://doi.org/10.3109/00365513.2013.803234
32. Footitt EJ, Clayton PT, Mills K, Heales SJ, Neergheen V, Oppenheim M, Mills PB (2013) Measurement of plasma B6 vitamer profiles in children with inborn errors of vitamin B6 metabolism using an LC-MS/MS method. J Inherit Metab Dis 36(1):139–145. https://doi.org/10.1007/s10545-012-9493-y
33. Albersen M, Bosma M, Jans JJ, Hofstede FC, van Hasselt PM, de Sain-van der Velden MG, Visser G, Verhoeven-Duif NM (2015) Vitamin B6 in plasma and cerebrospinal fluid of children. PLoS One 10(3):e0120972. https://doi.org/10.1371/journal.pone.0120972
34. Mathis D, Croft J, Chrastina P, Fowler B, Vianey-Saban C, Ruijter GJG (2022) The role of ERNDIM diagnostic proficiency schemes in improving the quality of diagnostic testing for inherited metabolic diseases. J Inherit Metab Dis 45(5):926–936. https://doi.org/10.1002/jimd.12523

5-Methyltetrahydrofolate

25

Rafael Artuch, Cristina Sierra,
and Thomas Opladen

Contents

Abbreviations

CFD	Cerebral folate deficiency
CSF	Cerebrospinal fluid
EC	Electrochemical detection
EDTA	Ethylenediaminetetraacetic acid
FD	Fluorescence detection
HPLC	High-performance liquid chromatography
5MTHF	5-Methyltetrahydrofolate
MTHFR	5,10-Methylen-tetrahydrofolate reductase
TRIS	Tris(hydroxymethyl)aminomethane

R. Artuch (✉) · C. Sierra
Clinical Biochemistry Department, Sant Joan de Déu Research Institute and CIBERER-ISCIII, Barcelona, Spain
e-mail: rafael.artuch@sjd.es

T. Opladen
Division for Child Neurology and Metabolic medicine, Centre for Paediatric and Adolescent Medicine, Heidelberg, Germany

25.1 Introduction

Folic acid is a water-soluble vitamin that functions as one-carbon donor in various metabolic cycles. It is involved in the biosynthesis of thymidylates and purines, methionine synthesis via homocysteine remethylation, the methylation of phospholipids and proteins, the serine and glycine interconversion, and the metabolism of histidine and formate, among other functions. It is therefore essential for the growth, reproduction, and maintenance of normal body function. The natural form is referred to as folate, it consists mainly of 5-methyltetrahydrofolate (5MTHF) and 10-formyltetrahydrofolate in their polyglutamate derivatives [1]. 5MTHF is an active one-carbon donor and it is transported through the blood–brain barrier to serve for a number of reactions in the central nervous system.

Serum folate deficiency is known to be associated with megaloblastic anemia, high blood

N. Blau, F. M. Vaz (eds.), *Laboratory Guide to the Methods in Biochemical Genetics*,
https://doi.org/10.1007/978-3-031-58819-8_25

levels of homocysteine, and neural tube defects in newborns as a result of folate shortage in the mother [2, 3]. Cerebral Folate Deficiency (CFD) is a recently recognized neurological disorder found in a number of children with psychomotor retardation, spastic paraplegia, cerebellar ataxia, and dyskinesia. These patients have very low 5MTHF in CSF, normal blood folate concentration [4, 5], and some patients display a high titer of folate receptor blocking auto-antibodies [6].

In this chapter, the analytical procedures for 5-methyltetrahydrofolate in Cerebrospinal Fluid (CSF) and serum are described. 5MTHF was initially measured by microbiological and radioisotope dilution assay [7, 8], and later by high-performance liquid chromatography (HPLC) using either electrochemical (EC), ultraviolet or fluorescence detection (FD) [9–11]. Compared to other methods, EC and FD detection are more sensitive. Recently, quantitation of 5MTHF in CSF using the liquid chromatography-electrospray tandem mass spectrometry method has been reported [12].

25.2 Clinical Significance

CFD is associated with the following inherited metabolic or genetic disorders: 5,10-methylentetrahydrofolate reductase (MTHFR) deficiency [13], 3-phosphoglycerate dehydrogenase deficiency [14], dihydropteridine reductase deficiency [15], Rett syndrome [16], and Aicardi-Goutières Syndrome [17]. Furthermore, folate deficiency may be associated with transport defects, such as congenital folate malabsorption (proton-coupled folate transporter deficiency), folate receptor alpha deficiency, some mitochondrial disorders (Kearns-Sayre syndrome, mitochondrial polymerase DNA gamma 1 deficiency) or with other conditions including severe malnutrition, some pharmacological treatments or formiminotransferase deficiency (FIGLU uria), among others.

25.3 Properties of 5MTHF

5MTHF is biologically synthesized from 5,10-methylene-tetrahydrofolic acid through the action of MTHFR. Commercially, it is available as a Ca-salt. 500 mg of 5MTHF calcium salt is soluble in 100 ml of water. A 1-mM solution in water has a pH of 5.9. Solutions are less stable than the solid form and should be made immediately before use. A 1-mM solution in water at room temperature for 2.5 h degrades by approximately 4% and within 4 h 10% has degraded. Dry powder open at room temperature for 3 months is relatively stable degrading by 0.5% over this time. The powder is hygroscopic absorbing 10% of its weight in water within 4 h (www.eprova.ch).

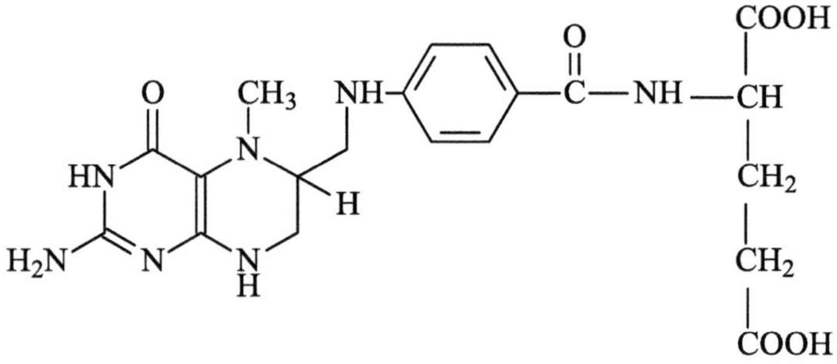

5-Methyltetrahydrofolic acid (5MTHF)

25.4 Methods

5MTHF is separated from other folates by reverse phase HPLC and can be quantified with EC and FD [18–20]. Prior to HPLC-EC serum was extracted with Sep-Pak C18 cartridges, while CSF was injected without pretreatment for either EC or FD [19, 20]. The two HPLC-EC systems described are very similar and both of them can be used for CSF or serum. HPLC-FD methods will be also detailed.

25.4.1 Pre-analytical

25.4.1.1 Specimen

Both CSF and serum samples should be immediately stored at −80 °C. CSF contaminated with blood should be immediately centrifuged (20–30% lower values).

Serum concentration: 400 µl serum is thawed immediately before measurement and mixed with 600 µl ice-cold 28 µmol/L ascorbic acid solution to protect against oxidation. After activation of the Sep-Pak C18 cartridge (Waters Chromatography) with 2 ml methanol and 10 ml of 14 µmol/L ascorbic acid solution the mixture is applied. The cartridge is washed with 3 ml of 14 µmol/L ascorbic acid solution and 5MTHF is eluted with 1.5 ml methanol in a light-protected tube. The eluates are evacuated under vacuum (Speed Vac, SVC 100, Savant) and dried samples are redissolved in 100 µl of 14 µmol/L ascorbic acid solution.

25.4.2 Regents and Chemicals

25.4.2.1 Chemicals for HPLC-EC System

- Water, HPLC quality.
- Sodium acetate (CH_3COONa, MW: 82.03) (Merck no. 6268)
- Acetic acid, anhydrous (CH_3COOH, MW: 60.05) (Sigma no. 338826)
- Ascorbic acid ($C_6H_8O_6$, MW: 176.12) (Fluka no. 95210)
- Ethylenediaminetetraacetic acid, EDTA ($(HO_2CCH_2)_2NCH_2CH_2N(CH_2CO_2H)_2$, MW: 292.24)
- Methanol, HPLC quality (Fluka no. 65541)
- 5MTHF-Ca ($C_{20}H_{23}N_7O_6$, MW 459.50) (Schircks Laboratories)
- Freon (Fridore AG)

Stock solutions

- 0.05% (w/v) ascorbic acid: Dissolve 500 mg of ascorbic acid in 500 ml of water. Store at 4 °C.
- 10% (w/v) EDTA: Dissolve 10 g of EDTA in 100 ml of water. Store at room temperature.
- 50 mM sodium acetate (Eluent): Dissolve 4.1 g of sodium acetate in 400 ml of water and add 0.25 ml of 10% EDTA, 5 ml of acetic acid, and 225 ml of methanol. pH is adjusted to 4.4–4.6 with 100% anhydrous acetic acid. Add water to mark 1000 ml. Degas in a vacuum. Store at room temperature.
- 28 µmol/L and 14 µmol/L ascorbic acid solution: Dissolve 5 mg of ascorbic acid in 1 ml of water. Dissolve 1:1000 for 28 µmol/L ascorbic acid solution and 1:2000 for 14 µmol/L ascorbic acid solution:
- Standard mixture: 200 µM 5MTHF. Dissolve 10.0 mg of 5MTHF-Ca in 100 ml of 0.05% ascorbic acid. Store 1 ml aliquots covered with Freon at −20 °C. Stable at least 1 year.
- 100 nM 5MTHF (CSF). Dilute 100 µl of 200 µM 5MTHF solution with 200 ml of 50 mM sodium acetate. Store 1 ml aliquots covered with Freon at −20 °C. Stable at least 1 year.
- 6.25–1600 nM (serum). Use 200 µM 5MTHF standard to prepare calibration standards (6.25, 12.5, 25, 50, 400, 800, 1600 nmol) by diluting with 50 mM sodium acetate.

25.4.2.2 Chemicals for HPLC-FD System

- Water, HPLC quality
- KH_2PO4 (Sigma, ref. P-5379)
- 85% Orthophosphoric acid (Merck, ref. 1.00573.2500)
- Ascorbic acid (Merck, ref. 1.00127.0100)
- Trizma base (TRIS) (Sigma, ref. T-6791)
- HCl (Merck, ref. 1.00317.1000)
- Acetonitrile for HPLC (Romil, ref. H049)
- 5MTHF calibrator (Sigma, ref. M-0132)

Stock solutions

- Reagent n°1: 5 mmol/L phosphate buffer, pH 2.3. Weigh 680 mg of KH_2PO_4 and dissolve it in 1 L of HPLC water and adjust to pH 2.3 by adding a few drops of 85% orthophosphoric acid.
- Reagent n° 2: Buffer Tris-HCl 0.1 mol/L. Weigh 1.00 g of ascorbic acid and 1.21 g of Trizma base (TRIS) and dissolve the two reagents in 100 mL of HPLC water. Adjust pH to 7.6 with HCl.
- Reagent n° 3: Acetonitrile.
- Standard mixture: 5-MTHF (external standard). Stock solution (3.5 mmol/L). Weigh 17.6 mg of 5-MTHF and dissolve it in 10 mL of TRIS-HCl buffer. Store at −80 °C. Stable for 6 months.
- Working solution (35 nmol/L). Make a 1/1000 dilution of the stock solution to reach 3.5 µmol/L concentration. It can be done in two steps: a first dilution with 20 µL of 5-MTHF 3.5 mmol/L and 1980 µL of TRIS-HCl buffer, and a second dilution with 100 µL of the previous dilution and 900 µL of TRIS-HCl buffer.
- Make a new 1/100 dilution of the previous solution (3.5 µmol/L) to reach the working concentration (35 nmol/L): 20 µL of 5-MTHF 3.5 µmol/L and 980 µL of TRIS-HCl buffer. Store at −80 °C. Stable for 6 months.

25.4.3 Instrumentation and Analytical Procedure

25.4.3.1 HPLC-EC

CSF Analysis

Modular isocratic HPLC with EC detection: HPLC pump 515 (Waters), Injector 7725i (Rheodyne), Pulsation Reduced (Portman Instruments), Coulochem Detector Model 5100A (ESA), Analytical Cell 5011 (ESA), and PC with chromatography software (SCPA). Run the system for 2–4 h before the first injection to allow equilibration of the EC detector, a critical issue in electrochemical analysis.

Analytical column:	ODS1, 250 × 4.6 mm, 5 µm (Stagroma)
Eluent:	50 mM sodium acetate
Flow:	1 ml/min
Injection:	Inject 10 uL of CSF

*Keep samples on ice.

Detector 1 settings:	Potential (V)	0.20
	Polarity	–
	Gain	0 × 0
	Response time	2 s
	Autozero	off
	Digital display	–
Detector 2 settings:	Potential (V)	0.00
	Polarity	+
	Gain	10 × 10
	Response time	2 s
	Autozero	OUT
	Digital display	CH 2 OUT

Serum Analysis

HPLC system Gold 128 (Beckman-Coulter) with autosampler (Midas, Spark) and electrochemical detection CLC-100 (Chromsystems, Munich, Germany) is used. Sample results are analyzed by 32 Karat Software (Beckman Coulter, Inc.).

Analytical column:	Spherisorb ODS-1, 5 µm, 250 mm × 4.6 mm (Stagroma)
Pre-column:	Spherisorb C8, 5 µm, 40 mm × 4.6 mm (Stagroma)
Eluent:	50 mM sodium acetate
Flow:	1 ml/min
Injection:	40 µL serum

EC detector settings: oxidation mode, potential of +300 mV.

25.4.3.2 HPLC-FD

HPLC Perkin Elmer series 200 is equipped with a quaternary pump, online degasser, autosampler, and Fluorescence Detector. Totalchrom Integration Program.

Analytical column: ODS pre-column cartridge (Teknokroma) and nucleosil 120 C18 5 μm 15 × 0.4 analytical column (Teknokroma).
Eluent: 5 mmol/L phosphate buffer, pH 2.3: Acetonitrile (95:5)
Flow: 1.3 ml/min
Injection: 30 μL
Retention time: 6 min
Total run: 20 min
Fluorescence detector:
Excitation: 295 nm.
Emission: 355 nm.

25.4.4 Calculation

External standard method for both systems.

25.4.5 Metrological Features

HPLC-EC

Linearity:	3–300 nmol/L (CSF)
	5–1600 nmol/L (serum)
Detection limit:	CSF 3 nmol/L
	Serum 4.5 nmol/L

HPLC-FD (only for CSF)

Linearity:	3–250 nmol/L
Detection limit:	1.5 nmol/L

Table 25.1 Reference values for CSF (the format of this table 25.21 is different from that of the other tables?. Can we unify it?.

Age (years)	HPLC-EC 5MTHF (nmol/L)	HPLC-FD 5MTHF (nmol/L)
0–0.49	64–182	66–203
0.5–1.99	64–182	66–203
2–4.99	63–111	59–126
5–10.99	41–117	59–126
11–15.99	41–117	32–124
>16	41–90	Not done

Table 25.2 Reference values for serum (HPLC-EC)

Age (years)	5MTHF (nmol/L)
0–0.49	38–90
0.5–1.99	36–96
2–4.99	22–71
5–10.99	14–77
11–15.99	12–63
>16	10–39

25.4.6 Post-analytical

25.4.6.1 Reference Values for CSF (Table 25.1)

25.4.6.2 Reference Values for Serum (HPLC-EC (Table 25.2)

25.4.6.3 Pathological Values

Figures 25.1, 25.2, and 25.3 show chromatogram samples for CSF and serum. Low 5MTHF in CSF is associated with a growing number of medical conditions, that have been extensively reviewed [21]. Five potential mechanisms (genetic and environmental) leading to CFD are proposed:

- Reduced transport of folates across the blood–brain barrier into CSF and across the ependymal barrier into the brain including disorders of folate receptor alpha (FRa) and PCFT, disorders of energy metabolism and ATP production and conditions damaging the transport

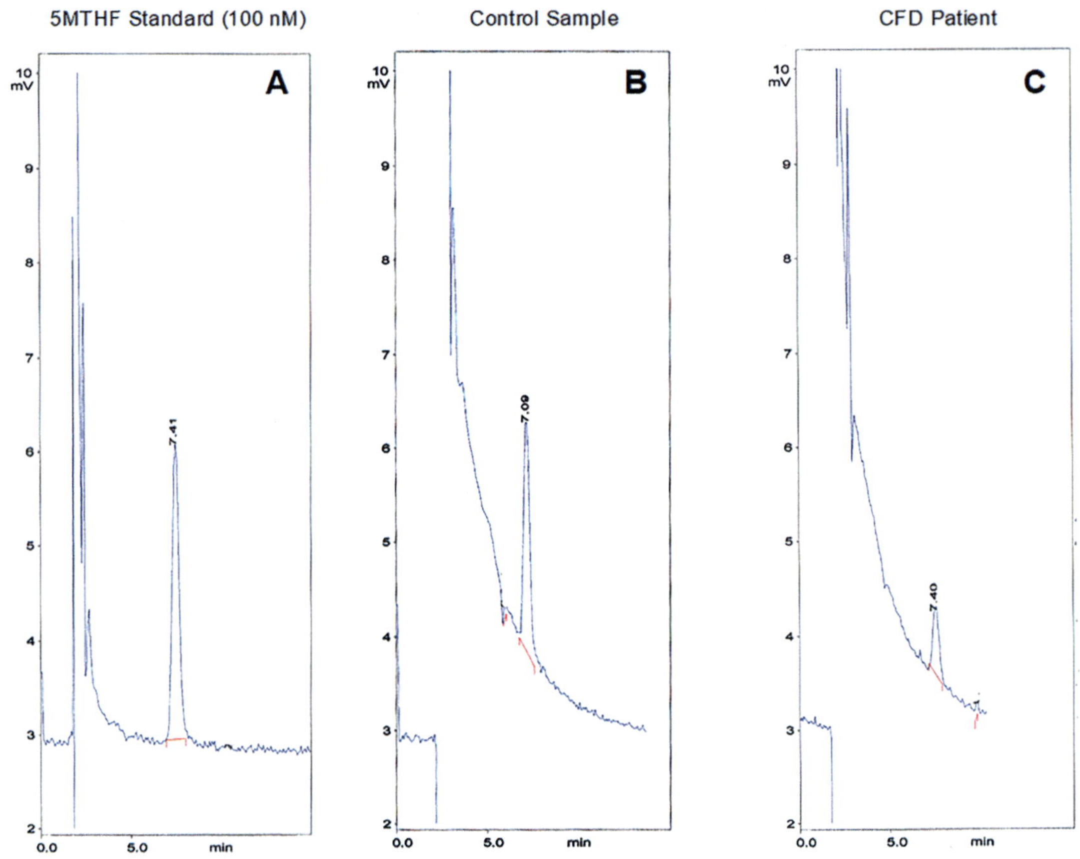

Fig. 25.1 Chromatogram of 5MTHF using electrochemical detection. (**a**) Standard, 100 nM; (**b**) Control CSF; (**c**) CSF from a patient with CFD

functions of brain-endothelial vessel walls and choroid epithelial cells.

- Decreased folate storage due to depletion of intracellular folyl-polyglutamate pool.
- Increased utilization of reduced folates in the brain, including hereditary conditions, iatrogenic conditions, infectious, parainfectious, and immune-mediated conditions.
- Increased catabolism of reduced folates within the nervous system including increased oxidative and/or nitrosative stress and inflammatory processes.
- Conditions affecting folate metabolism in the brain including hereditary conditions and hyperhomocysteinemia due to combined deficiencies of vitamin B2, B6, B9, and B12.

25.4.7 Quality Control

While internal and external quality control schemes are commonly available for total serum folate measurement in the clinical laboratory setting, no such schemes are currently available for 5MTHF. Regarding internal quality control, the use of real CSF samples stored at −80 °C can be used (pooling several remnant CSF samples is advisable). Regarding external quality control,

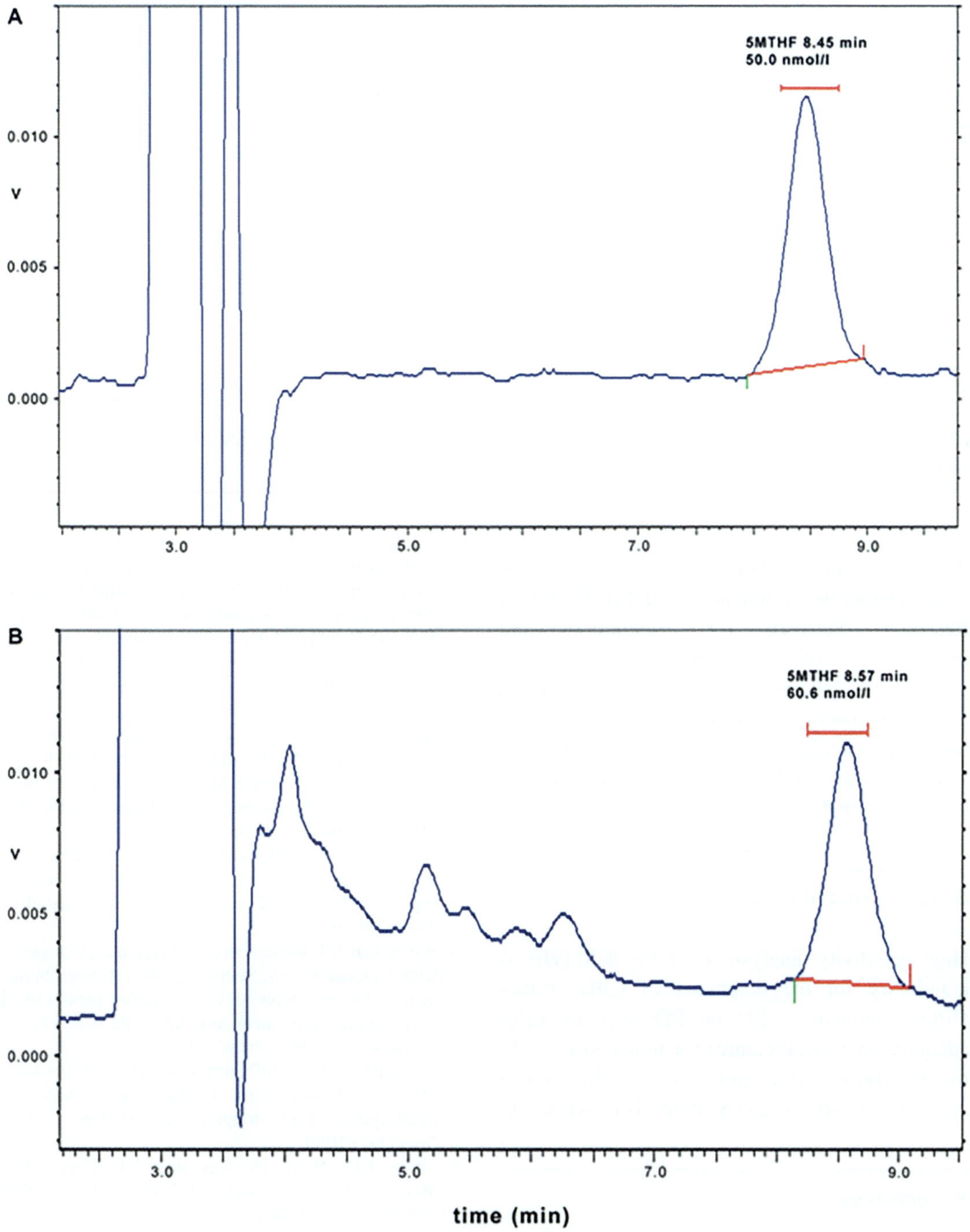

Fig. 25.2 Chromatogram of 5MTHF using electrochemical detection. (**a**) Standard, 50 nM; (**b**) Control serum

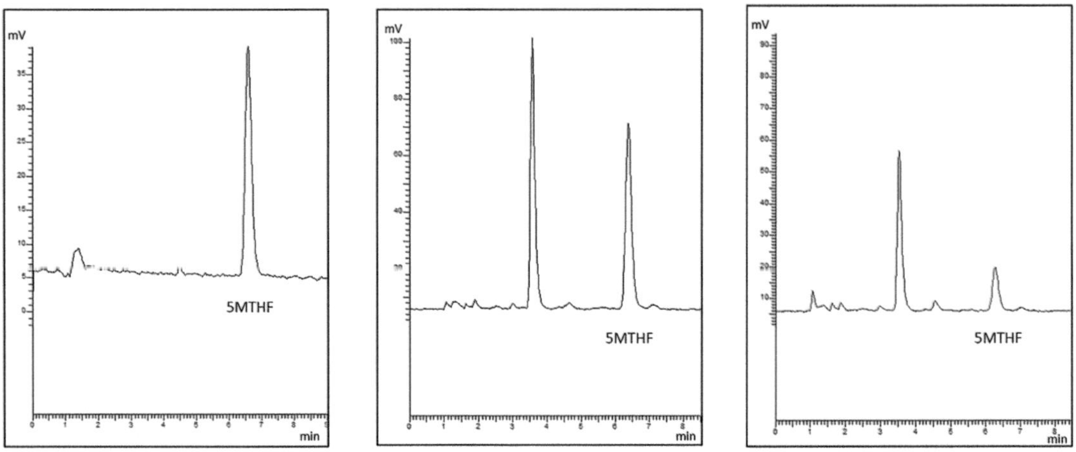

Fig. 25.3 Chromatogram of 5MTHF using fluorescence detection. (**a**) Standard, 100 nM; (**b**) Control CSF (normal value); (**c**) CSF from a patient with profound CFD

the analysis of CSF 5MTHF in the ERNDIM program is under development. ERNDIM (European Research Network for Evaluation and Improvement of Screening, Diagnosis, and Treatment of Inherited Disorders of Metabolism) is an independent not-for-profit foundation which has been providing External Quality Assurance schemes in the field of inborn errors of metabolism since 1994.

25.5 Conclusion

High-sensitivity analysis of CSF 5-MTHF is mandatory for the diagnosis of CFD. Either HPLC coupled to EC or FD systems offer adequate analytical features form this purpose. In the near future, development and application of tandem mass spectrometry methods is expected.

References

1. van der Put NMJ, van Straaten HWM, Trijbels FJM, Folate HJB (2001) Homocysteine and neural tube defects: an overview. Exp Biol Med 226(4):243–270
2. Yerby MS (2003) Clinical care of pregnant women with epilepsy: neural tube defects and folic acid supplementation. Epilepsia 44(Suppl 3):33–40
3. Moat SJ, Lang D, McDowell IF, Clarke ZL, Madhavan AK, Lewis MJ, Goodfellow J (2004) Folate, homocysteine, endothelial function and cardiovascular disease. J Nutr Biochem 15(2):64–79
4. Ramaekers VT, Blau N (2004) Cerebral folate deficiency. Develop Med Child Neurol 46:843–851
5. Ramaekers VT, Hausler M, Opladen T, Heimann G, Blau N (2002) Psychomotor retardation, spastic paraplegia, cerebellar ataxia and dyskinesia associated with low 5-methyltetrahydrofolate in cerebrospinal fluid: a novel neurometabolic condition responding to folinic acid substitution. Neuropediatrics 33(6):301–308
6. Ramaekers VT, Rothenberg SP, Sequeira J, Opladen T, Blau N, Quadros EV, Selhub J (2005) Autoantibodies against folate receptors of the choroid plexus block folate uptake and cause cerebral folate deficiency in childhood. NEJM 352:1985–1991
7. Dunnette J, Weinshilbourn RM (1977) Inheritance of low immunoreactive human plasma dopamine hydroxylase. Radioimmunoassay studies. J Clin Invest 609:1080
8. Thien KR, Blair JA, Leeming RJ, Cooke WT, Melikian V (1977) Serum folates in man. J Clin Pathol 30(5):438–448
9. Ghandour H, Bagley PJ, Shemin D, Hsu N, Jacques PF, Dworkin L, Bostom AG, Selhub J (2002) Distribution of plasma folate forms in hemodialysis patients receiving high daily doses of L-folinic or folic acid. Kidney Int 62(6):2246–2249

10. Burke PA, Luo ML, Zhu JW, Yaffe MB, Forse RA (1996) Injury induces rapid changes in hepatocyte nuclear factor-1-DNA binding. Surgery 120(2):374–381
11. Belz S, Frickel C, Wolfrom C, Nau H, Henze G (1994) High-performance liquid chromatographic determination of methotrexate, 7-hydroxymethotrexate, 5-methyltetrahydrofolic acid and folinic acid in serum and cerebrospinal fluid. J Chromatogr B Biomed Appl 661(1):109–118
12. Arning E, Bottiglieri T (2016) Quantitation of 5-methyltetrahydrofolate in cerebrospinal fluid using liquid chromatography-electrospray tandem mass spectrometry. Methods Mol Biol 1378:175–182
13. Skovby F (2005) Disorders of sulfur amino acids. In: Blau N, Duran M, Blaskovics M, Gibson KM (eds) Physician' guide to the laboratory diagnosis of metabolic disease. Springer, Heidelberg, pp 243–260
14. de Koning TJ, Duran M, Dorland L, Goosken R, Van Schaftingen E, Jaeken J, Blau N, Berger R, Poll-The BT (1998) Beneficial effect of L-serine and glycine in the management of seizures in 3-phosphoglycerate dehydrogenase deficiency. Ann Neurol 44(261–265):261
15. Ponzone A, Spada M, Ferraris S, Dianzani I, De Sanctis L (2004) Dihydropteridine reductase deficiency in man: from biology to treatment. Med Res Rev 24(2):127–150
16. Ramaekers VT, Steen-Ingemann H, Holm J, Opladen T, Senderek J, Häusler M, Heimann G, Fowler B, Maiwald R, Blau N (2003) Reduced folate transport to the brain in Rett syndrome. Neurology 61(4):506–514
17. Blau N, Bonafé L, Krägeloh-Mann I, Thöny B, Kierat L, Häusler M, Ramaekers VT (2003) Cerebrospinal fluid pterins and folates in Aicardi-Goutières syndrome: a new phenotype. Neurology 61:642–648
18. Hyland K, Surtees R (1992) Measurement of 5-methyltetrahydrofolate in cerebrospinal fluid using HPLC with coulometric electrochemical detection. Pteridines 3:149–150
19. Opladen T, Ramaekers VT, Heimann G, Blau N (2006) Analysis of 5-Methyltetrahydrofolate in serum of healthy children. Mol Genet Metab 87:61–65
20. Ormazabal A, García-Cazorla A, Pérez-Dueñas B, Gonzalez V, Fernández-Alvarez E, Pineda M, Campistol J, Artuch R (2006) Determination of 5-methyltetrahydrofolate in cerebrospinal fluid of paediatric patients: reference values for a paediatric population. Clin Chim Acta 371(1–2):159–162
21. Ramaekers VT, Quadros EV (2022) Cerebral folate deficiency syndrome: early diagnosis, intervention and treatment strategies. Nutrients 14(15):3096

α-Aminoadipic Semialdehyde and 6-Oxo-pipecolic Acid

26

Youssef Khalil, Peter T. Clayton, and Philippa B. Mills

Contents

Abbreviations

6-OH-PIP	6-Hydroxy-pipecolate
6-oxo-PIP	6-Oxo-pipecolic acid
AAA	Aminoadipic acid
AASA	Alpha-aminoadipic semialdehyde
AEA	Allysine ethylene acetal
ALDH7A1	α-Aminoadipic semialdehyde dehydrogenase
FMOC	9-Fluorenylmethyloxycarbonyl
LC-MS	Liquid chromatography-Mass spectrometry
MRM	Multiple reaction monitoring
P2C	Δ1-Piperideine-2-carboxylate
P2H2C	Piperidine-2-hydroxy-2-carboxylate
P5CR	Pyrroline-5-carboxylate reductase
P6C	Δ1-piperideine-6-carboxylic acid
PDE	Pyridoxine-dependent epilepsy
PLP	Pyridoxal 5′-phosphate

26.1 Introduction

L-Lysine is an essential amino acid that undergoes catabolism primarily in the liver via two routes; one is by formation of saccharopine and then alpha-aminoadipic semialdehyde (AASA) by the enzyme α-aminoadipic semialdehyde synthase (AASS) in the mitochondria. The other route is by the formation of pipecolic acid in the peroxisome which is then reduced by pyrroline-5-carboxylate reductase (P5CR) to Δ1-piperideine-6-carboxylic acid (P6C), the cyclic Schiff base form of AASA. AASA and

Y. Khalil · P. T. Clayton · P. B. Mills (✉)
Genetics and Genomic Medicine Department, UCL GOS Institute of Child Health, London, UK
e-mail: y.khalil@ucl.ac.uk; peter.clayton@ucl.ac.uk; p.mills@ucl.ac.uk

N. Blau, F. M. Vaz (eds.), *Laboratory Guide to the Methods in Biochemical Genetics*,
https://doi.org/10.1007/978-3-031-58819-8_26

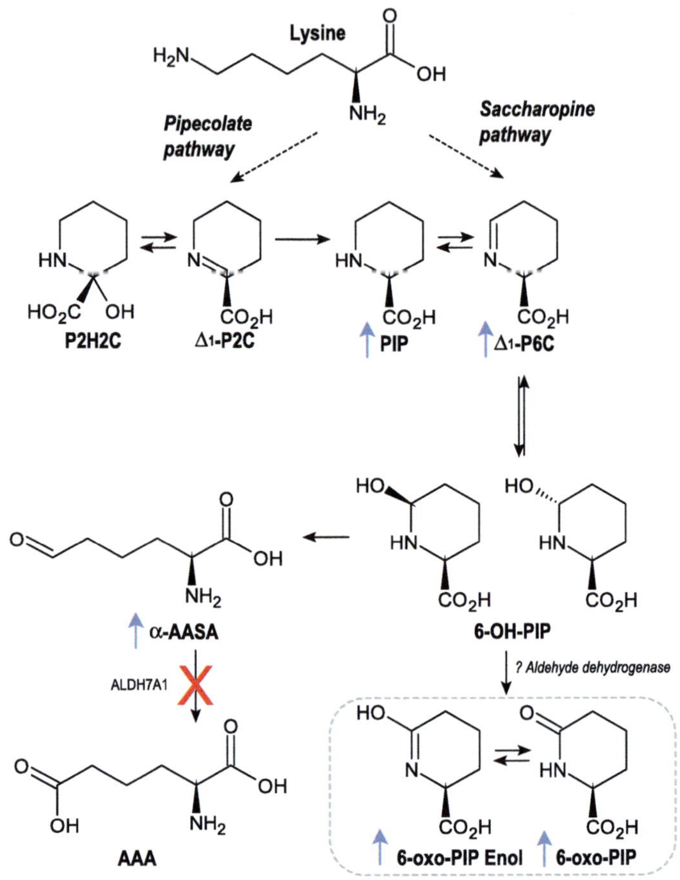

Fig. 26.1 Lysine oxidation pathway. The block in the pathway caused by ALDH7A1 deficiency is shown as a red X and results in the accumulation of metabolites indicated with a blue arrow. 6-Oxo-PIP and its enol form are formed by the oxidation of 6-OH-PIP, an α-AASA and P6C intermediate. PIP, pipecolate; Δ1-P6C, Δ1-piperideine-6-carboxylate; 6-OH-PIP, 6-hydroxy-pipecolate, 6-oxo-PIP, 6-oxo-pipecolate; Δ1-P2C, Δ1-piperideine-2-carboxylate; P2H2C, piperidine-2-hydroxy-2-carboxylate; AASA, α-aminoadipic semialdehyde; AAA, aminoadipic acid

P6C are in equilibrium. AASA continues down the degradation pathway where it is oxidised to aminoadipic acid by the enzyme α-aminoadipic semialdehyde dehydrogenase (ALDH7A1) (Fig. 26.1).

26.2 Clinical Significance

A deficiency of ALDH7A1, due to mutations in the corresponding gene (*ALDH7A1*), leads to accumulation of AASA-P6C in the blood, CSF, and urine. P6C binds to pyridoxal 5′-phosphate (PLP), the active form of vitamin B_6, leading to its depletion. This results in neonatal seizures as PLP plays an important role as a cofactor for more than 70 human enzymes, many of which are involved in neurotransmitter homeostasis. These seizures are partially or completely unresponsive to conventional anti-convulsant drugs but can be controlled by supraphysiological doses of pyridoxine (another form of vitamin B_6), hence the name of this disorder, pyridoxine-dependent epilepsy—ALDH7A1 (PDE-ALDH7A1) [1, 2]. A prompt diagnosis is important to ensure the best outcome can be achieved. Previously, when first described, the diagnosis was confirmed by carrying out a withdrawal trial of pyridoxine and seeing if the seizures reoccurred. It is no longer advised to do this as a diagnosis can be achieved genetically by sequencing *ALDH7A1*, or biochemically by measuring the concentration of AASA [3]. AASA can be measured whether the patient is receiving pyridoxine supplementation or not. Vitamin B_6 does not interfere with the analysis.

More recently, 6-oxo-pipecolic acid (6-oxo-PIP) has also been found to be elevated in the urine, plasma, and CSF of ALDH7A1-deficient patients [4]. 6-Oxo-PIP is derived from

6-hydroxypipecolate, an intermediate of the P6C and AASA equilibrium. As AASA accumulates, because of ALDH7A1 deficiency, 6-hydroxypipecolate is converted to 6-oxo-PIP by an unknown aldehyde dehydrogenase and accumulates. However, further research is needed on the validation of this biomarker before its use as a clinical assay.

26.3 Properties of the Metabolites

AASA (L-2-Amino-6-oxohexanoic acid) is an alpha amino acid that contains a primary amine and carboxylic acid on its alpha carbon and an oxo group on its epsilon carbon. It has a molecular weight of 145.15 Da. Its structure is shown in Fig. 26.1.

6-Oxo-PIP (*(S/R)*-2-Piperidone-6-carboxylic acid) is an imino acid derived from pipecolic acid with a molecular weight of 143.14 Da. Its structure is shown in Fig. 26.1.

26.4 Specimens for Analysis

AASA is generally analysed in urine but can also be analysed in plasma and CSF. AASA is unstable at room temperature, thus a sample must be frozen as soon as possible after collection and stored at −20 °C or ideally at −80 °C until analysis. 6-Oxo-PIP can be measured in urine, plasma, and CSF and has been shown to be stable in urine at room temperature for months [4].

26.5 Methods

26.5.1 Urine AASA Analysis by LC-MS

26.5.1.1 Principle

Following centrifugation of a urine sample, the AASA is derivatised with FMOC-chloride at pH 10.4 to form AASA-FMOC. As no stable isotope for AASA is available, a deuterated isotope of aminoadipic acid is used as an internal standard instead. AASA is analysed by LC-MS in negative ion mode using multiple reaction monitoring (MRM). AASA is not available as a reference standard but can be synthesised by deblocking L-allysine ethylene acetal (AEA) to form AASA:P6C (1:3) [5]. This mix can be used to create a calibration line. However, due to the equilibrium that exists between AASA and P6C and the lack of AASA internal standard this method is only semi-quantitative. The AASA concentration is determined as a ratio to the urine creatinine concentration.

26.5.1.2 Pre-analytical

A minimum sample volume of 100 µl of urine is needed. Due to difficulties in accurately quantifying AASA in extremely dilute urine samples, a minimum creatinine concentration of 0.5 mmol/l is necessary.

26.5.1.3 Reagents and Chemicals

1. Ammonium acetate
2. Acetonitrile
3. Acetone
4. 9-Fluorenylmethoxycarbonyl chloride (FMOC-chloride) (Merck)
5. Boric acid
6. Milli-Q water (Millipore)
7. L-Allysine ethylene acetal (AEA) (Merck)
8. Amberlyst-15 (dry) ion exchange resin (Merck)
9. D_3-Aminoadipic acid (D_3-AAA) (CDN Isotopes)
10. 25% Ammonium hydroxide
11. Sodium hydroxide
12. Eluent A: 4 mM ammonium acetate with formic acid, pH 5
13. Eluent B: Acetonitrile
14. 0.1 M Borate buffer pH 10.4: dissolve 620 mg boric acid in 100 ml H_2O. Adjust pH with 5 M NaOH
15. 5.8 mM FMOC-chloride solution: dissolve 15 mg FMOC-chloride in 10 ml acetone.
16. Internal standard D_3-AAA 10 µmol/L (1.64 mg in 1000 ml water)
17. 1 mM AASA Standard: Dissolve 5 mg AEA in 1 ml H_2O and add 15 mg Amberlyst beads then shake for 10 min at room temperature. Transfer the 1 ml solution to a 15-ml falcon

tube without the beads then wash the beads with 1 mL H_2O twice. Add these wash solutions to the falcon tube. Repeat the wash step with 0.5 ml 25% NH_4OH and combine this wash fraction with the others. Top up the falcon tube with 3.11 ml H_2O to make up a 6.61 ml solution of 1 mmol/L AASA. This solution will also contain 3 mmol/L P6C [6].
18. Make a 7-point AASA serial dilution with a range of 1–100 μmol/L (1, 2.5, 5, 10, 20, 50, 100).

26.5.1.4 Instrumentation

1. ACQUITY ultra-performance liquid chromatography (Waters)
2. Xevo TQ-S triple quadrupole mass spectrometer (Waters)
3. Discovery HS F5-5 column (5 cm × 2.1 mm, 5 μm) (Supelco)
4. Discovery HS F5 Supelguard™ Cartridge (2 cm × 2.1 mm, 5 μm) (Supelco)

26.5.1.5 Analytical Procedure

Using a refrigerated centrifuge, spin down 1 ml urine at 12,000 g for 10 min then pipette 10 μL urine into a 300 μl LC glass vial. Add 10 μl D_3-AAA internal standard (10 μmol/L) and 40 μl H_2O. Add 125 μL borate buffer followed by 125 μL FMOC-chloride solution then vortex briefly. For the calibration line, use 10 μl for each dilution point and the same volume for the rest of the solutions as the urine sample.

Samples are analysed immediately after FMOC-chloride derivatisation. Samples are kept at 10 °C in the ACQUITY ultra-performance liquid chromatography sample manager and 10 μl is injected onto the HS F5-5 column. The column is maintained at 30 °C and the samples are eluted from the column using the gradient described in Table 26.1. The mass spectrometry parameters are: Polarity = ES-; Capillary voltage = 1.50 kV; source temperature = 150 °C; Desolvation temperature = 350 °C. The MRM transitions used for monitoring FMOC-AASA and FMOC-D_3-AAA are listed in Table 26.2. The MRM parameters need to be optimised in each laboratory by tuning each compound using the reference standards.

26.5.1.6 Calculation

The AASA concentration is determined by integration of the AASA and D_3-AAA peaks and calculating the AASA/D_3-AAA peak area ratio. The concentration of AASA is then derived using the peak area ratio and the calibration line equation and divided by the urine creatinine concentration. The AASA level is reported in mmol/mol creatinine or the equivalent μmol/mmol creatinine.

Table 26.1 Liquid chromatography gradient for AASA analysis on a Discovery HS F5-5 column following FMOC derivatisation. Mobile phase A: 4 mM ammonium acetate pH 5; Mobile phase B: acetonitrile

Time (min)	Flow rate (mL/min)	%A mobile phase	%B mobile phase
0.00	0.50	95	5
1.90	0.50	95	5
2.00	0.25	80	20
10.00	0.25	20	80
12.00	0.50	0	100
16.00	0.50	0	100
18.05	0.65	95	5
20.00	0.65	95	5
20.01	0.50	95	5

Table 26.2 MRM parameters for the analysis of FMOC derivatised AASA and D_3-AAA in negative ion mode

	Precursor ion (m/z)	Product ion (m/z)	Cone voltage (V)	Collision Energy (V)
AASA-FMOC (quantitative)	366.09	144.1	4	6
AASA-FMOC (qualitative)	366.09	170.0	4	12
AASA-FMOC (qualitative)	366.09	126.1	4	20
D_3-AAA-FMOC	385.02	163.06	19	13

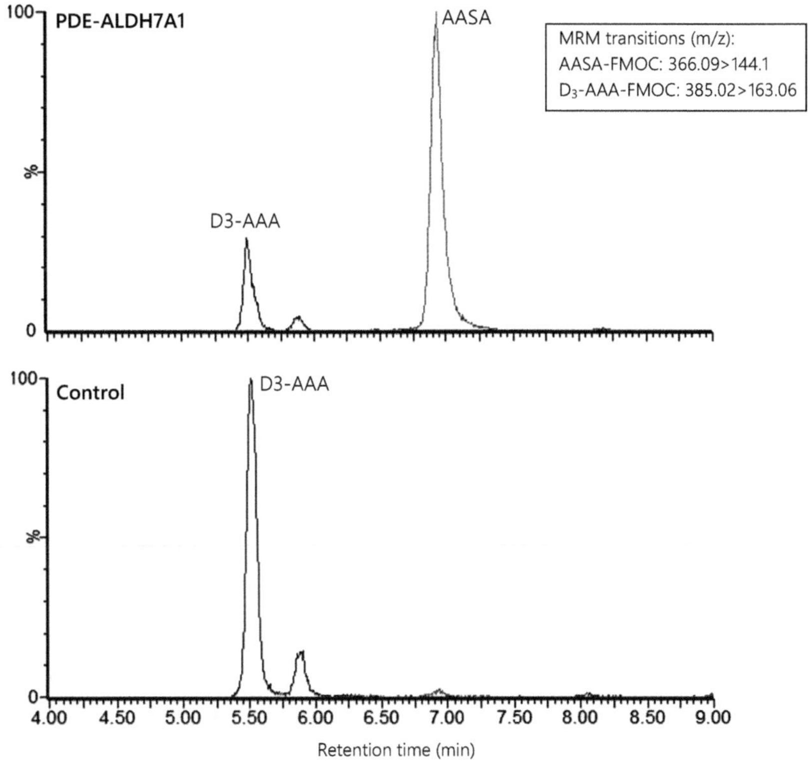

Fig. 26.2 Example chromatograms of AASA in the urine of a PDE-ALDH7A1 patient and a control following FMOC derivatisation. The FMOC-derivatised internal standard D_3-AAA is eluted from the Discovery HS F5-5 column before AASA-FMOC

26.5.1.7 Reference Values

Urine AASA reference levels are age-dependant with higher levels in newborns, but as this method is semi-quantitative, these ranges are likely to be lab specific. Therefore, the reference levels should be treated as a guideline. The levels are divided into three age groups: < 6 months: < 2 mmol/mol creatinine; 6–12 months: < 1 mmol/mol creatinine; and >12 months: < 0.5 mmol/mol creatinine.

26.5.1.8 Post-analytical

Interpretation

AASA concentrations above the control range suggest a deficiency in α-aminoadipic semialdehyde dehydrogenase. However, ALDH7A1 deficient patients on a lysine restricted diet and/or arginine supplementation can have normal urine AASA levels whilst Molybdenum Cofactor and Sulphite Oxidase deficient patients can have moderately elevated AASA levels [7]. Thus, elevated AASA levels should be followed by genetic sequencing of *ALDH7A1* to confirm the diagnosis.

Chromatograms

An example chromatogram of AASA in the urine of a PDE-ALDH7A1 patient and control following FMOC derivatisation can be seen in Fig. 26.2.

26.5.1.9 Quality Control

Every batch of samples analysed on the day contains a positive quality control (QC) sample. Low QC (LQC) and high QC (HQC) AASA standards with a concentration of 1 and 20 μmol/L, respectively, are used for intra-batch variability.

26.5.1.10 Stability

Quality control samples of normal urine spiked with varying concentrations of AASA-P6C have been used for testing the stability of AASA-P6C at room temperature and 4 °C [8]. When urine was spiked with a high concentration (65 μmol/L) of AASA-P6C there was a decrease of approximately 15% in the concentration after being left at room temperature for 24 h. This decrease was more marked in samples spiked with lower concentrations (15 and 25 μmol/L) of AASA-P6C, i.e. after 4 h at room temperature, there was up to a 35% reduction in AASA-P6C, and after 24 h more than a 55% decrease. When stored at 4 °C, the decrease was less marked after 24 h, i.e. samples spiked with high concentrations decreased by less than 10% whilst those of lower concentrations decreased by 15–20%. In a separate study of urine samples from confirmed ALDH7A1-deficient patients, samples were shown to be stable for up to 5 h (longest time looked at) at room temperature and up to 24 h when refrigerated [9]. Hence, it is recommended that urine samples be frozen within 4 h of collection and maintained at −20 or −80 °C until analysed. Samples have been shown to be stable for at least 1 year when stored at −80 °C and for at least 7 days at −20 °C [9].

26.5.2 Urine 6-oxo-PIP Analysis by LC-MS

26.5.2.1 Principle

Following centrifugation of a urine sample, 10 µl of urine is mixed with a solution of 6-oxo-PIP stable isotope, methanol, and acetonitrile. The sample is then analysed by LC-MS in positive ion mode using multiple reaction monitoring. A calibration line is created for quantitation and the 6-oxo-PIP concentration is determined as a ratio to the urine creatinine concentration.

26.5.2.2 Pre-analytical

A minimum sample volume of 100 µl of urine is needed. Due to the stability of 6-oxo-PIP, samples can be stored at room temperature before analysis.

26.5.2.3 Reagents and Chemicals

1. Methanol.
2. Acetonitrile.
3. Milli-Q water (Millipore).
4. Formic acid.
5. Ammonium acetate.
6. 6-Oxo-L-pipecolic acid (Merck).
7. Eluent A: 10 mM ammonium acetate with 0.1% Formic acid (add 385 mg ammonium acetate to 500 µl Milli-Q water then add 500 µl formic acid).
8. Eluent B: Methanol:Acetonitrile (1:1).
9. Internal standard D_3-6-oxo-PIP 10 µmol/l (1.46 mg in 1000 ml water). D_3-6-oxo-PIP can be synthesised as described by Wempe et al. (2019) [4].
10. 1 mM 6-oxo-PIP standard (14.3 mg in 100 ml water).
11. Make an 11-point 6-oxo-PIP serial dilution to cover a range of 1–500 µmol/l (1, 2, 5, 10, 20, 50, 100, 200, 300, 400, 500 µmol/l).

26.5.2.4 Instrumentation

1. ACQUITY ultra-performance liquid chromatography (Waters)
2. Xevo TQ-S triple quadrupole mass spectrometer (Waters)
3. UPLC column HSS T3 (1.8 µm, 2.1 × 150 mm) (Waters)

26.5.2.5 Procedure

Centrifuge 1 ml urine at 12,000 g for 10 min then pipette 10 µl urine into a 300 µl LC glass vial. Add 10 µl D_3-6-oxo-PIP internal standard (10 µmol/L), 20 µl Methanol, and 20 µl acetonitrile, then vortex. For the calibration line, use 10 µl for each serial dilution point and the same volume for the rest of the solutions as the urine sample. Ten microliters of the prepared sample or calibrants are subsequently injected onto the HSS T3 column and eluted with the gradient described in Table 26.3. The column temperature is set at 40 °C whilst the ACQUITY ultra-performance liquid chromatography sample manager temperature is set at 10 °C. The mass spectrometry parameters are: Polarity = ES+; Capillary voltage = 3.08 kV; Source temperature = 150 °C; Desolvation temperature = 450 °C. The MRM transitions used for monitoring 6-oxo-PIP and D_3-6-oxo-PIP are listed in Table 26.4.

26.5.2.6 Reference Values

Control urine 6-oxo-PIP concentrations are not significantly different across the three age groups established for the AASA reference range. The combined reference range for the age groups is 0–3.2 mmol/mol creatinine (previously unpublished).

Table 26.3 Liquid chromatography gradient for 6-oxo-PIP analysis on an HSS T3 column

Time (min)	Flow rate (mL/min)	%A mobile phase	%B mobile phase
0.00	0.40	99.0	1.0
2.50	0.40	99.0	1.0
4.50	0.40	5.0	95.0
5.00	0.40	5.0	95.0
5.10	0.40	99.0	1.0
8.00	0.40	99.0	1.0

Mobile phase A: 0.1% formic acid in 10 mM ammonium acetate; Mobile phase B: 1:1 methanol:acetonitrile

Table 26.4 MRM parameters for the analysis of 6-oxo-PIP in positive ion mode

	Precursor ion (m/z)	Product ion (m/z)	Cone voltage (V)	Collision energy (V)
6-Oxo-PIP (qualitative)	144.17	55.12	28	20
6-Oxo-PIP (qualitative)	144.17	70.15	28	18
6-Oxo-PIP (quantitative)	144.17	98.18	28	12
D_3-6-oxo-PIP	147.17	101.18	28	12

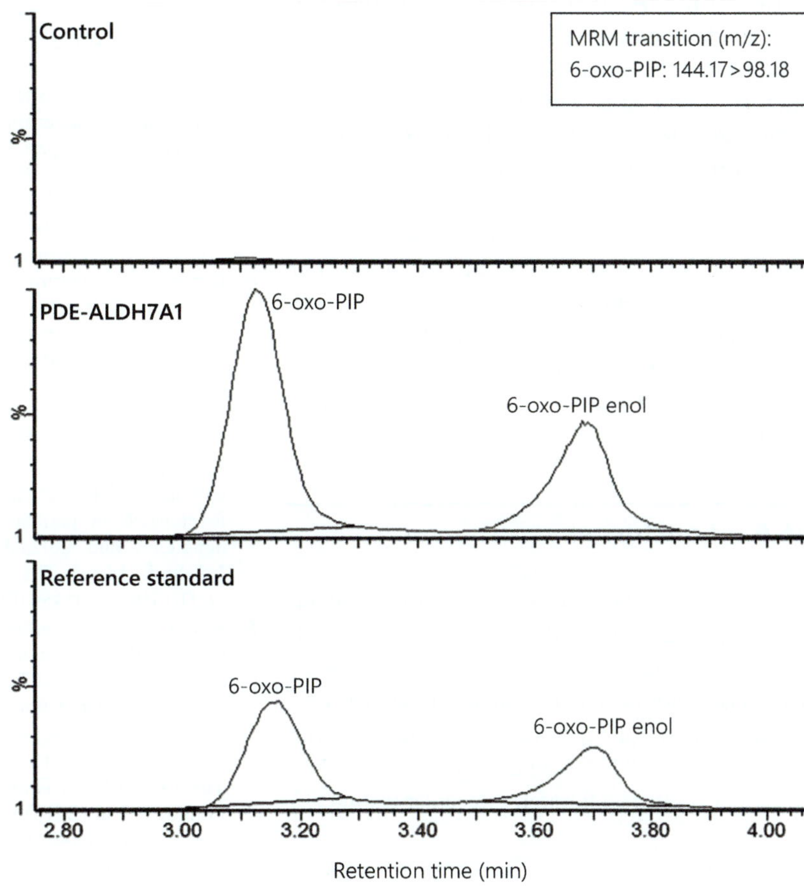

Fig. 26.3 Example chromatogram of 6-oxo-PIP in the urine of a PDE-ALDH7A1 patient and a control individual. Three transitions are monitored, two qualitative (not shown) and one quantitative (shown). The MRM transition monitored for quantitative analysis is *m/z* 144.17 > 98.18. 6-Oxo-PIP and its enol form are resolved and detected for both the PDE-ALDH7A1 patient sample and the reference standard. However, these are not normally detected in control samples

Chromatograms

An example of a chromatogram of 6-oxo-PIP in the urine of a PDE-ALDH7A1 patient and a control individual is shown in Fig. 26.3.

26.5.2.7 Pathological Values

Preliminary data (unpublished) shows that the urine 6-oxo-PIP levels are significantly more elevated in ALDH7A1-deficient patients above the age of 6 months (8.7–564 mmol/mol creatinine, median of 68 mmol/mol creatinine). The preliminary data also suggests some ALDH7A1 deficient newborns up to 6 months of age may show normal or borderline urinary 6-oxo-PIP levels even when AASA levels are elevated, however, these findings require further validation.

26.5.2.8 Quality Control

Every batch of samples analysed on the same day contains a positive QC sample. LQC and HQC 6-oxo-PIP standards are used for inter- and intra-batch variability.

26.5.2.9 Stability

The stability of urine 6-oxo-PIP has been tested in two ALDH7A1 deficient patient samples [4]. At room temperature, the 6-oxo-PIP levels were reduced by 5% from their initial concentration after 48 h, and up to 21% from their initial concentration after 1 week. At −15 °C and −70 °C, 6-oxo-PIP concentrations were reduced by ~10% and 15% from their initial concentrations, respectively, after 1 week. Notably, the 6-oxo-PIP concentration was reduced by up to 40% after 2 months at room temperature, whilst it was ~23% and ~25% less when stored at −15 °C and −70 °C, respectively. Thus, 6-oxo-PIP is more stable than AASA at room temperature, thereby facilitating easier transportation of samples.

26.6 Conclusion

Herein, we describe two methods for the detection of metabolic biomarkers associated with ALDH7A1 deficiency, and the pros and cons of using each biomarker. In tandem, these two biomarkers can provide a more accurate and specific diagnosis of ALDH7A1 deficiency which can then be further confirmed through molecular testing.

Acknowledgements All research at Great Ormond Street Hospital NHS Foundation Trust and UCL Great Ormond Street Institute of Child Health is made possible by the NIHR Great Ormond Street Hospital Biomedical Research Centre. The views expressed are those of the authors and not necessarily those of the NHS, the NIHR, or the Department of Health.

References

1. Mills PB, Footitt EJ, Mills KA, Tuschl K, Aylett S, Varadkar S et al (2010) Genotypic and phenotypic spectrum of pyridoxine-dependent epilepsy (ALDH7A1 deficiency). Brain 133(Pt 7):2148–2159. https://doi.org/10.1093/brain/awq143
2. Mills PB, Struys E, Jakobs C, Plecko B, Baxter P, Baumgartner M et al (2006) Mutations in antiquitin in individuals with pyridoxine-dependent seizures. Nat Med 12(3):307–309. https://doi.org/10.1038/nm1366
3. Coughlin CR 2nd, Tseng LA, Abdenur JE, Ashmore C, Boemer F, Bok LA et al (2021) Consensus guidelines for the diagnosis and management of pyridoxine-dependent epilepsy due to alpha-aminoadipic semialdehyde dehydrogenase deficiency. J Inherit Metab Dis 44(1):178–192. https://doi.org/10.1002/jimd.12332
4. Wempe MF, Kumar A, Kumar V, Choi YJ, Swanson MA, Friederich MW et al (2019) Identification of a novel biomarker for pyridoxine-dependent epilepsy: implications for newborn screening. J Inherit Metab Dis 42(3):565–574. https://doi.org/10.1002/jimd.12059
5. Sadilkova K, Gospe SM Jr, Hahn SH (2009) Simultaneous determination of alpha-aminoadipic semialdehyde, piperideine-6-carboxylate and pipecolic acid by LC-MS/MS for pyridoxine-dependent seizures and folinic acid-responsive seizures. J Neurosci Methods 184(1):136–141. https://doi.org/10.1016/j.jneumeth.2009.07.019
6. Rumbero A, Martin JF, Lumbreras MA, Liras P, Esmahan C (1995) Chemical synthesis of allysine ethylene acetal and conversion in situ into 1-piperideine-6-carboxylic acid: key intermediate of the alpha-aminoadipic acid for beta-lactam antibiotics biosynthesis. Bioorg Med Chem 3(9):1237–1240. https://doi.org/10.1016/0968-0896(95)00110-3
7. Mills PB, Footitt EJ, Ceyhan S, Waters PJ, Jakobs C, Clayton PT et al (2012) Urinary AASA excretion is elevated in patients with molybdenum cofactor deficiency and isolated sulphite oxidase deficiency. J Inherit Metab Dis 35(6):1031–1036. https://doi.org/10.1007/s10545-012-9466-1
8. Xue J, Wang J, Gong P, Wu M, Yang W, Jiang S et al (2019) Simultaneous quantification of alpha-aminoadipic semialdehyde, piperideine-6-carboxylate, pipecolic acid and alpha-aminoadipic acid in pyridoxine-dependent epilepsy. Sci Rep 9(1):11371. https://doi.org/10.1038/s41598-019-47882-2
9. Yuzyuk T, Liu A, Thomas A, Wilson JE, De Biase I, Longo N et al (2016) A novel method for simultaneous quantification of alpha-aminoadipic semialdehyde/piperideine-6-carboxylate and pipecolic acid in plasma and urine. J Chromatogr B Analyt Technol Biomed Life Sci 1017–1018:145–152. https://doi.org/10.1016/j.jchromb.2016.02.043

27 Pterins and Dihydropteridine Reductase

Alessio Cremonesi, Claudia Carducci, and Nenad Blau

Contents

Abbreviations

5-MTHF	5-methyl-tetrahydrofolate
6-MPH_4	6-methyltetrahydropterin
AR	Aldose reductase
BH_2	Dihydrobiopterin
BH_4	Tetrahydrobiopterin
Bio	Biopterin
CR	Carbonyl reductase
DBS	Dried blood spots
DHPR	Dihydropteridine reductase
DTPA	Diethylenetriaminepentaacetic acid
EDTA	Ethylenediaminetetraacetic acid
GEMO	Glyceryl-ether monooxygenase
GTP	Guanosine triphosphate
GTPCH	GTP cyclohydrolase I
HSDH	23α-hydroxysteroid dehydrogenase type 2
Neo	Neopterin
NOS	Nitric oxide synthase
PAH	Phenylalanine-4-hydroxylase
PCD	Pterin-4a-carbinolamine dehydratase
PKU	Phenylketonuria
Prima	Primapterin
Pte	Pterin
PTPS	6-pyruvoyl-tetrahydropterin synthase
RP	Reverse-phase

A. Cremonesi (✉)
Division of Clinical Chemistry and Biochemistry, University Children's Hospital Zürich, Zürich, Switzerland
e-mail: Alessio.Cremonesi@kispi.uzh.ch

C. Carducci
Department of Experimental Medicine, Sapienza University of Rome, Rome, Italy

N. Blau
Division of Metabolism, University Children's Hospital Zürich, Zürich, Switzerland

N. Blau, F. M. Vaz (eds.), *Laboratory Guide to the Methods in Biochemical Genetics*, https://doi.org/10.1007/978-3-031-58819-8_27

SR	Sepiapterin reductase
TCA	Trichloracetic acid
TH	Tyrosine-3-hydroxylase
TPH	Tryptophan-5-hydroxylase

27.1 Introduction

Tetrahydrobiopterin (BH_4) cofactor is essential for various enzyme activities, including phenylalanine-4-hydroxylase (PAH), tyrosine-3-hydroxylase (TH), tryptophan-5-hydroxylase (TPH), nitric oxide synthase (NOS), and glyceryl-ether mono-oxygenase (GEMO). The de novo biosynthesis pathway of BH_4 from guanosine triphosphate (GTP) involves GTP cyclohydrolase I (GTPCH), 6-pyruvoyl-tetrahydropterin synthase (PTPS), and sepiapterin reductase (SR). Three additional enzymes can also catalyze the last two steps of 6-pyruvoyl-tetrahydropterin reduction: aldose reductase (AR), carbonyl reductase (CR), and 3α-hydroxysteroid dehydrogenase type 2 (HSDH2). Cofactor regeneration requires pterin-4a-carbinolamine dehydratase (PCD) and dihydropteridine reductase (DHPR) [1].

The pteridines, a group to which BH_4 also belongs, constitute naturally occurring compounds with a base structure of pyrazino [2,3d] pyrimidine (Fig. 27.1). Pteridines with the structure 2-amino-4-oxo are designated by the term "pterins," and those with the structure 2,4-dioxo by the term "lumazines" (Fig. 27.1). Two groups of pterins can be distinguished: the first have p-aminobenzoate and glutamate attached to the pterin. These pteridines are designated as conjugated pterins (e.g., folic acid). The second group consists of unconjugated pterins, which contain neither of these two substitutions; instead, substitutions occur at the 6-position of the pterin ring nucleus by the aliphatic side chain. The most important blue-fluorescing unconjugated pterins are neopterin, monapterin, biopterin, isoxanthopterin, primapterin, and pterin (Fig. 27.1). Several

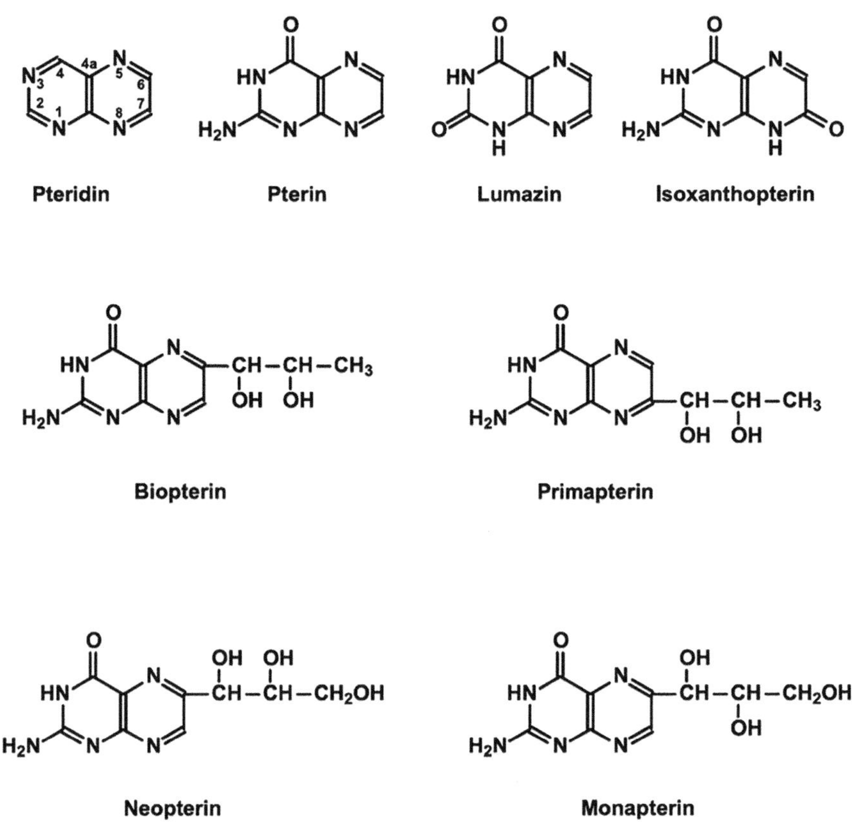

Fig. 27.1 The chemical structure of the most common blue-fluorescing pterins, including lumazin

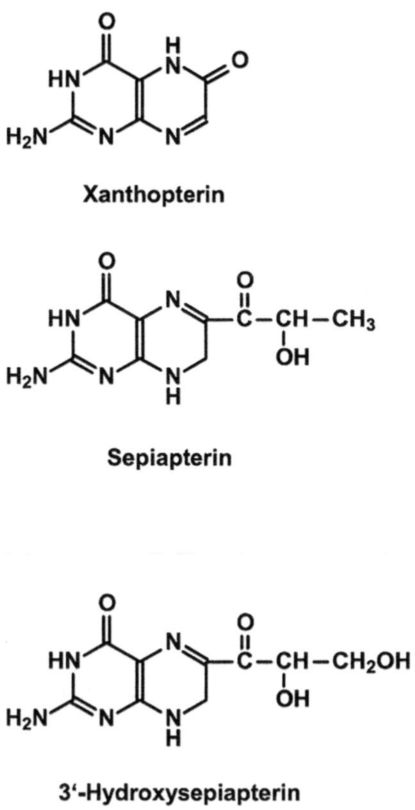

Fig. 27.2 The chemical structure of yellow-fluorescing pterins

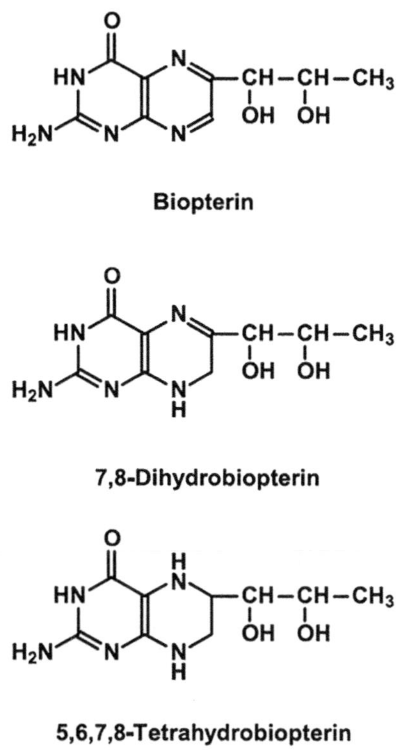

Fig. 27.3 The chemical structure of the three oxidation stages of biopterin

other pterins, like xanthopterin, sepiapterin, and 3′-hydroxy-sepiepterin, are yellow-fluorescing species (Fig. 27.2). Pterins can exist in different oxidation stages, but only fully reduced forms are biologically active. Biopterin is present in biological samples as biopterin, 7,8-dihydrobiopterin (BH_2), and 5,6,7,8-tetrahydrobiopterin (BH_4) (Fig. 27.3).

In man, BH_4 is degraded either nonenzymatically by side-chain cleavage to pterin or is enzymatically metabolized in the gastrointestinal tract to become a lumazine [2]. Pterin and dihydropterin are converted by xanthine dehydrogenase to isoxanthopterin and xanthopterin, respectively [3]. It is assumed, however, that most of the ingested BH_4 is used as a cofactor (mainly for PAH in the liver) and is catabolized to nonfluorescing compounds; it may even be degraded to CO_2 and ammonia.

27.2 Clinical Significance

Measurement of pterins in different biological fluids is the most common method for the screening and diagnosis of inborn errors of BH_4 metabolism. In addition, neopterin is widely used as a marker of T-helper cell 1-derived cellular immune activation [4]. Under physiological conditions, intracellular BH_4 level plays a pivotal role in the regulation of tyrosine and tryptophan hydroxylase, the initial and rate-limiting steps in the biosynthesis of the catecholamines and serotonin; disturbance of biogenic amine metabolism has been implicated as an etiological factor in a variety of neurological disorders, such as Parkinson's disease, familial dystonia, Alzheimer's disease, and endogenous depression.

BH_4 deficiency can be caused by variants in genes encoding the enzymes involved in its biosynthesis (GTPCH, PTPS, and SR) or regeneration (PCD/DCoH and DHPR). BH_4 deficiency due to autosomal recessive variants in BH_4-metabolizing enzymes (except SR) has been described as a cause of hyperphenylalaninemia (HPA). Depending on the enzyme defect and the mode of inheritance, patients are diagnosed by different analytical and biochemical approaches. Early diagnosis and immediate treatment are essential for a good outcome. Two forms of BH_4 deficiency may occur without HPA; the autosomal dominantly inherited form of GTPCH deficiency (dopa-responsive dystonia, DRD, initially described as Segawa disease) [5], and SR deficiency [6]. Patients presenting with HPA are usually detected through the neonatal screening programs for phenylketonuria (PKU), while those presenting without HPA are recognized either by their typical clinical signs and symptoms or by analysis of neurotransmitter metabolites and pterins in their cerebrospinal fluid (CSF).

Screening for a BH_4 deficiency should be done in all newborns with plasma phenylalanine levels higher than 120 μmol/L, as well as in older children with neurologic signs and symptoms [7]. The following tests are recommended: (1) analysis of pterins in urine or dried blood spots (DBS), (2) measurement of DHPR activity in DBS, (3) loading test with BH_4, (4) and analysis of pterins, folates, and neurotransmitter metabolites in CSF. The first two tests are essential and enable all BH_4 defects presenting with HPA to be differentiated. With some limitations, the BH_4-loading test is an additional, useful diagnostic tool for the rapid differentiation between classic PKU and BH_4 variants. This test alone cannot differentiate between some patients with a mild form of PKU/HPA and BH_4 variants. Analysis of neopterin, biopterin, 5-methyl-tetrahydrofolic acid (Chapter Folates), and the neurotransmitter metabolites 5-hydroxyindoleacetic acid and homovanillic acid (see Chap. 28), enables differentiation between severe and mild forms of BH_4 deficiencies.

Major diagnostic applications of neopterin measurements are, for example, monitoring of allograft recipients to recognize immunological complications early. Neopterin production provides prognostic information in patients with malignant tumor diseases and in HIV-infected individuals, high levels being associated with poorer survival expectations. Neopterin measurements are also useful to monitor therapy in patients with autoimmune disorders and in individuals with HIV infection. Screening of neopterin concentrations in blood donations allows the detection of acute infections in a nonspecific way and improves the safety of blood transfusions. As high neopterin production is associated with increased production of reactive oxygen species and with low serum concentrations of antioxidants like tocopherol, neopterin can also be regarded as a marker of reactive oxygen species formed by the activated cellular immune system. Therefore, measurement of neopterin not only allows estimation of the extent of cellular immune activation but also the extent of oxidative stress [4].

27.3 Properties of Pterins

Pterins are well known for their poor solubility in water and in most organic solvents. It has been postulated that this poor solubility in water results from possible hydrogen bonding of the 2-amino and 4-oxo groups between pterin molecules, and that this bonding is preferred to hydration. The nature of the substituents and their position of attachment to the pterin ring influence dramatically the specific solubility. The solubilities of xanthopterin and isoxanthopterin (Fig. 27.1) in water are one part in 40,000 and in 200,000, respectively. Fully oxidized pterins are either blue- or yellow-fluorescent compounds (Table 27.1), while the fluorescence of reduced derivatives is much weaker. Reduced pterins (e.g., BH_4) are extremely light and oxygen sensitive, and decompose rapidly to breakdown products like pterin, pterin-6-caboxylic acid, xanthopterin, and isoxanthopterin.

Table 27.1 Chemical and physical properties of the most common pterins

Compound	MW	Side Chain	Formula	Fluorescence λ_{ex}	Fluorescence λ_{em}
Pterin	163.1	–	$C_6H_5N_5O$	350	450
Xanthopterin	179.1	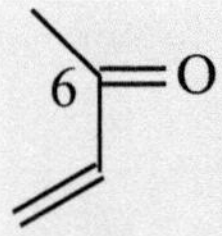	$C_6H_5N_5O_2$	425	530
Isoxanthopterin	179.1	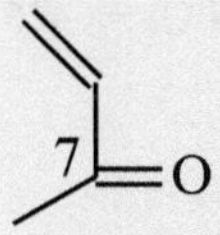	$C_6H_5N_5O_2$	365	480
Neopterin	253.2	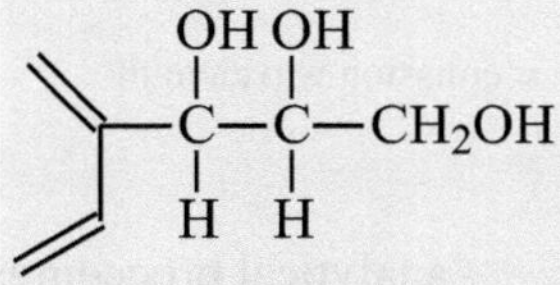	$C_9H_{11}N_5O_4$	350	450
7,8-Dihydroneopterin	255.2	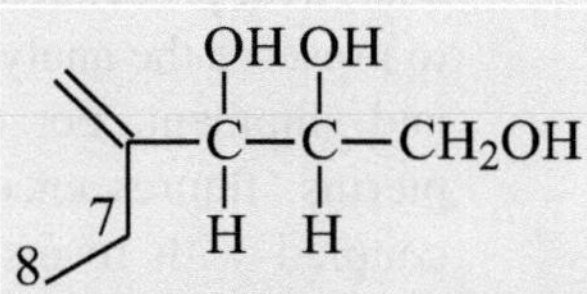	$C_9H_{13}N_5O_4$	–	–
Monapterin	253.2	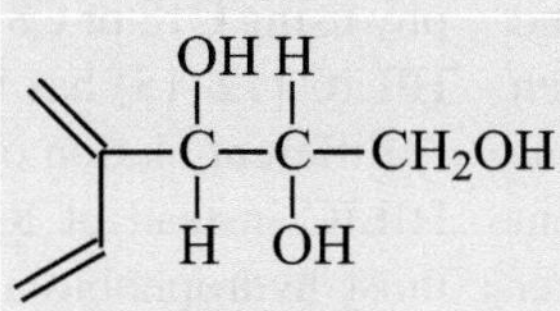	$C_9H_{11}N_5O_4$	350	450
Biopterin	237.2	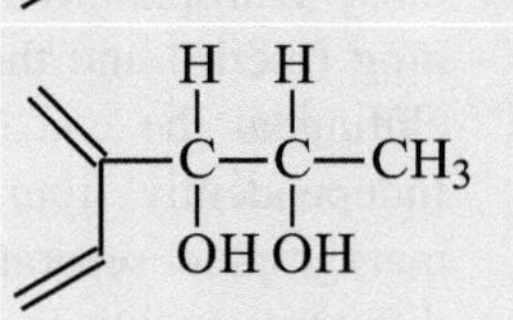	$C_9H_{11}N_5O_3$	350	450
7,8-Dihydrobiopterin	239.2	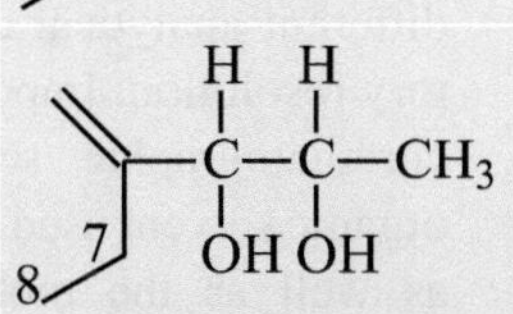	$C_9H_{13}N_5O_3$	–	–
5,6,7,8-Tetrahydrobiopterin	241.2	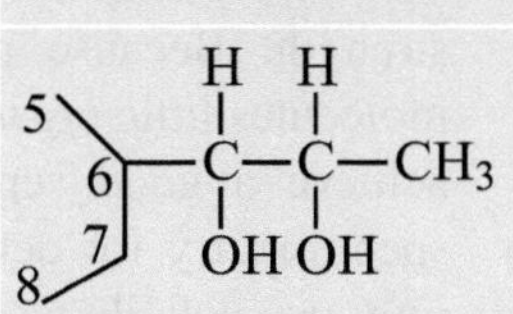	$C_9H_{15}N_5O_3$	–	–
Primapterin	237.2	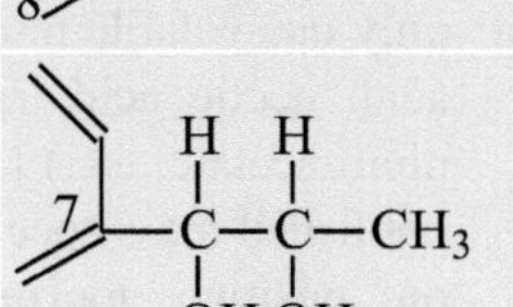	$C_9H_{11}N_5O_3$	350	450

(continued)

Table 27.1 (continued)

Compound	MW	Side Chain	Formula	Fluorescence λ_{ex}	Fluorescence λ_{em}
Sepiapterin	237.2	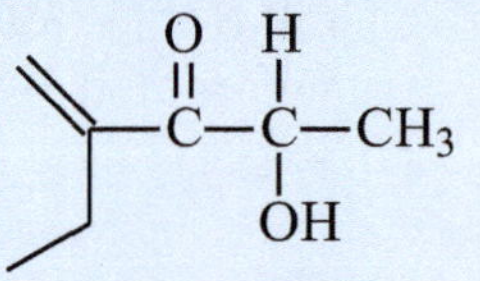	$C_9H_{11}N_5O_3$	425	530
3′-Hydroxysepiapterin	253.2	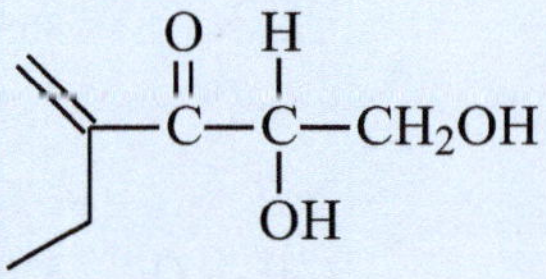	$C_9H_{11}N_5O_4$	425	530

MW Molecular weight, λ_{ex} = excitation wavelength, λ_{em} = emission wavelength

27.4 Methods

27.4.1 General Pre-analytical and Analytical Considerations

Due to the poor solubility of pterins in water and organic solvents, particular care must be taken when preparing highly concentrated solutions like stock solutions. Always follow the solvent recommendations provided by the manufacturer when preparing stock solutions. Since pterins are extremely sensitive to light and oxygen, pre-analytical conditions must be followed carefully to accurately measure these compounds. Oxidized urine samples (see below) are stable and can be sent at room temperature, whereas native urines must be sent on dry ice. CFS samples should be also sent on dry ice because other important CSF analytes besides pterins, e.g., neurotransmitter metabolites and 5-MTHF, are unstable at room temperature. In addition, sepiapterin is unstable in urine and CSF and these samples must be sent on dry ice as well. In contrast, DBS samples can be sent at room temperature since pterins in DBS are already fully oxidized and stable.

As already mentioned, most naturally occurring pterins are present in body fluids in reduced forms. Pterins need to be oxidized to highly fluorescent species when using fluorescence detection, while with mass spectrometry or electrochemical detection it is possible to measure both oxidized and reduced forms in native samples. When measuring native samples, pre-analytical procedures must be carefully followed to preserve the analytes during sample collection and shipment. For the quantitative analysis of pterins, fluorescence detection is most widely coupled with reverse-phase (RP) chromatography using C18 or C8 stationary phases [8–11] but HILIC [12–15] has been also successfully used. A different elution order is expected when using HILIC instead of RP chromatography with the most hydrophobic analyte eluting at the beginning (pterin) and the least hydrophobic analyte eluting at the end (neopterin) of the gradient. Independently from the detection mode, chromatographic separation must be optimized on different analytical columns to achieve satisfactory resolution. Important variables affecting the chromatographic separation are the type of organic solvent used to prepare the mobile phases as well as the mobile phases' pH and ionic strength. Because pterins are relatively polar molecules little organic modifier must be used to achieve a good separation. When using mass spectrometry for detection, it is fundamental to only use volatile mobile phase additives (formic acid, acetic acid, ammonium formate, ammonium acetate, etc.) [16–24]. The majority of the published mass spectrometric methods are based on positive ionization electrospray [16–18, 20–23] but a method using negative ionization electrospray also exists [19]. Both isocratic and gradient elution methods have been successfully used for pterins analysis. While in isocratic mode it is necessary to flush the columns after some injections to avoid contamination, with gradient elution it is possible to wash the column at the end of the gradient shortening the run time. When using fluorescence detection it is important to bear in mind that neopterin, monapterin, biopterin, primapterin, isoxanthopterin, and pterin are blue-fluorescing compounds, while sepiapterin and 3-hydroxysepiapterin are yellow-fluorescing compounds, which are detected using different excitation/emission wavelengths [25].

27.4.2 Analysis of Pterins in Urine, Plasma, Blood Spots, Amniotic Fluid, and CSF by HPLC and Fluorescence Detection

27.4.2.1 Pre-analytical Procedures

Pterins in Urine

Native urine samples should be protected from light, stored at −20 °C and shipped on dry ice, while oxidized urine samples can be stored and shipped at room temperature, but light protection is still recommended. Two procedures for the oxidation of urine (and other samples) are used: (1) oxidation with manganese dioxide (MnO_2) under acidic conditions, and (2) oxidation with iodine (iodine/potassium iodide, I_2/KI) under acidic and basic conditions. While the MnO_2 oxidation method is a routine method used to quantify total pterins (fully oxidized neopterin, monapterin, biopterin, primapterin, isoxanthopterin, and pterin), the I_2/KI method according to Fukushima and Nixon [26] is suited for the differential oxidation of pterins and quantification of BH_4. Total biopterin represents the sum of BH_4, BH_2, and fully oxidized biopterin. Under acidic conditions, BH_4 and BH_2 are oxidized to biopterin, while under basic conditions only BH_2 is oxidized to biopterin and BH_4 undergoes side-chain cleavage to form pterin. The difference in biopterin content between the two oxidations represents the actual BH_4 levels (see below).

Oxidation with MnO_2

Pipette 5 mL of fresh random urine into a centrifugation vial and adjust the pH to 1.0–1.5 with 200 μL of 6 M HCl. Add 100 mg of MnO_2 and shake for 5 min at room temperature. Centrifuge for 5 min at 3000 x*g*. Immediately transfer 1 mL of the clear supernatant into a vial. Wrap the vial in aluminum foil to protect it from light. The vial is then sent to the laboratory at ambient temperature. For collection of dried urine specimens, fresh random urine is processed as described for the liquid specimen (see above). Filter paper strips (3 × 5 cm, filter paper backing 165-0921, BioRad) are dipped into the clear supernatant of the oxidized urine up to 1 cm below the upper edge. Excess urine is wiped off and the filter paper is left to completely dry at room temperature in dim light. The filter strip is then sent to the laboratory in an envelope by express mail.

Oxidation with Iodine

1. Acidic pH oxidation: 150 μL of 0.2 M HCl are added to 100 μL of urine followed by the addition of 50 μL iodine solution (1% (w/v) I_2, 2% (w/v) KI). Samples are mixed and incubated for 1 h in the dark at room temperature. The reaction is stopped by adding 50 μL of 2% (w/v) ascorbic acid and 650 μL H_2O. The clear mixture is injected into the HPLC system.
2. Basic pH oxidation: 100 μL of 0.2 M NaOH are added to the same volume of urine, i.e., 100 μL, followed by addition of 50 μL iodine solution (1% (w/v) I_2, 2% (w/v) KI). Samples are mixed and incubated for 1 h in the dark at room temperature. The reaction is stopped by adding 200 μL of 2 M hydrochloric acid, 50 μL of 2% (w/v) ascorbic acid, and 500 μL H_2O. The clear mixture is injected into the HPLC system.

Pterins in Plasma

Blood is collected in ethylenediaminetetraacetic acid (EDTA) tubes (Sarstedt) and immediately centrifuged at 2000 x*g* for 10 min. Plasma is stored at −80 °C and, if necessary, shipped on dry ice and protected from light. When performing the analysis, plasma is thawed on ice before proceeding with the oxidation procedure.

Oxidation with MnO_2

Pipette 0.2 mL of plasma in an Ultrafree membrane filter unit with NMWL of 10 kDa (Merck),

30 µL of 30% (w/v) trichloroacetic acid (TCA), and 1 mg of MnO_2 and shake for 5 min at room temperature. Centrifuge for 10 min at 16,900 x*g*. The clear filtrate is injected into the HPLC system.

Oxidation with Iodine

1. Acidic pH oxidation: 20 µL of 1 M HCl are added to 100 µL of plasma followed by addition of 50 µL of iodine solution (1% (w/v) I_2, 2% (w/v) KI). Samples are mixed and incubated for 1 h in the dark at room temperature. The reaction is stopped by adding 10 µL of 5% (w/v) ascorbic acid and 20 µL H_2O. Oxidized samples are filtered using an Ultrafree membrane filter unit with NMWL of 10 kDa by centrifugation at 16,900 *xg* for 10 min at 4 °C. The clear filtrate is injected into the HPLC system.
2. Basic pH oxidation: 20 µL of 1 M NaOH are added to the same volume of plasma, i.e., 100 µL, followed by addition of 50 µL of iodine solution (1% (w/v) I_2, 2% (w/v) KI). Samples are mixed and incubated for 1 h in the dark at room temperature. The reaction is stopped by adding 10 µL of 5% (w/v) ascorbic acid and 20 µL of 2 M hydrochloric acid. Oxidized samples are filtered using an Ultrafree membrane filter unit with NMWL of 10 kDa by centrifugation at 16,900 x*g* for 10 min at 4 °C. The clear filtrate is injected into the HPLC system.

Pterins in Dried Blood Spot (DBS)

Pterins collected on DBS cards (Guthrie cards) are already fully oxidized and therefore these samples can be sent at room temperature and are processed without prior oxidation. For longer time storage DBS should be stored at −20 °C [27]. For every single measurement, 4 blood spots (6 mm diameter) are cut out from the Guthrie card and pterins are extracted with 500 µL of 0.02 M HCl and placed into an ultrasonic bath (Sonorex RK31, Bandelin) for 30 s. Extraction is continued for 10 min by incubation and vortexing the filter spots solution five times for 10 s at room temperature. The extract is centrifuged at 12,000 x*g* for 3 min at room temperature. A 60-µL aliquot of the clear supernatant is used for analysis of hemoglobin (Hb) on HemoCue® Plasma Low Hb Analyser (Hitado). The remaining supernatant is filtered using an Ultrafree membrane filter unit with NMWL of 10 kDa at 10,000 *xg* for 10 min at 4 °C. The clear filtrate is injected into the HPLC system.

Pterins in CSF and Amniotic Fluid

Pterins in CSF or amniotic fluid are unstable and these samples need to be shipped on dry ice (and protected from light). Before the analysis pterins need to be oxidized. For this, pipette 0.2 mL of CSF or amniotic fluid into an Ultrafree membrane filter unit with NMWL of 10 kDa, add 20 µL of 30% (w/v) TCA, and 1 mg of MnO_2. Shake the solution for 5 min at room temperature, followed by centrifugation for 10 min at 16,900 x*g* at 4 °C. The clear filtrate is injected into the HPLC system.

Sepiapterin in CSF and Urine

For sepiapterin determination in urine and in CSF, samples are analyzed without oxidation. Urine samples are protected from light and stored at −80 °C until analysis and, for shipment, samples are sent on dry ice. Before analysis, 360 µL urine is added together with 40 µL of 1% (w/v) ascorbic acid to an Ultrafree membrane filter unit with NMWL of 10 kDa, followed by centrifugation at 16,900 x*g* for 10 min at 4 °C. The clear filtrate is then pipetted into a new vial. CSF is injected directly without any sample preparation.

27.4.2.2 Reagents and Chemicals

Chemicals

1. Water, HPLC quality
2. HCl, 37% (Merck)
3. Methanol, HPLC grade (Sigma-Aldrich)
4. Isopropanol (Sigma-Aldrich)
5. Formic acid (MW 46.03; Merck)
6. NaOH (MW 40.00; Sigma-Aldrich)
7. Trichloroacetic acid (= TCA; Merck)
8. Activated MnO_2 (MW 86.94; Merck)
9. I_2 (MW 253.81; Sigma-Aldrich)
10. KI (MW 166.00; Sigma-Aldrich)
11. Ascorbic acid (MW 176.12; Sigma-Aldrich)

12. Dithioerythritol (= DTE; MW 154.25; Sigma-Aldrich)
13. Neopterin (= Neo; MW 253.20; Schircks Laboratories)
14. Monapterin (= Mon; MW 253.20; Schircks Laboratories)
15. Biopterin (= Bio; MW 237.20; Schircks Laboratories)
16. Primapterin (= Prima; MW 237.20; Schircks Laboratories)
17. Isoxanthopterin (= Ixa; MW 179.10; Schircks Laboratories)
18. Pterin (= Pte; MW 163.10; Schircks Laboratories)
19. Sepiapterin (= Sep; MW 237.2, LGC Standards)
20. Deoxysepiapterin (= Deoxy-Sep; MW 221.2; Schircks Laboratories
21. 3-Hydroxysepiapterin (= 3-OH-Sep; MW 253.2; Schircks Laboratories)

Reagents/Solutions

1. 1 M HCl: Dilute 8.264 mL 37% HCl and 100 mL H_2O, store at RT.
2. 0.1 M HCl: Dilute 10 mL 1 M HCl in 90 mL H_2O, store at RT.
3. 0.2 M HCl: Dilute 50 mL 1 M HCl in 200 mL H_2O, store at RT.
4. 0.05 M HCl: Dilute 10 mL 0.1 M HCl in 10 mL H_2O, store at RT.
5. 0.02 M HCl: Dilute 1 mL 0.2 M HCl in 90 mL H_2O, store at RT.
6. 1 M NaOH: Dissolve 20 g 1 M NaOH in 500 mL H_2O, store at RT.
7. 0.2 M NaOH: Dilute 50 mL 1 M NaOH in 200 mL H_2O, store at RT.
8. 0.1 M NaOH: Dilute 1 mL 1 M NaOH in 9 mL H_2O, store at RT.
9. 1% (w/v) ascorbic acid: Dissolve 0.1 g ascorbic acid in 10 mL H_2O; prepare each fresh (2 d stable).
10. 2% (w/v) ascorbic acid: Dissolve 0.2 g ascorbic acid in 10 mL H_2O; prepare freshly (2 d stable).
11. 5% (w/v) ascorbic acid: Dissolve 0.5 g ascorbic acid in 10 mL H_2O; prepare freshly (2 d stable).
12. 30% (w/v) TCA: dissolve 30 g TCA in 100 mL of H_2O. Store at RT.
13. 1% (w/v) I_2, 2% (w/v) KI: dissolve 1 g I_2 and 2 g KI in 100 mL of H_2O. Store at RT.
14. UHPLC Pterins mobile phase A (0.04% formic acid): add 40 μL formic acid to 1 L H_2O. Maximal storage at RT: 3 months.
15. UHPLC Pterins mobile phase B (100% methanol).
16. UHPLC Pterins mobile phase C (20% methanol): dilute 200 mL methanol with H_2O to a final volume of 1 L. Maximal storage at RT: 3 months.
17. UHPLC Pterins mobile phase D (50% methanol): dilute 500 mL methanol with H_2O to a final volume of 1 L. Maximal storage at RT: 3 months.
18. UHPLC Sepiapterin mobile phase A (0.02% formic acid): add 20 μL formic acid to 1 L H_2O. Maximal storage at RT: 3 months.
19. UHPLC Sepiapterin mobile phase B (80% methanol): dilute 800 mL methanol with H_2O to a final volume of 1 L. Maximal storage at RT: 3 months.

Stock Solutions and Standards (Pterins)

1. 1 mM Pterin stock solutions: Dissolve separately neopterin, monapterin, biopterin, primapterin, isoxanthopterin, or pterin in 1 mL 0.1 M NaOH. Keep for 30 s in an ultrasound bath. Add 9 mL 0.1 M HCl and keep for another 30 s in an ultrasound bath. The volumes of NaOH and HCl can be slightly adjusted to obtain a final concentration of 1 mM. Store samples as aliquots at −80 °C.
2. 10 μM Pterins working solutions: dilute 100 μL of each 1 mM Pterin stock solution with 9.9 mL 0.05 M HCl. Store samples as aliquots at −20 °C.
3. Urine pterins standard mixtures: pipette the different 10 μM pterins working solutions in a 10 mL flask and fill with 0.05 M HCl to the mark to prepare the calibration levels in Table 27.2. Store samples as aliquots at −20 °C.
4. Plasma/CSF/DBS/amniotic fluid pterins standard mixtures: pipette the different 10 μM

Table 27.2 Calibration range for pterins in urine

Pterins in urine (final concentrations)												
	Cal. 1			Cal. 2			Cal. 3			Cal. 4		
	Neo	Bio	Prima	Neo	Bio	Prima	Neo	Bio	Prima	Neo	Bio	Prima
µM	2	2	0.5	10	10	1	20	20	5	50	50	10

Table 27.3 Calibration range for pterins in Plasma/CSF/DBS/amniotic fluid

Pterins in plasma/CSF/DBS/amniotic fluid (final concentrations)										
	Cal. 1		Cal. 2		Cal. 3		Cal. 4		Cal. 5	
	Neo	Bio	Neo	Bio	Neo	Bio	Neo	Bio	Neo	Bio
nM	5	5	10	10	40	40	200	200	1000	500

Table 27.4 Calibration range for sepiapterin in urine

Sepiapterin in urine (final concentrations)					
	Cal. 1	Cal. 2	Cal. 3	Cal. 4	Cal. 5
µM	1	10	50	200	800

Table 27.5 Calibration range for sepiapterin in CSF

Sepiapterin in CSF (final concentrations)					
	Cal. 1	Cal. 2	Cal. 3	Cal. 4	Cal. 5
nM	0.5	1	5	20	100

pterins working solutions in a 10 mL flask and fill with 0.05 M HCl to the mark to prepare the calibration levels in Table 27.3. Store samples as aliquots at −20 °C.

Stock Solutions and Standards (Sepiapterin)

1. 1 mM Sepiapterin stock solutions: dissolve separately sepiapterin, 3-hydroxysepiapterin, deoxysepiapterin in H_2O to obtain a final concentration of 1 mM. Keep for 30 s in an ultrasound bath. Store samples as aliquots protected from light at −80 °C.
2. 10 µM Sepiapterin working solutions: dilute 100 µL of each 1 mM Sepiapterin stock solution with 9.9 mL H_2O. Store samples as aliquots protected from light at −20 °C.
3. Urine sepiapterin mixtures: pipette the 3 different 10 µM sepiapterin working solutions in a 10 mL flask and fill with 0.02% formic acid (mobile phase A) to the mark to prepare the calibration levels in Table 27.4. Store samples as aliquots protected from light at −20 °C.
4. CSF sepiapterin standard mixtures: pipette the 3 different 10 µM sepiapterin working solutions in a 10-mL flask and fill with 0.02% formic acid (mobile phase A) to the mark to prepare the calibration levels in Table 27.5. Store these samples as aliquots protected from light at −20 °C.

27.4.2.3 Instrumentation

Figures 27.4 and 27.5a–d describe the multivalve Agilent 1260–1290 Infinity II Prime LC-System used to analyze pterins and sepiapterin in urine, CSF, plasma, and DBS. To extend the lifetime of the analytical columns and to obtain clean baseline traces, trap columns are used. This is not necessary for the measurement of sepiapterin in CSF because this matrix is very clean and the excitation/emission wavelengths for this analyte are rather specific and thus the overall noise is low. For each matrix, different trap and analytical columns are used in order to prevent contaminations, carryover, or shifts in retention times. Using this configuration trap and analytical columns can be routinely used for up to 3000 and 8000 injections, respectively. For all matrices, the same type of trap column is used (Agilent Poroshell 120 SB-C8, 3 × 50 mm, 2.7 µm). For the measurements of pterins in CSF, plasma, and DBS the same type of analytical column is used (Agilent Poroshell 120 SB-C8, 3 × 100 mm, 2.7 µm), while for measuring pterins in urine a longer version of this same column (3 × 150 mm) is used due to the complexity of this matrix. The

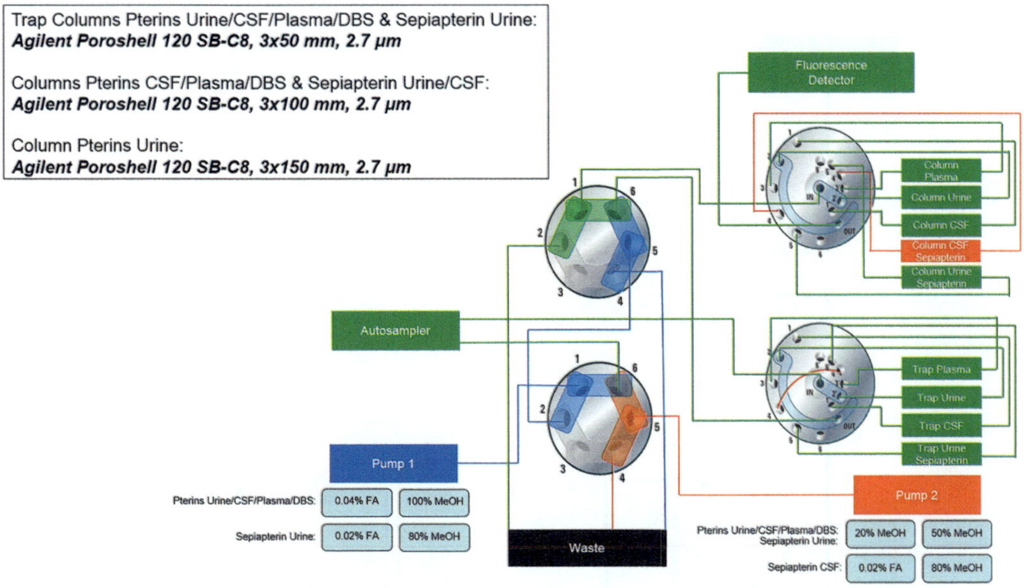

Fig. 27.4 Multivalve system to separate pterins and sepiapterin in different matrices. To improve the method robustness for each matrix a different trap column and analytical column is used with the exception of sepiapterin in CSF, for which no trap column is used. For all matrices the same type of trap column is used (Agilent Poroshell 120 SB-C8, 3 × 50 mm, 2.7 µm). For all matrices, the same type of analytical column is used (Agilent Poroshell 120 SB-C8) with different lengths (150 mm for urine and 100 mm for all other matrices)

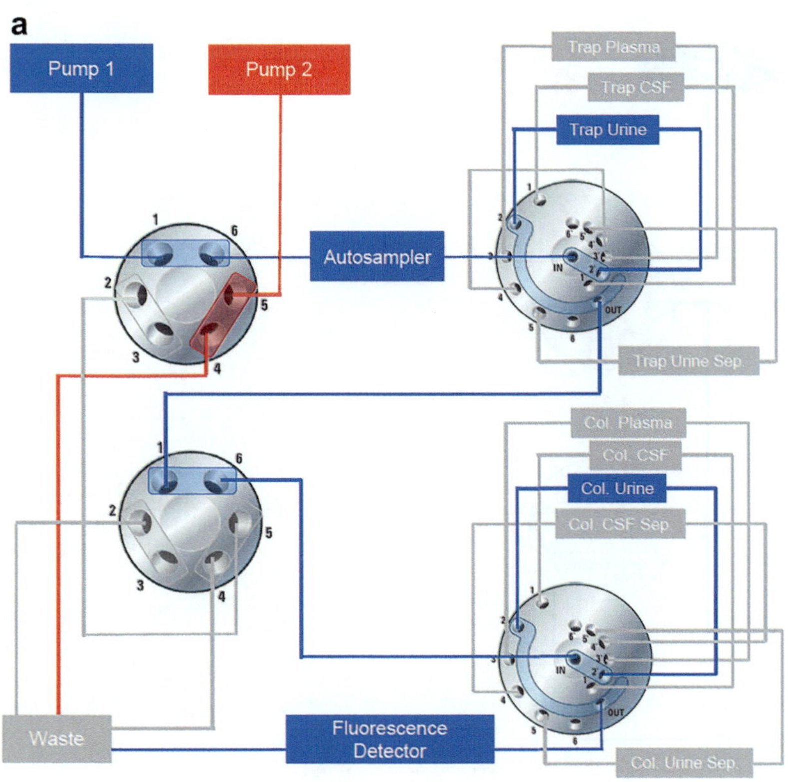

Fig. 27.5 (**a**) Valves configuration to load the analytes on the trap and analytical columns. (**b**) Valves configuration to separate the analytes on analytical columns and to wash the trap columns. (**c**) Valves configuration to wash the trap and analytical columns. (**d**) Valves configuration to analyze sepiapterin in CSF (without using a trap column)

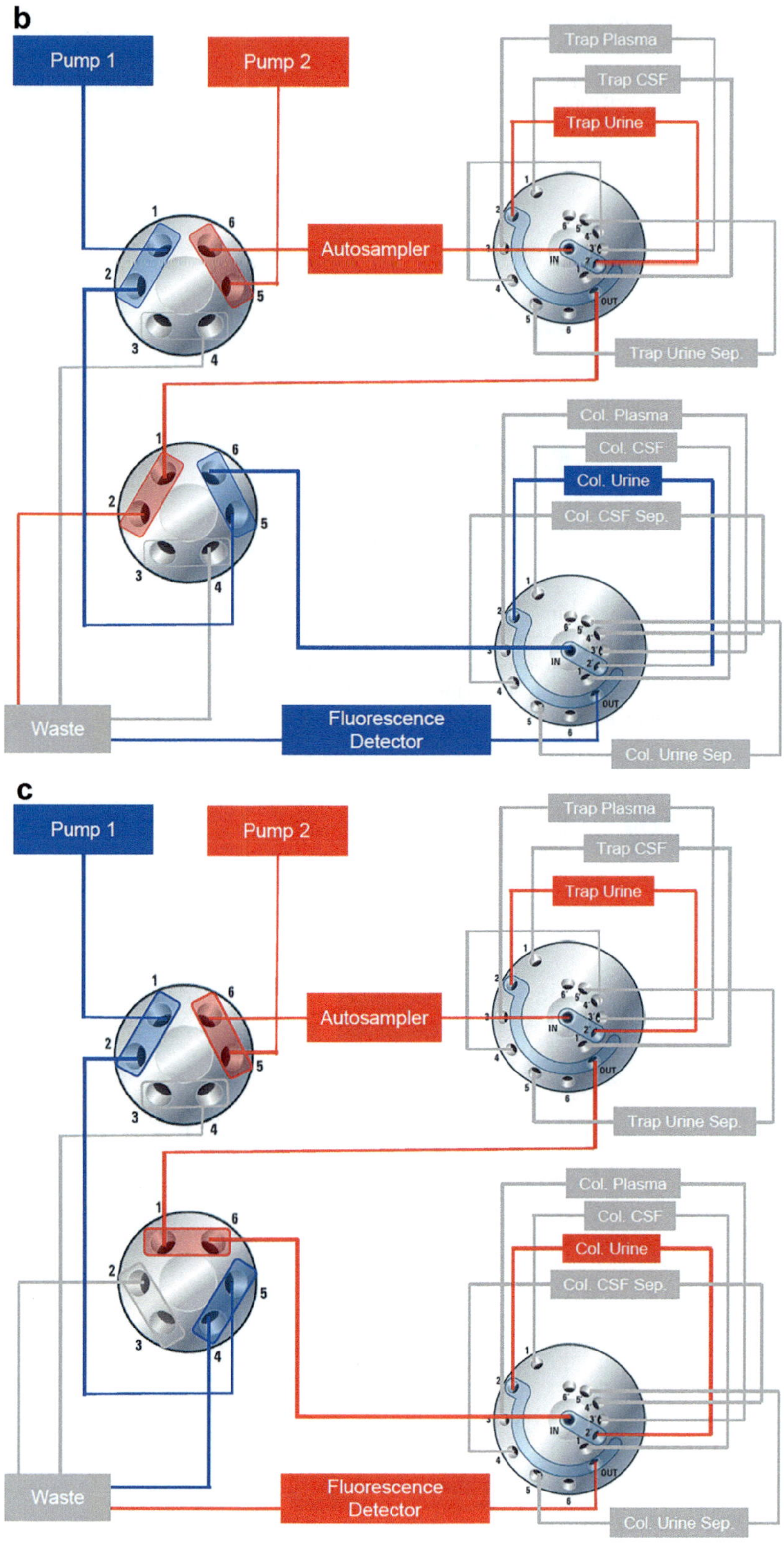

Fig. 27.5 (continued)

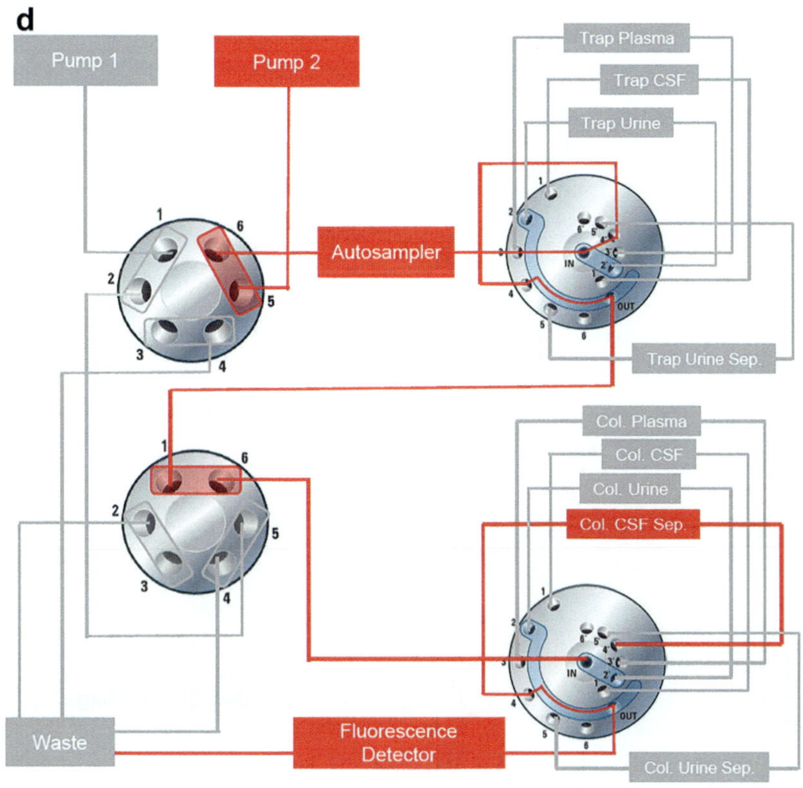

Fig. 27.5 (continued)

total run time is 10 min. Due to the different physiological concentrations of pterins in different matrices, different injection volumes are used: 20 μL for CSF, plasma, and DBS samples and 3 μL for urine samples. For the detection of blue-fluorescing pterins an excitation wavelength (λ_{Ex}) of 350 nm and an emission wavelength (λ_{Em}) of 450 nm are set, whereas to detect sepiapterin a λ_{Ex} of 425 nm and a λ_{Em} of 530 nm are used. For the quantification of sepiapterin a different method is used due to its chromatographic and fluorescing and properties.

More simple configurations and instrumentations can be also successfully used to measure these analytes, but it is highly recommended to wash thoroughly the columns after each application and, if possible, to use a trap column to extend the lifetime of the analytical column and to improve the method robustness. Since pterins are present at different physiological concentrations in the different matrices it is very important that clean baselines are obtained before injecting any samples. In general, a gradient elution should be favored over an isocratic elution, because with the latter hydrophobic contaminants can accumulate on the column over time. Moreover, the gradient and flow should be carefully optimized to reproducibly separate the structural isomers biopterin and primapterin (7-biopterin) from each other's.

Chromatograms of standard mixtures, healthy patients' samples as well as chromatograms of pterins from patients with different enzyme defects of BH_4 metabolism are shown in Fig. 27.6a–d. The patterns of pterins in plasma, dried blood, CSF, and urine are very similar for patients with BH_4 deficiency.

27.4.2.4 Calculations

A calibration curve is obtained from the standard solutions that are included in each sample series using the OpenLab CDS Software (Agilent).

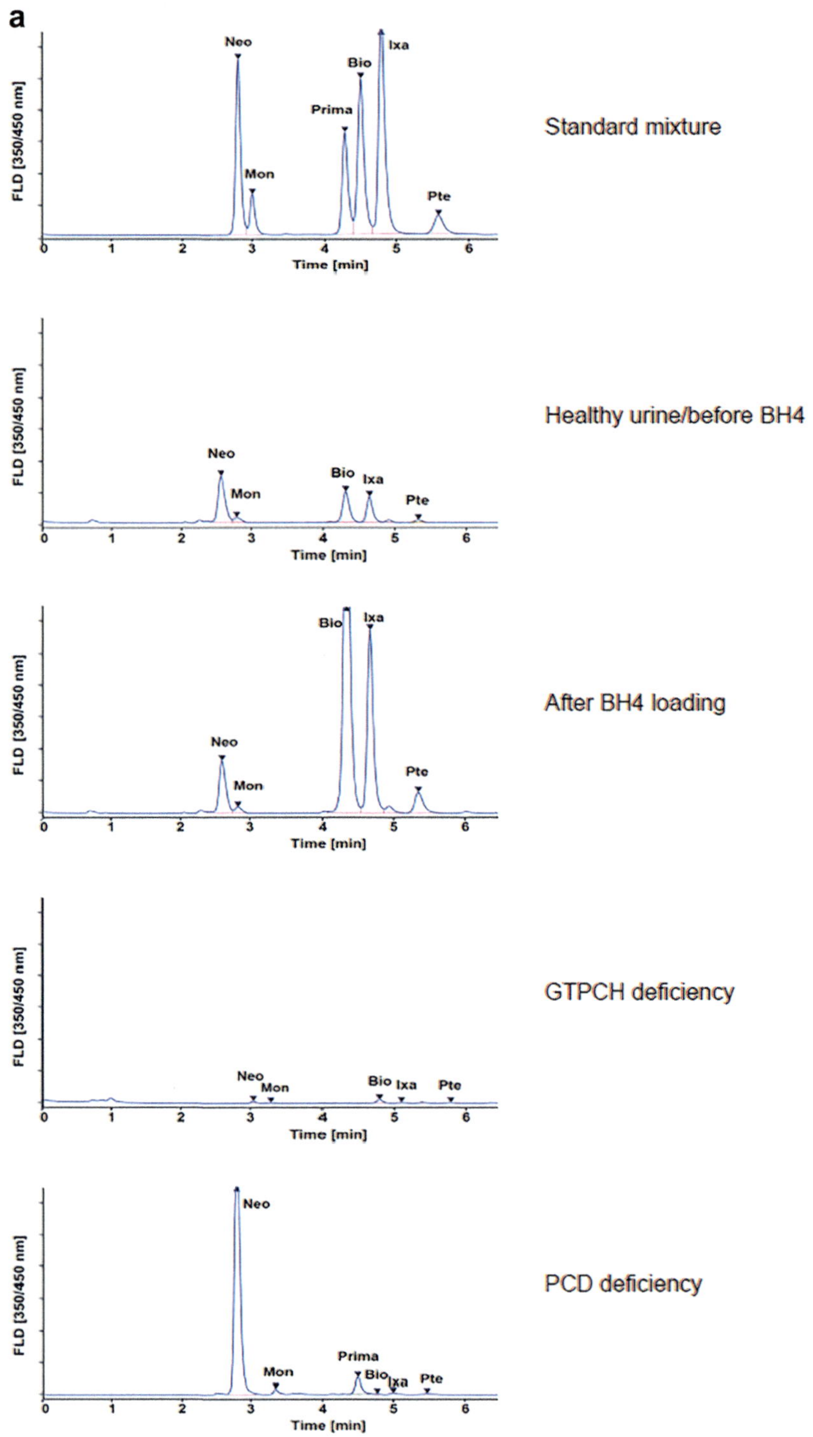

Fig. 27.6 (**a**) HPLC-FT of pterins in urines from a healthy individual as well as patients with known BH_4 defects. Neo = neopterin, Bio = biopterin, Mon = monapterin, Prima = primapterin, Ixa = isoxanthopterin, Pte = pterin. (**b**) HPLC-FT of pterins in CSF from a healthy individual as well as patients with known BH_4 defects and a patient with Aicardi-Goutières syndrome. Neo = neopterin, Bio = biopterin, Mon = monapterin, Prima = primapterin, Ixa = isoxanthopterin, Pte = pterin. (**c**) HPLC-FT of pterins in DBS from a healthy individual as well as a patient with PTPS deficiency. Neo = neopterin, Bio = biopterin, Mon = monapterin, Prima = primapterin, Ixa = isoxanthopterin, Pte = pterin. (**d**) HPLC-FT of sepiapterin in urine. Sep = sepiapterin, 3-OH-Sep = 3-hydroxysepiapterin, Deoxy-Sep = deoxysepiapterin

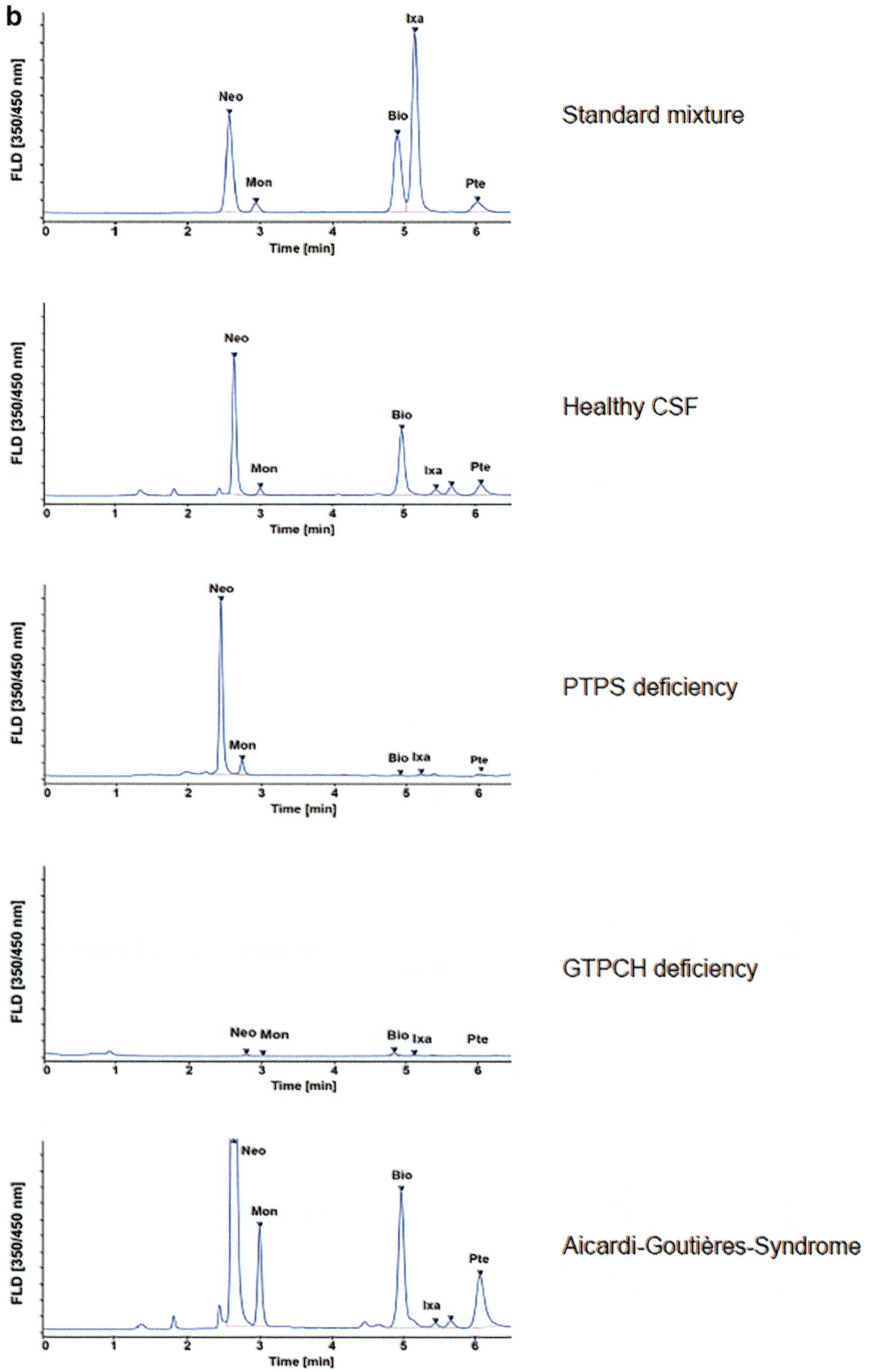

Fig. 27.6 (continued)

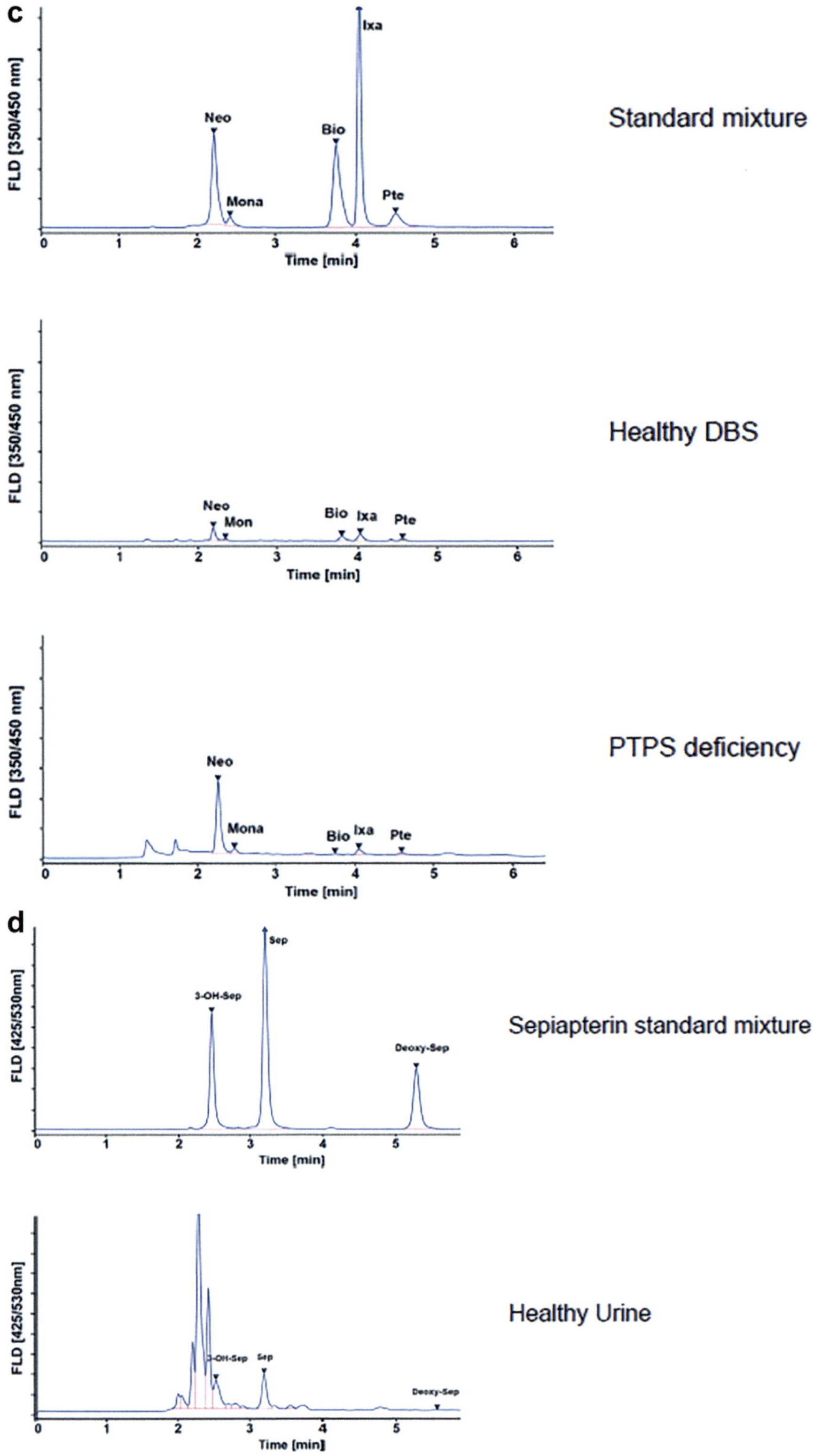

Fig. 27.6 (continued)

Table 27.6 Reference values for pterins in urine, serum, and cerebrospinal fluid

	Neo (U)	Bio (U)	Neo (S)	Bio (S)	Neo (CSF)	Bio (CSF)	Sep (CSF)
Age (years)	mmol/mol creatinine		nM				
Newborns	1.1–4.0	0.5–3.0	3–11	4–18	15–35	20–70	<1
0–1	1.1–4.0	0.5–3.0	3–11	4–18	12–30	15–40	<1
2–4	1.1–4.0	0.5–3.0	3–11	4–18	9–20	10–30	<1
5–10	1.1–4.0	0.5–3.0	3–11	4–18	9–20	10–30	<1
11–16	0.2–1.7	0.5–2.7	3–11	4–18	9–20	10–30	<1
>16	0.2–1.7	0.5–2.7	3–11	4–18	9–20	10–30	<1

Bio = biopterin, Neo = neopterin, Sep = sepiapterin, U = urine, S = serum, CSF = cerebrospinal fluid

Different calibration ranges are used for urine and plasma/CSF/DBS/amniotic fluid (see Tables 27.2, 27.3, 27.4 and 27.5). The calibration curves are calculated by linear regression analysis.

For an accurate clinical interpretation of the results, the concentrations of pterins in urine are reported as mmol/mol creatinine. Due to the relatively high concentrations of pterins in red blood cells, the concentrations of pterins in DBS can be reported as nmol/g Hb. Alternatively, they can be reported as nM.

When the actual BH_4 levels are measured, different oxidation of pterins is performed (see above). Under acidic conditions BH_4 and BH_2 are oxidized to biopterin, while under basic conditions only BH_2 is oxidized to biopterin and BH_4 undergoes side-chain cleavage to form pterin. The difference in biopterin content between the two oxidation procedures represents the actual BH_4 levels. To calculate the percentage of BH_4, the following equation can be used [26]:

$$\frac{\text{Biopterin}_{\text{acid}} - \text{Biopterin}_{\text{basic}}}{\text{Biopterin}_{\text{acid}}} * 100 = \%BH_4$$

27.4.2.5 Quality Control

A control urine, plasma, CSF, or DBS sample is analyzed in each series and serves as the internal quality control sample to assess the precision of the method. These samples are prepared by pooling different anonymized healthy samples. Since primapterin and sepiapterin are normally not present in any matrix, these analytes need to be spiked. External quality control is achieved by measuring urine samples obtained from the European Research Network for evaluation and improvement of screening, diagnosis, and treatment of Inherited disorders of Metabolism (ERNDIM) PTU Scheme, according to the scheme schedule (for more information visit www.erndimqa.nl).

Table 27.7 Reference values for pterins in amniotic fluid and dried blood from a Guthrie card

Neo (A)	Bio (A)	Neo (B)	Bio + Pte (B)	Bio + Pte (B)
nM		nmol/g Hb		%
16–40	6–21	0.31–4.45	0.15–2.91	13.9–78.3

A = amniotic fluid, B = blood, Hb = hemoglobin, Pte = pterin

Table 27.8 Pathological pterin values in plasma

Actual Phe[a]	Neo (S)	Bio (S)
μM	nM	nM
200–600	2–32	12–46
600–1200	9–27	24–39

[a] Serum neopterin and biopterin values depend strongly upon the actual hyperphenylalaninemia
Phe = phenylalanine, S = serum

27.4.2.6 Post-analytical

Figure 27.6a–d shows chromatograms of pterins in urine, CSF, and DBS from patients with different enzyme defects of BH_4 metabolism. The pattern of pterins in plasma, DBS, and CSF is similar to that in urine in patients with BH_4 deficiency. Only urine and DBS are suitable for screening.

Reference pterin values for urine, serum, and CSF are given in Table 27.6. Those for amniotic fluid and dried blood are given in Table 27.7. Pathological pterin values for plasma are given in

Table 27.9 Pathological pterin values in urine and CSF

Variant	Neo (U)	Bio (U)	Neo (CSF)	Bio (CSF)	Sep (CSF)
	mmol/mol creatinine		nmol/l		
GTPCH deficiency	<0.2	<0.2	0.05–3.0	1.5–7.5	n
PTPS deficiency (severe)	5.0–51.2	<0.5	47–402	1.0–16.0	n
PTPS deficiency (mild)	5.0–51.2	<0.5	25–230	13–56	n
DHPR deficiency (severe)	0.5–23.2	3.8–25.6	11–70	43–117	n
DHPR deficiency (mild)	0.5–23.2	3.8–25.6	11–70	43–117	n
PCD deficiency (benign)	4.1–22.5	0.7–1.5[a]	43–117	16–96	n
DRD	n	n	1.1–6.2	3.1–7.6	n
SR deficiency	n	n	14–51	72–102[b]	7–23

[a] Primapterin (7-biopterin) elevated
[b] 7,8-dihydrobiopterin elevated
DHPR = Dihydropteridine reductase, DRD = dopa-responsive dystonia, GTPCH = GTP cyclohydrolase I, n = normal, PCD = pterin-4a-carbinolamine dehydratase, PTPS = 6-pyruvoyltetrahydrobiopterin synthase, SR = sepiapterin reductase

Table 27.8, and those for urine and CSF are given in Table 27.9.

27.4.3 Analysis of Pterins in DBS and Urine by LC-MS/MS

27.4.3.1 Pre-analytical Procedure

Pterins in Dried Blood Spot (DBS)

Biopterin and neopterin in dried blood spots are stable for up to 7 days at room temperature. However, storage into a plastic bag with desiccant at 4–8 °C is recommended. For longer time storage DBS should be stored at −20 °C [27]. Two 6-mm diameter dots are punched from DBS into a vial. The DBS are eluted in 150 μL of DBS working extraction solution containing 50 nM $^{13}C_5$-neopterin, D_3-biopterin in 20 mM HCl and sonicated in an Elma Transsonic T/780 bath for 10 min. The extract solution is then filtered using Nanosep 10 K filters (PALL) at 15,000 *xg* for 15 min. The clear filtrate is injected into the LC-MS/MS system.

Pterins in Urine

Native urine samples should be protected from light, stored at −80 °C and shipped on dry ice. To prevent oxidation, a few drops of 6 M HCl can be added to urine until pH 1.5 is reached. The acidified urine samples can be stored at −20 °C. The LC-MS/MS analysis of total biopterin and total neopterin, obtained after the oxidation of reduced forms, is highly accurate and precise. The oxidation is similar to that reported in the Sect. 27.4.2. Briefly, 500 μL of fresh urine is adjusted to pH 1.0–1.5 with 6 M HCl. Oxidation of reduced pterins is performed by adding 110 μL of 0.5% (w/v) I_2, 1% (w/v) KI solution in 0.1 M HCl. The samples are then incubated for 45 min at room temperature and the reaction is stopped by the addition of 100 μL of 1% (w/v) ascorbic acid. Finally, 100 μL of internal standard solution containing 10 μM $^{13}C_5$-neopterin, D_3-biopterin are added. After the oxidation, a solid phase extraction (SPE) is recommended prior to MS/MS analysis. Cation exchange SPE columns (Varian) can be used. The various steps are: column activation with methanol and then with 0.1 M HCl, sample loading, washing step with 50 mM HCl, followed by water, and finally pterins elution with a solution of 70 mM potassium acetate/methanol (80:20, v/v).

27.4.3.2 Reagents and Chemicals

Chemicals

1. Water, HPLC quality
2. Methanol, LC-MS grade (Merck)
3. Formic acid (Merck)
4. Potassium acetate (Merck)
5. Neopterin (MW 253.2; Sigma-Aldrich)
6. Biopterin (MW 237.2; Sigma-Aldrich)

7. $^{13}C_5$-Neopterin (MW 258.2; Toronto Research Chemicals)
8. D_3-Biopterin (MW 240.2; Toronto Research Chemicals)

Regents/Solutions

1. UHPLC mobile phase A (0.2% formic acid): Add 2 mL formic acid to 998 mL H_2O.
2. UHPLC mobile phase B (methanol, 0.2% formic acid): Add 2 mL formic acid to 998 mL methanol.
3. 0.1 M HCl: Add 821 μL of 37% HCl to H_2O to reach a final volume of 100 mL.
4. 0.5% (w/v) I_2, 1% (w/v) KI in 0.1 M HCl: Dissolve 0.5 g I_2 and 1 g KI in 100 mL of 0.1 M HCl.
5. 1% (w/v) ascorbic acid: Dissolve 1 g ascorbic acid in 100 mL of H_2O.
6. 70 mM potassium acetate: Dissolve 686 mg potassium acetate in H_2O.
7. 70 mM potassium acetate/methanol (80:20, v/v): Dilute 80 mL 70 mM potassium acetate with 20 mL methanol.

Stock and Standard Solutions

1. 0.1 mM $^{13}C_5$-neopterin stock solution: dissolve 2.58 mg $^{13}C_5$-neopterin in 100 mL of 0.1 M HCl. Store 0.1 mL aliquots at −80 °C.
2. 0.1 mM D_3-biopterin stock solution: dissolve 2.40 mg D_3-biopterin in 100 mL of 0.1 M HCl. Store 0.1 mL aliquots at −80 °C.
3. 0.1 mM neopterin stock solution: dissolve 2.53 mg neopterin in 100 mL of 0.1 M HCl.
4. 0.1 mM biopterin stock solution: dissolve 2.37 mg biopterin in 100 mL of 0.1 M HCl
5. 10 μM $^{13}C_5$-neopterin, D_3-biopterin working internal standard solution for urine analysis: add 100 μL of each $^{13}C_5$-neopterin and D_3-biopterin stock solution to 800 μL of 0.1 M HCl.
6. 1 μM $^{13}C_5$-neopterin, D_3-biopterin working internal standard solution for DBS analysis: add 10 μL of each $^{13}C_5$-neopterin and D_3-biopterin stock solution to 980 μL of 20 mM HCl.
7. DBS working extraction solution containing 50 nM $^{13}C_5$-neopterin, D_3-biopterin: add 100 μL of 1 μM working internal standard solution to 1.9 mL of 20 mM HCl.

27.4.3.3 Instrumentation

Separation is obtained using an ACQUITY UPLC HSS T3, 2.1 × 50 mm, 1.8 μm column (Waters) maintained at 40 °C. The flow rate is set to 0.3 mL/min. Samples are kept at 10 °C in the autosampler and 10 μL of sample are injected. Gradient is reported in Table 27.10. The LC-MS/MS system consists of a Waters ACQUITY UPLC (composed of degasser, column oven, autosampler, and binary UPLC pumps) coupled with a Xevo TQMS mass spectrometer (Waters). The instrument is operated in positive electrospray ionization mode. The ionization source settings and the parameters for ion selection and collision activated fragmentation of the molecular ions are optimized by continuous infusion of pure standards (10 μM) at a flow rate of 10 μL/min to obtain the maximum signal of $[M+H]^+$ and the higher product ion signals (Table 27.11). The source operates at a capillary voltage of 0.7 kV and at a temperature of 150 °C. Nitrogen is used as desolvation and cone gas, while argon is used as collision gas. The desolvation gas is set at a flow of 900 L/h and a temperature of 600 °C, whereas collision cell gas flow is set to 0.15 mL/min. The MS/MS experiments are performed in multiple reaction monitoring mode and the data

Table 27.10 Optimized reverse-phase gradient

Time (min)	Flow rate (mL/min)	Mobile phase A (%)	Mobile phase B (%)
0.00	0.300	95.0	5.0
1.00	0.300	95.0	5.0
2.00	0.300	90.0	10.0
3.50	0.300	90.0	10.0
3.60	0.300	10.0	90.0
5.60	0.300	10.0	90.0
5.70	0.300	95.0	5.0
10.00	0.300	95.0	5.0

Table 27.11 MRM transitions and MS/MS parameters

Compound	Q1 (m/z)	Q3 (m/z)	Dwell time (ms)	Collision (V)	Cone (V)
Biopterin	238.2	220.2	78	16	26
Neopterin	254.2	206.0	78	18	28
D_3-Biopterin	241.2	223.1	78	16	26
$^{13}C_5$-Neopterin	259.0	210.1	78	18	0.28

Table 27.12 Reference values for pterins in DBS

Reference ranges (nM) Mean ± SD (range)			
Age	2–7 days	>7–30 days	>30 days
Bio	16.4 ± 4.7 (6.8–26.9)		
Neo	62.3 ± 19.0 (24.1–105.7)	34.8 ± 12.0 (15.8–64.2)	24.1 ± 10.6 (8.1–40.3)

Bio = biopterin, Neo = neopterin

is acquired and processed using the MassLynx V4.1 Software (Waters).

27.4.3.4 Calculations

A calibration curve is obtained from the standard solutions that are included in each sample series. The concentration ranges of the calibration standard are 0.2–10 μM for urine and 1–30 nM for DBS. The calibration curve is calculated by linear regression analysis, where peak area ratio is the independent variable and the concentration ratio is the dependent variable.

For an accurate clinical interpretation of the results, the concentrations of pterins in urine are reported as mmol/mol creatinine, while the concentrations of pterins in DBS are reported as nmol/g Hb as well as nM.

27.4.3.5 Quality Control

Quality control samples are included in each sample series. The control samples are processed and measured in the same way as the patient samples. Urine quality control samples are produced by pooling different anonymized healthy samples, aliquoted into vials and stored at −80 °C. DBS quality control samples are prepared by adding standard solutions to heparinized blood samples to obtain final concentrations of biopterin and neopterin of 100 nM. Then the blood is pipetted on filter papers, which are dried at room temperature and stored at −20 °C in plastic bag with a desiccant.

External quality control is achieved by measuring urine samples obtained from the European Research Network for evaluation and improvement of screening, diagnosis, and treatment of Inherited disorders of Metabolism (ERNDIM) PTU Scheme, according to the scheme schedule (for more information visit www.erndimqa.nl).

27.4.3.6 Post-analytical

Reference values of biopterin and neopterin in DBS are reported in Table 27.12. Biopterin DBS values do not seem to be influenced by age, while neopterin values are significantly higher in the first days of life. Patients affected by HPA deficiency, with high levels of phenylalanine, could have an increase in biopterin and neopterin values. The neopterin and biopterin analyses in DBS is very useful in the diagnosis of BH_4 deficiencies. The sensitivity is high for PTPS and GTPCH deficiencies, which are characterized by an increase of neopterin with a decrease of biopterin and a decrease of both analytes, respectively. In DHPR deficiency an increase of biopterin could be observed. Reference values of biopterin and neopterin in urine obtained in LC-MS/MS are not different from those obtained in HPLC-FT. Figure 27.7 shows chromatograms obtained from a DBS of a patient affected by PTPS deficiency and a patient affected by GTPCH deficiency.

27.4.3.7 Pitfalls

The use of a non-commercial analytical method requires an accurate optimization of the various phases that make up the analysis. The aim is to obtain a reliable sample preparation, efficient chromatographic separation, sensitive and selective detection, optimal data processing, and quantitative determination approach. In comparison with the other methods, using ESI-MS/MS a further variable must be considered, i.e., matrix effect. Matrix effect appears when a change in ionization efficiency of the analyte is induced by

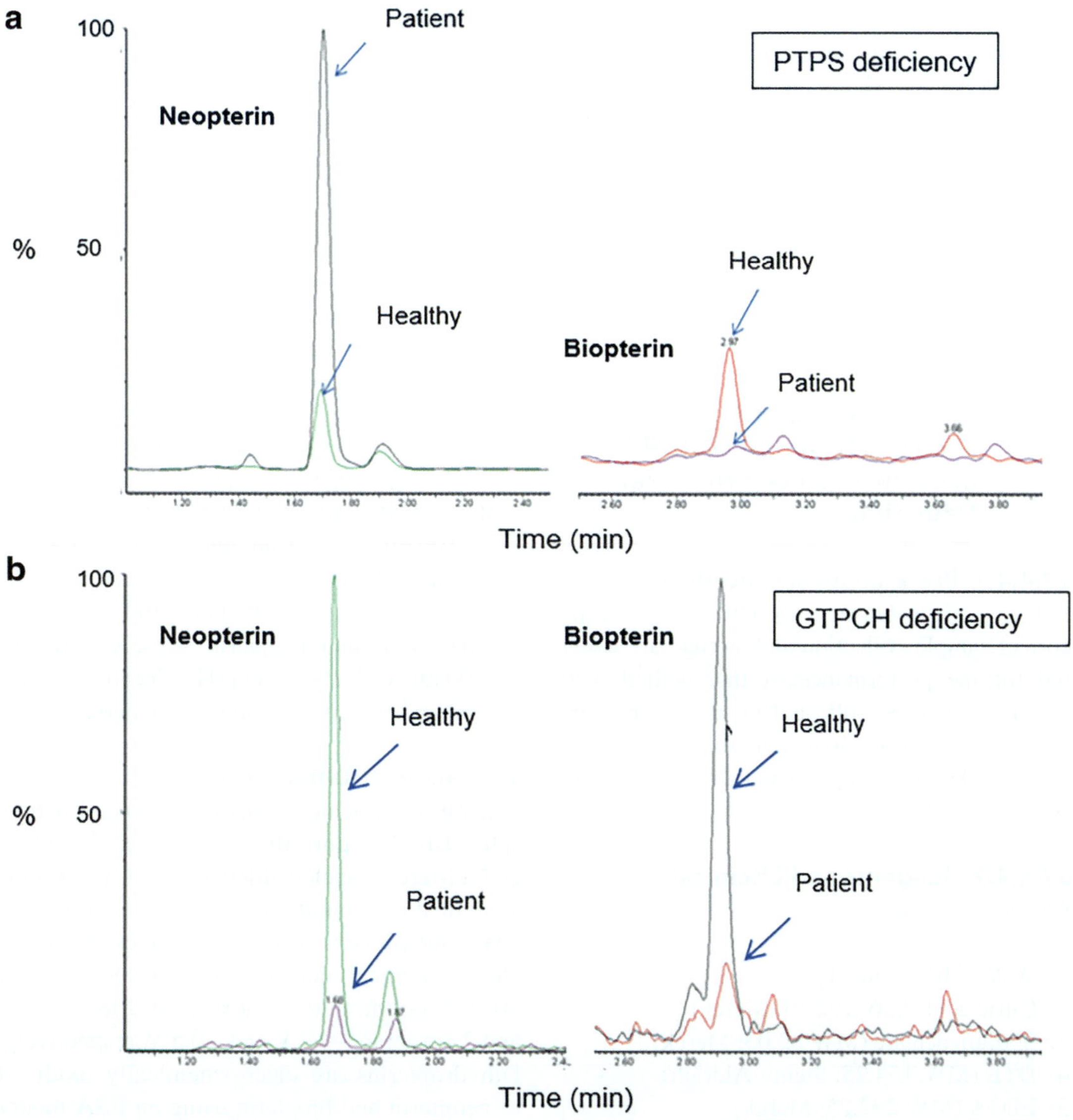

Fig. 27.7 Chromatograms obtained from a DBS of (**a**) a patient affected by PTPS deficiency (Neopterin 128.1 nM and Biopterin 0.8 nM) overlaid on a control sample (Neopterin 25.3 nM and Biopterin 23.7 nM) and (**b**) a patient affected by GTPGH deficiency (Neopterin 3.8 nM and Biopterin 5.2 nM) overlaid on a control sample (Neopterin 19.7 nM and Biopterin 17.7 nM)

coeluting compounds. These are considered the main obstacles of quantitative LC-MS/MS methods [28]. However, the use of stable isotopically labeled internal standards, such as in the case of biopterin and neopterin analysis, can compensate for the matrix effect. In the setting up of a new method, validation procedures must be followed (see ICH guideline M10 on bioanalytical method validation and study sample analysis, Committee for Medicinal Products for Human Use, EMA/CHMP/ICH/172948/2019, 25 July 2022; US Department of Health and Human Services, Food and Drug Administration, Guidance for Industry, Bioanalytical Method Validation, 2018) and even if the validity of the method has been previously demonstrated, it is necessary to confirm the reliability of the method in the specific laboratory in which the method will be applied.

Selectivity, sensitivity, linearity, accuracy, precision, matrix effects, and carryover are some of the variables that should be addressed in conformity with good laboratory practices. Regarding the analysis of biopterin and neopterin, particular attention should be paid to optimizing chromatographic separation and MS/MS parameters, in particular for DBS analysis, where high sensitivity is crucial.

27.4.4 Analysis of Tetrahydrobiopterin (BH_4) and Dihydrobiopterin (BH_2) in CSF with Electrochemical Detection

27.4.4.1 Pre-analytical Procedure

BH_4 in CSF is sensitive to auto-oxidation, and correct sample collection and storage are essential for the performance of the method. CSF (1 mL) should be collected in tubes containing 1 mg DTE and 1 mg diethylenetriaminepentaacetic acid (DTPA). Samples should be immediately frozen at −80 °C [27].

27.4.4.2 Reagents and Chemicals

Chemicals

1. Water, HPLC quality
2. Citric acid (MW 192.13; Merck)
3. Sodium acetate (MW 82.03; Merck)
4. DTE (MW 154.25; Sigma-Aldrich)
5. EDTA (MW 292.25; Merck)
6. DTPA (MW 393.34; Merck)
7. BH_4 dichloride (MW 314.2; Schircks Laboratories).
8. BH_2 (MW 239.2; Schircks Laboratories).

Regents/Solutions

1. HPLC mobile phase (50 mM sodium acetate, 5 mM citric acid, 48 μM EDTA, 160 μM DTE, pH 5.2): Dissolve 6.8 g sodium acetate, 1.05 g citric acid, 18 mg EDTA, and 25 mg DTE in 1 L of H_2O. The mobile phase is degassed with helium during the chromatographic run in order to remove as much air as possible from them.
2. 1% (w/v) DTE: dissolve 1 g DTE in 100 mL of H_2O.

Stock and Standard Solutions

1. 0.5 mM BH_4 stock solution: dissolve 15.7 mg BH_4 and 100 mg DTE in 100 mL of 0.1 M HCl. Store 1 mL aliquots at −80 °C.
2. 0.5 mM BH_2 stock solution: dissolve 12 mg BH_2 and 100 mg DTE in 100 mL of 0.1 M HCl. Store 1 mL aliquots at −80 °C.
3. 0.5 mM Neopterin stock solution: dissolve 12.7 mg neopterin and 100 mg DTE in H_2O. Add 10 μL of 10 M NaOH and add H_2O to 100 mL. Store 1 mL aliquots at −80 °C.
4. 100 nM Pterins (BH_4, BH_2, neopterin) working standard solutions: dilute 200 μL of each BH_4, BH_2 and neopterin stock solution to 100 mL with 1% (w/v) DTE. Keep in the dark and on ice (4 °C) for no more than 6 h.

27.4.4.3 Instrumentation

Separation is achieved using a Merck LiChrospher RP-18 5 μm (40 × 4 mm) guard column and a Merck Supelcosil RP-18 (250 Å ~ 4 mm) analytical column at a flow rate of 1.3 mL/min. BH_4 is measured with an ESA Coulochem 5100A electrochemical detector with a model ESA 5011 electrode in the redox mode. Electrodes 1 and 2 are set to −0.2 V and −0.5 V, respectively. Dihydropterins are electrochemically oxidized to neopterin and biopterin using an ESA model 5021 conditioning cell (+1.0 V). The fully oxidized pterins are detected by a Jasco FP-920 fluorescence detector at 350/450 nm.

Note that the actual voltages applied can vary between runs. Apply a voltage that gives a background current at electrode 1 of 0.5 μA and at electrode 2 of 0.05 μA. Get the system up and running, apply approximate voltages, then drop the flow rate to 0.3 mL/min and leave to run overnight. The next day, increase the flow rate to 1.3 mL/min and start changing the voltages to get the required background currents. Run the mobile phase to waste.

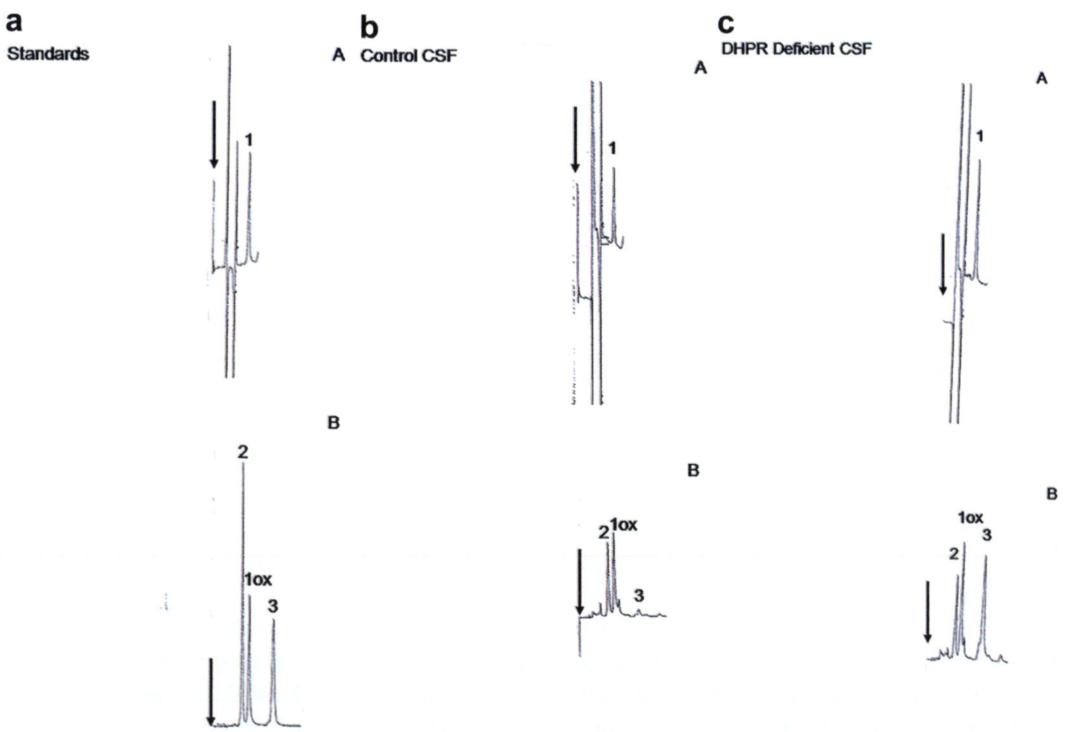

Fig. 27.8 Electrochemical detector (**A**) and fluorescence chromatograms (**B**), the latter generated following post-column oxidation. (**a**) Standards (50 nM). 1 = BH_4 (retention time = 5.12 min); 2 = dihydroneopterin (retention time = 4.17 min); 1ox = "oxidized BH_4" generated by electrochemical detector oxidation. N.B. This peak is variable in height/area and is not used for quantification; 3 = dihydrobiopterin (BH_2; retention time = 8.52 min). The arrow denotes the injection point. Figure courtesy of Dr. Simon Heales, London. (**b**) Control CSF from a 2-year-old male. (**c**) DHPR-deficient CSF in a 6-month-old male. Note the elevated BH_2. The shoulder on the peak is biopterin. Also, note that BH_4 concentration can be within the appropriate reference range in DHPR deficiency states [29]

27.4.4.4 Analytical Procedure

Run the standard mixture, quality control samples, and analytical samples using either an autosampler or with manual injection.

27.4.4.5 Calculation

A calibration curve is obtained from the standard solutions that are included in each sample series.

27.4.4.6 Post-analytical

Figure 27.8a–c shows chromatograms of reduced (electrochemical detection) and oxidized (fluorescence detection) pterins in the CSF of a 2-year-old control subject and a patient with DHPR deficiency. Elevated BH_2 is characteristic of CSF in patients with DHPR and SR deficiency. In both groups, BH_4 concentrations are subnormal and oxidized biopterin is elevated.

27.4.5 DHPR Activity in Erythrocytes

27.4.5.1 Principle

DHPR (EC 1.5.1.34) converts (or recycles) under in vivo conditions the quinonoid BH_2 in an NADH-dependent reaction to BH_4. Some patients and newborns with DHPR deficiency show completely normal blood or urinary pterin profiles and, therefore, the measurement of DHPR activity is essential in all patients with hyperphenylalaninemia, independently from pterin measurements.

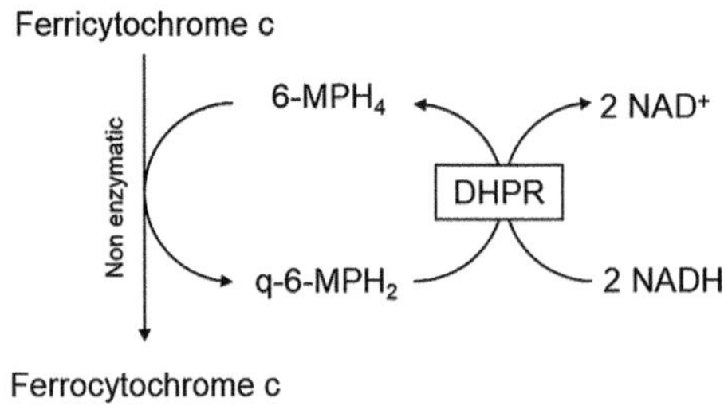

Fig. 27.9 Principle of the enzymatic method. For more details see text

Technically, DHPR activity can be assayed in different cells like erythrocytes, fibroblasts, amniocytes as well as in tissue samples (e.g., liver or brain lysates). However, for diagnostic purposes its measurement in dried blood from Guthrie cards can easily and reliably identify a DHPR deficiency. Therefore, this section will only focus on the measurement of DHPR activity in Guthrie cards. This is done by non-enzymatic coupling of the oxidation of 6-methyltetrahydropterin (6-MPH_4) to quinonoid 6-methyldihydropterin (q-6-MPH_4) to the reduction of ferricytochrome c to ferrocytochrome c. The formation of ferrocytochrome c over time, following an initial burst, proceeds proportionally to the amount of 6-MPH_4 that is reduced back from q-6-MPH_4 by DHPR in the presence of nicotinamide adenine dinucleotide (NADH). The formation of ferrocytochrome c can be monitored spectrophotometrically at 550 nm (Fig. 27.9). This assay is based on the method by Arai et al. [30, 31].

27.4.5.2 Pre-analytical Procedure

The DHPR activity from dried blood from Guthrie cards is stable and these samples can be sent at room temperature. For prolonged storage it is recommended to place them at 4 °C.

27.4.5.3 Reagents and Chemicals

Chemicals

1. Water, HPLC quality
2. HCl, 37% (Merck)
3. KCl (MW 74.6; Merck)
4. KOH (MW 56.1; Merck)
5. Trizma Base (MW 121.1; Sigma-Aldrich)
6. Ferricytochrome c (MW 12,384; Merck)
7. NADH, disodium salt (MW 709.4, Merck)
8. 6-MPH_4, dihydrochloride (MW 254.1; Schircks Laboratories)

Reagents/Solutions

1. 5 M HCl: to 41.5 mL 37% HCl and H_2O to a final volume of 100 mL.
2. 0.1 M HCl: dilute 4 mL 5 M HCl with H_2O to a final volume of 200 mL.
3. 0.01 M HCl: dilute 10 mL 0.1 M HCl with H_2O to a final volume of 100 mL.
4. 0.1 M KCl: dissolve 3.73 g KCl in 500 mL of H_2O.
5. 0.15 M KCl: dissolve 5.6 g KCl in 500 mL of H_2O.
6. 0.1 M KOH: dissolve 1.12 g KOH in 200 mL of H_2O.
7. 0.01 M KOH: dilute 20 mL 0.1 M KOH with H_2O to a final volume of 200 mL.
8. 40 mM Trizma Base Puffer, pH 7.6: dissolve 484 mg Trizma Base in 100 mL of 0.1 M KCl. Adjust to pH 7.6 with 5 M HCl.
9. 5 mM Trizma Base Puffer, pH 7.6, 100 mM KCl: dissolve 1.49 g KCl in 25 mL 40 mM Trizma Base Puffer and fill with H_2O to a final volume of 200 mL.
10. 1 mM Ferricytochrome c: dissolve 99.2 mg of ferricytochrome c in 8 mL of 5 mM Trizma Base Puffer, pH 7.6, 100 mM KCl.
11. 2 mM NADH: dissolve 1.4 mg NADH in 1 mL of 0.01 M KOH. Prepare freshly and store at 4 °C for maximally 1 day.
12. 20 mM 6-MPH_4: dissolve 5.1 mg of 6-MPH_4 in 1 mL 0.1 M HCl. Aliquots of 0.1 mL can be stored at −20 °C. Before use, an aliquot is thawed and diluted to 1 mM with 1.9 mL of 0.01 M HCl. This solution can be kept at 4 °C for maximally 2 days.

27.4.5.4 Instrumentation

Besides routine laboratory equipment, a Synergy HT Multi Detection Reader for Microplates for quantitative detection of ferrocytochrome c at 550 nm is required.

27.4.5.5 Analytical Procedure

To prepare hemolysates from Guthrie cards, 1 circle with a diameter of 6 mm is cut out from the filter paper and incubated for lysis in 0.2 mL of 0.15 M KCl for 30–40 min at room temperature in a 96-wells microplate. The DHPR activity is calculated in mU/mg Hb using the Hb concentration determined in g Hb/mL by standard laboratory methods. DHPR activity is assayed by measuring the linear increase in ferrocytochrome c monitored at 550 nm upon enzymatic incubation at room temperature.

For one sample prepare a mixture containing:
203.75 μL 40 mM Trizma Base Puffer, pH 7.6
12.5 μL 1 mM Ferricytochrome c solution
6.25 μL 2 mM NADH solution
2.5 μL 1 mM 6-MPH_4 substrate

Total volume: 225 μL

When measuring several samples, prepare a reaction mixture in a larger amount. This mixture is preincubated at 25 °C for 5 min before adding it to a microtiter plate, where 25 μL of samples (hemolysates, positive internal quality control, negative internal quality control) were previously added. The reaction is monitored by measuring the absorbance at 550 nm at room temperature over a period of 5 min at intervals of 30 s. According to this protocol, the starting hemolysates of 25 μL are diluted 10-fold.

27.4.5.6 Calculations

One unit of DHPR corresponds to the amount of DHPR enzyme able to produce 1 μmol of ferrocytochrome c per minute at 25 °C and pH 7.6. Thus, to calculate the DHPR activity in mU/mL it is necessary to determine the amount of ferrocytochrome c generated per minute at 25 °C in 1 mL. In the presented assay, the measured ΔA/min at 550 nm is multiplied by a correction factor of 2.606, which takes into account the reaction volume (250 μL), the sample volume (25 μL hemolysate), the optical path length (0.731 cm for a cylindrical well containing 250 uL solution) and $\Delta\varepsilon_{mM,550nm}$ (21.8 mM^{-1} cm^{-1}):

$$\frac{\Delta A/\min}{\Delta\varepsilon_{mM,550nm} * l * \mathrm{Vol}_{\mathrm{Reaction}}} * \frac{\mathrm{Vol}_{\mathrm{Reaction}}}{\mathrm{Vol}_{\mathrm{Sample}}} = \Delta A/\min * 2.606 = \mathrm{mU\,DHPR/mL}$$

$$\Delta A/_{\min} = A_{550nm}/\min_{\mathrm{Sample}} - A_{550nm}/\min_{\mathrm{Blank}}$$

$$\Delta\varepsilon_{mM,550nm} = \left(\varepsilon_{mM,550nm}\right)_{\mathrm{Ferrocytochrome}\,c} - \left(\varepsilon_{mM,550nm}\right)_{\mathrm{Ferricytochrome}\,c} = 21.8\,\mathrm{mM^{-1}cm^{-1}}$$

$$\text{Path length for a cylindrical well}\,(l) = \frac{V_{\mathrm{Reaction}}}{\pi * r^2} = \frac{250\,\mathrm{mm}^3}{\pi * 3.3^2\,\mathrm{mm}} = 7.31\,\mathrm{mm} = 0.731\,\mathrm{cm}$$

$$\mathrm{Vol}_{\mathrm{Reaction}} = 0.25\,\mathrm{mL}$$

$$\mathrm{Vol}_{\mathrm{Sample}} = 0.025\,\mathrm{mL}$$

The DHPR activity is then normalized by the amount of Hb (in mg) present in the hemolysate according to:

$$\frac{\frac{\mathrm{mU\,DHPR}}{\mathrm{mL}}}{\frac{\mathrm{mg\,Hb}}{\mathrm{mL}}} = \mathrm{mU\,DHPR/mg\,Hb}$$

27.4.5.7 Post-analytical

The reference range for DHPR activity in dried blood from Guthrie card is reported in Table 27.13.

Table 27.13 Reference range for DHPR activity in DBS

DHPR in DBS	1.1–4.2 mU/mg Hb

DHPR = dihydropteridine reductase

27.4.5.8 Quality Control

DHPR activity measurements are not available in external control programs and can only be compared by exchanging results with other laboratories (e.g., www.sciensano.be/en/rare-diseases). As positive internal quality control, we recommend using blood from healthy control subjects, whereas as negative internal quality control we recommend using blood collected from a DHPR-deficient patient.

27.5 Conclusion

This chapter outlines the protocols for quantifying pterins in urine, CSF, plasma, or DBS for the purpose of screening, diagnosing, or monitoring BH4 metabolism defects. The chapter extensively discusses three analytical techniques: reverse-phase HPLC coupled with fluorescence detection, LC-MS/MS analysis, and reverse-phase HPLC coupled with electrochemical detection. Special emphasis is placed on measuring oxidized pterin forms, as these methods are commonly employed in diagnostics and are highly reliable. When quantifying native pterin forms, additional adjustments are necessary. Due to the instability of pterins, meticulous sample preparation is crucial before analysis. The pre-analytical process is detailed before delving into each analytical method. These methods can be easily adapted for use in other laboratories following optimization steps; however, it is essential to verify the provided reference ranges in each laboratory. Additionally, a straightforward procedure for measuring DHPR activity in DBS is outlined. Enzymatic tests other than DHPR activity are currently utilized primarily for research purposes and have been supplanted by genetic analysis in diagnostic settings.

References

1. Opladen T, Hoffmann GF, Blau N (2012) An international survey of patients with tetrahydrobiopterin deficiencies presenting with hyperphenylalaninaemia. J Inherit Metab Dis 35(6):963–973
2. Kuster T, Matasovic A, Niederwieser A (1984) Application of gas chromatography-mass spectrometry to the study of biopterin metabolism in man. Detection of biolumazine and 2′-deoxysepialumazine. J Chromatogr 290:303–310
3. Blau N, de Klerk JB, Thony B, Heizmann CW, Kierat L, Smeitink JA et al (1996) Tetrahydrobiopterin loading test in xanthine dehydrogenase and molybdenum cofactor deficiencies. Biochem Mol Med 58(2):199–203
4. Murr C, Widner B, Wirleitner B, Fuchs D (2002) Neopterin as a marker for immune system activation. Curr Drug Metab 3(2):175–187
5. Ichinose H, Ohye T, Takahashi E, Seki N, Hori T, Segawa M et al (1994) Hereditary progressive dystonia with marked diurnal fluctuation caused by mutations in the GTP cyclohydrolase I gene. Nat Genet 8(3):236–242
6. Bonafe L, Thony B, Penzien JM, Czarnecki B, Blau N (2001) Mutations in the sepiapterin reductase gene cause a novel tetrahydrobiopterin-dependent monoamine-neurotransmitter deficiency without hyperphenylalaninemia. Am J Hum Genet 69(2):269–277
7. Blau N, Thony B, Spada M, Ponzone A (1996) Tetrahydrobiopterin and inherited hyperphenylalaninemias. Turk J Pediatr 38(1):19–35
8. Martin Tornero E, Duran Meras I, Espinosa-Mansilla A (2014) HPLC determination of serum pteridine pattern as biomarkers. Talanta 128:319–326
9. de Llanos AM, Espinosa-Mansilla A, Canada-Canada F, de la Pena AM (2011) Separation and determination of 11 marker pteridines in human urine by liquid chromatography and fluorimetric detection. J Sep Sci 34(11):1283–1292
10. Zurfluh MR, Giovannini M, Fiori L, Fiege B, Gokdemir Y, Baykal T et al (2005) Screening for tetrahydrobiopterin deficiencies using dried blood spots on filter paper. Mol Genet Metab 86(Suppl 1): 96–103
11. Antonozzi I, Carducci C, Vestri L, Pontecorvi A, Moretti F (1988) Rapid and sensitive method for high-performance liquid chromatographic analysis of pterins in biological fluids. J Chromatogr 459: 319–324

12. Novakova L, Kaufmannova I, Janska R (2010) Evaluation of hybrid hydrophilic interaction chromatography stationary phases for ultra-HPLC in analysis of polar pteridines. J Sep Sci 33(6–7):765–772
13. Guibal P, Leveque N, Doummar D, Giraud N, Roze E, Rodriguez D et al (2014) Simultaneous determination of all forms of biopterin and neopterin in cerebrospinal fluid. ACS Chem Neurosci 5(7):533–541
14. Koslinski P, Jarzemski P, Markuszewski MJ, Kaliszan R (2014) Determination of pterins in urine by HPLC with UV and fluorescent detection using different types of chromatographic stationary phases (HILIC, RP C8, RP C18). J Pharm Biomed Anal 91:37–45
15. Tomsikova H, Solich P, Novakova L (2014) Sample preparation and UHPLC-FD analysis of pteridines in human urine. J Pharm Biomed Anal 95:265–272
16. Burton C, Shi H, Ma Y (2016) Development of a high-performance liquid chromatography - Tandem mass spectrometry urinary pterinomics workflow. Anal Chim Acta 927:72–81
17. Burton C, Weng R, Yang L, Bai Y, Liu H, Ma Y (2015) High-throughput intracellular pteridinic profiling by liquid chromatography-quadrupole time-of-flight mass spectrometry. Anal Chim Acta 853:442–450
18. Burton C, Shi H, Ma Y (2013) Simultaneous detection of six urinary pteridines and creatinine by high-performance liquid chromatography-tandem mass spectrometry for clinical breast cancer detection. Anal Chem 85(22):11137–11145
19. Allegri G, Costa Netto HJ, Ferreira Gomes LN, Costa de Oliveira ML, Scalco FB, de Aquino Neto FR (2012) Determination of six pterins in urine by LC-MS/MS. Bioanalysis 4(14):1739–1746
20. Xiong X, Liu Y (2016) Chromatographic behavior of 12 polar pteridines in hydrophilic interaction chromatography using five different HILIC columns coupled with tandem mass spectrometry. Talanta 150:493–502
21. Arning E, Bottiglieri T (2016) LC-MS/MS analysis of cerebrospinal fluid metabolites in the pterin biosynthetic pathway. JIMD Rep 29:1–9
22. Jimenez Giron A, Martin-Tornero E, Hurtado Sanchez MC, Duran Meras I, Espinosa Mansilla A (2012) A simple HPLC-ESI-MS method for the direct determination of ten pteridinic biomarkers in human urine. Talanta 101:465–472
23. Yuan TF, Huang HQ, Gao L, Wang ST, Li Y (2018) A novel and reliable method for tetrahydrobiopterin quantification: benzoyl chloride derivatization coupled with liquid chromatography-tandem mass spectrometry analysis. Free Radic Biol Med 118:119–125
24. Santagata S, Di Carlo E, Carducci C, Leuzzi V, Angeloni A, Carducci C (2017) Development of a new UPLC-ESI-MS/MS method for the determination of biopterin and neopterin in dried blood spot. Clin Chim Acta 466:145–151
25. Carducci C, Santagata S, Friedman J, Pasquini E, Carducci C, Tolve M et al (2015) Urine sepiapterin excretion as a new diagnostic marker for sepiapterin reductase deficiency. Mol Genet Metab 115(4):157–160
26. Fukushima T, Nixon JC (1980) Analysis of reduced forms of biopterin in biological tissues and fluids. Anal Biochem 102(1):176–188
27. Howells DW, Hyland K (1987) Direct analysis of tetrahydrobiopterin in cerebrospinal fluid by high-performance liquid chromatography with redox electrochemistry: prevention of autoxidation during storage and analysis. Clin Chim Acta 167:23–30
28. Taylor PJ (2005) Matrix effects: the Achilles heel of quantitative high-performance liquid chromatography-electrospray-tandem mass spectrometry. Clin Biochem 38(4):328–334
29. Hyland K, Heales SJR (1993) Tetrahydrobiopterin and quinonoid dihydrobiopterin concentrations in CSF from patients with dihydropteridine reductase deficiency. J Inherit Metab Dis 16:608–610
30. Arai N, Narisawa K, Hayakawa H, Tada K (1982) Hyperphenylalaninemia due to dihydropteridine reductase deficiency: diagnosis by enzyme assays on dried blood spots. Pediatrics 70(3):426–430
31. Narisawa K, Arai N, Hayakawa H, Tada K (1981) Diagnosis of dihydropteridine reductase deficiency by erythrocyte enzyme assay. Pediatrics 68(4):591–592

Biogenic Amines

28

Simon J. R. Heales, Daniel Mackenzie,
Caroline Bhairo, and Simon A. S. Pope

Contents

Abbreviations

3-OMD/3-MD	3-methyldopa
5-HIAA	5-Hydroxyindoleacetic acid
AADC	Aromatic Amino Acid Decarboxylase
BH4	Tetrahydrobiopterin
COMT	Catechol-*O*-methyltransferase
DA	Dopamine
HVA	Homovanillic Acid
L-dopa	L-Dihydroxyphenylalanine
MAO + ALD	Monoamine oxidase and aldehyde dehydrogenase
MHPG	3-Methoxy-4-hydroxy-phenlglycol
PNPO	Pyridox(am)ine-5′-phosphate oxidase

S. J. R. Heales (✉) · S. A. S. Pope
Neurometabolic Unit, National Hospital, London, UK

UCL Great Ormond Street Institute of Child Health, London, UK
e-mail: simon.heales@nhs.net; simon.pope@nhs.net

D. Mackenzie · C. Bhairo
Neurometabolic Unit, National Hospital, London, UK

28.1 Clinical Significance

Whilst the term *biogenic amine* strictly encompasses all amines of biological origin, for the purpose of this chapter it will be employed to refer to the catecholamine (dopamine, noradrenaline) and serotonin group of neurotransmitters. These neurotransmitters are generated from the amino acid precursors, tyrosine and tryptophan, respectively, via the action of the tetrahydrobiopterin (BH4)-dependent tyrosine and tryptophan hydroxylases. Hydroxylation of the amino acid substrates leads to the formation of 3,4-dihydroxy-L-phenylalanine (L-dopa) and 5-hydroxytryptophan which are then decarboxylated via the pyridoxal phosphate-dependent aromatic amino acid decarboxylase (AADC) to yield dopamine and serotonin (Fig. 28.1a, b). In noradrenergic neurones, dopamine is further metabolised to noradrenaline through the action of dopamine-β-hydroxylase [1].

Within the central nervous system (CNS), dopamine, noradrenaline and serotonin have a diverse range of actions that include control of

N. Blau, F. M. Vaz (eds.), *Laboratory Guide to the Methods in Biochemical Genetics*,
https://doi.org/10.1007/978-3-031-58819-8_28

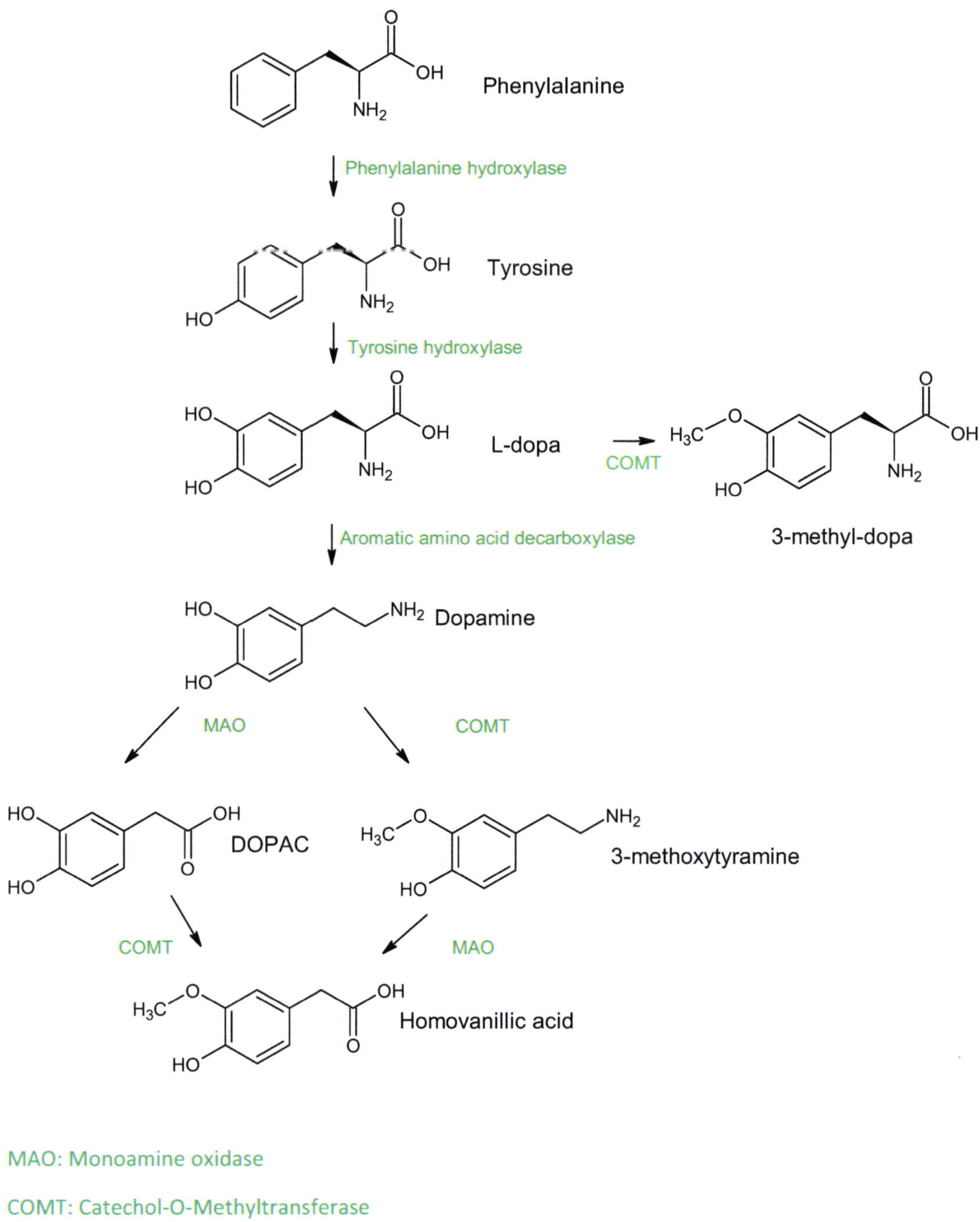

Fig. 28.1 (**a**) Biosynthesis and metabolism of dopamine. MAO: Monoamine oxidase. COMT: Catechol-O-Methyltransferase. (**b**) Biosynthesis and Metabolism of Serotonin

psychomotor function, regulations of motor coordination, processing of sensory input, arousal, emotional stability, temperature regulation, memory, appetite mood and sleep. The biogenic amines considered here, in common with other neurotransmitter groups, exert their biological action as a result of being released, in response to an action potential, from

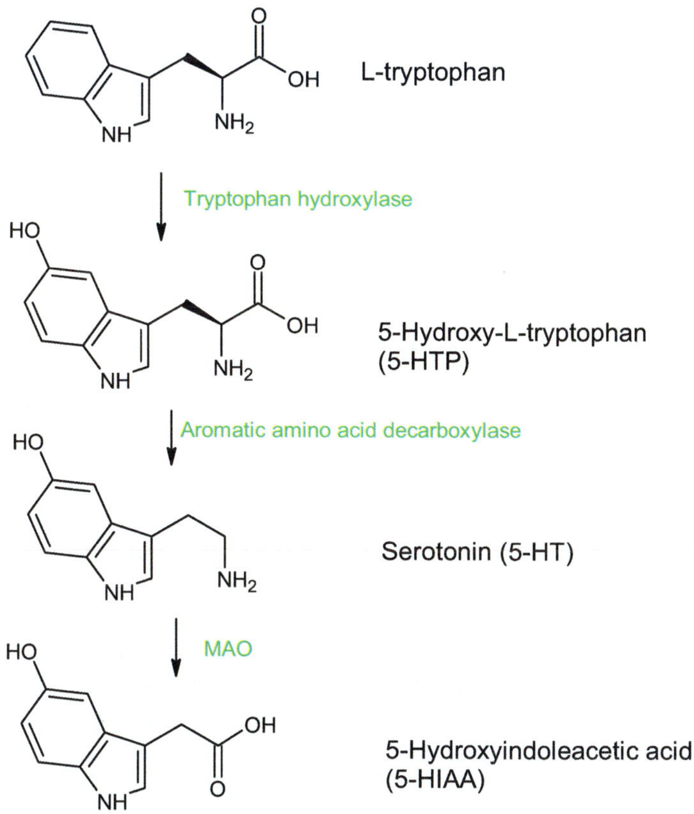

Fig. 28.1 (continued)

the presynaptic nerve terminal and binding to postsynaptic receptors where they elicit a biological response, e.g. excitation or inhibition of the target cell. The neurotransmitters are removed from the synaptic cleft and may be sequestered into vesicles and used again as neurotransmitters or are metabolised further. With regard to the latter, dopamine is acted upon by catechol-O-methyltransferase (COMT) and the monoamine oxidase/aldehyde dehydrogenase (MAO + ALD) system to form homovanillic acid (HVA). In a similar manner, MAO + aldehyde reductase and COMT form 3-methoxy-4-hydroxy-phenlglycol (MHPG) from noradrenaline. With regard to serotonin, MAO + ALD activity results in the generation of 5-hydroxyindoleacetic acid (5-HIAA). Determination of the concentration of these metabolites, particularly in cerebrospinal fluid (CSF), is considered to accurately reflect the integrity of catecholamine and serotonin metabolism within the CNS and has been instrumental in the identification and subsequent treatment of the disorders of biogenic amine metabolism, e.g. the inborn errors of BH4 metabolism, tyrosine hydroxylase deficiency, AADC deficiency and disorders of pyridoxal phosphate metabolism. Furthermore, with the advent of gene therapy, e.g. for AADC deficiency, inclusion of these disorders in newborn screening programmes is being proposed. However, this requires a good peripheral biomarker. For AADC, measurement of 3-methyl dopa in dried blood spots provides promising results [1, 2].

28.2 Properties of Analytes

As a result of a high index of clinical suspicion and, on occasion, supporting biochemical data from other investigations, one of the first specialist investigations to ascertain whether a patient has an inborn error of biogenic amine metabolism is, as mentioned above, analysis of the CSF concentrations of HVA and 5-HIAA. This is often performed in conjunction with the measurement of 3-methyldopa (3-OMD), also known as 3-methoxytyrosine. 3-OMD is formed from L-dopa via COMT activity and accumulates in conditions where aromatic amino acid decarboxylase activity is impaired (Fig. 28.1a).

The chemical properties of HVA, 5-HIAA, 3-OMD and 5-HTP make them amenable to reverse-phase high-performance liquid chromatography (HPLC) with electrochemical detection. Furthermore, the composition of CSF means that little, if any, sample preparation is required prior to analysis. However, the susceptibility of these metabolites to oxidation means that careful sample collection and storage are required in order to minimise analyte degradation.

Whilst HPLC has been the method of choice for CSF monoamine analysis, developments in sensitivity and accessibility of mass spectrometry are resulting in labs beginning to move to this platform. This approach will therefore also be discussed here.

As mentioned above, 3-OMD may also be measured in dried blood spots and used to identify patients with AADC deficiency. Indicative results will need to be followed up by enzymatic analysis, genetics (not discussed here) and invariably CSF profiling.

28.3 Methods

28.3.1 Specimen: CSF

Lumber CSF is required for this assay [3]. As there is a rostro caudal gradient for both HVA and 5-HIAA it is essential the first 0.5 ml of fluid from the spinal tap is collected to allow accurate evaluation and interpretation of results in comparison to the age-dependent reference range. The sample must be collected into a properly labelled screw-top microcentrifuge tube and frozen at the bedside in liquid nitrogen or dry ice. Samples should then be transported to the laboratory on dry ice and stored at −70 °C until analysis. Blood contamination will affect the results as iron released from red blood cells will degrade both HVA and 5-HIAA. If the CSF is contaminated with blood it should be placed on ice immediately and centrifuged within 5 min. The clear supernatant should then be frozen immediately and stored at −70 °C.

28.3.2 HPLC of Biogenic Amine Metabolites

28.3.2.1 Principle

HVA, 5-HIAA, 5-HTP and 3-OMD are amenable to separation by reverse-phase HPLC utilising an octadecyl silyl (ODS) column and detection by electrochemistry. The elution order of analytes on a reverse phase HPLC column is in order of increasing hydrophobicity, i.e. 5-HIAA elutes prior to HVA. The mobile phase is at an acidic pH to help suppress ionisation of the analytes. However, molecules with a catechol ring will remain positively charged even at pH 2. Consequently, an ion pairing agent, sodium octane sulphonic acid (OSS) is also added to the mobile phase to fully suppress the charge and hence optimise interaction with the stationary phase of the column. Methanol is also added in order to decrease the retention time of the analysis.

28.3.2.2 Chemicals

- Potassium dihydrogen orthophosphate (KH_2PO_4).
- 1-Octanesulfonic aid (OSS).
- Ethylenediaminetetraacetic acid (EDTA).
- Methanol (HPLC electrochemical detector grade).
- Hydrochloric acid (HCl).
- Homovanillic acid (HVA).
- 5-Hydroxyindoleacetic acid (5-HIAA)
- 3-Methoxy-L-tyrosine (3-Methyl-DOPA)
- 5-Hydroxy-L-trypyophan (5-HTP).

28.3.2.3 Instrumentation

Isocratic HPLC system with electrochemical detection (5021A and 5011A Electrodes, ESA Inc.,

Chelmsford, USA). 5021A Electrode, 50 mV, and E1 of the 5011A electrode set between 400–450 mV. N.B. The voltage of this electrode is an approximate value. As the electrode ages, the voltage may need to be increased. Signal is recorded from E1 of the 5011A electrode. A voltammogram (peak area for each standard vs. electrode 2 voltage) should be run at regular intervals to maintain linearity. A voltammogram must be performed when a new electrochemical cell is used.

Kya Tech Corporation HiQ sil C18W (4.6 mm × 250 mm) Column (Obtained from Kromatek Ltd) Column: C18, 5 μm, 25 cm × 4.6 mm.

Injection volume 50 μl.

Mobile Phase: 50 mM potassium dihydrogen phosphate (pH 2.750, adjust with conc. HCL), 1 mM OSS, 50 μM EDTA and 17% methanol. Ultrapure Milli-Q water (18.2 MΩ) must be employed throughout.

Flow rate 1.3 ml/min. Column was maintained at 35 °C by surrounding the column in a heating block.

28.3.2.4 Calibration

Standards

(a) *Stock solutions*

The stock standards of HVA and 5-HIAA and 3-Methyl-DOPA are made to a final concentration of 500 μM as follows:

HVA and 5-HIAA

Using a 100-ml volumetric dissolve 9.11 mg of HVA and 9.56 mg of 5-HIAA in 80 ml of ultrapure H_2O. Make up to 100 ml using Milli-Q H_2O and sonicate for 1 min. Mix thoroughly and aliquot immediately after the solid has dissolved. Aliquots of 0.5 ml should then be transferred into labelled Eppendorf tubes, frozen rapidly in liquid nitrogen and stored at −70 °C.

3-Methyl-DOPA

Using a 100-ml volumetric dissolve 10.56 mg 3-Methyl-DOPA in 80 ml of ultrapure H_2O. Make up to 100 ml using Milli-Q H_20, mix thoroughly and aliquot 0.5 ml into labelled Eppendorf tubes. Freeze rapidly in liquid nitrogen and store at −70 °C.

5-Hydroxytryptophan

Using a 100-ml volumetric dissolve 11.01 mg 5-hydroxy-L-Tryptophan in 80 ml of ultrapure H_2O. Make up to 100 ml using ultrapure H_2O, mix thoroughly and aliquot 0.5 ml into labelled Eppendorf tubes. Freeze rapidly in liquid nitrogen and store at −70 °C.

(b) *Working standard*

The working standard solutions of HVA and 5HIAA and 3-Methyl-DOPA are stable for a day if stored at 2–8 °C and are made to a final concentration of 500 nM as follows:

Remove stock solution from the −70 °C freezer, thaw and mix thoroughly. Add 3 drops of concentrated HCl to approximately 80 ml of Milli-Q H_20 and mix. Add 100 μL of HVA, 5-HIAA and 3-Methyl-DOPA stock standards and make up to 100 ml. Mix well and store at 2–8 °C.

Quality Control

An internal quality control (IQC) can be made by the pooling of clear, filtered CSF from previous patients with no evidence of a monoamine disorder. Mix thoroughly and aliquot and freeze immediately in labelled Eppendorf tubes at −70 °C. The control ranges should be established prior to analysis of patient samples by measuring the pooled CSF for a minimum of five occasions. For an analytical/diagnostic run to proceed, analysis of QC material must provide concentration values that are within two standard deviations (plus and minus) of the calculated mean for that particular QC. The construction of Levy-Jennings control charts provide historical information on overall performance and highlight potential deterioration in the performance of the system. The coefficient of variation should remain consistent, reflecting good system performance and providing additional confidence when reporting results.

An established external quality assurance (EQA) scheme for the monoamine metabolites has been available since 2016 run by the European Research Network for evaluation and improvement of screening, diagnosis and treatment of Inherited Disorders of Metabolism (ERNDIM). ERNDIM aims to educate and assess the ability of laboratories to diagnose inherited metabolic disease and operate quantitative, hybrid and qualitative schemes. The scheme for the monoamine metabolites is a 'hybrid scheme' which evaluates the ability of the participating laboratories to quantitatively analyse the concentrations of the analytes and to establish or exclude a specific diagnosis of an inborn errors in neurotransmitter metabolism. For the most up-to-date information visit: https://www.erndim.org.

Procedure

50 ul of standards, QC material or neat CSF are injected directly into the HPLC system. Blood-contaminated CSF should be filtered using a 10,000-molecular weight centrifugal filtration system—Pall Life Sciences, (nanosep with 10 K omega) prior to injection.

Calculation

The concentrations of each analyte can be calculated by relating the peak area in the sample chromatogram to peak area in the standard chromatogram as follows:

$$[\text{analyte}] = \frac{\text{peak area in sample}}{\text{peak area in standard}} \times 500\text{nM}$$

28.3.3 Interpretation

28.3.3.1 Chromatograms

Under normal conditions, a relatively clear chromatogram is generated following the injection of calibration standards or CSF sample (Fig. 28.2a, b). Peak identification and quantification of HVA,

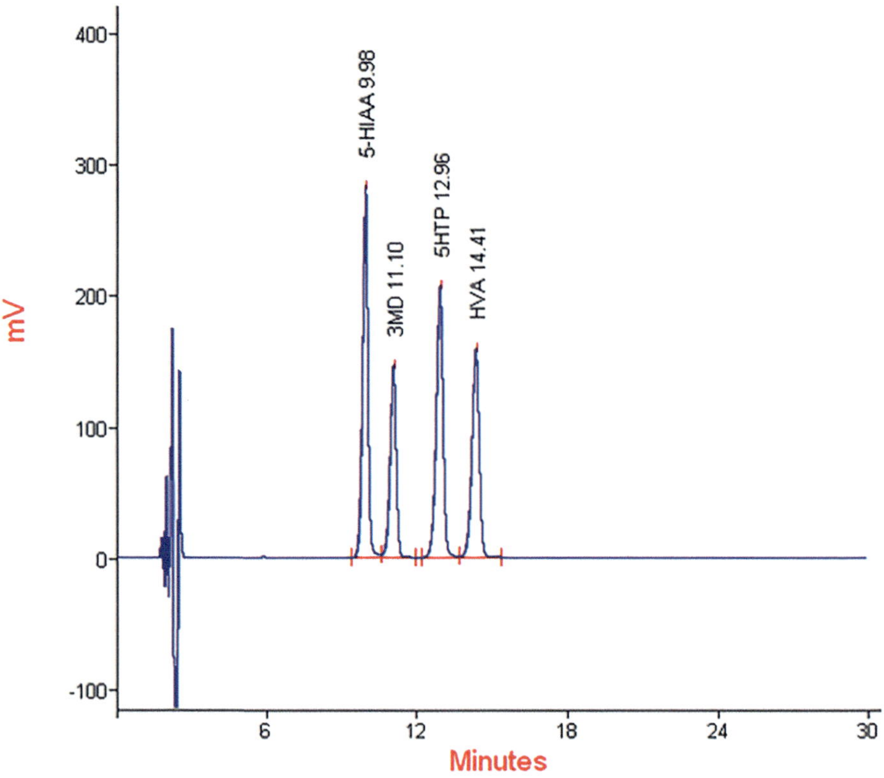

Fig. 28.2 (**a**) High-performance liquid chromatography (HPLC) with electrochemical detection (ECD) detection of neurotransmitter metabolites. **a** aqueous calibration standard mixture containing 500 nmol/L 5HIAA, 3MD, 5HTP and HVA. The numbers above the identity of the peak are the retention time for each analyte. (**b**) Cerebrospinal fluid (CSF) sample-internal quality control of pooled CSF showing no signs of a neurotransmitter disorder. (**c**) CSF sample-aromatic amino acid decarboxylase (AADC) deficient patient. Note, the poor resolution between 5HIAA and 3MD peak is due to the predominant 3MD concentration

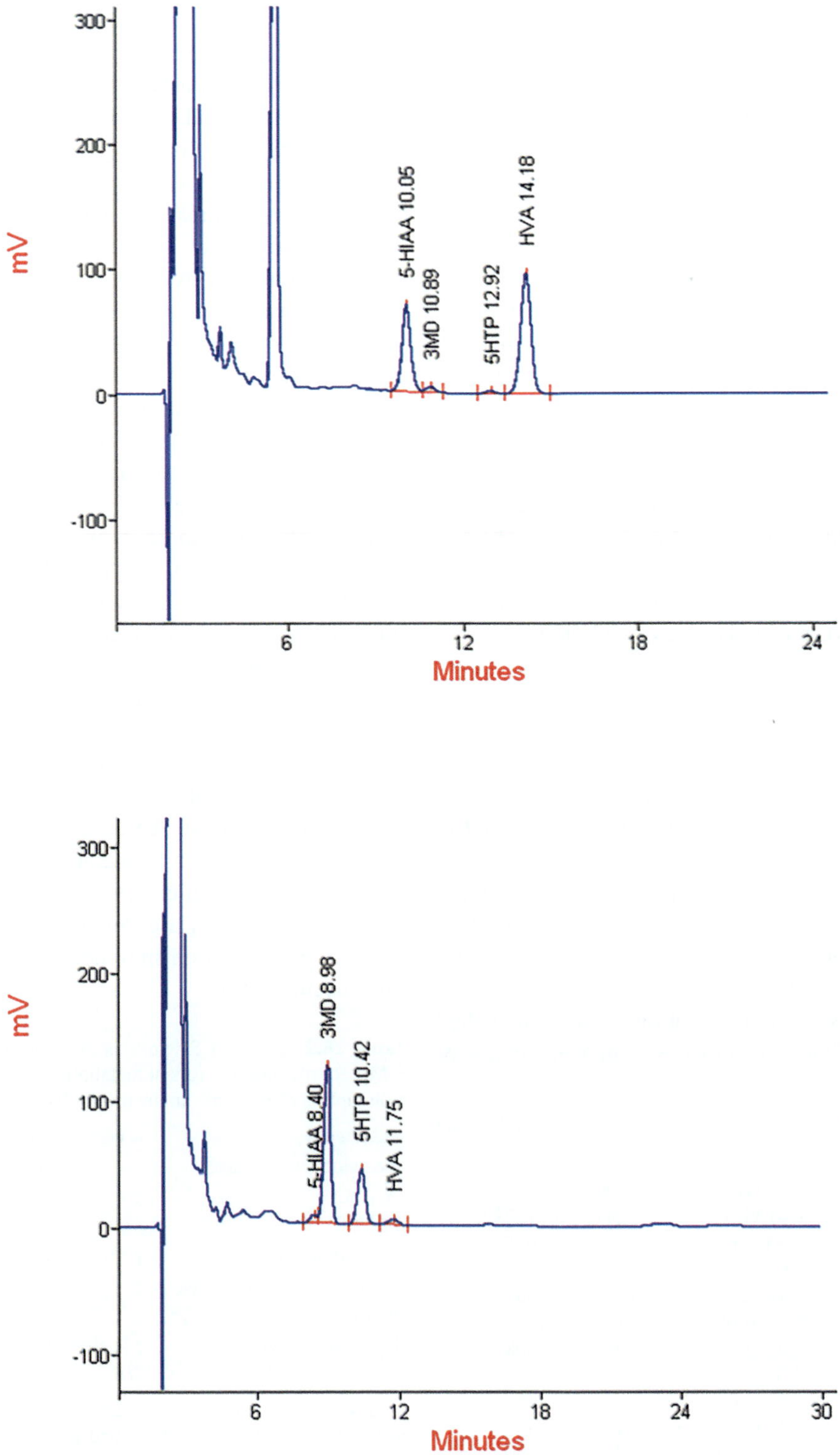

Fig. 28.2 (continued)

5-HIAA, 3-OMD and 5HTP (when present) is straightforward. The authenticity of a particular assigned peak can be checked further by spiking CSF samples with standards of HVA, 5-HIAA, 3-OMD and 5-HTP.

28.3.4 Reference Values

The data in Table 28.1, based on Hyland et al. [4] report reference ranges for HVA and 5-HIAA when the first 0.5 ml of CSF is collected for analysis. With regard to 3-OMD, <300 nmol/L is considered acceptable for children up to 6 months of age. Between 6 months and 2 years this declines to <100 nmol/L. For individuals 2 years and over the acceptable upper limit is 50 nmol/L. The reference interval for 5-HTP does not appear to be affected by age and with a concentration of <10 nmol/L not suggestive of pathology.

28.3.5 Diagnostic Profiles

Assignment of exact pathological values for the CSF biogenic amine metabolites encountered in the different known inborn errors of biogenic amine metabolism is not possible. Current experience suggests that the magnitude of the deficiency for a specific metabolite may vary considerably. However, the deviation of individual metabolite concentrations from established reference intervals provides profiles indicative of particular disorders. Table 28.2 provides a summary of such changes. However, these should be considered in conjunction with other investigations (see below) and the clinical picture before a suggested diagnosis is made and treatment regimens instigated.

Inspection of Table 28.2 shows that for tyrosine hydroxylase deficiency only the concentration of HVA is decreased in concentration. However, in cases of AADC deficiency, in view of the dual role this enzyme plays in dopamine and serotonin synthesis, both the HVA and 5-HIAA concentrations are decreased. Furthermore, deficiency of AADC will result in an accumulation of L-dopa which, as described above, is metabolised to 3-OMD. Hence in AADC deficiency, there is an accumulation of 3-OMD within the CSF (Fig. 28.2c). Whilst 5-HTP is not routinely recorded in our laboratory, this metabolite will also usually be elevated in AADC deficiency. AADC is an enzyme that has a requirement for the cofactor, pyridoxal phosphate. Consequently, decreased availability of pyridoxal phosphate (PLP), as occurs in pyridox(am)ine-5′-phosphate oxidase (PNPO) deficiency, can impair AADC activity and result in a CSF biogenic amine metabolite profile that mimics AADC deficiency. With regards to dopamine-β-hydroxylase deficiency an increase in HVA concentration may occur due to dopamine accumulation.

Table 28.1 Age related reference ranges for HVA and 5-HIAA

	Age (years)	Min	Max
HVA	0–0.33	324	1098
	0.34–0.66	362	955
	0.67–1.0	176	851
	1.1–5.0	154	867
	5.1–Adult	71	565
5HIAA	0–0.33	199	608
	0.34–0.66	63	503
	0.67–1.0	68	451
	1.1–5.0	89	367
	5.1–Adult	58	220

Values expressed as nmol/L. These reference ranges were established in the author's laboratory using the first 0.5 ml of lumbar CSF

Table 28.2 Typical CSF profiles of HVA, 5-HIAA and 3-MD for the inborn errors of metabolism associated with a disruption of biogenic amine metabolism

Deficiency	HVA	5-HIAA	3-MD
Tyrosine hydroxylase	↓	WR	WR
AADC	↓	↓	↑
PNPO	↓	↓	↑
Dopamine transporter	↑	WR	WR
Dopamine β Hydroxylase	↑	WR	WR
Pterin (recessive)	↓	↓	WR
Pterin (dominant)	↓ or WR	↓ or WR	WR
"Tryptophan Hydroxylase"	WR	↓	WR

A downward facing arrow indicates that a particular metabolite is below the established reference range. An upward facing arrow is indicative that a metabolite is above the established reference range. WR indicates that the concentration of the metabolite is likely to be within the reference range

BH4 is an obligatory cofactor for both tyrosine and tryptophan hydroxylase. Consequently, the inborn errors of BH4 metabolism are associated with impaired dopamine and serotonin turnover which is reflected by decreased concentrations of HVA and 5-HIAA in the CSF. Whilst such a pattern is particularly true for the autosomal recessive disorders of BH4 metabolism, an autosomal dominant disorder of BH4 metabolism, (autosomal dominant GTP cyclohydrolase deficiency) is not always associated with marked decreases in the CSF concentration of HVA and 5-HIAA.

To date, a fully documented case of tryptophan hydroxylase deficiency has not been described. However, such a deficiency would be predicted to have a characteristic CSF profile, i.e. an isolated decrease in CSF 5-HIAA concentration.

In addition, to the above inborn errors of metabolism, deficiencies in the catabolism of dopamine and serotonin would also be expected to result in decreased concentrations of HVA and 5-HIAA, e.g. MAO deficiency and COMT deficiency.

Marked elevations in the CSF concentration of HVA have been reported in a group of children with infantile parkinsonism-dystonia. This has been documented to arise from mutations in the dopamine reuptake transporter (SLC6A3).

28.3.6 Pitfalls

Apparent, perturbation of monoamine metabolism may arise as a consequence of an ongoing neurodegenerative process that results in the loss of dopamine and/or serotonin-containing neurones. Such loss could therefore produce a CSF metabolite profile characteristic of one of the inborn errors of metabolism documented above.

The effects of any medication a patient is receiving close to the time of lumbar puncture should also be considered. Ideally, these should be documented on the sample request form. Thus, patients receiving L-dopa will have an elevated concentration of 3-OMD in CSF which could, if the laboratory is not aware that the patient is receiving this drug, raise the possibility of AADC or PNPO deficiency. However, the 5-hydroxytryptophan concentration would be normal. Patients receiving ionotropic support in the form of dopamine will have markedly elevated levels of HVA which could, if the laboratory is ignorant of the dopamine administration, suggest a disruption of dopamine metabolism, i.e. a profile would be generated that would be comparable to that described for the dopamine transporter deficiency.

Other medication that could theoretically lead to the erroneous suggestion that a patient has an inborn error of biogenic amine metabolism inborn include serotonin reuptake inhibitors, MAO and COMT inhibitors.

In addition to the factors considered above, an apparent decrease in biogenic amine metabolite concentrations can arise as a result of poor sample handling or contamination of the CSF sample with blood. With regards to the former, a prolonged delay between obtaining a sample and freezing it can lead to a degradation of the metabolites to be analysed. This degradation is accelerated if the CSF is contaminated with blood. If blood contamination occurs, the sample should, if possible, be immediately centrifuged and the resulting clear supernatant stored frozen until analysis. If centrifugation is not possible, the CSF can be frozen. Upon thawing and prior to injection onto the HPLC system, the CSF can be centrifuged through a 10,000 molecular weight cut-off filter. The clear filtrate can then be analysed. Whilst this procedure will eventually remove the blood contamination, it is likely that some metabolite degradation will have occurred prior to filtration. Consequently, reports generated from samples processed this way should clearly make this point.

28.3.6.1 Liquid Chromatography-Mass Spectrometry (LC-MS) of Biogenic Amines

Although HPLC with electrochemical detection has been the 'gold standard' technique for measuring biogenic amines in CSF, many liquid chromatography-mass spectrometry (LC-MS) methods have also been described and are starting to be adopted by accredited clinical biochemistry laboratories. In the ERNDIM quality assurance scheme for 2022, approximately 20% of laboratories were using LC-MS to quantify CSF biogenic amines and this is expected to increase.

There is a wide variety of LC-MS methods for biogenic amines described in the literature and the method a laboratory chooses to use will depend on the equipment available, the sensitivity required and the volume of samples and turnaround times [5, 6].

Owing to the wide variety of sample preparation techniques, liquid chromatography systems, columns and mass spectrometers, it is difficult to give a 'standard' procedure. Therefore, two indicative methods are described below. One for the analysis of underivatised biogenic amines and one for benzoyl chloride derivatives of the biogenic amines [7].

The advantage of analysing biogenic amines without derivatising is that it allows simple sample preparation and avoids the use of toxic chemicals and the possible production of artefacts/degradation during the derivatisation process. However, many laboratories have faced ionisation/sensitivity problems when analysing underivatised biogenic amines, particularly with regards to homovanillic acid. Some of these problems can be overcome by increasing chromatographic separation/use of specialised columns, using volatile buffers to improve ionisation and using a high-sensitivity mass spectrometer. However, such high-sensitivity mass spectrometers are expensive and not available in all laboratories. Therefore, derivatisation methods have been developed using dansyl and benzoyl chloride. These derivatisation methods add benzoyl or dansyl moieties to amine and phenolic groups of the biogenic amines (Fig. 28.3) which improves chromatographic retention and separation and can also dramatically improve ionisation efficiency and sensitivity, particularly for homovanillic acid.

Table 28.3 gives details of the mass transitions and the mass spectrometer settings for both derivatised and underivatised methods. These settings will need to be optimised, depending on the specific LC-MS set-up being used.

Details of chromatographic conditions for both methods are also given in Table 28.4. The mobile phases and column are the same for both methods.

Mass spectrometer: Waters TQ-XS triple quadrupole mass spectrometer

LC system: Waters Acquity I-class

Column: Waters Acquity UPLC HSS PFP, 100A, 1.8 um, 2.1 mm × 100 mm (PN-186005967).

Mobile phases: A—Ultrapure water with 10 mM ammonium formate and 0.2% formic acid and B—Acetonitrile with 0.2% formic acid. All chemicals LC-MS grade.

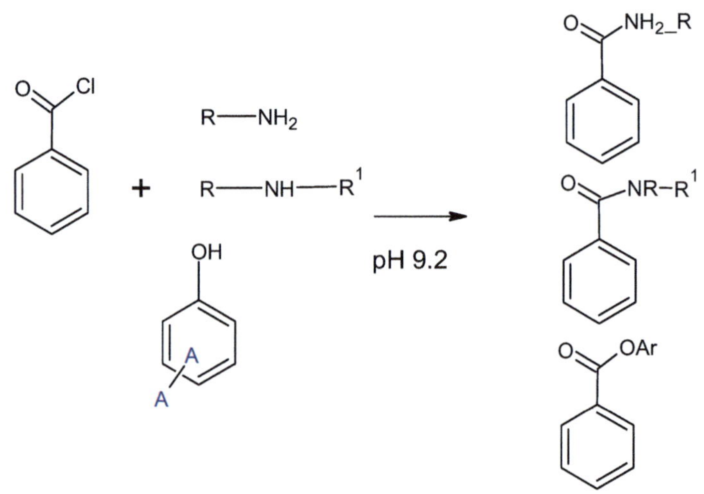

Fig. 28.3 Benzoyl chloride derivatisation of primary and secondary amines and phenols

Table 28.3 List of precursor and product ion *m/z* values for benzoyl chloride derivatised and underivatized biogenic amine metabolites

Compound	Adduct	Precursor ion (m/z)	Product ion (m/z)	Dwell (s)	Cone (V)	Collision energy (eV)	Internal standard	Polarity
Benzoyl derivatives								
5HIAA	5HIAA+Bz+NH4	313	105	0.032	25	15	5HIAA-D6	Pos
HVA	HVA+Bz+NH4	304	105	0.032	25	10	HVA-D5	Pos
3MD	3MD+Bz+H	420	105	0.032	25	15	3MD-D3	Pos
5HTP	5HTP+Bz+H	429	105	0.032	36	40	5HTP-D3	Pos
5HIAA-D6	5HIAA+Bz+NH4	319	105	0.032	25	15		Pos
HVA-D5	HVA+Bz+NH4	309	105	0.032	25	10		Pos
3MD-D3	3MD+Bz+H	423	105	0.032	25	15		Pos
5HTP-D3	5HTP+Bz+H	432	105	0.032	36	40		Pos
Underivatised								
5HIAA	5HIAA+H	192	146	0.032	30	14	5HIAA-D6	Pos
HVA	HVA-H	181	122	0.032	40	10	HVA-D5	Neg
3MD	3MD+H	212	153	0.032	30	14	3MD-D3	Pos
5HTP	5HTP+H	221	204	0.032	30	18	5HTP-D3	Pos
5HIAA-D6	5HIAA+H	198	152	0.032	30	14		Pos
HVA-D5	HVA-H	186	142	0.032	40	10		Neg
3MD-D3	3MD+H	215	156	0.032	30	14		Pos
5HTP-D3	5HTP+H	224	164	0.032	30	18		Pos

The table shows transitions used for quantitation but other transitions for each compound should also be used ('qualifiers') to confirm identity and accurate quantitation of each peak

Table 28.4 Chromatographic parameters for LC-MS methods

Derivatised method			
Time (min)	Flow rate ml/min	% A	% B
0	0.3	60	40
1	0.3	60	40
3.5	0.3	30	70
5.5	0.3	10	90
7	0.3	60	40
Total run time including equilibration = 15 min			
Underivatised method—isocratic			
Time (min)	Flow rate ml/min	% A	% B
0–15 min	0.3	85	15
Mobile phase A—Water with 0.2% formic acid and 10mM ammonium formate			
Mobile phase B—Acetonitrile with 0.2% formic acid			

Weak wash—5% Acetonitrile water; Strong wash—50% acetonitrile water; Seal wash—20% methanol.

Mass spectrometer settings: Desolvation temp—600 C, Gas flow—1000 L/h, Capillary—0.5 kV, Source temp—150C, Cone gas flow—150 L/h.

Centrifugal filters—Nanosep 10 K omega (PN—OD010C34).

Internal standards—Deuterated standards from CDN isotopes. Made up to a working concentration of 1000 nM for the underivatised method and 250 nM for the derivatised method.

Analysis of Underivatised Biogenic Amines

1. Take 150 μl of CSF and add 20 μl of IS mix. Mix and add to a 10-kDa centrifugal filter. Spin for 10 min at 13,000 g.
2. Transfer the filtered CSF to the LC-MS autosampler vial.
3. Inject 5–15 μl.

Analysis of Derivatives Biogenic Amines

1. Take 100 μl of CSF and add it to a 10-kDa centrifugal filter. Spin for 10 min at 13,000 g.
2. Take 25 μl of the filtered CSF, 25 μl of internal standard mixture, 25 μl of 100 mM sodium carbonate and finally 25 μl of benzoyl chlo-

ride (2% in Acetonitrile). Mix well and leave for 5 min. The derivatisation requires a basic pH so the reagents must be added in this order. The reaction goes to completion within 1 min.
3. Inject 5–15 µl.

28.3.7 Follow-up and Supporting Assays

Whenever a CSF profile is suggestive of a disruption of dopamine and/or serotonin metabolism, analysis of the pterin profile is extremely important in order to ascertain whether disruption of biogenic amine metabolism is directly related to alteration in BH4 availability.

Defects affecting dopamine and serotonin metabolism cannot be diagnosed with any confidence by metabolite profiling of peripheral fluids such as urine. Whilst, urinary profiles of neurotransmitter metabolites can support a proposed diagnosis, it is clear that such metabolites can be within normal reference ranges in disorders such as tyrosine hydroxylase deficiency. With regards to tyrosine hydroxylase, the enzyme is predominantly expressed in the brain and adrenal medulla which means that diagnostic enzymatic analysis is not an option. However, mutation analysis is now available in order to confirm or refute a finding from CSF analysis.

Analysis of dopamine-β-hydroxylase activity in plasma cannot be used to make a definitive diagnosis as approximately 4% of the population have very low plasma activities. Support for such a diagnosis can come from examining the noradrenaline:dopamine ratio in plasma. This is markedly reduced in this condition. Deficiencies of the MAO isoforms have been confirmed from enzymatic analysis of fibroblasts (MAO-A) or platelets (MAO-B).

With regards to AADC, vanillactic acid (a metabolite of 3-OMD) appears in the urine and can be detected by organic acid analysis. As predicted, an increase in urinary vanillactic acid has also been reported in PNPO deficiency. The pyridoxal phosphate concentration in CSF, which can be determined by HPLC, is also reported to be decreased in PNPO deficiency [8]. Mutation analysis can also now confirm or refute a suggestion of PNPO deficiency.

As AADC is expressed in plasma, enzymatic analysis to elucidate further upon a suggestive CSF profile can be performed.

28.3.8 AADC Activity

Principle
This assay measures the conversion of L-dopa into dopamine catalysed by AADC present in human plasma [9]. Dopamine is then detected by HPLC and electrochemical detection or by mass spectrometry [10].

Specimen
Lithium heparin plasma stored at −70 °C. Blood samples should be separated and frozen as soon as possible. A minimum of 250 µl of plasma is required for this assay. However, 0.5–1.0 ml of plasma is preferable.

Pitfall
L-dopa medication may interfere with this assay. Stop L-dopa treatment at least 3 days before blood sample collection.

Chemicals
- Disodium hydrogen orthophosphate anhydrous.
- Sodium dihydrogen orthophosphate 1 hydrate.
- Ethylenediaminetetraacetic acid dipotassium salt dihydrate for analysis (EDTA).
- 1,4-Dierythritol (DTE), minimum 99.0%.
- Pyridoxal 5′-phosphate (PLP) hydrate.
- 3,4-Dihydroxy-L-phenylalanine (L-dopa).
- Concentrated hydrochloric acid (~12 M).
- Perchloric acid ACS reagent 70%.
- (+)-Sodium-L-ascorbate.
- Dopamine hydrochloride.
- 1-Octanesulfonic acid sodium salt.
- 85% Orthophosphoric acid HPLC electrochemical grade.
- Methanol HiPerSolv for HPLC.
- Potassium hydroxide pellets.

Reagent Preparation

For Incubation

- *500 Mm Sodium Phosphate, 0.167 mM EDTA, pH 7.00*
  A. 500 mM Disodium hydrogen orthophosphate, 0.167 mM EDTA
     35.49 g Disodium hydrogen orthophosphate, 0.0675 g EDTA.K_2 dissolved in 400 ml of 18.2 MΩ H_2O. Make up 500 ml with 18.2 MΩ H_2O.
  B. 500 mM Sodium dihydrogen orthophosphate, 0.167 mM EDTA.
     34.5 g Sodium dihydrogen orthophosphate, 0.0675 g EDTA.K_2 dissolved in 400 ml of 18.2 MΩ H_2O. Make up to 500 ml with 18.2 MΩ H_2O.

  Titrate A with B to pH 7.00. Store at room temperature for up to 3 months.

- *Assay Buffer*

0.06 g dithioerythritol dissolved in 8 ml 500 mM sodium phosphate, 0.167 mM EDTA, pH 7.00, make up to 10 ml with 500 Mm sodium phosphate, 0.167 mM EDTA, pH 7.00. Prepare fresh on the day of assay.

- *0.7 mM Pyridoxal 5′-Phosphate (PLP)*

0.0017 g PLP.H_2O dissolved in 8 ml 18.2 MΩ H_2O, make up to 10 ml with 18.2 MΩ H_2O. Prepare fresh on the day of assay.

- *20 mM* l*-dopa*

0.0394 g L-dopa dissolved in 8 ml 18.2 MΩ H_2O + 10 µl 6 M HCl, make up to 10 ml with 18.2 MΩ H_2O. Prepare fresh on the day of assay.

- *0.8 M Perchloric acid*

12.1 ml 70% Perchloric acid to make up to 250 ml with 18.2 MΩ H_2O. Store at room temperature for up to 6 months.

For HLPC

- *6 M HCl*

49 ml of HCl sp. gr. 1.18 made up to 100 ml with 18.2 MΩ H_2O. Store at room temperature for up to 6 months.

- *500 µM Dopamine (DA) stock solution*

0.1 g (+)-sodium L-ascorbate, 0.005 g dopamine. HCl dissolved in 80 ml 18.2 MΩ H_2O, make up to 100 ml of 18.2 MΩ H_2O. Store in 1 ml aliquots at −70 °C for up to 12 months.

- *1000 nM DA working standard*

500 µl 6 M HCl, 40 ml 18.2 MΩ H_2O, 100 µL 500 µM DA stock solution, make up to 50 ml with 18.2 MΩ H_2O. Mix by inversion and store at +4 °C. Prepare fresh each day.

- *10% (w/v) EDTA*

10 g EDTA.K_2, dissolved in 80 ml 18.2 MΩ H_2O, make up to 100 ml with 18.2 MΩ H_2O. Store at room temperature for up to 3 months.

- *10 M Potassium Hydroxide*

56 g KOH dissolved in 60 ml 18.2 MΩ H_2O. Make up to 100 ml with 18.2 MΩ H_2O. Store at room temperature for up to 6 months.

- *Mobile Phase*

50 mM Sodium dihydrogen orthophosphate, 5 mM octanesulphonic acid, 0.0025% (w/v) EDTA, 0.2% (v/v) phosphoric acid and 23% (v/v) methanol in 18.2 MΩ H_2O, pH 3.6.

13.8 g Sodium dihydrogen orthophosphate, 2.164 g Octanesulphonic acid dissolved in 800 ml 18.2 MΩ H_2O, add 500 µL 10% EDTA, 5.6 ml orthophosphoric acid and 460 ml methanol. Make up to 1990 ml with 18.2 MΩ H_2O and titrate to pH 3.60 using 10 m KOH. Make up to 2 L with 18.2 MΩ H_2O.

When not attached to system store at 4 °C for up to 1 month.

- *50% Methanol*

500 ml methanol and 500 ml 18.2 MΩ H_2O mixed. Store at room temperature for up to 1 month.

Equipment

- Jasco PU-2080 Plus Intelligent HPLC pump.
- Jasco AS-2057 Plus Refrigerated Intelligent Autosampler.
- Degasys Populaire DP-Series Degasser.
- Jasco CO-2067 Plus Intelligent Column Oven.
- Kya Tech Corporation HiQ sil C18W Column from Kromatek Ltd.
- ESA Coulochem III Electrochemical Detector.
- ESA 5011A Analytical Cells (Voltages: E1 = 250–350 mV + E2 = 0).
- Thermo Scientific Dionex Conditioning Cell 5021A (Voltage: 20–25 mV).
- Eppendorf 5424 Benchtop Centrifuge.
- PC with Azur Data Capture and Analysis Software.

Procedure

1. Thaw clinical samples and QC. Prepare clinical samples and QC for incubation in accordance with the table below. A substrate blank must be performed for each clinical sample and QC. A plasma blank must be performed with each incubation.
2. Vortex all samples and incubate samples at +37 °C in a shaking water bath for 120 min.
3. Add 25 μl of 20 Mm L-dopa to full incubation samples and plasma blank. Do NOT add L-dopa to substrate blank. Vortex all samples. Incubate the samples at +37 °C in a shaking water bath for 90 min.
4. Add 250 μl 0.8 M perchloric acid to each sample and blank and incubate at room temperature for 10 min.
5. Centrifuge all samples and blanks at 12,000xg for 5 min at +4 °C. Transfer supernatants to fresh tubes and store at −70 °C until use. Discard pellets.

Sample	Assay buffer (μL)	0.7 mM PLP (μL)	Sample (μL)
Full incubation	150	25	50
Substrate blank	175	25	50
Plasma blank	200	25	0

Calculation

The concentration of dopamine (DA) produced during the incubation can be calculated using the equation:

$$(\text{Area of sample DA peak} / \text{Area of standard DA peak}) \times 1000$$

The result should be expressed in nM.

AADC activity should be calculated using the following equation:

$$(\text{DA concentration}(\text{nM}) \times 10) / 90$$

This result should be expressed as pmol/mL of plasma.

Reference range

Paediatric: 36–129 pmol/min/mL
Adult: 23.8–42.9 pmol/min/mL

28.3.9 3-*O*-Methyl Dopa in Dried Blood Spots

Measurement of 3-*O*-methyl dopa in dried blood spots is a useful tool for the diagnosis of aromatic amino acid decarboxylase (AADC) deficiency. This metabolite is typically highly elevated in blood spots of patients with AADC deficiency. A number of different methods have been described in the literature, some for newborn screening. Below we describe an indicative method, using the same LC-MS detection method for underivatised biogenic amines as described above. When analysing 3-*O*-methyl dopa in dried blood spots it is important to use appropriate age-related reference intervals, as the concentration typically falls with age in a similar way to the CSF metabolites. It is also important to be aware if the patient or the breast-feeding mother is taking L-dopa as this has been reported to give similar 3-*O*-methyl dopa levels to that observed in AADC deficiency.

1. Punch a 6-mm diameter dried blood spot into a 96-well plate or flat bottom tube. Add 20 ul

of water to rehydrate the blood spot. Leave for 5 min.

2. Add 20 ul of internal standard (100 nM of deuterated 3-*O*-methyl dopa) and 80 ul of water and mix well on an orbital shaker for 30 min at room temperature.
3. Spin down at 13,000 g for 5 min.
4. Filter the supernatant through a 0.2-micron centrifugal filter.
5. Inject 5–15 ul onto the LC-MS and analyse as above.

Acknowledgement SP is supported by a Medical Research Council (UK) Clinical Academic Research Partnership grant (MR/V03801X/1).

References

1. Blau N, Thony B, Cotton RGH, Hyland K (2001) Disorders of tetrahydrobiopterin and related biogenic amine. In: Scriver CR, Sly WS, Childs B, Beaudet AL, Valle D, Kinzler W, Vogelstein B (eds) The metabolic and molecular bases of inherited disease, 8th edn. McGraw Hill, New York, pp 1725–1776
2. Brautigam C, Wevers RA, Jansen RJT, Smeitink JAM et al (1998) Biochemical hallmarks of tyrosine hydroxylase deficiency. Clin Chem 44:1897–1904
3. Hyland K (2003) The lumbar puncture for diagnosis of paediatric neurotransmitter diseases. Ann Neurol 54:S13–S17
4. Hyland K, Surtees RAH, Heales SJR, Bowron A et al (1993) Cerebrospinal fluid concentrations of pterins and metabolites of serotonin and dopamine in a pediatric reference population. Pediatr Res 34:10–14
5. Carlo E, Santagata S, Sauro L, Tolve M, Manti F, Leuzzi V, Angeloni A, Carducci C (2021) Simultaneous determination of 5-hydroxytryptophan and 3-O-methyl dopa in dried blood spot by UPLC-MS/MS: a useful tool for the diagnosis of L-amino acid decarboxylase deficiency. J Chromatogr B 1185:122999
6. Galla Z, Raja C, Racz G, Gresco N, Barath A, Vecsei L, Bereczki C, Monostori P (2021) Simultaneous determination of 30 neurologically and metabolically important molecules: a sensitive and selective way to measure tyrosine and tryptophan pathway metabolites and other biomarkers in human serum and cerebrospinal fluid. J Chromatogr A 4(1635):461775
7. Wong J, Malec P, Mabrouk O, Ro J, Dus M, Kennedy R (2016) Benzoyl chloride derivatisation with liquid chromatography-mass spectrometry for targeted metabolomics of neurochemicals in biological samples. J Chromatogr A 1446:78–90
8. Mills PB, Surtees RAH, Champion MP et al (2005) Neonatal epileptic encephalopathy caused by mutations in the *PNPO* gene encoding pyridox(am)ine 5′-phosphate oxidase. Hum Mol Genet 14:1077–1086
9. Hyland K, Clayton P (1992) Aromatic amino acid decarboxylase deficiency: diagnostic methodology. Clin Chem 38:2405–2410
10. Hyland K, Surtees RAH, Rodeck C, Clayton PT (1992) Aromatic L amino acid decarboxylase deficiency: clinical features, diagnosis and treatment of a new inborn error of neurotransmitter amine synthesis. Neurology 42:1980–1988

Creatine and Its Metabolites

29

Desirée E. C. Smith, Mirjam M. C. Wamelink,
Gajja S. Salomons, and Jiddeke M. van de Kamp

Contents

Abbreviations

AGAT	L-Arginine:glycine amidinotransferase
Cr	Creatine
CSF	Cerebrospinal fluid
GA	Guanidinoacetic acid
GAMT	Guanidinoacetate methyltransferase
GC-MS	Gas chromatography-mass spectrometry
IS	Internal standard
LC-MS/MS	Liquid chromatography-tandem mass spectrometry
LLOD	Lower limit of detection
PFBBr	Pentafluorobenzylbromide
RT	Room temperature
SLC6A8	Creatine transporter

D. E. C. Smith · M. M. C. Wamelink (✉)
· G. S. Salomons
Amsterdam UMC location AMC, Laboratory Genetic Metabolic Diseases, Amsterdam, The Netherlands
e-mail: d.smith@amsterdamumc.nl; m.wamelink@amsterdamumc.nl; g.salomons@amsterdamumc.nl

J. M. van de Kamp
Department of Human Genetics, Amsterdam UMC location AMC, Amsterdam, The Netherlands
e-mail: jm.vandekamp@amsterdamumc.nl

29.1 Introduction

Creatine (Cr) plays an important role in energy transmission and storage in cells and tissues with high-energy demands. Creatine kinase catalyzes the reversible conversion of creatine and ATP to phosphocreatine and ADP. The creatine-phosphocreatine system serves as a cytosolic buffer for the regeneration of ATP and as a shuttle of high-energy phosphates. The Cr pool in the body is obtained from the diet and by de novo biosynthesis [1, 2]. Creatine is synthesized from arginine

N. Blau, F. M. Vaz (eds.), *Laboratory Guide to the Methods in Biochemical Genetics*,
https://doi.org/10.1007/978-3-031-58819-8_29

and glycine in a two-step reaction, using S-adenosylmethionine (SAM) as a methyl group donor. This two-step synthesis is mediated by L-arginine:glycine amidinotransferase (AGAT; EC 2.1.4.1) that catalyzes the reversible transamidination of the guanidino group from arginine to glycine, yielding guanidinoacetic acid (GA) and ornithine and guanidinoacetate methyltransferase (GAMT, EC 2.1.1.2) that uses SAM as a methyl donor, yielding Cr [3].

Many tissues contain a Na^+/Cl^--dependent Cr transporter (SLC6A8), which is responsible for Cr transport over the plasma membrane.

Clinical Significance

In humans, two inborn errors in Cr biosynthesis (AGAT and GAMT) and one in Cr transport (SLC6A8) are identified. SLC6A8 and GAMT deficiencies are the most common Cr deficiency syndromes. AGAT deficiency is rare. All three deficiencies lead to cerebral Cr deficiency which can be diagnosed by proton magnetic resonance spectroscopy of the brain (^{1}H-MRS), causing developmental delay and intellectual disability with prominent speech delay and frequent behavioral symptoms such as autistic features and short attention span. Severe extrapyramidal symptoms and intractable epilepsy are mainly associated with GAMT deficiency [4, 5] but epilepsy is also frequent in SLC6A8 deficiency [2, 4, 6] which can also be associated with mild extrapyramidal symptoms. In contrast, AGAT deficiency [4, 7] is associated with progressive myopathy and rarely with seizures.

AGAT and GAMT deficiencies are autosomal recessive conditions. SLC6A8 deficiency is X-linked causing (moderate to severe) intellectual disability and frequently epilepsy in all male patients while features in female heterozygotes are very variable from absence of symptoms or learning difficulties to mild-severe intellectual disability with or without epilepsy, depending on the X-inactivation pattern.

GA and Cr are decreased in the body fluids of AGAT-deficient patients. In GAMT deficiency GA is elevated and Cr decreased. Biochemical diagnosis of SLC6A8 deficiency in affected male patients is possible by analysis of Cr-to-creatinine ratio excretion in urine.

After finding aberrant metabolites in creatine metabolism or decreased cerebral creatine by H-MRS, the diagnosis of GAMT or AGAT deficiency can be confirmed by (1) enzymatic assays in various cell types, and SLC6A8 deficiency can be confirmed by a functional Cr-uptake assay in fibroblasts in males, and/or (2) by genetic analysis revealing mutation(s) in the *AGAT, GAMT, or SLC6A8*.

Prenatal diagnosis of all three defects is possible by genetic analysis.

29.2 Metabolites

29.2.1 Properties of Metabolites

Creatinine, Cr and GA are highly polar compounds. Their chemical structure and properties are depicted in Fig. 29.1 Creatinine is a breakdown product of Cr.

29.2.2 Determination of Creatinine

For completeness, the measurement of Creatinine is briefly described because of the relevance for biochemical diagnostics, but this is common practice in many labs and therefore not many details are provided. Creatinine in plasma, urine, and CSF is still commonly measured using the colorimetric Jaffé method where creatinine reacts with alkaline picrate and forms a red complex. The method is straightforward, but the non-specific nature of the reaction can cause falsely elevated concentration in the presence of among other things protein and glucose.

More recently, enzymatic methods using a creatine amidohydrolase or creatinine deaminase, HPLC and LC-MS/MS methods have been developed that are more specific and more suitable for this specific group of patients since creatinine concentrations in plasma samples of two patients with GAMT deficiency appeared signifi-

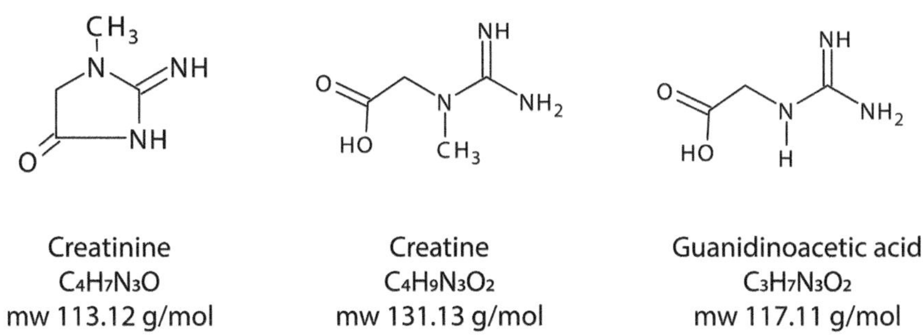

Fig. 29.1 Chemical structures of creatinine, creatine, and guanidinoacetic acid

cantly higher using the Jaffé method than when measured by HPLC [8].

The diagnostic value of creatinine measurements in the urine of patients with Cr biosynthesis defects is low since normal concentrations can be found in affected patients [9] although low plasma and CSF creatinine have been reported in a patient affected with GAMT deficiency [10].

Pitfalls

Bacteria in urine can degrade creatinine which can result in falsely low urine creatinine concentrations [11]. Depending on the method, non-specific interference due to factors like bilirubin levels or protein, can interfere with results [12]. Low urinary creatinine concentration and low 24-h urine creatinine excretion have been observed in patients with vanishing muscle mass (and thus a low body Cr pool), representing an unspecific finding in several myopathies and muscular dystrophies [13].

29.2.3 Determination of GA and Cr

To date, several methods are available for laboratories to determine GA and Cr in body fluids [14–19]. They vary between GC and LC approaches combined with mass spectrometry. Lower limits of detection (LLODs) do not differ greatly between the techniques, but sample preparation does.

Pre-analytical

For the determination of GA and Cr in urine (random urine samples are suitable), plasma, and cerebrospinal fluid (CSF) no special pre-analytical procedure needs to be carried out.

29.2.3.1 Gas Chromatography-Mass Spectrometry (GC-MS)

GA and Cr are determined using GC-MS following a derivatization of the guanidino group with hexafluoroacetylacetone and subsequent derivatization of the carboxylic groups with pentafluorobenzylbromide (PFBBr) [18, 20]. The reaction scheme is depicted in Fig. 29.2.

Reagents and Chemicals

Hexane p.a., toluene p.a., hydrochloric acid 37% and sodium bicarbonate (Merck, Darmstadt, Germany), hexafluoroacetylacetone, trimethylamine, PFBBr, Cr and GA (Sigma-Aldrich, Steinheim, Germany), acetonitrile Hipersolv (VWR, Poole, United Kingdom), [$^{13}C_2$]-GA (Amsterdam UMC location AMC, The Netherlands), and [$^{2}H_3$]-Cr (CDN Isotopes, Quebec, Canada).

Instrumentation

The measurement of GA and Cr are performed on an Agilent 5973 inert MSD system in selective-ion monitoring mode but can be performed on any GC-MS system that allows for negative chemical ionization.

Analytical Procedure

For urine:

- Add 50 μL of urine to a 1.5 mL glass vial.
- Add 50 μL of water (making a total volume of 100 μL).
- Add 50 μL internal standard (IS) containing 500 μM Cr and 250 μM GA.
- Add 50 μL of saturated aqueous sodium bicarbonate.
- Add 50 μL hexafluoroacetylacetone.
- Add 500 μL toluene.

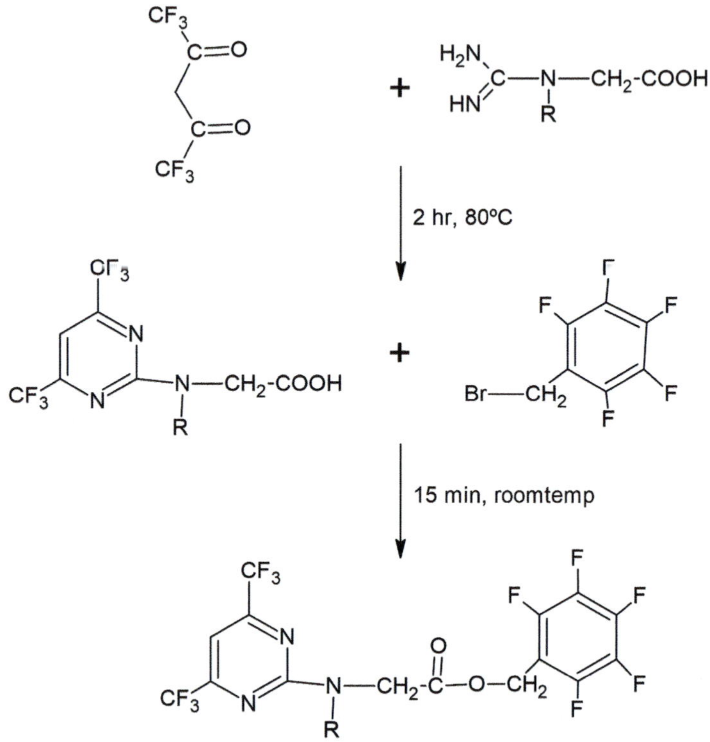

Fig. 29.2 Derivatization of creatine (R=CH_3) and guanidinoacetic acid (R=H) using hexafluoroacetylacetone and pentafluorobenzylbromide for measurement with GC-MS

Table 29.1 Calibration curves for guanidinoacetic acid (GA) in urine, plasma, and cerebrospinal fluid (CSF)

Standard	Urine	Plasma	CSF
	(nmol GA)	(nmol GA)	(nmol GA)
1	0	0	0
2	2.5	0.05	0.005
3	5	0.10	0.01
4	10	0.25	0.025
5	25	0.50	0.05
6	50	–	0.1

Table 29.2 Calibration curves for creatine (Cr) in plasma, cerebrospinal fluid (CSF), and urine

Standard	Plasma/CSF	Urine
	(nmol Cr)	(nmol Cr)
1	0	0
2	0.5	5
3	1.0	10
4	2.5	25
5	5.0	50
6	10	100
7	–	200

- The series is accompanied with a set of standards listed in Tables 29.1 and 29.2.
- Heat at 80 °C for 15 min under continuous stirring.
- Allow to cool.
- Transfer 100 μL of the upper toluene phase to another test tube.
- Evaporate the toluene at 40 °C using a nitrogen stream.
- *Add 100 μL 7% pentafluorobenzyl bromide in acetonitrile (PFBBr) (v/v) to the residue.
- Add 10 μL trimethylamine and mix.
- Derivatize at room temperature for 15 min.
- Add 200 μL 0.5 N HCl.
- Add 1 mL hexane and mix.
- Centrifuge 2 min at 830 g.
- Transfer the upper hexane layer to a vial and store until analysis at room temperature.

For plasma:

- Add 50 μL of plasma to a 1.5-mL glass vial.
- Add 50 μL internal standard (IS) containing 50 μM Cr and 5 μM GA.

- Add 50 μL of saturated aqueous sodium bicarbonate.
- Add 50 μL hexafluoroacetylacetone.
- Add 500 μL toluene.
- The series is accompanied with a set of standards listed in Tables 29.1 and 29.2.
- Heat at 80 °C for 15 min under continuous stirring.
- Allow to cool.
- Transfer 100 μL of the upper toluene phase to another test tube.
- Evaporate the toluene at 40 °C using a nitrogen stream.
- Proceed to make PFBBr derivatives as described with urine samples at the step denoted with an asterisk "*".

For CSF:

- Add 50 μL of plasma to a 1.5-mL glass vial.
- Add 50 μL internal standard (IS) containing 50 μM Cr and 0.5 μM GA.
- Add 50 μL of saturated aqueous sodium bicarbonate.
- Add 50 μL hexafluoroacetylacetone.
- Add 500 μL toluene.
- The series is accompanied with a set of standards listed in Tables 29.1 and 29.2.
- Heat at 80 °C for 15 min under continuous stirring.
- Allow to cool.
- Transfer 300 μL of the upper toluene phase to another test tube.
- Evaporate the toluene at 40 °C using a nitrogen stream.
- Proceed to make PFBBr derivatives as described with urine samples at the step denoted with an asterisk "*".

GC-MS Measurement

Separate 1 μL of the derivatized GA and Cr mixture using a SGE BPX-70 (25 m × 0.32 μm I.D., film thickness 0.25 μm) GC column or any other similar polar column, especially important for plasma and CSF samples using helium as carrier gas. The injection temperature is 250 °C. The oven temperature starts at 100 °C for 1 min, then rampes with 5 °C/min to 260°. The transfer line temperature is 260 °C. The negative chemical ionization mode is used with a mixture of 95% methane, 5% ammonia as reagent gas. For GA m/z –288 is used and for [$^{13}C_2$]-GA m/z –290. (IS) For Cr m/z –302 is used and for [$^{2}H_3$]-Cr m/z –305.

LLOD of GA in a 100 μl sample volume is 0.01 μmol/L.

Calculation

Construct a calibration curve from the standards using the (peak area analyte)/(peak area IS) ratio versus the amount (nmol) of Cr or GA in the standards. Cr and GA concentrations in the samples are subsequently calculated by interpolation of the (peak area analyte)/(peak area IS) ratio of the sample into the calibration curve. The concentration is calculated by dividing the nmol amount calculated by the amount of mL of the sample used to give a μmol/L concentration.

Quality control

Pools of the concerned sample type are run with each series.

Pitfalls

The stable-isotope dilution GC-MS measurement of GA and Cr in plasma, CSF, and urine is sensitive, robust, and reliable. The measurement of especially GA in CSF needs attention with regard to the chromatography. The analyses should be carried out on a GC column coated with a polar stationary phase to achieve proper separation of GA and interfering compounds. Longer derivatization with hexafluoroacetylacetone can result in conversion of creatinine to creatine.

29.2.3.2 Liquid Chromatography-Tandem Mass Spectrometry (LC-MS/MS)

Following protein precipitation, Cr and GA are butylated (Fig. 29.3) and subsequently separated using reversed-phase LC and detected by MS/MS [19]. A similar method for urine is also described where samples are diluted to a 1-mmol/L creatinine concentration before butylation [21]. Additional approaches involving underivatized metabolites are also published [22].

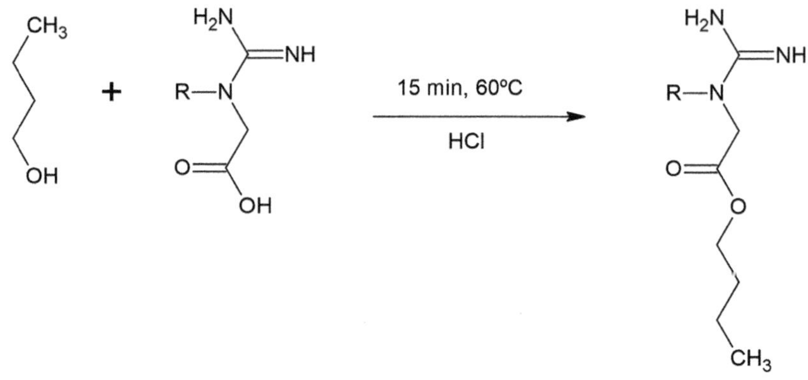

Fig. 29.3 Butylation of creatine (R=CH_3) and guanidinoacetic acid (R=H)

Reagents and Chemicals

Ammonium acetate (Merck, Darmstadt, Germany), HPLC grade acetonitrile and water (Romil Ltd., Cambridge, United Kingdom), formic acid 98–100% (VWR, Poole, United Kingdom), butanolic-HCl (Regis Technologies, Morton Grove, IL, USA), Cr and GA (Sigma-Aldrich, Steinheim, Germany), [$^{13}C_2$]-GA (CDN Isotopes, Canada) and 50 µmol/L [D_3]-Cr (Amsterdam UMC, The Netherlands).

Standard Curves

Every series is accompanied by an aqueous standard solutions series (0, 0.05, 0.5, 5, 10, 25, 50,100, and 200 µmol/L for Cr and 0, 0.02, 0.1, 1, 2, 5, 10, 20, and 40 µmol/L for GA).

Instrumentation

The measurement of GA and Cr are performed on an Agilent 1200 series coupled to a Sciex API3200 operating in the positive ion mode but can be performed on a similar LC-MS/MS system.

Analytical procedure

- Add 50 µL of EDTA-K2 plasma into a 1.5-ml Eppendorf.
- Add 25 µL internal standard (containing 10 µmol/L [$^{13}C_2$]-GA and 50 µmol/L [D_3]-Cr.
- Add 400 µL acetonitrile.
- Vortex 2 min.
- Centrifuge 5 min at 10,000 rpm.
- Transfer the supernatant to a vial.
- Dry sample at 40 °C using nitrogen.
- Add 80 µL butanolic-HCl to the residue.
- Derivatize at 65 °C for 15 min.
- Dry samples at 40 °C using nitrogen.
- Reconstituted residue in 200 µL acetonitrile-water 50:50 (v/v) containing 0.05% formic acid pH 4.

LC-MS/MS Measurement

The butylesters are separated on a Supelcosil LC-8 column (3 µm × 4.6 mm × 7.5 cm) using a mobile phase consisting of 5 mM ammonium acetate, 0.03% (v/v) formic acid, and 70% acetonitrile (v/v). Five µL is injected with a flow of 1 mL/min. Positive ion electrospray LC-MS/MS is used for detection. Collision energy for Cr is 27 V and for GA 21 V. The following transitions are monitored: m/z 174.1 > 101.0 for GA, 176.1 > 103.0 for [$^{13}C_2$]-GA, 188.1 > 90.0 for Cr and 191.1 > 93.0 for [D_3]-Cr.

LLOQ are 0.05 µmol/L for Cr (S/N ratio 12.3) and 0.02 µmol/L GA (S/N ratio 12.9).

Calculation

Construct a calibration curve from the standards using the (peak area analyte)/(peak area IS) ratio versus the amount (nmol) of Cr or GA in the standards. Cr and GA concentrations in the samples are subsequently calculated by interpolation of the (peak area analyte)/(peak area IS) ratio of the sample into the calibration curve. The concentration is calculated by dividing the nmol amount calculated by the amount of mL of the sample used to give a µmol/L concentration.

Pitfalls

Accurate separation between valine and GA needs to be obtained since they have identical mass transitions (molecular mass).

29.3 Interpretation of Metabolite Levels

The three inborn errors of creatine metabolism and transport can be distinguished based on creatine, creatinine, and guanidinoacetate measurements in urine and plasma.

AGAT deficiency can be diagnosed by the decreased concentrations of GA in plasma, urine, and CSF.

GAMT deficiency can be diagnosed by increased levels of GA in urine, plasma, and CSF.

In AGAT and GAMT deficiency, Cr is decreased or in the low–normal range in urine, plasma, and CSF. Creatinine in urine and plasma is decreased or in the low–normal range. This low urinary creatinine can result in the detection of increased concentrations of several apparent unrelated metabolites (e.g., amino acids, organic acids, and uric acid) when expressed per mol creatinine. This general increase can be a diagnostic hint for an error in creatine metabolism. Cr supplementation decreases GA in the plasma of GAMT-deficient patients but the levels do not normalize. GA remains also increased in cerebral H-MRS. Supplementation with Cr results in increased levels in both plasma and urine and partly replenishes cerebral creatine.

Male patients affected with the X-linked SLC6A8 defect have elevated urinary Cr concentrations only if this is expressed per mol creatinine, the Cr/creatinine ratio. In plasma, Cr concentrations are within the normal limits or mildly elevated. GA is within the reference range in both plasma and urine. Female patients/carriers of the defect may have elevated urinary Cr creatinine ratio. However, this is not a consistent finding in all carriers [23].

29.3.1 Reference Values

Reference values for both GA and Cr are age dependent [20, 24] and are listed in Tables 29.3 and 29.4. For urine, a separate reference range is necessary for newborns otherwise newborns with GAMT deficiency could be missed.

29.3.2 Pathological Values

Pathological values found for Cr deficiency syndromes are shown in Table 29.5 [6, 9, 13, 20, 24]. Pathological GA plasma values for three other children affected with AGAT deficiency have been described ranging from 0.01 to 0.04 μmol/l (controls 0.35–1.8) [25]. Other metabolites were listed in non-standard units. In urine samples, GA for these patients was more than ten times lower in comparison to controls. Cr levels for these patients are expressed as the sum of Cr and creatinine and are found to range from moderately lowered to low–normal.

Table 29.3 Reference values for GA

GA	*n*	Age (years)	Range
Urine (mmol/mol creatinine)	104	0–15	4–220[a]
	40	>15	3–78
Plasma (μmol/L)	33	0–15	0.35–1.8
	21	>15	1.0–3.5
CSF (μmol/L)	10	Not established	0.036–0.22

[a] Separate reference range is necessary for newborns otherwise newborns with GAMT deficiency could be missed

Table 29.4 Reference values for Cr

Cr	*N*	Age (years)	Range
Urine (mmol/mol creatinine)	44	0–4	6–1208
	36	4–12	17–721
	47	>12	11–244
Plasma (μmol/L)	33	0–10	17–109
	32	>10	6–50
CSF (μmol/L)	34	0–10	24–53
	7	>10	38–66

Table 29.5 Pathological values for Cr and GA in Cr disorders. L-Arginine:glycine amidinotransferase (AGAT), guanidinoacetate methyltransferase (GAMT), creatine transporter (SLC6A8) [6, 9, 13, 20]

	GA		Cr	
	Urine	Plasma	Urine	Plasma
Pathology	mmol/mol creatinine	µmol/L	mmol/mol creatinine	µmol/L
GAMT deficiency (*n* = 9)	>500[a]	>10	Normal	<12
SLC6A8 deficiency (*n* = 81)	Normal	Normal	Elevated[b]	Normal/mildly elevated
AGAT deficiency (*n* = 2)	<0.4	<0.05	Low-normal	<3

[a] GA in newborns with GAMT deficiency can be lower (190 mmol/mol creatinine), separate reference range is necessary for newborns

[b] Values obtained should always be related to the proper age-matched reference values and be judged as the ratio of creatine to creatinine. Variable values in females for this X-linked disorder

Table 29.6 GAMT activity in controls, heterozygotes, and GAMT-deficient patients

	GAMT activity (pmol/mg protein/h)		
	Lymphoblasts	Fibroblasts	Lymphocytes
Patients	<5 (*n* = 2)	<5 (*n* = 5)	n.d. (*n* = 3)
Heterozygotes	69, 130 (*n* = 2)	–	106, 115 (*n* = 2)
Controls	63–450 (*n* = 10)	60–243 (*n* = 7)	27–130 (*n* = 13)

29.4 Enzyme Assays

The diagnosis of GAMT deficiency, AGAT deficiency, and SLC6A8 deficiency can be confirmed by performing specific enzyme assays.

29.4.1 Guanidinoacetate Methyltransferase

GAMT activity is measured in homogenates of either lymphocytes (isolated using Histopaque™) or cultured lymphoblasts or fibroblasts [26, 27]. The assay is performed for 2 h at 37 °C in a 0.1-mol/L TRIS/HCl buffer pH 8.5 containing 1 mmol/L [$^{13}C_2$]-GA and 1.5 mmol/L [$^2H_3$]-S-adenosylmethionine. For lymphoblast, fibroblasts, or lymphocytes 0.5–1 mg, 0.1–0.4 mg, or 0.1 mg total protein was used, respectively. The reaction was stopped using methanol. Following the addition of 6.25 pmol of [$^2H_3$-Cr] as internal standard and a centrifugation step, the formed [$^{13}C_2$-$^2H_3$]Cr is measured by GC-MS procedure as described above. The amount of Cr formed divided by the amount of protein and incubation time represents GAMT activity (Table 29.6). A similar but radioactive approach was published by Ilas et al. [28].

GAMT activity in lymphocytes decreases when the isolation of lymphocytes from blood is delayed for more than a day [26].

29.4.2 L-Arginine: Glycine Amidinotransferase

AGAT activity can be determined in homogenates of either lymphocytes (isolated using Histopaque™) or lymphoblasts. In this procedure, 50 µg of total protein is incubated in 200 µL 0.1 M phosphate buffer pH = 7.5 containing 15 µmol/L [$^{15}N_2$]-L-arginine and 24 µmol/L [$^{13}C_2$-^{15}N]-glycine for 4–20 h at 37 °C [29]. The incubation was terminated by the addition of 30 µL of 4 M perchloric acid and subsequently neutralized by the addition of 20 µL of 6 M potassium hydroxide. After the addition of 1.25 nmoles of [1,2-$^{13}C_2$]-GA as internal standard, the formed [$^{13}C_2$,$^{15}N_3$]-GA, which is measured by the GC-MS procedure described above. The amount of [$^{13}C_2$,$^{15}N_3$]-GA formed is divided by the amount of protein used and the hours incubated to calculate AGAT activity (Table 29.7). An approach involving radioactive labels was described by Item et al. [30].

29.4.3 SLC6A8

The function of SLC6A8 is assayed using intact fibroblasts, which are cultured for 24 h in a medium that is enriched with either 25 μM or 500 μM Cr [31]. The intracellular concentration of Cr in the fibroblasts is determined using the GC-MS procedure described in detail in this chapter (Table 29.8).

Table 29.7 AGAT activity in controls and AGAT-deficient patients

	AGAT activity (pmol/mg protein/h)	
	Lymphoblasts	Lymphocytes
Patients	n.d. ($n = 10$)	Not performed
Controls	21–110 ($n = 15$)	1.3–8.5 ($n = 16$)

Table 29.8 Uptake of Cr by fibroblasts derived from controls and patients with SLC6A8 deficiency when exposed to either 25 μmol/L or 500 μmol/L for 24 h

	Cr uptake (pmol/μg protein/24 h)	
	25 μmol/L Cr	500 μmol/L Cr
Patients ($n = 13$)	0–3.2	2.1–10.5
Controls ($n = 13$)	18.8–38	31.9–51.7

29.5 Newborn Screening

29.5.1 Procedure

Several groups have shared their experience of the implementation and performance of their pilot studies or screening programs for GAMT [32–36]. Strategies are different but all start with flow injection FIA-MS/MS measurement of GA following butylation of the blood spots identical to the method described in the LC-MS/MS method section. False positive rates vary from 0.02% and 0.58%. Second-tier measurements, usually involving LC separation followed by MS/MS), lower these rates to <0.002%. Some groups apply *GAMT* gene sequencing as the third tier.

29.5.2 Reference and Pathological Values

In Table 29.9, the reference values for newborns for Cr, GA, and Cr/GA ratio in blood spots are shown. Especially Cr levels vary with age [36]

Table 29.9 Percentiles, minimum and maximum values, and average values for creatine (Cr) and guanidinoacetic acid (GA), expressed in μmol/L, obtained from blood spots from normal controls (controls) and patients with GAMT deficiency (GAMT) collected at <7 days of age and >7 days of age ($n = 3$). The ratio GA/creatine is also shown [36]

	Analyte	Age (days)	1%	50%	99.5%	99.9%	Range μmol/L	Mean (SD) μmol/L
Controls	Cr	<7	183	308	553	652	138–877	313 (80)
Controls	GA	<7	0.68	1.17	2.44	3.36	0.5–10.4	1.21 (0.35)
Controls	GA/Cr	<7	0.002	0.004	0.009	0.011	0.001–0.032	0.004 (0.001)
GAMT	Cr	<7					208–392	270 (83)
GAMT	GA	<7					12.1–28.7	18.8 (7.5)
GAMT	GA/Cr	<7					0.051–0.116	0.071 (0.030)
Controls	Cr	>7	75.9	158	346	430	38–512	172 (47)
Controls	GA	>7	0.66	1.08	3.48	5.96	0.4–12.8	1.18 (0.50)
Control	GA/Cr	>7	0.003	0.007	0.023	0.035	0.002–0.091	0.008 (0.004)
GAMT	Cr	>7					21–64	42 (21)
GAMT	GA	>7					6.0–13.5	10.1 (3.8)
GAMT	GA/Cr	>7					0.146–0.633	0.315 (0.275)

and therefore reference ranges for newborns up to 7 days and beyond 7 days were determined. Cr levels of newborns up to 7 days reflect more maternal supply than their own.

Only in December 2020, Utah U.S. reported the world's first true positive sample found in their GAMT deficiency newborn screening after the screening of over 250,000 samples (first-tier GA and Cr concentrations were 13.3 and 324 μmol/L, respectively). In 2021, New York followed (first-tier GA and Cr concentrations were 23.4 and 509 μmol/L, respectively) [37].

Cut-off values for GA range from 2.44 to 6.0 μmol/L [32–36]

29.6 Conclusions

In this chapter, we describe accurate methods for measuring Cr and GA in urine, plasma, and CSF for diagnosing inborn errors of metabolism in Cr biosynthesis and transport. Functional enzyme assays for confirmation of AGAT, GAMT, and Cr transporter deficiency are also described.

Acknowledgments The authors wish particularly to thank Eduard Struys, Nanda Verhoeven, and Cornelis Jakobs for their previous contribution to this chapter.

References

1. Salomons GS, Wyss M, Jakobs C (2005) Creatine. In: Coates PBM, Cragg G, Levine M, Moss J, White J (eds) Encyclopedia of dietary supplements. Marcel Dekker, pp 151–158
2. van de Kamp JM, Mancini GM, Salomons GS (2014) X-linked creatine transporter deficiency: clinical aspects and pathophysiology. J Inherit Metab Dis 37(5):715–733
3. Wyss M, Kaddurah-Daouk R (2000) Creatine and creatinine metabolism. Physiol Rev 80(3):1107–1213
4. Mercimek-Andrews S SG. Creatine Deficiency Disorders. 2009. In: GeneReviews® [Internet] [Internet]. Seattle: University of Washington, Seattle. Available from: https://www.ncbi.nlm.nih.gov/books/NBK3794/
5. Khaikin Y, Sidky S, Abdenur J, Anastasi A, Ballhausen D, Buoni S et al (2018) Treatment outcome of twenty-two patients with guanidinoacetate methyltransferase deficiency: an international retrospective cohort study. Eur J Paediatr Neurol 22(3):369–379
6. van de Kamp JM, Betsalel OT, Mercimek-Mahmutoglu S, Abulhoul L, Grunewald S, Anselm I et al (2013) Phenotype and genotype in 101 males with X-linked creatine transporter deficiency. J Med Genet 50(7):463–472
7. Stockler-Ipsiroglu S, Apatean D, Battini R, DeBrosse S, Dessoffy K, Edvardson S et al (2015) Arginine:glycine amidinotransferase (AGAT) deficiency: clinical features and long term outcomes in 16 patients diagnosed worldwide. Mol Genet Metab 116(4):252–259
8. Verhoeven NM, Salomons GS, Jakobs C (2005) Laboratory diagnosis of defects of creatine biosynthesis and transport. Clin Chim Acta 361(1–2):1–9
9. Verhoeven NM, Guérand WS, Struys EA, Bouman AA, van der Knaap MS, Jakobs C (2000) Plasma creatinine assessment in creatine deficiency: a diagnostic pitfall. J Inherit Metab Dis 23(8):835–840
10. Stöckler S, Marescau B, De Deyn PP, Trijbels JM, Hanefeld F (1997) Guanidino compounds in guanidinoacetate methyltransferase deficiency, a new inborn error of creatine synthesis. Metabolism 46(10):1189–1193
11. Seifter JL (2014) Urinary creatinine-splitting bacteria after ileal-loop diversion causing underestimate of glomerular filtration rate. Am J Med 127(12):e11–e12
12. Bargnoux A-S, Kuster N, Cavalier E, Piéroni L, Souweine J-S, Delanaye P et al (2018) Serum creatinine: advantages and pitfalls. J Lab Precision Med 3:3
13. Salomons GS, Johnston K, Plawner L, Cooper L, Barkovich J, Verhoeven NM et al (2005) The second family with Agat deficiency (creatine biosynthesis defect): diagnosis, treatment and the first prenatal diagnosis. J Inherit Metab Dis 28:224–224
14. Carducci C, Birarelli M, Santagata P, Leuzzi V, Carducci C, Antonozzi I (2001) Automated high-performance liquid chromatographic method for the determination of guanidinoacetic acid in dried blood spots: a tool for early diagnosis of guanidinoacetate methyltransferase deficiency. J Chromatogr B Biomed Sci Appl 755(1–2):343–348
15. Bodamer OA, Bloesch SM, Gregg AR, Stockler-Ipsiroglu S, O'Brien WE (2001) Analysis of guanidinoacetate and creatine by isotope dilution electrospray tandem mass spectrometry. Clin Chim Acta 308(1–2):173–178
16. Cognat S, Cheillan D, Piraud M, Roos B, Jakobs C, Vianey-Saban C (2004) Determination of guanidinoacetate and creatine in urine and plasma by liquid chromatography-tandem mass spectrometry. Clin Chem 50(8):1459–1461
17. Carducci C, Santagata S, Leuzzi V, Carducci C, Artiola C, Giovanniello T et al (2006) Quantitative determination of guanidinoacetate and creatine in dried blood spot by flow injection analysis-electrospray tandem mass spectrometry. Clin Chim Acta 364(1–2):180–187
18. Struys EA, Jansen EE, ten Brink HJ, Verhoeven NM, van der Knaap MS, Jakobs C (1998) An accurate stable isotope dilution gas chromatographic-mass

spectrometric approach to the diagnosis of guanidinoacetate methyltransferase deficiency. J Pharm Biomed Anal 18(4–5):659–665
19. Boenzi S, Rizzo C, Di Ciommo VM, Martinelli D, Goffredo BM, la Marca G et al (2011) Simultaneous determination of creatine and guanidinoacetate in plasma by liquid chromatography-tandem mass spectrometry (LC-MS/MS). J Pharm Biomed Anal 56(4):792–798
20. Almeida LS, Verhoeven NM, Roos B, Valongo C, Cardoso ML, Vilarinho L et al (2004) Creatine and guanidinoacetate: diagnostic markers for inborn errors in creatine biosynthesis and transport. Mol Genet Metab 82(3):214–219
21. Haas D, Gan-Schreier H, Langhans CD, Anninos A, Haege G, Burgard P et al (2014) Diagnosis and therapeutic monitoring of inborn errors of creatine metabolism and transport using liquid chromatography-tandem mass spectrometry in urine, plasma and CSF. Gene 538(1):188–194
22. Carling RS, Hogg SL, Wood TC, Calvin J (2008) Simultaneous determination of guanidinoacetate, creatine and creatinine in urine and plasma by un-derivatized liquid chromatography-tandem mass spectrometry. Ann Clin Biochem 45(Pt 6): 575–584
23. van de Kamp JM, Mancini GM, Pouwels PJ, Betsalel OT, van Dooren SJ, de Koning I et al (2011) Clinical features and X-inactivation in females heterozygous for creatine transporter defect. Clin Genet 79(3):264–272
24. Mørkrid L, Rowe AD, Elgstoen KB, Olesen JH, Ruijter G, Hall PL et al (2015) Continuous age- and sex-adjusted reference intervals of urinary markers for cerebral creatine deficiency syndromes: a novel approach to the definition of reference intervals. Clin Chem 61(5):760–768
25. Carducci C, Birarelli M, Leuzzi V, Carducci C, Battini R, Cioni G et al (2002) Guanidinoacetate and creatine plus creatinine assessment in physiologic fluids: an effective diagnostic tool for the biochemical diagnosis of arginine:glycine amidinotransferase and guanidinoacetate methyltransferase deficiencies. Clin Chem 48(10):1772–1778
26. Berends LM, Struys EA, Roos B, Holwerda U, Jansen EEW, Salomons GS et al (2017) Guanidinoacetate methyltransferase activity in lymphocytes, for a fast diagnosis. JIMD Rep 37:13–17
27. Verhoeven NM, Roos B, Struys EA, Salomons GS, van der Knaap MS, Jakobs C (2004) Enzyme assay for diagnosis of guanidinoacetate methyltransferase deficiency. Clin Chem 50(2):441–443
28. Ilas J, Mühl A, Stöckler-Ipsiroglu S (2000) Guanidinoacetate methyltransferase (GAMT) deficiency: non-invasive enzymatic diagnosis of a newly recognized inborn error of metabolism. Clin Chim Acta 290(2):179–188
29. Verhoeven NM, Schor DS, Roos B, Battini R, Stöckler-Ipsiroglu S, Salomons GS et al (2003) Diagnostic enzyme assay that uses stable-isotope-labeled substrates to detect L-arginine:glycine amidinotransferase deficiency. Clin Chem 49(5):803–805
30. Item CB, Stöckler-Ipsiroglu S, Stromberger C, Mühl A, Alessandrì MG, Bianchi MC et al (2001) Arginine:glycine amidinotransferase deficiency: the third inborn error of creatine metabolism in humans. Am J Hum Genet 69(5):1127–1133
31. Salomons GS, van Dooren SJ, Verhoeven NM, Cecil KM, Ball WS, Degrauw TJ et al (2001) X-linked creatine-transporter gene (SLC6A8) defect: a new creatine-deficiency syndrome. Am J Hum Genet 68(6):1497–1500
32. Pitt JJ, Tzanakos N, Nguyen T (2014) Newborn screening for guanidinoacetate methyl transferase deficiency. Mol Genet Metab 111(3):303–304
33. Sinclair GB, van Karnebeek CDM, Ester M, Boyd F, Nelson T, Stockler-Ipsiroglu S et al (2016) A three-tier algorithm for guanidinoacetate methyltransferase (GAMT) deficiency newborn screening. Mol Genet Metab 118(3):173–177
34. Mercimek-Mahmutoglu S, Sinclair G, van Dooren SJ, Kanhai W, Ashcraft P, Michel OJ et al (2012) Guanidinoacetate methyltransferase deficiency: first steps to newborn screening for a treatable neurometabolic disease. Mol Genet Metab 107(3):433–437
35. Mercimek-Mahmutoglu S, Pop A, Kanhai W, Fernandez Ojeda M, Holwerda U, Smith D et al (2016) A pilot study to estimate incidence of guanidinoacetate methyltransferase deficiency in newborns by direct sequencing of the GAMT gene. Gene 575(1):127–131
36. Pasquali M, Schwarz E, Jensen M, Yuzyuk T, DeBiase I, Randall H et al (2014) Feasibility of newborn screening for guanidinoacetate methyltransferase (GAMT) deficiency. J Inherit Metab Dis 37(2):231–236
37. Hart K, Rohrwasser A, Wallis H, Golsan H, Shao J, Anderson T et al (2021) Prospective identification by neonatal screening of patients with guanidinoacetate methyltransferase deficiency. Mol Genet Metab 134(1–2):60–64

Trimethylamine, Trimethylamine-N-oxide, N,N-Dimethylglycine and Dimethylated Sulfur Compounds

30

Shayne Mason and Marli Dercksen

Contents

Abbreviations

C	Degrees Celsius
^{1}H-NMR	Proton Nuclear Magnetic Resonance
cDNA	Complementary DNA
CH_2	Methylene
CH_3	Methyl
CSF	Cerebrospinal Fluid
DNA	Deoxyribonucleic Acid
DMG	Dimethylglycine
DMGU	Dimethylglycinuria
DMS	Dimethylsulfide
DMSO	Dimethylsulfoxide
$DMSO_2$	Dimethylsulfone
FMO3	Flavin-containing Monooxygenase 3
g	Gram
H_2O_2	Hydrogen Peroxide
H_2S	Hydrogen Sulfide
Hz	Hertz
IEM	Inborn Error of Metabolism
MAT	Methionine Adenosyltransferase
M	Molar

S. Mason (✉) · M. Dercksen
Biochemistry Department, Faculty of Natural and Agricultural Sciences, North-West University, Potchefstroom, South Africa
e-mail: marli.dercksen@nwu.ac.za

N. Blau, F. M. Vaz (eds.), *Laboratory Guide to the Methods in Biochemical Genetics*,
https://doi.org/10.1007/978-3-031-58819-8_30

mg	Milligram
mmol/L	Millimole Per Liter
MT	Methanethiol
MTO	Methanethiol Oxidase
OL	Oligosaccharide Lactulose
SAH	S-Adenosylhomocysteine
SAM	S-Adenosylmethionine
TMA	Trimethylamine
TMAO	Trimethylamine-N-Oxide
TMAU	Trimethylaminuria
TSP	Trimethylsilylpropanoic Acid
μL	Microliter
μmol/L	Micromole Per Liter
μs	Microsecond
USA	United States of America

30.1 Introduction

Various organic acidurias have been reported with unique body odors pending on the accumulating metabolites due to a specific enzyme deficiency. This chapter addresses inborn errors of metabolism (IEMs) associated with malodor with no other pathophysiological complications typically observed in other organic acidemias, such as isovaleric acidemia. Table 30.1 summarizes the conditions discussed in this chapter. Serendipitously, all these malodor disorders fall within the choline metabolism pathway, coupled with the methionine-homocysteine cycle, as illustrated in Fig. 30.1. Trimethylaminuria (TMAU) is certainly the most well-known inherited malodor syndrome known as "fishy odor" syndrome, whereby accumulating trimethylamine from dietary choline is unable/incompletely detoxified to trimethylamine-N-oxide. Renewed interest in unexplained causes of atypical malodor disorders has spiked, as evidenced by the recent description of methanethiol oxidase (MTO) deficiency due to phenotype-causing variants in the *SELENBP1* gene [1]. Additional metabolic causes for abnormal body fluid disorders include dimethylglycinuria (DMGU) and methionine adenosyltransferase (MAT) deficiency. The biochemical diagnosis of these disorders by the versatile diagnostic tool—proton nuclear magnetic resonance (^{1}H-NMR) spectroscopy—is showcased in this chapter.

Table 30.1 Biochemical summary of trimethylaminuria (TMAU), dimethylglycinuria (DMGU), methanethiol oxidase (MTO) deficiency, and methionine adenosyltransferase (MAT) deficiency

Name of disorder	Malodor	Altered protein	Gene	Diagnostic matrix	Metabolites of interest	OMIM
TMAU	Fish	FMO3	*FMO3*	Urine, serum	TMA, TMAO	602079
DMGU	Fish/ Cabbage	DMGDH	*DMGDH*	Urine, serum	DMG	605850
MTO deficiency	Cabbage	MTO	*SELENBP1*	Breath, urine, serum, CSF	H_2S, MT, DMS, DMSO, $DMSO_2$	604188
MAT deficiency	Cabbage	MAT I/III	*MAT1A*	Urine, plasma, CSF	Methionine, DMS, $DMSO_2$	250850

CSF = cerebrospinal fluid; FMO3 = flavin monooxygenase 3; TMA = trimethylamine; TMAO = trimethylamine-N-oxide; DMG = dimethylglycine; H_2S = hydrogen sulfide; MT = methanethiol; DMS = dimethylsulfide; DMSO = dimethylsulfoxide; $DMSO_2$ = dimethylsulfone

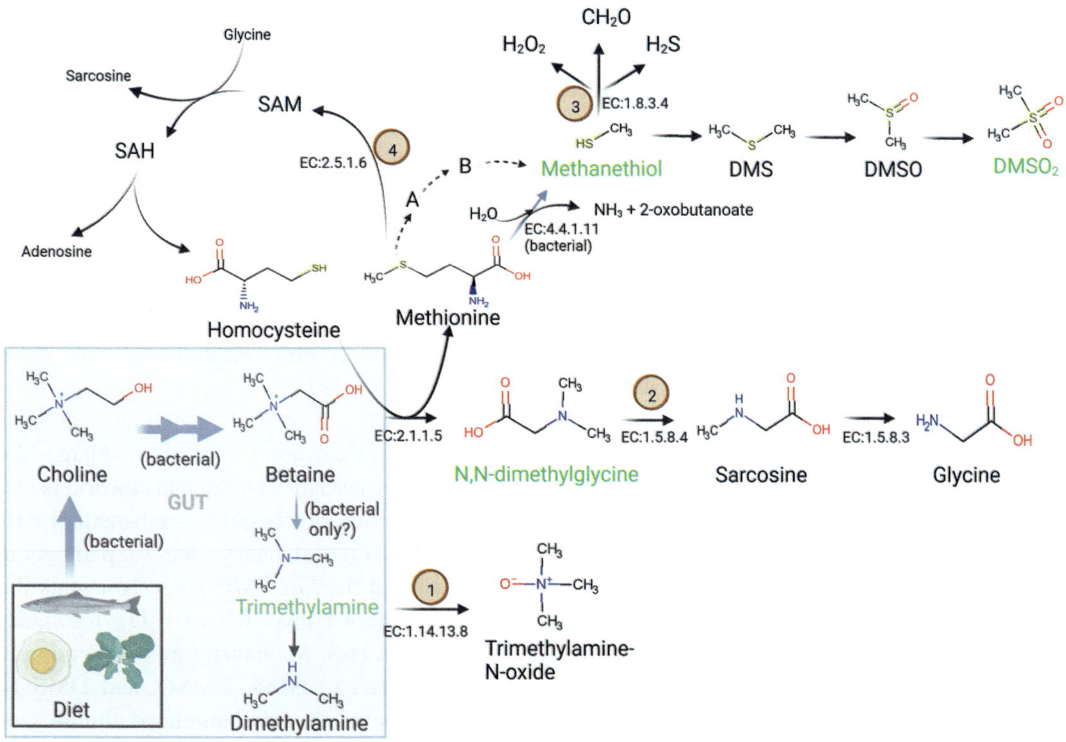

Fig. 30.1 Metabolic pathways indicating points of reported deficiencies resulting in malodor-related IEMs [① = trimethylaminuria (TMAU); ② = Dimethylglycinuria (DMGU); ③ methanethiol oxidase (MTO) deficiency; ④ = methionine adenosyltransferase (MAT) deficiency], and their increased volatile metabolites (highlighted in green). A secondary transamination pathway of methionine to methanethiol in methionine adenosyltransferase deficient individuals is indicated via dotted lines, with compound A (4-methylthio-2-oxobutyric acid) and compound B (3-methylthiopropionic acid). DMS = dimethylsulfide; DMSO = dimethylsulfoxide; $DMSO_2$ = dimethylsulfone; SAM = S-adenosylmethionine; SAH = S-adenosylhomocysteine. EC:1.14.13.8 = flavin-containing monooxygenase. EC:2.1.1.5 = betaine-homocysteine methyltransferase; EC:1.5.8.4 = dimethylglycine dehydrogenase; EC:1.5.8.3 = sarcosine dehydrogenase; EC:4.4.1.11 = methionine lyase (bacterial) EC:1.8.3.4 = methanethiol oxidase; EC:2.5.1.6 = methionine adenosyltransferase. NOTE: greyed box and arrows indicate gut bacterial pathways. Conversion of dietary intake to choline, and then to betaine, occurs primarily by gut microbiota; however, human choline dehydrogenase has been identified. The enzyme methionine lyase is also known to be bacterial. To our knowledge, there is no evidence of a human enzyme converting betaine to trimethylamine

30.2 Clinical Relevance

1. *Trimethylaminuria (TMAU)*

   Flavin-containing monooxygenase 3 (FMO3) deficiency results in trimethylaminuria (TMAU, also known as fish odor disease) and has been described as early as 1000–1400 BC in the Indian epic of the Bharata Dynasty [2, 3]. The disorder was clinically and biochemically characterized by Humbert et al. in 1970 [2]. The condition presents with body odor resembling that of decomposing/rotten fish due to excess excretion of trimethylamine (TMA) in the breath/saliva, sweat, urine, and reproductive fluids. According to the clinical gene card for TMAU, an estimated incidence of heterozygous carriers for severe/chronic TMAU, due to loss-of-function variants in the *FMO3* gene, is in the range of 0.5–1% with a frequency rate of 1 in 40,000 [4]. A large study in the USA revealed that one-third of a cohort with idiopathic malodor production tested positive for inherited TMAU on a metabolite level, leaving the remaining cohort with unexplained etiology [5].

Disease classification of TMAU includes: (1) primary trimethylaminuria due to pathogenic variants (>40) resulting in no residual FMO3 activity [6–8], (2) acquired TMAU, after a viral (hepatitis) infection, (3) transient childhood TMAU possibly due to enzyme immaturity (due to a delayed expression of the *FMO3* gene) which typically resolves with time, (4) episodic TMAU associated with menstruation (hormonal changes), and (5) precursor overload TMAU as a result of high choline and L-carnitine intake. Other researchers have confirmed that liver disease and gut dysbiosis, including helicobacter pylori infection, may also play a role in secondary TMAU [9, 10]. Less severe/intermittent/acquired forms (2–5) have been extensively researched, and a combination of missense or polymorphic variants in the *FMO3* gene and non-genetic factors (as mentioned above) may explain the less severe phenotype [11, 12]. Although debilitating physical symptoms are not observed in this condition, psychological consequences have been emphasized in various publications, motivating the need to test for TMAU [13–16].

Treatment strategies have been researched extensively and recently reviewed by Schmidt and Leroux [14]. Dietary intervention by limiting TMAU precursors, neutralizing the excreted TMAU by washing with low pH soap, as well as targeting gut metabolism via the use of antibiotics (metronidazole, neomycin, and rafaximin) and probiotics have demonstrated odor reduction in some patients. Although the use of sequestering agents such as activated charcoal and oligosaccharide lactulose (OL) reduced abnormal body odor, treatment with OL is non-specific and has resulted in diarrhea and subsequent dehydration [14].

2. *Methanethiol Oxidase (MTO) Deficiency*

The cause of blood-borne extraoral halitosis (10% of total halitosis cases reported) as a result of volatile sulfur compounds (mostly dimethylsulfide) may be due to systemic disease, a metabolic disorder such as hypermethioninemia or TMAU, or chronic consumption of food such as garlic, or medication intake [17]. An estimated prevalence of about 0.5–3% in the general population may suffer from blood-borne extraoral halitosis, and treatment strategies overlap with that of TMAU. In addition, a methionine-limiting diet, psychological therapy, and promoting oral hygiene are recommended in these patients [18, 19].

Recently, a novel metabolic disorder resulting in extraoral halitosis (cabbage-like odor) was described by Pol and coworkers [1]. As illustrated in Fig. 30.1, methanethiol oxidase (MTO) deficiency due to pathogenic variants in the *SELENBP1* gene prevents the methanethiol conversion to H_2O_2, formaldehyde, and H_2S. Methanethiol is subsequently metabolized to DMS, DMSO, and $DMSO_2$. Five patients in three unrelated families of Turkish and Portuguese descent (affected children in these two families had consanguineous parents) and as well as Dutch origin (no consanguinity reported). While the patients of the former two families did have additional clinical features (neurological involvement), the authors speculated that added genetic factors may be involved and, indeed, pathogenic variants in the *THAP4* gene were identified in one of the Turkish patients. The Dutch patients lacked any other clinical features and only presented with abnormal body odor.

Examining the breath and bodily fluids of these patients revealed an increase in DMS in the breath of the patients and DMSO and $DMSO_2$ in the blood, urine, and CSF. The correlation of methanethiol oxidase and reduced SELENBP1 protein expression was confirmed in fibroblast culture from two patients in separate families (Turkish and Dutch) mentioned [1]. In addition, reduced enzyme activity was noted in the erythrocytes in both patients. Interestingly, low expression of SELENBP1 has been observed in various cancer types and deficient methanethiol oxidase resulting in reduced H_2O_2 and H_2S may be relevant in cancer research [1, 20].

Evaluation of 15,000 exomes for biallelic pathogenic variants in the *SELENBP1* gene showed a frequency of approximately 1:90,000, translating to a carrier frequency of 1 in 300 individuals [1]. Expanded genome-wide association studies are required to conclusively determine the allele frequency worldwide. No additional case reports could be found about this disorder since it was first described in 2018.

3. *Dimethylglycinuria (DMGU)*

   Only one case of dimethylglycinuria due to deficient dimethylglycine dehydrogenase, a flavin-containing enzyme that requires folate as a cofactor, has been documented after a thorough review of the literature [21]. The clinical features on the one reported case included fish-/cabbage-like odor (likely due to increased dimethylglycine and methionine, respectively), chronic muscle fatigue, and increased serum creatine kinase. Sequencing of the *DMGDH* gene revealed a homozygous A → G substitution (326 A → G) in both the cDNA and genomic DNA of the patient. This mutation changed a His to an Arg (H109R). The result was decreased enzyme expression and inactivation of dimethylglycine dehydrogenase [22]. The current consensus is that dimethylglycinuria may have been an incidental finding of questionable clinical significance and not necessarily associated with the additional clinical features observed in this isolated case [23]. The conditions should be considered if the origin of the abnormal odor cannot be determined.

4. Methionine adenosyltransferase (MAT) I/III deficiency

   MAT I and III are two isoenzymes (first reaction in the transsulfuration pathway) that convert methionine to S-adenosylmethionine. The first case of deficient MAT resulting in hypermethioninemia in the liver of an infant was described in 1974 [24]. Thereafter, several cases with variable phenotypic features (asymptomatic to neurological involvement) have been documented. Typical methionine levels reported for MAT I/III deficiency range between 206 and 1394 μmol/L (normal plasma range < 50 μmol/L) [25]. The disorder may have both an autosomal dominant inheritance pattern (typically asymptomatic patients) or autosomal recessive inheritance pattern which most frequently accounts for clinical phenotype (cabbage-like odor with neurological component) [26]. The clinical significance of this enzyme deficiency remains questionable as many of these patients (often identified via newborn screening programs) present with no deleterious clinical features, except for a cabbage-like odor. A large cohort of 30 patients studied in the 1990s revealed that the majority of the patients (27) had no detrimental neurological deficits, but three patients did show neurological features including convulsions/episodic hypotonia/dystonia, as well as noted demyelination on the MRI [27]. A more recent study investigated 64 patients, with half of the group presenting with some form of neurological involvement (characterized by an abnormal brain MRI). It was suggested that a persistent methionine of above 800 μmol/L mostly presented with a neurological fallout [29]. A personalized approach consisting of regular clinical follow-ups in patients diagnosed via newborn screening is subsequently recommended due to the broad clinical consequence. Treatment of symptomatic patients with a methionine-limiting diet and S-adenosylmethionine showed clinical improvement [28].

   As a result of MAT I/III deficiency, accumulating methionine concentrations are redirected to the production of DMS to DMSO to $DMSO_2$, as illustrated in Fig. 30.1. Breath studies confirmed the increased presence of DMS in MATI/II-deficient patients [30–32]. Engelke et al., confirmed a raised $DMSO_2$ in plasma of patients with MAT I/III deficiency. This was however in patients with a plasma methionine concentration of greater than 300 μmol/L [33]. DMSO concentrations have not been reported for patients with the related deficiency. Most asymptomatic patients with MAT I/III deficiency present with some

Table 30.2 Chemical information and shifts for metabolites identified in ^{1}H-NMR spectra of urine samples with their associated IEMs. Multiplicity: s—singlet

Metabolite name	Chemical structure	Chemical formula	δ (ppm), multiplicity	Chemical group	# protons
Trimethylamine	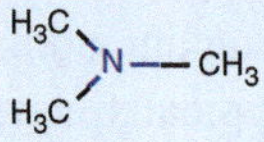	C_3H_9N	2.89 s	$(CH_3)_3$	9
Trimethylamine-N-oxide	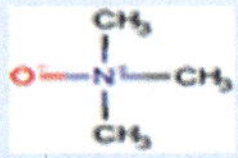	C_3H_9NO	3.26 s	$(CH_3)_3$	9
Dimethylamine	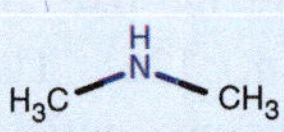	C_2H_7N	2.73 s	$(CH_3)_2$	6
N,N-Dimethylglycine	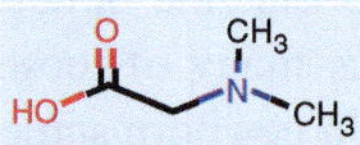	$C_4H_9NO_2$	2.93 s 3.72 s	$(CH_3)_2$ CH_2	6 2
Dimethylsulfone	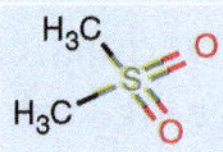	$C_2H_6O_2S$	3.14 s	$(CH_3)_2$	6
Dimethylsulfoxide	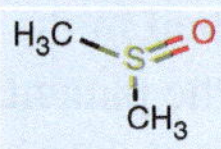	C_2H_6OS	3.13 s	$(CH_3)_2$	6
Dimethylsulfide		C_2H_6S	2.10 s	$(CH_3)_2$	6
Creatinine	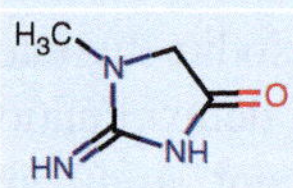	$C_4H_7N_3O$	4.06 s 3.05 s	CH_2 CH_3	2 3

degree of extraoral halitosis and cabbage-like body odor (persistent/transient) without neurological features. A methionine-limiting diet, together with the therapeutic options applied in TMAU, may provide relief [18].

30.3 Methods

The sample preparation method for the analysis of urine on ^{1}H-NMR is highly robust, repeatable, and reproducible. This is because there are very few steps that incorporate almost negligible analytical variation from an experienced analyst. Loading pure compounds of interest (TMA, TMAO, DMA, DMG, DMO, DMSO, $DMSO_2$) onto the ^{1}H-NMR to create spectral libraries allows for confident identification - ^{1}H-NMR is associated with high specificity. The major downside to ^{1}H-NMR, when compared to its mass spectrometry counterparts, is low sensitivity—a lower level of quantification of approx. 5 μmol/L. However, this limit in sensitivity is negated by the fact that the metabolites associated with malodor syndromes (shown in green in Fig. 30.1) arise from IEMs and thus occur in high abundance. Furthermore, the metabolites of interest here (Table 30.2) all have methyl protons that produce a sharp single ^{1}H-NMR peak, allowing for accurate quantification. Hence, ^{1}H-NMR spectroscopy is the ideal analytical platform for diagnosing malodor syndromes.

30.4 Pre-analytical

30.4.1 Preparation Before and After Sample Collection, and Sample Stability

For severe cases of malodor disease, analysis of random biological specimens will provide a biochemical confirmation of the disorder. For TMAU, a random urine sample (acidified after collection) is required for the detection of TMA and TMAO. Stability has been confirmed at −20 °C for 3 months [34]. Compound stability in various biological fluids for DMSO, $DMSO_2$, and dimethylglycine has not been completely addressed in literature.

A loading test is recommended for less severe/intermittent TMAU cases, where etiology is highly suggestive of genetic malodor condition, but baseline TMA concentrations are normal. The TMA and TMAO excretions are measured and the FMO3 oxidation capacity is calculated over a certain period (4–36 h, pending on loading protocol) to determine if the patient is affected. Pol et al. (2018) also showed the benefits of a methionine loading trail to confirm the involvement of MTO [1]. The documentation of odor index before and after a loading protocol has been encouraged in literature as this may assist in the differential diagnosis of a patient with TMAU or any other form of malodor condition [5]. The following TMAU loading protocols have been described in literature:

(a) Historically, and still at some centers today, choline (5–15 mg) loading tests are utilized as a diagnostic tool by measurement of TMA and TMAO before and after choline consumption; and subsequent determination of FMO3 oxidation capacity [35–39]. Human choline dehydrogenase [40], as well as the microbiome in the gut, may substantially differ in individuals and influence the formation of TMA and subsequent enzymatic conversion to TMAO. The possibility of inconsistent upstream contribution to TMA and subsequent TMAO production limits the positive predictive value of this approach [41].
(b) The Nijmegen protocol consists of eating a marine fish meal (300 g) to ensure adequate substrate (TMA) for conversion to TMAO. Subsequent collection of urine over a 6-hour period after the meal or sampling at 2–12 h after the meal has provided some patients with a positive diagnosis [34].
(c) Direct loading with TMA (600 mg orally) was found to be effective in the diagnosis of severe and mild cases of TMAU. This approach assures sufficient substrate to maximize flux through the oxidizing system [39]. TMA and TMAO are measured before and after (for 4–6 h) to determine the FMO3 oxidation capacity. A reduced dose of 300 mg was found to be too low while a higher dose of 900 mg saturated the N-oxidation system [16].

It is, however, important to note that some loading tests may fail or be borderline abnormal as adequate excretion of metabolic products is not always optimal. The reason for this may be the composition of the gut microbiome or due to delayed gastric emptying. For this reason, genetic testing is recommended to determine if a strong suspicion of primary TMAU is present/absent [16].

30.5 Reagents and Chemicals

A ^{1}H-NMR buffer solution is created using 1.5 M potassium dihydrogen phosphate (Sigma, CAS #: 7778-77-0) with 5.058 mM internal standard trimethylsilylpropanoic acid (TSP; Merck, CAS #: 5683-30-7) and 0.1% sodium azide (Sigma, CAS #: 26628-22-8) in deuterium oxide (Merck, CAS #: 7789-20-0). The final pH of the ^{1}H-NMR buffer solution is adjusted to precisely 7.40 using potassium hydroxide (Sigma, CAS #: 1310-58-3).

30.6 Instrumentation

A Bruker Avance III HD spectrometer with a 500-MHz magnetic field and a triple resonance probe head (optimized for ^{1}H observations) is used to analyze all urine samples at 300 K.

30.7 Analytical Procedure

The pH of each 1 mL aliquot of sample is adjusted to pH 7.00 ± 0.05 prior to sample preparation. Urine samples (approved by an ethics committee) are used here to represent various malodor disorders. Each 1 mL aliquot of urine sample is centrifuged at 12,000 *g* for 5 min and 540 μL supernatant transferred to a microcentrifuge tube containing 60 μL ^{1}H-NMR buffer solution. The sample is mixed by vortexing for 5 s and then centrifuged at 12000 *g* for 5 min. A 540-μL volume of supernatant is transferred to a 5-mm glass NMR tube, and the tube is capped. Each sample is loaded onto the NMR, the signal locked onto the deuterium signal, probe tuned and matched, magnet shimmed (magnetic field aligned), and radio frequency pulse calibrated. The following experimental parameters are used: spectral width = 12,000 Hz, acquisition time = 2.72 s, receiver gain = 64, and a NOESY pulse program suppresses the water signal at 4.72 ppm. A total of 128 scans are done via a 90° excitation pulse of 8 μs and a 4 s delay. Total run time per sample is 15 min. Fourier transformation converts the recorded free induction decay from a time domain to a frequency domain to generate a ^{1}H-NMR spectrum. Automatic processing calibrates the TSP peak to 0.00 ppm and performs phase and baseline corrections. All processing of NMR data is done using Bruker Topspin (V 4.6.5). Post-analytical processing (identifying metabolites of interest using pure compound spectral libraries, and quantification) is done using Bruker AMIX.

30.8 Post-analytical

Examples of ^{1}H-NMR spectra indicating malodor disorders are illustrated in Figs. 30.2, 30.3, 30.4, 30.5, and 30.6. The chemical shifts and chemical information of the corresponding metabolic markers of these IEMs are given in Table 30.2.

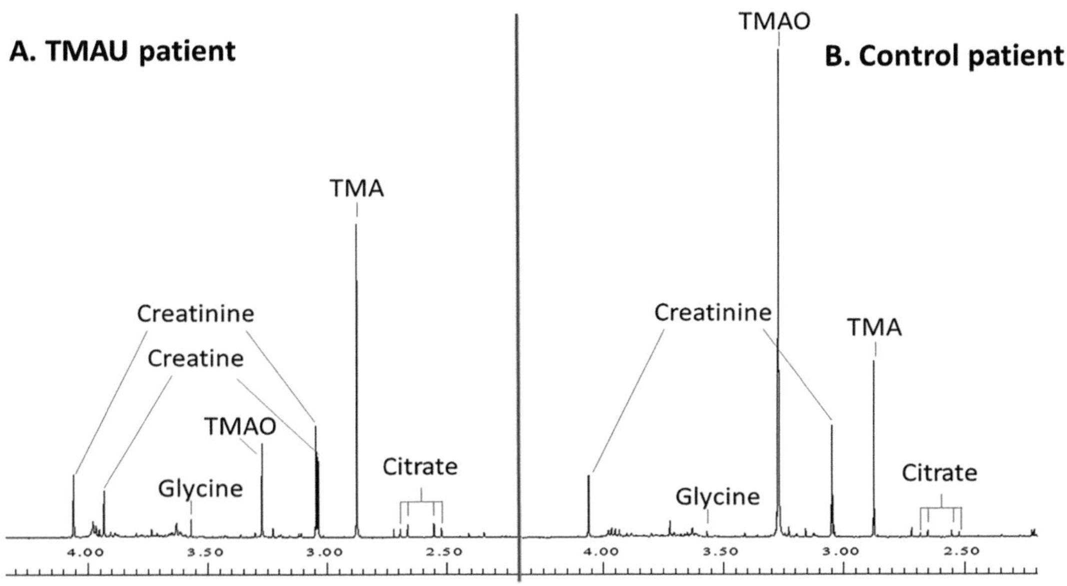

Fig. 30.2 ^{1}H-NMR representative spectra of urine from a TMAU patient (**a**; left) with a severe mutation, 2 h after TMA loading, and a control patient (**b**; right) 2 h after TMA loading. TMA is oxidized to TMAO under normal conditions, but for TMAU the TMA remains elevated and very little TMAO is formed

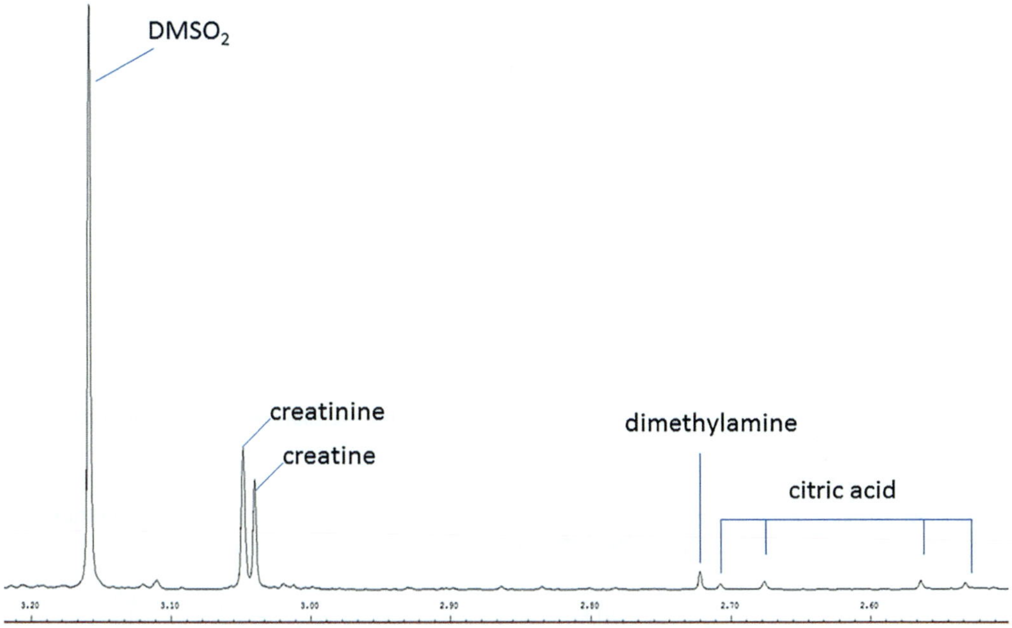

Fig. 30.3 ^{1}H-NMR representative spectrum of urine of a patient with highly elevated $DMSO_2$, likely due to reported supplementation

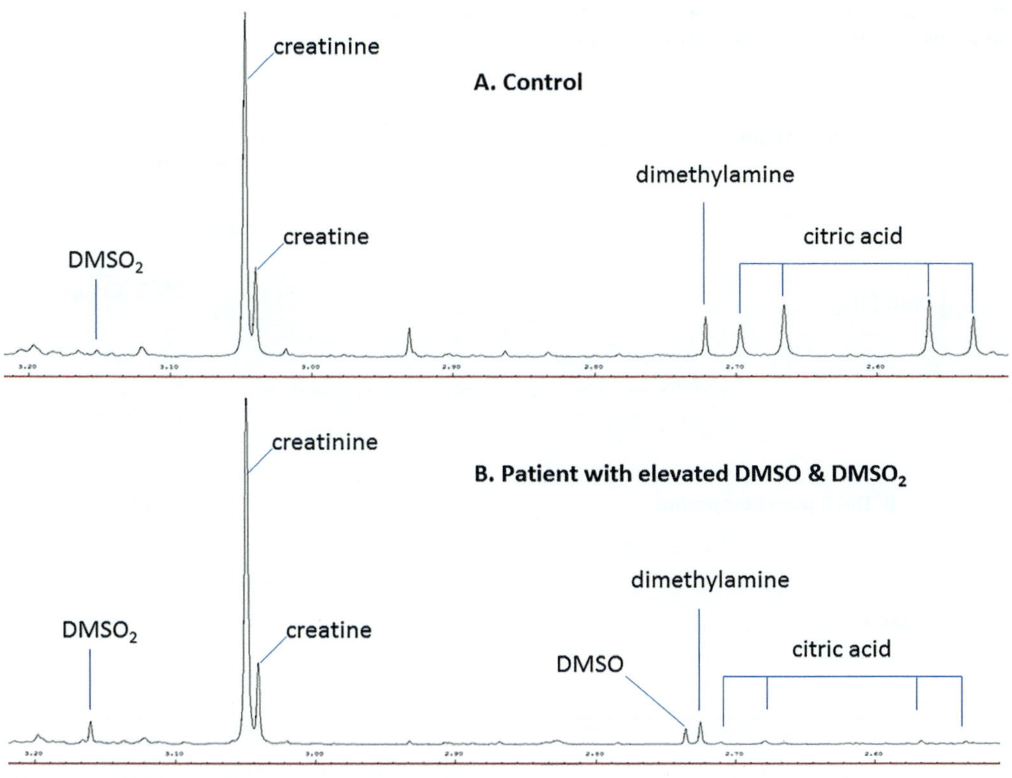

Fig. 30.4 ^{1}H-NMR representative spectra of urine from a control case (**a**; above), and a patient with elevated $DMSO_2$ and the presence of DMS (**b**; below)—likely a case of MAT I/III deficiency or MTO deficiency but requires genetic confirmation

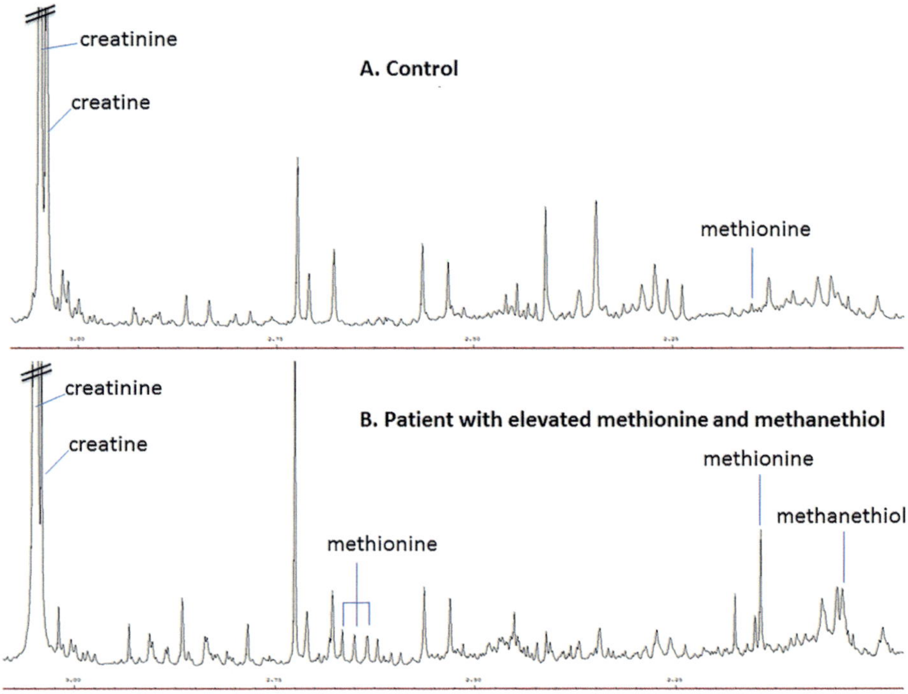

Fig. 30.5 ^{1}H-NMR representative spectra of urine from a control case (**a**; above), and a patient with elevated methionine and the presence of methanethiol (**b**; below)—likely a case of MAT I/III deficiency or MTO deficiency but requires genetic confirmation

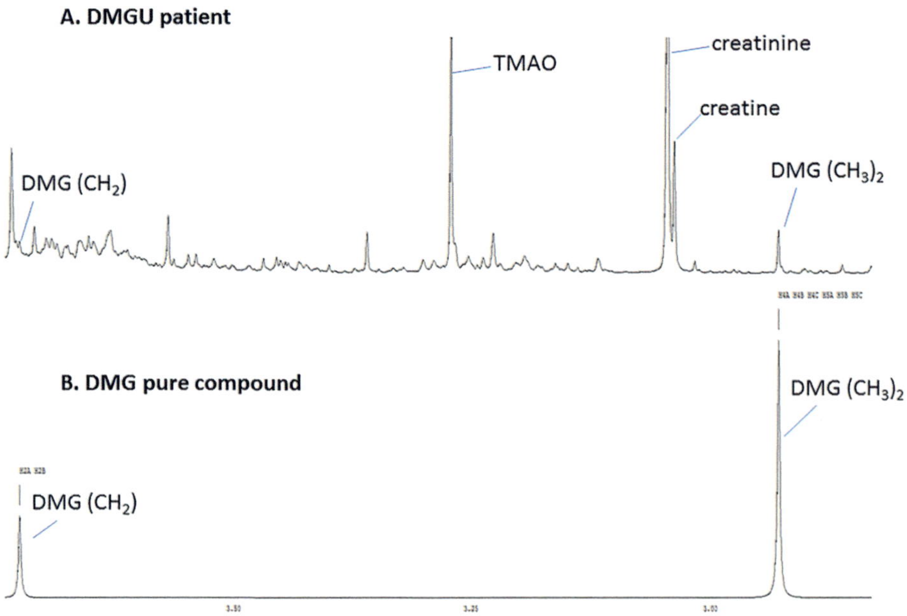

Fig. 30.6 ^{1}H-NMR representative spectrum of urine from a suspected DMGU patient (**a**; above), and a pure compound ^{1}H-NMR spectrum (**b**; below) indicating two singlets that represent the $(CH_3)_2$ and CH_2 moiety of DMG at 2.93 s and 3.72, respectively

30.9 Calculation

Bruker AMIX was used to calculate the integral (area under the peak) of each selected peak of interest. For metabolite quantification it is important to make a note of the number of protons each peak represents—see Table 30.2. All peak integral values are first divided by their respective number of protons to proton normalize all the integral values (i.e., removing the influence of the number of protons, so that all integrals are comparable). For urine samples, the integral of the peak of interest is divided by either the CH_2 or CH_3 moiety of creatinine and multiplied by 1000 to give a concentration of that metabolite as mmol/mol creatinine. For all other biofluids, divide the integral of the peak of interest by the TSP integral and multiply by the known amount of TSP in sample (0.5805 mmol/L in the protocol described here) to give absolute concentration (mmol/L; multiply by 1000 to give µmol/L).

Limited data, except for TMAU, are available on control reference ranges and pathological concentrations. Table 30.3 provides both although control groups remain small and do not reflect the true biological variation for all populations. The Human Metabolome Database (https://hmdb.ca) was also utilized as an additional source, although not indicated directly in Table 30.3.

30.10 Quality Control

External quality control schemes, such as ERNDIM (www.erndim.org), are not yet available for these specific metabolic disorders described in this chapter. Volatility of compounds under investigation, availability of specialized analytical resources over the world and unrecognition of disorders as clinically significant likely count against setting up an external quality control program. However, the NMR method described here has been validated for quantifying TMA and TMAO [43], and other methylamine metabolites [44] in urine.

30.11 Concluding Remark

We have shown in this chapter how ^{1}H-NMR spectroscopy can be utilized to identify and quantify metabolic markers of four malodor syndromes—TMAU (TMA and TMAO), DMGU (DMG), MTO deficiency (MT, DMSO $DMSO_2$), and MAT deficiency (methionine and $DMSO_2$). We have also described the clinical and biochemical background behind these IEMs and distinguished their various forms. However, this is not the full compendium of possible malodor syndromes in humans because various microbiota

Table 30.3 Reference ranges and pathological values from various referenced studies

Metabolite/measurement	Normal and disease	Serum*/plasma# (µmol/L)	CSF (µmol/L)	Urine (mmol/mol creat)
TMA (baseline)	Reference range [12]			18–20
	TMAU [34]			22–440
FMO3 metabolic capacity [16]	Reference			>92%
	TMAU severe			<84%
	TMAU mild			<92%
DMG	Reference range	1–5* [21]		1.6–10.4 [42]
	DMGU	221* (1 patient) [21]		441 and 457 (2 patients) [21]
$DMSO_2$	Reference range	1.5–16.1# [33]	5.1–17.5 [33]	1.3–49.0 [42]
	MAT deficiency	37.5–42.5# [33]		
	MTO deficiency [1]	ND–4# (5 patients)	41, 112 (2 patients)	28–60 (5 patients)
DMSO	Reference range [1]	<1	<1	<1
	MTO deficiency [1]	ND–6# (5 patients)	8 (2 patients)	4–31 (5 patients)

ND = Not detected, FMO3 metabolic capacity = TMAO/(TMAO+TMA) × 100, expressed as a percentage (%)

that inhabit our bodies are known to produce many different pungent chemicals [45]. Indeed, we are entering a (r)evolution in science whereby increased knowledge of our microbiota will yield insights into and of ourselves.

Acknowledgments Ron Wevers, Udo Engelke, and colleagues at the Radboud University Medical Centre, Nijmegen, The Netherlands are acknowledged for their contributions to the field of malodor syndromes. We dedicate this chapter to Dr. George Pretti (born: October 7, 1944; died: March 3, 2020 (age 75)), best known for his work on the chemistry of human body odors; from the non-profit Monell Chemical Senses Center in Philadelphia, USA.

References

1. Pol A et al (2018) Mutations in SELENBP1, encoding a novel human methanethiol oxidase, cause extraoral halitosis. Nat Genet 50(1):120–129
2. Humbert JA, Hammond KB, Hathaway WE (1970) Trimethylaminuria: the fish-odour syndrome. Lancet 2(7676):770–771
3. Mitchell SC (1996) The fish-odor syndrome. Perspect Biol Med 39(4):514–526
4. Shephard EA, Treacy EP, Phillips IR (2015) Clinical utility gene card for: trimethylaminuria - update 2014. Eur J Hum Genet 23(9). https://doi.org/10.1038/ejhg.2014.226
5. Wise PM, Eades J, Tjoa S, Fennessey PV, Preti G (2011) Individuals reporting idiopathic malodor production: demographics and incidence of trimethylaminuria. Am J Med 124(11):1058–1063
6. Hernandez D, Addou S, Lee D, Orengo C, Shephard EA, Phillips IR (2003) Trimethylaminuria and a human FMO3 mutation database. Hum Mutat 22(3):209–213
7. Shimizu M, Allerston CK, Shephard EA, Yamazaki H, Phillips IR (2014) Relationships between flavin-containing mono-oxygenase 3 (FMO3) genotype and trimethylaminuria phenotype in a Japanese population. Br J Clin Pharmacol 77(5):839–851
8. Phillips IR, Shephard EA (2020) Flavin-containing monooxygenase 3 (FMO3): genetic variants and their consequences for drug metabolism and disease. Xenobiotica 50(1):19–33
9. Yamazaki H, Fujieda M, Cashman JR, Kamataki T (2005) Mild trimethylaminuria observed in a Japanese cohort with liver damage. Am J Med 118(7):803–805
10. Mogilnicka I, Bogucki P, Ufnal M (2020) Microbiota and malodor-etiology and management. Int J Mol Sci 21(8). https://doi.org/10.3390/ijms21082886
11. Dolphin CT, Janmohamed A, Smith RL, Shephard EA, Phillips IR (1997) Missense mutation in flavin-containing mono-oxygenase 3 gene, FMO3, underlies fish-odour syndrome. Nat Genet 17(4):491–494
12. Mitchell SC, Smith RL (2001) Trimethylaminuria: the fish malodor syndrome. Drug Metab Dispos 29(4 Pt 2):517–521
13. Roddy D, McCarthy P, Nerney D, Mulligan-Rabbitt J, Smith E, Treacy EP (2021) Impact of trimethylaminuria on daily psychosocial functioning. JIMD Rep 57(1):67–75
14. Schmidt AC, Leroux J-C (2020) Treatments of trimethylaminuria: where we are and where we might be heading. Drug Discov Today 25(9):1710–1717
15. Khan SA, Shagufta K (2014) A rare case of fish odor syndrome presenting as depression. Indian J Psychiatry 56(2):185–187
16. Mackay RJ, McEntyre CJ, Henderson C, Lever M, George PM (2011) Trimethylaminuria: causes and diagnosis of a socially distressing condition. Clin Biochem Rev 32(1):33–43
17. Tangerman A, Winkel EG (2010) Extra-oral halitosis: an overview. J Breath Res 4(1):017003
18. Harvey-Woodworth CN (2013) Dimethylsulphidemia: the significance of dimethyl sulphide in extra-oral, blood borne halitosis. Br Dent J 214(7):E20
19. Izidoro C et al (2022) Revisiting standard and novel therapeutic approaches in halitosis: a review. Int J Environ Res Public Health 19(18). https://doi.org/10.3390/ijerph191811303
20. Elhodaky M, Diamond AM (2018) Selenium-binding protein 1 in human health and disease. Int J Mol Sci 19(11). https://doi.org/10.3390/ijms19113437
21. Moolenaar SH et al (1999) Defect in dimethylglycine dehydrogenase, a new inborn error of metabolism: NMR spectroscopy study. Clin Chem 45(4):459–464
22. Binzak BA et al (2001) Cloning of dimethylglycine dehydrogenase and a new human inborn error of metabolism, dimethylglycine dehydrogenase deficiency. Am J Hum Genet 68(4):839–847
23. Goodman SI (2022) Biochemical phenotypes of questionable clinical significance. Physician's Guide Diagnosis, Treatment, Follow-Up Inherited Metabolic Dis:1431–1447
24. Gaull GE, Tallan HH (1974) Methionine adenosyltransferase deficiency: new enzymatic defect associated with hypermethioninemia. Science 186(4158):59–60
25. Mudd SH (2011) Hypermethioninemias of genetic and non-genetic origin: a review. Am J Med Genet C Semin Med Genet 157C(1):3–32
26. Kim Y-M et al (2016) Determination of autosomal dominant or recessive methionine adenosyltransferase I/III deficiencies based on clinical and molecular studies. Mol Med 22:147–155
27. Mudd SH et al (1995) Isolated persistent hypermethioninemia. Am J Hum Genet 57(4):882–892
28. Surtees R, Leonard J, Austin S (1991) Association of demyelination with deficiency of cerebrospinal-fluid S-adenosylmethionine in inborn errors of methyltransfer pathway. Lancet 338(8782–8783):1550–1554

29. Chien Y-H et al (2015) Mudd's disease (MAT I/III deficiency): a survey of data for MAT1A homozygotes and compound heterozygotes. Orphanet J Rare Dis 10:99
30. Kaji H, Hisamura M, Saito N, Murao M (1981) Biochemical aspect of dimethyl sulfide breath test in the studies on methionine metabolism. Res Commun Chem Pathol Pharmacol 32(3):515–523
31. Kaji H et al (1979) Clinical application of breath analysis for dimethyl sulfide following ingestion of DL-methionine. Clin Chim Acta 93(3):377–380
32. Kaji H et al (1983) Nutritional aspect of methionine isomers studied by pulmonary exhalation of dimethyl sulfide and urinary excretion of alpha-keto-gamma-methiolbutyrate in humans. Jpn J Med 22(2): 106–111
33. Engelke UFH et al (2005) Dimethyl sulfone in human cerebrospinal fluid and blood plasma confirmed by one-dimensional (1)H and two-dimensional (1)H-(13) C NMR. NMR Biomed 18(5):331–336
34. Wevers RA, Engelke UFH Trimethylaminuria. Lab Guide Methods Biochem Genet:781–792. https://doi.org/10.1007/978-3-540-76698-8_34
35. Marks R, Greaves MW, Prottey C, Hartop PJ (1977) Trimethylaminuria: the use of choline as an aid to diagnosis. Br J Dermatol 96(4):399–402
36. Spellacy E, Watts RW, Goolamali SK (1980) Trimethylaminuria. J Inherit Metab Dis 2(4):85–88
37. Brewster MA, Schedewie H (1983) Trimethylaminuria. Ann Clin Lab Sci 13(1):20–24
38. Tjoa S, Fennessey P (1991) The identification of trimethylamine excess in man: quantitative analysis and biochemical origins. Anal Biochem 197(1):77–82
39. Al-Waiz M, Ayesh R, Mitchell SC, Idle JR, Smith RL (1989) Trimethylaminuria: the detection of carriers using a trimethylamine load test. J Inherit Metab Dis 12(1):80–85. https://doi.org/10.1007/bf01805534
40. Salvi F, Gadda G (2013) Human choline dehydrogenase: medical promises and biochemical challenges. Arch Biochem Biophys 537(2):243–252
41. Janeiro MH, Ramírez MJ, Milagro FI, Martínez JA, Solas M (2018) Implication of trimethylamine N-oxide (TMAO) in disease: potential biomarker or new therapeutic target. Nutrients 10(10). https://doi.org/10.3390/nu10101398
42. Bouatra S et al (2013) The human urine metabolome. PLoS One 8(9):e73076
43. Veyrat-Durebex C et al (2021) Quality consideration for the validation of urine TMA and TMAO measurement by nuclear magnetic resonance spectroscopy in fish odor syndrome. Anal Biochem 630:114330
44. Lee MB et al (2006) Validation of 1H NMR spectroscopy as an analytical tool for methylamine metabolites in urine. Clin Chim Acta 365(1–2):264–269
45. Mogilnicka I et al (2020) Microbiota and malodor—etiology and management. Int J Mol Sci 21(8):2886

mass spectrometers emerging in a specific pattern that may be matched to known compounds based on mass-to-charge (*m/z*), retention time, and fragment ion spectra. In contrast to isolated GC/MS/MS assays, metabolomic analyses usually involve the use of multiple "arms" or methods optimized for the capture of different targeted molecules. Protocols have been developed, for instance, for multi-arm assays based on hydrophilicity and charge [11, 13, 14], though theoretically, protocols may be designed to capture molecules in only a small group of metabolic pathways or only those of a particular class.

31.1.3.1 Mass Spectrometry-Based Metabolomics

MS-based metabolomic analyses are typically carried out on biofluids like plasma, cerebrospinal fluid (CSF), or urine though it is also possible to conduct analyses of dried blood spots and even tissue homogenates [15]. Protocols have, for example, been developed for the analysis of murine liver and brain samples through the use of dedicated homogenization and extraction steps [12]. Unless analyzed and processed immediately, samples must usually be quickly frozen and stored at −80 °C until it is time for processing. Thawed samples undergo a series of precipitation steps to remove proteins before the resultant extract is separated into aliquots and reconstituted with various solvent mixtures in an effort to isolate compounds based on charge and/or polarity. Aliquots can be created, for instance, to isolate positively charged, polar compounds by reconstitution with a solution of formic acid in water prior to placement on the appropriate column. Regardless of the class of compound being extracted, aliquots should always be spiked with an instrument performance standard to ensure consistency in injection steps, chromatography, and alignment for data processing [11].

31.1.3.2 Nuclear Magnetic Resonance Spectroscopy

The use of nuclear magnetic resonance spectroscopy (NMR) in metabolomics has grown exponentially in the past few decades—particularly in the areas of clinical research. NMR metabolomic data are generated through the use radio-frequency (RF) waves which generate an electromagnetic field around a biological sample. Variations in the electromagnetic field lead to differential energy absorption and re-emission resulting in a spectrum of emitted RF waves corresponding to each metabolite [16, 17]. Though less commonly used as a clinical assay, NMR metabolomics has the advantages of minimal sample preparation (though internal reference standards are still usually required), being able to easily measure metabolite concentrations and allowing for easier identification of compound structures. Both one- (1D-NMR) and two-dimensional (2D-NMR) spectra may be obtained for instance, with two-dimensional techniques incorporating protocols like correlation spectroscopy (COSY) or nuclear Overhauser effect spectroscopy (NOESY) to accurate identify specific chemical bonds and atomic orientations [18]. Both common, abundant nuclear isotopes like ^{1}H, ^{13}C, or ^{31}P as well as less common isotopes like ^{15}N, ^{19}F, and ^{17}O may be utilized [18]. Two-dimensional spectra may therefore be generated via homonuclear (e.g., ^{1}H-2D spectra) or heteronuclear (e.g., ^{1}H-^{13}C spectra) protocols with any number of theoretical combinations of isotopes being used depending on the specific metabolites being targeted [19].

NMR-based methods also predominantly examine biofluids like plasma, whole blood, and cerebrospinal fluids; however, the range of fluids that can be assayed is broad and may include saliva, bile, amniotic fluid, and exhaled breath condensate [17, 20, 21]. Tissue samples may also be examined as extracts or intact/en bloc, though this usually requires the use of high-resolution magic angle spinning (HRMAS) spectroscopy [22–24]. This technique involves the spinning of a sample at 54.7° (the so-called magic angle) relative to the applied magnetic field thereby reducing line-broadening effects and improving resolution. If not analyzed immediately, samples for NMR-metabolomics may be processed and stored in a similar fashion as those used for MS-based metabolomics.

31.1.4 Compound Identification and Molecule Libraries

Most MS-based metabolomic analyses will capture and measure compounds in the range of 50–1000 Daltons though the breadth of compounds in this range (both endogenous and exogenous) may make accurate identification challenging. This particular aspect of metabolomics has seen great advancements in the last few years with the improvements in machine learning and computational predictions [25–29]. Current methods usually involve the comparison of MS features (retention times, mass-to-charge ratios, and fragmentation patterns) to candidate molecules in publicly available databases like the Human Metabolome Database [30, 31], MassBank [32], Metabolomics Workbench [33], and PubChem [34] (Table 31.1). This approach, however, can prove limiting due to variability in spectral protocols, inconsistent library curation methodologies, or lack of knowledge surrounding fragmentation patterns for a large number of compounds. To circumvent this, predictive algorithms are being developed based on spectral features [39], the incorporation of collision-cross section measurements (e.g., differences in rotationally averaged collision area between a charged analyte and its neutral collision partner) [40] or network-based approaches wherein a particular analyte is annotated based on its expected relationship to other molecules in specific metabolic pathways [41].

The number of metabolites identified by NMR-based techniques has historically been viewed as more limited given difficulties with peak resolution (due largely to spectral overlapping) and a requirement for higher sample concentrations or quantities. The introduction of stronger magnetic fields and better analytical techniques, however, has begun to allow for less stringent sample requirements and an expansion in spectral libraries [42]. The use of protein extractions or coupling of NMR spectroscopy with an initial chromatography step can help to improve signal-to-noise and eliminate much of the interference observed in non-purified samples [43]. The complexity of NMR-metabolomic spectra has similarly necessitated the construction of various

Table 31.1 Public databases useful in the interpretation of metabolomic data

Database	Website URL	Organization	Information available
[1] **HMDB**—Human Metabolome Database	https://hmdb.ca/	The Metabolomics Innovation Centre (TMIC)	Comprehensive metabolomics resource
[2] **Metabolomics Workbench**	https://www.metabolomicsworkbench.org/	U.S. National Institutes of Health (NIH) Common Fund Metabolomics Program National Metabolomics Data Repository (NMDR)	National and international repository for metabolomics data and metadata; provides analysis tools and access to metabolite standards, protocols, tutorials, training, and more
[3] **MassBank**—High Quality Mass Spectral Database	https://massbank.eu/MassBank/	MassBank Consortium	Open-source mass spectral library for the identification of small chemical molecules of metabolomics, exposomics, and environmental relevance
[4] **PubChem**	https://pubchem.ncbi.nlm.nih.gov/	U.S. National Institutes of Health (NIH)	Comprehensive open chemistry database
[5] **KEGG**—Kyoto Encyclopedia of Genes and Genomes	https://www.genome.jp/kegg/pathway.html	Kyoto University Bioinformatics Center, Japan	Cell biological and metabolic pathway mapping that integrates genomic, chemical, and systemic functional information
[6] **Cytoscape**	https://cytoscape.org/	Cytoscape Consortium	Open-source software platform for visualizing molecular interaction networks and biological pathways and integrating these networks with annotations, gene expression profiles, and other state data
[7,8] **IEMbase**—Inborn Errors of Metabolism Knowledgebase	http://www.iembase.org/index.asp	IEMbase©	Comprehensive, curated hierarchical database of inborn errors of metabolism

[1][30]; [2][33]; [3][32]; [4][34]; [5][35]; [6][36]; [7][37]; [8][38]

spectral libraries as well as computational tools to allow for rapid and more accurate compound identification. Resources like ASICS [44], BATMAN [45, 46], or BAYESIL [47], for example, have been developed for more automated identification and quantitation of experimental plasma/blood spectra though additional resources continue to be developed for newer protocols based on urine and other fluids/tissues [48].

31.2 Untargeted Metabolomics in Clinical Practice

31.2.1 Data Handling and Approaches to Statistical Analysis

While both MS- and NMR-based metabolomic analyses have found utility in various research protocols, only MS-based protocols have been made broadly available in clinical practice. Currently, MS-metabolomic data may be generated on a sample-by-sample basis (N of 1 analysis) or as part of larger case-control cohorts. While case-control studies have the advantage of utilizing a minimal number of internal standards for calibration and qualification, N of 1 analyses may require more complex data handling steps for interpretation. A large reference population is often critical for the assessment of metabolomic data for N of 1 metabolomic screening [11, 15, 49, 50]. A lab-specific reference population, for instance, may be created by performing untargeted metabolomics on samples received from healthy individuals with special consideration given to include samples representative of as many biochemical pathways as possible. Raw data for each metabolite may then be median-scaled, $\log_2$-transformed, extreme outliers removed, and population z-scores generated based on the mean and standard deviation within this reference sample dataset [11, 15]. The z-score is the number of standard deviations that a data point differs from the reference population mean, representing the relative level of a given metabolite. Metabolites identified in each patient sample are then considered elevated or reduced when compared to this reference population.

31.2.2 Interpretation of Untargeted Metabolomic Data

By definition, targeted MS-based protocols examine only a small subset of analytes thereby making the interpretation of data relatively straightforward. The quantity and complexity of untargeted, MS-based metabolomic data, however, can be challenging [15, 49, 50]. Because expert knowledge is often limited to data obtained from currently available clinical tests, however, this knowledge base both enhances and hampers the analysis of such complex data sets. Untargeted metabolomics goes well beyond a targeted analyte panel, revealing the complexity of metabolism and highlighting connections, intersections, and dependencies of metabolic pathways that may not have been previously known and/or have not been traditionally considered in clinical testing.

31.2.2.1 Manual Review of Data

Metabolic perturbations may include single analytes, multiple molecules in a seemingly linear metabolic pathway, or profiles of altered molecules that were not previously considered related [8, 51–56]. Metabolomic profiles may also highlight pathways that span metabolic processes between organelles, alternate metabolic pathways impacted by blocks in metabolism, or may reveal even broader impacts on cellular biochemistry and metabolic relationships that challenge current knowledge. Much of the initial review of data, therefore, will rely on an in-depth knowledge of human metabolism and well-honed expertise in pattern recognition. Facilitating this review process are numerous well-curated databases that, taken together, can support the interpretation of untargeted metabolomic profiles. Multiple public databases, including those in Table 31.1, provide detailed knowledge of the chemistry and biochemistry of specific molecules, biochemical pathways, and metabolic disorders, as well as offer previously published data that may be interrogated to support interpretation of novel findings.

31.2.2.2 Automation, Machine Learning, and Discovery

Given the time and specialized expertise needed for accurate manual review of data, there have been efforts made to automate the interpretation of untargeted MS-metabolomic data. Analysis and interpretation of well-described metabolic abnormalities and metabolomic profiles that prove to be consistent across unrelated individuals with the same metabolic disorder can be readily automated for diagnostic purposes [57, 58]. Multiple analysis programs have been developed in recent years, and these resources are available to support data analysis [59–62]. Many of these existing resources and approaches to analysis of metabolomic data are linked to the databases and analytical tools listed in Table 31.1.

Machine learning or deep learning is a subset of artificial intelligence utilizing mathematical algorithms that improve learning through experience [63]. Here, a statistical model is constructed and then "trained" on a set of clinical data to build a model and make predictions. Following this, the algorithm may then be applied to a "testing" dataset with further fine-tuning to improve output [64, 65]. Algorithms for the interpretation of MS-metabolomic data are usually built on one of several methods like principal component analysis (PCA) [66], partial least squares (PLS) [67], random forest (RF) [68], kernel support vector machine (SVM) [69], and artificial neural networks (ANNs) [70]. Each method has its own inherent pros and cons with the final choice of method utilized based on factors like the size of the dataset, clarity in terms of the phenotype being examined, and whether or not the training of the algorithm is supervised or unsupervised. Supervision of an algorithm involves attempts to predict and classify a predetermined attribute in the output whereby unsupervised analyses aim to identify patterns without a specific or predetermined output attribute in mind [71]. No matter how a model is built, however, its performance must be evaluated via various different methods to ensure appropriateness and accuracy [64, 65, 72].

Given that untargeted metabolomic profiling will very likely identify molecules that have not been previously described in clinical medicine or show associations with known molecules that have not yet been linked to metabolic disease, as well as drug metabolites and metabolites and pathways altered due to treatment of disease, discovery is at the heart of these clinical studies. Machine learning further offers opportunities to identify associations that we are not able to visually observe or link in the manual interpretation of metabolomic data due to the limits of our existing knowledge [73, 74]. Various studies have shown that training systems in the assessment of profiles obtained from individuals with known disorders have revealed new molecules and pathways altered either due to the metabolic disease or as a consequence of treatment [73–77]. Furthermore, the sensitivity of metabolomic profiling in combination with machine learning may confirm metabolic disease in cases that are subtle, such as was shown by Thistlethwaite et al. [73] in the diagnosis of an adult with PEX6 deficiency [73].

31.2.3 Correlating Metabolomic Data with Phenotypic Data

Untargeted metabolomic analysis generates large amounts of data on the metabolites present in a sample (Fig. 31.2). The metabolites detected are the product of the interaction between the individual, the individual's microbiome and the environment. Phenotype data, which refers to the observable characteristics or traits of an individual, is critical in the interpretation of untargeted

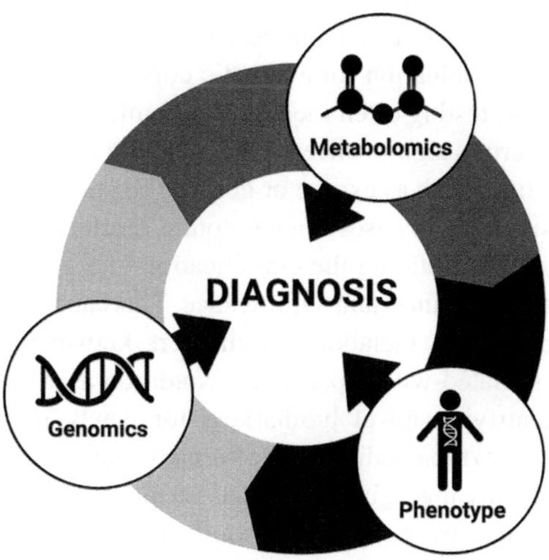

Fig. 31.2 Integration of untargeted metabolomics and genomics in clinical practice. Integrated analysis of metabolomic profiling with genomic findings and clinical phenotype leads to diagnosis (Created with BioRender.com)

metabolomic results. Phenotype data can include information about the physiological state, disease status, age of onset of the phenotype, sex, and other relevant factors that can influence metabolism. In addition, phenotype data can also be used to identify biomarkers that are associated with specific diseases or conditions. By correlating metabolomic data with phenotype data, researchers can identify novel diagnostic analytes or patterns associated with disease states. Overall, incorporating phenotype data into metabolomic analysis can provide valuable insights into the underlying biology of metabolic changes and help to identify potential targets for therapeutic intervention.

31.2.4 Correlating Metabolomic Data with Genomic Data

In the analytic hierarchy, metabolomic data is the most proximate to the phenotype or physiologic effect(s), whereas proteomics, RNA expression, and DNA sequence are respectively progressively more remote to the physiology or pathophysiology. Incorporating genomic and metabolomic data can provide valuable insights into biological systems and their functions. There are several ways in which genomic and metabolomic data can be integrated (Fig. 31.2). As genomic sequencing has become more widely available, this is often a first-line test for someone undergoing evaluation for a genetic condition. In such cases, testing often identifies genomic variants of uncertain significance, especially with untargeted testing, such as exome or genome sequence analysis. In such cases, metabolomic results may be useful in refining the classification of a genomic variant of uncertain significance. This can include analysis of metabolic biomarkers known to be associated with a particular condition as well as identifying novel biomarkers for newly identified or rare conditions. Conversely, metabolomic analysis may give equivocal results that can be clarified by genetic or genomic testing. This can be due to challenges of teasing apart the effects of the environment (including medications, supplements, and diet) and the individual's microbiome from the individual as well as challenges in distinguishing affected individuals from carriers of various autosomal recessive conditions. In such cases, genetic or genomic analysis may be helpful in understanding the origins or potential consequences of metabolic alterations identified on metabolic analysis. Therefore, integrating genomic and metabolomic data can provide a more complete view of biological systems and their functions, which can lead to new insights and applications in medicine.

31.2.5 Correlation of Metabolomic Data with Other Clinical Variables

31.2.5.1 Diet

Currently, most clinical metabolomic applications are focused on screening for inborn errors of metabolism; however, untargeted metabolomics is equally capable of identifying environmental or exogenous biochemicals. The most obvious application of this is in pharmacological research, but increasingly metabolic analysis has been employed to identify biomarkers that can reveal an individual's dietary intake. The benefit of employing metabolomics in population-based studies is to remove the reporting bias of dietary intake. Nutrimetabolomic profiles have been identified for a wide variety of foods/food groups which could streamline research into the interaction of diet and human physiology and the impact on disease in large population-based studies [78]. In the future, metabolomic analysis may be employed to make specific dietary recommendations, based on the current diet and interaction between the individual, their diet, and their microbiome.

31.2.5.2 Microbiome

The microbiome refers to the diverse communities of microorganisms that live on and in the human body, including bacteria, viruses, fungi, and other microbes. These microorganisms can have a profound impact on shaping an individual's metabolism and their metabolomic profile. For example, the microorganisms that compose the intestinal microbiome produce a wide range of metabolites, including fatty acids, amino acids, and neurotransmitters that can affect host metabolism. In addition, the microorganisms can influence the absorption and metabolism of nutrients from the diet, modulate immune function, and impact the regulation of various physiological processes. A dysbiotic microbiome, characterized by an imbalance of microbial populations, has been linked to various diseases, including obesity, type 2 diabetes mellitus, and inflammatory bowel diseases. For individuals with inborn errors of metabolism, we overall have a poor understanding of how the gastrointestinal microbiome may affect therapies commonly employed to treat diseases. It is likely that we significantly underestimate the effects of the microbiome on the therapies that are commonly employed [79].

31.2.5.3 Medications

In interpreting metabolic results for an individual, a list of their medications is important to include, along with the phenotype data. It is obvious that drugs employed in the treatment of common diseases, such as diabetes mellitus and hypercholesterolemia, will impact various metabolomic analytes related to glucose and lipid metabolisms. In addition, medications that alter the activity of an enzyme, such as methotrexate or allopurinol, will likely impact the levels of purines and pyrimidines in a biological sample. Medications that are intended to replace an endogenously synthesized biochemical, such as dopamine agents employed in the treatment of movement disorders and Parkinson's disease, or therapies that alter the production of neurochemicals, such as GABA can produce metabolomic results that may closely mimic human diseases, such as aromatic amino acid decarboxylase deficiency or GABA transaminase deficiency [8, 80]. It is also important to be mindful that medications, particularly antibiotics, can also alter the intestinal microbiome, which in turn can lead to dysbiosis as has been previously mentioned.

31.2.5.4 Assessing Mechanisms of Disease and Response to Therapy

Metabolomic analysis has also been used to better understand the pathophysiology of various conditions and assess response to therapeutic interventions. Serine deficiency disorders are one example of this. L-serine plays an essential role in a broad range of cellular processes, including protein synthesis, neurotransmission, folate metabolism, and synthesis of complex lysophospholipids. The phenotype of serine deficiency disorders is primarily neurological; however, the exact mechanism of the neurodevelopmental phenotype had been poorly understood. Metabolomic analyses indicated that an imbalance in sphingomyelins, phosphatidylcholines, and phosphatidylethanolamines contributed to the phenotype, and serine therapy provided partial correction of these imbalances [6, 81]. Metabolomics can also be used to understand the response to investigational therapies as in the case of glycogen storage disease type Ib. Neutropenia is commonly seen in this condition and is secondary to the toxic effects of 1,5-anhydroglucitol on neutrophils. Therapies that target renal sodium-glucose cotransporters, such as empagliflozin, have been employed as an off-label use to treat neutropenia in glycogen storage disease type Ib. Clinical analysis of 1,5-anhydroglucitol is not clinically available but can be detected on many metabolomic platforms. In individuals with glycogen storage disease type Ib treated with empagliflozin, metabolomic analysis demonstrated a reduction in 1,5-anhdyroglucitol that correlated with improvements in neutropenia and neutrophil dysfunction [82].

31.3 Summary

Accurate diagnosis of metabolic disease is the primary goal, and untargeted metabolomics allows affordable high-throughput disease screening that can be translated to the clinic. A breadth of metabolic pathways may be interrogated in a single biological sample (Table 31.2), facilitating diagnosis. While untargeted metabolomics offers an opportunity for discovery both inside and outside clinical settings, its use in the clinical lab supports broad screening for IEMs well beyond those identified in the newborn screen, offers a means to monitor disease management, and when integrated with genomics, provides a precision medicine approach to diagnosis of rare disease.

Table 31.2 Metabolomic findings in inborn errors of metabolism. Relative alterations in specific metabolic conditions are shown for the biofluids indicated. Inborn errors of metabolism are categorized based on the nosology proposed in the International Classification of Inherited Metabolic Disorders (ICIMD) [83]

ICIMD Nosology Group	ICIMD Nosology Number	Disorder	Key disease-related metabolite(s) identified		
			Plasma	Urine	Cerebrospinal fluid
Urea cycle disorders and inherited hyperammonemias	1.1.02.01	Carbamoyl phosphate synthetase I deficiency[1,2]	Glutamine [↑]		
			Pyroglutamine [↑]		
			Citrulline [↓]		
	1.1.03.01	Ornithine transcarbamolyase deficiency[1–4]	Glutamine [↑]		
			Citrulline [↓]		
			Ornithine [↑]		
			Uridine [↑]		
			Ororate [↑]		
	1.1.04.01	Argininosuccinate synthetase deficiency[1–6]	Citrulline [↑]	Citrulline [↑]	
			Homocitrulline [↑]	3-Ureidopropionate [↑]	
			N-Acetylcitrulline [↑]	4-Ureidobutyrate [↑]	
			3-Ureidopropionate [↑]	N-Carbamoylaspartate [↑]	
			Uridine [↑]	Ororate [↑]	
				Ornithine [↓]	
	1.1.05.01	Argininosuccinate lyase deficiency[1–3,5–8]	Argininosuccinate [↑]	Argininosuccinate [↑]	Argininosuccinate [↑]
			Citrulline [↑]	Citrulline [↑]	Citrulline [↑]
			Cytidine [↑]	N6-Succinyladenosine [↑]	Orotidine [↑]
			N-Acetylcitrulline [↑]		Arginine [↑]
	1.1.06.01	Arginase I deficiency[1–4]	Arginine [↑]		
			4-Guanidinobutanoate [↑]		
			N-Acetylarginine [↑]		
			Ororate [↑]		
	1.1.07.01	Mitochondrial ornithine transporter deficiency[2]	Homocitrulline [↑]		
			Ornithine [↑]		
	1.1.08.01	Citrin deficiency[2,9]	Citrulline [↑]	Citrulline [↑]	
			4-Hydroxyphenylpyruvate [↑]	4-Hydroxyphenylpyruvate [↑]	
			Galactitol [↑]	Galactitol [↑]	
			Arginine [↑]	4-Hydroxyphenyllactate [↑]	
			Phenyllactate [↑]	Galactose [↑]	
				Ornithine [↑]	

Organic acidurias	1.2.01.01	3-Methylcrotonyl-CoA carboxylase 1 deficiency[1,3,10]	3-Hydroxyisovalerate [↑]	3-Hydroxyisovalerate [↑]	
			3-Hydroxyisovalerylcarnitine [↑]	3-Hydroxyisovalerylcarnitine [↑]	
			3-Methylcrotonylglycine [↑]	3-Methylcrotonylglycine [↑]	
	1.2.02.01	Methylmalonic aciduria due to methylmalonyl-CoA epimerase deficiency[6]	Propionylcarnitine		
	1.2.05.01	Glutaryl-CoA dehydrogenase deficiency[1,3,6,11]	3-Hydroxyglutarate [↑]	3-Hydroxyglutarate [↑]	
			Glutarate [↑]	Glutarate [↑]	
			Glutarylcarnitine [↑]	Glutarylcarnitine [↑]	
			Glutarylglycine [↑]		
	1.2.06.01	Succinate-hydroxymethylglutarate-CoA transferase deficiency (GA3)[2]	Glutarate [↑]		
			Arachidate (20:0) [↑]		
			Octadecadienedioate (C18:2-DC) [↑]		
			Octadecenedioate (C18:1-DC) [↑]		
			Hexadecanedioate (C16-DC) [↑]		
	1.2.07.01	Isovaleryl-CoA dehydrogenase deficiency[1,3,6,10,12]	Isovalerate [↑]	Isovalerate [↑]	Isovalerylglycine [↑]
			Isovalerylglycine [↑]	Isovalerylglycine [↑]	Isovalerylcarnitine [↑]
			Isovalerylcarnitine [↑]	Isovalerylcarnitine [↑]	Beta-hydroxyisovalerate [↓]
			Beta-hydroxyisovalerate [↓]		
	1.2.08.01	Isobutyryl-CoA dehydrogenase deficiency[10]	Butyrylcarnitine [↑]		
	1.2.09.01	2-Methylbutyryl-CoA dehydrogenase deficiency[6]	2-Methylbutyrylcarnitine [↑]		
			2-Methylbutyrate [↑]		
	1.2.13.01	Mitochondrial short-chain enoyl-CoA hydratase 1 deficiency[2]	Beta-hydroxyisovalerate [↑]		
			3-Hydroxy-3-methylglutarate [↑]		
	1.2.14.01	3-Hydroxyisobutyryl-CoA hydrolase deficiency[2]	3-Hydroxyisobutyrate [↓]		
	1.2.18.01 and 1.2.19.01	Propionic acidemia due to propionyl-CoA carboxylase subunit alpha/beta deficiency[1,3,6,8,10,11]	Propionylcarnitine [↑]		Propionylcarnitine [↑]
			2-Methylcitrate [↑]		2-Methylcitrate [↑]
			3-Hydroxypropioate [↑]		3-Hydroxypropioate [↑]
			Glycine [↑]		Glycine [↑]
			Propionylglycine [↑]		Propionylglycine [↑]
			Carnitine [↓]		Carnitine [↓]
	1.2.21.01	Methylmalonic aciduria due to methylmalonyl-CoA mutase deficiency[1,3,5,6,10,11]	2-Methylcitrate [↑]	2-Methylcitrate [↑]	
			Methylmalonylcarnitine [↑]	Methylmalonylcarnitine [↑]	
			Methylmalonate [↑]		
			Propionate [↑]		
			Propionylcarnitine [↑]		
	1.2.23.01	Malonyl-CoA decarboxylase deficiency[1]	Malonate [↑]		
			Malonylcarnitine [↑]		

(continued)

Table 31.2 (continued)

ICIMD Nosology Group	ICIMD Nosology Number	Disorder	Key disease-related metabolite(s) identified		
			Plasma	Urine	Cerebrospinal fluid
Disorders of branched-chain amino acid metabolism	1.3.02.01 and 1.3.03.01	Branched-chain ketoacid dehydrogenase E1 alpha/beta deficiency[1–3,6,11]	Allo-isoleucine [↑]		
			Isoleucine [↑]		
			Leucine [↑]		
			Valine [↑]		
			2-Hydroxy-3-methylvalerate [↑]		
			3-Methyl-2-oxovalerate [↑]		
			Beta-hydroxyisovalerate [↑]		
Disorders of phenylalanine and tyrosine metabolism	1.4.01.01	Phenylalanine hydroxylase deficiency[1,3,6,11,13–15]	Phenylalanine [↑]	Phenylalanine [↑]	
			Phenylacetate [↑]	Phenylacetate [↑]	
			Phenyllactate [↑]	Phenyllactate [↑]	
			Phenylpyruvate [↑]	Phenylpyruvate [↑]	
	1.4.02.01	Tyrosine aminotransferase deficiency (tyrosinemia type II)[11]	Tyrosine [↑]		
	1.4.05.01	Homogentisic acid oxidase deficiency (alkaptonuria)[1,6,16]	Homogentisate [↑]		
			Tyrosine [↑]		
	1.4.07.01	Fumarylacetoacetase deficiency (tyrosinemia type I)[1,2,5,6,8,11]	Tyrosine [↑]	Tyrosine [↑]	Tyrosine [↑]
			3-(4-Hydroxyphenyl)lactate [↑]	3-(4-Hydroxyphenyl)lactate [↑]	3-(4-Hydroxyphenyl)lactate [↑]
			4-Hydroxyphenylacetate [↑]	4-Hydroxyphenylacetate [↑]	4-Fumarylacetoacetic acid [↑]
			Succinylacetone [↑]		
			Phenyllactate [↑]		
			Phenylpyruvate [↑]		
			N-Acetyltyrosine [↑]		
Disorders of sulfur amino acid and sulfide metabolism	1.5.01.01	Mitochondrial sulfur dioxygenase deficiency (ethylmalonic encephalopathy)[2,6,17]	2-Methylbutyrylglycine [↑]		
			Ethylmalonate [↑]		
			Methylsuccinate [↑]		
			Butyrylcarnitine [↑]		
			Isobutyrylcarnitine [↑]		
	1.5.02.01	Methionine adenosyltransferase I/III deficiency[6,8]	Methionine [↑]		Methionine [↑]
			Glutamylmethionine [↑]		S-Adenosylmethionine [↓]
			N-(1-Deoxy-1-fructosyl)methionine [↑]		
	1.5.06.01	Cystathionine beta-synthase deficiency[1–3,6,10]	Homocysteine [↑]		
			Methionine [↑]		
			Methionine-sulfoxide [↑]		
			S-Adenosylmethionine [↑]		
	1.5.13.01	Methionine synthase deficiency (cblG)[2]	S-Adenosylhomocysteine [↑]		
			Methionine [↓]		

Disorders of glycine and serine metabolism	1.6.01.01	Nonketotic hyperglycinemia due to glycine decarboxylase deficiency (glycine encephalopathy)[2,8]	Glycine [↑]		Glycine [↑]
	1.6.01.02	3-phosphoglycerate dehydrogenase deficiency[18,19]	Glycine [↓]		
			Serine [↓]		
			Glycerophosphocholine [↓]		
			Glycerophosphoethanolamine [↓]		
	1.6.02.01	Phosphoserine aminotransferase deficiency[18,19]	Glycine [↓]		
			Serine [↓]		
			Glycerophosphocholine [↓]		
			Glycerophosphoethanolamine [↓]		
Disorders of proline and ornithine metabolism	1.7.01.01	Ornithine aminotransferase deficiency[1,6]	3-Amino-2-piperidone [↑]		
			Guanidinoacetate [↑]		
			Ornithine [↑]		
			Proline [↑]		
	1.7.14.01	Pyrroline-5-carboxylate synthetase deficiency (spastic paraplegia type 9A)[2]	Proline [↓]		
			Ornithine [↓]		
			Citrulline [↓]		
Disorders of lysine metabolism	1.8.01.01	Alpha-aminoadipic semialdehyde synthase deficiency (hyperlysinemia)[6]	Lysine [↑]		
			N-Acetyl-L-lysine [↑]		
			Pipecolate [↑]		
Disorders of glutamate/glutamine and aspartate/asparagine metabolism	1.9.01.03	Aspartoacylase deficiency (Canavan disease)[6]	N-Acetylaspartate [↑]		
Disorders of histidine metabolism	1.10.01.01	Histidine ammonia-lyase deficiency (histidinemia)[6]	Histidine [↑]		
			Imidazole lactate [↑]		
	1.10.02.01	Urocanase deficiency[2,20]	*cis*-Urocanate [↑]	*cis*-Urocanate [↑]	
			trans-Urocanate [↑]	*trans*-Urocanate [↑]	
			Imidazole propionate [↑]	Imidazole propionate [↑]	
				Hydantoin-5-propionate [↓]	
Disorders of amino acid transport	1.11.06.01	Lysinuric protein intolerance[1,3,6]	Arginine [↓]	Glutarylcarnitine [↓]	
			Lysine [↓]	N6-Trimethyllysine [↓]	
			Ornithine [↓]	N6-Acetyllysine [↑]	
			Glutamine [↑]		

(continued)

Table 31.2 (continued)

ICIMD Nosology Group	ICIMD Nosology Number	Disorder	Key disease-related metabolite(s) identified		
			Plasma	Urine	Cerebrospinal fluid
Other disorders of amino acid metabolism	1.12.02.01	Aminoacylase I deficiency[1,6]	N-Acetylglutamate [↑]		
			N-Acetylglutamine [↑]		
			N-Acetylmethionine [↑]		
			N-Acetylthreonine [↑]		
Disorders of methylamine metabolism	2.3.01.01	Dimethylglycine dehydrogenase deficiency[6]	Dimethylglycine [↑]		
Disorders of galactose and fructose metabolism	3.1.01.02	Galactose-1-phosphate uridyltransferase deficiency[2,6]	Galactonate [↑]		
			Galactitol [↑]		
	3.1.02.03	Aldolase B deficiency	Fructose [↑]		
Disorders of gluconeogenesis	3.2.01.01	Glycerol kinase deficiency, isolated	Glycerol [↑]		
Disorders of glycogen metabolism	3.4.01.02 and 3.4.04.01	Glucose-6-phosphatase deficiency (GSD1a) and Glucose-6-phosphate transporter deficiency (GSD1b)[2,21,22]	Pyruvate [↑]		
			α-Ketoglutarate [↑]		
			Creatine [↑]		
			Succinate [↓]		
Disorders of pentose metabolism	3.5.03.01	Transaldolase deficiency[2,6,23]	Sedoheptulose [↑]	Sedoheptulose [↑]	
			Ribitol [↑]	Ribitol [↑]	
			Arabitol/xylitol [↑]	Arabitol/xylitol [↑]	
			Erythritol [↑]	Erythritol [↑]	
			Erythronate [↑]	Erythronate [↑]	
			Inosine [↑]	Xanthosine [↑]	
	3.5.04.01	Transketolase deficiency[23]	Sedoheptulose [↑]	Sedoheptulose [↑]	
			Ribitol [↑]	Ribitol [↑]	
			Arabitol/xylitol [↑]	Arabitol/xylitol [↑]	
			Erythritol [↑]	Erythritol [↑]	
			Erythronate [↑]	Erythronate [↑]	
			Inosine [↑]	Xanthosine [↑]	
Disorders of carnitine metabolism	4.1.01.01	Primary carnitine deficiency[1,2]	Carnitine [↓]		
	4.1.03.01	Carnitine palmitoyltransferase 2 deficiency[1]	Hexadecanoylcarnitine [↑]		
			Oleoylcarnitine [↑]		
			Oleoylcarnitine [↑]		
	4.1.05.01	ε-N-Trimethyllysine hydroxylase deficiency[2,3]	N6-Trimethyllysine [↑]		
			Deoxycarnitine [↑]		
			Acetylcarnitine [↓]		
			Carnitine [↓]		

Disorders of mitochondrial fatty acid oxidation	4.2.01.01	Short-chain acyl-CoA dehydrogenase deficiency[2]	Ethylmalonate [↑]		
			Butyrylcarnitine [↑]		
			Methylsuccinate [↑]		
	4.2.02.01	Medium chain acyl-CoA dehydrogenase deficiency[1–3,6,10]	Hexanoate [↑]	5-Hydroxyhexanoate [↑]	
			Octanoate [↑]	Adipate [↑]	
			Hexanoylcarnitine [↑]	Hexanoylglycine [↑]	
			Hexanoylglycine [↑]	Octanoylcarnitine [↑]	
			N-octanoylglycine [↑]		
			Octanoylcarnitine [↑]		
			Suberate [↑]		
			Sebacate [↑]		
	4.2.03.01	Very-long-chain acyl-CoA dehydrogenase deficiency[1–3,6,11]	Hexadecanoylcarnitine [↑]		
			Octadecanoylcarnitine [↑]		
			Oleoylcarnitine [↑]		
			Tetradecanoylcarnitine [↑]		
			Tetradecenoylcarnitine [↑]		
			Tetradecadienoylcartinine [↑]		
	4.2.05.01	Long-chain 3-hydroxyacyl-CoA dehydrogenase deficiency[1,2,6]	3-Hydroxyhexadecanoylcarnitine [↑]		
			3-Hydroxyoctadecenoylcarnitine [↑]		
			3-Hydroxytetradecanoylcarnitine [↑]		
	4.2.06.01 and 4.2.07.01 and 4.2.08.01	Multiple acyl-CoA dehydrogenase deficiency (glutaric aciduria types IIA/B/C)[1,2,6,11]	Glutarate [↑]		
			Decanoylcarnitine [↑]		
			Glutarylcarnitine [↑]		
			Hexanoylcarnitine [↑]		
			Hexadecenoylcarnitine [↑]		
			Tetradecenoylcarnitine [↑]		
			Tetradecanoylcarnitine [↑]		
Disorders of ketone body metabolism	4.3.16.01	3-Hydroxy-3-methylglutaryl-CoA lyase deficiency[3,6]	2-Methyl-3-hydroxybutyrate [↑]		
			3-Hydroxyisovalerate [↑]		
			3-Hydroxymethylglutarate [↑]		
			3-Methylglutarylcarnitine [↑]		
			3-Methylglutaconate [↑]		
	4.3.13.01	Mitochondrial acetoacetyl-CoA thiolase deficiency (beta-ketothiolase deficiency)[1,6,10]	2-Methyl-3-hydroxybutyrate [↑]		
			Tiglate [↑]		
			Tiglylcarnitine [↑]		
			Tiglylglycine [↑]		

(continued)

Table 31.2 (continued)

ICIMD Nosology Group	ICIMD Nosology Number	Disorder	Key disease-related metabolite(s) identified		
			Plasma	Urine	Cerebrospinal fluid
Disorders of pyruvate metabolism	5.1.01.01	Pyruvate dehydrogenase E1 alpha deficiency[2]	Lactate [↑]		
			Pyruvate [↑]		
			Alanine [↑]		
	5.1.05.01	Dihydrolipoyl dehydrogenase deficiency[2]	3-Ethyl-2-oxobutyrate [↑]		
			2-Hydroxyglutarate [↑]		
			3-Methyl-2-oxovalerate [↑]		
			4-Methyl-2-oxopentanoate [↑]		
			Lactate [↑]		
			Pyruvate [↑]		
Disorders of the Krebs cycle	5.2.09.02	Plasma membrane citrate transporter deficiency[2,24]	Citrate [↑]	16α-Hydroxy DHEA 3-sulfate [↑]	Citrate [↑]
					2-Methylcitrate [↑]
	5.2.05.01	GTP-specific succinyl-CoA ligase α subunit deficiency[2]	Succinylcarnitine [↑]		
			Propionylcarnitine [↑]		
			Methylmalonate [↑]		
			Hexanoylcarnitine [↑]		
			Butyrylcarnitine [↑]		
			Malate [↓]		
Disorders of creatine metabolism	5.3.03.01	Guanidinoacetate methyltransferase deficiency[2,3,6,8]	Guanidinoacetate [↑]	Guanidinoacetate [↑]	
			Creatinine [↓]		
			Creatine [↓]		
Disorders of mtDNA-encoded tRNA and rRNA	6.2.11.01	Mitochondrial tRNA(Leu) 1 deficiency (mitochondrial encephalopathy, lactic acidosis, and stroke-like episodes (MELAS))[25–27]	Alanine [↑]	2-Hydroxyglutarate [↑]	
			Lactate [↑]	4-Pentenoate [↑]	
			Pyruvate [↑]	Glycolate [↑]	
			Arginine [↓]		
Disorders of complex V subunits and assembly factors	7.5.01.01	Transmembrane protein 70 deficiency[2]	3-Methylglutaconate [↑]		
			3-Methylglutarylcarnitine [↑]		
			Alanine [↑]		
			Lactate [↑]		
Disorders of lipoic acid and iron-sulfur metabolism	8.2.03.01	Lipoyltransferase 1 deficiency[2,28]	β-alanine [↑]		
			4-Guanidinobutanoate [↑]		
			Oxalate [↓]		
			Malonate [↓]		
	8.2.04.01	NFU1 deficiency[2,29]	Glycine [↑]		

Disorders of mitochondrial nucleotide pool maintenance	9.1.08.01	Thymidine phosphorylase deficiency (mitochondrial neurogastrointestinal encephalopathy)[1–3]	Deoxyuridine [↑]
			Thymidine [↑]
			Thymine [↑]
			Uracil [↑]
			5,6-Dihydrothymine [↓]
Disorders of mtDNA replication and maintenance	9.2.01.04	Mitochondrial DNA polymerase gamma catalytic subunit deficiency (POLG)— (progressive external ophthalmoplegia, mitochondrial recessive ataxia syndrome, MIRAS)[25]	Sorbitol [↑]
			Alanine [↑]
			Cystathionine [↑]
			Glucuronate [↑]
			Carnosine [↓]
Disorders of mitochondrial transcript processing and modification	10.1.12.01	tRNA 5-methylaminomethyl-2-thiouridylate-methyltransferase deficiency[2]	Glycochenodeoxycholate [↑]
			Glycohyocholate [↑]
			Lactate [↑]
			Pyruvate [↑]
			Taurocholate [↑]
	10.1.15.01	17-beta-hydroxysteroid dehydrogenase type 10 deficiency[2]	2-Hydroxy-3-methylvalerate [↑]
			2-Methyl-3-hydroxybutyrate [↑]
			3-Hydroxyisovalerate [↑]
			Tiglylcarnitine [↑]
			Tiglylglycine [↑]
Disorders of mitochondrial metabolite repair	12.1.02.01	L-2-hydroxyglutarate dehydrogenase deficiency[2]	2-Hydroxyglutaric acid [↑]
	12.1.22.01	Acetyl-CoA-synthase 3 deficiency[6]	Malonate [↑]
			Methylmalonate [↑]
Disorders of peroxisomal fatty acid oxidation	14.2.03.01	D-bifunctional protein deficiency[2,30]	1-Lignoceroyl-GPC (24:0) [↑]
			Pipecolate [↑]
			Docosadioate [↑]
			Sphingomyelin [↓]
	14.2.06.01	Phytanoyl-CoA hydroxylase deficiency (classic Refsum disease)[6]	Phytanate [↑]
Disorders of sterol biosynthesis	14.7.01.02	Mevalonate kinase deficiency[1]	Mevalonate [↑]
			Mevalonolactone [↑]
	14.7.14.01	Smith-Lemli-Opitz syndrome[2]	7-Dehydrocholesterol [↑]
			Cholesterol [↓]

(continued)

Table 31.2 (continued)

ICIMD Nosology Group	ICIMD Nosology Number	Disorder	Key disease-related metabolite(s) identified		
			Plasma	Urine	Cerebrospinal fluid
Disorders of bile acid metabolism	14.8.04.02	Sterol 27-hydroxylase deficiency (cerebrotendinous xanthomatosis)[6]	Cholestane-pentol [↑]		
	14.8.05.02	α-Methylacyl-CoA racemase deficiency[1,2]	Dihydroxycholestanoate[↑]		
			Phytanate [↑]		
			Phytanoyl-carnitine [↑]		
			Pristanoyl-carnitine [↑]		
			Trihydroxycholestanoate [↑]		
Disorders of pyrimidine metabolism	16.1.03.01	Uridine monophosphate synthase deficiency[6]	Dihydroororate [↑]		
			Orotate [↑]		
	16.1.03.02	Beta-ureidopropionase deficiency[2,6]	3-Aminoisobutyrate [↑]		
			3-Ureidopropionate [↑]		
			3-Ureido-isobutyrate [↑]		
			5,6-Dihydrothymine [↑]		
			5,6-Dihydrouracil [↑]		
Disorders of purine metabolism	16.2.03.01	Adenylosuccinate lyase deficiency[1,5,6,31]	SAICA riboside [↑]	2-Deoxyinosine [↑]	
			Succinyladenosine [↑]	Adenine [↑]	
				N6-succinyladenosine [↑]	
				Xanthosine [↑]	
	16.2.11.01	Hypoxanthine guanine phosphoribosyltransferase deficiency (Lesch-Nyhan syndrome)[2,5]	Inosine [↑]	Hypoxanthine [↑]	
			Uracil [↑]	Inosine [↑]	
				Urate [↑]	
				Xanthosine [↑]	
	16.2.21.01	AICAR transformylase-IMP cyclohydrolase deficiency[2]	Succinyladenosine [↑]		
Disorders of multiple glycosylation pathways	18.4.03.02	NANS-CDG (N-acetylneuraminic acid synthase deficiency)[2,6]	N-acetylglucosamine / N-acetylgalactosamine [↑]		
			N-acetylmannosamine [↑]		
			N-acetylneuraminate [↓]		
Peroxisomal biogenesis disorders	19.3.01.01 and 19.3.05.01	Peroxin 1 deficiency, Peroxin 6 deficiency[2,30]	1-Lignoceroyl-GPC (24:0) [↑]		
			Pipecolate [↑]		
			7-HOCA [↑]		
			Docosadioate [↑]		
			Palmitoyl sphingomyelin [↓]		
			Oleoyl sphingomyelin [↓]		
			Sphingomyelin [↓]		

Disorders of glycoprotein degradation	20.3.04.01	Beta-mannosidase deficiency[1]	GlcNAc-Man [↑]		
Disorders of riboflavin metabolism	21.3.03.01	Riboflavin transporter 3 deficiency (Brown-Vialetto-Van Laere syndrome type 2)[2,32]	Hexanoylcarnitine [↑]		
			Octanoylcarnitine [↑]		
			Decanoylcarnitine [↑]		
			2-Hydroxyglutarate [↑]		
			Kynurenate [↓]		
			Picolinate [↓]		
			Methionine sulfone [↓]		
Disorders of pyridoxine metabolism	21.6.02.02	Alpha-amino adipic semialdehyde (AASA) dehydrogenase deficiency[6,33]	6-Oxopiperidine-2-carboxylate [↑]	6-Oxopiperidine-2-carboxylate [↑]	6-Oxopiperidine-2-carboxylate [↑]
			2-Oxopropylpiperidine-2-carboxylate [↑]	2-Oxopropylpiperidine-2-carboxylate [↑]	2-Oxopropylpiperidine-2-carboxylate [↑]
			Alpha-aminoadipic semialdehyde [↑]	Alpha-aminoadipic semialdehyde [↑]	Alpha-aminoadipic semialdehyde [↑]
			Tetrahydropiperidine-2-carboxylate [↑]	Tetrahydropiperidine-2-carboxylate [↑]	Tetrahydropiperidine-2-carboxylate [↑]
			Pipecolate [↑]	Pipecolate [↑]	Pipecolate [↑]
Disorders of biotin metabolism	21.7.02.01	Holocarboxylase synthetase deficiency[3,5]	2-Methylcitrate [↑]	2-Methylcitrate [↑]	
			3-Hydroxypropanoate [↑]	2-Methylmalonylcarnitine [↓]	
			3-Methylcrotonylglycine [↑]	3-Hydroxypropanoate [↑]	
			Propionylcarnitine [↑]	Succinate [↓]	
			Propionylglycine [↑]	Succinylcarnitine [↓]	
	21.7.17.01	Sodium-dependent multivitamin transporter deficiency[2]	Pantothenate [↓]		
			Carnitine [↓]		
Disorders of folate metabolism	21.8.03.01	5,10-methylenetetrahydrofolate reductase deficiency[2]	Betaine [↓]		
			Methionine [↓]		
			Methionine sulfone [↓]		
			Methionine-sulfoxide [↓]		
	21.8.06.01	Formimidoyltransferase cyclodeaminase deficiency[1,6,10]	Formiminoglutamate [↑]		
			Hydantoin-5-propionate [↑]		
Disorders of cobalamin metabolism	21.9.09.01	Methylmalonic aciduria and homocystinuria, cblC type[2]	3-Hydroxy-3-methylglutaric acid [↑]		
			Methylmalonate [↑]		
			Propionylcarnitine [↑]		
			Methionine [↓]		

(continued)

Table 31.2 (continued)

ICIMD Nosology Group	ICIMD Nosology Number	Disorder	Key disease-related metabolite(s) identified		
			Plasma	Urine	Cerebrospinal fluid
Disorders of molybdenum cofactor metabolism	21.10.01.01	Cyclic pyranopterin monophosphate synthase deficiency (molybdenum cofactor deficiency A)[5,6]	4-Pyridoxate [↑]	Cysteine S-sulfate [↑]	
			Xanthine[↑]	Xanthine [↑]	
			Xanthosine[↑]		
			Urate [↓]		
	21.10.04.01	Molybdenum cofactor sulfurase deficiency (xanthinuria, type II)[6]	Xanthine [↑]		
			Xanthosine [↑]		
			4-Pyridoxate [↓]		
			5-Hydroxyisourate [↓]		
			N1-Methyl-2/4-pyridone-5-carboxamide [↓]		
			Urate [↓]		
Monoamine neurotransmission	23.1.02.01	Aromatic L-amino acid decarboxylase deficiency[2,34]	3-Methoxytyrosine (3-O-methyldopa) [↑]	3-Methoxytyrosine (3-O-methyldopa) [↑]	3-Methoxytyrosine (3-O-methyldopa) [↑]
			3-Methoxytyramine sulfate [↓]	Vanillate [↑]	
			Dopamine 3-O-sulfate [↓]	Vanillylmandelate [↓]	
			5-HIAA [↓]	Vanilloylglycine [↑]	
			Homovanillate [↓]		
			Vanillylmandelate [↓]		
Gamma-aminobutyric acid neurotransmitter disorders	23.2.05.01	GABA transaminase deficiency[2,35,36]	2-Pyrrolidinone [↑]	2-Pyrrolidinone [↑]	2-Pyrrolidinone [↑]
			Succinamate [↑]	Succinimide [↑]	Succinimide [↑]
			4-Guanidinobutanoate [↑]		Homocarnosine [↑]
	23.2.06.01	Succinic semialdehyde dehydrogenase deficiency SSADH[2,37]	2-Pyrrolidinone [↑]	2-Pyrrolidinone [↑]	
			4-Guanidinobutanoate [↑]	4-Guanidinobutanoate [↑]	
			Argininate [↑]		
Disorders of steroid metabolism	24.2.04.01	3-Beta-hydroxysteroid dehydrogenase deficiency[6]	3β,7α-Dihydroxy-5-cholestenoate [↑]		
			3β,7α-Dihydroxy-5-cholenoate [↑]		
			Dihydroxycholenoyl glycine [↑]		

[1][84]; [2][4]; [3][15]; [4][76]; [5][49]; [6][85]; [7][50]; [8][86]; [9][87]; [10][88]; [11][89]; [12][90]; [13][91]; [14][92]; [15][93]; [16][94]; [17][95]; [18][81]; [19][6]; [20][7]; [21][52]; [22][82]; [23][9]; [24][96]; [25][97]; [26][98]; [27][99]; [28][100]; [29][101]; [30][102]; [31][103]; [32][104]; [33][105]; [34][80]; [35][106]; [36][8]; [37][107]

References

1. Almannai M, Marom R, Sutton VR (2016) Newborn screening: a review of history, recent advancements, and future perspectives in the era of next generation sequencing. Curr Opin Pediatr 28:694–699
2. Richards S, Aziz N, Bale S, Bick D, Das S, Gastier-Foster J, Grody WW, Hegde M, Lyon E, Spector E, Voelkerding K, Rehm HL, Acmg Laboratory Quality Assurance Committee (2015) Standards and guidelines for the interpretation of sequence variants: a joint consensus recommendation of the American College of Medical Genetics and Genomics and the Association for Molecular Pathology. Genet Med 17:405–424
3. Meng L, Pammi M, Saronwala A, Magoulas P, Ghazi AR, Vetrini F, Zhang J, He W, Dharmadhikari AV, Qu C, Ward P, Braxton A, Narayanan S, Ge X, Tokita MJ, Santiago-Sim T, Dai H, Chiang T, Smith H, Azamian MS, Robak L, Bostwick BL, Schaaf CP, Potocki L, Scaglia F, Bacino CA, Hanchard NA, Wangler MF, Scott D, Brown C, Hu J, Belmont JW, Burrage LC, Graham BH, Sutton VR, Craigen WJ, Plon SE, Lupski JR, Beaudet AL, Gibbs RA, Muzny DM, Miller MJ, Wang X, Leduc MS, Xiao R, Liu P, Shaw C, Walkiewicz M, Bi W, Xia F, Lee B, Eng CM, Yang Y, Lalani SR (2017) Use of exome sequencing for infants in intensive care units: ascertainment of severe single-gene disorders and effect on medical management. JAMA Pediatr 171:e173438
4. Liu N, Xiao J, Gijavanekar C, Pappan KL, Glinton KE, Shayota BJ, Kennedy AD, Sun Q, Sutton VR, Elsea SH (2021) Comparison of untargeted metabolomic profiling vs traditional metabolic screening to identify inborn errors of metabolism. JAMA Netw Open 4:e2114155
5. Mastrokolias A, Pool R, Mina E, Hettne KM, van Duijn E, van der Mast RC, van Ommen G, t Hoen PA, Prehn C, Adamski J, van Roon-Mom W (2016) Integration of targeted metabolomics and transcriptomics identifies deregulation of phosphatidylcholine metabolism in Huntington's disease peripheral blood samples. Metabolomics 12:137
6. Glinton KE, Benke PJ, Lines MA, Geraghty MT, Chakraborty P, Al-Dirbashi OY, Jiang Y, Kennedy AD, Grotewiel MS, Sutton VR, Elsea SH, El-Hattab AW (2018) Disturbed phospholipid metabolism in serine biosynthesis defects revealed by metabolomic profiling. Mol Genet Metab 123:309–316
7. Glinton KE, Levy HL, Kennedy AD, Pappan KL, Elsea SH (2019) Untargeted metabolomics identifies unique though benign biochemical changes in patients with pathogenic variants in UROC1. Mol Genet Metab Rep 18:14–18
8. Kennedy AD, Pappan KL, Donti T, Delgado MR, Shinawi M, Pearson TS, Lalani SR, Craigen WE, Sutton VR, Evans AM, Sun Q, Emrick LT, Elsea SH (2019) 2-Pyrrolidinone and Succinimide as clinical screening biomarkers for GABA-transaminase deficiency: anti-seizure medications impact accurate diagnosis. Front Neurosci 13:394
9. Shayota BJ, Donti TR, Xiao J, Gijavanekar C, Kennedy AD, Hubert L, Rodan L, Vanderpluym C, Nowak C, Bjornsson HT, Ganetzky R, Berry GT, Pappan KL, Sutton VR, Sun Q, Elsea SH (2020) Untargeted metabolomics as an unbiased approach to the diagnosis of inborn errors of metabolism of the non-oxidative branch of the pentose phosphate pathway. Mol Genet Metab 131:147–154
10. Alaimo JT, Glinton KE, Liu N, Xiao J, Yang Y, Reid Sutton V, Elsea SH (2020) Integrated analysis of metabolomic profiling and exome data supplements sequence variant interpretation, classification, and diagnosis. Genet Med 22:1560–1566
11. Ford L, Kennedy AD, Goodman KD, Pappan KL, Evans AM, Miller LAD, Wulff JE, Wiggs BR, Lennon JJ, Elsea S, Toal DR (2020) Precision of a clinical metabolomics profiling platform for use in the identification of inborn errors of metabolism. J Appl Lab Med 5:342–356
12. Want EJ, Masson P, Michopoulos F, Wilson ID, Theodoridis G, Plumb RS, Shockcor J, Loftus N, Holmes E, Nicholson JK (2013) Global metabolic profiling of animal and human tissues via UPLC-MS. Nat Protoc 8:17–32
13. Gika HG, Theodoridis GA, Vrhovsek U, Mattivi F (2012) Quantitative profiling of polar primary metabolites using hydrophilic interaction ultrahigh performance liquid chromatography-tandem mass spectrometry. J Chromatogr A 1259:121–127
14. Lioupi A, Virgiliou C, Walter TH, Smith KM, Rainville P, Wilson ID, Theodoridis G, Gika HG (2022) Application of a hybrid zwitterionic hydrophilic interaction liquid chromatography column in metabolic profiling studies. J Chromatogr A 1672:463013
15. Miller MJ, Kennedy AD, Eckhart AD, Burrage LC, Wulff JE, Miller LA, Milburn MV, Ryals JA, Beaudet AL, Sun Q, Sutton VR, Elsea SH (2015) Untargeted metabolomic analysis for the clinical screening of inborn errors of metabolism. J Inherit Metab Dis 38:1029–1039
16. Bothwell JH, Griffin JL (2011) An introduction to biological nuclear magnetic resonance spectroscopy. Biol Rev Camb Philos Soc 86:493–510
17. Wishart DS, Lewis MJ, Morrissey JA, Flegel MD, Jeroncic K, Xiong Y, Cheng D, Eisner R, Gautam B, Tzur D, Sawhney S, Bamforth F, Greiner R, Li L (2008) The human cerebrospinal fluid metabolome. J Chromatogr B Analyt Technol Biomed Life Sci 871:164–173
18. Larive CK, Barding GA Jr, Dinges MM (2015) NMR spectroscopy for metabolomics and metabolic profiling. Anal Chem 87:133–146
19. Marchand J, Martineau E, Guitton Y, Dervilly-Pinel G, Giraudeau P (2017) Multidimensional NMR approaches towards highly resolved, sensitive and high-throughput quantitative metabolomics. Curr Opin Biotechnol 43:49–55

20. Nagana Gowda GA, Pascua V, Raftery D (2021) Extending the scope of (1)H NMR-based blood metabolomics for the analysis of labile antioxidants: reduced and oxidized glutathione. Anal Chem 93:14844–14850
21. Wishart DS (2008) Metabolomics: a complementary tool in renal transplantation. Contrib Nephrol 160:76–87
22. Beckonert O, Coen M, Keun HC, Wang Y, Ebbels TM, Holmes E, Lindon JC, Nicholson JK (2010) High-resolution magic-angle-spinning NMR spectroscopy for metabolic profiling of intact tissues. Nat Protoc 5:1019–1032
23. Dietz C, Ehret F, Palmas F, Vandergrift LA, Jiang Y, Schmitt V, Dufner V, Habbel P, Nowak J, Cheng LL (2017) Applications of high-resolution magic angle spinning MRS in biomedical studies II-human diseases. NMR Biomed 30
24. Gowda GAN (2018) Profiling redox and energy coenzymes in whole blood, tissue and cells using NMR spectroscopy. Metabolites 8
25. Chen G, Walmsley S, Cheung GCM, Chen L, Cheng CY, Beuerman RW, Wong TY, Zhou L, Choi H (2017) Customized consensus spectral library building for untargeted quantitative metabolomics analysis with data independent acquisition mass spectrometry and MetaboDIA workflow. Anal Chem 89:4897–4906
26. Dehaven CD, Evans AM, Dai H, Lawton KA (2010) Organization of GC/MS and LC/MS metabolomics data into chemical libraries. J Cheminform 2:9
27. Duhrkop K, Fleischauer M, Ludwig M, Aksenov AA, Melnik AV, Meusel M, Dorrestein PC, Rousu J, Bocker S (2019) SIRIUS 4: a rapid tool for turning tandem mass spectra into metabolite structure information. Nat Methods 16:299–302
28. Tada I, Tsugawa H, Meister I, Zhang P, Shu R, Katsumi R, Wheelock CE, Arita M, Chaleckis R (2019) Creating a reliable mass spectral-retention time library for all ion fragmentation-based metabolomics. Meta 9
29. Zhou Z, Luo M, Chen X, Yin Y, Xiong X, Wang R, Zhu ZJ (2020) Ion mobility collision cross-section atlas for known and unknown metabolite annotation in untargeted metabolomics. Nat Commun 11:4334
30. Wishart DS, Guo A, Oler E, Wang F, Anjum A, Peters H, Dizon R, Sayeeda Z, Tian S, Lee BL, Berjanskii M, Mah R, Yamamoto M, Jovel J, Torres-Calzada C, Hiebert-Giesbrecht M, Lui VW, Varshavi D, Varshavi D, Allen D, Arndt D, Khetarpal N, Sivakumaran A, Harford K, Sanford S, Yee K, Cao X, Budinski Z, Liigand J, Zhang L, Zheng J, Mandal R, Karu N, Dambrova M, Schioth HB, Greiner R, Gautam V (2022) HMDB 5.0: the human metabolome database for 2022. Nucleic Acids Res 50:D622–DD31
31. Wishart DS, Knox C, Guo AC, Eisner R, Young N, Gautam B, Hau DD, Psychogios N, Dong E, Bouatra S, Mandal R, Sinelnikov I, Xia J, Jia L, Cruz JA, Lim E, Sobsey CA, Shrivastava S, Huang P, Liu P, Fang L, Peng J, Fradette R, Cheng D, Tzur D, Clements M, Lewis A, De Souza A, Zuniga A, Dawe M, Xiong Y, Clive D, Greiner R, Nazyrova A, Shaykhutdinov R, Li L, Vogel HJ, Forsythe I (2009) HMDB: a knowledgebase for the human metabolome. Nucleic Acids Res 37:D603–D610
32. Horai H, Arita M, Kanaya S, Nihei Y, Ikeda T, Suwa K, Ojima Y, Tanaka K, Tanaka S, Aoshima K, Oda Y, Kakazu Y, Kusano M, Tohge T, Matsuda F, Sawada Y, Hirai MY, Nakanishi H, Ikeda K, Akimoto N, Maoka T, Takahashi H, Ara T, Sakurai N, Suzuki H, Shibata D, Neumann S, Iida T, Tanaka K, Funatsu K, Matsuura F, Soga T, Taguchi R, Saito K, Nishioka T (2010) MassBank: a public repository for sharing mass spectral data for life sciences. J Mass Spectrom 45:703–714
33. Sud M, Fahy E, Cotter D, Azam K, Vadivelu I, Burant C, Edison A, Fiehn O, Higashi R, Nair KS, Sumner S, Subramaniam S (2016) Metabolomics workbench: an international repository for metabolomics data and metadata, metabolite standards, protocols, tutorials and training, and analysis tools. Nucleic Acids Res 44:D463–D470
34. Kim S, Chen J, Cheng T, Gindulyte A, He J, He S, Li Q, Shoemaker BA, Thiessen PA, Yu B, Zaslavsky L, Zhang J, Bolton EE (2023) PubChem 2023 update. Nucleic Acids Res 51:D1373–D1D80
35. Kanehisa M, Furumichi M, Sato Y, Kawashima M, Ishiguro-Watanabe M (2023) KEGG for taxonomy-based analysis of pathways and genomes. Nucleic Acids Res 51:D587–DD92
36. Shannon P, Markiel A, Ozier O, Baliga NS, Wang JT, Ramage D, Amin N, Schwikowski B, Ideker T (2003) Cytoscape: a software environment for integrated models of biomolecular interaction networks. Genome Res 13:2498–2504
37. Lee JJY, Wasserman WW, Hoffmann GF, van Karnebeek CDM, Blau N (2018) Knowledge base and mini-expert platform for the diagnosis of inborn errors of metabolism. Genet Med 20:151–158
38. Ferreira CR, van Karnebeek CDM, Vockley J, Blau N (2019) A proposed nosology of inborn errors of metabolism. Genet Med 21:102–106
39. McEachran AD, Chao A, Al-Ghoul H, Lowe C, Grulke C, Sobus JR, Williams AJ (2020) Revisiting five years of CASMI contests with EPA identification tools. Metabolites 10
40. Zhou Z, Luo M, Zhang H, Yin Y, Cai Y, Zhu ZJ (2022) Metabolite annotation from knowns to unknowns through knowledge-guided multi-layer metabolic networking. Nat Commun 13:6656
41. Chen L, Lu W, Wang L, Xing X, Chen Z, Teng X, Zeng X, Muscarella AD, Shen Y, Cowan A, McReynolds MR, Kennedy BJ, Lato AM, Campagna SR, Singh M, Rabinowitz JD (2021) Metabolite discovery through global annotation of untargeted metabolomics data. Nat Methods 18:1377–1385
42. Edison AS, Colonna M, Gouveia GJ, Holderman NR, Judge MT, Shen X, Zhang S (2021) NMR: unique strengths that enhance modern metabolomics research. Anal Chem 93:478–499

43. Nagana Gowda GA, Raftery D (2017) Recent advances in NMR-based metabolomics. Anal Chem 89:490–510
44. Lefort G, Liaubet L, Canlet C, Tardivel P, Pere MC, Quesnel H, Paris A, Iannuccelli N, Vialaneix N, Servien R (2019) ASICS: an R package for a whole analysis workflow of 1D 1H NMR spectra. Bioinformatics 35:4356–4363
45. Hao J, Astle W, De Iorio M, Ebbels TM (2012) BATMAN – an R package for the automated quantification of metabolites from nuclear magnetic resonance spectra using a Bayesian model. Bioinformatics 28:2088–2090
46. Hao J, Liebeke M, Astle W, De Iorio M, Bundy JG, Ebbels TM (2014) Bayesian deconvolution and quantification of metabolites in complex 1D NMR spectra using BATMAN. Nat Protoc 9:1416–1427
47. Ravanbakhsh S, Liu P, Bjorndahl TC, Mandal R, Grant JR, Wilson M, Eisner R, Sinelnikov I, Hu X, Luchinat C, Greiner R, Wishart DS (2015) Accurate, fully-automated NMR spectral profiling for metabolomics. PLoS One 10:e0124219
48. Vignoli A, Ghini V, Meoni G, Licari C, Takis PG, Tenori L, Turano P, Luchinat C (2019) High-throughput metabolomics by 1D NMR. Angew Chem Int Ed Engl 58:968–994
49. Kennedy AD, Miller MJ, Beebe K, Wulff JE, Evans AM, Miller LA, Sutton VR, Sun Q, Elsea SH (2016) Metabolomic profiling of human urine as a screen for multiple inborn errors of metabolism. Genet Test Mol Biomarkers 20:485–495
50. Kennedy AD, Pappan KL, Donti TR, Evans AM, Wulff JE, Miller LAD, Reid Sutton V, Sun Q, Miller MJ, Elsea SH (2017) Elucidation of the complex metabolic profile of cerebrospinal fluid using an untargeted biochemical profiling assay. Mol Genet Metab 121:83–90
51. Hermans ME, van Weeghel M, Vaz FM, Ferdinandusse S, Hollak CEM, Huidekoper HH, Janssen MCH, van Kuilenburg ABP, Pras-Raves ML, Wamelink MMC, Wanders RJA, Welsink-Karssies MM, Bosch AM (2022) Multi-omics in classical galactosemia: evidence for the involvement of multiple metabolic pathways. J Inherit Metab Dis 45:1094–1105
52. Mathis T, Poms M, Kofeler H, Gautschi M, Plecko B, Baumgartner MR, Hochuli M (2022) Untargeted plasma metabolomics identifies broad metabolic perturbations in glycogen storage disease type I. J Inherit Metab Dis 45:235–247
53. Pandi A, Diehl C, Yazdizadeh Kharrazi A, Scholz SA, Bobkova E, Faure L, Nattermann M, Adam D, Chapin N, Foroughijabbari Y, Moritz C, Paczia N, Cortina NS, Faulon JL, Erb TJ (2022) A versatile active learning workflow for optimization of genetic and metabolic networks. Nat Commun 13:3876
54. Pappan KL, Kennedy AD, Magoulas PL, Hanchard NA, Sun Q, Elsea SH (2017) Clinical metabolomics to segregate aromatic amino acid decarboxylase deficiency from drug-induced metabolite elevations. Pediatr Neurol 75:66–72
55. Peters TMA, Merx J, Kooijman PC, Noga M, de Boer S, van Gemert LA, Salden G, Engelke UFH, Lefeber DJ, van Outersterp RE, Berden G, Boltje TJ, Artuch R, Pias-Peleteiro L, Garcia-Cazorla A, Baric I, Thony B, Oomens J, Martens J, Wevers RA, Verbeek MM, Coene KLM, Willemsen M (2023) Novel cerebrospinal fluid biomarkers of glucose transporter type 1 deficiency syndrome: implications beyond the brain's energy deficit. J Inherit Metab Dis 46:66–75
56. Zhou G, Li S, Xia J (2020) Network-based approaches for multi-omics integration. Methods Mol Biol 2104:469–487
57. Beger RD, Dunn W, Schmidt MA, Gross SS, Kirwan JA, Cascante M, Brennan L, Wishart DS, Oresic M, Hankemeier T, Broadhurst DI, Lane AN, Suhre K, Kastenmuller G, Sumner SJ, Thiele I, Fiehn O, Kaddurah-Daouk R, Medicine for "Precision, and Initiative Pharmacometabolomics Task Group"-Metabolomics Society (2016) Metabolomics enables precision medicine: "a white paper, community perspective". Metabolomics 12:149
58. Speyer CB, Baleja JD (2021) Use of nuclear magnetic resonance spectroscopy in diagnosis of inborn errors of metabolism. Emerg Top Life Sci 5:39–48
59. Haijes HA, van der Ham M, Hcmt Prinsen MH, Broeks, van Hasselt PM, de Sain-van der Velden MGM, Verhoeven-Duif NM, Jans JJM (2020) Untargeted metabolomics for metabolic diagnostic screening with automated data interpretation using a knowledge-based algorithm. Int J Mol Sci 21
60. Li X, Milosavljevic A, Elsea SH, Wang CC, Scaglia F, Syngelaki A, Nicolaides KH, Poon LC (2021) Effective aspirin treatment of women at risk for preeclampsia delays the metabolic clock of gestation. Hypertension 78:1398–1410
61. Shin SY, Fauman EB, Petersen AK, Krumsiek J, Santos R, Huang J, Arnold M, Erte I, Forgetta V, Yang TP, Walter K, Menni C, Chen L, Vasquez L, Valdes AM, Hyde CL, Wang V, Ziemek D, Roberts P, Xi L, Grundberg E, Consortium Multiple Tissue Human Expression Resource, Waldenberger M, Richards JB, Mohney RP, Milburn MV, John SL, Trimmer J, Theis FJ, Overington JP, Suhre K, Brosnan MJ, Gieger C, Kastenmuller G, Spector TD, Soranzo N (2014) An atlas of genetic influences on human blood metabolites. Nat Genet 46:543–550
62. Yazdani A, Yazdani A, Elsea SH, Schaid DJ, Kosorok MR, Dangol G, Samiei A (2019) Genome analysis and pleiotropy assessment using causal networks with loss of function mutation and metabolomics. BMC Genomics 20:395
63. Hamet P, Tremblay J (2017) Artificial intelligence in medicine. Metabolism 69S:S36–S40
64. Galal A, Talal M, Moustafa A (2022) Applications of machine learning in metabolomics: disease modeling and classification. Front Genet 13:1017340

65. Xi B, Gu H, Baniasadi H, Raftery D (2014) Statistical analysis and modeling of mass spectrometry-based metabolomics data. Methods Mol Biol 1198:333–353
66. Nyamundanda G, Brennan L, Gormley IC (2010) Probabilistic principal component analysis for metabolomic data. BMC Bioinform 11:571
67. Gromski PS, Muhamadali H, Ellis DI, Xu Y, Correa E, Turner ML, Goodacre R (2015) A tutorial review: metabolomics and partial least squares-discriminant analysis – a marriage of convenience or a shotgun wedding. Anal Chim Acta 879:10–23
68. Melo C, Navarro LC, de Oliveira DN, Guerreiro TM, Lima EO, Delafiori J, Dabaja MZ, Ribeiro MDS, de Menezes M, Rodrigues RGM, Morishita KN, Esteves CZ, de Amorim ALL, Aoyagui CT, Parise PL, Milanez GP, do Nascimento GM, Freitas ARR, Angerami R, Costa FTM, Arns CW, Resende MR, Amaral E, Junior RP, Ribeiro-do-Valle CC, Milanez H, Moretti ML, Proenca-Modena JL, Avila S, Rocha A, Catharino RR (2018) A machine learning application based in random forest for integrating mass spectrometry-based metabolomic data: a simple screening method for patients with Zika virus. Front Bioeng Biotechnol 6:31
69. Mahadevan S, Shah SL, Marrie TJ, Slupsky CM (2008) Analysis of metabolomic data using support vector machines. Anal Chem 80:7562–7570
70. Mendez KM, Broadhurst DI, Reinke SN (2019) The application of artificial neural networks in metabolomics: a historical perspective. Metabolomics 15:1–14
71. Alloghani M, Al-Jumeily D, Mustafina J, Hussain A, Aljaaf AJ (2020) A systematic review on supervised and unsupervised machine learning algorithms for data science. In: Supervised and unsupervised learning for data science
72. Trainor PJ, DeFilippis AP, Rai SN (2017) Evaluation of classifier performance for multiclass phenotype discrimination in untargeted metabolomics. Metabolites 7:30
73. Thistlethwaite LR, Li X, Burrage LC, Riehle K, Hacia JG, Braverman N, Wangler MF, Miller MJ, Elsea SH, Milosavljevic A (2022) Clinical diagnosis of metabolic disorders using untargeted metabolomic profiling and disease-specific networks learned from profiling data. Sci Rep 12:6556
74. Thistlethwaite LR, Petrosyan V, Li X, Miller MJ, Elsea SH, Milosavljevic A (2021) CTD: an information-theoretic algorithm to interpret sets of metabolomic and transcriptomic perturbations in the context of graphical models. PLoS Comput Biol 17:e1008550
75. Bravo-Merodio L, Williams JA, Gkoutos GV, Acharjee A (2019) Omics biomarker identification pipeline for translational medicine. J Transl Med 17:155
76. Burrage LC, Thistlethwaite L, Stroup BM, Sun Q, Miller MJ, Nagamani SCS, Craigen W, Scaglia F, Sutton VR, Graham B, Kennedy AD, Ucdc Members of the, Milosavljevic A, Lee BH, Elsea SH (2019) Untargeted metabolomic profiling reveals multiple pathway perturbations and new clinical biomarkers in urea cycle disorders. Genet Med 21:1977–1986
77. Milosavljevic S, Glinton KE, Li X, Medeiros C, Gillespie P, Seavitt JR, Graham BH, Elsea SH (2022) Untargeted metabolomics of Slc13a5 deficiency reveal critical liver-brain axis for lipid homeostasis. Metabolites 12
78. Wang Y, Hodge RA, Stevens VL, Hartman TJ, McCullough ML (2020) Identification and reproducibility of plasma metabolomic biomarkers of habitual food intake in a US diet validation study. Meta 10
79. Miller MJ, Bostwick BL, Kennedy AD, Donti TR, Sun Q, Sutton VR, Elsea SH (2016) Chronic oral L-carnitine supplementation drives marked plasma TMAO elevations in patients with organic acidemias despite dietary meat restrictions. JIMD Rep 30:39–44
80. Atwal PS, Donti TR, Cardon AL, Bacino CA, Sun Q, Emrick L, Reid Sutton V, Elsea SH (2015) Aromatic L-amino acid decarboxylase deficiency diagnosed by clinical metabolomic profiling of plasma. Mol Genet Metab 115:91–94
81. Ferreira CR, Goorden SMI, Soldatos A, Byers HM, Ghauharali-van der Vlugt JMM, Beers-Stet FS, Groden C, van Karnebeek CD, Gahl WA, Vaz FM, Jiang X, Vernon HJ (2018) Deoxysphingolipid precursors indicate abnormal sphingolipid metabolism in individuals with primary and secondary disturbances of serine availability. Mol Genet Metab 124:204–209
82. Tallis E, Karsenty CL, Grimes AB, Karam LB, Elsea SH, Sutton VR, Rawls-Castillo BL, Liu N, Soler-Alfonso C (2022) Untargeted metabolomic profiling in a patient with glycogen storage disease Ib receiving empagliflozin treatment. JIMD Rep 63:309–315
83. Ferreira CR, Rahman S, Keller M, Zschocke J, Icimd Advisory Group (2021) An international classification of inherited metabolic disorders (ICIMD). J Inherit Metab Dis 44:164–177
84. Bonte R, Bongaerts M, Demirdas S, Langendonk JG, Huidekoper HH, Williams M, Onkenhout W, Jacobs EH, Blom HJ, Ruijter GJG (2019) Untargeted metabolomics-based screening method for inborn errors of metabolism using semi-automatic sample preparation with an UHPLC- Orbitrap-MS platform. Metabolites 9
85. Coene KLM, Kluijtmans LAJ, van der Heeft E, Engelke UFH, de Boer S, Hoegen B, Kwast HJT, van de Vorst M, Huigen M, Keularts I, Schreuder MF, van Karnebeek CDM, Wortmann SB, de Vries MC, Janssen MCH, Gilissen C, Engel J, Wevers RA (2018) Next-generation metabolic screening: targeted and untargeted metabolomics for the diagnosis of inborn errors of metabolism in individual patients. J Inherit Metab Dis 41:337–353
86. Haijes HA, van der Ham M, Gerrits J, van Hasselt PM, Hcmt Prinsen, de Sain-van der Velden MGM,

Verhoeven-Duif NM, Jans JJM (2019) Direct-infusion based metabolomics unveils biochemical profiles of inborn errors of metabolism in cerebrospinal fluid. Mol Genet Metab 127:51–57
87. Kuhara T, Ohse M, Inoue Y, Cooper AJ (2011) A GC/MS-based metabolomic approach for diagnosing citrin deficiency. Anal Bioanal Chem 400:1881–1894
88. Scolamiero E, Cozzolino C, Albano L, Ansalone A, Caterino M, Corbo G, di Girolamo MG, Di Stefano C, Durante A, Franzese G, Franzese I, Gallo G, Giliberti P, Ingenito L, Ippolito G, Malamisura B, Mazzeo P, Norma A, Ombrone D, Parenti G, Pellecchia S, Pecce R, Pierucci I, Romanelli R, Rossi A, Siano M, Stoduto T, Villani GR, Andria G, Salvatore F, Frisso G, Ruoppolo M (2015) Targeted metabolomics in the expanded newborn screening for inborn errors of metabolism. Mol BioSyst 11:1525–1535
89. Jacob M, Malkawi A, Albast N, Al Bougha S, Lopata A, Dasouki M, Abdel Rahman AM (2018) A targeted metabolomics approach for clinical diagnosis of inborn errors of metabolism. Anal Chim Acta 1025:141–153
90. Peters TMA, Engelke UFH, de Boer S, van der Heeft E, Pritsch C, Kulkarni P, Wevers RA, Maap Willemsen MM, Verbeek, Coene KLM (2020) Confirmation of neurometabolic diagnoses using age-dependent cerebrospinal fluid metabolomic profiles. J Inherit Metab Dis 43:1112–1120
91. Janeckova H, Hron K, Wojtowicz P, Hlidkova E, Baresova A, Friedecky D, Zidkova L, Hornik P, Behulova D, Prochazkova D, Vinohradska H, Peskova K, Bruheim P, Smolka V, Stastna S, Adam T (2012) Targeted metabolomic analysis of plasma samples for the diagnosis of inherited metabolic disorders. J Chromatogr A 1226:11–17
92. Xiong X, Sheng X, Liu D, Zeng T, Peng Y, Wang Y (2015) A GC/MS-based metabolomic approach for reliable diagnosis of phenylketonuria. Anal Bioanal Chem 407:8825–8833
93. Blasco H, Veyrat-Durebex C, Bertrand M, Patin F, Labarthe F, Henique H, Emond P, Andres CR, Antar C, Landon C, Nadal-Desbarats L, Maillot F (2017) A multiplatform metabolomics approach to characterize plasma levels of phenylalanine and tyrosine in phenylketonuria. JIMD Rep 32:69–79
94. Davison AS, Norman BP, Ross GA, Hughes AT, Khedr M, Milan AM, Gallagher JA, Ranganath LR (2019) Evaluation of the serum metabolome of patients with alkaptonuria before and after two years of treatment with nitisinone using LC-QTOF-MS. JIMD Rep 48:67–74
95. Lim J, Shayota BJ, Lay E, Elsea SH, Bekheirnia MR, Mary Elizabeth M, Tessier SF, Kralik GM, Rice CS-A, Scaglia F (2021) Acute strokelike presentation and long-term evolution of diffusion restriction pattern in ethylmalonic encephalopathy. J Child Neurol 36:841–852
96. Bainbridge MN, Cooney E, Miller M, Kennedy AD, Wulff JE, Donti T, Jhangiani SN, Gibbs RA, Elsea SH, Porter BE, Graham BH (2017) Analyses of SLC13A5-epilepsy patients reveal perturbations of TCA cycle. Mol Genet Metab 121:314–319
97. Buzkova J, Nikkanen J, Ahola S, Hakonen AH, Sevastianova K, Hovinen T, Yki-Jarvinen H, Pietilainen KH, Lonnqvist T, Velagapudi V, Carroll CJ, Suomalainen A (2018) Metabolomes of mitochondrial diseases and inclusion body myositis patients: treatment targets and biomarkers. EMBO Mol Med 10
98. Esterhuizen K, Lindeque JZ, Mason S, van der Westhuizen FH, Rodenburg RJ, de Laat P, Smeitink JAM, Janssen MCH, Louw R (2021) One mutation, three phenotypes: novel metabolic insights on MELAS, MIDD and myopathy caused by the m.3243A > G mutation. Metabolomics 17:10
99. Sharma R, Reinstadler B, Engelstad K, Skinner OS, Stackowitz E, Haller RG, Clish CB, Pierce K, Walker MA, Fryer R, Oglesbee D, Mao X, Shungu DC, Khatri A, Hirano M, De Vivo DC, Mootha VK (2021) Circulating markers of NADH-reductive stress correlate with mitochondrial disease severity. J Clin Invest 131
100. Stowe RC, Sun Q, Elsea SH, Scaglia F (2018) LIPT1 deficiency presenting as early infantile epileptic encephalopathy, Leigh disease, and secondary pyruvate dehydrogenase complex deficiency. Am J Med Genet A 176:1184–1189
101. Birjiniuk A, Glinton KE, Villafranco N, Boyer S, Laufman J, Mizerik E, Scott D, Elsea SH, Galambos C, Varghese NP, Scaglia F (2020) Multiple mitochondrial dysfunctions syndrome 1: an unusual cause of developmental pulmonary hypertension. Am J Med Genet A 182:755–761
102. Wangler MF, Hubert L, Donti TR, Ventura MJ, Miller MJ, Braverman N, Gawron K, Bose M, Moser AB, Jones RO, Rizzo WB, Sutton VR, Sun Q, Kennedy AD, Elsea SH (2018) A metabolomic map of Zellweger spectrum disorders reveals novel disease biomarkers. Genet Med 20: 1274–1283
103. Donti TR, Cappuccio G, Hubert L, Neira J, Atwal PS, Miller MJ, Cardon AL, Sutton VR, Porter BE, Baumer FM, Wangler MF, Sun Q, Emrick LT, Elsea SH (2016) Diagnosis of adenylosuccinate lyase deficiency by metabolomic profiling in plasma reveals a phenotypic spectrum. Mol Genet Metab Rep 8:61–66
104. Pillai NR, Amin H, Gijavanekar C, Liu N, Issaq N, Broniowska KA, Bertuch AA, Sutton VR, Elsea SH, Scaglia F (2020) Hematologic presentation and the role of untargeted metabolomics analysis in monitoring treatment for riboflavin transporter deficiency. Am J Med Genet A 182:2781–2787
105. Engelke UFH, Van Outersterp RE, Merx J, Van Geenen FAMG, Van Rooij A, Berden G, Huigen MCDG, Kluijtmans LAJ, Peters TMA, Al-Shekaili

HH (2021) Untargeted metabolomics and infrared ion spectroscopy identify biomarkers for pyridoxine-dependent epilepsy. J Clin Invest 131
106. Koenig MK, Bonnen PE (2019) Metabolomics profile in ABAT deficiency pre- and post-treatment. JIMD Rep 43:13–17
107. Brown M, Turgeon C, Rinaldo P, Pop A, Salomons GS, Roullet J-B, Michael Gibson K (2020) Longitudinal metabolomics in dried bloodspots yields profiles informing newborn screening for succinic semialdehyde dehydrogenase deficiency. JIMD Rep 53:29–38

Index

N. Blau, F. M. Vaz (eds.), *Laboratory Guide to the Methods in Biochemical Genetics*,
https://doi.org/10.1007/978-3-031-58819-8

C

D

E

Q

R

S